Modern Analytical Methods in Art and Archaeology

CHEMICAL ANALYSIS

A SERIES OF MONOGRAPHS ON ANALYTICAL CHEMISTRY
AND ITS APPLICATIONS

Editor
J. D. WINEFORDNER

VOLUME 155

WILEY-INTERSCIENCE

A JOHN WILEY & SONS, INC., PUBLICATION

New York • Chichester • Weinheim • Brisbane • Singapore • Toronto

Modern Analytical Methods in Art and Archaeology

ENRICO CILIBERTO AND GIUSEPPE SPOTO

Editors

Dipartimento di Scienze Chimiche
Università di Catania
Catania, Italy

WILEY-INTERSCIENCE

A JOHN WILEY & SONS, INC., PUBLICATION

New York • Chichester • Weinheim • Brisbane • Singapore • Toronto

This book is printed on acid-free paper.⊗

Published simultaneously in Canada.

For ordering and customer service, call 1-800-CALL-WILEY.

Library of Congress Cataloging-in-Publication Data:

Modern analytical methods in art and archaeology / E. Ciliberto and G. Spoto, editors.
p. cm. — (Chemical analysis ; v. 155)
Includes index.
ISBN 0-471-29361-X (alk. paper)
1. Art—Expertising. 2. Chemistry, Analytic—Methodology. 3. Archaeological chemistry. I. Ciliberto, E. (Enrico) II. Spoto, G. (Giuseppe) III. Series.
N8558.M63 2000
702′.8′8—dc21 99-057435

Printed in the United States of America.
10 9 8 7 6 5 4 3 2 1

CONTENTS

CONTRIBUTORS

Jorge A. Ascencio, Instituto Nacional de Investigaciones Nucleares, Amsterdam 46-202, Hipódromo Condesa, 06100 México, D.F., México

Mauro Bacci, Istituto di Ricerca sulle Onde Elettromagnetiche, CNR "Nello Carrara", Via Panciatichi 64, 50127, Firenze, Italy

Mike J. Baxter, Department of Mathematics, Statistics and Operational Research, Nottingham Trent University, Clifton Campus, Nottingham, United Kingdom, NG11 8NS

Alex von Bohlen, Institut für Spektrochemie und Angewandte Spektroskopie, ISAS, Bunsen-Kirchhoff-Str. 11, D-44139 Dortmund, Germany

Silvia Bruni, Dipartimento di Chimica Inorganica, Metallorganica e Analitica, Università degli Studi di Milano, Via G. Venezian 21, 20133 Milano, Italy

Caitlin E. Buck, School of History and Archaeology, University of Wales, PO BOX 909, Cardiff, Wales, CF1 3XU, United Kingdom

Thomas Calligaro, Laboratoire de Recherche des Musées de France, CNRS UMR 171, 6, rue des Pyramides 75041 Paris Cedex 01 France

Raúl J. Cano, Environmental Biotechnology Institute and Biological Sciences Department, California Polytechnic State University, San Luis Obispo, California 93407 U.S.A.

Franco Cariati, Dipartimento di Chimica Inorganica, Metallorganica e Analitica, Università degli Studi di Milano, Via G. Venezian 21, 20133 Milano, Italy

Enrico Ciliberto, Dipartimento di Scienze Chimiche, Università di Catania, Viale Andrea Doria 6, 95125 Catania, Italy

Jean-Claude Dran, Laboratoire de Recherche des Musées de France, CNRS UMR 171, 6, rue des Pyramides 75041 Paris Cedex 01 France

Richard P. Evershed, School of Chemistry, University of Bristol, Cantock's Close, Bristol, BS8 1TS United Kingdom

Noel H. Gale, Isotrace Laboratory, Oxford University, Keble Road, OX1 3RH Oxford, United Kingdom

Rainer Grün, Research School of Earth Sciences, Australian National University, Canberra ACT 0200, Australia

Ronald G. V. Hancock, SLOWPOKE-2 Facility, Department of Chemistry and Chemical Engineering, Sawyer Building, Royal Military College, 11 General Crerar Crescent, Kingston, Ontario, Canada, K7K 7B4

Robert E. M. Hedges, Oxford Radiocarbon Accelerator Unit, Research Laboratory for Archaeology and the History of Art, Oxford University, 6 Keble Road, OX1 3QJ, Oxford, United Kingdom

Miguel José-Yacamán, Instituto Nacional de Investigaciones Nucleares, Amsterdam 46-202, Hipódromo Condesa, 06100 México, D.F., México and Instituto de Fisica, Universidad Nacional Autónoma de México, Apdo. Postal 20-364, Delegación Alvaro Obregón, 01000 México, D.F., México

Luc Moens, University Ghent, Laboratory of Analytical Chemistry, Proeftuinstraat 86, B-9000 Ghent, Belgium

Hector Neff, Missouri University Research Reactor Center, University of Missouri, Columbia, Missouri, U.S.A.

Marianne Odlyha, Birkbeck College, School of Biological and Chemical Sciences, University of London, Gordon House, 29 Gordon Square, London WC1H OPP, United Kingdom

A. Mark Pollard, Dept. of Archaeological Science, University of Bradford, Bradford, West Yorkshire, BD7 1DP, United Kingdom

Richard G. Roberts, Department of Earth Sciences, La Trobe University, Melbourne, Victoria 3083, Australia

Joseph Salomon, Laboratoire de Recherche des Musées de France, CNRS UMR 171, 6, rue des Pyramides 75041 Paris Cedex 01 France

Giuseppe Spoto, Dipartimento di Scienze Chimiche, Università di Catania, Viale Andrea Doria 6, 95125 Catania, Italy

Zofia Stos-Gale, Isotrace Laboratory, Oxford University, Keble Road, OX1 3RH Oxford, United Kingdom

Sebastiano O. Troja, Department of Physics and Astronomy, University of Catania, Corso Italia 57, 95129 Catania, Italy

Peter Vandenabeele, University Ghent, Laboratory of Analytical Chemistry, Proeftuinstraat 86, B-9000 Ghent, Belgium

Guenther A. Wagner, Forschungsstelle Archaeometrie der Heidelberger Akademie der Wissenschaften, am Max-Planck-Institut fuer Kernphysik, Postfach 103980, Heidelberg, Germany

Suzanne M. M. Young, Archaeometry Laboratories, Harvard University, Peabody Museum, 11 Divinity Avenue, Cambridge, Massachusetts 02138 U.S.A.

PREFACE

The use of analytical methods in studies concerning works of art and archaeological artifacts has contributed in various ways to the growth of many scientific and historical areas. Knowing as much as possible about the material constituting objects of artistic and archaeological value is fundamental for the preservation of their heritage. The composition of the materials used in the past, their age, and the methods used for their production facilitate our understanding of historical events, ancient culture, and usage. The interaction of various areas such as analytical chemistry, history of art, and archaeology has led to the development of a variety of methods used in archaeological and conservation chemistry. These methods are used to date objects, to understand the provenance of ancient materials, to reconstruct ancient usage, and to identify the degradation processes that affect the integrity of works of art. This gives us a brief insight into the fascinating path that we have chosen to follow in the editing of this book. The aim of the book is to offer an updated overview of state-of-the-art analytical methods in archaeology and art and to provide archaeologists, conservators, scientists, and students interested in the field with an introduction to some of these methods.

Archaeologists, conservators, and students with a scientific background at an M.Sc. or Ph.D. level will find brief explanations of the basic principles of the analytical methods discussed and indications as to relevant information these provide in the study of artistic and archaeological objects. The potential and limitations of these methods and the status of the main scientific topics related to science, art, and archaeology will be also addressed.

Scientists both with and without previous experience in the field of interest will find (i) assistance in the mastery of new analytical methods or traditional analytical methods, together with an indication of new applications in the field of art and archaeology; (ii) information on state-of-the-art research in the field; and (iii) sources for updated reviews on recent applications. We have tried to achieve these goals by sectioning the book into four parts: elemental analysis, molecular and structural analysis, isotope analysis, dating, and provenance methods, and data handling and statistical analysis. Each part, with the exception of data handling and statistical analysis, includes several chapters and is prefaced by an introductory chapter that attempts to provide an overview of the methods discussed and to guarantee cohesion between the chapters. The readers with nonspecific experience in the analytical methods used in art and archaeology should preliminarily refer to these chapters. The first chapter of the book provides reasons why art, archaeology, and analytical chemistry are placed together in this book.

The chapters of the book, written by authors with proven experience in the field, are self-contained. This provides easy access to suit the requirements and interests of individual readers. In order to be self-contained, each chapter is prefaced by a general introduction which briefly provides the theoretical principles of the discussed method(s) or technique(s). One or more paragraphs in which the most important applications of the method(s) or technique(s) to art and archaeology follow. An up-dated reference list closes each chapter.

We hope that the present book will offer an opportunity to further consolidate links between science, art, and archaeology.

ACKNOWLEDGMENT

We would like to acknowledge each author for time, effort, and concern devoted to the creation of this book. We also acknowledge Emilio Castellucci, Giuseppe Condorelli, Domenico Grasso, Ronald G. V. Hancock, Salvatore Ingrassia, Giuseppe Pappalardo, Goffredo Petrone, John W. Steeds, Giuseppe Vecchio, and Güenther A. Wagner for their suggestions and proofreading of parts of this book.

Finally we thank the Series Editor and the staff of John Wiley & Sons for their suggestions, hard work, and endless patience.

Enrico Ciliberto
Giuseppe Spoto
Catania, Italy

CHEMICAL ANALYSIS

A SERIES OF MONOGRAPHS ON ANALYTICAL CHEMISTRY AND ITS APPLICATIONS

J. D. Winefordner, *Series Editor*

Vol. 1. **The Analytical Chemistry of Industrial Poisons, Hazards, and Solvents**. *Second Edition*. By the late Morris B. Jacobs

Vol. 2. **Chromatographic Adsorption Analysis**. By Harold H. Strain (*out of print*)

Vol. 3. **Photometric Determination of Traces of Metals**. *Fourth Edition*
Part I: General Aspects. By E. B. Sandell and Hiroshi Onishi
Part IIA: Individual Metals, Aluminum to Lithium. By Hiroshi Onishi
Part IIB: Individual Metals, Magnesium to Zirconium. By Hiroshi Onishi

Vol. 4. **Organic Reagents Used in Gravimetric and Volumetric Analysis**. By John F. Flagg (*out of print*)

Vol. 5. **Aquametry: A Treatise on Methods for the Determination of Water**. *Second Edition* (*in three parts*). By John Mitchell, Jr. and Donald Milton Smith

Vol. 6. **Analysis of Insecticides and Acaricides**. By Francis A. Gunther and Roger C. Blinn (*out of print*)

Vol. 7. **Chemical Analysis of Industrial Solvents**. By the late Morris B. Jacobs and Leopold Schetlan

Vol. 8. **Colorimetric Determination of Nonmetals**. *Second Edition*. Edited by the late David F. Boltz and James A. Howell

Vol. 9. **Analytical Chemistry of Titanium Metals and Compounds**. By Maurice Codell

Vol. 10. **The Chemical Analysis of Air Pollutants**. By the late Morris B. Jacobs

Vol. 11. **X-Ray Spectrochemical Analysis**. *Second Edition*. By L. S. Birks

Vol. 12. **Systematic Analysis of Surface-Active Agents**. *Second Edition*. By Milton J. Rosen and Henry A. Goldsmith

Vol. 13. **Alternating Current Polarography and Tensammetry**. By B. Breyer and H. H. Bauer

Vol. 14. **Flame Photometry**. By R. Herrmann and J. Alkemade

Vol. 15. **The Titration of Organic Compounds** (*in two parts*). By M. R. F. Ashworth

Vol. 16. **Complexation in Analytical Chemistry: A Guide for the Critical Selection of Analytical Methods Based on Complexation Reactions**. By the late Anders Ringbom

Vol. 17. **Electron Probe Microanalysis**. *Second Edition*. By L. S. Birks

Vol. 18. **Organic Complexing Reagents: Structure, Behavior, and Application to Inorganic Analysis**. By D. D. Perrin

Vol. 19. **Thermal Analysis**. *Third Edition*. By Wesley Wm. Wendlandt

Vol. 20. **Amperometric Titrations**. By John T. Stock

Vol. 21. **Reflectance Spectroscopy**. By Wesley Wm. Wendlandt and Harry G. Hecht

Vol. 22. **The Analytical Toxicology of Industrial Inorganic Poisons**. By the late Morris B. Jacobs

Vol. 23. **The Formation and Properties of Precipitates**. By Alan G. Walton

Vol. 24. **Kinetics in Analytical Chemistry**. By Harry B. Mark, Jr. and Garry A. Rechnitz

Vol. 25. **Atomic Absorption Spectroscopy**. *Second Edition*. By Morris Slavin

Vol. 26. **Characterization of Organometallic Compounds** (*in two parts*). Edited by Minoru Tsutsui

Vol. 27. **Rock and Mineral Analysis**. *Second Edition*. By Wesley M. Johnson and John A. Maxwell

Vol. 28. **The Analytical Chemistry of Nitrogen and Its Compounds** (*in two parts*). Edited by C. A. Streuli and Philip R. Averell

Vol. 29. **The Analytical Chemistry of Sulfur and Its Compounds** (*in three parts*). By J. H. Karchmer

Vol. 30. **Ultramicro Elemental Analysis**. By Günther Tölg

Vol. 31. **Photometric Organic Analysis** (*in two parts*). By Eugene Sawicki

Vol. 32. **Determination of Organic Compounds: Methods and Procedures**. By Frederick T. Weiss

Vol. 33. **Masking and Demasking of Chemical Reactions**. By D. D. Perrin

Vol. 34. **Neutron Activation Analysis**. By D. De Soete, R. Gijbels, and J. Hoste

Vol. 35. **Laser Raman Spectroscopy**. By Marvin C. Tobin

Vol. 36. **Emission Spectrochemical Analysis**. By Morris Slavin

Vol. 37. **Analytical Chemistry of Phosphorus Compounds**. Edited by M. Halmann

Vol. 38. **Luminescence Spectroscopy in Analytical Chemistry**. By J. D. Winefordner, S. G. Schulman, and T. C. O'Haver

Vol. 39. **Activation Analysis with Neutron Generators**. By Sam S. Nargolwalla and Edwin P. Przybylowicz

Vol. 40. **Determination of Gaseous Elements in Metals**. Edited by Lynn L. Lewis, Laben M. Melnick, and Ben D. Holt

Vol. 41. **Analysis of Silicones**. Edited by A. Lee Smith

Vol. 42. **Foundations of Ultracentrifugal Analysis**. By H. Fujita

Vol. 43. **Chemical Infrared Fourier Transform Spectroscopy**. By Peter R. Griffiths

Vol. 44. **Microscale Manipulations in Chemistry**. By T. S. Ma and V. Horak

Vol. 45. **Thermometric Titrations**. By J. Barthel

Vol. 46. **Trace Analysis: Spectroscopic Methods for Elements**. Edited by J. D. Winefordner

Vol. 47. **Contamination Control in Trace Element Analysis**. By Morris Zief and James W. Mitchell

Vol. 48. **Analytical Applications of NMR**. By D. E. Leyden and R. H. Cox

Vol. 49. **Measurement of Dissolved Oxygen**. By Michael L. Hitchman

Vol. 50. **Analytical Laser Spectroscopy**. Edited by Nicolo Omenetto

Vol. 51. **Trace Element Analysis of Geological Materials**. By Roger D. Reeves and Robert R. Brooks

Vol. 52. **Chemical Analysis by Microwave Rotational Spectroscopy**. By Ravi Varma and Lawrence W. Hrubesh

Vol. 53. **Information Theory As Applied to Chemical Analysis**. By Karl Eckschlager and Vladimir Stepanek

Vol. 54. **Applied Infrared Spectroscopy: Fundamentals, Techniques, and Analytical Problem-solving**. By A. Lee Smith

Vol. 55. **Archaeological Chemistry**. By Zvi Goffer

Vol. 88. **Applied Geochemical Analysis**. By C. O. Ingamells and F. F. Pitard

Vol. 89. **Detectors for Liquid Chromatography**. Edited by Edward S. Yeung

Vol. 90. **Inductively Coupled Plasma Emission Spectroscopy: Part 1: Methodology, Instrumentation, and Performance; Part II: Applications and Fundamentals**. Edited by J. M. Boumans

Vol. 91. **Applications of New Mass Spectrometry Techniques in Pesticide Chemistry**. Edited by Joseph Rosen

Vol. 92. **X-Ray Absorption: Principles, Applications, Techniques of EXAFS, SEXAFS, and XANES**. Edited by D. C. Konnigsberger

Vol. 93. **Quantitative Structure-Chromatographic Retention Relationships**. By Roman Kaliszan

Vol. 94. **Laser Remote Chemical Analysis**. Edited by Raymond M. Measures

Vol. 95. **Inorganic Mass Spectrometry**. Edited by F. Adams, R. Gijbels, and R. Van Grieken

Vol. 96. **Kinetic Aspects of Analytical Chemistry**. By Horacio A. Mottola

Vol. 97. **Two-Dimensional NMR Spectroscopy**. By Jan Schraml and Jon M. Bellama

Vol. 98. **High Performance Liquid Chromatography**. Edited by Phyllis R. Brown and Richard A. Hartwick

Vol. 99. **X-Ray Fluorescence Spectrometry**. By Ron Jenkins

Vol. 100. **Analytical Aspects of Drug Testing**. Edited by Dale G. Deustch

Vol. 101. **Chemical Analysis of Polycyclic Aromatic Compounds**. Edited by Tuan Vo-Dinh

Vol. 102. **Quadrupole Storage Mass Spectrometry**. By Raymond E. March and Richard J. Hughes

Vol. 103. **Determination of Molecular Weight**. Edited by Anthony R. Cooper

Vol. 104. **Selectivity and Detectability Optimization in HPLC**. By Satinder Ahuja

Vol. 105. **Laser Microanalysis**. By Lieselotte Moenke-Blankenburg

Vol. 106. **Clinical Chemistry**. Edited by E. Howard Taylor

Vol. 107. **Multielement Detection Systems for Spectrochemical Analysis**. By Kenneth W. Busch and Marianna A. Busch

Vol. 108. **Planar Chromatography in the Life Sciences**. Edited by Joseph C. Touchstone

Vol. 109. **Fluorometric Analysis in Biomedical Chemistry: Trends and Techniques Including HPLC Applications**. By Norio Ichinose, George Schwedt, Frank Michael Schnepel, and Kyoko Adochi

Vol. 110. **An Introduction to Laboratory Automation**. By Victor Cerdá and Guillermo Ramis

Vol. 111. **Gas Chromatography: Biochemical, Biomedical, and Clinical Applications**. Edited by Ray E. Clement

Vol. 112. **The Analytical Chemistry of Silicones**. Edited by A. Lee Smith

Vol. 113. **Modern Methods of Polymer Characterization**. Edited by Howard G. Barth and Jimmy W. Mays

Vol. 114. **Analytical Raman Spectroscopy**. Edited by Jeanette Graselli and Bernard J. Bulkin

Vol. 115. **Trace and Ultratrace Analysis by HPLC**. By Satinder Ahuja

Vol. 116. **Radiochemistry and Nuclear Methods of Analysis**. By William D. Ehmann and Diane E. Vance

Vol. 117. **Applications of Fluorescence in Immunoassays**. By Ilkka Hemmila

Vol. 118. **Principles and Practice of Spectroscopic Calibration**. By Howard Mark

Vol. 119. **Activation Spectrometry in Chemical Analysis**. By S. J. Parry

Vol. 120. **Remote Sensing by Fourier Transform Spectrometry**. By Reinhard Beer

Vol. 121. **Detectors for Capillary Chromatography**. Edited by Herbert H. Hill and Dennis McMinn

Vol. 122. **Photochemical Vapor Deposition**. By J. G. Eden

Vol. 123. **Statistical Methods in Analytical Chemistry**. By Peter C. Meier and Richard Zünd

Vol. 124. **Laser Ionization Mass Analysis**. Edited by Akos Vertes, Renaat Gijbels, and Fred Adams

Vol. 125. **Physics and Chemistry of Solid State Sensor Devices**. By Andreas Mandelis and Constantinos Christofides

Vol. 126. **Electroanalytical Stripping Methods**. By Khjena Z. Brainina and E. Neyman

Vol. 127. **Air Monitoring by Spectroscopic Techniques**. Edited by Markus W. Sigrist

Vol. 128. **Information Theory in Analytical Chemistry**. By Karel Eckschlager and Klaus Danzer

Vol. 129. **Flame Chemiluminescence Analysis by Molecular Emission Cavity Detection**. Edited by David Stiles, Anthony Calokerinos, and Alan Townshend

Vol. 130. **Hydride Generation Atomic Absorption Spectrometry**. By Jiri Dedina and Dimiter L. Tsalev

Vol. 131. **Selective Detectors: Environmental, Industrial, and Biomedical Applications**. Edited by Robert E. Sievers

Vol. 132. **High Speed Countercurrent Chromatography**. Edited by Yoichiro Ito and Walter D. Conway

Vol. 133. **Particle-Induced X-Ray Emission Spectrometry**. By Sven A. E. Johansson, John L. Campbell, and Klas G. Malmqvist

Vol. 134. **Photothermal Spectroscopy Methods for Chemical Analysis**. By Stephen E. Bialkowski

Vol. 135. **Element Speciation in Bioinorganic Chemistry**. Edited by Sergio Caroli

Vol. 136. **Laser-Enhanced Ionization Spectrometry**. Edited by John C. Travis and Gregory C. Turk

Vol. 137. **Fluorescence Imaging Spectroscopy and Microscopy**. Edited by Xue Feng Wang and Brian Herman

Vol. 138. **Introduction to X-Ray Powder Diffractometry**. By Ron Jenkins and Robert L. Snyder

Vol. 139. **Modern Techniques in Electroanalysis**. Edited by Petr Vanýsek

Vol. 140. **Total Reflection X-Ray Fluorescence Analysis**. By Reinhold Klockenkamper

Vol. 141. **Spot Test Analysis: Clinical, Environmental, Forensic, and Geochemical Applications, Second Edition**. By Ervin Jungreis

Vol. 142. **The Impact of Stereochemistry on Drug Development and Use**. Edited by Hassan Y. Aboul-Enein and Irving W. Wainer

Vol. 143. **Macrocyclic Compounds in Analytical Chemistry**. Edited by Yury A. Zolotov

Vol. 144. **Surface-Launched Acoustic Wave Sensors: Chemical Sensing and Thin-Film Characterization**. By Michael Thompson and David Stone

Vol. 145. **Modern Isotope Ratio Mass Spectrometry**. Edited by T. J. Platzner

Vol. 146. **High Performance Capillary Electrophoresis: Theory, Techniques, and Applications**. Edited by Morteza G. Khaledi

Vol. 147. **Solid Phase Extraction: Principles and Practice**. By E. M. Thurman

Vol. 148. **Commercial Biosensors: Applications to Clinical, Bioprocess and Environmental Samples**. Edited by Graham Ramsay

Vol. 149. **A Practical Guide to Graphite Furnace Atomic Absorption Spectrometry**. By David J. Butcher and Joseph Sneddon

Vol. 150. **Principles of Chemical and Biological Sensors**. Edited by Dermot Diamond

Vol. 151. **Pesticide Residue in Foods: Methods, Technologies, and Regulations**. By W. George Fong, H. Anson Moye, James N. Seiber, and John P. Toth

Vol. 152. **X-Ray Fluorescence Spectrometry**. *Second Edition.* By Ron Jenkins

Vol. 153. **Statistical Methods in Analytical Chemistry**. *Second Edition.* By Peter C. Meier and Richard E. Zünd

Vol. 154. **Modern Analytical Methodologies in Fat and Water Soluble Vitamins**. Edited by Won O. Song, Gary R. Beecher, and Ronald R. Eitenmiller

Vol. 155. **Modern Analytical Methods in Art and Archaeology**. Edited by Enrico Ciliberto and Giuseppe Spoto

Modern Analytical Methods in Art and Archaeology

CHAPTER

1

ANALYTICAL METHODS IN ART AND ARCHAEOLOGY

ENRICO CILIBERTO

Dipartimento di Scienze Chimiche, Università di Catania, Catania, Italy

1.1. WHY ART, ARCHAEOLOGY, AND ANALYTICAL CHEMISTRY TOGETHER?

The numerous problems faced when studying archaeological artifacts and art objects from a scientific and analytical point of view require careful attention. To begin with, it would be worth considering what art, archaeology, and an experimental science such as analytical chemistry, three essentially different aspects of human culture, may have in common. This bring us into the realms of a lengthy philosophical and epistemological debate. One thing is, however, certain. The scientists who work in these fields and others besides are frequently led to consider the ways in which their studies relate to the real-life human context of their work. Two examples taken from the past may better serve to illustrate this point. In a letter to Fortunio Liceti in 1640, Galileo Galilei wrote: "Experience is to be preferred over all manner of discourse as being amongst the surest ways of reaching Truth." Preceding Galileo by more than 100 years, the great scientist and artist Leonardo da Vinci is also reported as having repeatedly observed that "wisdom comes of experience." The experimental method is still fundamental in science today. The historical and archaeological sciences along with those more commonly labeled as "experimental" must thus openly confront the universe of signs and symbols around us if they are to continue to progress and flourish.

If it may be confidently said that archeology shows how humans reach back to the events of the past in an attempt to reconstruct the cultural heritage of humanity, then science can be defined as our constant endeavor to unfathom reality through coded systems and methods. Apart from the

Modern Analytical Methods in Art and Archaeology, Edited by Enrico Ciliberto and Giuseppe Spoto. Chemical Analysis Series, Vol. 155
ISBN 0-471-29361-X

objects studied and the models and language systems employed to that purpose, both archaeology and science share the same desire for knowledge and use similar methods to attain it. Art, on the other hand, enjoys a freer status. The artist tends to see himself or herself as a bridge between reality, which is sometimes perceived intuitively and is rarely openly expressed, and common people. Beholding a work of art, we discern a universal message that goes beyond the constraints of space and time. Though they may do so in different ways, science, history, and art thus unitedly express the basic human desire for knowledge and our constant search for universality along the road to progress.

Let us now turn to what we mean by the term *cultural heritage*. What this term, in fact, indicates is the whole of human cultural patrimony, that is, everything that refers to the history of civilization. This includes all that is considered of interest from a variety of points of view—archaeological, historical, and artistic. It also embraces all works and documents of value contained in archives or preserved in book collections. The various ways in which this cultural knowledge takes shape and is preserved may also be defined as cultural heritage. This is exactly how art, archaeology, and the experimental sciences are brought to touch on one another. In a single object it is often possible to pick out aspects pertaining to all three fields. Figure 1.1 shows a well-known archaeological find, commonly known as the Painted Sarcophagus found in H. Triada, Crete (Late Minoan, IIIA.), now kept in the archaeological museum of Eraklion. Apart from being a find of immense archaeological importance, it is also undoubtedly a great work of art. Scientific studies and investigations, involving scanning electron microscopy and X-ray diffraction, were carried out by the Italian School of Archaeology in Athens (Militello, 1998). These studies provided information concerning the variety of pigments used by the ancient Minoans, and we can now say which techniques were used in the frescoes. An indication was also given as to the general degree of technology attained by the Minoans. This in turn has contributed to a greater understanding of the level of civilization these people achieved.

Analytical investigations are of value for numerous other reasons. The systematic analysis of works of art throughout history has enabled us to compile criteria against which forgeries can be detected and compared. If, for example, Prussian Blue (a pigment in use from the first half of the eighteenth century) is found in a painting believed to be an Italian Renaissance original, questions will inevitably be raised as to the possibility of it being a forgery or whether the work was touched up in more recent times (Sheare et al. 1983). Enquiries of a historical nature should also be included in studies of this kind. Research on ancient sources is often useful in analytical studies concerning works of art and archaeological finds. S. M. Vinaz's extremely

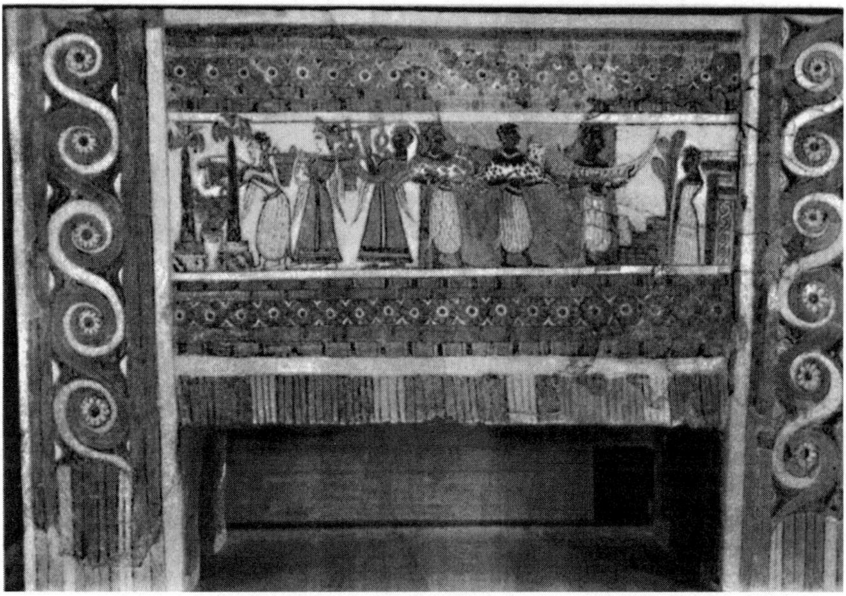

Figure 1.1. Painted Sarcophagus (Late Minoan IIIA) from Haghia Triada (Crete): Archaeological remains or work of art? (Reproduced with permission of Scuola Archeologica Italiana di Atene.)

helpful review of collected sources with reference to medieval painting techniques is one work worth mentioning in this context (Vinaz, 1998). Readers will also find a selection of important documents for wall painting techniques from the Latin era to modern times in the appendix of an interesting book by Mora, Mora, and Phillipot (Mora et al. 1984).

The study of degradation indicators is yet another area in which analytical investigation proves invaluable. All things, man-made or not, are subject to processes that lead to the progressive degradation of the materials involved, although the rate at which these degrade may vary. Knowing which processes cause degradation and identifying their indicators allows us to find suitable methods for slowing the course of damage in order to ensure that the pieces concerned will endure for as long as possible. This leads us to consider another fundamental issue in cultural heritage—conservation. The question is often raised as to how large amounts of money and significant human resources can justifiably be expended for the conservation of cultural heritage to the detriment of other more deserving social problems such as poverty and unemployment. Some reply that our cultural patrimony should be preserved for posterity. Others believe that conservation is an essential part of the world tourist trade or see cultural heritage in terms of economic investment. The personal opinion put forward here is that people cannot build a future without looking to their roots. It is in this context that the conservation of cultural heritage should be acknowledged. In a historic lecture at the British Museum on the 23 of November, 1977, Harold J. Plenderleith gave a unique personal insight into the important role that the conservation of cultural heritage has come to play in the twentieth century (Plenderleith, 1998). The pressing necessity for the preservation of the world's cultural patrimony has also been recognized by authoritative bodies such as UNESCO. UNESCO has compiled a list of objects and places of cultural value whose protection should be given priority. The list currently stands at 582 items, but additions are steadily being made. A host of little-known or "hidden" art objects and archaeological artifacts in fact deserve equal attention. Much of the world's culture risks being lost irretrievably, and this threat will persist should conservation projects not be undertaken at a local level. It should by now be clear that accurate, detailed knowledge of the material used in the creation of objects of cultural value is the primary requisite for conservation and restoration procedures of all types and kinds.

1.2. MODERN ANALYTICAL METHODS

People's move away from simply basing their deduction on the feelings evoked by an object's form, color, or smell toward a more systematic, scientific approach to observation was a giant cognitive leap forward. We enlarged upon our knowledge and extended it beyond the limits of our sense organs. We might be tempted to consider this extension of our knowledge as being the essence of science itself. Looking at it this way, scientific instruments are nothing more than artificial extensions of our sense organs. The microscope and spectroscope may thus be considered as the extension of the power of sight. This, however, is only one of the many facets of science. Another essential, and by no means less important, element of scientific investigation can be found in what may be termed as its *intensive* nature. This comes into play when we search for basic units that go beyond what we are able to perceive. From an *extensive* point of view, scientific instruments may increase our understanding of details not ordinarily perceptible by human sense alone. Intensive investigation instead involves organizing data in a different, more consistent way. It is no longer a matter of simply listing things as they are perceived. Science's extension of our knowledge of things must necessarily be accompanied by parallel improvements in our ability to penetrate objects and phenomena at an intensive level. Both aspects are required for scientific investigation to prove fruitful. When looking at the complexities and variability of analytical data, for example, the need to set these within a logical framework becomes obvious. It is necessary to look closely at how data are obtained and to examine how they relate to the context in question. Indeed, this may serve as an indication of how crucial it is to distinguish between experimental techniques and analytical methods. When the object for analysis comes from the field of archaeology or the artistic sphere, this distinction is of the utmost importance. The chapters of this book indicate quite clearly that no analytical technique may be used successfully without first answering a number of fundamental questions as to the type of information required, how significant this information may be in an art history context, and whether the proposed technique may lead to the formation of models that are necessary for the subsequent interpretation of the data obtained. The experimental techniques employed to obtain the required data are clearly intrinsic to the analytical method used. Experimental procedures carried out for archeological objects or works of art also require further aspects to be taken into account.

1.2.1. Sampling

The sampling procedure is one of the most delicate aspects of the analytical approach. The following are just some of the various questions that should be addressed: What amount of specimen is required to achieve an acceptable degree of accuracy in the results? Will sampling be possible to a reasonable extent, ensuring a representative specimen of the object under analysis? Are there reasons to suppose that the areas sampled have been altered in any way? Will damage sustained by the object be within reasonable limits? In connection with this last question, a distinction is often made between destructive and microdestructive sampling techniques. However, leaving aside any formal definition regarding sampling techniques and procedures, this is a complex area where a great number of problems need to be tackled. It is worth mentioning that a frequently indispensable preliminary step consists in the visual survey of the specimen. In the case of microspecimens, this requires the use of the optical and/or electron microscope. Manipulating samples only tens of micrometers in dimension is never straightforward and should be carried out with due care. Information in the literature concerning sampling techniques and related areas is scant. A useful contribution to the field can, however, be found in Tsang and Cunningham's article on the techniques developed at the Conservation Analytical Laboratory of the Smithsonian Institution for the preparation of thick cross sections (50 μm and up) and microtomed cross sections (10 μm) of small samples from works of art (Tsang and Cunningham, 1991). The importance of sampling procedures as outlined above leads logically to the next area for consideration—the analytical techniques themselves.

1.2.2. Analytical Techniques

A wide variety of analytical techniques are available for the study of archaeological specimens and those taken from works of art. In the past, these techniques were based predominantly on assay methods, some of these being extremely sensitive and highly selective. For the past 40 years, however, instrumental techniques have had the upper hand, many of these having recently been reviewed by Pollard and Heron (1996). A great number of these techniques may be used in the context of elemental analysis (see Chapter 2), while others find their place in molecular and structural methods (see Chapter 7). Some are used specifically in the context of isotope analysis, dating, and provenance methods (see Chapter 15). The various techniques in use at present are dealt with amply in the chapters to follow. However, a few points regarding methodology should be made.

It should always be remembered that the choice of technique will depend greatly on the nature of the problem it intends to solve. At times, the technique employed fails to be the most suitable solution for the observation of a given chemical or physical property. The required degree of accuracy may not be attained, thus hindering the solution of the problem in question. Techniques may be chosen for financial reasons or on the basis of equipment availability. However, once the field of analysis has been established, it is advisable to try to find a balance between the necessity for accuracy and the need for economy.

One particularly delicate matter should be mentioned—calibration. All scientists know that analytical results depend on what can be experimentally measured. Observable properties are generally correlated with the concentrations of chemical species that are to be determined. This relation is usually linear. It is thus simply necessary to determine experimentally the parameters modulating the mathematical function. This must be done using analytical standards that are representative for actual samples and compatible with the technique employed. Before being applied to real, specific cases, analytical procedures should be tested on standard samples in the selected field of concentration. When this is done accurately, it is also possible to compare various analytical techniques and evaluate their limits and potential. An excellent article by Hatcher et al. (1995) provides further examples of how this can be done. The authors compare various techniques for the elementary analysis of ceramic samples on the basis of measurements taken for certified standards.

A lively debate has sprung up over the past few years surrounding nondestructive analytical techniques. Such techniques allow analytical information to be obtained with no damage whatsoever to the sample or in some cases the object in question. All visible alterations are avoided, and the object remains aesthetically unimpaired. The ability to study works of art in this way is of enormous value when sampling is not feasible or when fragments used for analysis have to be put back in their original location at the end of the study. This is, however, not so crucial where archeological objects are concerned. Numerous fragments and leftover pieces are often available and the use of specimens for analysis is therefore less restricted. Of the non-destructive techniques of major interest, mention should be made of spectroscopic imaging (Bacci et al. 1997; Bacci and Picollo 1996; infra Chapter 12). This involves the use of wavelengths between 400 and 2400 nm—a range totally harmless for the majority of materials. These imaging techniques provide a wealth of information concerning pigments. It is also possible to see where objects have been touched up or have sustained degradation. Other nondestructive techniques use impinging X-ray or bombarding particles. The high sensitivity of today's X-ray photon detectors makes it possible to use low doses of bombarding radiation and keep exposure time to a minimum. This limits the possibility of degradation due to ionizing radiation.

High spatial resolution techniques may be termed as a compromise between destructive and nondestructive analytical techniques. They provide information for micrometer-sized areas and can be employed when dealing with extremely small specimens. High spatial resolution techniques may also be used to look at the spatial distribution of the chemical species in the fragment of material under study (Ciliberto et al. 1995; Allen et al. 1995). This is particularly important since it is thus possible to use the "physical" image of a fragment of material to select the areas of analytical relevance. Analytical techniques such as these are becoming increasingly relevant. Most of today's major museums usually have on-site equipment for high spatial resolution analyses.

1.2.3. Data Handling

The way in which the data obtained from analysis are subsequently elaborated is an extremely important point for consideration, so much so that an entire chapter of this book has been dedicated to this subject alone (see Chapter 20). The first thing to remember is that all quantitative analysis data should be interpreted in line with the degree of accuracy afforded by the technique used. This is essential if the reliability of the analytical data is to be established. Unreliable data is of no use whatsoever. However, other problems need to be addressed specifically in the field of studies involving archaeological finds and works of art. Various questions arise when one is looking at an archaeological object, such as: What kind of object is it? Where does it come from? How old is it? What was it used for? For works of art, questions run along the following lines: Who executed it? How was it executed? What may have caused it to degrade? When looking for the answers to important questions such as these, a descriptive approach to the analytical data often proves inadequate. It is thus often necessary to construct mathematical models representing homogenous samples of the analytical data. Orton's key work of the 1980s, and the numerous articles it inspired, provides further information on this subject (Orton, 1980). An increasing number of studies are now performed using a statistical approach for the interpretation of data. Terms such as principal component analysis, cluster analysis, Bayesian analysis, Mahalanobis distances are now common in the language of modern analytical chemistry—a field that requires an increasingly high degree of interpretation acumen.

1.3. FURTHER READING AND CONCLUDING REMARKS

Many authors have competently illustrated the points made above. Zvi Goffer's (1980) much acclaimed book certainly deserves being singled out. Goffer adeptly outlines not only many aspects of analytical investigations but also provides description of various materials such as construction materials, ceramics, metals, glasses, and pigments. Aspects of degradation and the conservation of ancient objects are also discussed. For readers interested in the physical methods used in archaeology and art history, Ulrich Leute's (1987) introductory book is recommended. Topics include the characterization and dating of objects and modern prospecting methods, an area of great importance in the field of archaeology. Pollard and Heron's (1996) work on archeological chemistry provides an uncommonly competent introduction to this field and is ample proof of the authors' vast experience in the sector. One of the founding fathers of the new scientific approach to archaeometry, M. J. Aitken has together with R. E. Taylor recently edited one of the most up-to-date books on dating methods in archaeology (Taylor and Aitken, 1997).

By virtue of his extensive experience in the application of science in the field of cultural heritage, J. B. Lambert has achieved great success with his introductory book on the subject (Lambert, 1997). Lambert places the accent on materials, although historical and geographical aspects are also discussed. The Glossary is extremely well researched and is particularly useful in its promotion of interdisciplinary terminology—something that has long been felt to be missing in the fields of chemistry and physics as in those of history and art history.

Chemistry symposia on the subjects treated in this book have also proved extremely useful and productive. Those organized by the History of Chemistry Division of the American Chemical Society have led to the publication of five volumes treating a variety of up-to-date topics, including biomolecular analysis. The volumes cover a period extending from 1974 to 1996 and thus allow the reader to follow the progress of techniques and methods in this area of science. Readers may, for example, trace the chronology of studies carried out for the Shroud of Turin and see that many of the questions posed about this relic remain as yet unanswered (Carter 1984; Jumper et al. 1984; Kouznetsov et al. 1996; Jull et al. 1996).

The authors of the present volume are hopeful that the variety of references provided in each chapter may serve as an up-to-date reference aid for those wishing to undertake analytical studies of archaeological artefacts or works of art. It is also sincerely hoped that the following chapters, written by experts and researchers in the fields discussed, will allow new generations of scientists to break away from the rigid constraints of excessive specialization and breathe in the fresh, new air of collaboration to work together on some of the most fascinating aspects of human culture.

Acknowledgments

I would like to acknowledge J. D. Winefordner for his encouragement in the preparation of this work and Vincenzo La Rosa for his discussions on wide-ranging archaeological topics. Scuola Archeologica Italiana di Atene is also acknowledged for photographic materials.

REFERENCES

Allen, G. C., I. T. Brown, E. Ciliberto, and G. Spoto (1995), *Eur. Mass Spectrom.* **1**, 496.

Bacci, M., and M. Picollo (1996), *Stud. Cons.* **41**, 136.

Bacci, M., S. Baronti, A. Casini, F. Lotti, M. Picollo, and S. Porcinai (1997), in *Materials Issues in Art and Archaeology*, Vol. 5, Materials Research Society Symposium Proceecings Vol. 462, P. B. Vandiver, J. R. Druzik, J. F. Merkel, and J. Stewart, Eds., Materials Research Society, Warrendale, PA, p. 39.

Carter, G. F. (1984), in *Archaeological Chemistry*, Vol. 3, J. B. Lambert, Ed., Advances in Chemistry Series 205, American Chemical Society, Washington, DC, pp. 425–446.

Ciliberto, E., I. Fragalà, N. A. Mancini, and G. Spoto (1995), *Micros. Microanal. Microstruct.* **6**, 533.

Goffer, Z. (1980), *Archaeological Chemistry*, John Wiley & Sons, New York.

Hatcher, H., M. S. Tite, and J. N. Walsh (1995), *Archaeometry* **37**(1), 83.

Jumper, E. J., A. D. Adler, J. P. Jackson, S. F. Pellicori, J. H. Heller, and J. R. Druzik (1984), in *Archaeological Chemistry*, Vol. 3, J. B. Lambert, Ed., Advances in Chemistry Series 205, American Chemical Society, Washington, DC, pp. 447–476.

Jull, A. J. T., D. J. Donahue, and P. E. Damon (1996), in *Archaeological Chemistry*, Vol. 5, M. V. Orna, Ed., Advances in Chemistry Series 625, American Chemical Society, Washington, DC, pp. 248–253.

Kouznetsov, D. A., A. A Ivanov, and P. R. Valetsky (1996), in *Archaeological Chemistry*, Vol. 5, M. V. Orna, Ed., Advances in Chemistry Series 625, American Chemical Society, Washington, DC, pp. 229–247.

Lambert, J. B. (1997), *Traces of the Past: Unraveling the Secrets of Archaeology Through Chemistry*, Addison-Wesley, Reading, MA.

Leute, U. (1987), *Archaeometry*, VCH, New York.

Mora, P., L. Mora, and P. Phillipot (1984), *Conservation of Wall Paintings*, Butterworths, London.

Militello, P. (1998), *Gli Affreschi Minoici di H. Triada*, Scuola Archeologica di Atene, Ausilio, Padova.

Orton, C. (1980), *Mathematics in Archaeology*, William Collins Sons, Glasgow.

Plenderleith, H. J. (1998), *Stud. Cons.* **43**, 129.

Pollard, A. M., and C. Heron (1996), *Archaeological Chemistry*, Royal Society of Chemistry, Cambridge.

Sheare, J. C., D. C. Peters, G. Hoepfner, and T. Newton (1983), *Anal. Chem.* **55**, 875A.

Taylor, R. E., and M. J. Aitken (1997), *Chronometric Dating in Archaeology*, Plenum, New York.

Tite, M. S., M. Bimson, and M. R. Cowellin (1984), *Archaeological Chemistry*, Vol. 3, J. B. Lambert, Ed., Advances in Chemical Series 205, American Chemical Society, Washington, DC, pp. 215–242.

Tsang, J., and R. H. Cunningham (1991), *J. Am. Inst. Cons.* **30**(2), 163.

Vinaz, S. M. (1998), *Stud. Cons.* **43**, 114.

CHAPTER

2

ELEMENTAL ANALYSIS

RONALD G. V. HANCOCK*

SLOWPOKE Reactor Facility and Department of Engineering and Applied Chemistry, University of Toronto, Toronto, Ontario, Canada

2.1. INTRODUCTION

Elemental analysis serves many disciplines, helping answer questions related to materials, old and new. Sometimes elemental analysis works alone and sometimes it is best employed in conjunction with physical analysis techniques. As such, elemental analysis, and each specific elemental analytical technique, is like a tool in a toolbox, designed for a specific purpose or set of purposes, but with its eventual usefulness based on the imagination and skill of its user.

Much of what is written below applies more to materials from the world of archaeology rather than from the world of art history. But, since many archaeological artifacts are works of art, if not art works, many thoughts on the elemental analysis of materials from archaeology should be translatable to the materials of art history.

The primary function of elemental analysis in archaeology and art is the characterization of ancient and historic materials. By measuring the concentrations of a number of elements in a sample of a particular material, the analyst may establish an elemental fingerprint for the sample, thereby identifying the sample. Using elemental fingerprints, numbers of samples may be compared to establish similarities or differences. The quality of each elemental fingerprint depends on the number of elements chosen and on the natures of the elements themselves.

Using elemental fingerprints, researchers may begin to answer simple questions about the material under study: What was it made of? Is this

* SLOWPOKE-2 Facility, Department of Chemistry and Chemical Engineering, Royal Military College, Kingston, Ontario, Canada

Modern Analytical Methods in Art and Archaeology, Edited by Enrico Ciliberto and Giuseppe Spoto. Chemical Analysis Series, Vol. 155
ISBN 0-471-29361-X

sample made of the same material as that sample? Where was it made? How was it made? When was it made? Once researchers can answer these sorts of questions, they may then progress to the larger questions of trade and other interactions of peoples.

The rationale of comparing fingerprints appears to be clear. Samples that physically look the same (nonchemical evidence) and have identical elemental fingerprints were made from the same source material(s). Samples that physically look the same and have similar elemental fingerprints were probably made from the same source material(s). Samples that physically look the same but have different elemental fingerprints were probably made from different source material(s). Samples that look physically different but have similar elemental fingerprints were probably made from similar or the same source material(s). Note that in the last case, because of my personal biases, the chemistry tends to override the physical descriptors. Note also that in the last three cases, the level of certainty is variable. This follows the invaluable warning by Harbottle (1982) that unequivocal sourcing of most objects is often impossible.

2.2. DESCRIPTIONS

An elemental fingerprinting operation is somewhat akin to forming a description of a person. If appropriate numbers and types of parameters are chosen for quantification or description, the descriptive system should be valuable. If, however, incorrect numbers and/or types of parameters are chosen for quantification or description, the descriptive system may range in value from adequate to useless.

For example, to uniquely describe a person we need to employ a number of different parameters. Some parameters may involve qualitative descriptions while others may involve semiquantitative and quantitative measurements. Of the latter, different length measurements help define the relative proportions of the person: total body length (height); arm span; leg length; size of head; shoulder width; chest, waist, and hip circumferences. Add to these the body mass (weight), finger and footprints, dental records, and DNA data and one has a wonderful scientific description of the naked individual that is great for a coroner. But this information may or may not be helpful to someone waiting to meet the described individual at an international airport. In this case, if the person to be identified is part of a confusing mass of similarly sized bodies, other parameters are needed for the accurate differentiation of the individual from a large crowd of people.

For our normal, visual, people identification system, we use some of the above parameters (mainly height and mass) but also add other characteristics such as age, gender, skin coloring, hair coloring and styling, along with identifying clothing, glasses or lack thereof, and other accessories. This more general type of physical description may be fine for finding someone in a crowd but does not tell what the person is like. For this latter type of description we must seek other parameters such as cultural tastes and habits, political leanings, sense of humor, and self-confidence.

The same sorts of levels of descriptive power are available in elemental analyses. Interestingly, but understandably, we are collectively better at describing other people than we are at generating or interpreting diagnostic elemental fingerprints.

2.3. DIAGNOSTIC ELEMENTAL FINGERPRINTS

Three factors are of primary importance in creating diagnostic elemental fingerprints: the number of elements analyzed, the types of elements analyzed, and the quality of each elemental analysis.

2.3.1. Number of Elements

The selection of appropriate elements to be analyzed may be vital to the answering of specific questions. But, more often than not, analysts have access only to a single particular analytical technique (or, perhaps, to several), and the technique restricts the maximum number of elements that may be quantified in a particular material (e.g., Hughes et al. 1976).

Analysis of a single element is not usually going to be enough to identify a sample, although in some special cases, such as the identification of metal scraps or of pigments in paintings, one element might be adequate, at least on the simple "what is it?" level. On average, though, the larger the number of elements analyzed, the better the individual sample is described, and the clearer the fingerprint should become. But it is not quite as simple as this since the types of elements analyzed are also important.

2.3.2. Types of Elements

The types of elements chosen to establish an elemental fingerprint will depend on the nature of the material being tested. For ceramics, made from natural sediments, and for lithics, the primary anions are silicates, carbonates, and sulfates, with smatterings of hydroxides, halides, oxides, sulfides, and perhaps oxy-anions of heavier elements. So, very low concentrations of elements in the bulk matrix may be associated with trace concentrations of each element in different major mineral phases and/or with low concentrations of specific element-rich minerals, mixed within the matrix material. For example, chromium tends to be present in low (10s to 100s of ppm) in many minerals (and sediments) but on occasion is found as tiny particles of chromite, or other chromium-rich minerals, scattered within a sediment. Also, gold is notorious for its presence as tiny elemental nuggets.

Researchers talk of ceramics being made from sediments (or clays) and often assume that all sediments at a specific location are homogeneous. In many cases, this is a reasonably valid assumption, especially in regions where large river systems provide good sediment mixing. But it is known from physical geography studies that sediments from different horizons at a specific location may be mineralogically either homogeneous or heterogeneous (e.g., Mahaney and Hancock 1991), and hence their chemistries will be similar or variable. For ceramics and lithics, the diagnostic elements include the major elements, along with minor and trace elements that are primarily associated with silicates and carbonates (e.g., Volterra and Hancock 1994).

Metals are a whole different story. Depending on the minerals used in their production (e.g., sulfide ores or oxide ores), the types of elements that will be included in the final product metal, either as alloying trace elements or trapped in inclusions, will depend on the ores' source(s) and on the exact nature of the metal production process. For example, seventeenth-century smelted European copper tends to attract sulfide-forming elements such as gold, silver, arsenic, antimony, indium, selenium, cobalt, and nickel (Hancock et al. 1991), while zinc tends to be clean except for low levels of cadmium. On the other hand, iron tends to include traces of other transition metals such as cobalt, manganese, vanadium, and chromium, while lead retains traces of the sulfide-forming elements found in copper.

Since many metallic elements selectively form compounds with different anions, some minor to trace elements may be useful discriminators in a wide range of materials, while others are specifically useful to a limited range of materials. For example, gold is often useless in ceramic analysis because of the nugget effect, may be sometimes useful in the separations of lithics, but appears to be vital in the seventeenth-century world of European smelted copper.

Hence, a knowledge of the concentrations of the halogen elements (e.g., chlorine, bromine, and iodine) in a sample may represent one aspect of the sample description, while the alkali (sodium, potassium, rubidium, and cesium) and alkali–earth (magnesium, calcium, strontium, and barium) element concentrations represent two more aspects, transition element contents represent another aspect, and the lanthanide concentrations yet another, with other groups of elements providing more information. Any of these elemental subgroups may provide diagnostic information that may be useful in answering the question(s) posed about the material under study, but in some cases some of the elemental information may be irrelevant.

Analysis of a wise and representative mixture of these different types of elements is therefore desirable and necessary to form an elemental fingerprint that is as discriminating as possible. The aim of analysts is not to under- or overdescribe a sample. In practice, this is often difficult, especially with materials that are unfamiliar.

Inorganic pigments form a special case since they usually need only major element analyses, and there is usually prior knowledge of the diagnostic elements. Minor and trace elements tend to become important only if the geological sources of the pigments need to be established. Analysts must be careful in advocating the use of elemental analyses to solve all problems. In many cases of pigment identification, mineralogical analyses are equally, if not more, useful than elemental analyses (e.g., Laver 1997; Wainwright et al. 1986). So, being aware of the most appropriate physical or chemical analytical approach to be applied to a particular problem is the crux of a successful analysis.

2.3.3. Environmental Storage Effects

An added point of interest in analyzing archaeological or historical materials, rather than newly made materials, is the possibility of changes generated by time and storage environment. Depending on the natures of the materials under study, elements may be leached out or new minerals may even crystallize in pores and cavities within artifacts. For example, archaeological bone is a wonderful host to many unusual elements and minerals (e.g., El-Kammar et al. 1989).

2.4. CHAIN OF EVENTS FOR A SUCCESSFUL ANALYSIS

Analysts tend to assume that their elemental fingerprints give them information about the individual objects just after (or even before) their time of production. But, in reality, the elemental analysis is only part of a chain of processes to which the object has been subjected, each link of which must be considered before the elemental analysis can be assured of successfully characterizing an archaeological artifact.

2.4.1. Production

The chain starts with the production of each object. Knowledge of the effects of the production process is especially important for processed materials such as ceramics, metals, and glass, primarily because the act of producing these objects may alter some elemental concentrations found in the raw materials. If the questions being asked relate to the sources of processed raw materials, analysis of artifacts that are the final products of these raw materials may be a waste of time.

2.4.2. Use

The use to which an object is put may also effect its elemental composition. While funerary items may be expected to represent the "as-new" object, household objects may have been subjected to the vagaries of human use, including repeated washing, leaching by stored liquids, and even the effects of weather.

2.4.3. Long-Term Storage

Once the object is discarded and is eventually interred in the ground, environmental storage effects may produce significant changes in elemental compositions, especially the near-surface mineralogical and hence elemental compositions of an artifact. With leaching and redeposition effects, trace elemental contents of porous materials such as bones may bear no resemblance to what they were at the time of death of their host.

2.4.4. Recovery

The next insult to which the object, now an artifact, may be subjected may occur during its recovery. Acidic and other cleaning processes may not only clean the surfaces of an artifact but may penetrate the body, altering the elemental chemistry by dissolving more soluble components.

2.4.5. Elemental Analyses

Depending on availability, a large number of different analytical techniques may be applied to a particular problem. Many of these techniques have been thoroughly reviewed by Pollard and Heron (1996: Chapter 2). Usually, though, analysts have access to a restricted number of techniques and must address archaeological questions and materials with this limitation in mind, as displayed in the following chapters.

Analytical techniques for elemental analysis tend to fall into two categories: surface or near-surface elemental analysis techniques and bulk elemental analysis techniques. Different techniques are appropriate for addressing different questions, with no single technique being universal. Each technique has advantages and disadvantages relative to the other techniques and widely varying sensitivities for the detection of specific elements in different materials. No matter the technique, they all have imposing or impossible titles that are usually abbreviated into acronyms.

Surface or near-surface elemental analysis techniques are best for pigment analysis, corrosion analysis, and even for bulk elemental analysis if the material is homogeneous. Analytical techniques in this general category include total reflection X-ray fluorescence spectrometry (TRXRF; see Chapter 4), particle-induced X-ray emission spectrometry (PIXE; see Chapter 6), and electron microprobe analysis (EMPA).

Bulk elemental analysis techniques may be subdivided into those that require the analytical sample to be in solution and those that can deal with solid samples. Dissolution techniques (see Chapter 3) include atomic absorption spectroscopy (AAS or AA), optical emission spectroscopy (OES), inductively coupled plasma–atomic emission spectrometry (ICP–AES), inductively coupled plasma–mass spectrometry (ICP–MS), and X-ray fluorescence spectrometry (XRF) if solid glassy solutions are included. A primary potential problem with dissolution techniques is that the analyst needs to be extremely careful to dissolve all of the sample, since the analytical equipment analyses only what is in solution.

Techniques that may be used to analyze solid samples include X-ray fluorescence spectrometry (XRF), inductively coupled plasma–mass spectrometry with laser ablation (ICP–MS–LA; see Chapter 3), and instrumental neutron activation analysis (INAA or NAA; see Chapter 5). ICP–MS–LA is a developing technique that as of 1998 has not proven reliable on a wide range of elements in broadly differing materials, although its proponents have great expectations for it in the near future.

Many analytical techniques are matrix sensitive so analytical standards tend to be made from materials of similar composition that are processed in similar fashions. The expertise of the analyst in choosing standards is vital in avoiding the later pitfalls of overinterpreting the multielemental analytical data generated from specific sets of samples.

Hardly any elemental analytical techniques are truly nondestructive. With care, the surface analysis techniques produce minimal damage. But of the bulk analysis techniques, only INAA, using elements that produce short-lived radioisotopes, is nondestructive and then only for small samples.

Analytical problems may occur with a number of poor choices, most of which are caused by analytical/material inexperience and/or a lack of adequate analytical funding. "Fast and exact" is an analytical aim that is not usually achieved.

The first potentially poor choice can occur when the appropriate sample size to analyze is not selected. This is not usually important for homogeneous materials but is often a source of confusion with more heterogeneous (e.g., Bromund et al. 1976) or with more heavily weathered materials. The next poor choice comes when the most appropriate analytical procedure to employ (see above) is not selected. Then comes the actual elemental analysis with all of its hidden subtleties.

The use of in-house or international analytical standards, in association with interlaboratory comparisons, tends to ensure analytical accuracy as long as analytical procedures are followed faithfully and competently.

Preanalysis sample preparation may significantly affect the elemental fingerprints obtained. Some analysts prefire ceramics to a high temperature prior to analysis. This may severely alter the absolute value of the element concentrations obtained, especially for highly calcareous ceramics that were previously low fired and where, after manufacture, much of the calcium remains in the form of carbonate. Thus, the analytical heating process may obscure potentially valuable secondary information associated with tempering materials or with environmental storage effects.

Some analytical procedures require that each sample be analyzed in solution. Materials that are not readily dissolved (or, what is worse in the case of heterogeneous materials, some components of the sample that are not readily dissolved) may create anomalous elemental fingerprints, immaterial of the quality of the analyses that follow the sample dissolution step.

2.4.6. Data Interpretation

Given that the problems of question asking, artifact selection, subsampling, and analysis have all been competently addressed, and a collection of analytical data has been assembled, the final hurdle is to unravel the material and archaeological interpretation of the data. There is a range of approaches that may be employed for this task (see Chapter 20 for an in-depth discussion). The simplest first step is to arrange the data in such a way that it can be visually inspected. Even with quite large data collections, this can be a valuable procedure, especially when students are involved. If truly anomalous data are identified, the primary analyst can be asked to assess each anomaly, and the overall reliability of the data set can be established. This approach deals with trivial analytical errors that all analysts deny ever happened to them once the final report or publication is ready to print.

Sometimes, during this visual inspection of the data, the differences among elemental fingerprints become apparent and the task of data interpretation becomes easy. But, for very large data sets, or for intractable data sets that do not separate readily into their various component groupings, much more sophisticated computer-based data analysis programs must be brought to bear on the interpretive process (e.g., Bishop and Neff 1989; and see Chapter 20).

Whatever approach is chosen, the weaknesses and strengths of the chosen approach should be considered so that researchers err on the side of underinterpretation rather than overinterpretation. False associations of samples with established groupings of samples, and false rejections of samples from established groupings of samples (producing outliers) are not creditworthy achievements.

In cases where long-distance trade is not an issue, and given that the data are trustworthy, the finding of sample outliers by the application of a particular approach often points to the failure of the user of the approach to use it appropriately, rather than to the finding of unusual samples. The old adage that "a bad workman blames his tools" may very well translate into "an inexpert data interpreter finds outliers."

In all cases, the final interpretation should always make sense in terms of the specific material under investigation. That is to say, the underlying chemistry (and perhaps geochemistry) must be sensible.

The final quirk that may cause embarrassment in data interpretation comes from the use of published data from the literature, without realizing that the data may contain typographical errors that were missed before final publication. This may be one of the reasons why database compilers and keepers range from being very careful and concerned to being neurotic and obsessive about the validity of their data.

2.5. CONCLUDING REMARK

Analysts must always keep in mind that while the artifacts that they study are real, their analyses, elemental fingerprints, and interpretations all hinge on potentially biased descriptions of the artifacts, and so may or may not reflect the reality of the artifacts, depending on their skills, intuition, and good luck.

REFERENCES

Bishop, R. L., and H. Neff (1989), in *Archaeological Chemistry*, Vol. 4, R. O. Allen, Ed., Advances in Chemistry Series 220, American Chemical Society, Washington, DC, p. 57.

Bromund, R. H., N. W. Bower, and R. H. Smith (1976), *Archaeometry* **18**, 218.

El-Kammar, A., R. G. V. Hancock, and R. O. Allen (1989), in *Archaeological Chemistry*, Vol. 4, R. O. Allen, Ed., Advances in Chemistry Series 220, American Chemical Society, Washington, DC, p. 333.

Hancock, R. G. V., L. A. Pavlish, R. M. Farquhar, R. Salloum, W. A. Fox, and G. C. Wilson (1991), *Archaeometry*, **33**, 69.

Harbottle, G. (1982), in *Contexts for Prehistoric Exchange*, J. E. Ericson and T. K. Earle, Eds., Academic, New York, p. 13.

Hughes, M. J., M. R. Cowell, and P. T. Craddock (1976), *Archaeometry*, **18**, 19.

Laver M. E. (1997), in *Artist's Pigments: A Handbook of Their History and Characteristics*, Vol. 3, E. W. FitzHugh, Ed., National Gallery of Art, Washington, DC, p. 295.

Mahaney, W. C., and R. G. V. Hancock (1991), *J. Radioanal. Nucl. Chem. Articles* **148**, 81.

Pollard, A. M., and C. Heron (1996), *Archaeological Chemistry*, Royal Society of Chemistry, Cambridge, p. 375.

Volterra, V., and R. G. V. Hancock (1994), *J. Radioanal. Nucl. Chem. Articles* **180**, 37.

Wainwright, I. N. M., J. M. Taylor, and R. D. Harley (1986), in *Artist's Pigments: A Handbook of Their History and Characteristics*, Vol. 1, R. Feller, Ed., National Gallery of Art, Washington, DC, p. 219.

CHAPTER

3

ATOMIC SPECTROSCOPY AND SPECTROMETRY

SUZANNE M. M. YOUNG

Archaeometry Laboratories, Harvard University, Peabody Museum, Cambridge, Massachusetts, U.S.A.

A. MARK POLLARD

Department of Archaeological Science, University of Bradford Bradford, West Yorkshire, United Kingdom

3.1. INTRODUCTION

Compositional data are an important adjunct to the study of ancient extractive and fabrication technology and can be a component of trade studies. For the most part in archaeological sciences, standard instrumentation is used—originally optical emission spectroscopy (OES) and neutron activation analysis (NAA), followed by atomic absorption spectrometry (AAS), X-ray fluorescence spectrometry (XRF), and electron microprobe (EMP) as these techniques were developed. More recently, these have gradually been replaced by inductively coupled plasma–atomic emission spectroscopy (ICP–AES), inductively coupled plasma–mass spectrometry (ICP–MS), and other forms of mass spectroscopy. After a discussion of sampling techniques (Section 3.2), this chapter focuses on AAS (3.4), ICP–AES (3.5), and ICP–MS (3.6). The final section (3.7) discusses what is perhaps the greatest challenge—in the light of increasing analytical sophistication, how should projects be designed that integrate analytical information with other data sources in an archaeologically meaningful manner.

3.1.1. Elemental Analysis in Archaeology

Instrumental methods of compositional analysis have been used in the study of antiquities since the 1930s. Optical emission spectroscopic analysis of

---

Modern Analytical Methods in Art and Archaeology, Edited by Enrico Ciliberto and Giuseppe Spoto. Chemical Analysis Series, Vol. 155
ISBN 0-471-29361-X

Early Bronze Age museum collections was used to group metal artifacts, assuming that similarity in minor and trace element impurity patterns in artifacts reflected common centers of production that (conveniently!) used consistent sources of raw materials and processes. This approach developed into large-scale analytical programs by the 1950s. Realizing the naiveté of some of these assumptions, work relating the composition of metal artifacts to their parent ore bodies also commenced in the 1950s. Although this interpretation of chemical composition is highly problematic for several reasons (including the thermodynamics of the processes and the questions of mixing and recycling), this work continues today and has been extended to a whole range of other inorganic materials, particularly ceramics and glass. Although only small samples are required for most instrumental methods of analysis, the prospect of even limited destruction of artifacts is understandably a problem for most curators and places a premium on limited or nondestructive sampling wherever possible.

Using a solid spark source and photographic recording of the emission spectrum, OES was the standard method of analysis for archaeological pottery, obsidian, faience, and metals almost up the 1980s. It was replaced by AAS, but only gradually because the need for liquid samples required a more complicated sampling process, sometimes involving aggressive dissolution techniques and difficulties in keeping some elements in solution. Moreover, only one element is measured at a time, giving rise to the problem that "unexpected" elements might be missed. The 1980s saw an explosion in the range of analytical techniques available to archaeometrists, and the pace of instrumental development was very rapid. Old techniques were upgraded, and numerous new ones, such as ICP–AES and ICP–MS, became commercially available. With the ability to measure up to 20 elements (effectively) simultaneously, ICP–AES replaced AAS for samples that can easily be brought into solution and became the technique of choice for large-scale analytical programs. The transition was again gradual, however, because the instruments are substantially more expensive than AAS and require more skill to operate and maintain.

ICP–MS has now earned widespread acceptance. It can analyze more elements simultaneously and often at greater sensitivity than the typical ICP–AES used by archaeometrists (not many have access to the new high-sensitivity ICP–AES). In addition to compositional analysis, ICP–MS allows analysts to monitor concentrations of individual isotopes or measure isotope ratios. The precision, although currently lower than that of thermal ionization mass spectrometry (TIMS) or the new ICP–magnetic sector–multicollector MS (ICP–MS–MS), can be sufficient to address questions of archaeological importance. It is certainly a much more rapid and inexpensive technique than TIMS and is likely to replace it as a routine tool for isotopic ratio determination in the next few years. For compositional analysis in archaeology, ICP–MS and ICP–AES now complement each other and together dominate the field.

Philosophical changes in the 1980s attempted to move the discipline of archaeometry toward the twin goals of continuing to carrying out serious scientific work underpinned by adequate theoretical structures, but with increased relevance to archaeology. Unfortunately, the blind purchase of scientific services by archaeologists continues to happen, perhaps on the mistaken basis that scientific appendices improve the value of the work. Choice of technique is often based upon access to instruments and cost rather than the suitability of the resulting data to solve the formulated question. Archaeological science must still strive to get the scientific data used appropriately and the implications understood by scholars in the broader field. Now, with so many options available, analytical decisions increasingly should be based on what kind of data is needed to answer the specific question, although this decision is still constrained by restricted access to unique material.

3.2. SAMPLES FROM ARCHAEOLOGY AND ART

Although the analytical instrumentation used is normally standard, archaeometrists must always consider the unique nature of archaeological objects. There is always a necessity to minimize or simplify sample preparation. More stringently, there is often a requirement to leave artifacts intact. Techniques such as ICP–MS, which require samples of less than a few milligrams (mg), are by far the most attractive to archaeologists, art historians, and curators. This constraint has to be balanced against the likelihood of elemental inhomogeneity within the sample.

ICP–MS is now attracting considerable attention as more methods of sample introduction for solid-state analysis, such as slurry nebulization and laser ablation, are introduced. Laser ablation–ICP–MS (LA–ICP–MS), with its potential for spatially resolved microanalysis similar to EMP but with much greater sensitivity and potential for isotopic analysis, is extremely attractive to archaeological scientists. In situations where artifacts are small enough to fit into the sample chamber, no sample need be extracted at all, and the only damage to the object is a crater of a few tens of micrometers where the laser has “burned out” the material for analysis. Although this will undoubtedly be the preferred choice in the future, the following sections assume that a sample is to be taken from the object, either for solid-state analysis or for dissolution, since this is currently the most common situation.

3.2.1. Sampling

Artifacts tend to be inhomogenous and corroded. They may have changes caused by corrosion during burial or artificial patination, inclusions from unreacted or residual raw materials, inhomogeneity from manufacturing techniques, or glazed surfaces. Sometimes it is this very heterogeneity that is the subject of the analytical investigation. Often, however, it is an overall "bulk" chemical composition that is required, and this is the most common archaeological application of spectrometry. In such cases it is necessary to select a representative sample for analysis, avoiding areas of corrosion, plating, glazing, or excessive material segregation. The most obvious area to avoid is, of course, the corroded surface, since this is palpably unrepresentative.

Before samples are taken, a small area of the artifact surface should be cleaned (perhaps using an abrasive wheel) to expose unaltered material and then a quantity of this is removed using a small drill, usually of hardened steel. Samples of less than 50 mg can be analyzed by any of the techniques described here, but sometimes larger samples are needed simply to give a more representative sample. The most sensitive ICP instruments can actually analyze samples of about 1 mg—as small as can be handled preparatively—but there is a severe risk of being misled by inhomogeneity. The sample is then inspected under a microscope or powerful hand lens to make sure that it is free from visible contamination. Some samples should be bathed in acid to ensure that there is no contamination from the sampling, with the choice and concentration of acid determined by the nature and size of the sample. It is usually better, however, if possible, to cut a small section for both microscopic examination and chemical analysis. This will always give more information than taking a drilling, but drilled samples minimize the visual impact of the process. There is also a greater risk of contamination than with cut samples.

3.2.2. Preparation of Sample Solution

Small, contamination-free samples must be extracted and completely dissolved in acid, preferably nitric acid (HNO_3). However, not all samples can be dissolved or kept in this solution, and sometimes the use of hydrochloric acid (HCl) for some metals, bone, or paint samples, or hydrofluoric acid (HF), for mineral, ceramic, or glass samples, is necessary. There can be problems with the use of these acids. In general, when HF is used, samples should be dried down and the residue dissolved in other acids. Some archaeomaterials are difficult to dissolve and careful choices must be made in the selection of preparation method. There are as many "optimal" preparations as there are materials and analysts—it is always best to try a series of preparations to arrive at the one that most adequately meets the need.

Metals are generally the most simple to dissolve in acid. Iron, iron alloys, copper, and copper alloys (including Cu–As, Cu–Sn, and Cu–Sn tertiary and quaternary alloys) are usually dissolved in an excess of concentrated HNO_3. This procedure allows most of the metals present to be taken into solution. Tin-rich alloys, lead and high-lead alloys, pewter, gold, and silver need to be dissolved in a mixture of concentrated acids (25 or 50% v/v HNO_3 in HCl). Where copper- or iron-rich alloys contain low, but significant, levels of tin, lead, or precious metals it is not always clear which acid is best used for the initial dissolution. In general, such elements in trace concentrations will dissolve in concentrated HNO_3 acid. However, even where these elements are present at the 1% concentration level, it may be necessary to use combinations of acids for copper- and iron-based artifacts. Samples containing significant amounts (above about 1%) of Au or Sn should remain in concentrated acid for 48 hr prior to analysis to allow complete dissolution. It should, however, be remembered that the use of combined HNO_3 and HCl is not always appropriate. While chlorine ions help to dissolve some elements, they interfere with others in ICP–AES and –MS analysis—for example, chlorine ions cause Ni to be removed from the sample stripper cones in an irregular manner that cannot be corrected.

Where samples contain tin there is a danger that it may precipitate from solution before analysis. This tendency is lessened by adding a small quantity of HF to the solution, but this introduces additional complications in that Teflon labware must be used to avoid the leaching of glassware. Furthermore, the amount of HF must be kept below 0.1% in the analyte to avoid etching any quartz components in the instrument. In practice, especially when samples containing more than a few percent of tin are to be prepared, it is better to dilute sample solutions to final analyte no more than a day or two prior to analysis. Where samples are of gold matrix and the need is to analyze the platinum group elements (PGEs), glass labware or Teflon must be used for all stages of preparation and analysis since PGEs absorb onto the walls of the standard polymer bottles at a rate as high as 70% loss in 2 hr from standard 5% acid solutions.

Metal slags, glasses, minerals, ceramics, and porcelain are generally much more difficult to prepare, and require more aggressive techniques. They may be dissolved using elevated temperature and pressure digestions in concentrated acids or by using greater amounts of HF. This can be achieved by heating tightly sealed vessels of Teflon or bombs containing the samples and acids. In this case the HF must then be evaporated off before taking the sample up into a solution of HNO_3 for analysis. Samples of marble or bone

Table 3.1. Methods of Elemental Analysis

Technique	Sensitivity	Accuracy
AAS—flame	1–10 ppm	1–5%
AAS—graphite furnace	1–10 ppb	1–5%
OES/AES	100 ppm	7–10%
ICP–AES	0.5–100 ppb	3–5%
ICP–MS	50–1000 ppt	0.1–1.0%
LA ICP MS	1–100 ppm	5–10%

usually dissolve better with the use of organic acids in conjunction with a strong acid such as HNO_3.

All sample solutions must be diluted to give concentrations within the optimal detection range of the instrument (see Table 3.1). It is always best to dilute just before analysis to minimize the possibility of selected elements "falling out" of solution as the pH rises. Internal spikes can be added to each sample solution to correct for drift and flame or beam fluctuations during measurement. This internal standard must be carefully selected because, while it is desirable to have a spike as close as possible to the element of interest, it obviously precludes the measurement of the spike element. When the composition of the sample is truly unknown, a quick semiquantitative analysis should be carried out to select the best sample dilutions and internal standards.

Acid blanks—samples prepared in exactly the same way as the unknowns, but without sample—should always be analyzed and if necessary subtracted from sample solutions, especially for quantitative techniques. They also help correct or at least monitor some of the matrix effects caused by being in dilute solutions. Whenever possible, the best standards to use are those in the same matrix as the samples that can then undergo identical procedures. Matrix is the term for the major component of the sample—for example, silica in glass or calcium phosphate in bone. Matrix effects plague all techniques and can only be minimized by removing the matrix elements using column or electroplating chemistry or by analyzing known materials with the same matrix. Further details for particular techniques are discussed in the following sections.

3.3. IDENTIFICATION AND QUANTIFICATION TECHNIQUES

The techniques discussed here—AAS, OES, ICP–AES, and ICP–MS—all involve the study of atomic species rather than molecules or ions. For each technique, atomization, the freeing of atoms from molecules in which they are bound, must occur. This dissociation is achieved by volatilizing liquids or gases or ablating solid samples. Atomization of the released fragments is completed using heat energy. The conversion efficiency of sample to atomic vapor depends on temperature—too high and the atoms will ionize; too low and some elements will not atomize. Ions absorb radiation at different wavelengths than atoms of the same element and are therefore not "seen" at atomic wavelengths. However, high temperatures also minimize matrix effects. Some of the newer techniques actually use ionized samples. The right balance must therefore be found for the required elements and the machine being used. All the techniques described here can achieve identification of elements present in the sample, but not all can be adequately quantified for every element.

3.4. ATOMIC ABSORPTION SPECTROSCOPY

3.4.1. Theory

Radiant electromagnetic energy can be absorbed by atoms causing electrons to move up to orbitals of higher energy. This absorption occurs at characteristic frequencies because different elements have different energy levels and therefore characteristic energy differences between their orbitals. Each element has a unique set of these resonance frequencies, corresponding to each possible electron transition. Light of these specific frequencies is therefore diminished in power as it passes through an atomic vapor because of selective absorption. In atomic absorption spectroscopy, the sample is irradiated with light of a particular frequency (corresponding to a resonance frequency for a particular element), the loss of power is measured, and this is used to quantify the amount of that element in the sample. However, only one, or at best a few, element(s) can be measured at the same time since each requires a different frequency.

Absorption of radiant energy by matter is quantified by Beer's law, which states that the diminution of power is proportional to the number of absorbing atoms in the path of a beam of radiation of a single wavelength. The more radiation absorbed, the higher is the concentration of the absorbing material. Mathematically,

$$A = abC$$

where A is absorbance; a is absorptivity (a proportionality constant that is different for each element) in liters per gram-centimeters; b is the optical path length through which the radiation passes in centimeters; and C is molar concentration. A plot of absorbance versus concentration is linear with slope equal to ab. Absorbance is best measured as the following ratio:

$$A = \log(P_{\text{blank}}/P_{\text{sample}})$$

where P_{blank} is the radiant power of unabsorbed (transmitted) radiation emerging from a blank solution, and P_{sample} is the radiant power of transmitted radiation emerging from a sample solution.

Usually the electron transition between ground state and the first excited state is the line with the strongest absorptivity, and this therefore is usually the line of preference for quantitative analysis. Beer's law assumes that the concentration of analyte atoms in the sample (C) is constant. In AAS, the sample is in the form of a liquid introduced into a flame at some flow rate and atomized. In order for these laws to work, the sample flow rate must be uniform and the liquid sample of constant concentration. Fluctuations in flow rate are a source of deviation from Beer's law and will lead to analytical inaccuracy. Other sources of deviation are discussed below.

3.4.2. Instrumentation

Figure 3.1 shows a schematic diagram of a typical AAS. A sample containing the element(s) to be determined is dissolved into an aqueous solution (see

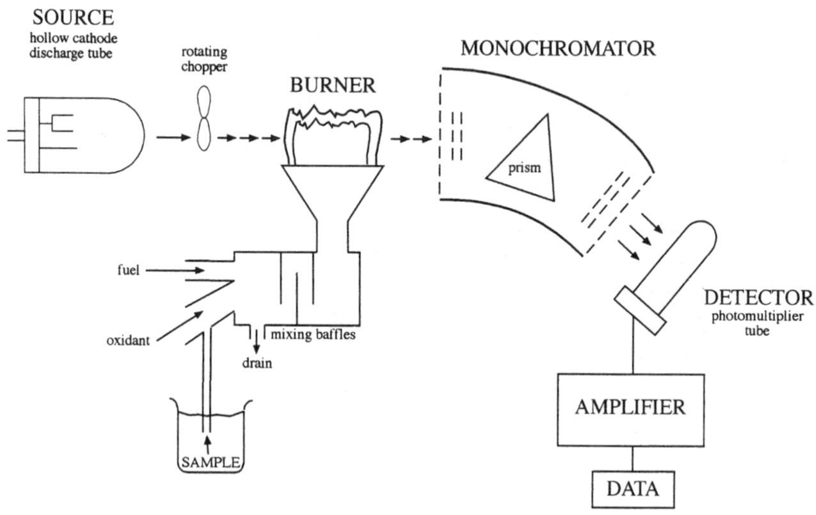

Figure 3.1. Schematic diagram of Atomic Absorption Spectroscopy

above). The sample solution must be dissociated to yield metal atoms at the point where absorbance is measured. There are a variety of methods available. In a typical flame most atoms are in the ground state, making the use of a flame to atomize samples very attractive and the most common method in AAS. The sample solution is introduced to the flame as an aerosol (very tiny droplets). These desolvate in the flame into tiny specks of solid matter that then atomize. These particles are carried through the flame at very high velocity, so not all of the sample is atomized, but the flow rates can be adjusted so that enough atomic material is present for the measurements to be made.

There are two types of burners for instruments using flames: the premixed nebulizer burner and the total consumption integral aspirator burner. In the former, an oxidant gas flows past a capillary tip drawing sample solution up from a container into a chamber. The speed of this process automatically causes the sample solution to fragment into tiny droplets. Fuel gas also flows into this same chamber. The stream of gas and liquid aerosol passes through an appropriate filter system to remove the large droplets that would fail to atomize. A "fishtail" burner is then used to give a long flat flame. The light from the light source passes through the long axis of the flame, giving the maximum optical path (usually 5 to 10 cm) for absorption; the longer the path, the greater the instrument sensitivity. Gas pressure, fuel flow rate, and oxidant flow rates all regulate the flow of the sample through the burner, essential for reproducible analysis of standard and sample. Typically, less than 1% of the sample is actually atomized and in the optical path for measurement. The flame temperature is dependent upon the choice of fuel and oxidant. For example, coal gas and air produce a flame of 1800°C, natural gas and air produce a flame of 1700°C, acetylene and air produce a flame of 2200°C, and acetylene and nitrous oxide produce a flame of 3000°C. The flame temperature must be constant for maximum reproducibility.

The total consumption integral aspirator burner has a constricted tip through which the sample solution is drawn by air instead of filtering out the big droplets. The sheer force of the moving air disperses the droplets. This combines with the fuel and goes straight into the flame. This method consumes much less sample solution, but the sample flow tends to be much less uniform and the path length through the flame is shorter. This drastically reduces both sensitivity and reproducibility and so is much less commonly used.

There is a flameless way to provide an atomized sample using an electrically heated graphite furnace. A few milliliters of sample solution are put on a tantalum ribbon, graphite rod, or tiny graphite crucible. The solution is evaporated under an infrared (IR) lamp outside the furnace or with a current inside the furnace. Temperature is increased rapidly (within a few seconds) and in a controlled manner to 2000 to 3000°C to form a cloud of atomic vapor. This vapor enters a stream of Ar gas to both prevent oxidation of furnace components and prevent formation of oxides from the sample. This gas stream is swept through a graphite tube, through which the beam of radiation travels. Such furnaces are low in background noise and have higher sensitivity than flame methods because a larger fraction of the sample is atomized. However, matrix effects can be much more severe with a furnace, and analytical precision is often worse (5 to 10%). Some elements are difficult to measure this way, for example, volatile hydrides of As, Ge, Sb, Se, Sn, and Te can form, requiring a more complicated analytical protocol. In general, graphite furnaces are much more cumbersome to use than flame systems.

The most common source of radiation used in AAS is a hollow cathode discharge tube, consisting of an anode and hollow cylindrical cathode in inert gas (Ne or Ar if Ne emits close to the resonance line of analysis) at low pressure. The tube is sealed with a quartz window through which the beam is transmitted. The energy spectrum emitted by the tube is entirely dependent upon the material from which the cathode is made. A voltage as applied to the lamp (usually up to 300 V) and atoms of gas are ionized in the resulting electronic discharge. These ions (Ne^+ or Ar^+) accelerate toward the negative cathode and collide with it dislodging atoms of the cathode metal—a process known as sputtering. The dislodged metal atoms become excited through further collisions with the gaseous ions. These excited metal atoms decay back to their ground state once they reach cooler regions inside the cathode tube, emitting a wavelength characteristic of the cathode metal. The dependency of the output spectrum on the cathodic material is why only one metal can be analyzed per analysis. AAS systems require different lamps for each element, which must be changed each time a new element is to be measured. The larger AAS instruments have several different lamps on a rotating turret that can be moved into position sequentially without disrupting the entire system. Some alloys can be used as cathodes in the lamps to emit lines for several metals, but rarely do they perform as well as single element lamps. Lamps are currently available for 65 elements, not including C, N, O, S, Cs, Lu, Os, Th, U, radioactive elements, noble gases, or halogens.

Electrodeless discharge lamps are available for 17 of the more volatile elements. This consists of a bulb—a small amount of pure metal and inert gas at low pressure in a quartz tube—with a ceramic cylinder with an RF (radio frequency) coil on it. The RF field ionizes the inert gas and its energy vaporizes the elements and excites the atoms, which emit a spectrum of radiation. Electrodeless discharge lamps, when available, are brighter and offer better sensitivity, that is, better precision and lower detection limits, than conventional hollow cathode lamps.

To reduce the effect of radiative noise (or background radiation) from the flame or furnace, a rotating "chopper" is placed between the source of the beam and the flame or furnace, which makes the beam "flash" at a set frequency as it passes through the sample. By synchronizing the detector with the chopper, it is possible to measure the background radiation from the flame or furnace while the light source is blocked, and subtract it from the signal. This type of background correction (Smith-Hiefte) is essential for any quantitative work.

The optical system allows light from the source to pass through a particular region of the flame at a certain distance above the burner. Having passed through the sample, a monochromator isolates a narrow wavelength band from the transmitted radiant energy. In most instruments this is a set of slits on either side of a prism or a diffraction grating. The prism or diffraction grating disperses the radiation according to wavelength, and the slits select a certain wavelength, the resonance line, for measurement by the detector. It filters out any other resonance lines from the cathode spectrum. Usually lines for analysis can be found that are easily isolated from other lines so AAS monochromators need not be highly sophisticated. Most lines of interest are in the ultraviolet (UV) and visible part of the spectrum and are detected with a photomultiplier tube. The signal is amplified and the data collected and analyzed using a computerized digital processor.

When in a magnetic field, each line of radiation will split into at least three lines: one unchanged line polarized perpendicularly to one line of lengthened wavelength and one line of shortened wavelength. This is the Zeeman effect. A polarizer can separate these lines. Radiation polarized perpendicularly to the field is not absorbed by the atomic vapor. Radiation polarized parallel to the field is absorbed by the atomic vapor as normal. The background for both is identical. Thus, by subtracting the perpendicular radiation lines from the absorbed parallel radiation lines, backgroundless absorption spectra can be obtained. This is the most expensive method of background correction. Two types of instruments are sold: direct Zeeman atomic absorption (ZAA) with the magnet positioned around the radiation source, and inverse-ZAA with the magnet positioned around the atomizer. Inverse-ZAA is superior because there is generally greater background drift in ZAA due to difficulties associated with operating a cathode ray tube fully inside a magnetic field.

3.4.3. Additional Considerations

Any deviations from the assumptions of Beer's law cause deviation in the linear relationship between sample concentration and absorbance, which obviously reduces the accuracy of the method. Instrumental deviations are caused by the actual bandwidth characteristics of filters in monochromators. Beer's law assumes strictly monochromatic radiation, and anything that increases bandwidth is a problem. Every effort must be made to avoid band broadening at all stages in the process. Chemically induced deviations are caused by shifts in the position of a chemical or physical equilibrium, which includes the absorbing species, causing it to change in concentration during analysis. For this reason, the pH and ionic strength of a sample solution must be kept constant.

Spectral methods of analysis are actually remarkably free of interference because there are always multiple lines on which to make measurements. When there is interference from the matrix, radiation source, or flame, it is usually possible to select a different line to analyze. The linear behavior over the range of sample concentration of any line to be used must first be checked with standards of differing concentrations. This is rarely a problem since sample solutions can be made to any concentration—usually the more dilute, the better the linear response.

There are a number of problems specific to particular elements. A few species do have very close preferred absorption lines (e.g., Ni at 231.096 nm and Sb at 231.147 nm), which can be difficult to separate if both are present. Rare earth elements still cannot be analyzed well by AAS. Ti, Al, and V oxidize in flames to form refractory oxides that have very different absorption spectra and so can go unmeasured. To measure any of these one must use the hottest possible flame (acetylene with N_2O) if the atomization is done by flame. Sample solutions of different viscosity are not comparable—viscosity affects the efficiency of aspiration of the sample, and so two solutions with the same concentration of a metal, but different viscosities, will produce very different instrument readings.

Matrix effects plague almost all analytical chemical methods and are certainly the greatest problem in AAS. The matrix can effect the atomization process. Some elements may be dissociated into atoms, excited, and gone before other elements begin to atomize. Standards as close to the composition of the unknown sample solutions as possible should be used to monitor and correct any such effects. The matrix can also contain absorbing molecular species, which can cause problems if they pass into the solution and the flame temperature is not high enough to dissociate them. Molecules absorb over much wider spectral ranges than atoms and can mask or give false positives in some cases (e.g., Ba is seriously affected by the presence of Ca, which forms a hydroxide species).

3.5. EMISSION SPECTROSCOPY (OPTICAL AND ICP)

3.5.1. Method

Emission spectroscopy is concerned with the transition from the atom's excited state back to its ground state, resulting in the emission of energy.

Excited atoms will deactivate or revert to a lower energy state, spontaneously, emitting radiation of a wavelength characteristic of that element. Alkali and alkaline earth elements emit very strongly in the visible region with just the heat from a flame—the flame provides the energy to excite the atoms, and transfer to a cooler part of the flame allows the atoms to de-excite, emitting characteristic colors. This is the basis of the simple flame tests for these elements. With more powerful excitation all metals and some nonmetals will emit radiation. The emission spectrum of an element can be as simple as a few lines, such as that of Na, all the way up to thousands of lines, such as that of U, but are always distinct and reproducible. Emission resonance lines are usually in the UV region but can also include visible and even near IR. Just as with absorption spectra, some lines from different elements may lie close to each other despite the uniqueness of electronic energy levels for each element, but it is usually possible to find lines for each element without overlap. Major resonance lines of elements of interest in archaeological materials are given in Table 3.2.

In emission spectroscopy, the sample is atomized and excited thermally. The radiation is recorded as the excited atoms revert to lower energy. The fraction of sample atoms in an electronically excited state, and the number of excited states attainable, depends on the temperature. The lower the temperature, the less the number of excited states that can be populated, thus producing simpler spectra that can be more easily resolved into resonance line without interference. One major advantage of emission spectroscopy is that, with standards spread throughout the emission spectrum, calibration can be made over a greater concentration range than any other method. Another advantage over AAS is that multiple elements can be measured simultaneously, and unexpected elements can be detected. With appropriate standards emission spectroscopy can be calibrated over the entire periodic table.

3.5.2. Instrumentation

Figure 3.2 shows a schematic diagram of a generalized emission spectrometer. As with AAS, samples for ES must be completely dissolved (if they are not to be laser ablated from a solid). Sources of atomization and excitation for liquid samples include flames, furnace, electric arcs and sparks, and plasmas. Flame sources are exactly as described above for AAS, but other fuel and oxidant combinations are frequently used, including hydrogen with air to produce 2000°C, hydrogen/oxygen to 2650°C, acetylene/oxygen (3200°C), and C_2N_2/oxygen to produce 4800°C. Although lower temperatures give simpler spectra, in general higher temperatures are desired in order to excite a large population of the atoms in a sample. The sample is introduced as an aerosol into the flame, which dissociates the atoms. The atoms

Table 3.2. Most Sensitive or Most Commonly Used Resonance Lines for Elements of Interest to Archaeologists

Elements	AAS (nm)	ES–Atom (nm)	ES–Ions (nm)
Mg	285.21	285.21	279.55
Al	396.15	396.15	266.92
Si	—	288.16	283.16
P	—	253.57	—
	—	213.62	
	—	178.28	
S	180.73	180.73	—
Ca	422.67	422.67	393.37
V	318.54	437.92	309.31
Cr	425.44	425.43	267.72
	—	399.86	
Mn	279.48	403.08	257.61
Fe	248.33	317.99	259.94
Co	240.73	240.73	228.62
Ni	232.00	232.00	231.60
Cu	324.75	278.02	—
	327.40	193.76	—
Zn	213.86	213.86	202.55
As	228.80	278.02	—
	234.90	193.76	—
Se	—	203.99	—
Rb	780.02	780.02	—
Sr	460.73	460.73	421.55
Rh	343.49	343.49	—
Pd	244.80	363.47	—
	—	360.96	—
	—	340.46	—
Ag	328.07	328.07	—
	338.29	—	—
Cd	228.80	326.11	214.44
Sn	235.48	284.00	356.83
	—	224.60	—
Sb	217.58	231.15	—
	—	206.83	—
Te	214.27	214.27	—
Re	—	346.05	197.31
Os	—	263.71	225.58
Ir	—	208.88	—
Pt	265.95	306.47	214.42
	—	265.95	—
Au	242.80	242.80	—
	267.60	—	—

Table 3.2. (Continued)

Elements	AAS (nm)	ES–Atom (nm)	ES–Ions (nm)
Hg	253.65	253.65	194.23
Pb	217.00	405.78	220.35
	283.31	–	–
Bi	223.06	223.06	–
Th	–	–	283.73
U	–	–	385.96

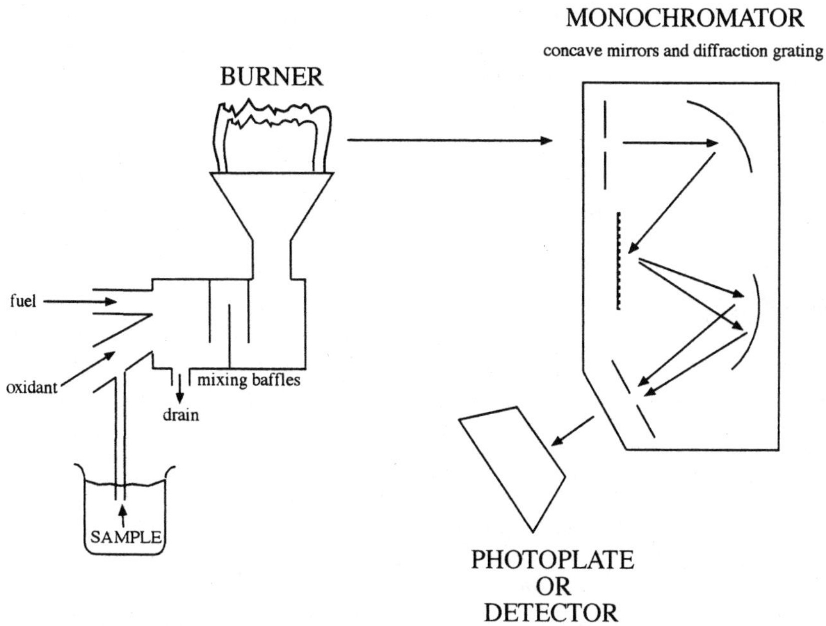

Figure 3.2. Schematic diagram of Emission Spectroscopy

acquire enough energy through collision to become electronically excited, and subsequently emit when the energy falls.

Simple OES produces a spectrum of resonance lines in the visible region of the spectrum, separates them with a system of concave mirrors, a diffraction grating, and slits, and then allows either viewing or a photographic plate to record the spectrum. This method is only semiquantitative at best, partly because of problems with reproducibility of the photographic plates, but the advantage of OES is that it can measure elements in high concentrations.

Electrical excitation methods (flameless sources) can achieve higher temperatures than flame source methods—up to 5000°C with a continuous or interrupted arc. Plasmas (very hot ionized gases) can produce temperatures of about 8000°C. With sparks discharged through a graphite or coal electrode, temperatures as high as 10,000°C can be achieved. With these very high temperatures, ions, rather than excited atoms, tend to be produced that will reduce the analytical sensitivity unless special techniques are employed.

In older OES machines, small solid samples, as powder, were placed into graphite cups, and the samples excited by striking a spark between an electrode and the cup. Samples dried onto electrodes and excited by electrical heating may also be used. In direct current (DC) arc source emission spectroscopy, an electric arc is struck between two graphite rod electrodes. One electrode holds a drop of sample solution in a small depression. The drop is oven dried or dried under an IR lamp prior to use. When an arc is struck, this sample residue volatilizes, atomizes, and excites the atoms in the sample. There are major problems with quantification—even with suitable standards and blanks, reproducibility is low. Not all elements volatilize uniformly. Electric arcs also wander and flicker across the surface of the electrode, causing different areas of the deposited sample to be excited in an uncontrolled fashion. Mixing the sample with pure graphite powder improves the stability of the arc due to the increased electrical conductivity.

Spark sources also discharge a voltage between two electrodes, one of which contains a dried drop of the sample solution. The other electrode is pointed and is positioned just above the sample electrode. A higher voltage than used in DC arcing is produced by a Tesla coil that steps up the voltage in very short (millisecond) bursts. This causes a spark to discharge between the two electrodes, vaporizing a minute amount of the sample. This method is more stable and reproducible than the DC arc, and the higher voltage means that the problem of selectively unvolatilized elements is minimized.

If a metal is to be analyzed, samples, if large enough and regarded as disposable, may be machined into the shape of an electrode and used directly. Sufficient amounts of sample can also be mixed with copper or arsenic and pressed into pellets that can be clamped in and act as the sample electrode. Powdered samples can also be mixed with graphite for use as an electrode. When the sample itself becomes the electrode, the method is call glow discharge.

Emission spectroscopy can be carried out with the same instrumentation as absorption spectroscopy—normally, absorption machines can be operated in either mode. Flames and furnaces identical to those described for AAS can be used as sources in atomic emission spectroscopy for small volumes of sample. With the hollow cathode discharge method, the sample is made into a disk or cathode in a lamp similar to the hollow cathode source described above. This is a very efficient method of excitation and therefore has lower detection limits than the other sources. It does, however, require considerable machining resources to carry out.

Plasma torches, rather then flame burners, are now routinely used as emission sources for aspirated sample solutions. The plasma is typically argon (Ar), which therefore requires a significant energy input to generate conditions for self-sustaining plasma. This initial energy can be supplied in different ways, for example, direct current plasma (DCP), capacitively couple microwave plasma (CMP), or inductively coupled plasma (ICP). DCP is created by a DC current passed between two electrode in a heated gas. CMP is created by microwave power, created by an energized magnetron, transferred through electrodes to the gas stream. ICP is created by igniting flowing gas with Tesla sparks making the gas electrically conductive as it flows through a radio frequency induction coil. All these methods developed as the need for simultaneous multielement analysis increased. If temperatures are sufficiently high and atomic and ionic species dominate the emission spectrum, lines can be found for elements of widely differing ionization potentials. Plasma sources offer greater temporal and spatial stability than other high-temperature sources. The DCP method consumes Ar at an extremely high rate and is, therefore, the least cost-effective method of analysis.

Plasma source methods are the best for quantitative multielement determination. The source is very reproducible and consistent. For an ICP, the combination of a high-velocity argon flow heated by electrical induction and initiated by an electrical spark produces a plasma fireball of about 8000°C at the mouth of the torch. The sample is aspirated into the torch, where it is both atomized and ionized with the aim of producing singly charged ions. Ions are excited and an emission spectrum measured. Excited ions have unique and elementally characteristic patterns, as do excited atoms, but as noted above the ionic spectrum is different from the atomic spectrum for any given element.

More recently, a new method for converting a solid sample into a vapor suitable for emission analysis has become of great importance. This uses a laser (light amplification by stimulated emission of radiation), which produces radiation of high intensity, directionality, and monochromaticity. A laser beam can carry sufficient energy to remove a tiny sample from a solid specimen and also vaporize, atomize, excite, and ionize this sample. Laser ablation devices can vaporize a hole of a few micrometers in diameter from a solid sample and send it directly into the ICP source, eliminating the need for a sample solution. This saves considerable preparation time dissolving and diluting samples and is extremely attractive to archaeologists because it is considerably less destructive. The most commonly used lasers are the ruby laser, which consists of crystals of Al_2O_3 doped with Cr_2O_3 (emission frequency 694 nm), and the Nd:YAG (Nd^{3+} in $Y_3Al_5O_{12}$ garnet crystals—emission frequency 1064 nm, which can also be filtered to 266 nm if it is preferable to ablate within the UV range rather than in IR).

In all emission spectrometers, light from the source is filtered by a scanning monochromator that can be used to measure emissions of many different wavelengths, thus analyzing for multiple elements in the sample. For alkali metal analysis only, low-temperature sources may be used and inexpensive filters can replace the monochromator to give a qualitative analysis (the flame photometer).

Older optical methods recorded the emission spectrum on photographic film giving a pattern of dark lines on the negative. The photographic plate has now been replaced with a detector, usually a photomultiplier since detection within the UV and visible ranges is required. With arc and spark sources, there is only transient emission and not enough time for a single detector to scan through all the wavelengths. With multiple photomultiplier tubes positioned behind slits along the focal curve with one tube for an internal standard element and the other tubes for other element, this problem can be overcome, but it is still difficult to quantify. Data produced by these instruments may be very precise, but it cannot be assumed to be accurate, since matrix, sample preparation, and standards all effect the accuracy of the data.

Flame and plasma sources produce more stable signals, providing the flow rate of the sample into the source is constant. Although multiple detectors can be used at fixed wavelengths to monitor predetermined elements, it is most common to have a single photodetector that is scanned under computer control through the UV and visible emission spectrum, either recording the whole spectrum or monitoring the intensity of selected lines. Figure 3.3 shows a typical system for ICP–AES, with separate UV and visible detectors for greatest sensitivity. Some instruments measure the radiation in two different directions from the flame, allowing direct comparison of radiant power emitted by elements in the same sample. In all cases, the detected signal can, of course, now be amplified and digitized to yield computerized data.

3.5.3. Additional Considerations

Conventional OES as carried out on archaeological material before the 1980s is now regarded largely as a qualitative technique—elements present can be identified but not quantified with sufficient precision to give useful data. The magnitude of the emission signal produced is related to the power input from the thermal excitation and to the number of excited atoms in the optical path at the moment of measurement, so later emission spectros-

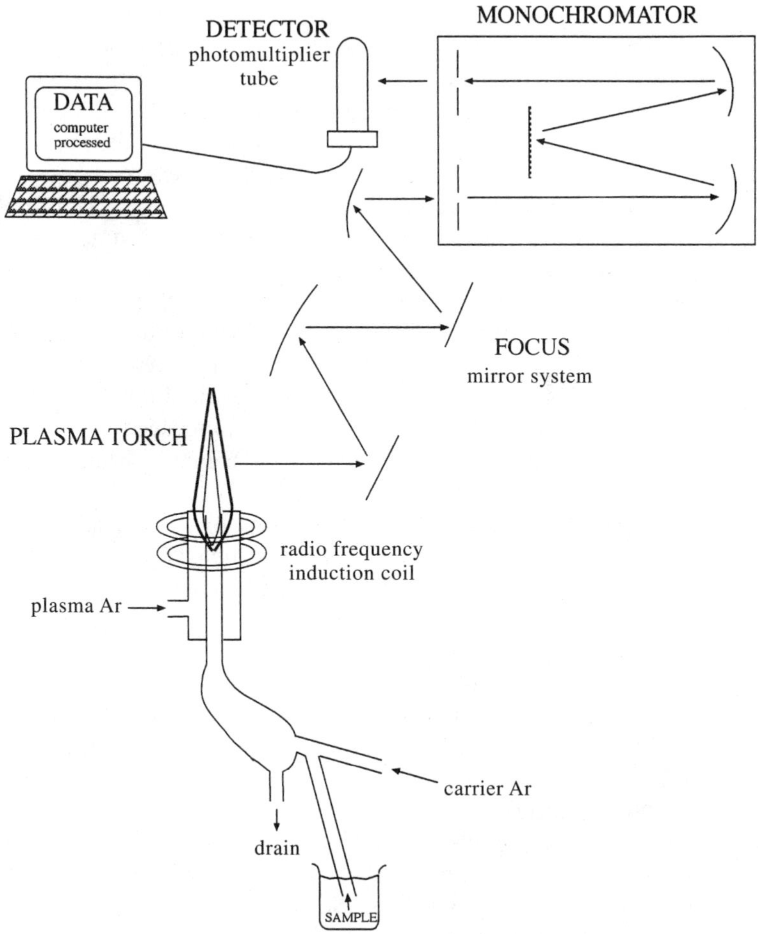

Figure 3.3. Schematic diagram of Atomic Emission Spectroscopy

copies, with better regulated and stable thermal sources, became quantitative or at least semiquantitative. However, a complete mathematical theory for this relationship does not exist. Ideally, the optical density will equal $kP\,\Delta T$, where P is the radiant power and ΔT is the exposure time; k should be constant, but this is true only over a short range. Calibration of concentration can only be done empirically through comparison of standards and unknowns. True quantification without standards is impossible. Calibration is best with internal standards so that unknown lines within a sample can be compared to lines of a similar element of known composition within the same sample. Internal standards can already be in a sample and simply known from another method of analysis or may be added in known amounts to the sample as a spike. Standard lines should be as close as possible to the unknowns in both concentration and wavelength in order to minimize any nonlinearity. The greatest problem with arc and spark methods is differences in volatility of elements within the sample. The differences can be so great that some elements can volatize and even burn out completely before other elements have been heated sufficiently to appear in the spectrum.

Alkali metals do not analyze well by high-temperature methods such as plasma source spectroscopy. They ionize quickly, so there is no atomic emission, but their ionic emission is close to invisible because, once ionized, they behave as noble gases. Basic flame methods are often best for alkali metals. By contrast, high-resolution detectors are required for the analysis of the rare earth elements, U, and Th, or sample solutions need to be stripped of the major element matrix of the sample by column chromatography before analysis.

Although it is the least destructive method, laser ablation of solid samples must occur inside a vacuum chamber so that the sample vapor can be carried into the plasma source. These chambers are usually about 4 cm in diameter, implying that most artifacts (other than small coins, jewelry, or broken objects) will not fit in the chamber without sampling. However, to avoid cutting pieces of objects, small diamond-coated pads have been developed for sampling the surface by abrasion, as well as new experimental work with O-ring seals pressed against whole objects to create a small *in situ* vacuum chamber.

Both IR and UV lasers can be used for ablation. Both ablate uniform craters in amorphous materials, but IR wavelengths can propagate along crystal edges and cause craters to be uneven. For this reason UV lasers are better suited for crystalline materials. Standardization of laser ablation sources is difficult. Most work is still semiquantitative at best because materials ablate at different rates, sending different amounts of sample into the vapor state and on into the plasma. This makes matrix matched samples and standards essential, but very few actually exist, especially for trace elements in complex materials, which need to be homogenous at the appropriate level.

As with absorption methods, spectral overlaps do still exist, and oxide or hydroxide formation can cause spectral interference as well as removing that element from the measurement. With plasma methods, doubly charged species can form, also removing those species from measurement or causing spectral overlap. Interferences are a much greater problem with all AES methods than with AAS methods. Matrix effects cause the greatest problems by increasing background spectral features close to the elements of interest. Although the mechanisms are not fully understood, anions in the sample solution can cause interferences or reduced signals. For example, sulfate and phosphate in the solution lowers the emission of calcium.

In spite of all these possible complications, AES can generally measure up to 40 elements; ICP–AES can do even better, measuring up to 70 elements if the detector is suitable for rare earth elements. Very low backgrounds give a high signal-to-noise ratio, resulting in very low detection limit—in the range of parts per billion for the ICP–AES method.

3.6. ATOMIC MASS SPECTROSCOPY (ICP–MS AND LA–ICP–MS)

3.6.1. Method

Mass spectrometry is not concerned with energy levels or radiation. The sample is separated according to mass and the number of atoms and molecules of each mass is counted, after ionization, giving a mass spectrum. With ICP–MS, the mass spectrum is detected by a quadrupole mass spectrometer that measures the motion of charged ions in an electromagnetic field.

For proper quantification, count rate in the detector must be directly or indirectly related to sample concentration. As with emission spectroscopy, this relationship must be determined empirically (by calibration) since analytical conditions can change rapidly, and this is most easily done with liquid samples. To obtain quantitative concentration data, standard solutions of known concentration must be analyzed with the unknown samples. It is good practice to prepare standard solutions to give at least three different concentrations of each of the elements of interest to bracket the range expected in the analyte. These three analyses are then used to produce a calibration for each selected isotope that can be examined for linearity between concentration and observed count rate. Internal standard isotope spikes are also used in every sample to check on the calibration of the mass scale. Acid blanks are always analyzed and subtracted from sample solutions.

3.6.2. Instrumentation

A typical instrument configuration is shown in Figure 3.4. The technique shares a number of features with other spectrometric techniques in that samples are either introduced as aspirated liquids or laser ablated solids, and subsequently atomized and ionized at elevated temperatures. As with ICP–AES, in ICP–MS an inductively coupled plasma is the source of ionization.

The key differences between ICP–AES and ICP–MS are in the detection systems. Whereas ICP–AES has a spectrophotometer to measure the light output, ICP–MS passes ions from the plasma into a quadrupole mass spec-

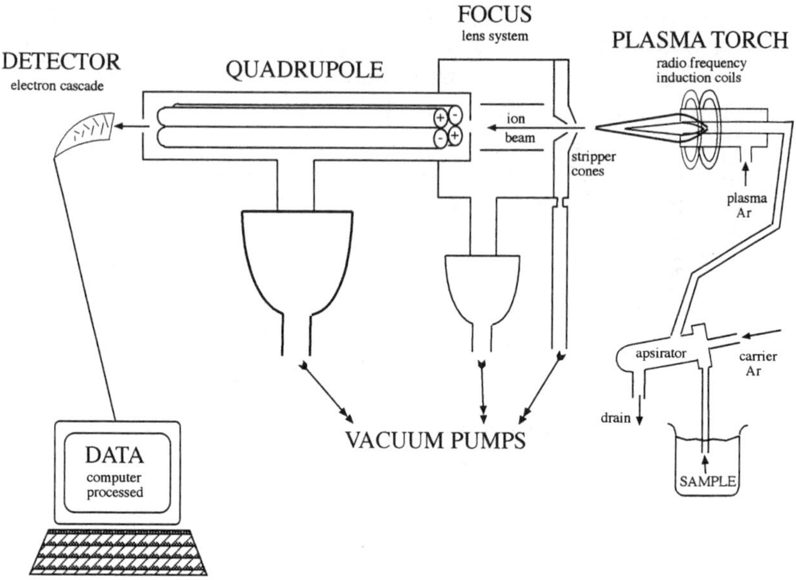

Figure 3.4. Schematic diagram of ICP-Quadrupole Spectrometry

trometer. The quadrupole utilizes two pairs of oppositely charged metal rods, each superimposing a radio frequency alternating current (AC) signal onto a DC voltage. This combination of RF and DC fields imparts a narrowing spiral path onto ions in the beam so that, for any set of operating conditions, only one particular mass value can pass through the quadrupole into the detector. Channel electron multipliers (electron cascade) and pulse counters are common detectors. All other ions are deflected out of the beam and pumped away by the vacuum system. As each isotope of any given element has a different atomic mass, the quadrupole has the effect of separating individual isotopes for separate measurement.

In order to enable quantitative elemental analysis, care must be taken to ensure that the sample dilution chosen gives concentrations within the optimal detection range of the instrument for the element of interest. Increasing dead time as the count rate increases leads to a lowering of efficiency so that, at high count rates, the detector counts fewer ions than actually exist. In practice major isotopes of the most abundant elements may yield nonlinear calibrations as the count rate rises toward detector saturation. Only isotopes in the linear range may be used for concentration determinations, and the calibration data are used to select those isotopes to be measured. If isotopic concentrations for a given element are at natural abundance, then elemental compositions may be calculated from the measurement of a single isotope. If isotopic fractionation is suspected, it is desirable to measure other isotopes of the same element as a check. This is obviously necessary if the measurement of isotopic fractionation is the object of the exercise.

Ratios of different isotopes of the same element can also be determined by ICP–MS but are far more difficult than simple chemical analysis. However, ICP–MS can measure isotope ratios up to three significant figures. The most widely used isotope system in archaeometallurgy is currently that of lead. With careful work, the precisions of the ratios as measured by ICP–MS are 0.2% for $^{206}Pb/^{207}Pb$, 0.5% for $^{208}Pb/^{207}Pb$, and 0.7 to 1.5% for $^{206}Pb/^{204}Pb$, but the latter may be much worse if the ^{204}Pb signal drops below the optimal instrument detection range, as may happen if the more abundant isotopes are within range. This cannot be avoided as it is neither wise nor recommended to measure the solution twice in different concentrations to give the ratio. This lower precision of quadrupole ICP–MS for lead isotope ratios suggests that thermal ionization mass spectrometry (TIMS) remains a better option for discriminating between sources of similar isotopic composition, at least until the new ICP–magnetic sector multicollectors (discussed below) become more widely available. However, it has often been suggested that not every archaeological problem requires measurements of the highest precision, and the much higher sample throughput of ICP–MS still appears to have something to offer in some areas of archaeological research.

3.6.3. Additional Considerations

Theoretically the mass range of ICP–MS covers the isotopes of all elements in the periodic table with the exception of H, He, C, N, O, F, Ne, and Ar. In practice, for any particular analysis there are matrix effects, as well as chemical and isotopic interferences including the presence of Ar molecular species, and atmospheric interferences that have to be taken into account in order to accurately interpret the data. Despite these complications, detection limits are typically much lower than for many other analytical methods, typically of the order of 1 ppb for light elements, and 50 parts per trillion for heavy elements. A new generation of high sensitivity ICP–MS instruments with better detectors are now available giving detection limits even lower than this, although pretreatment by column chemistry to remove major elements is required. Precision in compositional analysis is of the order of ± 1 to 3% for most elements. Measurement precisions as high as 0.2% can be achieved for some isotope ratio measurements. It is important to remember, however, that it is not possible to separate different isotopes of identical mass. Where two elements are present in the sample that have isotopes of the same mass (isobaric interference) other steps must be taken.

The lightest elements (those with atomic mass below 4.5) can be detected, but with greatly reduced accuracy compared with those at the higher end of the mass spectrum because of quadrupole instability. The mass range 11.5 through 23.5 is subject to interferences from atmospheric gases (C, O, and N) and from Ne in the carrier gas. Masses 24.5 through 42.5 also exhibit problems—S, Cl, and Si cannot be measured because molecular ions from O_2 and N_2 from the atmosphere interfere. ^{40}Ar within the plasma itself coincides with the major isotope of calcium (^{40}Ca), but ^{44}Ca can be measured instead. $^{40}Ar^{16}O$ formed by the oxidation of the carrier gas interferes with ^{56}Fe (92% natural abundance) and requires the analysis of minor iron isotopes at significantly reduced precision. Although there are ways of improving the performance of Fe determination by ICP–MS, in fact ICP–AES is capable of more accurate and more precise measurement of Fe. As a final specific example, $^{40}Ar^{35}Cl$, which forms when HCl has been used in the dissolution process and excess chlorine ions remain in solution, interferes with ^{75}As. In such cases, the data gathered at mass 75 reflects the amount of Cl present in solution, not the concentration of As in the original sample. In fact, although this list seems extensive, the only elements above 4.5 amu that are not excluded because of being atmospheric gases and that are subject to interferences with no alternative isotopes are fluorine and phosphorus.

Interference problems can be particularly significant for specific elements in certain sample matrices. For example, the major isotope of zinc (^{64}Zn, 48.9%) falls within the mass range normally skipped for copper to avoid the signal from ^{63}Cu and ^{65}Cu. When measuring zinc in a copper matrix, calculations have to be based on the less abundant zinc isotopes (^{66}Zn, ^{67}Zn, ^{68}Zn, and ^{70}Zn). Even so, the second most abundant isotope of zinc (^{66}Zn) may still be overcounted due to excessive copper in solution. Similar considerations apply to antimony in tin-rich samples. Both isotopes of antimony (^{121}Sb and ^{123}Sb) fall within the atomic mass range normally skipped in the analysis of tin-based materials. If the region of overlap between the isotopes of Sb and the heavier isotopes of Sn is scanned there is a tendency to overestimate Sb values. It is very important to map out such problems of matrix interference before every particular analysis.

ICP–MS has a shorter linear calibration range than ICP–AES. Because of this, one cannot analyze major, minor, and trace elements simultaneously in the same sample because only one concentration level will be in optimum detection range. To achieve this, the sample must be analyzed at two concentrations—first, a determination made on a dilution optimized for the major and minor elements, and subsequently, a more concentrated sample solution analyzed for trace elements with the instrument set to omit particular masses corresponding to the major elements. For similar reasons, in ICP–MS isotope ratio determinations difficulties may be encountered if isotopes with significantly different abundances are being compared.

Another limitation of IR–ICP–MS (isotope ratio–ICP–MS) is that quadrupole systems are less tolerant of a fluctuating ion beam or high-voltage instability than a slit-resolved system such as TIMS, which is capable of making isotope ratio measurements to five significant figures. This and poor peak shape does not permit a precision of better than 0.1% for IR–ICP–MS. In practice, this precision may never actually be reached—this figure actually applies to the measurement of silver isotopes at natural abundance (^{107}Ag at 51.355% and ^{109}Ag at 48.655%). This close match in natural abundance is optimal for ICP–MS. As the difference in abundance increases, the difference in count rate of the isotopes increases, and the expected precision of the ratio measurement decreases.

Although there is no isotopic fractionation in solution ICP–MS (as there is in TIMS as a result of ionization, volatility, and space charge effects), there is a preference for heavy isotopes in the mass spectrometric separation. This mass bias must be monitored with standards and subsequently normalized. Ideally ICP–MS standards should have a ratio close to that of the unknown samples and be in solutions with concentrations matched to the samples so that count rate rates are as close as possible.

In spite of these problems, IR–ICP–MS is still an important tool for archaeological scientists because large numbers of measurements can be performed without extensive sample preparation, using small samples and, most importantly, at very low cost.

3.7. FUTURE DIRECTIONS IN INSTRUMENTAL ANALYSIS AND ARCHAEOLOGY

It is clear from the above that the instrumentation available for chemical analysis in archaeology has advanced at a remarkable pace over the past 20 years. For almost all applications, conventional instrumentation can be used, since the sampling requirements have reduced considerably. The advent of laser ablation for solid samples is likely to see further advances in the analysis of museum and archaeological objects, but enthusiasm must be tempered with caution in recognition of the increased importance of inhomogeneity as sample size falls.

The biggest advance of all is likely to come from the increasing availability of high-precision isotopic ratio measurements for a wide range of elements. Although TIMS produces higher precision data than quadrupole ICP–MS, an inductively coupled plasma–magnetic sector multiple collector–mass spectrometer (ICP–MS–MS) has now been developed that, for some elements, provides the highest precision yet attainable. This equipment has recently been used in studies of lead isotopes and the results indicate comparable, if not superior, resolution to TIMS. In the magnetic sector–ICP–

MS, the quadrupole detector is replaced by a higher resolution magnetic sector mass spectrometer with multiple detectors. Several isotopes can be detected simultaneously, allowing direct measurement of isotopic ratios and eliminating variation due to plasma stability and the like. This machine combines the best attributes of ICP–MS and TIMS, and appears to offer a clear way forward in inorganic isotopic analysis.

We should not be deluded, however, into believing that better analytical data automatically results in better understanding. Since the 1950s and increasingly through 1970s, it became routine for archaeometrists to gather compositional data on a wide range of archaeological materials using the available analytical techniques. Much of these data was reported either as journal articles or as appendices to archaeological reports, irrespective of whether additional insights were provided by an analysis of these data. Some archaeometric work did little to add to our understanding of prehistory. The quality of information (as opposed to data) provided by any analytical program depends critically on a number of factors—having a well-designed hypothesis with clear milestones for refutation, how well the project is designed to test this hypothesis, how suitable the samples are, and how reliable the analytical data are in the context of the problem set. Perhaps the key question for any researcher is "how would I know if my hypotheses was wrong?" In this context, it is essential to know if the analytical data generated are sufficiently precise to resolve between competing hypotheses. It is, however, also generally a waste of money to gather data that is considerably more precise than necessary. It is to be hoped that archaeologists will embrace these new technological innovations for the analysis of ancient materials with enthusiasm but apply them to the resolution of problems of matching subtlety and sophistication. Only then will we truly be in a position to know more about our human past.

BIBLIOGRAPHY

GENERAL

Dean, J. A. (1995), *Analytical Chemistry Handbook*, McGraw-Hill, New York.

ATOMIC ABSORPTION SPECTROSCOPY

Angino, E. E., and G. K. Billings (1967), *Atomic Absorption Spectrometry in Geology*, Elsevier, New York.

Barnes, R. (1976), *Emission Spectroscopy*, Halsted, New York.

Boumans, P. W. J. M. (1987a), *Inductively Coupled Plasma Emission Spectroscopy, Part I: Methodology, Instrumentation, and Performance*, John Wiley & Sons, New York.

Boumans, P. W. J. M. (1987b), *Inductively Coupled Plasma Emission Spectroscopy, Part II: Applications and Fundamentals*, John Wiley & Sons, New York.

Christian, G. D., and F. J. Feldman (1970), *Atomic Absorption Spectroscopy: Applications in Agriculture, Biology, and Medicine*, John Wiley & Sons, New York.

Dean, J. A. (1960), *Flame Photometry*, McGraw-Hill, New York.

Dean, J. A. (1995), *Analytical Chemistry Handbook*, McGraw-Hill, New York.

Dean, J. A., and T. C. Rains (1971a), *Flame Emission and Atomic Spectrometry: Theory*, Vol. 1, Marcel Dekker, New York.

Dean, J. A., and T. C. Rains (1971b), *Components and Techniques*, Vol. 2, Marcel Dekker, New York.

Dean, J. A., and T. C. Rains (1971c), *Elements and Matrices*, Vol. 3, Marcel Dekker, New York.

Dean, J. A., and T. C. Rains (1975), *Flame Emission and Atomic Absorption Spectrometry*, Vols. 1–3, Marcel Dekker, New York.

Dedina, J. (1995), *Hydride Generation Atomic Absorption Spectrometry*, John Wiley & Sons, New York.

Ebdon, L. (1982), *An Introduction to Atomic Absorption Spectroscopy—a Self Teaching Approach*, Heydon, Philadelphia.

Ewing, G. W. (1985), *Instrumental Methods of Chemical Analysis*, McGraw-Hill, New York.

Fitch, A., Y. Wang, S. Mellican, and S. Macha (1996), Lead lab: Teaching instrumentation with one analyte, *Anal. Chem.* **68**(23), 727A.

Hall, G. E. M., and J. L. Bouvier (1988), *A Procedure to Lower the Limits of Detection for Silver, Cadmium and Lead in the Analysis of Geological Materials by Atomic Absorption Spectrometry*, Canadian Government Pub. Centre, Ottawa.

Horton, B. (1995), Spectroscopic analysis, *Nature* **377**(6544), 87.

Hughes, M. J., M. R. Cowell, and P. T. Craddock (1976), Atomic absorption techniques in archaeology, *Archaeometry* **18**(1), 19.

Jackson, K. W., and G. Chen (1996), Atomic absorption, atomic emission, and flame emission spectrometry, *Anal. Chem.* **68**(12), 231R.

Jackson, K. W., and T. M. Mahmood (1994), Atomic absorption, atomic emission, and flame emission spectrometry, *Anal. Chem.* **66**(12), 252H.

Jenkins, F. A., and H. E. White (1976), *Fundamentals of Optics*, McGraw-Hill, New York.

Kirkbright, G. F. (1974), *Atomic Absorption and Fluorescence Spectroscopy*, Academic, New York.

Lajunen, L. H. J. (1992), *Spectrochemical Analysis by Atomic Absorption and Emission*, Royal Society of Chemistry, Cambridge.

Lov, B. V. (1970), *Atomic Absorption Spectrochemical Analysis*, Adam Hilger, London.

Mallard, W. G., G. R. Dalton, and J. W. Gallagher (developers) (1991), *NIST Spectroscopic Properties of Atoms and Atomic Ions Database (Computer File)*, NIST, Gaithersburg, MD.

Mayer, B. (1982), *Guidelines to Planning Atomic Spectrometric Analysis*, Elsevier, New York.

Metcalfe, E. (1987), *Atomic Absorption and Emission Spectroscopy*, published on behalf of ACOL by John Wiley & Sons, New York.

Miller, J. M., and K. G. Strothkamp (1995), Instruments for atomic spectroscopy in elementary chemistry laboratories, *J. College Sci. Teaching* **24**(4), 264.

Montaser, A., and D. Golightly (1992), *Inductively Coupled Plasmas in Analytical Atomic Spectroscopy*, New York.

Mottana, A., and F. Burragato (1990), *Absorption Spectroscopy in Mineralogy*, Elsevier, New York.

Ottaway, J., and A. Ure (1983), *Practical Atomic Absorption Spectroscopy*, Pergamon, New York.

Pinta, M. (1975), *Atomic Absorption Spectrometry*, Adam Hilger, London.

Price, W. J. (1979), *Spectrochemical Analysis by Atomic Absorption.* Heyden, London.

Pyle, S. M., J. M. Nocerino, S. N. Deming, J. A. Palasota, J. M. Palasota, E. L. Miller, D. C. Hillman, C. A. Kuharic, W. H. Cole, P. M. Fitzpatrick, M. A. Watson, and K. D. Nichols (1996), Comparison of AAS, ICP-AES, PSA, and XRF in determining lead and cadmium in soil, *Environ. Sci. Technol.* **30**(1), 204.

Robinson, J. W. (1966), *Atomic Absorption Spectroscopy*, Marcel Dekker, New York.

Sacks, R. (1981), *Treatise on Analytical Chemistry*, Part I, Vol. 7, John Wiley & Sons, New York.

Slavin, M. (1978), *Atomic Absorption Spectroscopy*, John Wiley & Sons, New York.

Van Loon, J. C. (1980), *Analytical Atomic Absorption Spectroscopy: Selected Methods*, Academic, New York.

Wach, F. (1995), GI-AAS: A mature technique for a modest investment, *Anal. Chem.* **67**(1), 51A.

Walsh, M., and M. Thompson (1983), *A Handbook of inductively coupled plasma spectroscopy*, Methuen, New Jersey.

EMISSION SPECTROSCOPY

Alcock, N. W. (1995), Flame, flameless, and plasma spectroscopy, *Anal. Chem.* **67**, 503R.

Alkemade, C. T. J., and R. Herrmann (1979), *Fundamentals of Analytical Flame Spectroscopy*, Adam Hilger, London.

Barnard, T. W. (1993), Solid-state detector for ICP-OES, *Anal. Chem.* **65**, 123.

Barnes, R. M. (1978), in *McGraw-Hill Yearbook of Science and Technology*, McGraw-Hill, New York, p. 342.

Borman, S. (1993), Near-field spectroscopy: Technique images single molecules (near-field scanning optical microscopy) (news of the week), *Chem. Eng. News* **71**(48), 6.

Boumans, P. W. J. M. (1987), *Inductively Coupled Plasma Emission Spectroscopy, Part I, Methodology, Instrumentation and Performance*, John Wiley & Sons, New York.

Dean, J. A. (1960), *Flame Photometry*, McGraw-Hill, New York.

Dean, J. A., and T. C. Rains (1969), *Flame Emission and Atomic Absorption Spectrometry*, Vol. 1, McGraw-Hill, New York.

Dean, J. A., and T. C. Rains (1971), *Components and Techniques*, Vol. 2, McGraw-Hill, New York.

Dean, J. A., and T. C. Rains (1975), *Elements and Matrices*, Vol. 3, McGraw-Hill, New York.

Epstein, S. (1977), *Chemical Analysis by Emission Spectroscopy*, Franklin, NJ.

Gill, R. J. (1993), AAS or ICP-OES: Are they competing techniques? *Am. Lab.* **24-F**.

Komolov, S. A. (1992), *Integralnnapieaea Vtorichno-Elektronnapieaea Spektroskopipeiaea Poverkhnosti* (Total Current Spectroscopy of Surfaces), Gordon and Breach Science, Philadelphia.

Koropchak, J. A. (1982), Liquid sample introduction to ICP spectrometries, *Spectroscopy* **8**, 20.

McCartney, E. J. (1983), *Absorption and Emission by Atmospheric Gases: The Physical Processes*, John Wiley & Sons. New York.

Montaser, A., and D. W. Golightly (1992), *Inductively Coupled Plasmas in Analytical Atomic Spectrometry*, 2nd ed., VCH, New York.

Nachtrieb, N. H. (1950), *Principles and Practice of Spectrochemical Analysis*, McGraw-Hill, New York.

Noble, D. (1994), ICP-AES from fixed to flexible, *Anal. Chem.* **66**, 105A.

Olesik, J. W., J. A. Kinzer, and B. Harkleroad (1994), Inductively coupled plasma optical emission spectrometry using nebulizers with widely different sample consumption rates, *Anal. Chem.* **66**, 2202.

Pinta, M. (1966), *Detection and Determination of Trace Elements; Absorption Spectrophotometry, Emission Spectroscopy, Polarography*, Ann Arbor Science, Ann Arbor.

Seiler, H. G., A. Sigel, and H. Sigel (1994), *Handbook on Metals in Clinical and Analytical Chemistry*, Marcel Dekker, New York.

Trautman, J. K., J. J. Macklin, L. E. Brus, and E. Betzig (1994), Near-field spectroscopy of single molecules at room temperatures, *Nature* **369**, 40.

U.S. National Bureau of Standards, Metallurgy Division (1974), *Soft X-ray emission spectra of metallic solids: critical review of selected systems and annotated spectral index*, U.S. Department of Commerce, National Bureau of Standards.

Weber, A. P. (1992), Sensitivities of inductively coupled plasma optical emission spectrometry for dry and wet aerosols, *Anal. Chem.* **64**, 672.

Walsh, J. N., and R. A. Howie (1986), Recent developments in analytical methods: Uses of inductively coupled plasma source spectrometry in applied geology and geochemistry, *Appl. Geochem.* **1**, 161.

INDUCTIVELY COUPLED PLASMA–MASS SPECTROMETRY

Aggarwal, J. K. (1996), Determination of the rare earth elements in aqueous samples at sub-ppt levels by inductively coupled plasma mass spectrometry and flow injection ICPMS, *Anal. Chem.* **68**, 4418.

Alcock, N. W. (1993), Flame, flameless, and plasma spectroscopy (clinical chemistry), *Anal. Chem.* **65**, 463R.

Alves, L. C. (1993), Measurement of vanadium, nickel, and arsenic in seawater and urine reference materials by inductively coupled plasma mass spectrometry with cryogenic desolvation, *Anal. Chem.* **65**, 2468.

Alves, L. C. (1992), Reduction of polyatomic ion interferences in inductively coupled plasma mass spectrometry by cryogenic desolvation, *Anal. Chem.* **64**, 1164.

Arrowsmith, P. (1987), Laser ablation of solids for elemental analysis by inductively coupled plasma mass spectrometry, *Anal. Chem.* **59**, 1437.

Beary, El. S., P. J. Paulsen, L. B. Jassie, and J. D. Fassett (1997), Determination of environmental lead using continuous-flow microwave digestion isotope dilution inductively coupled plasma mass spectrometry, *Anal. Chem.* **69**, 758.

Beary, E. S., and P. J. Paulsen, (1993), Selective application of chemical separations to isotope dilution inductively coupled plasma mass spectrometric analyses of standard reference materials, *Anal. Chem.* **65**, 1602.

Beauchemin, D. (1995), On-line standard addition method with ICPMS using flow injection. (inductively coupled plasma mass spectrometry), *Anal. Chem.* **67**, 1553.

Byrdy, F. A., and J. A. Caruso (1994), Elemental analysis of environmental samples: using chromatography coupled with plasma mass spectrometry, *Environ. Sci. Technol.* **28**, 528A.

Clemons, P. S., M. G. Minnich, and R. S. Houk (1995), Attenuation of metal oxide ions in inductively coupled plasma mass spectrometry with a graphite torch injector, *Anal. Chem.* **67**, 1929.

Cromwell, E. F., and P. Arrowsmith (1995), Semiquantitative analysis with laser ablation inductively coupled plasma mass spectrometry, *Anal. Chem.* **67**, 131.

Date, A. R., and A. L. Gray (1989), *Applications of Inductively Coupled Plasma Mass Spectrometry*, Blackie, Glasgow.

Dziewatkoski, M. P., L. B. Daniels, and J. W. Olesik (1996), Time-resolved inductively coupled plasma mass spectrometry measurements with individual, monodisperse drop sample introduction, *Anal. Chem.* **68**, 1101.

Evans, E. H., W. Pretorious, L. Ebdon, and S. Rowland (1994), Low-pressure inductively coupled plasma ion source for molecular and atomic mass spectrometry, *Anal. Chem.* **66**, 3400.

Gray, A. L. (1988), in *Inorganic Mass Spectrometry*, F. Adams, R. Gijbels, R. Van Grieken, Eds., John Wiley & Sons, New York, p. 257.

Halliday, A. N., D.-C. Lee, J. N. Christensen, A. J. Walder, P. A. Freedman, C. E. E. Jones, C. M., Hall, W. Yi, and D. Teagle (1995), Recent developmens in inductively coupled plasma magnetic sector multiple-collector mass spectrometry, *Int. J. Mass Spectrom. Ion Proc.*, **146/147**, 21.

Hiroyasu, I., M. Linscheid, M. Morita, and K. Ostuka (1994), Resolution of interelement spectral overlaps by high-resolution inductively coupled plasma mass spectrometry, *Anal. Chem.* **66**, 1588.

Holland, G., and A. N. Eaton (1993), *Applications of Plasma Source Mass Spectrometry*, Vol. 2, Royal Society of Chemistry, Cambridge.

Huang, Y., Y. Shibata, and M. Morita (1993), Instrumentation and performance of micro laser ablation system, part 1 (Micro laser ablation-inductively coupled plasma mass spectrometry), *Anal. Chem.* **65**, 2999.

Hutton, R. C. (1986), Application of inductively coupled plasma source mass spectrometry (ICP-MS) to the determination of trace metals in organics, *J. Anal. Atom. Spectrom.* **1**, 259.

Jarvis, K. E., and A. L. Gray (1990), *Plasma Source Mass Spectrometry*, Special Publication 85, Royal Society of Chemistry, London.

Jarvis, I., and K. E. Jarvis (1985), Rare-earth element geochemistry of standard sediments: A study using inductively coupled plasma spectrometry, *Chem. Geol.* **53**, 335.

Jarvis, I., and K. E. Jarvis (1992), Plasma spectrometry in the earth sciences: Techniques, applications and future trends, *Chem. Geol.* **95**, 1.

Jarvis, K. E., A. L. Gray, and R. S. Houk (1991), *Handbook of Inductively Coupled Plasma Mass Spectrometry*, Blackie, Glasgow.

Jenner, G. A., H. P. Longerich, S. E. Jackson, and B. J. Fryer (1990), ICP-MS—A powerful tool for high-precision trace-element analysis in earth sciences: Evidence from analysis of selected U. S. G. S. reference samples, *Chem. Geol.* **83**, 133.

Ketterer, M. E. (1996), Inductively coupled and microwave induced plasma sources for mass spectrometry, *Anal. Chem.* **68**, 486A.

Ketterer, M. E., and D. A. Biddle (1992), Effect of changes in abundances of interfering polyatomic ions (multivariate calibration in inductively coupled plasma mass spectrometry, part 2), *Anal. Chem.* **64**, 1819.

Lamoureux, M. M., D. C. Gregoire, C. L. Chakrabarti, and D. J. Goltz (1994), Characterization (Modification of a commercial electrothermal vaporizer for sample introduction into an inductively coupled plasma mass spectrometer, part 1), *Anal. Chem.* **66**, 3208.

Lee, D. C., and A. N. Halliday (1995), Precise determinations of the isotopic compositions and atomic weights of molybdenum, tellurium, tin and tungsten using

ICP magnetic sector multiple-collector mass spectrometry, *Int. J. Mass Spectrom. Ion Proc.* **146/147**, 35.

Lichte, F. E. (1995), Determination of elemental content of rocks by laser ablation inductively coupled plasma mass spectrometry, *Anal. Chem.* **67**, 2479.

Liu, Y., V. Lopez-Avila, J. J. Zhu, D. R. Wiederin, and W. F. Beckert (1995), Capillary electrophoresis coupled on-line with inductively coupled plasma mass spectrometry for elemental speciation, *Anal. Chem.* **67**, 2020.

Longerich, H. P., G. A. Jenner, B. J. Fryer, and S. E. Jackson (1990), Inductively coupled plasma–mass spectrometric analysis of geologic samples: A critical evaluation based on case studies. *Chem. Geol.* **83**, 105.

Machuron-Mandard, X., and J. C. Birolleau (1993), in *Applications of Plasma Source Mass Spectrometry*, Vol. 2 G. Holland, A. N. Eaton, Eds., Royal Society of Chemistry, Cambridge, p. 196.

Moor, H. C., T. Schaller, and M. Sturm (1996), Recent changes in stable lead isotope ratios in sediments of Lake Zug, Switzerland, *Environ. Sci. Technol.* **30**, 2928.

Newman, A. (1996), Elements of ICPMS (inductively coupled plasma mass spectrometry) (evaluation), *Anal. Chem.* **68**, 46A.

Peters, G. R. (1993), Characterization of an interface allowing either nebulization or gas chromatography as the sample introduction system in ICPMS, *Anal. Chem.* **65**, 97.

Sang-Ho, N. M., R. L. Wellington, and A. Montaser (1993), Investigation of helium inductively coupled plasma–mass spectrometry for the detection of metals and nonmetals in aqueous solutions, *Anal. Chem.* **65**, 2784.

Shum, S. C. K. (1993), Elemental speciation by anion exchange and size exclusion chromatography with detection by inductively coupled plasma mass spectrometry with direct injection nebulization, *Anal. Chem.* **65**, 2972.

Smith, R. G. (1993), Determination of mercury in environmental samples by isotope dilution/ICPMS (inductively coupled plasma mass spectrometry), *Anal. Chem.* **65**, 2485.

Stetzenbach, K. J., M. Amano, D. K. Kreamer, and V. F. Hodge (1994), Testing the limits of ICP–MS: Determination of trace elements in ground water at the part-per-trillion level (inductively coupled plasma mass spectrometer), *Ground Water* **32**, 976.

Taylor, H. E., J. R. Garbarino, D. M. Murphy, and R. Beckett (1992), Inductively coupled plasma–mass spectrometry as an element-specific detector for field-flow fractionation particle separation, *Anal. Chem.* **64**, 2036.

Tykot, R. H., and S. M. M. Young (1996), in *Archaelogocal Chemistry: Organic, Inorganic and Biochemical Analysis*, M. V. Orna, Ed., American Chemical Society, Washington DC, p. 116.

Vanhaecke, F., L. Moens, R. Dams, J. Papadakis, and P. Taylor (1997), Applicability of high-resolution ICP–mass spectrometry for isotope ratio measurements (inductively coupled plasma), A*nal. Chem.* **69**, 268.

Walder, A. J., and N. Furuta (1993), High-precision lead isotope ratio measurement by inductively coupled plasma multiple collector mass spectrometry, *Anal. Sci.* **9**, 675.

Yi, Y. V., and A. Masuda (1996), Simultaneous determination of ruthenium, palladium, iridium, and platinum at ultratrace levels by isotope dilution inductively coupled plasma mass spectrometry in geological samples, *Anal. Chem.* **68**, 1444.

Young, S. M. M., and R. C. Pflaum (1995), in *Pittsburgh Spectroscopy Conference*, March 5–10, 1995, New Orleans, H. A. Barrel and M. B. Perry, Eds., Pittsburgh, p. 945.

CHAPTER

4

X-RAY FLUORESCENCE

LUC MOENS

University Ghent, Laboratory of Analytical Chemistry, Ghent, Belgium

ALEX VON BOHLEN

Institut für Spektrochemie und Angewandte Spektroskopie, ISAS, Dortmund, Germany

PETER VANDENABEELE

University Ghent, Laboratory of Analytical Chemistry, Ghent, Belgium

4.1. INTRODUCTION

X-ray fluorescence (XRF) analysis allows the concentrations of most elements of the periodic table to be determined. The method has many advantages that make it very suited for the analysis of artifacts. It allows solid materials to be analyzed directly without necessitating dissolution or any other time-consuming or destructive sample preparation. Small objects can be brought to the XRF instrument and in many cases analyzed as such. The method is consumption free and, apart from occasional radiation damage, will leave the artifact unaltered and suited for analysis with other methods. Mobile XRF instruments allow larger objects to be studied on site. Obviously, it is also possible to sample the artifacts if such is ethically acceptable. A further advantage of the method is its capability to determine major, minor, and trace element concentrations. Finally, XRF is a well-established method, the possibilities and limitations of which are known and well understood. The operation costs are low and the analysis times are short.

Accurate quantitative analysis of unprepared artifacts may be difficult because XRF (except total reflection XRF, TXRF) suffers from matrix effects for which it is not easy to correct. Also, the concentrations measured with

Modern Analytical Methods in Art and Archaeology, Edited by Enrico Ciliberto and Giuseppe Spoto. Chemical Analysis Series, Vol. 155
ISBN 0-471-29361-X

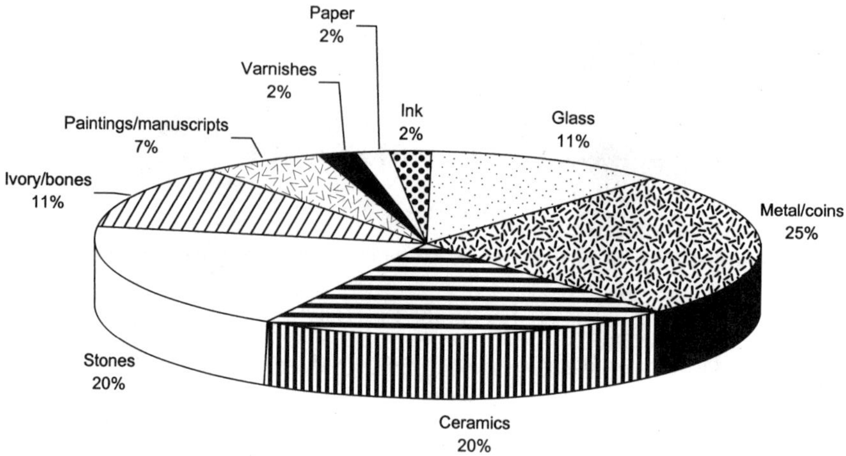

Figure 4.1. Relative importance (1995–1998) of different XRF application fields in archaeology and art history.

XRF mainly reflect the composition of the outer layers of the material. For weathered objects this composition may considerably deviate from that of the unaltered material below the patina. These limitations are obviously alleviated when samples are taken from the artifact and sample preparation methods aiming at accurate quantification are applied. TXRF, under optimal circumstances, is free from matrix effects and quantification can be simple.

A literature search for the period 1995 to 1998 revealed that XRF analysis is still widely used to identify and characterize a variety of archaeological and art historical materials. Figure 4.1 depicts the different application fields and their respective importance (period 1996 to 1998). Metallic objects (including coins) and potsherds make more than 50% of the applications, but stone material and bones and ivory have been studied as well. Also painted surfaces and glass were found to have been the subject of XRF analysis.

In this chapter the XRF techniques used will be discussed, and the application field will be reviewed via a limited number of typical applications. Rather than emphasizing the analytical details, the archaeometric potential of the method will be illustrated.

4.2. NATURE OF X-RAYS AND X-RAY FLUORESCENCE

X-rays are part of the electromagnetic spectrum (cf. Fig. 4.2) and were discovered and described by W. C. Röntgen in 1895 (Bertin 1975; Broll 1996). Only a part of their spectral range between 0.1 and 100 keV or according to

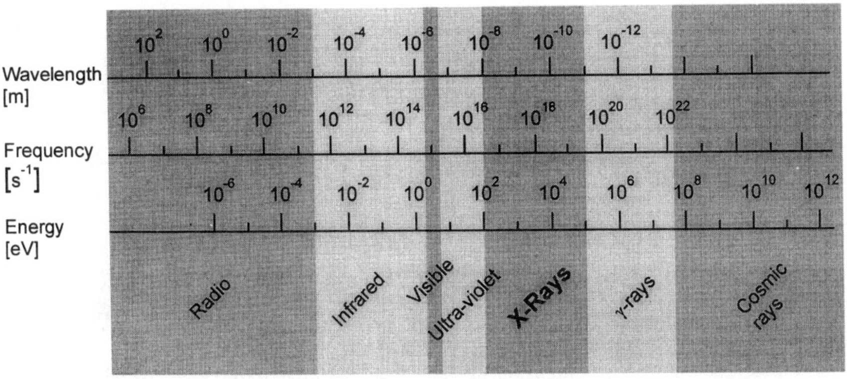

Figure 4.2. Electromagnetic spectrum.

the relationship $E = hc/\lambda$, between 10 and 0.01 nm is used in X-ray spectral analysis for material sciences. The X-rays are generated by bombardment of a target by means of charged particles, electrons, or ions leading to X-ray emission or by high-energy radiation leading to X-ray fluorescence. The observed X-ray spectra are composed of a continuum with wide spectral range superimposed by well-defined X-ray lines.

When applying sufficient kinetic energy, both, radiation and particles are able to penetrate into the atoms and to remove electrons from their inner shells. After the ejection of electrons, the atoms are in an energetically unstable configuration, and electrons from outer shells fill up the vacancies. The energy difference between the shells involved in the transition process of electrons is emitted—in the case of inner electrons shells—as X-ray photons. The selection rules of quantum theory dictate which electron transitions are allowed. The nomenclature of emitted X-rays is based on conventional rules, as illustrated in Figure 4.3. For instance, in $K_{\beta 1}$, the capital K refers to the K shell (Bohr's atom model) from which initially an electron was expelled. Greek letters and digits are used to further denote the X-rays in a conventional and nonsystematic way; thereby reference is made to the orbitals from which an electron leaves to fill up the originally created vacancy and to the relative intensities of the X-rays. The wavelength of the emitted X-rays depends on the atomic number Z of the excited atoms. For a given type of X-ray (e.g., K_α or $K_{\beta 1}$) the dependence of its wavelength on Z is described by Moseley's law:

$$\lambda = k_1/(Z - k_2)^2$$

where λ is the wavelength and k_1 and k_2 are constants. Each element emits X-rays with specific wavelengths that characterize the element. This char-

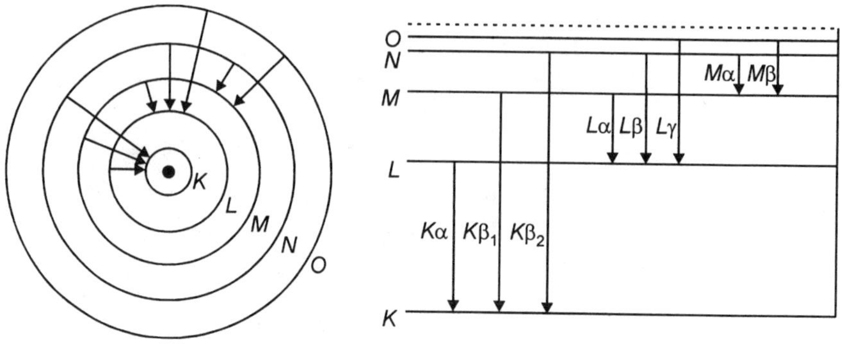

Figure 4.3. Electron transitions for different shells involved in the emission of X-rays.

acteristic radiation can be detected and is graphically represented by a characteristic X-ray spectrum. In contrast to the atomic emission lines in the ultraviolet (UV) and visible part of the spectrum, the characteristic X-ray spectra are simple: whereas Fe has alone 6000 UV–vis emission lines, not more than 600 X-ray lines are relevant in X-ray spectrometry and are sufficient to characterize all the elements. In general, no distinction can be made between characteristic X-rays from isotopes of the same element or from atoms of the same element present in different chemical compounds. Except for the two first elements of the periodic table of the elements, hydrogen and helium, that do not produce X-ray spectra, all elements are accessible to a qualitative and quantitative X-ray analysis. Thereby the wavelengths or energies of the emitted X-rays identify the elements contained in a sample, while the intensities of individual signals are proportional to the concentrations of these elements. Main constituents up to 100% as well as minor constituents or even traces can thus be determined in a great variety of samples.

For detailed information concerning the physical nature, the instrumentation, the quantification, and classical applications of X-rays and X-ray spectrometry several monographs are available: Bertin (1975), Williams (1987), Jenkins (1988), and Klockenkämper (1996).

4.3. X-RAY FLUORESCENCE SPECTROMETRY

Being a form of electromagnetic radiation, X-rays can be described in terms of wave phenomena or as corpuscles. In particular, X-rays show interference effects that can be explained effectively by wave theory (Bertin 1975). One of these effects is coherent scatter (Williams 1987) usefully interpreted as a

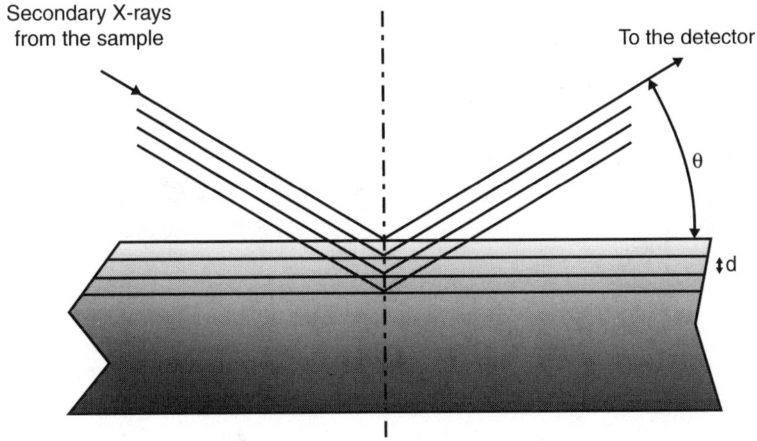

Figure 4.4. Diffraction of X-rays by a crystal according to Bragg; λ = wavelength, θ = Bragg angle of diffraction, d = interplanar spacing of diffracting planes.

wave phenomenon by Bragg, who considers it as a reflection on a stack of crystal planes (cf. Fig. 4.4). When a collimated X-ray beam of wavelength λ is directed at an angle θ onto a crystal with crystal planes of spacing d, a reinforced scatter of the radiation can be observed (Williams 1987) under the angle of reflection on condition that Bragg's law

$$n\lambda = 2d \cdot \sin\theta$$

is fulfilled for an integer number n.

4.3.1. Wavelength-Dispersive X-Ray Fluorescence Spectrometry

This so-called Bragg reflection is fully used in wavelength-dispersive X-ray fluorescence spectrometry (WDXRF) to select the characteristic X-rays of interest originating from a sample, by selecting the corresponding angle θ. A schematic representation of a WDXRF spectrometer is shown in Figure 4.5.

In WDXRF samples are irradiated by the primary X-rays of an X-ray source. The induced secondary X-ray radiation consists of different wavelengths originating from the different elements of the sample and reflecting its elemental composition. The secondary (or fluorescence) radiation originating from the sample is guided through a collimator that allows only rays with parallel propagation to pass. The X-rays reach a Bragg reflector, that is, a crystal, and are reflected according to Bragg's law. Scanning the angle θ by rotating the crystal and simultaneous displacement of the detector with a

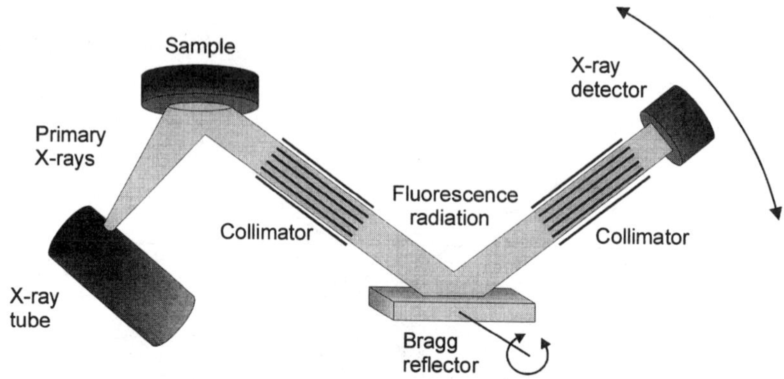

Figure 4.5. Schematic representation of a wavelength dispersive X-ray fluorescence spectrometer (WDXRF).

high mechanical accuracy (better than 0.001°) make it possible to sequentially detect X-rays of different wavelength. The radiation selected by the Bragg reflector will pass a second collimator before reaching the detector. By the aid of a computer, the signals are recorded and composed to a spectrum or when only some elements are of interest, the intensity of the X-rays of the corresponding wavelengths are reported.

Several excitation sources are used in X-ray spectrometry. The most common are X-ray tubes producing a broad beam spanning a large solid angle. These are built as end- or side-window tubes that are operated at high voltages up to 100 kV; the current (milliampere) is adjusted to a maximum load of 2 or 3 kW. Also fine-focus tubes (power 0.6 to 3 kW) or tubes with rotating anodes (power up to 18 kW) are used in WDXRF spectrometers.

Since it is impossible to select all wavelengths using a single Bragg reflector crystal, most X-ray spectrometers are equipped with several crystals that can be used according to the needs of the experiment. The choice will depend, for instance, on the elements to be detected, the required spectral resolution, or the stability of the crystal. A list of crystals regularly used in WDXRF and their specifications is given in Table 4.1. Detailed information can be found in the literature (Bertin 1975).

In WDXRF two detectors with different characteristics are normally used. These are arranged in a tandem geometry and are switched on depending on the wavelength of the radiation to be recorded. The first detector of this arrangement is a gas flow proportional counter for the registration of long-wave (low-energy) radiation. The second is a scintillation counter for the detection of the hard X-rays having higher energies. Specifications and working methods are described in detail elsewhere (Bertin 1975; Heinrich

Table 4.1. Bragg Reflectors Regularly Used in WDXRF[a]

		Lightest element detectable (Z) at max $\theta = 70°$	
Crystal	$2d$ (Å)	Using K Lines	Using L Lines
Lithium fluoride, LiF (220)	2,848	V (23)	Pr (57)
Lithium fluoride, LiF (200)	4,208	K (19)	In (49)
Germanium, Ge (111)	6,532	S (16)	Zr (40)
Pentaerytrol PET (002)	8,742	Si (14)	Rb (37)
Thailium acid phtalate, TIAP (1010)	25,75	O (8)	V (23)
Lead stearate decanoate, PbSD	100	B (5)	Ca (20)
Multilayer, W/C	120	B (5)	

[a]The crystallographic orientation is given in brackets.

1981; Klockenkämper 1980). The energy resolution of these types of detectors is poor and therefore a preselection by means of diffraction (Bragg reflector) is necessary.

4.3.2. Energy-Dispersive X-Ray Fluorescence Spectrometry

The development of semiconductor materials had consequences in X-ray spectrometry. Indeed the energy resolution of lithium-drifted silicon [Si(Li)] detectors is sufficiently high to allow X-ray spectrometry without previous dispersion with a Bragg reflector. X-ray photons entering the Si(Li) crystal interact primarily by photoelectric absorption, producing electron–hole pairs. The number of pairs produced is proportional to the energy of the photon, thus making X-ray spectrometry possible. The produced charge is swept from the semiconductor diode by the applied voltage and produces an electric pulse. The signals produced by the X-ray photons are very weak and have to be amplified by a low-noise amplifier transistor (field effect transistor, FET) (Heinrich 1981). For optimal operation of the system and to freeze the lithium drifting process, effective cooling with liquid nitrogen is indispensable. By processing the signals with a fast multichannel analyzer (MCA), energy-dispersive X-ray fluorescence spectrometry (EDXRF) will produce X-ray spectra with a spectral resolution sufficient for most applications. Though its spectral resolution is poor in comparison to WDXRF, EDXRF has the advantage of being simpler and cheaper and to allow truly simultaneous observation of all X-rays.

As in WDXRF, in EDXRF different X-ray tubes are applied as primary X-ray source. Additionally X-ray emitting radioisotope sources such as cadmium-109 (^{109}Cd), americium-241 (^{241}Am), and others are used instead of X-ray tubes. Radioisotope sources, among others, are used in compact portable spectrometers that are applied in field work for fast screening of the elemental composition of a material.

Recently, synchrotrons (Haller and Knöchel 1996) have been used as extremely intense X-ray sources that make XRF analysis with a superior sensitivity possible (synchrotron radiation XRF, SRXRF). The high initial intensity available at the source moreover allows microbeams to be used, offering excellent spatial resolution (μ-SRXRF) and good sensitivity.

4.3.3. Samples and Sample Preparation

In general, no restrictions exist concerning the samples or their preparation for either WD- or EDXRF. Conducting or isolating solids, powdered materials, and liquids can be subjected to X-ray fluorescence analysis. One of the most prominent properties of XRF analysis is that the samples can be recovered completely after the analysis. Unfortunately, some samples show alterations caused by the irradiation with X-rays and a change in color or a change in mechanical properties caused by radiolysis is occasionally observed. Some minerals, glasses, plastics, and organic materials are affected more seriously than others. The sample chambers of the spectrometers in general contain the end of the X-ray tube, parts of the spectrometer and of the crystal changer (WDXRF), detectors or parts of these, collimators, and the sample stage. Usually the sample chamber is evacuated down to a pressure of the order of 10 Pa. Nevertheless, XRF instruments can be operated at normal atmospheric conditions or filled with helium depending on the sample and/or the analytical tasks. Especially in the latter cases, the size of the sample chamber can be sufficiently large to accommodate small archaeological objects (jewels, statuettes, etc.). The sample chamber among others is designed to prevent X-rays from escaping the instrument and thus is important for safe operation. Nevertheless, open-beam instruments have been used, especially for *in situ* analysis of large artifacts (e.g., paintings). Safety in this case will depend on keeping an appropriate distance during operation and on the use of protection shields.

4.3.4. Quantification

Unfortunately, WD- and EDXRF operated in conventional modes are affected by matrix effects that cannot be neglected. These are caused by absorption of X-rays in the sample and by fluorescence enhancements when the radiation is crossing the sample. Also effects related to the surface texture, to the inhomogeneity of the sample, and to the particle size have been observed. All these effects lead to curved instead of straight calibration lines and have to be taken into account if quantitative evaluation of data and accurate determination of the elemental composition of a sample are the goals. Many calibration and correction procedures are being applied in XRF analysis (Bertin 1975):

1. *Standard Addition and Dilution Methods.* The pure sample itself is measured as well as at least one sample to which a known amount of the elements to be determined (analytes) is added. Sample preparation is necessary to make the two specimens as similar in density, surface condition, and the like as possible. In this method, the sample actually provides its own standard(s) in its own matrix.
2. *Thin-Film Methods.* The specimens (samples and standards) are made so thin that absorption/enhancement effects become irrelevant.
3. *Matrix Dilution Methods.* The matrix of all specimens is diluted with a suited material so that the effect of the matrix is determined by the material used for the dilution.
4. *Comparison Standard Methods.* The X-ray intensities from the samples are compared with those of standards, having the same form as the sample(s) and as nearly as possible the same matrix and analyte concentration.
5. *Internal Standardization.* The comparison standard method is improved by quantitative addition to all specimens of an internal standard element. The calibration function involves the intensity ratios of the analyte and internal standard lines.
6. *Standardization with Scattered X-rays.* The intensity of the primary X-rays scattered by the specimen is used to correct for absorption/enhancement effects.
7. *Experimental Correction.* Various other experimental techniques have been described to minimize or compensate matrix effects.
8. *Mathematical Correction.* Absorption/enhancement effects are corrected for mathematically by use of experimentally derived parameters.

The most prominent methods are the fundamental parameter approaches based on standards or on programs for the standardless analysis (Lachance and Claisse 1995). Monte Carlo calculations (Vince et al. 1995a, 1995b) and chemometric techniques (Adams and Allen 1998; Luo et al. 1998) are applicable as well. It is beyond the scope of this book to go into further detail on the different calibration methods. Suffice it to say that the development of powerful hardware and software in combination with new mathematical and computational techniques improved XRF in the last decades from a semi-quantitative to a fully quantitative method.

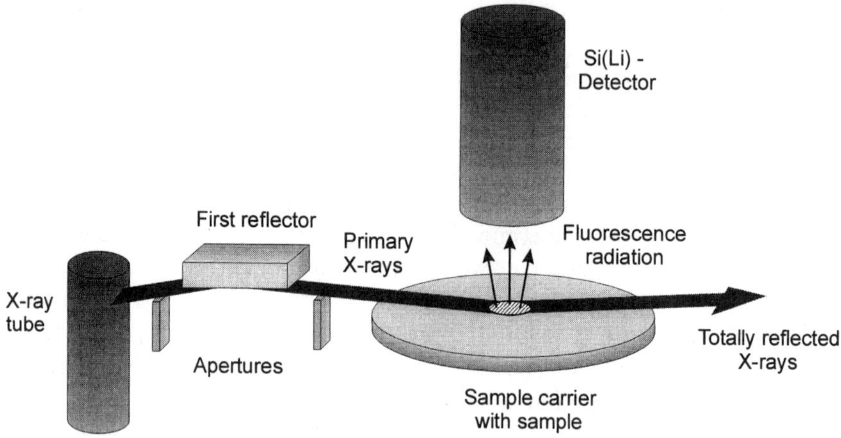

Figure 4.6. Schematic representation of a total reflection X-ray fluorescence spectrometer (TXRF).

4.3.5. Total Reflection X-Ray Fluorescence Spectrometry

Total reflection X-ray fluorescence spectrometry (TXRF) is a newly developed method that differs from conventional energy dispersive X-ray fluorescence (EDXRF) by the excitation geometry (Klockenkämper and von Bohlen 1992; Prange and Schwenke 1992; Klockenkämper et al. 1992; Klockenkämper 1996). In TXRF a flat, sheetlike primary X-ray beam is filtered and collimated when passing a first reflector (see Fig. 4.6). Then it is directed at a small glancing angle $<0.1°$ to a very flat and smooth surface of a sample carrier. The incident beam is totally reflected at the sample carrier's surface. It therefore hardly interacts with the carrier but effectively excites the elements of the sample. The fluorescence radiation emitted by the sample is recorded by a Si(Li) detector positioned directly above it. Usually TXRF instruments operate at normal pressure in the sample chamber. Absorption of X-rays by the gas in the chamber is minimized since the sample detector distance is very small.

The special excitation arrangement leads to a substantial decrease of the spectral background originating from the sample carrier. If applying only small amounts of sample (microgram or microliter size), a very effective excitation to fluorescence of the sample and an effective collection of the secondary radiation is observed. The detectable amount of a great number of elements is in the order of a picogram (10^{-12} g). Moreover, if a critical sample mass (upper limit), depending on the matrix of the sample, is not exceeded, no matrix effects occur and multielement analysis with a simple and reliable quantification becomes possible (Klockenkämper and von Bohlen 1988).

In TXRF spectra the intensities I of the X-rays are recorded. If the above-mentioned conditions concerning the sample are fulfilled, no interelement effects are observed and the recorded intensities are proportional to the mass fractions of the corresponding elements. The intensities have only to be divided by their respective relative sensitivities S. The sensitivities are constants of proportionality that have to be determined only once by analyzing the residue after drying of aqueous standard solutions. By the absence of matrix effects these sensitivity factors can be applied to all kinds of samples.

If the sample mass is too low to be determined accurately, only relative mass fractions (m_x) of the detected elements can be calculated by the following equation (von Bohlen et al. 1994; Klockenkämper 1996):

$$m_x = \frac{I_x/S_x}{\sum_k I_k/S_k} \times 100\%$$

Thus, relative mass fractions with respect to the sum of the masses of all detected elements (k) can be calculated. However, a full quantification becomes possible if the sample mass is known and the method of internal standardization is applied. For this purpose, a known quantity (m_i) of only one standard element, which is not contained in the sample, is added to the sample. After recording the spectrum, the absolute masses of the detected elements (m_x) are calculated from the following equation (von Bohlen et al. 1994; Klockenkämper 1996):

$$m_x = \frac{I_x/S_x}{I_{\text{int}}/S_{\text{int}}} \times m_{\text{int}}$$

and when relating the masses m_x to the total sample mass, mass fractions (e.g., in micrograms per gram) can be obtained. As mentioned above, the TXRF sample chamber is normally filled with air, so that the characteristic X-ray radiation of elements with atomic number ≤ 13 is absorbed and these elements are not detectable.

TXRF instrumentation is, compared to other modern techniques, relatively simple. An excitation unit containing a high-voltage generator and an X-ray tube with housing, a total reflection module, and a detection unit composed of a Si(Li) detector and a multichannel analyzer are needed. The operating and maintenance costs are low.

TXRF proved to be a versatile tool for chemical analyses, especially for extreme trace and microanalyses and for direct analyses of solid samples as was reviewed by several authors (Reus and Prange 1993; Klockenkämper et al. 1992; Prange and Schwenke 1992; Klockenkämper 1996; Moens et al. 1994).

4.4. APPLICATIONS

The examples cited here are far from a complete overview and should only show the versatility and the variety of applications of X-ray fluorescence analysis.

4.4.1. Analysis of Ancient Coins

Metal analysis is one of the most prominent applications of XRF in archaeometry. In particular, the analysis of ancient coins proved to be very useful. In general, coins were produced from well-controlled alloys by a known mint and often were provided with a date of issue. Many references concerning the technology, the typology, and the chronology can be found in historical documents and in modern studies. Coins are, except for the rough surface, ideal objects for XRF analysis. They can be investigated without any sample preparation and nearly all elements of interest present as major or minor constituents, as well as some traces can be detected in coins of all forms and sizes. Further advantages are their favorable physical properties allowing nondestructive multielement analysis in a reasonable time. Only layers of patina, bleaching, or plating have to be considered when discussing the results of elemental analyses. Indeed, the observed fluorescence radiation originates from the top layer of the sample, resulting in a "depth of information" of 1 to some 100 μm, depending on the applied technique, the alloy, and the elements in question. Therefore XRF results do not necessarily represent the composition of the bulk of the coin material.

Until now, the elemental composition of hundreds or even thousands of ancient coins, predominantly of Roman origin, has been determined by XRF analysis. Very different approaches have been used for this purpose, and it is notable that XRF techniques often have been used in combination with other methods.

Condamin and Picon (1964) were the first to examine corrosion effects on dinarii and concluded that the surface silver content will be markedly higher than the silver content of the alloy within the coin. The notable discrepancy between the silver contents in Roman coins induced several approaches going hand in hand with the development of techniques to study problems of bleaching, plating, corrosion, and wear off by circulation. Walker (1976, 1977, 1978) presented an extended work related to the metrology of Roman silver coinage. Several other works dealing with the examination of Roman coins (e.g., Carter 1978; Butcher et al. 1997) and related analytical problems (e.g., Lutz and Pernicka 1996) have been published. Some of these are presented below.

In his approach, Klockenkämper (1978) uses WDXRF and a scanning electron microscope with energy dispersive X-ray detection (SEM-EDX) for the characterization of more than 200 coins. The combination of the two techniques with different penetration depth of the primary X-rays and of electrons, respectively, into the material provided insight into the silver enrichment at the surface of dinarii relative to the silver content in deeper layers. The results showed the decay of values in the Roman Empire: Less silver was used for coinage alloys, but the surface of the coins was enriched starting from about the year A.D. 200. Similar results are reported for Roman coins (dinarii, quintarii, etc.) found in Switzerland, Augusta Raurica, and analyzed by Zwicky-Sobczyk and Stern (1997). They combined the results of EDXRF analysis using silver-K and silver-L lines and of density measurements to characterize more than 600 coins. The intensity ratio of silver X-ray lines in modern silver alloys is, theoretically, a clue to the presence of plating, but with most ancient silver coins this ratio is influenced by the surface morphology and by corrosion. However, the silver-K intensities of massive and of uncorroded plated silver coins are clearly different.

Though most work has been done on Roman coins, other coins of very different regions and ages were analyzed using XRF as well. Ancient dirhams (silver coins) from the Abbasid period 158–218 Hirji (A.D. 775 to 833) of the Great Islamic Empire were analyzed by Al-Kofahi et al. (1997). The dirhams were found to have silver (about 41 to 84%) as a major constituent and minor quantities of 1% or more of silicon, copper, mercury, gold, and lead, of 0.1 to 1% of aluminium, phosphorus, iron, and tungsten and less than 0.1% of titanium, nickel, and zinc. The results of the analyses show that the dirhams coined in the capital of the state, Madinat Essalam, are of good quality, having a silver concentration of at least 80%, whereas those fabricated in locations far from the capital have significantly lower silver concentrations (41 to 54%).

Genuine and counterfeit German *Reichsgoldmünzen* dated 1872 to 1914 made of gold were studied by Klockenkämper et al. (1990). A gold content of about 90% and a silver content of about 0.4% were found to be reliable criteria of genuineness. Counterfeits frequently show more gold and often more or less silver. Additionally, several counterfeits have a gold enrichment at the surface.

Early Russian platinum roubles, of the beginning of the eighteenth century, showing gold inclusions and lower fineness than reported by the mint were examined by Auer et al. (1998). In earlier times platinum ores contain-

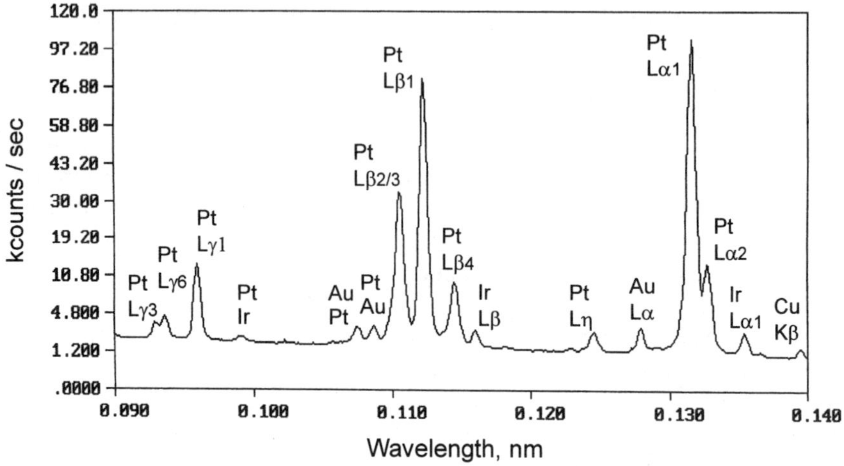

Figure 4.7. WDXRF spectrum of a platinum rouble of the year 1838 showing gold and iridium lines next to those of platinum; analyzing crystal LIF (220); rhodium X-ray tube 100 kV, 30 mA.

ing ca. 64% platinum found at the Ural could be refined only by a complex chemical procedure. The refined platinum was minted in a powder metallurgical process to coins of 2, 3, or 6 roubles. These should have 96 to 99% of platinum and up to 0.5% of ruthenium, 0.25% of palladium, 1.2% of iridium, and traces of chromium, iron, and copper. The analysis of 2 rouble coins revealed the presence of additional elements like manganese, zinc, silver, and gold. An XRF spectrum of a platinum rouble with gold inclusions is shown in Figure 4.7. The detection of manganese, iron, nickel, copper, zinc, palladium, silver, iridium, and gold suggests a mixing of refined platinum and original platinum ores for the coinage.

The transfer of gold and silver originating from the New World starting about the year 1500 and the propagation in the Old World and later mixing with European, Russian, and Ottoman gold used in coinage was studied by Guerra (1995, 1998). Typical element traces, for example, arsenic in copper, indium in silver, rhodium, palladium, and platinum in gold, present in coinage metals of different provenance were analyzed by a combined procedure including methods of neutron activation, laser ablation mass spectromerty, proton-induced X-ray fluorescence, and XRF.

4.4.2. Pottery and Porcelain

Among the most important archaeological finds are ceramic materials, and it should not come as a surprise that numerous XRF studies deal with pottery or porcelain. Small samples are usually taken and either analyzed as such

(LaBreque et al. 1998) or prepared for analysis (Punyadeera et al. 1997) following standard XRF procedures. Pottery samples can often be linked to a common provenance based on petrographic and chemical characteristics originating from the clay matrix or the temper. The interpretation of the analytical information in most cases will be based on multivariate statistical methods. XRF has been used for the determination of major, minor, and trace elements characterizing pottery of known or unknown provenance. For instance, LaBreque et al. (1998) have used radioisotope (cadmium-109) XRF to study the origin of 12 Majolica ceramics, found in the Americas, by comparing the concentrations of lead, rubidium, strontium, and zirconium in these with the concentrations of the same elements in 29 samples of known provenance. The lead–tin enamel of the shards was removed with a diamond saw, and the remaining clay fraction was prepared for analysis by grinding and sieving of the obtained powder. A fraction of the powder was transferred to a sample holder and measured as such. Principal component analysis was used to establish the provenance of the unknown samples.

Yu and Miao (1996) used EDXRF for the determination of trace elements in Chinese blue and white porcelains of the Ming dynasty, in imitations, and in a modern porcelain object. The porcelain objects were analyzed as such. For calibration, samples of broken objects from the Ming and Qing dynasties were taken and analyzed with EDXRF and inductively coupled plasma atomic emission spectroscopy (ICP-AES). For the latter analysis the outer 1-mm layer of the shards was taken as a sample. The material was ground and sieved prior to dissolution and ICP-AES analysis. The quantitative results were used for principal component analysis. Using the concentrations of 13 trace elements, this allowed porcelain from different periods during the Ming dynasty to be distinguished and imitations to be separated from authentic artifacts.

Punyadeera et al. (1997) used EDXRF for a provenance study of Iron Age pottery from South Africa. They compared the amounts of 10 transition metals in 107 potsherds originating from four archaeological sites. The shards were ground and the powder was homogenized, mixed with a binder, and compressed to a pellet of standardized dimensions and with a well-polished flat surface. Pellets prepared from blank samples and from four geological samples of known composition were used for calibration purposes. The data obtained were subjected to correspondence analysis to distinguish between groups. The groupings were interpreted in terms of social and cultural interactions between sites.

4.4.3. Obsidian

Obsidian was widely used and traded in prehistoric times as raw material for tool making. Provenance determination of obsidian found at archaeological sites therefore is an interesting tool for studying trade patterns and intercultural contacts. Among others, Vázquez and Escola (1995) have used WDXRF to determine the concentration of five trace elements (rubidium, zirconium, strontium, titanium, and manganese) and one major element (iron) in obsidian samples from archaeological and geological sites. Obsidian flakes were ground and the powder was sieved and compressed to a pellet using cellulose as a backing. Synthetic standards, mimicking obsidian, were made by mixing known amounts of the elements of interest with basic components (mainly silicon dioxide) that reflect the major element composition of obsidian. It was found that rubidium, zirconium, and strontium are very reliable indicators for the provenance determination of the obsidian found at the site of Casa Chavez Monticulos (450 B.C. to A.D. 650) in Argentina. The obsidian deposit of Ona, at a distance of 80 to 90 km from the site, turned out to have been the source of the material used at the site.

4.4.4. Ivory

The analysis of mammoth ivory was the subject of a study (though not a purely archaeological study) by Shimoyama et al. (1998). The Convention on International Trade in Endangered Species (CITES), while protecting elephants has also created a market for smugglers. Archaeological mammoth ivory is obviously not meant in the CITES, but it has been discovered that elephant ivory is transported together with and as mammoth ivory. Visual distinction of these materials is extremely difficult and therefore a method was designed whereby nondestructive XRF analysis of the dentine is used to distinguish mammoth ivory from modern elephant ivory. Use was made of glancing angle XRF for the determination of the intensity ratios of the K_α X-rays of strontium and calcium and of bromine and phosphorus. These ratios are sufficiently different for both types of ivory to allow mammoth ivory clearly to be distinguished from smuggled African elephant ivory.

4.4.5. *In Situ* Element Mapping with EDXRF

Element mapping of small objects can be performed with a small diameter X-ray beam and by stepwise moving the object, at each position performing XRF analysis. EDXRF equipment can be made to be portable or at least displaceable. Thus it becomes feasible to transfer the equipment to a museum or any site where artifacts need to be studied. Moreover the equipment can be attached to a mechanical system that allows it to be positioned at will with respect to large artifacts. The primary X-ray beam is aimed at a selected spot on the surface of the object and the characteristic fluorescence X-rays are measured. Schreiner et al. (1992) have used such a system for *in situ* mapping of elements in large objects such as paintings. In one such application, they performed element mapping of Indian miniature paintings (seventeenth century) in the Millionenzimmer at the Schönbrunn palace in Vienna (Austria). The spatial resolution of the mapping was 1 mm and the measuring time per spot was 30 sec. This led to a long total analysis time (up to hours), but it was possible to identify most of the pigments used: red lead, vermilion, white lead, a green copper pigment, azurite, and ochre. In addition silver and gold were found to be present in metallic form. Some pigments could not be identified and were assumed to exclusively consist of light elements that escape XRF analysis under the conditions used (e.g., ultramarine or organic pigments). The mapping allowed original sections to be distinguished from later additions.

In situ analysis of paintings is hampered by the fact that the observed X-rays originate not only from the pigments under study but also from superficial contamination, varnish, underlying paint layers, or even the carrier. Critical interpretation of this mixed information is therefore necessary and in most cases sufficient to come to relevant conclusions.

4.4.6. Glass

Glass Analysis with SRXRF. Janssens et al. (1996) and Adams et al. (1997) have used synchrotron radiation for the analysis of glass. These authors studied 90 Roman glass samples (about A.D. 4 to 68) originating from Qumrân (Jordan) and determined the concentrations of trace elements with μ-SRXRF. Unlike the major element composition, the trace element pattern showed considerable variation and allowed the objects to be divided in two different groups (Fig. 4.8). It was concluded that objects from each of both groups originated from the same batch of glass, suggesting a local production or import from the same source. The same authors also used the method for trace element mapping, making line scans over glass sections in a direction perpendicular to the surface. Thus the alteration of the glass could be studied.

Glass Analysis with TXRF. The characterization of valuable glass objects exhibited in museums becomes possible when no samples need to be taken or when only a few grains of material are required. Samples of ca. 100 µg can be taken with a diamond drill from the bottom of the artifacts (Wegstein 1997). The technique was applied to sample artifacts in the Hessisches Landesmuseum Kassel, Germany. The small samples were stored in plastic containers to avoid loss or contamination during transport to the laboratory.

Subsamples of only one or two grains were used to perform semiquantitative analyses of a Bohemian *Reichsadlerhumpen* dated back to the year

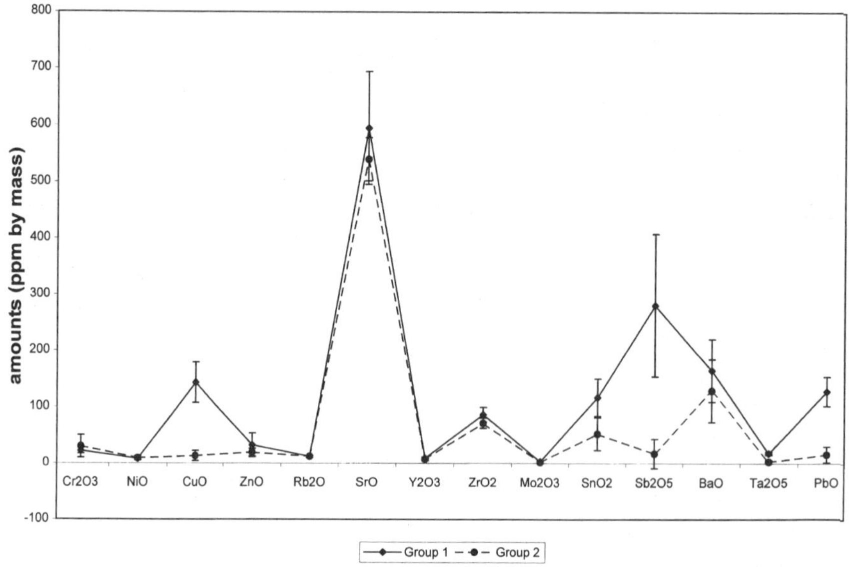

Figure 4.8. Concentrations determined by SRXRF in two different types of Roman glass from Qurnrân (Jordan).

1572. Relative amounts referring to the amount of iron (~100 arbitrary mass units) were determined with TXRF: potassium = 2160, calcium = 2260, manganese = 147. Minor constituents and trace elements were titanium, zinc, rubidium, strontium, barium, and lead. In the glass of this tumbler all elements found in the ashes of native plants (potassium, calcium, manganese, iron, zinc, rubidium, strontium, barium), sands (titanium, iron), and lime are present. The presence of other traces (e.g., lead) can be explained by the usual recycling of old glass. Furthermore, potassium- and manganese-rich components must have been added to the glass because the amounts present cannot be explained by the inclusion of native plant ashes alone. For comparison, glass fragments from old Prague (sixteenth century) and from the glassworks in the Vogler (a forest in Lower Saxony) were analyzed. Nearly identical element ratios of potassium : calcium of 1 : 1 and of manganese : iron of 2 : 1 were found as in the Bohemian *Reichsadlerhumpen*. In contrast to these results, the analysis of recent glass from glass-making regions in Lower Saxony and Thuringa shows a potassium:calcium ratio of 1 : 5 and less manganese than iron.

4.4.7. Analysis of Artist's Pigments

The analysis of valuable works of art such as paintings, miniatures, or polychrome sculptures is ethically acceptable only if no damage is inflicted to the artefact (as in direct XRF without sampling) or if the damage is negligible. For TXRF only microsamples (less than 1 μg) are required, and a gentle microsampling method was developed, consisting of rubbing a dry cotton wool bud (Q-tip) over the painted surface. This will remove a minute amount of material, without causing any visible damage. A fraction of the material can next be transferred to the TXRF sample carrier.

TXRF was extensively used for identification and fingerprinting purposes. Apart from being practically nondestructive, the method proved to be advantageous because samples can easily be taken by restorers or art historians, the response time is short (analysis time of a few minutes), and the costs are very low. However, sampling is possible only when the pigment layer is not covered by varnish. For most old paintings, therefore, samples can only be taken during restoration. Also, the sampling only concerns the surface of the object, and the analysis consequently does not reveal any information on deeper paint layers.

TXRF proved to be an excellent method to identify artists' pigments via the detection of key elements, a principle also applied in the *in situ* XRF pigment analysis outlined before. Especially ancient pigments are predominantly of mineral origin and can indeed be characterized by the presence of one to three typical elements (Klockenkämper et al. 1993; Moens et al. 1995). In addition, TXRF allows the mixing ratios of different pigments, used to produce a paint of a particular hue, to be determined. Finally, it is also possible to determine minor and trace element concentrations that reflect the characteristics of the minerals used, of the refinement process, and of the actual paint-making recipe used in the artist's workshop. This concentration pattern is often a unique result of various parameters and therefore can be used, with the necessary precautions, to tell different workshops apart.

Apart from this, pigment identification and characterization can provide data for various other investigations: general art historical or archaeological studies on the materials used, conservation and restoration, detection of forgeries, and relative dating. The latter two aims are within reach because some pigments are known to have appeared (dating *post quem*) on artists' pallets and/or to have disappeared (dating *ante quem*) at a certain time.

Apart from the detection of anachronistic pigments, the detection of pigments that were not found in any of an artist's authentic works can also reveal falsification. For instance, in 1994 a privately owned painting claimed to be painted by Modigliani was offered to the organizers of a comprehensive exhibition on the work of this master. The pallet of Modigliani has been studied by Delburgo et al. (1981), who analyzed 15 authentic oil paintings with *in situ* XRF. It was reported that in those paintings Modigliani only used cadmium and chrome yellow, ochre, vermilion, chromium green, Prussian blue, white lead, zinc white, and organic pigments. In the painting discussed here, cerulean blue and emerald green where detected via TXRF analysis of microsamples, which could indicate falsification (Devos 1996). Further radiographic research revealed anomalies in the painting technique as well, and the painting was rejected by the exhibition.

4.4.8. Analysis of Ink

Medieval manuscripts are often illuminated with miniatures and decorative motives (Van Hooydonk et al. 1998, Vandenabeele et al., 1999). The pigments used have been studied among others with TXRF. Also writing ink, in most cases gall ink, was found to yield interesting information on the writer. Gall ink was made by adding iron sulfate to gall extract. A complex between iron and gallic acid is formed that, only after the writing, turns into a black ink by oxidation of the iron (Wunderlich 1994). In the scriptorium new ink had to be made quite regularly, and it is likely that the impurity pattern was slightly different for each new batch and quite different for different scriptoria. This allowed constructing a chronology of undated letters and notes by Galileo based on impurity patterns determined with proton-induced X-ray emission (PIXE) and comparison of these patterns with those of inks on dated bills and diaries (Giuntini et al. 1995). Recently, ink analysis was used to study a series of manuscripts from the collection of Raphael de Mercatellis (1437 to 1508), native son of Philip the Good, Duke of Burgundy (Derolez 1979). One of the manuscripts (Decretum Gratiani) consists of three volumes from which microsamples where taken from the ink on several pages. The relative concentrations of a number of elements, as determined by TXRF are shown in Figure 4.9. All samples show a similar pattern, except one, a sample from a colophon, mentioning the acquisition date (1505). The

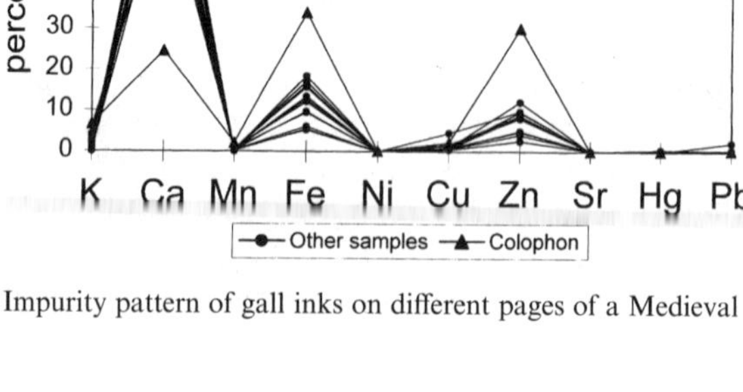

Figure 4.9. Impurity pattern of gall inks on different pages of a Medieval manuscript.

analysis suggests that the colophon was not written at the same time or the same place as the manuscript.

4.4.9. Violin Varnishes

Historical violin varnishes are mixtures of natural products in which the main components are organic substances such as drying oils, essential oils, resins, waxes, organic colorants, spirit, and others. Beside these major components some inorganic substances were added to control the properties of the varnish. It is indeed possible to influence the color, the hue, the transparency, the hardness, and also to reduce drastically the drying time of oil-based varnishes. A great variety of old formulas describing how to produce and how to process varnishes for violinmakers are known, but unfortunately those of early Italian masters were lost. Especially these varnishes, applied by prominent violinmakers and families of violinmakers remain unattainable. The particular beauty of these varnishes and their capability to preserve the wood and thus the sound quality of the instruments can therefore not be reproduced.

Small flakes of varnishes were collected from stringed instruments that were under restoration. The flakes were removed carefully from the pre-cleaned surface of the varnishes by the aid of a scalpel and deposited in small paper bags for transport and storage. Fractions of such samples with masses below 20 μg were analyzed directly by TXRF without any additional preparation except the deposition on Plexiglas carriers by means of wooden toothpicks (von Bohlen 1999; von Bohlen and Meyer 1996). The multielement analyses were performed in a counting time of 100 to 300 sec. More than 20 elements (silicon, phosphorus, sulfur, chlorine, potassium, calcium, titanium, chromium, manganese, iron, cobalt, nickel, copper, zinc, arsenic, bromine, rubidium, strontium, barium, lead, potassium, calcium) were detected simultaneously and relative detection limits of the order of 10 pg/μg for elements with atomic number $Z \geq 24$ were estimated. In Figure 4.10 a TXRF spectrum of an Andrea Guarneri varnish of a violin of ca. 1660 is shown. Elements such as iron, arsenic, and lead could be related to pigments added to the varnishes, whereas manganese, cobalt, and lead are related to siccatives accelerating the drying process of the oil-based varnishes. Occasionally elements are detected (e.g., silver or tin) that seem to have been introduced accidentally.

Unfortunately, the masses of the analyzed flakes were too low to be weighed accurately, and, consequently, it was impossible to calculate the mass fractions of the detected elements in relation to the total sample mass. However, a possibility for quantification is given when relative mass fractions of the detected elements are calculated with respect to the sum of all

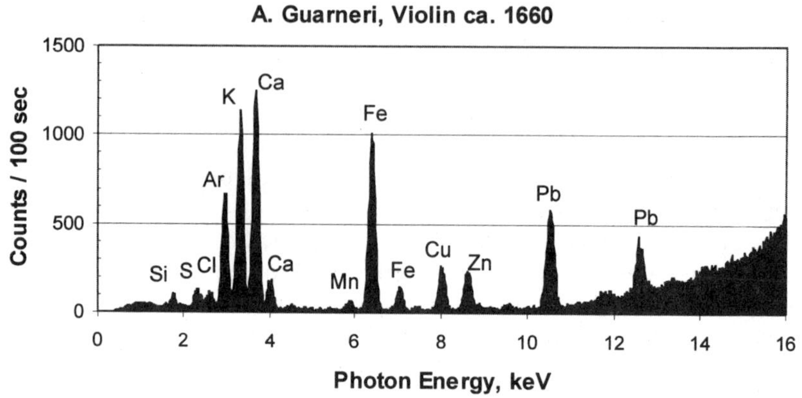

Figure 4.10. TXRF spectrum of an Andrea Guarneri varnish obtained from a violin of about 1660. A sample with a mass of less than 20 μg was used for the analysis. Excitation: molybdenum X-ray tube, 50 kV, 38 mA and spectrum acquisition time of 100 sec.

detected elements (see before). After normalization of the data, a generation and visualization of multielement patterns was performed by a graphical representation of TXRF results using star plots (cf. Fig. 4.11). This combination makes an easy classification of element pattern in violin varnishes possible. It can be used to characterize unknown historical varnishes, to distinguish retouches made on historical stringed instruments using modern varnishes, and for general classification of the elemental distribution in varnishes.

4.5. CONCLUSION

XRF-based methods, versatile and various as they are, will continue to be of great use to archaeometric research. For qualitative analysis the non-destructive character and the speed of analysis allow the element composition of an artifact to be known instantly and without damaging the object. This information is often sufficient to determine the nature of the material used for making the object. Quantification may require expert knowledge, but numerous examples show that accuracy and precision are within reach. The useful information present in the vast amount of data, generated by the simultaneous multielement analysis, often is extracted from the data set with the aid of statistical analysis.

In recent years the means have been created to perform microanalysis and spatially resolved analysis with good resolution. Microsamples, as required for TXRF, can be taken from most artifacts without inflicting any visible damage and can easily be analyzed quantitatively. Microbeams, as generated

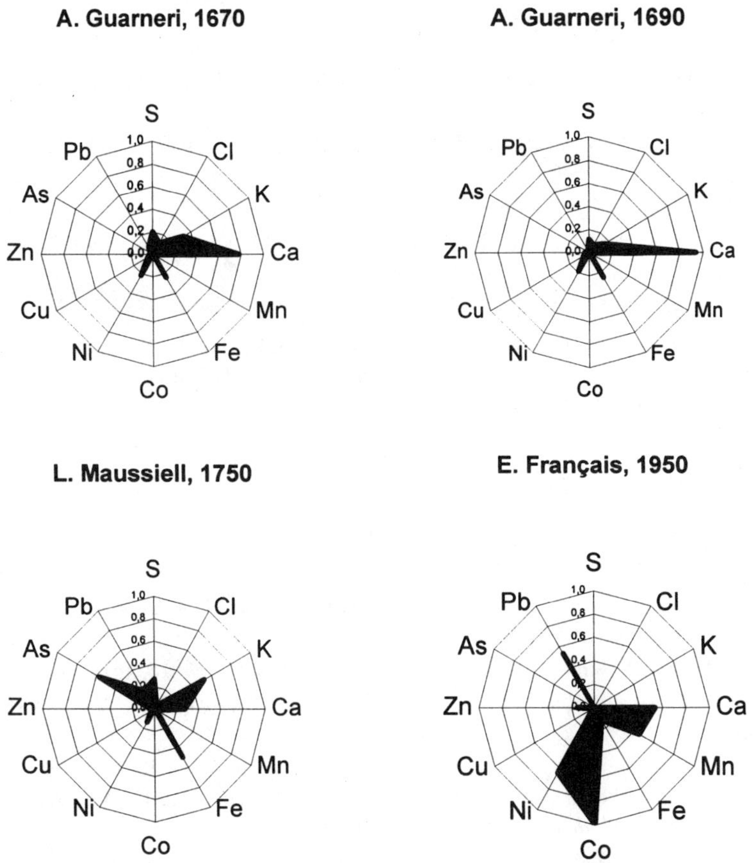

Figure 4.11. Radial plots for the characterization of elemental pattern of varnishes of two different celli made by A. Guarneri, of a varnish of a cello made by L. Maussiell, and of a modern violin varnish made by E. Français. Relative concentrations were obtained by TXRF analysis of small varnish flakes.

by special X-ray source setups and synchrotrons, allow element mapping at submilimeter scale. It is expected that new applications of these new possibilities will continue to be developed, making XRF an even more attractive analytical tool than it is today.

REFERENCES

Adams, M. J., and J. R. Allen (1998), *J. Anal. Atom. Spectrom.* **13**, 119.

Adams, F., A. Adriaens, A. Aerts, I. De Raedt, K. Janssens, and O. Schalm (1997), *J. Anal. Atom. Spectrom.* **12**, 257.

Al-Kofahi, M. M., K. F. Al-Tarawneh, and J. M. Shobaki (1997), *X-Ray Spectrom.* **26**, 10.

Auer, E., Th. Rehren, A. von Bohlen, D. Kirchner, and R. Klockenkämper (1998), *Metalla* **5**, 71.

Bertin, E. P. (1975), *Principles and Practice of X-Ray Spectrometry*, 2nd ed., Plenum, New York.

Broll, N. (1996), *J. Phys. IV* **6**, 583.

Butcher, K., M. Ponting, and G. Chandler (1997), *AJN Second Series* **9**, 17.

Carter, G. F. (1978), *Numismatic Chronicle* **19**, 67.

Condamin, J., and M. Picon (1964), *Archaeometry* **7**, 98.

Delbourgo, S., and L. Faillant-Dumas (1981), L'Etude au laboratoire de Recherche des musées de France, in *Amadeo Modiglini, 1884–1920*, Musée d'Art Moderne de la Ville de Paris, Paris.

Derolez, A. (1979), *The Library of Raphael de Mercatellis*, Story Scientia, Ghent.

Devos, W. (1996), Ph.D. thesis, Laboratory of Analytical Chemistry, University of Ghent.

Guerra, M. F. (1995), *Appl. Radiat. Isot.* **46**, 583.

Guerra, M. F. (1998), *X-Ray Spectrom.* **27**, 73.

Giuntini L., F. Lucarelli, P. A. Mandò, W. Hooper, and P. H. Barker (1995), *Nucl. Instrum. Methods Phys. Res. B* **95**, 389.

Haller, M., and A. Knöchel (1996), *J. Trace Microprobe Tech.* **14**, 461.

Heinrich, K. F. J. (1981), *Electron Beam X-Ray Microanalysis*, Van Nostrand Reinhold, New York.

Janssens, K., A. Aerts, L. Vincze, F. Adams, C. Yang, R. Utui, K. Malqvist, K. W. Jones, M. Radtke, S. Garbe, F. Lechtenberg, A. Knöchel, and H. Wouters (1996), *Nucl. Instrum. Methods Phys. Res. B* **109/110**, 690.

Jenkins, R. (1988), *X-Ray Fluorescence Spectrometry*, Wiley-Interscience, London.

Klockenkämper, R. (1978), *Fresenius Z. Anal. Chem.* **290**, 212.

Klockenkämper, R. (1980), in *Ullmans Encyklopädie der technischen Chemie*, Verlag Chemie GmbH, Weinheim, p. 501.

Klockenkämper, R. (1996), *Total Reflection X-Ray Fluorescence Analysis*, John Wiley & Sons, New York.

Klockenkämper, R., and A. von Bohlen (1988), *Spectrochim. Acta B* **44**, 461.

Klockenkämper, R., and A. von Bohlen (1992), *J. Anal. Atom. Spectrom.* **7**, 273.

Klockenkämper, R., M. Becker, and H. Otto (1990), *Spectrochim. Acta B* **45**, 1043.

Klockenkämper, R., J. Knoth, A. Prange, and H. Schwenke (1992), *Anal. Chem.* **64**, 1115A.

Klockenkämper, R., A. von Bohlen, L. Moens, and W. Devos (1993), *Spectrochim. Acta B* **48**, 239.

LaBrecque, J. J., J. E. Vaz, J. M. Cruxent, and P. A. Rosales (1998), *Spectrochim. Acta B* **53**, 95.

Lachance, G. R., and F. Claisse (1995), in *Quantitative X-ray Fluorescence Analysis: Theory and Application*, G. R. Lachance and F. Claisse, Eds., Wiley, New York.

Lutz, J., and E. Pernicka (1996), *Archaeometry* **38**, 313.

Luo, L., A. Ji, G. Ma, and C. Guo (1998), *X-Ray Spectrom.* **27**, 17.

Moens, L., W. Devos, R. Klockenkämper, and A. von Bohlen (1994), *TRAC Trend. Anal. Chem.* **13**, 198.

Moens, L., W. Devos, R. Klockenkämper, and A. von Bohlen (1995), *J. Trace Microprobe Tech.* **13**(2), 119.

Prange, A., and H. Schwenke (1992), *Adv. X-Ray Anal.* **35**, 899.

Puyandeera, C., A. E. Pillay, L. Jacobson, and G. Whitelaw (1997), *X-Ray Spectrom.* **26**, 249.

Reus, U., and A. Prange (1993), *Spectrosc. Eur.* **5**, 25.

Schreiner, M., M. Mantler, F. Weber, R. Ebner, and F. Mairinger (1992), *Adv. X-Ray Anal.* **35**, 1157.

Shimoyama, M., T. Nakanishi, Y. Hamagana, T. Ninomiya, and Y. Ozaki (1998), *J. Trace Microprobe Tech.* **16**, 175.

Vandenabeele, P., B. Wehling, L. Moens, B. Dekeyzer, B. Cardon, A. von Bohlen, and R. Klockenkämper (1999), *The Analyst* **194**, 169.

Van Hooydonk, G., M. De Reu, L. Moens, J. Van Aelst, and L. Milis (1998), *Eur. J. Inorg. Chem.* **5**, 639.

Vázquez, C., and P. Escola (1995), *J. Radioanal. Nucl. Chem. Lett.* **200**, 373.

Vince, L., K. Janssens, F. Adams, M. L. Rivers, and K. W. Jones (1995a), *Spectrochim. Acta B* **50**, 127.

Vince, L., K. Janssens, F. Adams, M. L. Rivers, and K. W. Jones (1995b), *Spectrochim. Acta B* **50**, 1481.

von Bohlen, A. (1999), *J. Trace Microprobe Tech.* **17**, 177.

von Bohlen, A., and F. Meyer (1996), *Restauro* **7**, 472.

von Bohlen, A., P. Rechmann, J. L. Tourmann, and R. Klockenkämper (1994), *J. Trace Elem. Electrolytes Health Dis.* **8**, 37.

Walker, R. D. (1976), *Brit. Archaeol. Rep. Suppl. Ser.* **5**, 1.

Walker, R. D. (1977), *Brit. Archaeol. Rep. Suppl. Ser.* **22**, 1.

Walker, R. D. (1978), *Brit. Archaeol. Rep. Suppl. Ser.* **40**, 1.

Wegstein, M., H. Urban, P. Rostam-Khani, A. Wittershagen, and B. O. Kolbesen (1997), *Spectrochim. Acta B* **52**, 1057.

Williams, K. L. (1987), *An Introduction to X-Ray Spectrometry*, Allen & Unwin, London.

Wunderlich, C. H. (1994), *Restauro* **6**, 414.

Yu, K. N., and J. M. Miao (1996), *X-Ray Spectrom.* **25**, 281.

Zwicky-Sobczyk, C. N., and W. B. Stern (1997), *Archaeometry* **39**, 392.

CHAPTER

5

NEUTRON ACTIVATION ANALYSIS FOR PROVENANCE DETERMINATION IN ARCHAEOLOGY

HECTOR NEFF

University of Missouri Research Reactor Center, University of Missouri, Columbia, Missouri, U.S.A.

5.1. INTRODUCTION

Neutron activation analysis is a technique for characterizing the elemental composition of virtually any material. Its potential as an archaeological tool was recognized not long after the dawn of the nuclear age, when J. Robert Oppenheimer suggested its use to R. W. Dodson and E. V. Sayre of Brookhaven National Laboratory in 1954 (Harbottle 1976). Experimental work was then undertaken by Sayre, and the results were reported at a meeting of archaeologists and chemists at the Institute for Advanced Study at Princeton in 1956 (Harbottle 1976; Sayre and Dodson 1957). Around the same time, a group at Oxford started employing neutron activation analysis in provenance investigations of pottery and coins (Emeleus 1958, 1960). Early attempts to use the technique for archaeological purposes were hampered by the poor resolution of available detection systems (e.g., Bennyhoff and Heizer 1965), but the advent of the lithium-drifted germanium (Ge[Li]) detector in the 1960s (Sayre 1966, cited in Harbottle 1976) prompted a flurry of archaeological applications (Harbottle 1976). Perlman and Asaro (1969, 1971) provided a definitive description of standard-comparator instrumental neutron activation analysis (INAA) as applied to archaeological provenance determination at the Lawrence-Berkeley Laboratory (LBL). Techniques similar to those described by Perlman and Asaro were used at Brookhaven National Laboratory (BNL) beginning in the late 1960s (Sayre et al. 1971; Abascal-M et al. 1974). During the 1970s and 1980s, archaeologists turned more and more frequently to INAA to help determine the sources of pottery,

Modern Analytical Methods in Art and Archaeology, Edited by Enrico Ciliberto and Giuseppe Spoto. Chemical Analysis Series, Vol. 155
ISBN 0-471-29361-X

obsidian, chert, and other materials (Hughes et al. 1991a; Kuleff and Djingova 1990; Neff 1992; Neff and Glascock 1995). By the early 1990s, INAA was regarded as the technique of choice for such applications (Bishop et al. 1990; Gilmore 1991).

My aims in this chapter are to describe how INAA works and to review briefly its potential and actual contributions to archaeological science. The three basic topics to be covered are technique, methodology, and example applications. Under "technique" the reader will find a discussion of neutron activation and basic procedures of gamma ray measurement. This discussion highlights some of the desirable properties of INAA vis á vis other techniques as well as some of the technical pitfalls that can hinder its effective application in archaeology. "Methodology" concerns the means by which the analytical measurements produced by INAA can be used to determine the sources of archaeological artifacts. This section touches on some issues that arise in the quantitative analysis of INAA data, but the nitty-gritty statistical details are left for Baxter and Buck (Chapter 20). The obsidian and ceramic provenance studies discussed briefly in the final section are chosen to illustrate how appropriate methods lead, by several routes, from analytical measurements to source assignments.

5.2. INSTRUMENTAL NEUTRON ACTIVATION ANALYSIS

Neutron activation analysis depends on the fact that neutrons in the environment of a sample will interact with the nuclei of atoms in the sample in such a way that different atoms (isotopes) are produced. Many of the product isotopes are radioactive, and their decay can be monitored. Although neutrons can be supplied by charged-particle accelerators or radioisotopic sources, such as ^{252}Cf, the neutrons most commonly used in neutron activation analysis come from the fissioning of ^{235}U during the chain reaction inside nuclear reactors. It is probably safe to say that virtually all applications of neutron activation analysis in archaeology have utilized reactor-supplied neutrons (e.g., Kuleff and Djingova 1990). Typically, a package of samples is placed inside the reactor core or somewhere in the reflector region around the reactor, where there is a flux of neutrons sufficient to activate a variety of nuclei.

Details of INAA procedures differ from laboratory to laboratory and depend on such things as neutron flux, sample introduction facilities, counting facilities, and choices made by key personnel during the development of each program (Gilmore 1991; Glascock 1992; Harbottle 1976; Hughes et al. 1991a; Kuleff and Djingova 1990). Although the emphasis in the following discussion is on the phenomena and principles underlying INAA, exemplifying its use requires reference to the specific procedures used at one or a few research facilities. Procedures used in archaeometric research at the University of Missouri Research Reactor Center (MURR), which are used for exemplification in the following discussion, have been described previously by Glascock (1992; Glascock et al. 1994b). For general principles of INAA, the following account relies heavily on publications by Glascock (1991, 1994, 1998) and Parry (1991) and secondarily on works by Alfassi (1990), Gilmore (1991), Hughes et al. (1991a), Kuleff and Djingova (1990), Kushelevsky (1994), Landsberger (1994), Perlman and Asaro (1969; 1971), Tsechanski (1994), and Williams and Wall (1991). A chart of the nuclides (Walker et al. 1989) provides key information used in the following discussion as well. Much of the primary literature on nuclear phenomena and activation analysis techniques appears in the *Journal of Radioanalytical and Nuclear Chemistry, Articles*.

5.2.1. Nuclear Reactions Involving Neutrons

Interactions between neutrons and nuclei include scattering reactions on the one hand and capture reactions on the other hand (Kushelevsky 1994). The former phenomena can be exploited to investigate the structure of crystalline materials using techniques similar to X-ray diffraction. Neutron capture reactions, in which an incident neutron is captured by the nucleus of an atom, are the phenomena of interest in a discussion of neutron activation analysis.

In part, the kind of reaction that takes place when a neutron strikes a nucleus depends on the energy of the neutron. Neutrons emitted by the fission of ^{235}U have energies of around 2 MeV and are called "fast neutrons." Fast neutrons lose energy rapidly through elastic scattering from collisions with moderator materials, such as water or graphite. In the energy range 0.5 eV to 1 MeV, they are referred to as "epithermal neutrons." Below 0.5 eV they are referred to as "thermal neutrons." The reactions of greatest utility for neutron activation analysis are those that involve thermal (or "slow") neutrons. Most of these are (n, γ), that is, "n–gamma," reactions, an example of which is the interaction of ^{23}Na with a thermal neutron:

$$ {}^{1}_{0}n + {}^{23}_{11}Na \rightarrow {}^{24}_{11}Na + \gamma $$

In this case, a neutron (atomic mass N of 1 and atomic number Z of 0) strikes a sodium nucleus ($N = 23$, $Z = 11$), the neutron is absorbed by the nucleus, and a "prompt" gamma ray is emitted from the nucleus. In actuality, there is an intermediate step in the reaction, when a compound nucleus exists for an extremely short period of time before the emission of the prompt gamma ray.

Prompt gamma rays can be exploited for neutron activation analysis if samples are placed in a neutron beam outside a reactor and gamma rays are counted during irradiation (Glascock 1994; see Glascock et al. 1984 for an archaeological application). More commonly, however, the decay of radioactive nuclei formed by the emission of the prompt gamma ray is monitored sometime after the sample is removed from the reactor, and the technique thus is more precisely termed delayed instrumental neutron activation analysis (Glascock 1994; Landsberger 1994). In accord with standard practice, however, the more widely used acronym "INAA" will be used to refer in the rest of this chapter to DINAA.

In the case of the radioactive nucleus ^{24}Na produced by the (n, γ) reaction discussed above, radioactive decay yields the stable end product ^{24}Mg together with an electron and a neutrino, as follows:

$$^{24}_{11}\mathrm{Na} \rightarrow {}^{24}_{12}\mathrm{Mg} + \beta^- + \nu$$

In addition, the decay of ^{24}Na, which takes place with a half-life of about 15 hr, gives off gamma rays, which can be counted for the quantification of Na in neutron activation analysis.

Some of the radioactive isotopes whose gamma rays are counted in INAA are not produced by (n, γ) reactions. Cobalt-58 is produced from ^{58}Ni via a (n, p) reaction involving fast neutrons, as follows:

$$^{1}_{0}\mathrm{n} + {}^{58}_{28}\mathrm{Ni} \rightarrow {}^{58}_{27}\mathrm{Co} + {}^{1}_{1}\mathrm{p}$$

Gamma rays emitted by the decay of ^{58}Co, which has a half-life of about 71 days, can then be counted for determination of Ni in the sample.

Another departure from direct determination of concentrations from isotopes produced by (n, γ) reactions involves the determination of Th. In this case, the first step is the activation of ^{232}Th:

$$^{1}_{0}\mathrm{n} + {}^{232}_{90}\mathrm{Th} \rightarrow {}^{233}_{90}\mathrm{Th} + \gamma$$

Thorium-233 then decays with a half-life of 22.3 min as follows:

$$^{233}_{90}\mathrm{Th} \rightarrow {}^{233}_{91}\mathrm{Pa} + \beta^-$$

Protactinium-233 then decays with a half-life of 27 days as follows:

$$^{233}_{91}\mathrm{Pa} \rightarrow {}^{233}_{92}\mathrm{U} + \beta^-$$

The latter decay yields a gamma ray that can be counted after an extended

(3- to 4-week) decay period for the quantification of thorium in the sample. Uranium can be determined via a similar reaction involving activation of ^{238}U and subsequent counting of a gamma ray given off during decay of a daughter isotope. The first two columns of Table 5.1 list, in order of increasing half-life, radioactive isotopes used in INAA at MURR and the stable targets from which they are produced.

5.2.2. Natural Isotopes and Isotopic Abundance

The number of atoms of a given element activated by exposure to a flux of neutrons depends, among other things, on how many atoms of that element are present in the sample. However, a given element (with a fixed atomic number) may be associated naturally with a range of numbers of neutrons, and the different natural isotopes of the element will be affected differently by interaction with neutrons. Only a few or perhaps none of the natural isotopes may produce a radioactive isotope that is useful in neutron activation analysis.

As an example, potassium atoms, which have 19 protons, may have 20, 21, or 22 associated neutrons, and the atomic masses of the three nuclei are therefore 39, 40, and 41. Potassium-39, -40, and -41 have natural abundances of 93.2581, of 0.0117, and 6.7302%, respectively. These abundances alone might suggest that ^{39}K would be the most likely isotope on which to focus in trying to determine the amount of potassium in a sample. However, activation of ^{39}K produces ^{40}K, and activation of ^{40}K produces ^{41}K, both of which are stable.* Only the activation of ^{41}K yields a radioactive isotope (^{42}K) that is useful in neutron activation analysis. The number of potassium nuclei available for activation, then, is the total number of potassium nuclei in the sample multiplied by 0.067302, the abundance of the ^{41}K isotope.

The second column of Table 5.1 shows natural isotopes and their abundance for the target isotopes used in INAA at MURR.

5.2.3. Activation Rate

The activation rate for a given element in a sample depends not just on the number of target nuclei present but also on neutron flux and the probability of radiative capture of a neutron across the whole neutron energy range (Landsberger 1994: 124). Capture probability is customarily expressed as a unit of area called the barn ($=10^{-24}$ cm^2). The relation is thus expressed as

* Potassium-40 is actually a naturally occurring radioactive isotope of potassium, but its half-life is about 1250 million years. Natural decay of ^{40}K to ^{40}Ar is the basis for potassium–argon dating, a technique that is especially useful for dating volcanic tephras.

Table 5.1. Radioactive Isotopes Used in Delayed Instrumental Neutron Activation Analysis at MURR, Arranged by Half-life[a]

Radioactive Isotope and Production Mode	Target Isotope and Isotopic Abundance	Cross Section for Thermal Neutron Capture (barns)	Resonance Integral (barns)	Half-life	Irradiation Length (Short or Long)[b] and Typical MURR Decay Time
^{28}Al (n, γ)	^{27}Al (100%)	0.226	0.16	2.24 min	Short, 25 min
^{52}V (n, γ)	^{51}V (99.75%)	4.79	2.63	3.75 min	Short, 25 min
^{51}Ti (n, γ)	^{50}Ti (5.4%)	0.171	0.115	5.76 min	Short, 25 min
^{49}Ca (n, γ)	^{48}Ca (0.187%)	1.12	0.50	8.72 min	Short, 25 min
^{139}Ba (n, γ)	^{138}Ba (71.7%)	0.405	0.36	84.63 min	Short, 25 min
^{165}Dy (n, γ)	^{164}Dy (28.2%)	1697	424	2.33 hr	Short, 25 min
^{56}Mn (n, γ)	^{55}Mn (100%)	13.2	13.9	2.58 hr	Short, 25 min
^{42}K (n, γ)	^{41}K (6.73%)	1.45	1.41	12.36 hr	Short, 25 min
^{24}Na (n, γ)	^{23}Na (100%)	0.513	0.0303	14.96 hr	Short, 25 min
^{76}As (n, γ)	^{75}As (100%)	3.86	52.5	26.32 hr	Long, 7–8 d
^{140}La (n, γ)	^{139}La (99.91%)	9.34	11.6	40.27 hr	Long, 7–8 d
^{153}Sm (n, γ)	^{152}Sm (26.7%)	220	3168	46.27 hr	Long, 7–8 d
^{239}Np (n, γ)[c]	^{238}U (99.27%)	2.75	284	23.47 min + 2.36 d	Long, 7–8 d
^{175}Yb (n, γ)	^{174}Yb (31.8%)	128	58.9	4.19 d	Long, 7–8 d
^{177}Lu (n, γ)	^{176}Lu (2.6%)	2100	1160	6.71 d	Long, 7–8 d
^{147}Nd (n, γ)	^{146}Nd (17.19%)	1.45	2.9	10.98 d	Long, 7–8 d
^{86}Rb (n, γ)	^{85}Rb (72.17%)	0.494	7.31	18.66 d	Long, 21–28 d
^{233}Pa (n, γ)[c]	^{232}Th (100%)	7.26	83.7	22.3 min + 27.0 d	Long, 21–28 d
^{51}Cr (n, γ)	^{50}Cr (4.35%)	15.2	8.1	27.7 d	Long, 21–28 d
^{141}Ce (n, γ)	^{140}Ce (88.48%)	0.575	0.48	32.5 d	Long, 21–28 d
^{181}Hf (n, γ)	^{180}Hf (35.1%)	13.5	34.0	42.39 d	Long, 21–28 d

^{59}Fe (n, γ)	^{58}Fe (0.28%)	1.31	1.28	44.5 d	Long, 21–28 d
^{124}Sb (n, γ)	^{123}Sb (42.7%)	4.09	118	60.2 d	Long, 21–28 d
^{95}Zr (n, γ)	^{94}Zr (17.38%)	0.053	0.268	64.02 d	Long, 21–28 d
^{85}Sr (n, γ)	^{84}Sr (0.56%)	0.69	9.14	64.84 d	Long, 21–28 d
^{58}Co (n,p)	^{58}Ni (68.27%)	113[d]	[d]	70.82 d	Long, 21–28 d
^{160}Tb (n, γ)	^{159}Tb (100%)	23.8	426	72.3 d	Long, 21–28 d
^{46}Sc (n, γ)	^{45}Sc (100%)	26.3	11.3	83.81 d	Long, 21–28 d
^{182}Ta (n, γ)	^{181}Ta (99.98%)	20.4	679	114.5 d	Long, 21–28 d
^{65}Zn (n, γ)	^{64}Zn (48.6%)	0.726	1.42	243.9 d	Long, 21–28 d
^{134}Cs (n, γ)	^{133}Cs (100%)	30.7	390	2.06 y	Long, 21–28 d
^{60}Co (n, γ)	^{59}Co (100%)	37.13	74	5.27 y	Long, 21–28 d
^{152}Eu (n, γ)	^{151}Eu (47.8%)	5900	5564	13.33 y	Long, 21–28 d

[a] All data are from Glascock (1991).

[b] The short irradiation for pottery, obsidian, and chert used at MURR is 5 sec. The long irradiation for pottery and most geological materials is 24 hr, while the time used for obsidian and chert is 70 hr.

[c] As described in the text, uranium and thorium are determined from the decay of daughter isotopes produced by decay of isotopes produced initially by the activation of these elements.

[d] Average cross section in a ^{235}U fission neutron spectrum is given for ^{58}Ni, which activates to ^{58}Co via a (n, p) reaction.

$$R = \sigma\phi n$$

where R is the activation rate, σ is the cross section in barns, ϕ is the neutron flux in neutrons per square centimeter per second, and n is the number of target nuclei (Glascock 1994; Parry 1991).

For thermal neutrons, the most probable velocity is 2200 m/s (0.0253 eV), so thermal neutron capture probabilities are usually given as barns at this velocity. Neutrons in the epithermal energy range (above 0.5 eV) may also be captured by nuclei. For these higher energy neutrons, however, the resonance integral is used to express the probability of radiative capture. Considering reactions in both the thermal and epithermal energy regions, the activation rate R for a sample may be expressed as

$$R = \sigma\phi_{\text{th}}n + I\phi_{\text{epi}}n$$

where σ is the thermal neutron cross section in barns, ϕ_{th} is the flux of thermal neutrons (neutrons $\times$ cm^{-2} $\times$ s^{-1}), I is the resonance integral for all neutron energies above thermal, ϕ_{epi} is the flux of epithermal neutrons, and n is the total number of target nuclei (Glascock 1994; Landsberger 1994; Parry 1991). Glascock (1991: Table 1) has compiled thermal neutron cross sections and resonance integrals for target isotopes useful in neutron activation analysis. Cross sections and resonance integrals are reproduced here (Table 5.1) for the 33 isotopes determined in analysis of archaeological ceramics and other geological materials at MURR.

It turns out that for so-called $1/v$ absorbers (elements for which thermal neutron capture cross sections decrease linearly with increasing neutron velocity, typically the lighter elements (see Table 5.1), σ and I are about the same order of magnitude (Parry 1991: 13). However, since the flux of thermal neutrons is about 50 times the flux of epithermal neutrons in research reactors (Glascock 1994,81), thermal neutron activation generally accounts for most of the induced activity for most elements.

5.2.4. Decay Rate

Radioactive isotopes produced by neutron activation start to undergo radioactive decay with a characteristic half-life. The decay rate depends on the original number of radioactive nuclei present (N) scaled by a constant value λ that is unique to each radionuclide (Glascock, 1998; Parry 1991: 13–15):

$$\frac{dN}{dt} = -\lambda N$$

Integrating from the beginning of the decay, when there are N_0 atoms of the nuclide in question, to some later time t yields

$$\ln\left(\frac{N_t}{N_0}\right) = -\lambda t$$

and the number of radionuclide atoms present at the later time becomes

$$N_t = N_0 e^{-\lambda t}$$

Since activity is proportional to the number of radionuclides present (N), this equation also gives the level of activity produced by a given radionuclide in a sample after the decay period.

The "half-life" of a radionuclide can be obtained by setting $N_t = \frac{1}{2} N_0$. In this case,

$$\ln\left(\frac{(1/2)N_0}{N_0}\right) = -\lambda t_{1/2}$$

or

$$\ln\left(\frac{1}{2}\right) = -\lambda t_{1/2}$$

The half-life, therefore, is related to the decay constant as follows:

$$t_{1/2} = \frac{\ln(2)}{\lambda} = \frac{0.693}{\lambda}$$

(Glascock, 1998). The last two columns of Table 5.1 show the half-lives of 33 isotopes measured in INAA of most archaeological and geological samples at MURR along with the decay times typically allowed before gamma counting.

5.2.5. Activity at Count Time

The activity in a sample at the time it is counted is proportional to the number of radionuclide atoms present, which is determined by (1) the combined growth and decay of the radionuclide during the time it was being irradiated, which determines the number of radionuclides present at the end of irradiation, and (2) radioactive decay of the nuclide between the end of irradiation and the start of the count.

Taking the definitions of the activation rate and decay rate given above, the production rate during irradiation, which is the difference between the activation and decay rates, becomes

$$\frac{dN}{dt} = \sigma\phi n - \lambda N$$

(Glascock, 1998; Parry 1991: 15). From this it follows that the number of radionuclide atoms present at the end of the irradiation of length T can be determined from

$$N_{\text{eoi}} = \frac{\sigma\phi n}{\lambda}(1 - e^{-\lambda T})$$

The activity present at the end of irradiation is thus dependent on the half-life of the nuclide in question. Glascock (1998) points out that, if irradiation is short compared to the nuclide half-life, the parenthetical expression on the right side of this equation approximates to λT, so that the number of nuclides is approximately the production rate multiplied by the irradiation time. On the other hand, as irradiation time is lengthened compared to nuclide half-life, the growth of N levels off, and induced activity for a given neutron flux approaches saturation (Glascock, 1998; Parry 1991).

Decay between the end of irradiation and the beginning of the count (decay time T_d) means that the number of nuclides present at count time (N_{soc}) will be

$$N_{\text{soc}} = N_{\text{eoi}} e^{-\lambda T_d}$$

As illustrated by the last two columns of Table 5.1, isotopes with short half-lives are counted after relatively short decay times, when activity is still easily detectable above the background in a gamma ray spectrum. Conversely, longer lived isotopes are counted following a longer decay period, after most of the activity from shorter lived isotopes is gone and there are fewer spectral interferences (see below).

5.2.6. Decay Schemes

In INAA, it is the counting of gamma rays emitted by the decay of radio-isotopes created by neutron activation that yields the basic data leading to quantification of elemental abundances. Most activated nuclei decay by emission of a beta particle. However, daughter nuclei produced by beta emission are often in excited states initially, and additional energy transitions are necessary before they arrive at a ground state. The latter transitions are accompanied by the emission of gamma rays at different energies, and it is these gamma rays that are normally counted in INAA.

Decay schemes range from simple to quite complex. A simple decay scheme is exemplified by the decay of ^{28}Al, which is produced by the neutron activation of ^{27}Al (e.g., Parry 1991: Figure 4.2). The activated nucleus decays with a half-life of 2.25 min via emission of a β^- particle with a maximum energy of 2.88 MeV. Initially, the ^{28}Si daughter is in an excited state at 1.779 MeV, but a transition to the ground state takes place almost immediately, and this is accompanied by emission of a 1.779-MeV gamma ray. This gamma ray can be used for quantification of aluminum by a measurement after a short irradiation and short decay. More complex decay schemes involve β^- emission to several excited states of the daughter nucleus, followed by emission of gamma rays of several different energies via transitions of the excited daughters. Diagrams of various decay schemes are given by Parry (1991: 42–49) and Glascock (1998). Branching ratios give the abundance of each transition and its associated gamma ray energy, which is one criterion to consider for selecting gamma rays for quantification in INAA. In the case of ^{28}Al, for instance, all decays produce an excited ^{28}Si nucleus at the 1.779-MeV level, so the abundance associated with the 1.779-MeV gamma ray is 100%. Alfassi (1990) and Glascock (1991: Table 3) give gamma rays and their percentage abundances for a number of isotopes useful in INAA. In sum, a sample containing multiple elements, once irradiated, will consist of numerous radioactive isotopes whose decay may yield one or more gamma rays.

5.2.7. Gamma Spectra

Gamma rays interact with matter in several ways, one of which (the photoelectric effect) is an absorption process in which all of the energy of the gamma ray photon is transferred to a photoelectron inside the detector (Glascock, 1998; Kushelevsky 1994; Parry 1991; Tsechanski 1994). This process is used for their detection and measurement by various semiconductor detectors, the most useful of which for INAA is the high-purity germanium detector. Typically, gamma ray emissions are counted across a range of energies up to about 3.2 MeV, yielding a gamma spectrum whose characteristics are thus determined in some sense by the makeup of the sample. An example gamma ray spectrum produced following the irradiation of an archaeological basalt sample is shown in Figure 5.1. Gamma ray spectroscopy consists of various techniques that are used to extract qualitative and quantitative information about sample characteristics from gamma ray spectra such as the one shown in Figure 5.1.

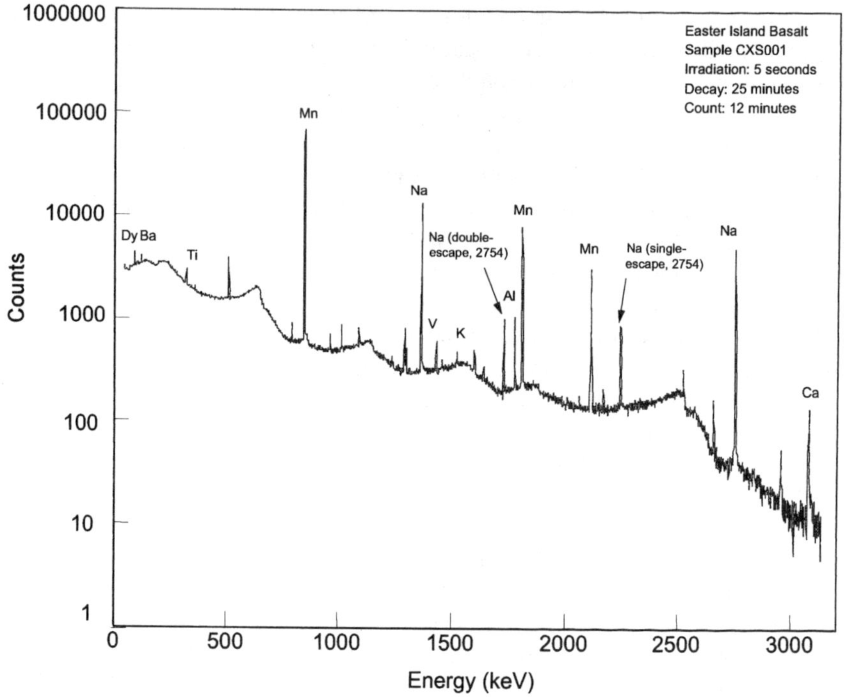

Figure 5.1. Gamma ray spectrum obtained during a 12-min count after short irradiation and decay of an archaeological basalt sample at MURR.

Unfortunately, the characteristics of gamma ray spectra are determined not just by the various decaying radioisotopes in the sample and their associated gamma rays, but also by other interactions between gamma rays and matter that take place in and around the detector. One such interaction is Compton scattering, in which some of the energy of a gamma ray is lost to an electron by inelastic scattering and a scattered gamma ray of some lower energy is produced (Kushelevsky 1994; Tsechanski 1994). If the lower energy gamma ray escapes from the detector rather than losing the rest of its energy through the photoelectric effect, then an event of lower energy than the incident gamma ray is recorded by the detector. Thus, the result of Compton scattering is an elevated background energy continuum that ranges up to the "Compton edge," which is the maximum energy that can be transferred to an electron by a gamma ray of given energy.

Another interaction is pair production, which is the creation of an electron–positron pair from a gamma ray of greater than 1.02 MeV, with the remaining energy above 1.02 MeV shared between the electron and positron as kinetic energy (Kushelevsky 1994; Tsechanski 1994). Annihilation by the positron of an electron in the absorber produces two 0.511-MeV photons (0.511 MeV is the rest mass energy of an electron), with the remaining energy above 1.02 MeV shared between the two. Pair production may record the full energy of the incoming gamma ray (neither photoelectron escapes the detector), a single escape event (one photon escapes the detector) at 0.511 MeV less than the energy of the original gamma ray, or a double-escape event (both photons escape) at 1.022 MeV less than the original gamma ray. Thus, for example, 2.754-MeV gamma rays from decay of ^{24}Na produce the 2.754 full-energy peak, a single-escape peak at 2.23 MeV, and a double-escape peak at 1.732 MeV (Fig. 5.1).

Besides Compton scattering and pair production, other effects that shape gamma ray spectra include β^- emission from the sample, X-ray emission from the photoelectric effect of gamma rays absorbed by lead shielding, and sum peaks from two gamma rays that are emitted almost simultaneously from the sample (Glascock, 1998). Like Compton scattering and pair production, these effects either raise the background or add spurious peaks to the gamma ray spectrum.

A typical gamma spectroscopy setup in use today consists of a detector connected to a preamplifier, amplifier, and analog to digital converter (ADC), which records a count in the appropriate channel when a gamma ray interacts via the photoelectric effect with the detector crystal (Parry 1991; Tsechanski 1994). Most analyzers used today consist of 8192 channels. Thus, if a spectrum is collected for the energy range from 0 to 3300 keV, each channel represents 0.403 keV. As the number of counts recorded in a given channel increases during a count, the entire spectrum at that channel is raised. As shown in Figure 5.1, the end result can be depicted as a histogram on which the x axis represents channel number (which is converted into an energy scale through an energy calibration) and the y axis represents number of counts recorded.

5.2.8. Identification of Peaks

Given a gamma ray spectrum such as that shown in Figure 5.1, one goal is to identify the nuclides responsible for the various peaks. Initial identification can be accomplished by consulting tables of gamma ray energies, the most useful of which are arranged by half-life (e.g., Glascock 1991: Tables 4 and 5). In Figure 5.1, for instance, a number of peaks are identifiable by their characteristic gamma ray energies. Peaks in a spectrum will be different depending on irradiation and decay times. For instance, a peak at 320.1 keV in a gamma ray spectrum accumulated 25 min after irradiation (e.g., Fig. 5.1) would be identified as ^{51}Ti, which has a 5.76-min half-life (Table 5.2),

Table 5.2. Gamma Rays Used for INAA at MURR

Determined Element	Activated Product and Half-Life	Recommended Gamma Ray (keV)[a]	Other Gamma Rays (keV)[a]	Isotopes Interfering with Recommended Gamma Ray, Half-lives, and Interfering Energy (keV)[a]
^{27}Al	^{28}Al (2.24 min)	1778.99	None	None
^{51}V	^{52}V (3.75 min)	1434.08	None	None
^{50}Ti	^{51}Ti (5.76 min)	320.08	928.64	None
^{48}Ca	^{49}Ca (8.72 min)	3084.54	4072.00	None
^{138}Ba	^{139}Ba (84.63 min)	165.85	None	None
^{164}Dy	^{165}Dy (2.33 hr)	94.7	279.76, 361.67, 633.43, 715.33	^{233}Th, 22.3m, 94.7
^{55}Mn	^{56}Mn (2.58 hr)	1810.72	846.76, 2113.05	None
^{41}K	^{42}K (12.36 hr)	1524.58	None	None
^{23}Na	^{24}Na (14.96 hr)	1368.60	2754.00	^{123}Sb, 60.2d, 1368.2
^{75}As	^{76}As (26.32 hr)	559.1	563.23, 657.05, 1212.92, 1216.08, 228.52	^{114m}In, 49.51d, 558.4
^{139}La	^{140}La (40.27 hr)	1596.21	328.76, 432.49, 487.02, 751.64, 815.77, 867.85, 919.55, 925.19	None
^{152}Sm	^{153}Sm (46.27 hr)	103.18	69.67	^{153}Gd, 241.6d, 103.2; ^{239}Np, 2.36d, 103.7 (X-ray)
^{238}U	^{239}Np (23.47 min + 2.36 d)[b]	228.18	106.12, 209.75, 277.60, 334.32	^{132}Te, 78.2h, 228.3; ^{182}Ta, 114.5d, 229.3
^{174}Yb	^{175}Yb (4.19 d)	396.33	113.81, 282.52	None
^{176}Lu	^{177}Lu (6.71 d)	208.36	112.95	^{239}Np, 2.36d, 209.8
^{146}Nd	^{147}Nd (10.98 d)	91.1	319.41, 531.01	None
^{75}Rb	^{86}Rb (18.66 d)	1076.6	None	None
^{232}Th	^{233}Pa (22.3 min + 27.0 d)[b]	312.01	75.34, 86.65, 300.18, 340.59, 38.66, 415.93	^{42}K, 12.36h, 312.4

^{50}Cr	^{51}Cr (27.7 d)	320.08	None	^{147}Nd, 10.98d, 319.4; ^{177m}Lu, 160.9d, 319.0
^{140}Ce	^{141}Ce (32.5 d)	145.44	None	^{183}Ta, 5.1d, 144.1; ^{175}Yb, 4.19d, 144.9
^{180}Hf	^{181}Hf (42.39 d)	482.18	136.28, 345.94	^{194m}Ir, 171d, 482.7
^{58}Fe	^{59}Fe (44.5 d)	1099.25	142.65, 192.35, 1291.60	None
^{123}Sb	^{124}Sb (60.2 d)	1690.98	602.73, 645.86, 722.79, 1368.16, 2090.94	None
^{94}Zr	^{95}Zr (64.02 d)	756.73	724.20	^{154}Eu, 8.59y, 756.9
^{84}Sr	^{85}Sr (64.84 d)	514.00	None	None
^{58}Ni	^{58}Co (70.82 d)	810.77	None	^{152}Eu, 13.33y, 810.8
^{159}Tb	^{160}Tb (72.3 d)	879.38	86.79, 197.04, 215.65, 298.58, 962.32, 966.17, 1177.96, 1271.88	None
^{45}Sc	^{46}Sc (83.8 d)	889.28	1120.55	None
^{181}Ta	^{182}Ta (114.5 d)	1221.41	84.68, 100.11, 113.67, 116.42, 152.43, 156.39, 179.39, 198.35, 222.11, 229.32, 264.08, 1121.30, 1189.05, 1231.02, 1257.42, 1289.16	None
^{64}Zn	^{65}Zn (243.9 d)	1115.55	None	^{46}Sc, 83.81d, 1120.6; ^{152}Eu, 13.33y, 1112.1; ^{160}Tb, 72.3d, 1115.1; ^{182}Ta, 114.5d, 1113.4
^{133}Cs	^{134}Cs (2.06 yr)	795.85	563.23, 569.32, 604.70, 801.93, 1365.15	^{232}Th, 1.4E + 10y, 795.0
^{59}Co	^{60}Co (5.27 yr)	1173.24	1332.50	None
^{151}Eu	^{152}Eu (13.33 yr)	1408.00	121.78, 244.69, 344.29, 411.12, 443.89, 778.92, 867.38, 964.11, 1085.89, 1112.08	None

[a] Recommended gamma rays, other gamma rays, and spectral interferences are from Glascock (1991); also see Alfassi (1990).

[b] See Table 5.1 for notes on determination of uranium and thorium.

whereas a peak at the same energy in a gamma ray spectrum collected four weeks after an irradiation would be identified as the long-lived isotope (27.7 days) ^{51}Cr (Table 5.2). Table 5.2 shows gamma rays for various elements determined at MURR together with potential spectral interferences on the main gamma ray used for quantification.

5.2.9. Quantification of Elemental Concentration from Gamma Ray Spectra

In order to turn a gamma ray spectrum into a "chemical fingerprint" of elemental concentrations in an unknown sample, the activity represented by peaks selected for quantification must first be calculated. Since peak area is determined directly by the number of counts recorded in the channels under that peak, measured activity (or count rate) is just the peak area divided by the count time. Peak area is estimated by calculating the total area under the peak and subtracting an estimated background, which is obtained by selecting several channels on either side of the peak to get an average background in that region, then summing over the number of channels under the peak. Spectral interferences, as discussed below, can complicate background subtraction.

The variance in determination of peak area is usually calculated as total counts in the signal peak plus twice the subtracted background, so the percent standard deviation, which is how the error is normally reported, becomes

$$\%\sigma = \frac{100(S + 2B)^{1/2}}{S}$$

where S is the net counts in the peak of interest and B is the background subtracted out (Gilmore 1991; Glascock, 1998; Parry 1991). Gilmore (1991: Appendix 2) gives a variance calculation appropriate if multiple channels on either side of the peak are used to estimate background and demonstrates that error diminishes substantially as more channels are used for the background calculation.

Absolute quantification of elemental abundances from the activities determined from gamma ray spectra requires control over all nuclear parameters (isotopic abundance, atomic weight, neutron capture cross section and resonance integral, intensity of the measured gamma ray, decay constant) and measurement conditions (thermal and epithermal neutron fluxes, efficiency of the detector, sample weight, irradiation time, decay time, and count time). In the vast majority of archaeometric applications, however, use of a calibration standard dramatically simplifies the calculations. As Glascock (1998) and Landsberger (1994) point out, if all measurement conditions, including sample weights, fluxes, irradiation, decay, and count times, are identical for an unknown and a standard (with a known concentration of a given element), then the concentration, C_{unk}, in the unknown is

$$C_{\text{unk}} = C_{\text{std}} \frac{A_{\text{unk}}}{A_{\text{std}}}$$

where C_{std} is the concentration in the standard, A_{unk} is the count rate in the unknown, and A_{std} is the count rate in the standard. Since decay times and sample weights typically vary, the unknown concentrations are usually found by

$$C_{\text{unk}} = C_{\text{std}} \left(\frac{A_{\text{unk}}}{A_{\text{std}}}\right) \left(\frac{(e^{\lambda T_d})_{\text{unk}}}{(e^{\lambda T_d})_{\text{std}}}\right) \left(\frac{W_{\text{std}}}{W_{\text{unk}}}\right)$$

where λ is the decay constant, T_d is the decay time, W_{std} is the standard weight, and W_{unk} is the unknown weight.

The vast majority of laboratories undertaking INAA for archaeological provenance determination utilize a multielement calibration standard. The basic procedure is to collect gamma ray spectra for both unknowns and standards under conditions that are as closely similar as possible, then to use one of the two above equations to calculate concentrations of various elements whose peaks appear in the gamma spectra. These calculations are, of course, computerized. Quality control or check standards are commonly included in each irradiation in order to provide an independent check on the quality of the data and in order to identify mistakes, such as weighing errors in the standards. Table 5.3 lists a number of standards currently in use or used in the recent past.

5.2.10. Spectral Interferences

Table 5.2 shows that many of the gamma rays used for INAA at MURR are close to gamma rays produced by other radioactive isotopes. These nearby gamma rays can complicate the identification and quantification of the area under a peak, since some of the counts from the nearby peak can add to the area of the peak of interest. The high-purity germanium detectors in use today resolve gamma ray peaks much more clearly than sodium iodide detectors available prior to the mid-1960s. As mentioned in the introduction (and see Harbottle 1976), it was the invention of such high-resolution detectors that sparked a surge of interest in the use of INAA for archaeological provenance determination.

Table 5.3. Some Standards Used in INAA-Based Archaeometric Characterization[a]

Standard	Source	Laboratories	Purposes
SRM1633 (coal fly ash)	NIST	Smithsonian/NIST	Calibration, medium, long
SRM1633a (coal fly ash)	NIST	MURR, Smithsonian/NIST, Texas A&M, Budapest	Calibration, short, medium, long; QC (Budapest)
SRM1633b (coal fly ash)	NIST	MURR	To replace SRM1633a when supply runs out
SRM688 (basalt rock)	NIST	MURR, Texas A&M	Calibration of Ca, short; QC (Texas A&M)
SRM278 (obsidian)	NIST	MURR	Calibration, short, medium, long (obsidian unknowns); QC (pottery, long)
Old Ohio Red Clay	Resco Products	MURR, Smithsonian/NIST, BNL, Texas A&M	QC
New Ohio Red Clay	Resco Products	MURR, Texas A&M	QC
USGS Rocks	USGS	BNL, Demokritos	Calibration
Perlman-Asaro standard Pottery	LBL	LBL, Hebrew University, Demokritos, Bonn	Calibration, short, medium, long
British Museum standard Pottery	British Museum Research Lab	British Museum	Calibration, medium and long
Bonn standard pottery	Bonn	Bonn	Calibration, short, medium, and long

[a] *References:* Glascock (1992), Harbottle (1982), Hughes et al. (1991a), Mommsen et al. (1988), Perlman and Asaro (1969, 1971), Yellin et al. (1978). *Abbreviations*: QC = quality control (or check standard); NIST = National Institute of Standards and Technology, formerly National Bureau of Standards; MURR = Missouri University Research Reactor Center; BNL = Brookhaven National Laboratory; LBL = Lawrence Berkeley Laboratory; USGS = United States Geological Survey.

Even with high-resolution detectors, some peaks are so close together that they cannot be resolved. Sometimes the result is a distorted peak that obviously has contributions from two underlying gamma rays. Even worse, in some cases peaks may appear Gaussian despite being determined by decays from two distinct isotopes. The latter possibility makes it imperative that tables of possible interferences be consulted (e.g., Glascock 1991: Tables 6–8) in selecting peaks for quantification (Gilmore 1991; Williams and Wall 1991). Other things being equal, it is preferable to select a peak without interferences if one is available for the isotope of interest. For instance, for the quantification of manganese at MURR, the 1810.7-keV gamma ray from ^{56}Mn is used (Table 5.2; Figure 5.1) rather than the 846.76 gamma ray because of interference on the latter by the 843.8 gamma ray from ^{27}Mg. Without this interference, the 846.76 line would probably be chosen because of its higher branching ratio abundance (Glascock 1991: Table 3). This spectral interference can preclude detection of ^{27}Mg, since the low branching ratio abundance of the alternate, interference-free line at 1014.43 keV coupled with low neutron capture cross section and low (11%) isotopic abundance of the parent ^{26}Mg may submerge this peak below the background. More importantly, however, ^{27}Mg is produced by a (n, p) reaction from ^{27}Al, so any value for magnesium would have to be corrected for the contribution from the fast neutron reaction (see discussion of nuclear interferences below).

If a peak with potential spectral interferences must be used, choice of counting time can minimize contributions from interfering gamma rays. For instance, by waiting 4+ weeks after irradiation to determine ^{233}Pa, which has a 27-day half-life and is used to determine ^{232}Th, interferences from the much shorter lived isotope ^{42}K (12.36 hr) is minimized (Table 5.2). Contributions to peak area from interfering peaks can be removed by determining the number of counts in a peak of the interfering isotope that is itself free of interferences, then calculating the contribution to the peak of interest from the ratio of the abundances of the two gamma rays of the interfering isotope (Gilmore 1991; Williams and Wall 1991). For the isotopes listed in Table 5.2, the contributions from interfering gamma rays is so small as to be practically negligible with the decay times used at MURR, and corrections for interference generally are not carried out.

5.2.11. Nuclear or Primary Interferences

One kind of nuclear interference is the production of isotopes of interest via fast neutron reactions involving elements with an atomic number one or two higher than the element of interest (Glascock, 1998). The most important such interference in archaeological applications, that involving production of

^{27}Mg from ^{27}Al, has already been mentioned. Additional contribution to ^{27}Mg comes from a (n, α) reaction of ^{30}Si. The element of interest is magnesium, and the isotope produced by thermal neutron activation of ^{26}Mg does, in fact, have two measurable gamma rays. However, with such high concentrations of aluminum in some archaeological samples, particularly pottery, the fast neutron fluxes of most reactors are sufficient to produce a significant contribution to ^{27}Mg from the fast neutron reaction. Glascock et al. (1985: Table 6) report that uncorrected magnesium concentrations on a key standard reference material, SRM-1633a (coal fly ash), were too high by a factor of 3.77. Magnesium is not reported routinely in archaeometric applications at MURR, largely because of this problem.

Another kind of nuclear interference involves the production of isotopes by fission of ^{235}U caused by neutron bombardment of the sample. As reported by Glascock et al. (1986), the fission products can make important contributions to lanthanum, cerium, neodymium, zirconium, and barium concentrations, with zirconium and one commonly analyzed isotope of barium being the most severely affected. These authors give correction factors by which to obtain fission yields of the various elements given a measured uranium content; the fission yields can be subtracted from the affected elements to determine the correct elemental concentration. The problem is, of course, most severe when uranium concentrations are high: at uranium concentrations of 125 ppm, the pattern of light rare earth element concentrations in a shale reference material was clearly distorted, barium was overestimated by a factor of 3.3, and zirconium was overestimated by a factor of 34 (Glascock et al. 1986: 4). Fortunately, uranium content in most materials likely to be the subject of archaeological provenance investigations averages about 2 ppm or less. The exception is some pottery from the Southwest U.S., in which uranium values close to 50 ppm have been observed, and significant corrections, especially to zirconium concentrations, are thus required.

5.2.12. Sample Preparation

One of the main advantages of INAA for archaeometric applications is that solid geological samples can be analyzed, and this greatly facilitates sample processing. The solid sample analysis capability arises from (1) the low neutron capture cross sections of the most abundant elements in geological samples (silicon, aluminum, and oxygen), which permits neutrons to penetrate such matrices with a minimum of attenuation, and (2) the fact that most geological materials are ineffective at stopping the high-energy gamma rays given off by radioactive decay, which minimizes self-shielding. Still, Landsberger (1994) points out that self-shielding can be a problem with high concentrations of elements with high neutron capture cross sections (e.g., boron, cadmium, gold, and silver) and that concentrations of high-atomic-number elements can attenuate the detection of low-energy gamma rays. Such effects would seem most worrisome in the analysis of metal artifacts, especially alloys containing gold, silver, and tin.

Despite the ease of sample preparation relative to other techniques, such as inductively coupled plasma–atomic emission spectroscopy (ICP–AES) and ICP–mass spectrometry (MS) (Chapter 3), care must nonetheless be taken in preparing solid samples for INAA. Contamination from metal tools used in processing is one potential problem. For instance, the tungsten carbide drill bits often used to obtain powdered material from ceramics may add cobalt, manganese, and iron, especially if the ceramic fabric is hard and/or the drill bit is dull. A distinct ^{187}W line at 487 keV is a good mark of drill-bit contamination. To minimize contamination of ceramic pastes at MURR, samples are processed by burring all surfaces with a silicon carbide burr, then washing the burred sample in deionized water, then crushing the burred and cleaned sherd in a mortar and pestle. Lithic materials, which sometimes have to be prepared with a geological saw and/or broken in a press, are also washed thoroughly prior to weighing.

Weighing error, which contributes to overall analytical error, should be minimized by standard but careful laboratory procedures. Weights should be obtained on a high-quality analytical balance. Powdered samples, such as samples from ceramics, should be dried in a drying oven then kept in a dessicator until they are weighed. In addition, any processing of samples between weighing and irradiation should be undertaken with sufficient care that no sample material is inadvertently lost.

In most cases, samples will be placed on a gamma ray detector in the same container in which they were irradiated. This means that the containers must have the lowest possible concentrations of the various trace elements to be measured. At MURR, high-density polyethylene vials are used for the short irradiation, and high-purity quartz is used for the long irradiation. The quartz vials are washed in aqua regia following the irradiation to remove contaminant isotopes from preirradiation handling and reactor pool water. If an unknown new container is being used, blanks should be analyzed to verify suitability.

Finally, encapsulated samples and standards should be as close as possible to the same size and shape and must be placed at the same distance from the detector. This is to ensure that a gamma ray emitted from one sample has the same probability of reaching the detector as that emitted by another sample and thereby to maximize the comparability of unknown and standard spectra. Similar size and shape at MURR are achieved by placing approximately the same amount of sample in the bottom of each quartz vial. Also, obtaining sufficient activity (through longer irradiation times and/or higher neutron fluxes) permits sample-to-detector distances to be increased, so that the sample "looks" more like a point source to the detector. Additional control on sample geometry can be introduced by rotating the sample while it is being counted, so that the detector "sees" a cylinder even though the sample may actually be nonuniformly distributed in its container. Another technique originally recommended by Perlman and Asaro (1969) is to press powdered samples into uniform pellets.

5.2.13. Strengths of INAA

The advantages of the INAA technique described above for characterizing archaeological artifacts are (1) extremely high precision, high accuracy, and high sensitivity for many major, minor, and trace elements; (2) capability to analyze very small samples (less than 100 mg); (3) relative immunity from matrix effects and consequent ease of sample preparation (see above); (4) simultaneous measurement of 30 or more elements with a fully instrumental technique; and (5) demonstrated potential for intercalibrating measurements between laboratories (e.g., Glascock 1992; Hughes et al. 1991b).

Landsberger (1994) has discussed the concepts of precision (or reliability) and accuracy as it applies to analytical measurements in general and INAA in particular. Precision refers to the extent to which a measurement on the same sample can be replicated. On this score, INAA rivals other highly reliable techniques, such as ICP–MS, ICP–AES, atomic absorption spectrometry (AAS), and high-precision X-ray fluorescence spectrometry (XRF), all of which can claim precisions in the 1 to 5% range for certain elements under realistic conditions (Glascock 1992; Blackman et al. 1993; Pollard and Heron 1996; Tykot and Young 1996; Garcia Heras et al. 1997). Sample preparation procedures can, however, dramatically attenuate precision, as demonstrated by comparison of bulk INAA data on ceramic pastes with ICP–AES data from the same samples treated by a weak hydrochloric acid extraction process to obtain a liquid suitable for ICP (Neff et al. 1996; Burton and Simon 1993 are the original proponents of this approach). In INAA, inattention to weighing precision, sample geometry and other procedural details, either with standards or with unknowns, will reduce the level of precision.

Accuracy refers to the extent to which a measurement approximates the "true value" of some quantity of interest. In the case of analytical measurements by INAA or some other method, the quantities of interest are elemental concentrations, and accuracy can be evaluated by analyzing certified reference materials (Landsberger 1994). In standard-comparator INAA, concentrations are normally obtained by comparison of activities with activities in standard reference materials (SRMs) (see above), so accuracy depends in large part on the accuracy of the elemental concentrations used for the standard. Glascock (1991) presents certified as well as literature average values for a number of SRMs; at MURR, the literature average values are normally used rather than the certified values. Bishop et al. (1990) suggest that the important thing in provenance research is not so much to ensure that measurements are accurate but to ensure that they can be calibrated to one another, so that databases from multiple laboratories and, conceivably, multiple techniques can be combined in a single study.

Another scale on which INAA can be compared to other techniques is sensitivity. This refers to the smallest detectable amount of an element that can be measured with a given tool, such as INAA. INAA has achieved its status as the "technique of choice" in provenance investigations partly because of its sensitivity to many trace elements and the resultant potential for detecting source-related chemical differences along multiple dimensions of major, minor, and trace element variation (Bishop et al. 1990). In the past decade and a half, however, techniques based on the use of an (ICP) torch, including ICP–AES and ICP–MS, have rivaled INAA in sensitivity for a range of sample matrices (Pollard and Heron 1996). In particular, ICP–MS appears currently to wear the "most sensitive" crown for multielement characterization of silicate materials (Parry 1991: Table 1.2; Tykot and Young 1996; Chapter 3, this volume).

Although INAA is a destructive technique in that analysis involves removal and pulverization of sample material, the amount that must be removed is often so small that even very valuable museum pieces can be analyzed without any visible harm to the object. As little as 30 mg of material can simply be removed by scraping, drilling, or burring in an inconspicuous place. Although sensitivity remains high, even with very small quantities, there is some loss in precision, and the issue of whether the small sample is representative of the bulk object composition becomes more problematic. Because of the latter issue, much larger quantities are normally homogenized for analysis at MURR, as mentioned above. With the tons of potsherds that often result from archaeological excavations, however, a few grams of powdered sherd material seems a small price to pay for detailed information on provenance.

The relative immunity from matrix effects and consequent ease of sample processing in INAA has already been discussed. It is these qualities that have enabled INAA to retain its title of "technique of choice" in provenance investigations despite equal or better precision with ICP–AES and better sensitivity with ICP–MS. INAA remains the best technique for achieving high precision and high sensitivity with easy, relatively error-free sample processing.

Determination of multiple elements is, other things being equal, a necessity in provenance investigations simply because sources often are very similar on many elements, and the potential for discriminating sources increases with number of elements determined. The multielement capability of INAA should be obvious from the above discussion. Spectra representing peaks from numerous isotopes are collected simultaneously, and computer programs subsequently identify and quantify the peaks of interest, then calculate elemental concentrations via comparison with standards. Depending on how many irradiations and gamma counts are undertaken, one may expect somewhere between 20 and 40 elements to be determined from INAA. The 33 elements reported in pottery analyses at MURR have proven extremely useful in a wide range of provenance investigations. Here again, however, it is worthwhile to point out that ICP–AES and, especially, ICP-MS rival or surpass INAA in the number of elements that can be routinely determined. An archaeometric laboratory that combined INAA with one of these other techniques would set a new benchmark in terms of multielement artifact characterization.

A final advantage of INAA is that analysts and archaeologists have recognized the importance of and potential for intercalibrating data and have worked toward this goal for some time (Bishop et al. 1990; Harbottle 1982; Hughes et al. 1991a; Kuleff and Djingova 1990; James et al. 1995; Neff and Glascock 1995; Yellin et al. 1978). If two laboratories use the same calibration standard, then, as long as the same values are assumed for the standard, the data should be comparable. If two laboratories use different standards (see Table 5.3 for some common standards), then, to achieve data compatibility, the calibration standards or quality control standards used in one laboratory must be analyzed as unknowns in the other laboratory. Ideally, multiple replicates of several different reference materials will be analyzed in order to maximize the compatibility of the data. Reanalyzing some of the unknowns from the other laboratory is an additional check on the intercalibration. Despite these steps, different neutron fluxes, decay and counting times, and other variable procedural details will yield a different set of "good" elements in different laboratories, and careful consideration must be given to which elements will be retained in the combined analysis. Finding that a subgroup distinction recognized in compositional data corresponds well to a laboratory difference should motivate strong skepticism regarding the validity of the calibration.

5.2.14. Disadvantages of INAA

Today, as in the past, the main perceived disadvantage of INAA is that it can only be undertaken where there is a source of neutrons available, normally a research reactor. Partly as a result of the limited availability of neutron sources, the commercial cost of INAA (perhaps $100 to $150 per sample for archaeological artifacts) is beyond the reach of most archaeological research budgets.

Because of limited availability and high cost, INAA in the service of archaeological research was developed originally by a few research groups at well funded government laboratories where researchers were encouraged to seek new applications for nuclear techniques. In North America, as already mentioned, the main early R&D in archaeometric INAA was done at BNL (Sayre and Dodson 1957; Harbottle 1976) and LBL (Perlman and Asaro 1969, 1971). Because of the limited number of research reactors, application of INAA to archaeological research never spread to more than a few tens of laboratories around the world [e.g., see Neff and Glascock (1995) for a North American summary]. One way that the limited access to neutrons can be overcome is by having a batch of samples irradiated as a service at a research reactor, then having them shipped back to a home laboratory for gamma counting (e.g., Hughes et al. 1991a). In North America, a National Science Foundation (NSF) funded program at MURR has subsidized the cost of INAA for archaeologists since 1988, and this has dramatically increased the number and diversity of archaeological applications of INAA over the past decade (summarized on the MURR World Wide Web page, http://www.missouri.edu/~glascock/archlab.htm). In this light, a recent charge that INAA is the domain of a few "inbred" practitioners (Rice 1996: 166–167) is fallacious.

The high per-sample cost of INAA can be calculated from the cost of neutrons, costs for disposal of radioactive waste, and costs of all supplies and labor consumed. In government programs, such as the collaborative program between the Smithsonian's Center for Materials Characterization and Analysis and the National Institute of Standards and Technology, these costs are absorbed as operating costs, so that the analysts and archaeologists served by the programs do not need to generate funds to cover them. Similarly at MURR, NSF funding covers the costs of everything except supplies consumed in the analysis, so that collaborating archaeologists pay far less than the true per-sample cost of INAA. For unsubsidized applications, however, including cultural resource management, per-sample costs remain in the $100 to $150 range, and this continues to restrict the use of INAA. Short-lived isotope analysis at small reactors may cost substantially less than this (Hancock 1998).

Besides limited availability and high cost, another disadvantage of INAA sometimes mentioned is that it "contributes to the world's inventory of radioactive garbage" (Mallory-Greenough et al. 1998: 94). Although this charge is true, the extremely small mass of most samples irradiated for INAA (50 to 200 mg) and the fact that they consist of short-lived isotopes make their contribution to nuclear waste disposal problems inconsequential.

There is another way in which nuclear waste issues compromise the long-term prospects of INAA, however. The problems associated with storage and disposal of spent reactor fuel have increased the cost of reactor construction and operation to such an extent that reactor decommissioning is likely to outpace new construction in the near future, if it has not already happened. That means that facilities where INAA can be undertaken will inevitably become even less common during the coming decades. In light of this prediction, it seems prudent for archaeologists interested in provenance research and their analyst colleagues to begin building a bridge to a non-nuclear future. For one thing, to prevent the obsolescence of the huge INAA databases built up over the past three decades, efforts should be initiated now to develop protocols for new techniques, particularly ICP–MS, that will yield data that can be calibrated to the existing databases (Bryan et al. 1996).

Despite the likelihood that INAA will eventually disappear from the analytical scene, it remains at present one of the most versatile tools available for characterizing archaeological artifacts. One of the main reasons for characterizing artifacts, whether by INAA or other chemical characterization techniques, is to help determine source. The following section concerns the methodological means by which analytical units measured in the laboratory can be mapped onto the geographical coordinate units in terms of which source locations may be described.

5.3. METHODOLOGY

Inclusion of a section on methodology of provenance investigation in an overview chapter on the use of INAA in archaeology is warranted by the close association that INAA has with provenance issues. This association dates to the very first discussions of what activation analysis might offer archaeology (Sayre and Dodson 1957). Moreover, much of the explicit development of provenance determination methods has taken place in the context of INAA-based studies. But, whereas development of the INAA technique has been the domain of physicists and chemists, development of methods for determining provenance using INAA data has been a joint endeavor by technical experts (physicists and chemists) and those with an interest in applying the technique (archaeologists) (Harbottle 1976: 42).

Before proceeding, it is worthwhile to remark that the fascination of archaeologists with provenance issues is not too surprising considering that the archaeological record is recognizable as the product of hominid activity in large part because of the spatial displacement of materials from their natural settings. A scientific account of this record inevitably requires a description of the kinds and amounts of materials that were moved and the directions and distances over which they were moved. Provenance determination permits archaeologists to investigate such diverse topics as mobility patterns of paleoindian hunters in North America (Hoard et al. 1992; Tankersley 1998), prehistoric migrations of Puebloan people in the North American Southwest (Triadan 1997), and patterns of economic and political interaction in the Aztec empire (e.g., Hodge et al. 1993; Neff and Hodge, in press), to name just a few applications.

5.3.1. Provenance Postulate

Although the basic proposition underlying chemistry-based provenance determination was clearly understood by the early 1970s (e.g., Abascal-M et al. 1974; Bieber et al. 1976; Harbottle 1976; Olin and Sayre 1971; Perlman and Asaro 1969, 1971; Sayre et al. 1971), it was not until 1977 that Weigand, Harbottle, and Sayre (1977: 24) recognized explicitly that efforts to trace artifacts to sources depends on the postulate "that there exist differences in chemical composition between different natural sources that exceed, in some recognizable way, the differences observed within a given source." Weigand and his colleagues named this the "Provenience Postulate"; I will refer to it as the "provenance postulate" in order to avoid confusion arising from the fact that the term "provenience" is also used as a synonym for archaeological context. These authors add that the provenance postulate "has not been very often tested by the exhaustive sampling and analysis of materials from single sources" (Weigand et al. 1977: 24), so it is clear that the postulate is not an assumption about conditions in the natural world, but rather a statement of the conditions that must be met in order for source determination to be possible. Specifically, the provenance postulate states that artifact raw material sources can be determined by chemical characterization as long as between-source chemical differences exceed within-source variation.

Although Weigand et al. (1977) intended the provenance postulate to apply to quantitative variation encountered in chemistry-based provenance investigations, this is an unnecessarily narrow formulation. Mineralogy-based sourcing rests on the same basic principle, and the postulate applies to qualitative as well as quantitative differences. So a more general statement of the fundamental principle underlying provenance research is thus something like the following: Sourcing is possible as long as there exists some qualitative or quantitative chemical or mineralogical difference between natural sources that exceeds the qualitative or quantitative variation within each source.

The provenance postulate can be thought of as a bridge between units measured in two distinct domains (Neff 1998). One domain is that of the individual analyzed specimen, in which the units are imposed by the analytical technique, for example, ppm of the various elements measured by INAA or some other analytical techniques, presence/absence of some unusual mineral, and proportion of common minerals in the total nonplastic assemblage. The other domain is geographic space, in terms of which "source" may be described. The units in this second domain are conventional geographic coordinates [e.g., latitude and longitude or Universal Transverse Mercator (UTM) coordinates]. As I have suggested elsewhere (Neff 1998), this perspective makes it clear that reliability (precision) and validity (accuracy) in provenance investigations are not solely properties of the analytical technique (e.g., INAA), but rather these properties also depend on raw material properties and on the methods used to span the gap between analytical units and geographic coordinate units. Analytical considerations certainly play a role; for example, if chemical data are so noisy that intersource differences are invisible, then reliability is severely attenuated (Neff et al. 1996). However, reliability and validity also depend on the adequacy of sampling both raw materials and artifacts, on natural variation in raw materials, on whether artifact raw materials were processed in some way by humans, on the potential for postmanufacturing compositional change in the artifacts, and on whether supplementary analytical techniques are brought to bear to tease apart various sources of compositional variation.

5.3.2. Sampling of Raw Materials and Artifacts

Logically, source determination efforts based on the provenance postulate can follow one of two separate paths (Fig. 5.2). If sources are localized and relatively easy to identify, as in the case of obsidian flows, the known sources can be characterized and artifacts of unknown provenance can then be compared to the range of variation of the known groups. If sources are widespread, as is true especially in the case of ceramic raw materials, sampling and characterizing most or all of the possible sources are impractical. As a result, ceramic provenance research generally involves an alternative approach, in which reference groups are created from the unknowns rather than knowns. In this more common approach to ceramic source determination, elemental concentration units are mapped onto geographic coordinates by comparing individual known samples (raw materials) to the unknown ceramic reference groups. Sampling requirements differ depending on which of the two approaches summarized in Figure 5.2 is adopted.

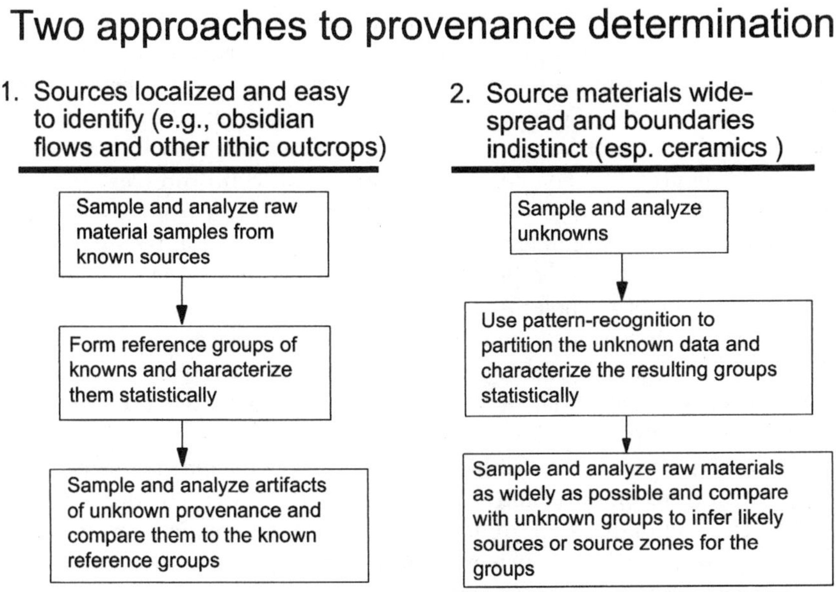

Figure 5.2. Two approaches to provenance determination.

5.3.2.1. Sampling When Natural Sources Are Localized

For the case of localized sources, the goal must be to sample the sources intensively enough to characterize the range of natural variation within each. Highly homogeneous materials, the epitome of which is obsidian, often exhibit clear differentiation between sources on many different elements, so that multivariate statistical discrimination of sources (see Chapter 20) is seldom necessary. In such cases, 5 to 10 samples from each source will often provide an adequate basis for discrimination. More variable materials, such as chert, impose greater sampling demands because individual elements or combinations of just a few elements will not reveal intersource difference, and multivariate statistical discrimination becomes essential. In either case, once potential sources are characterized, individual artifacts (unknowns) may be compared to the characterized sources in order to determine where on the ground the artifact originated.

5.3.2.2. Sampling When Natural Sources Are Widespread

For the case of widely available raw materials (the right side in Fig. 5.2), epitomized by ceramics, sampling design is more problematic. A given collection of ceramics contains representatives of an unknown number of natural sources in unknown proportions. When such a collection is sampled for chemical characterization, some sources may end up not being represented in the sample, and some sources may be represented so sparsely that their existence will not be revealed by standard pattern recognition techniques (see Chapter 20). The best general approach to this problem is to increase sample size: With a larger random sample drawn from a given collection, it becomes more likely that the sources recognized in the sample will approximate in number and proportion the natural sources actually present in the collection as a whole. A variation on this approach is to stratify the collection, say, based on decorative criteria, and to focus on selected sampling strata that are of particular interest. Stratified sampling designs also can be used to ensure that very unusual artifact categories are not overlooked in the sample.

Inevitably, some of the sampled unknowns will remain unassigned following pattern recognition and statistical analysis. Baxter and Buck (Chapter 20) refer to samples that are inconsistent with the remainder of the data as outliers. Some such inconsistent samples may indeed be statistical outliers, that is, specimens that diverge from one of the recognized groups because of chance or some other effect. Others may pertain to groups that are so lightly represented in the sample that they cannot be recognized as groups. With ceramics, paste preparation effects may contribute to both kinds of outliers, since preparation may make a specimen diverge from group compositional norms (Arnold et al. 1991; Bishop and Neff 1989; Neff et al. 1988b, 1989) or, in more extreme cases, it may create an entirely new compositional profile (Neff 1998; Neff et al. 1988a). Paste preparation effects are considered in more detail below.

Once reference groups are formed from the unknowns (and outliers are identified), the strategy on the right side of Figure 5.2 calls for sampling of raw materials. Although it is sometimes argued in the case of ceramics that raw material sampling should be based on consultation with modern potters (e.g., Rice 1987: 319), an exclusive reliance on modern potters is misguided because it precludes sampling raw materials that might have been exploited in the past but are currently not perceived as suitable or are unknown to modern potters. Rice's (1977, 1978a) failure to identify source clays for Formative white wares from the Valley of Guatemala (Neff et al. 1990) illustrates the fallibility of this approach to raw material sampling. A systematic approach (e.g., Neff 1995a; Neff and Bove, 1999; Neff et al. 1992), in which an attempt is made to sample the range of available raw materials as widely as possible, provides a more reliable basis for linking reference groups with locations on the ground.

5.3.2.3. Samples from Production Sites

When samples of production debris from kiln sites are available, the method of ceramic provenance research approximates the method used for materials whose sources are highly localized (i.e., the left side of Fig. 5.2). In effect, the kiln can be conceived as a localized source, much like a lithic outcrop, and the group to which production byproducts (wasters) from the kiln site are found to belong can be used as a reference group for that manufacturing location. In practice, kiln or workshop samples have been quite useful in ceramic provenance studies in the Old World (e.g., Mommsen et al. 1988; Pollard and Heron 1996: Chapter 4), but their use in the New World has been rather limited. Exceptions include several projects in coastal Peru (Chapdelaine et al. 1995; Hayashida 1995; Shimada et al. 1994). Elsewhere in the New World, production debris from apparent waster dumps, even though not directly associated with kiln sites, has sometimes proved useful in establishing provenance (Neff and Bishop 1988; see Section 5.4.2 below).

Although manufacturing location is not identical to the raw material source whose location is sought in the standard approach to ceramic sourcing (right side of Fig. 5.2), it is rare for clay sources to be more than a few kilometers from kiln sites (Arnold 1985). Nonetheless, the possibility that clays might be brought in from some distance and that clay from more than one source might be used at a single kiln must always be borne in mind when using samples from production sites to form reference groups. More generally, it is not always obvious whether a kiln sample is representative of the range of compositional variation of that kiln's output (Pollard and Heron 1996: 144).

Another stumbling block in the use of kiln debris to tie reference groups to locations in geographic space is that some kiln sites may remain undiscovered or may not have been included in the analyzed sample. As a result, while some reference groups may be tied to specific locations because of their inclusion of kiln wasters, other groups remain without clear provenance. As one example, although Mommsen et al. (1988) formed one of their reference groups around kiln wasters, a second group had no such linkage to a location on the ground, and provenance thus had to be inferred on other grounds (see below).

5.3.2.4. Criterion of Abundance

In practice, ceramic compositional investigations are very often undertaken without any sampling of raw materials or kiln debris. And, as just pointed out, some studies that do include kiln debris may not have all of the relevant kilns represented. Such sampling limitations preclude establishing direct linkages between ceramic reference groups and geographic locations. In order to circumvent these limitations, ceramic provenance researchers have adopted the "criterion of abundance" (Bishop et al. 1982), which extends to compositional research the commonsense assumption that things are most frequent closest to where they originate. That is, a homogeneous compositional group of unknown derivation is assumed to come from the site or region where it is most heavily represented archaeologically. For example, although Mommsen et al. (1988) were able to link one group of Mycenaean pottery from the Argolid to a location on the ground through analysis of kiln wasters, a production zone for the second group had to be inferred based on the criterion of abundance.

The problem with the criterion of abundance is that one can imagine numerous conditions under which it would be a poor guide to the origin of ceramics fitting a particular composition. Peculiarities of past consumption patterns may cause a particular kind of pottery to peak in frequency at some distance from its production center. Or, due to vagaries of archaeological sampling, relative frequencies in excavated assemblages may not accurately represent variation in the frequency of occurrence of pottery derived from some source. And vagaries of sampling the assemblages for analysis may distort the relative frequencies of various compositional groups. All in all, too much faith in provenance assignments based on the criterion of abundance would seem to be ill-advised.

5.3.3. Natural Variation in Raw Materials

5.3.3.1. Geological Process and Regional Geological History

Arguments about source locations, whether based on sampling of raw materials, samples from production sites, or the "criterion of abundance," can often be strengthened by considering features of regional geology. This stands to reason, of course, since the artifacts subjected to provenance investigation typically start out as natural geological materials. Obsidian and other lithics are unaltered geological material; ceramics start out as soils or sediments containing variable quantities of clay minerals, nonplastic sand- and silt-sized mineral grains and rock fragments, and organic matter; and metals start out as ores. Geological maps and highly detailed descriptions of both bedrock geology and soils, often including both mineralogical and chemical variation, are available for many regions of the world and constitute an extremely valuable resource for provenance researchers.

Geological process considerations are highlighted in several recent provenance studies of lithic material. Malyk-Selivanova et al. (1998), in a study of chert formations in the Brooks Range of Alaska, show that geochemical signals of the sedimentary environment in which the chert formed, diagenetic processes of chert formation, and geological origin of the chert-bearing sediments all prove useful in achieving discrimination of the chert-bearing formation of this region. In a study of steatite quarries from eastern North America, Truncer et al. (1998) observe that transition metal elements are particularly effective in differentiating the quarries and speculate that this arises from differences in the parent rock, the transition metals tending to be relatively immobile in the weathering of ultramafic rock.

Recent chemistry-based ceramic provenance projects have incorporated considerations of regional geology on several spatial scales. Steponaitis et al. (1996), for instance, have shown that groups of Mississippian pottery pertaining to broad regions of the southeast and midwest U.S. differ chemically in precisely the ways expected based on differences in clay mineralogy among the different drainage basins. At a smaller geographical scale, in the northern Valley of Guatemala, a localized outcrop of granodiorite helps explain the peculiar chromium–cobalt profile of modern pottery from the area and helps eliminate it as a source of certain regionally distributed prehistoric ceramics (Neff et al. 1994). At a scale intermediate between these two extremes, Bronze Age White Slip II pottery from Cyprus is argued to derive from clays weathering from ultrabasic rocks of the Troodos Mountains in part because of high observed concentrations of iron, scandium, and other transition metals coupled with relatively low concentrations of light rare earth elements (Gomez et al. 1995; Neff et al., in press b; Rautman et al. 1993). In one tantalizing recent study involving chemical characterization of Nile Valley pottery by ICP–MS, Mallory-Greenough et al. (1998) have argued that climatic variation in the headwaters regions of the Nile cause changes in vegetative cover and sediment load make-up that are detectable in the chemical composition of the Nile alluvium and the pottery produced from it.

Pollard and Heron (1996: Chapter 4) have recently stressed the importance of geological processes in chemistry-based ceramic provenance research, arguing that geochemical considerations should be incorporated into such studies much more explicitly than has been done in the past. Their argument applies not just to the formation of clay deposits through deposition and weathering but also to paste preparation and postdepositional change, which are the topics of a later section of this chapter.

5.3.3.2. Raw Material Variation and the Issue of Reliability

Prior to their procurement by humans, the bulk composition of artifact raw materials is determined by parent rock geochemistry and weathering history.

Initially, then, composition is necessarily a *valid* reflection of source in the sense that the composition of the material carried away from the source by a human is a sample drawn from the range of natural compositional variation of that source. On the other hand, whether composition is a *reliable* monitor is an empirical question, the answer to which depends on the total variation in measurements that could be obtained on the source material.

Whether two sources can be distinguished reliably depends on the extent to which the ranges of chemical variation in the two sources overlap in multielement concentration space, which in turn depends on the magnitude of the difference between two centroids (between-source difference) and the dispersion of data points around each centroid (within-source variation; see Chapter 20 for a discussion of statistical group discrimination). The latter parameter, S_T^2, the total group variance, may be partitioned into S_N^2 (natural variance), S_S^2 (variance due to sampling the matrix for analysis), and S_A^2 (variance due to analytical error) (Harbottle 1976). Procedural steps to maximize reliability thus entail using samples large enough to ensure that they are representative of the bulk composition and refining the analytical techniques so as to minimize analytical error. The remaining determinant of reliability, S_N^2, is raw material variation itself, which must be considered on a case-by-case basis in relation to the scale of geographical resolution required to answer the archaeological questions of interest.

As already mentioned, obsidian epitomizes a sourceable material because obsidian flows so clearly meet the criteria of within-source homogeneity and between-source difference specified by the provenance postulate as necessary for source determination to be possible. Another way of stating this observation is to say that chemical characterization of obsidian is a highly *reliable* method for matching obsidian artifacts to sources. Other methods, such as visual sourcing, might (and have) been suggested (e.g., Braswell et al. 1994), but their reliability does not match the near-100% success rate achievable by chemical means. Even in the case of obsidian, however, geological processes can attenuate the reliability of matching artifacts to sources on the ground. For instance, mechanical weathering can distribute obsidian cobbles far from a source vent and can cause the geographic distributions of chemical "sources" to overlap (Hughes 1998). Further, some obsidian sources have multiple flows, between which there may be only subtle chemical differences. In such cases, reliable resolution at the individual flow level (rather than the source level) may require a highly precise multielement technique such as INAA along with a greater investment in sampling to ensure representation of the various chemical fingerprints (Glascock et al. 1998).

More obvious reliability problems affect efforts to source lithic materials that have longer and more complex geological histories. In the study of steatite quarry samples from the east coast of North America, mentioned above, Truncer et al. (1998) found posterior classification success rates between 40 and 96% using cross-validated Mahalanobis distances from the group centroids (see Chapter 20). In some cases, posterior classification was reliable at the regional level even though numerous specimens were misassigned at the individual quarry level, thus demonstrating that reliability may depend on the scale of geographical resolution required.

Chert is another lithic material for which chemical characterization may not always yield reliable intersource discrimination. Hoard et al. (1992, 1993) have successfully differentiated Oligocene-age chert outcrops in the Great Plains, some of which were utilized prehistorically, and the study by Malyk-Selivanova et al. (1998), cited above, achieved reliable separation of Brooks Range chert-bearing formations. On the other hand, efforts to discriminate cherts from different parts of northern Belize have not produced reliable source distinctions, the whole area being essentially a single chert source (Cackler et al., 1999). Clearly the geographical extent and geological context of chert "sources" are crucial determinants of whether chert provenance analysis will yield answers to the archaeological questions being posed.

Clays are so ubiquitous and their geological histories are so varied that the reliability of distinguishing between natural clay deposits varies widely. In general, if the parent rock contributions and weathering histories of two argillaceous soils or sediments are sufficiently distinct, then the provenance postulate will apply, that is, chemical variation between the two deposits will exceed variation within each, and sourcing pottery to one deposit or the other is possible in principle. In geologically diverse areas, the chemical distinctiveness of ceramics derived from different sources may approach that of obsidian source groups. For instance, raw material and locally produced pottery from four contemporary pottery-making towns in the Valley of Oaxaca, Mexico, are easily and clearly distinguished from one another on the basis of just two elemental concentrations (Fig. 5.3). Therefore, using pottery and raw materials from the modern production centers as reference groups, points of origin (sources) could be determined for "unknowns" from the modern Valley of Oaxaca with near-100% reliability.

Unfortunately, geological processes of clay formation often do not create discrete, chemically homogeneous clay sources such as those illustrated in Figure 5.3 but instead produce extensive deposits that vary in composition both vertically and horizontally. A homogeneous "source" of argillaceous sediments may be so widespread that reliable distinctions at a geographical scale useful to archaeological investigation may be impossible. A case in point is the quaternary alluvium of the Mississippi River valley, which can be considered a single source from the confluence of the Ohio and Mississippi Rivers south to the confluence of the Arkansas and Mississippi Rivers

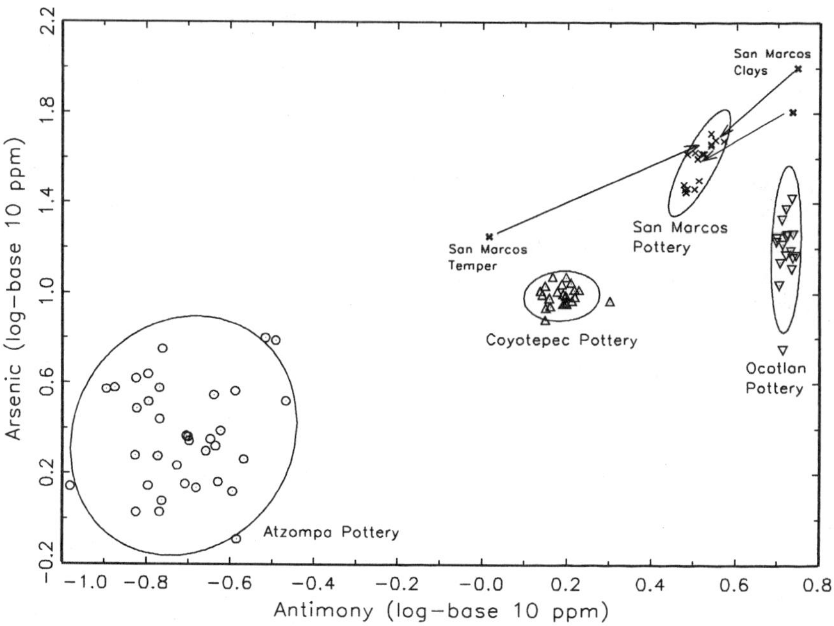

Figure 5.3. Antimony and arsenic log-concentrations in pottery produced in four pottery-making towns in the Valley of Oaxaca, Mexico. Concentrations of a temper and two clays obtained in the town of San Marcos are also shown. Ellipses represent 90% confidence levels for membership in the four groups.

(O'Brien et al. 1995), a situation that clearly precludes identifying movement of pottery north and south within the Mississippi Valley.

More generally in sedimentary clay deposits, proportions of material weathered from different source rocks may vary gradually across horizontal space, while textural variation as well as *in situ* chemical weathering may create vertical differentiation within a single stratigraphic column. Extensive raw material sampling in the quaternary sediments of the Guatemalan Pacific coastal plain suggests that geographic variation in the composition of the region's argillaceous sediments approximates this model (Neff and Bove, 1999; Neff et al. 1992; see Section 5.4.2 below). In such an environment, compositional variation in ceramics may reflect procurement location, but geographically discrete, chemically homogeneous source clays like those from the modern communities in the Valley of Oaxaca (Fig. 5.3) may be difficult if not impossible to identify. Clearly, sampling discrete sources is impossible under these conditions. Also, such conditions highlight the importance of explicitly identifying the geographic scale at which the source is conceived.

5.3.4. Artifact Life History

In the case of some materials, especially ceramics and metals, modifications during an artifact's life history may attenuate both the reliability and the validity of chemical analysis as a monitor of source differences. I focus here on ceramics, which have been the subject of a large number of INAA-based provenance investigations, but similar problems, albeit more extreme, affect trace element provenance investigations of metals (Pollard and Heron 1996: 303–306).

Like lithics, ceramic life histories encompass (1) residence in the geological environment prior to use as implements by humans, (2) procurement of raw materials and preparation by humans, and (3) use and residence in the burial environment after discard. A key difference between the life histories of ceramics and lithic artifacts such as obsidian is that ceramics may undergo considerable chemical alteration during manufacture, use, and burial. One possibility considered by Neff et al. (1988b, 1989) is that paste preparation or other postprocurement changes may obscure source-related differences by increasing the chemical heterogeneity of ceramic groups; this is a reduction in the *reliability* of chemistry-based provenance investigation (Neff 1998). But these changes also may cause the chemical profile of the analyzed ceramics to diverge *systematically* from that of the clay originally carried away from the source by the potter. This latter effect is an attenuation of *validity*: Since ceramic composition does not just reflect the composition of the clay source, compositional groups within a database of analyzed ceramics may no longer reflect clay–source differences alone, but rather may also reflect differences engendered by paste preparation, manufacturing, and/or diagenesis. The special problems that paste preparation and diagenesis might create for chemistry-based sourcing have been appreciated for some time (Sayre and Dodson 1957).

5.3.4.1. Paste Preparation Effects

Two broad categories of processes, additive and subtractive, may contribute singly or jointly to enrichment or dilution of elemental concentrations in ceramic pastes during manufacture. Additive processes are the best studied (e.g., Arnold et al. 1991; Neff et al. 1988b, 1989); they include addition of temper, colorants, water, clay from a separate source, or any other intentional or accidental addition of material to a ceramic paste. The effect of additions during manufacture is intuitively obvious: the bulk profile of the finished product is a weighted linear expression of the profiles of components mixed during manufacture (Arnold et al. 1991; Bishop and Neff 1989; Bishop et al. 1982; Neff et al. 1988a, 1988b, 1989). Figure 5.3 illustrates this effect for components (clay and temper) mixed during the manufacture of pottery in San Marcos, Oaxaca: as indicated by the arrows in the figure, natural concentrations of arsenic and antimony in the clays are diluted by the addition of temper, creating a pottery profile that is intermediate between clays and temper on both elements.

Subtractive processes include sifting, levigation, and manual removal of particular size ranges of nonplastics. What is left after subtractive processing expresses the original composition as modified by the selective removal of constituents. For instance, Blackman (1992) has documented elemental enrichment–dilution effects of size sorting in clays from the Khabur drainage, Syria; some elements were found to be concentrated in the sand fraction, some in the silt fraction, and some in the clay fraction. The compositional profile of the finished pot thus diverges from the unprocessed clay, but the specific nature of the divergence depends on how much of what size fractions were removed by processing.

Elemental enrichment and dilution during the manufacturing process leads to three related complications for ceramic provenance research. First, the compositional resemblance of finished pots to their constituent raw materials may be obscured, so successful provenance determination may require comparison of finished products to processed (e.g., tempered or levigated) rather than raw source materials (Arnold et al. 1991). Second, differential paste preparation might create two or more compositional groups among pots derived from a single clay source; for example, Formative period Fine Red ware from southern Guatemala apparently subdivides into two compositional groups because two different volcanic ash tempers were added (Neff et al. 1988a, 1994). These first two complications primarily concern the *validity* of composition as a measure of geographic space: in the first case, no raw clay source falls within the range of compositional variation of the analyzed ceramics, whereas in the second case, the existence of multiple compositional groups reflects not the use of multiple clay sources but rather the use of multiple paste preparation recipes.

The third potential complication of paste preparation has more to do with *reliability* of composition as a measure of geographic space. Additive and subtractive preparation may obscure compositional differences between clay sources by moving the centroids of the groups derived from those sources closer together and by broadening their natural ranges of chemical variation. While tempering in particular has been considered a serious problem in compositional studies (Arnold et al. 1978; Rice 1978b), a series of simulation experiments (Neff et al. 1988b, 1989) suggests that attenuation of group distinctiveness as a result of addition of temper may not be particularly significant unless the added temper is highly variable and elemental concentrations in the temper approach those in the clay for most elements involved in the group separation.

Firing has been considered a further possible source of chemical change during manufacture. However, recent experiments (Cogswell et al. 1996a; also see Kilikoglou et al. 1988) have failed to identify any changes in concentration of a large number of elements determined by neutron activation analysis through a range of firing temperatures up to 1100°C. The only detected loss was bromine, apparently due to volatilization, but this is of little concern because bromine is seldom determined by INAA. [Bromine can, however, be an important indicator of burial in a marine environment, so its detection may be of interest in some studies (Hancock 1998).]

5.3.4.2. Use and Diagenesis

After manufacture, both use of a vessel and postdepositional processes (diagenesis) may modify the bulk chemistry of a ceramic fabric. Although neither use nor diagenesis has been well studied, particularly for many of the elements commonly used in provenance research, it is reasonable to suspect that ceramic paste composition will not remain stable but instead will change gradually as a function of chemistry and temperature in the various environments to which the ceramic is subjected after manufacture.

Use-related compositional changes that may affect provenance studies are poorly documented. Phosphorus is perhaps the only element whose concentration has been explicitly linked to use. Duma (1972) has suggested that the concentration and distribution of phosphorus in a fabric may be affected by cooking. Dunnell and Hunt (1990) question the usefulness of phosphorous concentrations as a marker for use and suggest instead that original clay composition and diagenesis are as important or more important determinants of phosphorous concentrations and distribution in the ceramic fabric. Walter and Besnus (1989) reach similar conclusions. In any case, however, phosphorus has rarely if ever been used in provenance research. Cogswell et al. (1996b) report a pattern of zinc enrichment near the interior surfaces of Mississippian cooking vessels that appears to be due to cooking of organ meat.

Diagenetic alteration of ceramic composition is widely documented. Obvious examples of diagenetic effects on ceramic composition are shell-tempered Mississippian sherds from which most of the shell has been removed by leaching in the burial environment. Myers et al. (1992) have demonstrated a similar enrichment–dilution effect on calcareous majolica pottery deposited in mildly acidic burial environments. Experiments with powdered test tiles indicate that concentrations of many elements determined by neutron activation analysis are affected by a 24-hr bath in 10% HCl (Cogswell et al. 1998). Conversely, several studies have documented post-depositional crystallization of calcite (e.g., García Heras 1993), and fixation of other elements during diagenesis has been suggested as well (e.g., Dunnell and Hunt 1990; Lemoine and Picon 1982; Picon 1985, 1987; Walter and Besnus 1989). However, experiments by Franklin and Vitali (1985) imply that compositional changes during diagenesis are expected mainly to affect the surface layers of sherds. Clearly, more laboratory research is needed before generalizations can be advanced about the importance of diagenesis as a confounding effect in provenance studies.

5.3.4.3. Supplementary Analyses

Based on the foregoing discussion, multiple compositional groups of ceramics generally cannot be assumed to represent different clay sources. Rather, this is just one hypothesis. The groups also could arise from differences in paste preparation, different use or diagenetic histories, or some combination of these effects. These are competing hypotheses about the meaning of the groups. If raw materials have not been sampled or the available sample does not demonstrate a direct connection between the ceramic groups and specific locations on the ground, then the various competing hypotheses cannot be ruled out based on the bulk chemical data. In such cases, supplementary analytical approaches can be used to test the competing hypotheses. Combining several techniques in ceramic provenance research was explicitly advocated long ago by Shepard (1965) and more recently by Bishop et al. (1982) and Rice (1987: 375).

Supplementary analyses may range from relatively low-technology examination with a low-power microscope to microprobe chemical characterization of various components of the clay matrix. The general premise underlying the approach is that determining precisely *where* particular elements are localized within a ceramic fabric (clay matrix, temper particles, edges of pore spaces, near the vessel surface, etc.) provides a tool for determining *when* (i.e., at what stage during manufacture or use) the concentrations of those elements were modified. The enrichment of zinc in certain Mississippian cooking pots (Cogswell et al. 1996b), mentioned above, is one example of a pattern discovered in bulk INAA data that was elucidated by microprobe analysis: consistent enrichment of zinc near interior surfaces permitted the inference that a use-related effect, such as cooking of high-zinc organ meats, produced the enrichment observed in the bulk chemical data. In another study (Neff et al., 1999), microprobe analyses of specimens in two compositional groups of cream paste pottery from southern Mesoamerica showed that both the volcanic glass particles (temper) and clay matrices were distinct, a finding that is inconsistent with the hypothesis that the bulk chemical difference was created by different tempering practices.

Sensitivity is a limitation of the wavelength-dispersive X-ray spectrometer used in the studies just cited. There is a new technique, however, laser ablation ICP–MS, that makes it possible to carry out spot analyses of ceramic paste components with sensitivities for a wide range of trace elements at sub-ppm levels. The coupling of laser ablation ICP–MS with bulk analysis via INAA or another high-precision technique (e.g., ICP–AES or ICP–MS) may be the wave of the future in chemistry-based provenance investigation.

5.4. EXAMPLES

In order to exemplify the use of INAA in provenance research, two case studies involving artifacts from southern Mesoamerica are presented briefly in this section. The first example, which involves obsidian artifacts from the Pacific coast of Guatemala, illustrates a case in which artifacts are easily linked to a small number of localized, chemically homogeneous sources. The second example, which involves a widely traded Mesoamerican ceramic ware known as Plumbate, illustrates some of the complexities that arise in ceramic provenance investigations.

5.4.1. Obsidian from Pacific Coastal Guatemala

After ceramics, obsidian is probably the most common artifact found on archaeological sites on the Pacific coast of Guatemala. In order to understand shifts in obsidian source usage through time, Frederick J. Bove and his collaborators initiated an effort to determine sources for a large number of excavated obsidians from various chronological contexts. Since there are only three main sources, and they are to some extent visually distinct, one approach was to rely on visual identification (Carpio 1993). However, development of an abbreviated INAA protocol for southern Mesoamerican obsidian (Glascock et al. 1994b) provided the opportunity to obtain more reliable determination of sources for a statistically useful sample of artifacts.

Sampling of the three Guatemalan sources (El Chayal, San Martin Jilotepeque, and Ixtepeque) was undertaken by Robert H. Cobean (Cobean et al. 1991). As described by Glascock et al. (1994b) the abbreviated approach to obsidian source determination relies on elements determined by a 5-sec irradiation, 25-min decay, and 12-min gamma count. Sodium and manganese prove to be the most useful elements for source discrimination in most cases, with barium sometimes providing additional discrimination. Because the three Guatemalan sources are completely discriminated by sodium and manganese alone, attribution of artifacts to sources can be accomplished by visual inspection of bivariate plots.

Since 1993, nearly 1600 obsidian artifacts from Pacific coastal Guatemalan contexts have been analyzed by abbreviated INAA at MURR. The sample considered here consists of 117 artifacts found in Early Formative contexts (1800 B.C.–800 B.C.) on the lower coastal plain of the departments of Escuintla (Arroyo 1994) and Suchitepequez (Arroyo and Neff 1996; Neff and Arroyo, in press). The obsidian sources are in the volcanic mountain chain that lies immediately north of the coastal plain. Thus, obsidian source determination identifies inland regions that were known and exploited by different groups of Early Formative coastal people.

Inspection of a sodium–manganese plot of the source sample 90% confidence ellipses together with data points for the Early Formative artifacts (Fig. 5.4) readily reveals the sources of the 117 artifacts. Several observations about Early Formative raw material exploitation are immediately

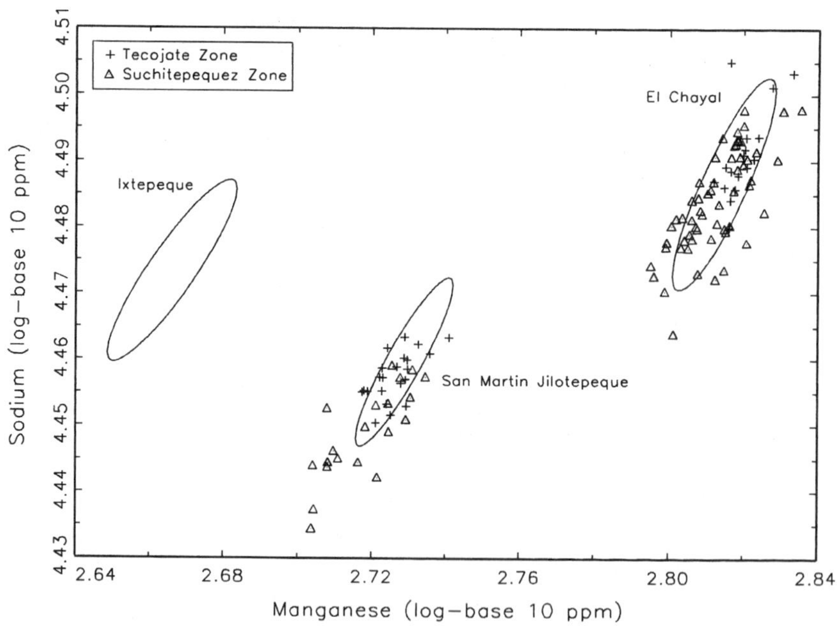

Figure 5.4. Manganese and sodium log-concentrations in obsidian artifacts from Early Formative contexts on the Pacific coast of Guatemala. Ellipses represent 90% confidence levels for groups of source samples (individual source sample data not shown). The scatter of data points outside of the confidence ellipses suggests that the range of variation sampled by the Early Formative people was somewhat different than that sampled by the archaeologists at the three sources.

apparent. For one thing, the southeasternmost source, Ixtepeque, was not exploited at all by Early Formative groups at the sampled locations. Second, a majority of the artifacts come from El Chayal. Third, representation of the two exploited sources in the two archaeological zones differs, with El Chayal clearly dominant in the Suchitepequez zone (58 vs. 17 from San Martin Jilotepeque) and San Martin Jilotepeque slightly dominant (22 vs. 17 from El Chayal) in the Tecojate zone. The implications of these observations for southern Mesoamerican prehistory are outside the scope of this chapter, but it is clear that the obsidian provenance study yields concrete data on variability in how Early Formative coastal people exploited resources of the adjacent highlands.

This example is not atypical of obsidian provenance projects. In regions where all or most of the sources have been sampled, such as Guatemala, the success rate for INAA-based provenance determination is near 100%, even with the abbreviated approach used here. The few artifacts that cannot be securely linked to any of the sampled sources by the abbreviated approach can be subjected to a complete analysis in order to ascertain whether they are outliers from the known sources on the elements determined by the abbreviated analysis, representatives of extraregional sources, or representatives of sources not yet in the databank. The key to this approach is obtaining representative samples of as many of the sources as possible. The MURR obsidian databank now affords highly reliable sourcing not only for Guatemala but also for the rest of Mesoamerica (e.g., Cobean et al. 1991; Glascock et al. 1998), large parts of western North America (e.g., Ambroz et al. submitted), and Andean South America (Brooks et al. 1998; Glascock et al. 1994a).

5.4.2. Plumbate Pottery

Plumbate is a distinctive class of pottery found in Late Classic and Early Postclassic contexts across a wide area of Mesoamerica (Neff 1984, 1995b; Neff and Bishop 1988; Shepard 1948). The Late Classic variant, San Juan Plumbate, is common on the Pacific coast of southern Mexico, Guatemala, and El Salvador and in neighboring highland regions. Tohil Plumbate, the Early Postclassic variant, occurs as far south as Panama, in western and central Mexico, and in northern Yucatan. Whether there was a single production zone or multiple zones for each Plumbate variety, whether the two varieties originated in the same zone(s), and where the production zone(s) was/were located are key questions about long-distance interaction in Late Classic and Early Postclassic Mesoamerica.

Research prior to 1980 had indicated that there were two types of Plumbate paste, suggesting at most two production zones (Shepard 1948), and that Plumbate frequencies in archaeological collections peaked on the Pacific coast of southern Mesoamerca, near the present border between Mexico and Guatemala (Shook 1965). The Mexico–Guatemala border region was therefore hypothesized to contain the source of both Plumbate varieties despite the low frequency of vessels in the Early Postclassic Tohil decorative style (Shepard 1948, 1952; Shook 1965). INAA-based chemical characterization research initiated in 1980 (Neff 1984) demonstrated that the paste composition associated with Tohil decoration outside the presumed production region also characterizes a sizable minority of simpler vessels in the San Juan style within the production region. Thus, because (1) Plumbate is so frequent in archaeological assemblages in the Mexico–Guatemala border region and (2) compositional analysis shows that sherd collections from this region contain both of the two main Plumbate paste compositional variants, the "criterion of abundance" clearly favors the hypothesis that all Plumbate originated on the Pacific coast, near the present Mexico–Guatemala border.

The INAA study carried out in the early 1980s also produced more direct evidence of local production of San Juan Plumbate. Possible Plumbate wasters and several amorphous clay lumps interpreted as ceramic production byproducts were collected at a littoral zone site in extreme southwestern Guatemala, where Plumbate pottery constituted the vast majority of recovered artifacts (Neff 1984, 1989a; Neff and Bishop 1988). Many of the wasters fell within the range of variation of the San Juan Plumbate reference group. In addition, one of the amorphous clay lumps, while statistically outside the range of variation of San Juan Plumbate, was similar enough chemically to the group to suggest that it was, in fact, a byproduct discarded during preparation of San Juan Plumbate paste. Based on these findings, Plumbate of San Juan composition is linked to the Mexico–Guatemala border region not only by the criterion of abundance but also by inclusion in the group of probable wasters as well as paste preparation byproducts.

Until recently, attribution of Tohil Plumbate to the hypothesized production region remained indirect, being based on the criterion of abundance. Tohil wasters and/or byproducts were not identified at the dumps that yielded San Juan wasters. Further, Tohil compositions were found to be low in frequency at all sites on the Guatemalan side of the border. These distributional patterns suggested that Tohil clay sources might have been located north and west of the lower coastal region where San Juan Plumbate seemed to be concentrated and probably originated (Neff 1984, 1989b; Neff and Bishop 1988). Unfortunately, fieldwork was not undertaken further west, at possible Plumbate production sites within Mexico.

Recently, an opportunity arose to test the foregoing hypotheses about Plumbate provenance directly by comparison of raw materials to Plumbate reference groups. The reference groups in this case are comprised of Plumbate analyzed recently at MURR (Neff 1995b) and do not include the earlier

BNL analyses, for which many fewer elements were determined. The raw clay database consists of 288 samples collected during raw material surveys on the Guatemalan coast (Neff et al. 1992; Neff and Bove, 1999). A sample of 29 sands from the coastal plain is also relevant, since microscopic examination indicates that Plumbate paste had fine volcanic sand or crushed volcanic rock added as temper (e.g., Shepard 1948). Although the raw material sample is best for the central coastal departments of Escuintla and Suchitepequez, sampling during 1995–1997 extended the coverage east to the El Salvadoran border and west to the Mexican border (sampling locations are indicated in Fig. 5.5).

To model the ceramic pastes that could be made at different locations on the coastal plain, each raw clay sample was paired with the nearest sampled sand, and the compositional profile of a 30% mixture of the sand with the clay was calculated. The resulting 288 "ceramic pastes" were then projected against the MURR Plumbate reference groups by calculating Mahalanobis distances of the paste from the reference group centroids (Chapter 20 discusses Mahalanobis distance). Hotelling's T^2 probabilities derived from the 288 Mahalanobis distances provide the basis for estimating a probability surface, on which peaks and plateaus indicate likely zones where Plumbate raw materials might have been procured.

Probability surfaces for the San Juan and Tohil reference groups are shown in Figure 5.5. Although probabilities do not rise above 1% on either map, it is clear that the only high spots are near the Mexico–Guatemala border, in the region inferred on the basis of other evidence to be the Plumbate source region. That is, of all the clays sampled across the entire Guatemalan Pacific coast, only a few sampled from the border region thought to be the source of Plumbate fall close enough to the range of variation defined by the Plumbate reference groups to be considered potential source clays. Considering the low intensity of raw material sampling coverage in this region, it is, perhaps, remarkable that such a clear pattern is discernible in the raw materials data. Perhaps even more remarkable, comparison of the San Juan and Tohil maps (Fig. 5.5) accords with the inference based on distributional data that the Tohil Plumbate source may be located slightly north and west of the San Juan source. Further raw material sampling both on the Guatemalan side of the border and on the Mexican side would certainly help to increase the specificity of these provenance inferences.

The main value of the foregoing update of the Plumbate study is that it exemplifies all of the basic approaches to ceramic provenance determination discussed in Section 5.3 above. At first, attribution of Plumbate to the Mexico–Guatemala border region was based entirely on the criterion of abundance. Later, probable wasters and pottery-making by-products found within the suspected source zone were linked to one of the main Plumbate

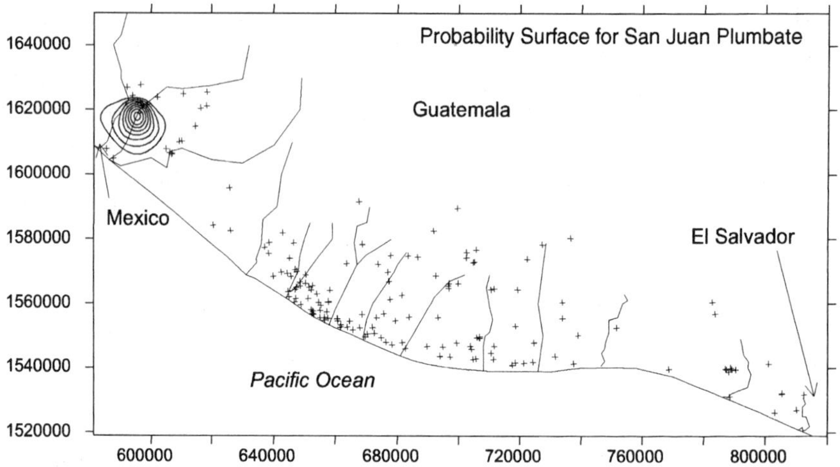

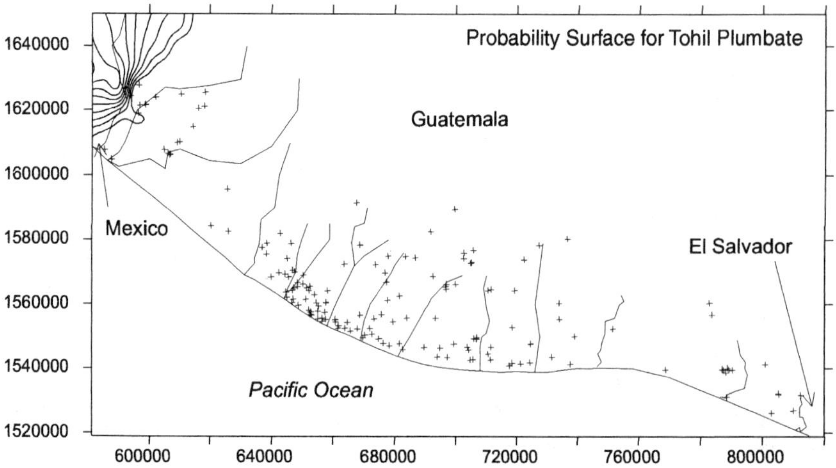

Figure 5.5. Maps representing variation in the Mahalanobis distance-based probability that raw materials sampled at various locations on the Pacific coast of Guatemala could be members of the San Juan Plumbate reference group (top) or the Tohil Plumbate reference group (bottom). Pluses indicate clay sampling locations. Clay elemental concentrations were corrected for the effect of temper by factoring in the effect of adding 30% of the nearest sand temper (locations not shown). Maximum probability for San Juan, 0.021%; maximum probability for Tohil, 0.097%. Axis units are UTM coordinates. More intensive sampling of clays and potential tempers on both sides of the Mexico–Guatemala border would be required to obtain a closer approximation of the Plumbate reference group compositions. However, the raw materials data clearly accord with the criterion of abundance and by-product characterization, all of which point to the international border region as the likely Plumbate production zone.

reference groups. And now, with the new raw materials data discussed here, both Plumbate reference groups can be linked directly to geographic locations within the hypothesized source zone.

5.5. CONCLUSION

After more than three decades of successful application, the reliability of INAA-based provenance determination is now firmly established. Numerous publications reporting the results of such studies appear in regional and international archaeological journals and in more technical journals such as *Archaeometry* and the *Journal of Archaeological Science*. Besides obsidian and ceramics, which are emphasized in the above discussion, INAA has been used successfully to investigate the provenance of metals (Anselmi et al. 1997; Hancock et al. 1995), glass (Djingova and Kuleff 1991, 1992; Kenyon et al. 1995), jade (Lange 1993), turquoise (Harbottle and Weigand 1992; Weigand et al. 1977), chert (Malyk Selivanova et al. 1998; Hoard et al. 1992, 1993), steatite (Truncer et al. 1998), and other materials.

The major limitation on archaeological applications of INAA, namely the small number of research reactors, has been successfully surmounted by programs that specialize in collaborative research with archaeologists. The key to success of such programs is the merging of technical expertise in INAA with careful attention to the methodological issues discussed in Section 5.3 above. Use of appropriate quantitative methods to evaluate patterns of resemblance in the data is essential as well (see Chapter 20).

Acknowledgments

I am indebted to a number of people who have taught me aspects of how INAA can be used in provenance research. At the top of the list are Ron Bishop, Jim Blackman, Mike Glascock, Gar Harbottle, and Ed Sayre, to whom I offer my most sincere thanks. I especially appreciate the time Mike Glascock took to comment on an earlier draft of this chapter. Ron Hancock also provided useful comments on an earlier draft.

I have had the good fortune to interact with a number of individuals who have helped me to refine my ideas about provenance methodology and to gain a better appreciation of the archaeological research potential of reliable provenance information. Among these are Dean Arnold, Ron Bishop, Fred Bove, Basil Gomez, Mary Hodge, Dan Larson, Ann Ramenofsky, Marcus Rautman, Daniela Triadan, and Nieves Zedeño.

Many individuals have made concrete contributions to the studies summarized here. Jim Cogswell and Lou Ross have carried out microprobe work to demonstrate the feasibility of supplementing bulk chemical analysis with spot analysis. Individuals involved in the Guatemalan obsidian provenance project in one way or another include Barbara Arroyo, Fred Bove, Mike Brewer, Edgar Carpio, Francisco Estrada Belli, Mike Glascock, and Taryn Johnson. And recent progress on the Plumbate study has depended on contributions by Barbara Arroyo, Fred Bove, Kelly Chakov, Jessica Coats, Mike Glascock, and Mike Love.

Financial support for many of the studies cited here was provided by the U.S. National Science Foundation through grant Nos. BNS89-11580, BNS88-01707, SBR91-02016, SBR95-03035, and SBR96-00334.

REFERENCES

Abascal-M., R., G. Harbottle, and E. V. Sayre (1974), in *Archaeological Chemistry*, C. W. Beck, Ed., American Chemical Society, Washington, DC, p. 81.

Alfassi, Z. B. (1990), in *Activation Analysis*, Vol. 2, Z. B. Alfassi, Ed., CRC Press, Boca Raton, FL, p. 3.

Ambroz, J. A., M. D. Glascock, and C. E. Skinner (submitted), *J. Archaeol. Sci.*

Anselmi, L. M., M. A. Latta, and R. G. V. Hancock (1997), *Northeast Anthropol.* **53**, 47.

Arnold, D. E. (1985), *Ceramic Theory and Culture Process*, Cambridge University Press, New York.

Arnold, D. E., P. M. Rice, W. A. Jester, W. N. Deutsch, B. K. Lee, and R. I. Kirsch (1978), in *The Ceramics of Kaminaljuyu*, R. K. Wetherington, Ed., Pennsylvania State University Press, University Park, PA, p. 543.

Arnold, D. E., H. Neff, and R. L. Bishop (1991), *Am. Anthropol.* **93**, 70–90.

Arroyo, B. (1994), The Early Formative in Southern Mesoamerica: An Explanation for the Origins of Sedentary Villages, Ph.D. thesis, Vanderbilt University.

Arroyo, B., and H. Neff (1996), Informe Tecnico del Proyecto Suchitepequez, Temporada 1996, unpublished report submitted to the Instituto de Antropología e Historia de Guatemala.

Bennyhoff, J. A., and R. F. Heizer (1965), *Am. Antiq.* **30**, 348.

Bieber, A. M. Jr., D. W. Brooks, G. Harbottle, and E. V. Sayre (1976), *Archaeometry* **18**, 59.

Bishop, R. L., and H. Neff (1989), in *Archaeological Chemistry*, Vol. 4, R. O. Allen, Ed., Advances in Chemistry Series 220, American Chemical Society, Washington, DC, p. 57.

Bishop, R. L., R. L. Rands, and G. R. Holley (1982), in *Advances in Archaeological Method and Theory*, Vol. 5, M. B. Schiffer, Ed., Academic, New York, p. 275.

Bishop, R. L., V. Canouts, P. L. Crown, and S. P. De Atley (1990), *Am. Antiq.* **55**, 537.

Blackman, M. J. (1992), in *Chemical Characterization of Ceramic Pastes in Archaeology*, H. Neff, Ed., Prehistory Press, Madison, W

I, p. 113.

Blackman, M. J., G. J. Stein, and P. B. Vandiver (1993), *Am. Antiq.* **58**, 60.

Braswell, G. E., E. W. Andrews V, and M. D. Glascock (1994), *Ancient Mesoamerica* **5**, 173.

Brooks, S. O., M. D. Glascock, and M. Giesso (1998), *Nature* **386**, 449.

Bryan, N. D., L. J. Holmes, and S. M. A. Hoffman (1996), Updating Ancient Ceramic Elemental Concentration Databases: Bring in the New Technique, But Without Throwing Out the Old Data, poster presented at the 1996 International Symposium on Archaeometry, Urbana, IL, May 20–24, 1996.

Burton, J. H., and A. W. Simon (1993), *Am. Antiq.* **58**, 45.

Cackler, P. R., M. D. Glascock, H. Neff, H. Iceland, K. A. Pyburn, D. Hudler, T. R. Hester, and B. M. Chiarulli (1999), *J. Archaeol. Sci.* **26**, 389.

Carpio R., E. H. (1993), in *The Balberta Project: The Terminal Formative—Early Classic Transition on the Pacific Coast of Guatemala*, F. J. Bove, S. M. Medrano B., B. Lou P., and B. Arroyo L., Eds., University of Pittsburgh Memoirs in Latin American Archaeology No. 6, Pittsburgh, PA, p. 84.

Chapdelaine, C., G. Kennedy, and S. U. Castillo (1995), *Bull. Intitut Francais d'Etudes Andines* **24**(2), 183.

Cobean, R. H., J. R. Vogt, M. D. Glascock, and T. L. Stocker (1991), *Latin Am. Antiq.* **2**, 69.

Cogswell, J. W., H. Neff, and M. D. Glascock (1996a), *J. Archaeol. Sci.* **23**, 283.

Cogswell, J. W., L. M. Ross, Jr., M. J. O'Brien, H. Neff, and M. D. Glascock (1996b), Postmanufacture Effects on the Chemical Characterization of Prehistoric Pottery: Evidence from the Central Mississippi River Valley, paper presented at the 1996 International Symposium on Archaeometry, Urbana, IL, May 20–24.

Cogswell, J. W., H. Neff, and M. D. Glascock (1998), *Am. Antiq.* **63**, 63.

Djingova, R., and I. Kuleff (1991), in *Neutron Activation and Plasma Emission Spectrometric Analysis in Archaeology: Techniques and Applications*, M. J. Hughes, M. R. Cowell, and D. R. Hook, Eds., British Museum Occasional Papers No. 82, British Museum, London, p. 121.

Djingova, R., and I. Kuleff (1992), *Archaeometry* **34**, 53.

Duma, G. (1972), *Curr. Anthropol.* **13**, 127.

Dunnell, R. C., and T. L. Hunt (1990), *Curr. Anthropol.* **31**, 330.

Emeleus, V. M. (1958), *Archaeometry* **1**, 6.

Emeleus, V. M. (1960), *Archaeometry* **3**, 16.

Franklin, U. M., and V. Vitali (1985), *Archaeometry* **27**, 3.

García Heras, M. (1993), Deposiciones invisibles: Micro-procesos de calcitización postdeposicional en cerámicas celtibéricas, in *Arqueología Espacial* **16–17**, IV Coloquio Internacional de Arqueología Espacial, Procesos Post-deposicionales, Teruel, p. 391.

García Heras, M., R. Fernandez-Ruiz, and J. D. Tornero (1997), *J. Archaeol. Sci.* **24**, 1003.

Gilmore, G. R. (1991), in *Neutron Activation and Plasma Emission Spectrometric Analysis in Archaeology: Techniques and Applications*, M. J. Hughes, M. R. Cowell, and D. R. Hook, Eds., British Museum Occasional Papers No. 82, British Museum, London, p. 1.

Glascock, M. D. (1991), *Tables for Neutron Activation Analysis*, University of Missouri, Columbia.

Glascock, M. D. (1992), in *Chemical Characterization of Ceramic Pastes in Archaeology*, H. Neff, Ed., Prehistory Press, Madison, WI, p. 11.

Glascock, M. D. (1994), in *Chemical Analysis by Nuclear Methods*, Z. B. Alfassi, Ed., John Wiley & Sons, New York, p. 75.

Glascock, M. D. (1998), in *Instrumental Multielement Chemical Analysis*, Z. B. Alfassi, Ed., Kluwer Academic Publishers Group, Dordrecht, Motherlands, p. 93.

Glascock, M. D., T. G. Spalding, J. C. Biers, and M. F. Cornman (1984), *Archaeometry* **26**, 96.

Glascock, M. D., W. Z. Tian, and W. D. Ehman (1985), *J. Radioanal. Nucl. Chem.* **92**(2), 379.

Glascock, M. D., P. I. Nabelek, D. D. Weinrich, and R. M. Coveney, Jr. (1986), *J. Radioanal. Nucl. Chem.* **99**(1), 121.

Glascock, M. D., M. Giesso, R. L. Burger, and E. Salazar (1994a), South American Obsidian Studies at the Missouri University Research Reactor, paper presented at the 1994 SAA Meetings, Anaheim, CA, April 20–24.

Glascock, M. D., H. Neff, K. Stryker, and T. N. Johnson (1994b), *J. Radioanal. Nucl. Chem.* **180**(1), 29.

Glascock, M. D., G. E. Braswell, and R. H. Cobean (1998), in *Archaeological Obsidian Studies: Method and Theory*, Vol. 3, M. S. Shackley, Ed., Advances in Archaeological and Museum Science, Plenum, New York, p. 15.

Gomez, B., M. Rautman, H. Neff, and M. D. Glascock (1995), *Report of the Department of Antiquities, Cyprus* 1995, 113.

Hancock, R. G. V., L. A. Pavlish, W. A. Fox, and M. A. Latta (1995), *Archaeometry* **37**, 339.

Harbottle, G. (1976), in *Radiochemistry*, Vol. 3, G. W. A. Newton, Ed., The Chemical Society, London, p. 33.

Harbottle, G. (1982), in *Archaeological Ceramics*, J. S. Olin and A. D. Franklin, Eds., Smithsonian Institution, Washington, DC, p. 67.

Harbottle, G., and P. C. Weigand (1992), *Sci. Am.*, February, 78.

Hayashida, F. (1995), State Pottery Production in the Inca Provinces, Ph.D. dissertation, Department of Anthropology, University of Michigan, Ann Arbor.

Hoard, R. J., S. R. Holen, M. D. Glascock, H. Neff, and J. M. Elam (1992), *J. Archaeol. Sci.* **19**, 655.

Hoard, R. J., J. R. Bozell, S. R. Holen, M. D. Glascock, H. Neff, and J. M. Elam (1993), *Am. Antiq.* **58**, 698.

Hodge, M. G., H. Neff, M. J. Blackman, and L. D. Minc (1993), *Latin Am. Antiq.* **4**, 130.

Hughes, R. E. (1998), in *Measuring Time, Space, and Material: Unit Issues in Archaeology*, A. F. Ramenofsky and A. Steffen, Eds., University of Utah Press, Salt Lake City, p. 115.

Hughes, M. J., M. R. Cowell, and D. R. Hook (1991a), in *Neutron Activation and Plasma Emission Spectrometric Analysis in Archaeology: Techniques and Applications*, M. J. Hughes, M. R. Cowell, and D. R. Hook, Eds., British Museum Occasional Papers No. 82, British Museum, London, p. 29.

Hughes, M. J., M. R. Cowell, and D. R. Hook (1991b), in *Neutron Activation and Plasma Emission Spectrometric Analysis in Archaeology: Techniques and Applications*, M. J. Hughes, M. R. Cowell, and D. R. Hook, Eds., British Museum Occasional Papers No. 82. British Museum, London, p. IX.

James, W. D., R. L. Brewington, and H. J. Shafer (1995), *J. Radioanal. Nucl. Chem.*, Articles **192**(1), 109.

Kenyon, I., R. G. V. Hancock, and S. Aufrieter (1995), *Archaeometry* **37**, 323.

Kilikoglou, V., Y Maniatis, and A. P. Grimanis (1988), *Archaeometry* **30**, 37.

Kuleff, I., and R. Djingova (1990), in *Activation Analysis*, Vol. 2, Z. B. Alfassi, Ed., CRC Press, Boca Raton, FL, p. 427.

Kushelevsky, A. P. (1994), in *Chemical Analysis by Nuclear Methods*, Z. B. Alfassi, Ed., John Wiley & Sons, New York, p. 3.

Landsberger, S. (1994), in *Chemical Analysis by Nuclear Methods*, Z. B. Alfassi, Ed., John Wiley & Sons, New York, p. 121.

Lange, F. (1993), *Precolumbian Jade: New Geological and Cultural Interpretations*, University of Utah Press, Salt Lake City.

Lemoine, C., and M. Picon (1982), *Revue d'Archéométrie* **6**, 101.

Mallory-Greenough, L. M., J. D. Greenough, and J. V. Owen (1998), *J. Archaeol. Sci.* **25**, 85.

Malyk-Selivanova, G. M. Ashley, R. Gal, M. D. Glascock, and H. Neff (1998), *Geoarchaeology* **13**, 673.

Mommsen, H., E Lewandowski, J. Weber, and Ch. Podzuweit (1988), in *Proceedings of the Twenty-Sixth Archaeometry Symposium*, R. M. Farquhar, R. G. V. Hancock, and L. A. Pavlish, Eds., Archaeometry Laboratory, Department of Physics, University of Toronto, Canada, p. 165.

Myers, J. E., J. S. Olin, and M. J. Blackman (1992), Archaeological Implications of the Leaching of Calcium Compounds from Calcareous Ceramics at Sites in the Southeastern United States and Carribean, poster presented at the 1992 SAA meetings, Pittsburgh, PA.

Neff, H. (1984), The Developmental History of the Plumbate Pottery Industry in the Eastern Soconusco Region, A.D. 600 Through A.D. 1250, Ph.D. dissertation, Department of Anthropology, University of California, Santa Barbara.

Neff, H. (1989a), in *Ancient Trade and Tribute: Economies of the Soconusco Region of Mesoamerica*, B. Voorhies, Ed., University of Utah Press, Provo, p. 175.

Neff, H. (1989b), in *Ancient Trade and Tribute: Economies of the Soconusco Region of Mesoamerica*, B. Voorhies, Ed., University of Utah Press, Provo, p. 249.

Neff, H. (1992), in *Chemical Characterization of Ceramic Pastes in Archaeology*, H. Neff, Ed., Prehistory Press, Madison, WI, p. 1.

Neff, H. (1995a), in *Evolutionary Archaeology: Methodological Issues*, P. Teltser, Ed., University of Arizona Press, Tucson, p. 69.

Neff, H. (1995b), *JOM* **47**(3), 52.

Neff, H. (1998), in *Measuring Time, Space, and Material: Unit Issues in Archaeology*, A. F. Ramenofsky and A. Steffen, Eds., University of Utah Press, Salt Lake City, p. 115.

Neff, H. and B. Arroyo (in press), in *Proceedings of the 1996 Chacmool Conference*, University of Calgary.

Neff, H., and R. L. Bishop (1988), *Am. Antiq.* **53**, 505.

Neff, H., and F. J. Bove (1999), in *Proceedings of the 1996 Urbana International Symposium on Archaeometry*, J. Henderson, H. Neff, and Th. Rehren, Eds., special Issue, *J. Archaeol. Sci.* **16**, 1037.

Neff, H., and M. D. Glascock (1995), *J. Radioanal. Nucl. Chem.* **196**(2), 273.

Neff, H., and M. G. Hodge (in press), in *Place of Jade: Society and Economy in Ancient Chalco*, M. G. Hodge, Ed., University of Pittsburgh Memoirs in Latin American Archaeology.

Neff, H., R. L. Bishop, and D. E. Arnold (1988a), *J. Field Archaeol.* **15**, 339.

Neff, H., R. L. Bishop, and E. V. Sayre (1988b), *J. Archaeol. Sci.* **15**, 159.

Neff, H., R. L. Bishop, and E. V. Sayre (1989), *J. Archaeol. Sci.* **16**, 57.

Neff, H., R. L. Bishop, and D. E. Arnold (1990), *Ancient Mesoamerica* **1**, 171.

Neff, H., F. J. Bove, B. Lou, and M. F. Piechowski (1992), in *Chemical Characterization of Ceramic Pastes in Archaeology*, H. Neff, Ed., Prehistory Press, Madison, WI, p. 59.

Neff, H., F. J. Bove, E. Robinson, and B. Arroyo (1994), *Latin Am. Antiq.* **5**, 333.

Neff, H., M. D. Glascock, R. L. Bishop, and M. J. Blackman (1996), *Am. Antiq.* **61**, 389.

Neff, H., J. W. Cogswell, L. J. Kosakowsky, F. Estrada Belli, and F. J. Bove (1999), *Latin Am. Antiq.* **10**, 281.

Neff, H., M. Rautman, and B. Gomez (in press b), in *Excavation Techniques and Treatment of Finds: Balancing the Constraints of Excavation with Conservation and Analytical Potential*, S. Pike, Ed., selected papers from a workshop sponsored by the Malcolm H. Wiener Laboratory, American School of Classical Studies at Athens in cooperation with the Corinth Excavations, March 29, 1997, Athens, Greece, Wiener Laboratory Publication Number 4.

O'Brien, M. J., J. W. Cogswell, R. C. Mainfort, Jr., H. Neff, and M. D. Glascock (1995), *Southeastern Archaeology* **14**(2), 181.

Olin, J. S., and E. V. Sayre (1971), in *Science and Archaeology*, R. H. Brill, Ed., MIT Press, Cambridge, p. 196.

Parry, S. J. (1991), *Activation Spectrometry in Chemical Analysis*, John Wiley & Sons, New York.

Perlman, I., and F. Asaro (1969), *Archaeometry* **11**, 21.

Perlman, I., and F. Asaro (1971), in *Science and Archaeology*, R. H. Brill, Ed., MIT Press, Cambridge, p. 182.

Picon, M. (1985), *Revue d'Archéometrie* **9**, 27.

Picon, M. (1987), *Revue d'Archéometrie* **11**, 41.

Pollard, A. M., and C. Heron (1996), *Archaeological Chemistry*, Royal Society of Chemistry, Cambridge.

Rautman, M., B. Gomez, H. Neff, and M. D. Glascock (1993), *Report of the Department of Antiquities, Cyprus*, **1993**, 233.

Rice, P. M. (1977), *J. Field Archaeol.* **4**, 221.

Rice, P. M. (1978a), in *The Ceramics of Kaminaljuyú, Guatemala*, R. K. Wetherington, Ed., Pennsylvania State University Monograph Series on Kaminaljuyú, University Park, PA, p. 401.

Rice, P. M. (1978b), in *The Ceramics of Kaminaljuyú, Guatemala*, R. K. Wetherington, Ed., Pennsylvania State University Monograph Series on Kaminaljuyú, University Park, PA, p. 511.

Rice, P. M. (1987), *Pottery Analysis: A Sourcebook*, University of Chicago Press, Chicago.

Rice P. M. (1996), *J. Archaeol. Res.* **4**(3), 165.

Sayre, E. V. (1966), in *Comptes Rendus 7th International Congress on Glass, Brussels, 1965*, Gordon & Breach, New York, Paper No. 220.

Sayre, E. V. (1975), Brookhaven Procedures for Statistical Analysis of Multivariate Archaeometric Data, unpublished manuscript, Brookhaven National Laboratory.

Sayre, E. V., and R. W. Dodson (1957), *Am. J. Archaeol.* **61**, 35.

Sayre, E. V., L. H. Chan, and J. A. Sabloff (1971), in *Science and Archaeology*, R. H. Brill, Ed., MIT Press, Cambridge, p. 165.

Shepard, A. O. (1948), *Plumbate: A Mesoamerican Tradeware*, Carnegie Institution of Washington, Washington, DC, Pub. No. 573.

Shepard, A. O. (1952), Personal letter to Philip Drucker dated October 13, 1952, letter on file, University of Colorado Museum, Boulder.

Shepard, A. O. (1965), *Ceramics for the Archaeologist*, 5th printing, Carnegie Institution of Washington, Washington, DC, Pub. No. 609 (first published 1956).

Shimada, I., V. Chang, C. G. Elera, H. Neff, M. D. Glascock, U. Wagner, and R. Gebhard (1994), in *La Tecnología y la Organización de Producción de las Ceramicas en los Andes Prehispánicos*, I. Shimada, Ed., Fondo Editorial, Pontificia Universidad Católica del Perú, Lima p. 67.

Shook, E. M. (1965), in *Handbook of Middle American Indians*, Vol. 2, G. R. Willey, Ed., University of Texas Press, Austin, p. 180.

Steponaitis, V. P., M. J. Blackman, and H. Neff (1996), *Am. Antiq.* **61**, 555.

Tankersley, K. B. (1998), *Am. Antiq.* **63**, 7.

Triadan, D. (1997), *Ceramic Commodities and Common Containers: Production and Distribution of White Mountain Red Ware in the Grasshopper Region, Arizona*, Anthropological Papers of the University of Arizona, No. 61, Tucson.

Truncer, J., M. D. Glascock, and H. Neff (1998), *Archaeometry* **40**, 23.

Tsechanski, A. (1994), in *Chemical Analysis by Nuclear Methods*, Z. B. Alfassi, Ed., John Wiley & Sons, New York, p. 23.

Tykot, R. H., and S. M. M. Young (1996), in *Archaeological Chemistry: Organic, Inorganic, and Biochemical Analysis*, M. V. Orna, Ed., American Chemical Society Symposium Series 625, American Chemical Society, Washington, DC, p. 116.

Walker, F. W., J. R. Parrington, and F. Feiner (1989), *Nuclides and Isotopes*, 14th ed., General Electric Company, Nuclear Energy Operations, San Jose, CA.

Walter, V., and Y. Besnus (1989), *Revue d'Archéométrie* **13**, 55.

Weigand, P. C., G. Harbottle, and E. V. Sayre (1977), in *Exchange Systems in Prehistory*, T. K. Earle and J. E. Ericson, Eds., Academic, New York, p. 15.

Williams, C. T., and F. Wall (1991), in *Neutron Activation and Plasma Emission Spectrometric Analysis in Archaeology: Techniques and Applications*, M. J. Hughes, M. R. Cowell, and D. R. Hook, Eds., British Museum Occasional Papers No. 82, Britich Museum, London, p. 105.

Yellin, J., I. Perlman, F. Asaro, H. V. Michel, and D. F. Mosier (1978), *Archaeometry* **20**, 95.

CHAPTER

6

PARTICLE-INDUCED X-RAY EMISSION

JEAN-CLAUDE DRAN, THOMAS CALLIGARO, and JOSEPH SALOMON

Laboratoire de Recherche des Musées de France, Paris, France

6.1. INTRODUCTION

Particle-induced X-ray emission (PIXE) is an ion beam analytical technique using a particle accelerator, generally an electrostatic one, single-ended or tandem with a voltage of a few megavolts, and allowing the determination of the concentration of the elements contained in a material by means of the emission of characteristic X-rays. X-ray production by a beam of charged particles was reported for the first time by Chadwick (1912), who bombarded various targets with alpha particles. The analytical use of this phenomenon had to wait more than half a century and the growth of semiconductor detectors to enable a true multielemental analysis by energy-dispersive X-ray spectrometry (EDS). This was initiated by Johansson et al. (1970), who gave its name to the technique. PIXE rapidly gained a great popularity because of its high sensitivity allowing trace element analysis down to about the 10-ppm level for most elements, thus some three orders of magnitude better than the electron microprobe. However, the quantification of the analyses was not straightforward, particularly for thick targets, due to complex matrix effects (ion energy loss, X-ray self-absorption, secondary fluorescence).

Sophisticated software programs are now able to take these effects into account so that an accuracy of about 10% can be routinely reached (Campbell et al. 1993).

PIXE applications to art and archaeology rapidly grew for two additional reasons:

1. PIXE is recognized as nondestructive and therefore meets the constraints linked to the uniqueness and precious character of museum

Modern Analytical Methods in Art and Archaeology, Edited by Enrico Ciliberto and Giuseppe Spoto. Chemical Analysis Series, Vol. 155
ISBN 0-471-29361-X

objects; this is due to the large cross section of the underlying phenomenon of inner shell ionization and thus to the low beam intensity and total ion dose necessary to the analysis.

2. The technique can be easily implemented in the external beam mode, that is, in air or preferably in a helium atmosphere, because it can stand some energy dispersion since the above-mentioned cross section varies smoothly with energy; external beam configuration allows the *in situ* analysis of large or fragile objects without the need of sampling or vacuum.

The scope of this chapter is to provide the basic knowledge for applications in the field of art and archaeology. For further information, the reader is referred to two extensive textbooks (Johansson and Campbell 1988; Johansson and al. 1995), book sections (Cohen and Clayton 1989), and numerous review articles (Mitchell and Barfoot 1981; Bird et al. 1983; Breese et al. 1992; Cahill 1980; Demortier 1991; Sciuti and Suber 1991; Swann 1997).

6.2. PHYSICAL PRINCIPLES

6.2.1. Production of Characteristic X-Rays

The PIXE technique relies on the emission of X-rays induced by the interaction of energetic light ions (most frequently protons of a few mega-electron-volts) with constituent atoms of materials. This interaction involves three successive steps:

1. An electron is expelled from an inner shell of the atom (inner shell ionization, e.g., the K shell, which is the deepest one containing two electrons).
2. The electronic vacancy is filled by an electron originating from a more external shell (e.g., the L shell, containing up to eight electrons distributed into three subshells L_1, L_2, and L_3).
3. The resulting excited state releases the excess electron binding energy $E_K - E_L$ by the emission of either an X-ray of this energy, characteristic of the target atom, or an Auger electron from a third electronic shell (e.g., M) with a kinetic energy $E_K - E_L - E_M$. The relative probability of X-ray emission is called the fluorescence yield ω and for the K lines varies from 0 to 1 when increasing the atomic number.

The emitted X-rays correspond to the electronic transitions (Fig. 6.1) predicted by quantum mechanics and follow the so-called selection rules based

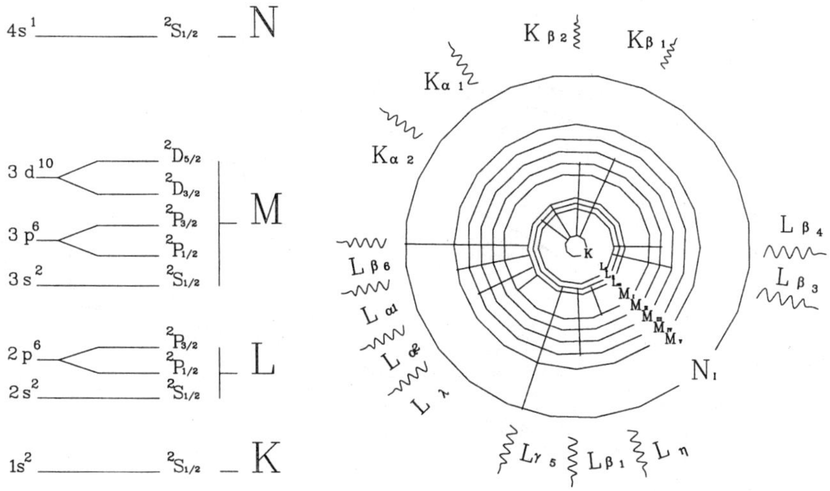

Figure 6.1. Scheme of electronic shell structure of Cu atom and associated transitions following primary ionization and leading to PIXE spectrum.

on the atom quantum numbers (*n* principal, *l* orbital angular momentum, *j* total angular momentum. These selection rules state that $\Delta n > 0$, $\Delta l = \pm 1$, $\Delta j = 0, \pm 1$. The multiplicity of the allowed transitions induces several discrete lines in the spectrum, grouped in series of decreasing energy: *K* lines, *L* lines, and so on, corresponding to the shells where occurred the primary ionization. Each series contains several components, for example, K_α and K_β (Siegbahn notation) of increasing order of energy but of decreasing intensity. The energy values of the components and the intensity ratios provide the fingerprint pattern that identifies the element. For a given series the X-ray energy steadily increases with the atomic number *Z* of the target atom, according to Moseley's law: $E = C(Z - s)^2$, where *C* is a constant different for each spectral series and *s* the shielding constant close to 1. Only *K* and *L* lines are used for analytical purposes. In EDS based on a silicon detector, the detector efficiency limits the range of useful X-rays from 1 to 40 keV. Therefore the useful energy range of the *K* lines extends from about 1 keV for sodium ($Z = 11$) to 37 keV for neodymium ($Z = 60$), and that of the *L* lines from 5 keV for neodymium to 13 keV for uranium ($Z = 92$).

6.2.2. Cross Section of X-Ray Production

For a given single-element target of atomic number *Z*, the cross section of the production of a given X-ray line is $\sigma^x = \sigma_z \omega_z b_z$, where σ_z is the cross section of primary inner shell ionization, ω_z the fluorescence yield, and b_z the

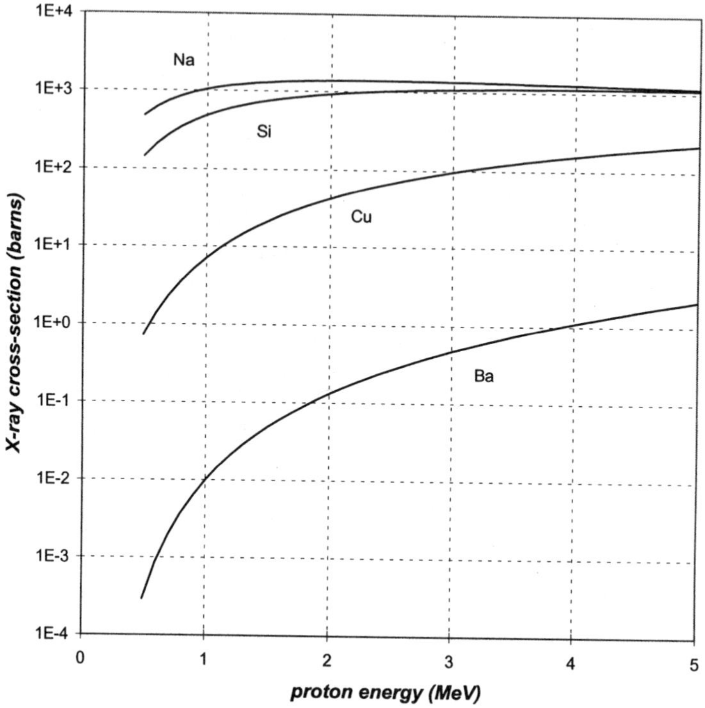

Figure 6.2. Variations of cross section of KX-ray emission with proton energy on several targets.

relative intensity of the line; ω_z and b_z are characteristic features of the atom and at least for the *K* lines do not depend on the ionization process, whereas σ_z is directly related to it. Particle-induced ionization follows roughly the same law as electron-induced ionization, namely the speed matching law, which states that σ^i is maximum when the velocity of the incident particle matches the orbital velocity of the electron in a given shell. Consequently for a given target, σ_z and σ^x increase with the energy *E* of the incident ion up to a maximum and then slowly decrease (Fig. 6.2). With protons and for most atoms except the lightest ones, the value of the maximum exceeds the practical energy range (2 to 3 MeV) set by background radiation (see the next section). At a given incident energy σ_z and σ^x for both the *K* and *L* lines rapidly fall when increasing the target atomic number *Z* (Fig. 6.3). For intermediate and heavy elements, *L* lines are clearly more favorable than *K* lines by several orders of magnitude. Ions heavier than protons must have the same speed as protons and thus the same energy per nucleon to have the same ionizing power; consequently, an energy of 48 MeV is needed for ^{16}O

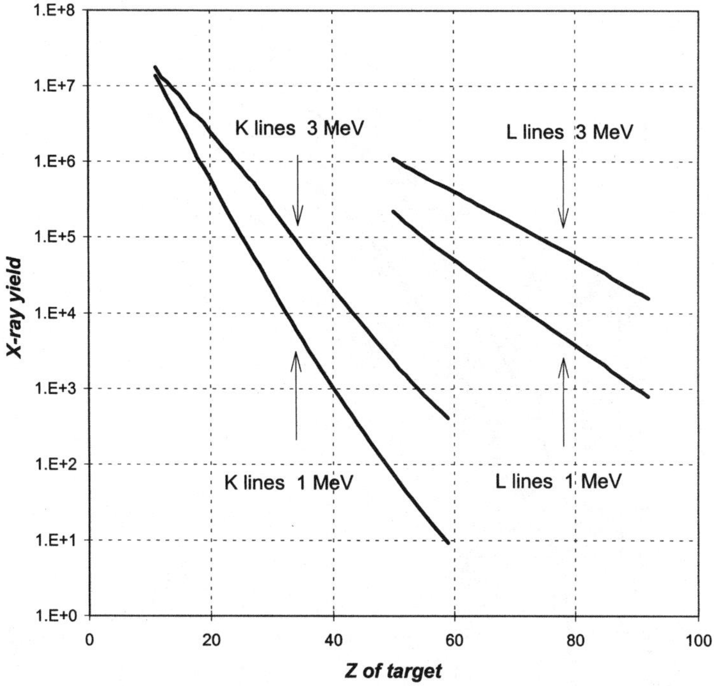

Figure 6.3. Variations of KX-ray yield with atomic number for thin targets of 1 μg/cm² thickness bombarded with 1 μc of 1- and 3-MeV proton beams.

ions, which is out of reach with small accelerators. However, heavy ions are sometimes used to induce multiple ionization, which can provide information on the chemical state of the target atom through satellite lines in the X-ray spectrum.

A universal scaling law for the cross section of *K*-shell primary ionization (Fig. 6.4) was successfully inferred from theoretical calculations, based on two models, the impulse or binary encounter approximation (BEA; Garcia et al. 1973) and the planar wave Born approximation (PWBA). The first model, based on classical mechanics, describes the ionization process as a direct collision between the projectile nucleus and the bound electron. The second model relies on the perturbation theory of quantum mechanics and considers the phenomenon as a transition from an initial state (plane-wave projectile and bound atomic electron) to a final one (plane-wave projectile and ejected electron). The PWBA model was notably improved and transformed into the ECPSSR model by Brandt and Lapicki (1981), who introduced several correction factors linked to the effects of the energy loss (E), the nuclear Coulomb field (C), the perturbation of the stationary atomic

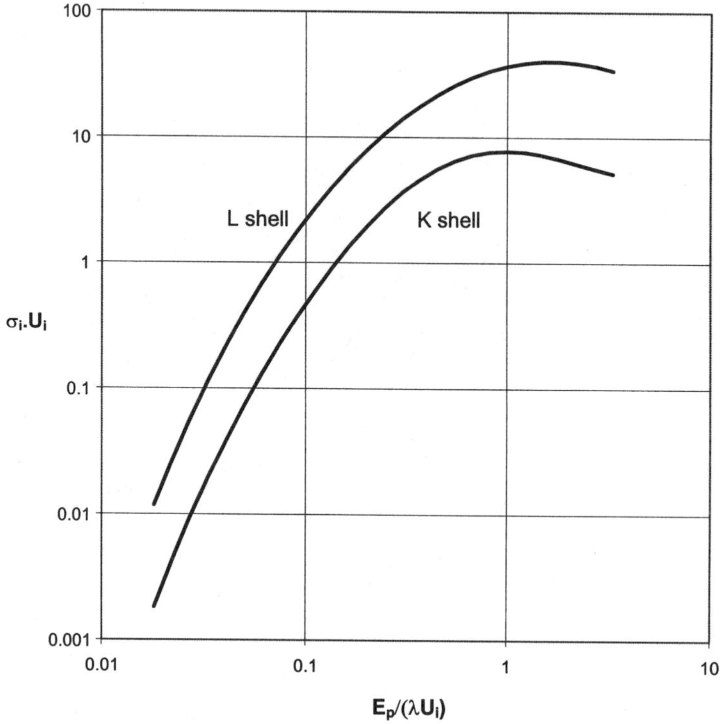

Figure 6.4. Universal scaling law for *K*- and *L*-shell ionization of elements calculated using polynomial fit of Johansson and Johansson (1976).

state (PSS) as well as relativistic effects (R). If one considers a projectile of nuclear charge Z_1, mass M, and energy E (in kilo-electron-volts), expelling a K-shell electron of binding energy U_K with a cross section σ_K, Figure 6.4 expresses the variations of $U_K{}^2\sigma_K/Z_1{}^2$ as a function of $E/\lambda U_K$ with $\lambda = M/m$ the ratio of the projectile mass to that of the electron. There is a good agreement between the theoretical calculations and a large amount of experimental data. One can also note a maximum for $E/\lambda U_K = 1$, which is another expression of the speed matching quoted above.

6.2.3. Background Radiation

As in other spectrometric techniques, the detection limit is set by the amplitude of the background. X-ray spectra induced by electrons exhibit, in addition to characteristic lines, a continuum due to the bremsstrahlung radiation originating from the slowing down of incident electrons during Coulomb interactions with nuclei. This continuum extends to the energy of the incident electron, and its intensity is maximum at half this value. In the case of X-rays induced by protons, the bremsstrahlung directly produced by these particles is considerably reduced (the intensity being roughly inversely proportional to the square of the projectile mass). The only important source of bremsstrahlung is that due to secondary electrons. As the maximum energy transferred to electrons is $4Em/M$, where m and M are respectively the electron and projectile masses, the extension of this bremsstrahlung is directly related to the incident particle energy E. Thus, for a proton energy of 3 MeV, the background extends up to 6 keV. However, in the case of insulating targets, especially when analyzed under vacuum, charging effects can produce voltages of several tens of kilovolts with a subsequent acceleration of secondary electrons and extension of the background. As shown in the next section, the background radiation is not the sole limitation to the sensitivity of the technique; several other phenomena that can be called instrumental artifacts occur that can markedly blur the spectra. They are directly related to X-ray interaction with the semiconductor detector, most often a Si(Li) detector.

6.3. X-RAY DETECTION

6.3.1. Detector Characteristics

PIXE analysis generally relies on EDS using semiconductor detectors (most often a silicon crystal in which lithium has drifted) that convert the X-ray photons into electron–hole pairs, resulting in voltage pulses of height proportional to the photon energy. However, the efficiency of this process is not constant over the entire energy range, due mostly to the detector structure and dimensions. A schematic cross section of a Si(Li) detector is shown in Figure 6.5, taken from Potts (1987). The crystal is connected to a liquid nitrogen cryostat via a copper cold finger and is mounted in a special housing hold under high vacuum and terminated by a very thin window made of a low-Z material, usually a Be foil a few micrometers thick. The presence of such a window restricts the energy of detectable X-rays to a minimum of about 1 keV corresponding to the K X lines of Na. On the other side of the energy range, the detector thickness (from 2 to 5 mm) is not sufficient to stop all photons and thus its efficiency drops steeply (Fig. 6.6).

Another important parameter is detector resolution, which characterizes its ability to separate X-ray lines of very close energy values. It is usually expressed as the full width at half maximum (FWHM) of a reference photopeak, namely that corresponding to the Mn K_α line at 5.9 keV emitted by a

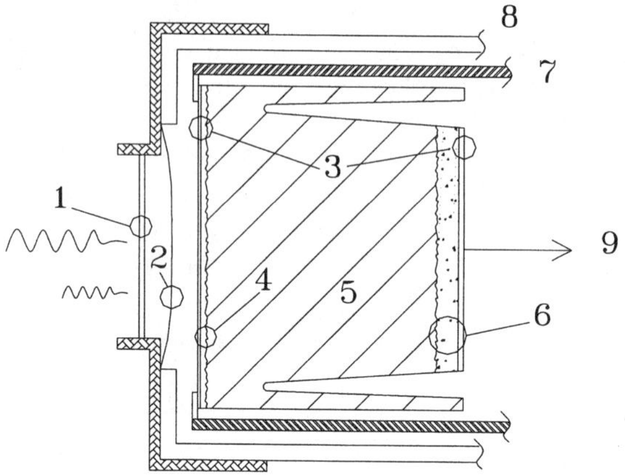

Figure 6.5. Schematic view of typical Si(Li) detector (Potts 1987): (1) filter (various materials and thicknesses); (2) thin window (generally about 10 μm Be); (3) Au front and rear contact layers (0.02 μm); (4) Si dead layer (0.03–0.1 μm); (5) detector crystal (3–5 mm); (6) Li diffusion layer (300 μm); (7) support connected to Cu cold finger; (8) detector housing; (9) signal to field effect transistor.

^{55}Fe radioactive source. Current Si(Li) detectors have a resolution in the range 140 to 160 eV.

6.3.2. Detection Artifacts

In a typical PIXE spectrum as that represented in Figure 6.7, one notes beside the *K* and *L* lines of copper, the following features purely instrumental (from the low- to the high-energy sides):

1. *Escape Peaks.* Small peaks distant from the corresponding *K* or *L* lines of an energy difference of 1.74 keV (exactly the energy of the Si *K* X-rays). These peaks derive from incomplete conversion of the X-photons into electron–hole pairs. Indeed, these photons undergo the photoelectric absorption within the detector, most often by interaction with an electron of the Si *K* shell. The expelled electron collides with other Si atoms producing further electron–hole pairs. The original Si atom deexcites by emitting either a Si *K* X-ray or an Auger electron, which are both likely to be absorbed within the detector active volume in such a way that the full energy of the primary photon is deposited in the detector. However, when the primary ionization occurs close to the detector surface, there is a significant probability that the Si *K* X-ray escapes without further ionization. The amplitude of such escape peaks is usually lower than 1% of the main peak.

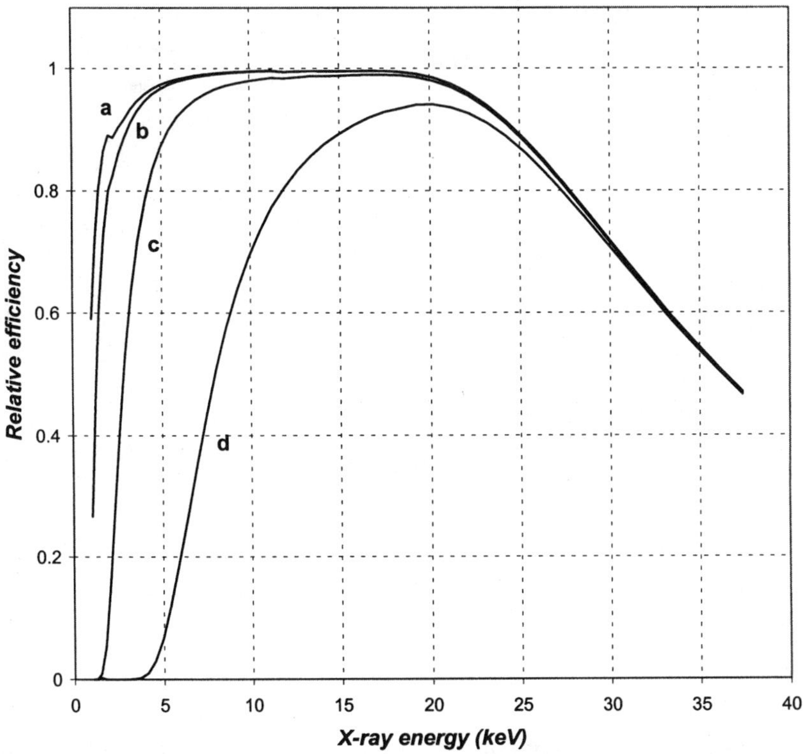

Figure 6.6. Variations of efficiency of a Si(Li) detector with X-ray energy: (*a*) Si(Li) crystal alone with 0.01-μm Au contact layer and 0.1-μm dead layer; (*b*) addition of 8-μm Be window; (*c*), (*d*) addition of either 125-μm Be or 50-μm Al filter.

2. *Pileup Peaks.* These peaks arise from the inability of the electronic system to separate coincident pulses of distinct energies E_1 and E_2. If the time interval between two incoming X-rays is shorter than the pulse shape of the amplifier, the two corresponding pulses are summed together, yielding a continuum from E_1 to E_2 up to $E_1 + E_2$. If the pulse processor is able to reject pairs of pulses in such a situation, it will still remain a small peak at energy $E_1 + E_2$, corresponding to the simultaneous detection of the two X-rays. The count rate of the pileup peak is $C_p = 2\tau C_1 C_2$, where C_1 and C_2 are the count rates of the two X-rays and τ the resolving time of the amplifier.

3. *Compton Continuum.* Such a continuum mostly originates from the Compton effect undergone on the sample and the Si detector by the gamma rays induced by proton interaction with light nuclei, mostly Na, Al, and Si and to a lesser extent F. This last feature also explains the limitation to 3

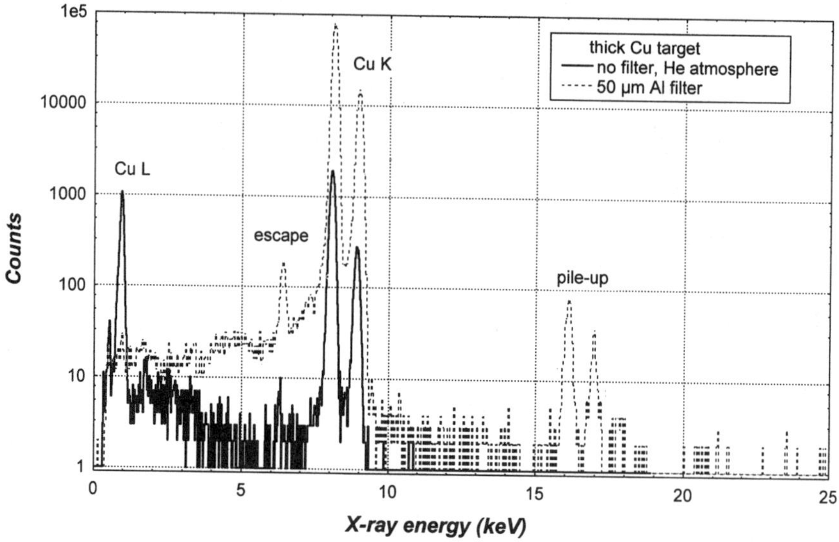

Figure 6.7. PIXE spectrum of copper thick sample obtained with Si(Li) detector and showing various detection artifacts.

MeV of the incident proton energy: above this value the probability of nuclear reactions with gamma ray emission is notably increased.

4. *Peak Tailing*. The response of a Si(Li) detector to a monochromatic X-ray source is modeled by a Gaussian peak, and on the low-energy side by one or two exponential tails and a flat plateau (see Section 6.7). The low-energy tail affects the detection limit of elements with X-ray lines of lower energy than the matrix ones. For example, this is the case of metallic alloys containing high concentration of heavy elements such as Hg, Au, and Pb. Moreover, the peak tailing features are characteristic of each detector and must be accurately measured at regular time intervals because of possible evolution, in order to perform a real quantitative analysis.

6.4. YIELD–CONCENTRATION RELATIONSHIP

Quantitative analysis requires us to establish a reliable correlation between the number of counts Y_z in a particular X-ray line under given experimental conditions and the number of atoms per cubic centimeter, N_z, of the corresponding element of atomic number Z assumed to be uniformly distributed. The simplest situation occurs for a thin target for which both the energy loss of the incident protons (and the subsequent variation of the ionization cross section) and the absorption of the X-rays of interest are negligible. This corresponds to a thickness of the order of 1 μm in light matrices leading to an attenuation of the Na K_α X-rays lower than 10%.

6.4.1. Thin Targets

The relationship between Y_z and N_z is quite straightforward:

$$Y_z = \frac{N_z t \sigma_z \omega_z b_z Q \varepsilon_z \Omega}{4\pi \cos\theta_i}$$

where t is the thickness of the target, Q the number of incident protons, ε_z the detector efficiency for the corresponding X-ray, Ω the detector solid angle, and θ_i the angle between the beam and the sample normal. Considering that

$$N_z = \frac{\rho C_z N_a}{A_z}$$

where ρ is the sample specific gravity, C_z the weight concentration of element Z, N_a the Avogadro number, and A_z the mass number, it comes

$$Y_z = \frac{\rho C_z t N_a \sigma_z \omega_z b_z Q \varepsilon_z \Omega}{4\pi \cos\theta_i A_z}$$

Putting $k_z = N_a \sigma_z \omega_z b_z \varepsilon_z \Omega / (4\pi \cos\theta_i A_z)$, which is a parameter independent of both the element content and the number of incident protons, defined as the thin target sensitivity for element of atomic number Z, the previous equation becomes

$$Y_z = k_z Q \rho C_z t = k_z Q M_z$$

M_z being the mass per unit area (in microgram per square centimeter). Therefore, knowing k_z and Q (generally expressed as the integrated proton charge in microcoulomb), the mass per unit area M_z can be deduced from the measured yield Y_z. Such an analysis is absolute and does not need the use of a standard. The accuracy can be better than 10% when using K lines due to the accurate knowledge of K-shell ionization cross sections but poorer with L lines. This method can be used to get an X-ray yield calibration curve, Y versus Z, characteristic of the analytical system. This is performed by means of a series of thin samples, monoelemental or compounds, vacuum deposited on a polymer foil. When an unknown element is detected and its X-ray yield measured, the use of the yield curve allows the determination of the mass per unit area of the corresponding element. Once such a calibration curve has been established, it is valid as long as the analytical system is not modified.

6.4.2. Thick Targets

Two phenomena intervene to complicate the yield–concentration relationship. First, the incident protons progressively slow down within the sample from their initial energy E_0 to zero at the end of their range. Consequently, the cross section regularly decreases as a function of the depth x. The depth–energy relationship

$$x = \frac{1}{\rho} \int_{E_0}^{0} \frac{dE}{S(E)} \tag{6.1}$$

is inferred from the energy loss variations $dE/dx = -\rho S(E)$ in which $S(E)$ is the matrix stopping power. In addition, the X-rays produced at that depth are attenuated by the overlayer of material. The calculation is performed by dividing the sample into successive thin layers and adding their contribution to the global X-ray production. The overall yield is expressed as

$$Y_z = \frac{\Omega Q \varepsilon_z \omega_z b_z N_a}{4\pi A_z} \int_{E_0}^{0} \frac{\sigma_z(E) T_z(E)}{S(E)} dE \tag{6.2}$$

where the integral corresponds to a correction factor that takes into account the depth variation of the cross section and the X-ray self-absorption, the transmission factor at a depth $x(E)$ being

$$T_z(E) = \exp\left[-\left(\frac{\mu}{\rho}\right)_z \frac{\cos\theta_i}{\cos\theta_o} \int_{E_0}^{E} \frac{dE}{S(E)}\right]$$

where θ_o is the detection angle with respect to the normal to the sample surface.

6.5. OPTIMIZATION PROCEDURES

The Si(Li) detectors and their associated electronics on which are based most PIXE facilities in operation suffer from intrinsic limitations:

1. a limited energy resolution;
2. relatively long duration of the pulses, which limits the count rate to about two thousand counts per second at most in order to avoid important pile-up;
3. intense peaks due to major elements that exhibit a tail on the decreasing energy side, which interferes with small peaks of trace elements of lower atomic numbers; and
4. a sensitivity to backscattered charged particles that tends to lower the resolution.

The above features can markedly reduce the detection limit, particularly in the case of relatively high atomic number matrices like metallic alloys. Different attempts have been made to overcome such difficulties, which are based on two principles:

1. Attenuate matrix X-rays by an adequate filter.
2. Reduce or even avoid the emission of matrix X-rays.

6.5.1. Filters

The presence of a filter in front of the Si(Li) detector is already needed as a protection against backscattered particles. As mentioned above, its role is also to attenuate the matrix X-rays in order to reduce both the count rate detrimental to the detector resolution and the background on the low-energy side of the corresponding peaks. However, the simultaneous attenuation of the lower energy X-rays emitted by light elements like Na, for example, cannot be avoided, but this can be minimized by choosing a selective filter, that is, containing an element with an X-ray absorption edge just below the X-ray energy of interest, generally one or two units of atomic number below that of the emitting matrix element. In some cases, the attenuation by the filter is excessive and does not allow the accurate measurement of the matrix element content together with those of minor and trace elements. To cope with this situation, a small hole is made on the filter in order to allow the partial detection of nonattenuated matrix X-rays. This type of filter configuration is called a pinhole, or funny, filter (Harrison and Eldred 1973). It should also be stressed that the use of a filter can induce unwanted interfering radiation: the X-ray lines of the filter constituent element. A secondary filter can then be superimposed to the first one in order to suppress them, leading to a multilayer structure for the filter (Swann and Fleming 1990).

6.5.2. Selection of Beam Energy

Early PIXE operations most often used two different proton energies and different filter thicknesses to optimize the analytical conditions. The following procedure was applied:

1. The matrix elements of usually low Z are first analyzed with 1-MeV protons using a filter of appropriate thickness to stop the backscattered particles (about 15 μm Be including the intrinsic detector window).
2. High-energy (3-MeV) protons are used for the analysis of high-Z elements, frequently at trace level concentration, with a sufficient detection limit. A thicker filter (about 50 μm Al) is then necessary, which also totally absorbs the X-rays emitted by matrix elements.

Considerable progress has been made these last years in order to enhance the analytical performance and reduce the number of steps necessary for a complete measurement. This is linked with the development of external beams (see Section 6.8.1) and the use of two X-ray detectors to cover the whole range of X-ray energy. Details will be given in Section 6.8.3 devoted to a short description of the external beam line of the accelerator facility of the Research Laboratory of the Museums of France in Paris (Fig. 6.8).

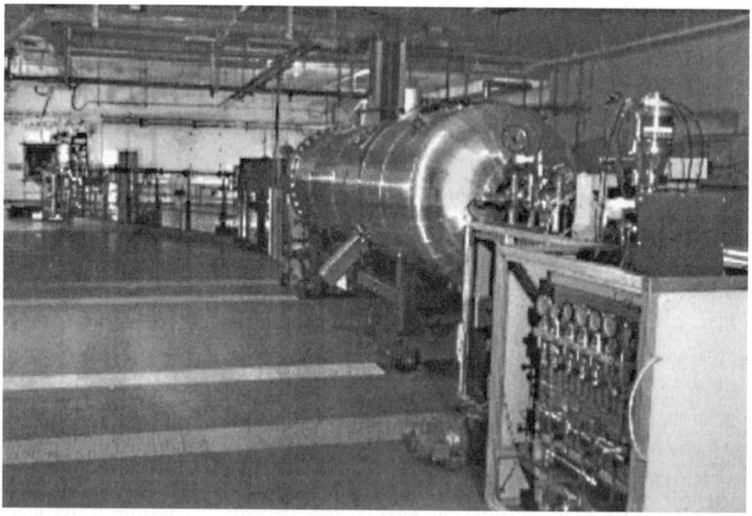

Figure 6.8. View of AGLAE IBA facility of Research Laboratory of the Museums of France.

6.5.3. Peculiar Setups

For some particular samples such as metallic alloys, current EDS setups are sometimes unable to separate interfering X-ray lines, whatever the number of beam energy values and the number of detectors. One solution would be to turn to wavelength-dispersive X-ray spectrometry (WDS), based on a Bragg crystal that provides a much better resolution than the Si(Li) detector for low-energy (1 to 5-keV) X-rays (Maeda et al. 1998), on the order of 10 eV for conventional crystals and even lower for special crystals. However, the technique would then lose its multielemental character, since each element must be measured separately. In addition, the count rate is considerably reduced, and consequently the duration of the measurements is markedly increased with subsequent greater risk of damage. Another way to avoid interferences it to selectively excite the X-ray line of the trace element of interest while avoiding the emission of matrix X-rays of higher energy. This favorable situation can be reached by irradiating a primary target of atomic number intermediate between those of the trace element and the matrix element and using the emitted X-rays to induce the trace element X-ray by fluorescence (Peisach 1987; Zeng et al. 1993); for example, a primary target of Ge produces K_α X-rays of 9.88 eV, sufficient to excite the K_α X-rays of Cu and Zn present at trace level in a lead matrix but not the interfering L_α lines of Pb (10.55 and 10.45 keV).

6.6. SPECTRUM PROCESSING BY COMPUTER CODES AND QUANTITATIVE ANALYSIS

6.6.1. Computer Codes

The use of sophisticated computer codes, such as GUPIX, GEOPIXE, PIXAN, and WitsHEX, is the key toward quantitative analysis by PIXE. These codes are the result of many years of effort and are being continuously improved, by incorporating the most recent theoretical developments (cross sections), accurate measurements of basic parameters (stopping powers for protons, X-ray attenuation coefficients, fluorescence yields), as well as the various instrumental artifacts (pile-up peaks, escape peaks, peak lineshape). We will briefly describe below the characteristics of the GUPIX program (Maxwell et al. 1988), which is now the most widely used.

6.6.2. Spectrum Synthesis

The response of the Si(Li) detector to a monochromatic X-ray is modeled by the sum of a Gaussian and at the low-energy side of an exponential tail and a plateau (Campbell et al. 1987). Thus a monochromatic line ideally centered at channel c will give birth to a more complex structure where the content of any channel i can be numerically calculated. Five parameters, $a_1, \ldots, a_5$, defined hereafter, are characteristic of the detector. The first three are defined by the channel–energy relationship:

$$i(E) = a_1 + a_2 E + a_3 E^2$$

and the two others by the expression of the detector resolution:

$$\kappa(E) = \sqrt{a_4 + a_5 E}$$

The escape peaks are calculated channel by channel using the following empirical formula (Johansson 1982):

$$\mathrm{esc}(E - 1.742\ \mathrm{keV}) = y(E) C_1 \left[1 - T \log\left(1 + \frac{1}{T}\right)\right] \qquad T = C_2 E^{C_3}$$

with $C_1 = 0.01896$, $C_2 = 0.009255$, $C_3 = 2.9662$, $1.84\ \mathrm{keV} < E < 17.5\ \mathrm{keV}$. The parameters $a_1, \ldots, a_5$ will be adjusted during the fitting procedure.

The overall theoretical spectrum of the sample is obtained by summing the contributions of all X-ray lines for each element and then for all elements detectable in the sample. Consequently, the content of each channel i of the spectrum is expressed by

$$y_i^{\mathrm{sim}} = \sum_{z=\mathrm{elements}} a_z \left[\sum_{c=K \text{ or } L \text{ lines}} R_{zc}\, y(i, c) \right]$$

The energies of the c lines are calculated from the binding energies of the electronic shells; R_{zc} is the relative height of the c line with respect to the main line (i.e., K_α or L_α) for element Z and a_z is a parameter proportional to the element concentration. Other spectrum features such as continuum background and pile-up peaks are also simulated with special numerical procedures.

6.6.3. Fitting Procedure

The processing of PIXE spectra is ruled by the same principle whatever the software package used. Assuming a given elemental composition, the program simulates a theoretical spectrum that is compared to the experimental one. The simulation parameters, in particular the sample composition, are optimized by iteration in order to reduce the gap between the two spectra.

6.6.4. Calculation of the Elemental Concentrations

To convert the peak heights a_z obtained above into concentrations C_z, it is necessary to know the yields Y_z under the experimental conditions and for a given matrix (proton energy, target composition, etc.). The theoretical relationship giving the yield Y_z per unit of weight concentration of element Z in a thick target is given by Eq. (6.1) and (6.2) of Section 6.4.

The data necessary for the calculation of this yield are the following:

Tabulated Data

ionization cross sections σ_z of K and L shells;
fluorescence yields ω_z of K and L lines;
X-ray attenuation coefficients μ_z; and
stopping powers $S(E)$.

Experimental Data

overall detector efficiency ε_z, including filters;
detector solid angle Ω; and
integrated charge Q.

The parameters ε_z and Ω are obtained by means of thin or thick standards or calibrated radioactive sources, while Q must be measured with a current integrator.

The concentration of element Z in the sample C_z^{spl} can then be deduced from the measured area of the main peak a_z:

$$C_z^{spl} = \frac{a_z}{Y_z}$$

This procedure is absolute if all the Y_z can be calculated, which requires us to know ε_z, Ω, and Q. In practice, these instrumental parameters are rather difficult to accurately assess. Consequently, one relies on procedures that bypass these parameters and reduce the sources of errors. These procedures are presented next.

Internal Standard. If one element of atomic number Z' present in the spectrum has a known concentration, the concentrations of all other elements can be deduced from the relation

$$C_z^{spl} = C_{z'}^{spl} \frac{a_z^{spl}}{a_{z'}^{spl}} \frac{Y_{z'}}{Y_z}$$

In the ratio of the yields Y the instrumental parameters Q and Ω vanish, and the matrix effects are those of the same matrix for different elements. However, it remains the ratio of the values of the detector efficiency for different elements. This procedure can be applied to samples that have been previously analyzed with a complementary technique [X-ray fluorescence spectrometry (XRF), scanning electron microscope (SEM), inductively coupled plasma (ICP)], which is seldom the case.

External Standard. This approach consists in using an external standard [National Institute for Standards and Technology (NIST), U.S. Geological Survey (USGS), or a sample previously analyzed] and comparing its PIXE spectrum to that of the sample. The concentration of element Z in the unknown sample is related to that in the standard by the formula

$$C_z^{spl} = C_z^{std} \frac{a_z^{spl}}{a_z^{std}} \frac{Y_z^{std}}{Y_z^{spl}}$$

This time, all the instrumental parameters vanish from the yield ratio. The main drawback of this method is the need to accurately know the proton dose Q received by both the standard and the sample. In order to apply this method to art and archaeology, it has been necessary to develop a reliable technique of beam monitoring in external beam mode.

Summing Up the Components to 100%. The concentrations of all components satisfy to the relation

$$C_z^{spl} = \frac{a_z / Y_z}{\sum_z a_z / Y_z}$$

which implies $\sum_z C_z^{spl} = 100\%$. Such a constraint substitutes to the measurements of Q and Ω, which is quite advantageous with an external beam for which Q is difficult to measure. The key point in this approach is the assessment of the detector efficiency. The method is checked a posteriori by applying it to standards. It is only applicable to samples for which all constituent elements appear in the spectrum. It is for example the case of metallic objects made of bronze, brass, or gold. It can be extended to samples based on oxides (e.g., stones, ceramics) by assuming that the sum of oxides is 100%.

6.7. CAPABILITY OF THE TECHNIQUE

The popularity of PIXE in various research fields stems from its analytical performances combined with the ease of operation and its nondestructive character. The main constraint is obviously the need of sophisticated and costly equipment and a well-trained technical staff. Let us recall the most important qualities of PIXE:

1. *Multielemental Character.* Almost all elements of the periodic table can be analyzed simultaneously with the exception of the lightest ones (H to Ne). However, PIXE is frequently combined with another ion beam analytical technique, called PIGE (particle-induced gamma ray emission) for the measurement of some light elements (F, Na, Al, Si).

2. *Sensitivity.* As stated earlier, PIXE is a very sensitive technique suitable for trace element analysis. The sensitivity depends on the experimental conditions and the sample chemical composition in major elements. In the best case, for thick targets, the detection limit of elements in the range $20 < Z < 30$ is of the order of 1 ppm. This limit is governed by the signal-to-background ratio of Si(Li) detectors. The mean detection limit (MDL) for each element is defined by

$$\text{MDL} = 3.29\sqrt{B}$$

where B is the value of the background taken within a region extending at three standard deviations at each side of the Gaussian. This MDL expressed in count number is converted into concentration by means of the yield Y.

3. *Accuracy.* Sources of errors are linked to each step of the spectrum processing. During the spectrum acquisition, the content of each channel has a statistical uncertainty and the dose measurement has a few percent uncertainty. The spectrum fitting is also the source of errors, but it is during the complex numerical calculations of the yields that the risk of errors is important. Whereas the theoretical and tabulated data are known with an accuracy better than 5% and the error linked to the numerical integration is negligible, the assessment of the overall detector efficiency is an important source of uncertainty. The filter transmission is hardly known with good accuracy, and at low energy the intrinsic efficiency of the detector is hardly measurable. This makes the analysis of light elements ($11 < Z < 15$) particularly difficult. Consequently, the concentrations measured on a thick sample have an accuracy of the order of 5 to 10%. This accuracy is generally sufficient for the study of objects of art and archaeology, which usually do not have a homogeneity better than 5 to 10%.

4. *Nondestructive Character.* This property is of particular significance when dealing with precious objects of cultural heritage. Under proper conditions of exposure to the proton beam (intensity below a few tens nanoamperes per square centimeter for a fixed millimeter-sized beam), and particularly in an external beam mode, most materials of interest stand

irradiation without any visible damage. Concern still remains for organic materials like paper, parchment, leather, or painting (Cahill et al. 1984; Zeng et al. 1990; Cheng et al. 1998), particularly with microbeams (Cookson 1988; McColm and Cahill 1991; Cholewa et al. 1991), but reducing the intensity and cooling the sample with a helium flow generally avoid any risk. Another beam effect should also be mentioned, namely the possible mobilization of sodium and other alkalis in silicate crystals and glasses with the risk of erroneous measurements (Campbell et al. 1997).

5. *Rapidity and Ease of Measurement.* It usually takes between 3 and 15 min to collect a PIXE spectrum. The processing time is of the same order. However, it should be stressed than the interpretation of data for archaeological purpose can be much longer.

On the other hand, the following limitations are worth mentioning:

1. *Lack of Information on the Chemical State.* PIXE is strictly elemental and for instance will provide the amount of Fe without the distinction between Fe^{2+} and Fe^{3+}.
2. *Analysis of Near-Surface Layer.* The maximum probed depth is a few tens of micrometers variable with the elements of interest and depending on the matrix composition. This fact can constitute a serious problem whenever one wishes to determine nondestructively the bulk composition of an altered archaeological object.

6.8. EXPERIMENTAL SETUPS

6.8.1. External Beam

A unique advantage of the PIXE technique is its easy implementation under nonvacuum conditions, that is, in air or even better in a helium atmosphere. The main reason is that contrary to others ion beam analytical techniques [Rutherford backscattering (RBS) or nuclear reaction analysis (NRA)], it does not require high-energy definition and can stand a significant energy straggling linked to the beam interaction with both the exit window and the atmosphere without losing its analytical performances. This is due to the smooth variation of the X-ray production cross section with energy. Therefore, very shortly after its invention, PIXE was applied in the external beam mode, originally for the analysis of biological samples but very soon also for that of archaeological objects (Seaman and Shane 1975; Katsanos et al. 1976; Deconninck and Bodart 1978; Williams 1984; Mando 1994). Beam extraction into air is achieved through an exit window designed to withstand the atmospheric pressure. Such a window is made of a thin foil of various materials selected on the following grounds (Raisanen 1984; Anttila et al. 1985; Rauhala and Raisanen 1985):

1. Beside its good mechanical properties, it should not be sensitive to radiation damage.
2. The background radiation due to beam interaction with the window should be minimized.
3. Energy loss and straggling should be minimized.

Besides the obvious easy sample handling and positioning in air, the use of an external beam provides several notable advantages:

1. Objects of large size and complex structure and shape can be analyzed *in situ* without the need of sampling, a fact particularly valuable for art objects.
2. It is possible to study materials containing volatile compounds that could otherwise not stand vacuum.
3. The risk of damage due to heating is considerably reduced because of efficient cooling by the air or the helium flow.
4. Charging of insulating materials is totally avoided without the need of the thin conductive coating required when operating under vacuum. Indeed, charging effects can produce potentials up to several tens of kilo-electron-volts and accelerate accordingly secondary electrons with a subsequent extension of the bremsstrahlung background to higher energy. The detection limit would then be markedly increased. Moreover, deposition of a conductive coating on a precious museum object would of course be prohibited.

The main difficulty when operating in the external beam mode is the precise determination of the integrated dose of particles that is needed for accurate analysis. Several means have been employed for beam monitoring: a beam chopper inside or outside the vacuum, the Ar X K_α line produced along the beam path in air, the X-ray or RBS signal of a major component of the target used as an internal standard, or a specific signal emitted by the exit foil, which can be gamma rays, X-rays, or backscattered particles. The latter seems to be the most reliable technique.

6.8.2. Microbeam

A common feature of analytical tools based on charged particles, whether they are electrons or ions, is the ability to focus the beam by means of elec-

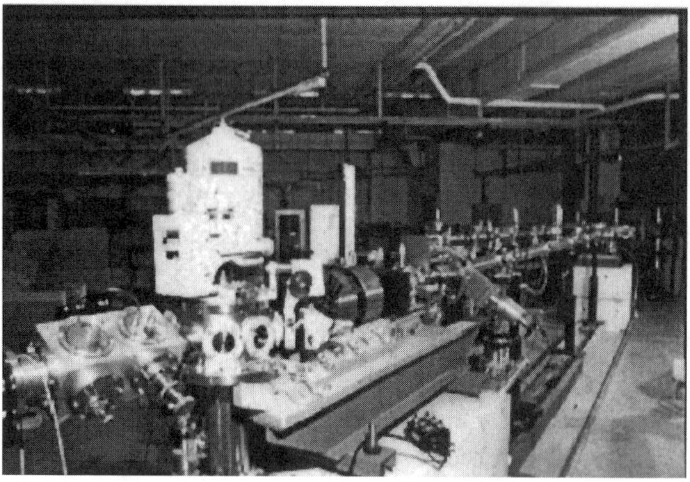

Figure 6.9. View of nuclear microprobe.

trostatic or magnetic lenses down to very small diameters, in the micrometer range for ions. The resulting instruments, which rely on mega-electron-volt ions, and a set of analytical techniques that include PIXE are called nuclear microprobes (Fig. 6.9), whereas those based on sputtering by kilo-electron-volt ions and mass spectrometry are named ion microprobes. The beam can be scanned over the sample in providing elemental micromapping with a high lateral resolution and even a three-dimensional distribution when using depth profiling techniques. The first nuclear microprobe was built at the end of the 1960s (Cookson et al. 1972), at the same time as the development of PIXE. Because of its high sensitivity, PIXE is well suited to the microprobe mode, but the relatively large probed depth (over 10 μm in most thick samples) excludes the need of beam focusing below 1 μm for the analysis of thick samples. Another limitation to the use of very narrow beams stems from the high density of energy deposited and the subsequent risk of damage. Nuclear microprobes are usually operated under vacuum, a fact that limits their usefulness for the analysis of objects of cultural heritage. However, it is generally admitted that applications of nuclear microprobes in art and archaeology do not require very high lateral resolution (Swann 1997) and that focused external beams of size around 50 μm are well adapted to this field. This type of set-up has been developed on the ion beam analysis facility of the Research Laboratory of the Museums of France in Paris (Calligaro et al. 1998a), which is named AGLAE, an acronym for Accélérateur Grand Louvre d'Analyse Elémentaire. The focusing system consists of three magnetic quadrupoles; previously this system was equipped with a microprobe built by the Oxford proton microprobe unit (Watt and Grime 1987). It is worth describing below a few details of the set-up (Calligaro et al. 1996).

6.8.3. AGLAE External Beam Line

The beam conditions in the PIXE mode are currently a proton energy of 3 MeV, intensity of 1 nA (which can be reduced down to 50 pA for the study of paperlike materials), and diameter from about 0.1 to 1 mm for routine work. Two Si(Li) detectors are used to cover the entire X-ray energy range. They are set at 45° with respect to the beam direction, the one dedicated to major light elements (low-energy X-rays) in an horizontal plane and the second for trace elements (high-energy X-rays) in a vertical plane. Their characteristics are the following:

Low-Energy Detector. Active surface 10 mm^2, window made of 0.25-μm-thick boron nitride, filter of 0.5-μm C foil to stop the light; fitted with a 50-mm-long magnet to deflect backscattered protons; helium flow to fill the 70-mm long path between the beam spot and the detector.

High-Energy Detector. Active surface 50 mm^2; 5-μm-thick Be window; Al filter of thickness depending on the experiment, designed to stop backscattered protons and matrix X-rays; the distance to the spot can be as low as 15 mm. The detector is shielded against X-rays emitted by the exit window by means of a 1-mm-thick lead cover.

The beam line is terminated by a conical nozzle of about 120 mm length containing the exit window (Fig. 6.10). The nozzle can be accurately positioned with respect to the lens axis by means of a x–y micrometric stage, so that the beam hits the center of the exit window. The characteristics of the exit window depend on the type of analysis performed. Metallic foils have been preferred to previously used Kapton windows because of their much greater stability to radiations. For millimeter-sized beams, the exit window has an area of about 4 mm^2 and is made of a 10-μm Al foil. For micro-PIXE operation, we have successively obtained a beam spot of about 20 to 30 μm with a 0.75-μm-thick Al foil and of roughly 10 μm with a 0.1-μm-thick Si_3N_4 foil.

The beam monitoring usually relies on the detection of particles backscattered by the exit foil, which is tilted at 45° with respect to the beam and directly faces the surface barrier Si detector. Such a geometry permits an accurate monitoring, as the detector collects the entire signal even when the beam spot on the window is not well centered. However, when producing an external microbeam, the small thickness of the Al exit window and the subsequent extremely low count rate of backscattered particles exclude this

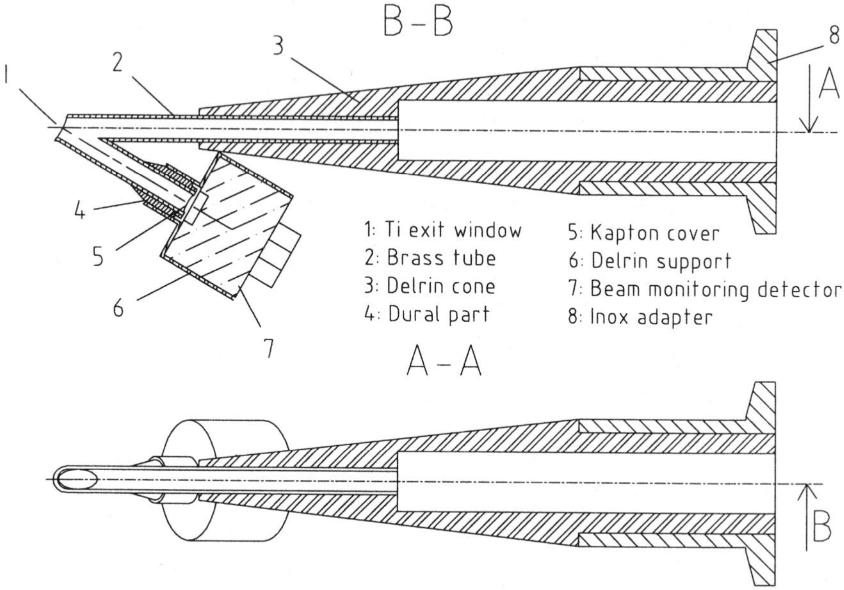

Figure 6.10. Cross section of exit nozzle of AGLAE external beam line.

monitoring method. The Al X-rays emitted by the window are more favorable, but the small dimensions of the nozzle do not fit a classical liquid-nitrogen-cooled Si(Li) detector and require a more compact detector cooled by Peltier effect.

The sample is placed as close as possible to the exit window in order to reduce the spot broadening due to the angular straggling induced by the window crossing. The minimum distance that allows X-ray detection with external Si(Li) detectors is about 3 mm. The sample position is remotely controlled with an accuracy of 0.1 mm by a set of laser beams and a video camera. The beam profiling has been performed by using a reference sample made of a cross section of a 2-μm Zr foil imbedded in aluminum and submitted to a linear scan with steps depending on the beam size. The Si(Li) detector dedicated to high-energy X-rays is used to record the K_α and K_β Zr lines.

6.9. EXAMPLES OF APPLICATIONS

Applications of PIXE in the field of art and archaeology started almost since the early development of the technique with the analysis of a pottery shard

(Gordon and Kraner 1972). They were quickly extended to all types of objects of the cultural heritage. However, because of the features described in Section 6.7, PIXE is more appropriate to some materials than to others. It is particularly suitable for the *in situ* analysis of thin layers of high-Z compounds on top of low-Z substrates. This ideal situation is encountered, for example, in drawings and illuminated manuscripts (Cahill et al. 1981; Kusko and Schwab 1987). In contrast, it is much more difficult to analyze easel paintings *in situ* without sampling, due to both the usual presence of a thick varnish coating and the complex structure of the paint layer. Recent efforts have been made to overcome this difficulty by using multiple-energy PIXE, which permits in some cases to identify the pigments and layer arrangement (Brissaud et al. 1996; Neelmeijer et al. 1996). Other materials are difficult to analyze *in situ*: metallic alloys because of the much higher detection limit than in light matrices and glasses because of the formation of an altered surface layer. It is out of the scope of the present chapter to provide a complete description of the numerous works published in the literature. Rather we will concentrate on the most recent ones dating from 1990 with particular emphasis on those performed in our laboratory. Instead of drawing a comprehensive list of applications classified according to the types of materials, as done elsewhere, we will stress the underlying issue of archaeology, art history, and conservation.

6.9.1. Materials Identification

Materials identification is the basic objective of the scientific characterization of museum objects and provides important clues for the knowledge of the fabrication procedure and the authentication of the work of art. This task is quite easily performed by PIXE in the external beam mode. This approach is illustrated by several investigations (Table 6.1) performed on manuscripts, miniatures, and drawings and intended to identify the nature of inks, metal points, and pigments. Important information was inferred on the use of pigments in medieval manuscripts: for instance, the very precious ultramarine blue, derived from lapis lazuli, has been surprisingly found on twentieth-century every-day mass books (McArthur et al. 1990). On Renaissance drawings by Pisanello, the use of different metal points was clearly evidenced (Duval 1997): lead or silver–mercury alloy on unprepared paper or parchment and pure silver on a preparation made of bone white and calcium carbonate. An interesting application of ink/pigment analysis on manuscripts concerns the enhanced visibility of faded characters through elemental mapping (Lovestam and Swietlicki 1990). Among numerous other works, we can also quote the identification of components of prestigious statues (Calligaro et al. 1998b) or jewels (Querré et al. 1996) kept in the Louvre in

Table 6.1. Materials Identification by PIXE Analysis of Major Elements

Artistic or Archaeological Objects	Component of Interest	Reference
Antique papyrus	Ink	Lovestam and Swietliki 1990
Medieval miniatures	Pigments	McArthur et al. 1990 Bussotti et al. 1997
Medieval Copt manuscripts	Inks and pigments	McArthur 1995
Medieval and Renaissance manuscripts	Inks	Del Carmine et al. 1993 Cambria et al. 1993
Historical manuscripts	Inks	Giuntini et al. 1995 Del Carmine et al. 1996
Chinese ancient paper money	Seal ink	Wu et al. 1993
Renaissance book engravings	Pigments	Neelmeijer et al. 1995
Renaissance drawings	Metal points	Duval 1997
Egyptian pectorals	Gems	Bouquillon et al. 1992
Antique jewels	Gems	Querré et al. 1996
Antique statuette	Gems	Calligaro et al. 1998b

Paris. The first study, devoted to a Parthian statue representing the goddess of love Ishtar, allowed the authors to identify the inlays forming the eyes and navel as rubies and not colored glasses as previously assumed by the museum curators.

6.9.2. Materials Provenance

This issue is probably the most frequently addressed in archaeological works because it is the key toward the knowledge of ancient trade routes and links between populations. For many years it was almost the monopoly of instrumental neutron activation analysis (INAA) due to the high sensitivity of this technique for trace elements that constitute the fingerprint of the material source (see Chapter 5). Now PIXE and to a lesser extent XRF (X-ray fluorescence) have replaced INAA in the great majority of studies. The current methodology combines trace element analysis of artifacts and samples of well-defined geological sources and multivariate statistical methods that first identify the discriminant elements and permit to ascribe to each artifact a source with a calculated probability. This approach has been applied to numerous materials (see Table 6.2), including stones (obsidians, flints), gems (rubies, emeralds), and ceramics. An example is provided by the determination of the provenance of the ruby inlays of the above-mentioned statuette. PIXE spectra for both major and trace elements are reported in Figure 6.11. A large collection of rubies of well-known provenance has been submitted to

Table 6.2. Materials Provenancing by PIXE Analysis of Trace Elements

Artistic or Archaeological Objects	Material of Interest	Reference
Tool or arrowhead	Obsidian	Poupeau et al. 1996
Inlays of antique statuette	Ruby	Calligaro et al. 1998b
Medieval jewels	Garnet	Farges 1998
Potsherds from South Africa	Clay	Peisach et al. 1990
		Jacobson et al. 1994
		Bollong et al. 1997
Islamic and Chinese potteries	Porcelain	Fleming and Swann 1992
Italian medieval potteries	Clay	Zucchiatti et al. 1993
		Zucchiatti et al. 1998
Byzantine ceramics	Paste and glaze	Waksman et al. 1994

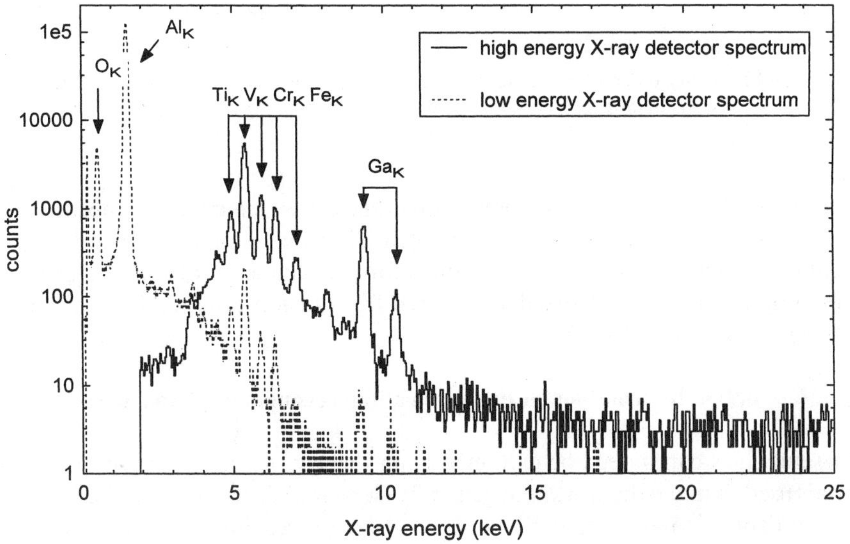

Figure 6.11. PIXE spectrum of rubies of Parthian statuette.

PIXE analysis under the same experimental conditions. Multivariate statistical processing of all data of elemental analysis was performed and led to the conclusion that the most likely provenance of the statuette rubies is Burma. The binary diagram in Figure 6.12 illustrates this statement.

In the case of ceramics, *in situ* analyses are seldom done because of potential bias due to their heterogeneous character. Therefore, most studies imply sampling a small amount of material that is ground and transformed

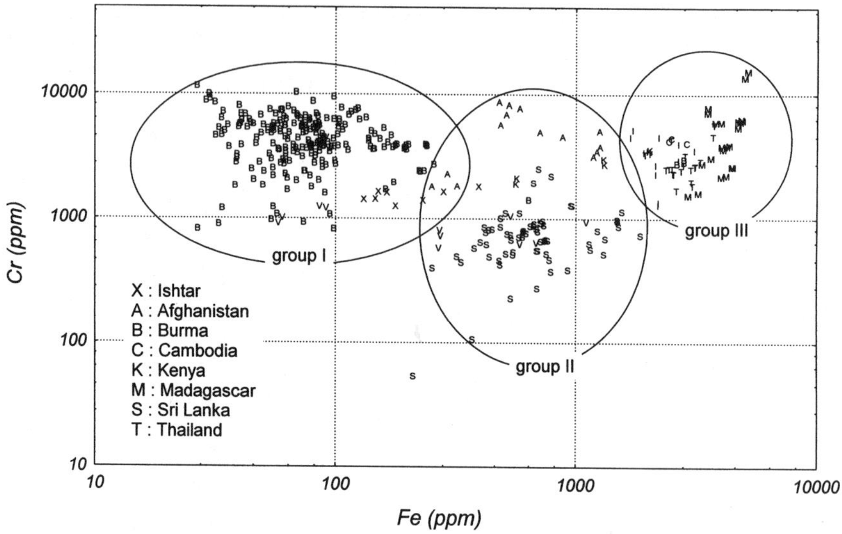

Figure 6.12. Binary diagram of Fe versus Cr contents in rubies of various provenances. Note that rubies of statuette of Fig. 6.11 fall inside first of three distinct groups.

into a flat pellet for measurement. Sometimes, the elemental composition is not sufficient to discriminate among potential sources, and thus an additional parameter can be helpful: for instance, for provenancing obsidian artifacts, the solidification age of obsidian inferred from fission track dating was decisive (Poupeau et al. 1996).

6.9.3. Information on the Fabrication Technology (Table 6.3)

Useful clues can indeed be extracted from the simple identification of materials used to make the work, but generally elemental distribution, either lateral or in-depth, is required. PIXE with good lateral resolution using a focused beam is thus preferable. A good illustration of the wealth of information that can be provided by PIXE analysis is the work of Demortier and co-workers on the soldering technology of gold jewelry of various origins (Demortier 1992, 1996).

6.10. CONCLUSION

PIXE has now reached its mature state and appears to be a very reliable analytical technique suited to almost all areas of research and technology. It

Table 6.3. Works Dealing with Fabrication Technology

Artistic or Archaeological Objects	Component of Interest	Reference
Panamanian gold artifacts	Au–Cu gilding	Fleming et al. 1990
Roman gold jewelry	Soldering joints	Weber et al. 1990
Antique and medieval gold artifacts	Soldering joints	Demortier 1992, 1996
Gold artifacts	Au–Cu alloy	Demortier and Rubalcaba-Sil 1996; Rubalcaba-Sil and Demortier 1996, 1997

combines several features, which explains its great popularity in art and archaeology: high sensitivity, nondestructive character, and rapidity and flexibility of implementation, particularly in the external beam mode. In fact, it is probably the most useful ion beam analysis technique in this field and served as the strongest argument for the construction of the AGLAE facility, which up to now is the only one in the world entirely devoted to the study of objects of cultural heritage. However, this technique has limitations and should not be considered a panacea. For the study of museum objects, its use should be restricted to investigations needing both high sensitivity and good lateral resolution. Otherwise, X-ray fluorescence could be substituted, as an alternative technique.

Acknowledgments

We thank J. P. Mohen for his constant support and interest in our work. We are also grateful to several of our colleagues: B. Moignard for his contribution to the improvement of the experimental set-up, A. Bouquillon and G. Querré for their pertinent comments and suggestions, and D. Bagault and D. Vigears for the photographs of the AGLAE facility.

REFERENCES

Anttila, A., J. Raisanen, and R. Lappalainen (1985), *Nucl. Instrum. Methods B* **12**, 245.

Bird., J. R., P. Duerden, and D. J. Wilson (1983), *Nucl. Sci. Applications B* **1**, 357.

Bollong, C. A., L. Jacobson, M. Peisach, C. A. Pineda, and C. G. Sampson (1997), *J. Archaeol. Sci.* **24**, 319.

Bouquillon, A., G. Querré, and J. P. Poirot (1992), *Proceedings of the Third International Conference on Non-Destructive Testing Microanalytical methods and environmental evaluation for study and conservation of Works of Art*, M. Marabelli and P. Santopadre, eds. ICR-AIPhD-Beta Gamma, Rome, Italy.

Brandt, W., and G. Lapicki (1981), *Phys. Rev. A* **23**, 1717.

Breese, M. B. H., G. W. Grime, and F. Watt (1992), *Ann. Rev. Nucl. Part. Sci.* **42**, 1.

Brissaud, I., G. Lagarde, and P. Midy (1996), *Nucl. Instrum. Methods B* **117**, 179.

Bussotti, L., M. P. Carboncini, E. Castellucci, L. Giuntini, and P. A. Mando (1997), *Stud. Cons.* **42**, 83.

Cahill, T. A. (1980), *Ann. Rev. Nucl. Part. Sci.* **30**, 211.

Cahill, T. A., B. Kusko, and R. N. Schwab (1981), *Nucl. Instrum. Methods* **181**, 205.

Cahill, T. A., B. H. Kusko, R. A. Eldred, and R. N. Schwab (1984), *Archaeometry* 26, **3**.

Calligaro, T., J. D. MacArthur, and J. Salomon (1996), *Nucl. Instrum. Methods B* **109/110**, 125.

Calligaro, T., J. C. Dran, H. Hamon, B. Moignard, and J. Salomon (1998a), *Nucl. Instrum. Methods B* **136–138**, 339.

Calligaro, T., A. Mosmann, J.-P. Poirot, and G. Querré (1998b), *Nucl. Instrum. Methods B* **136–138**, 846.

Cambria, R., P. Del Carmine, M. Grange, F. Lucarelli, and P. A. Mando (1993), *Nucl. Instrum. Methods B* **75**, 488.

Campbell, J. L., A. Perujo, and B. M. Millman (1987), *X-ray Spectrom.* **16**, 195.

Campbell, J. L., D. Higushi, J. A. Maxwell, and W. J. Teesdale (1993), *Nucl. Instrum. Methods B* **77**, 95.

Campbell, J. L., G. K. Czamanske, L. MacDonald, and W. J. Teesdale (1997), *Nucl. Instrum. Methods B* **130**, 608.

Chadwick, J. (1912), *Philos. Mag.* **24**, 594.

Cheng, H., D. Yanfang, H. Wenquan, and Y. Fujia (1998), *Nucl. Instrum. Methods B* **136–138**, 897.

Cholewa, M., G. Brench, B. J. Kirby, and G. J. F. Legge (1991), *Nucl. Instrum. Methods B* **54**, 101.

Chu, W., J. Liu, Z. Zhang, and K. B. Ma (1994), in *Materials Science and Technology, Characterization of Materials*, Vol. 2B., E. Lifshin, Ed., VCH, Weinheim, p. 423.

Cohen, D. D., and E. Clayton (1989), in *Ion Beam for Materials Analysis*, J. R. Bird and J. S. William, Eds., Academic, Marrickville, NSW Australia, p. 209.

Cookson, J. A. (1988), *Nucl. Instrum. Methods B* **30**, 324.

Cookson, J. A., A. T. G. Ferguson, and F. D. Pilling (1972), *J. Radioanal. Chem.* **12**, 39.

Deconninck, G., and F. Bodart (1978), *Nucl. Instrum. Methods* **149**, 609.

Del Carmine, P., F. Lucarelli, P. A. Mando, and A. Pecchioli (1993), *Nucl. Instrum. Methods B* **75**, 480.

Del Carmine, P., L. Giuntini, W. Hooper, F. Lucarelli, and P. A. Mando (1996), *Nucl. Instrum. Methods B* **113**, 354.

Demortier, G. (1991), *Nucl. Instrum. Methods B* **54**, 334.

Demortier, G. (1992), *Nucl. Instrum. Methods B* **64**, 481.

Demortier, G. (1996), *Nucl. Instrum. Methods B* **113**, 347.

Demortier, G., and J. L. Rubalcaba-Sil (1996), *Nucl. Instrum. Methods B* **118**, 352.

Duval, A. (1997), *Proceedings of the Fifth International Conference on Non-Destructive Testing Microanalytical methods and environmental evaluation for study and conservation of Works of Art*, M. Marabelli and C. Parisi, eds., AIPhD-ICR, Rome, Italy.

Farges, F. (1998), *Am. Mineral.* **83**, 323.

Fleming, S. J., and C. P. Swann (1992), *Nucl. Instrum. Methods B* **64**, 528.

Fleming, S. J., C. P. Swann, P. E. McGovern, and L. Horne (1990), *Nucl. Instrum. Methods B* **49**, 293.

Garcia, J. D., R. J. Fortner, and T. M. Kavanagh (1973), *Rev. Mod. Phys.* **45**, 111.

Giuntini, L., F. Lucarelli, P. A. Mando, W. Hooper, and P. H. Barker (1995), *Nucl. Instrum. Methods B* **95**, 389.

Gordon, G. M., and H. W. Kraner (1972), *J. Radioanal. Chem.* **12**, 181.

Jacobson, L., C. A. Pineda, D. Morris, and M. Peisach (1994), *Nucl. Instrum. Methods B* **85**, 901.

Janssens, K., A. Aerts, L. Vincze, F. Adams, C. Yang, R. Utui, K. Malmqvist, K. W. Jones, M. Radtke, S. Garbe, F. Lechtenberg, A. Knochel, and H. Wouters (1996), *Nucl. Instrum. Methods B* **109/110**, 690.

Johansson, G. (1982), *X-ray Spectrom.* **11**, 194.

Johansson, S. A. E., and J. L. Campbell (1988), *PIXE: A Novel Technique for Elemental Analysis*, Wiley, New York.

Johansson, S. A. E., and T. B. Johansson (1976), *Nucl. Instrum. Methods* **137**, 473.

Johansson, S. A. E., J. L. Campbell, and K. G. Malqvist (1995), *Particle Induced X-ray Emission Spectrocopy (PIXE)*, John Wiley & Sons, New York.

Johansson, T. B., K. R. Akselsson, and S. A. E. Johansson (1970), *Nucl. Instrum. Methods* **84**, 141.

Katsanos, A., A. Xenoulis, A. Hadjhantoniou, and R. W. Fink (1976), *Nucl. Instrum. Methods* **137**, 119.

Kusko, B. H., and R. N. Schwab (1987), *Nucl. Instrum. Methods B* **22**, 401.

Lovestam, N. E. G., and E. Swietlicki (1990), *Nucl. Instrum. Methods B* **45**, 307.

Maeda, K., K. Hasegawa, H. Hamanaka, and K. Ogiwara (1998), *Nucl. Instrum. Methods B* **134**, 418.

Mando, P. A. (1994), *Nucl. Instrum. Methods B* **85**, 815.

Maxwell, J. A., J. L. Campbell, and W. J. Teesdale (1988), *Nucl. Instrum. Methods B* **43**, 218.

McArthur, J. D. (1995), *Techne* **2**, 68.

McArthur., J. D., P. Del Carmine, F. Lucarelli, and P. A. Mando (1990), *Nucl. Instrum. Methods B* **45**, 315.

McColm, D. W., and T. A. Cahill (1991), *Nucl. Instrum. Methods B* **54**, 91.

Mitchell, I. V., and K. M. Barfoot (1981), *Nucl. Sci. Applications* **1**, 99.

Neelmeijer, C., W. Wagner, H. P. Schramm, and U. Thiel (1995), *Nucl. Instrum. Methods B* **99**, 390.

Neelmeijer, C., W. Wagner, and H. P. Schramm (1996), *Nucl. Instrum. Methods B* **118**, 338.

Peisach, M. (1987), *J. Radioanal. Nucl. Chem.* **110**, 461.

Peisach, M., C. A. Pineda, and L. Jacobson (1990), *Nucl. Instrum. Methods B* **49**, 309.

Potts, P. J. (1987), in *A Handbook of Silicate Rock Analysis*, Blackie, Glasgow, p. 286.

Poupeau, G., L. Bellot-Gurlet, O. Dorighel, T. Calligaro, J.-C. Dran, and J. Salomon (1996), *C. R. Acad. Sci. Paris* **323**, 443.

Querré, G., A. Bouquillon, T. Calligaro, M. Dubus, and J. Salomon (1996), *Nucl. Instrum. Methods B* **109/110**, 686.

Raisanen, J. (1984), *Nucl. Instrum. Methods B* **3**, 220.

Rauhala, E., and J. Raianen (1985), *Nucl. Instrum. Methods B* **12**, 321.

Rubalcaba-Sil, J. L., and G. Demortier (1996), *Nucl. Instrum. Methods B* **113**, 275.

Rubalcaba-Sil, J. L., and G. Demortier (1997), *Nucl. Instrum. Methods B* **130**, 297.

Scuti, S., and G. Suber (1991), *Rivista Nuovo Cimento* **14**, 1.

Seaman, G. G., and K. C. Shane (1975), *Nucl. Instrum. Methods* **126**, 473.

Swann, C. P. (1997), *Nucl. Instrum. Methods B* **130**, 289.

Swann, C. P., and S. J. Fleming (1990), *Nucl. Instrum. Methods B* **49**, 65.

Waksman, S. Y., A. Pape, and Ch. Heitz (1994), *Nucl. Instrum. Methods B* **85**, 824.

Watt, F., and G. W. Grime (1987), *Principles and Applications of High-Energy Ion Microbeams*, Adam Hilger, Bristol.

Weber, J., Th. Beier, U. Diehl, D. Lambrecht, H. Mommsen, and F. J. Pantenburg (1990), *Nucl. Instrum. Methods B* **50**, 221.

Williams, E. T. (1984), *Nucl. Instrum. Methods B* **3**, 211.

Wu, X., X. Zeng, and F. Yang (1993), *Nucl. Instrum. Methods B* **75**, 458.

Zeng, X., X. Wu, Q. Shao, J. Tang, and F. Yang (1990), *Nucl. Instrum. Methods B* **47**, 143.

Zeng, X., X. Wu, H. Yao, F. Yang, and T. A. Cahill (1993), *Nucl. Instrum. Methods B* **75**, 99.

Zucchiatti, A., H. J. Annegarn, M. A. Kneen, and C. Varaldo (1993), *Nucl. Instrum. Methods B* **75**, 463.

Zucchiatti, A., F. Cardoni, P. Prati, F. Lucarelli, P. A. Mando, and G. P. Martino (1998), *Nucl. Instrum. Methods B* **136–138**, 893.

CHAPTER

7

MOLECULAR AND STRUCTURAL ANALYSIS

GIUSEPPE SPOTO

Dipartimento di Scienze Chimiche, Università di Catania, Catania, Italy

7.1. INTRODUCTION

A molecule is the smallest unit of a compound that can exist by itself and retains its chemical properties. Molecules involve the bonding of at least two atoms and are sometimes the smallest distinct neutral aggregates to be found in matter. Nature provides a huge variety of molecules. Some of these contain very few atoms, as in the case of oxygen, which has two atoms. However, molecules made up of thousands of atoms are not uncommon, for example, hemoglobin and myoglobin, two important molecules belonging to the category defined as proteins, which help to carry and fix the oxygen in the body.

Molecular analytical techniques may be defined as those that identify and characterize molecules. This definition may, however, be extended to include the formula unit, which is the smallest neutral unit that reproduces the ratio between the elements in ionic compounds. Ions are chemical species that retain positive or negative electric charge(s).

It should be pointed out that molecular analysis does not indicate whether the characterized species exist as finite and independent molecules, as is the case for gases such as oxygen or nitrogen. Nor does this type of analysis show whether the molecules are present as aggregates involving bonds of varying strength.

Molecular techniques differ from elemental analytical techniques (compare Chapter 2). Elemental analysis identifies and often quantifies elements that may also be found in molecules or ionic compounds. However, no clues are provided as to the different kind of molecule or unit formula in which they may be found. Thus, if SiO_2 is analyzed using a suitable elemental

Modern Analytical Methods in Art and Archaeology, Edited by Enrico Ciliberto and Giuseppe Spoto. Chemical Analysis Series, Vol. 155
ISBN 0-471-29361-X

analytical technique, only evidence of the presence of silicon and oxygen is obtained. An adequate molecular analytical technique instead provides a more or less explicit indication of the presence of SiO_2 as a whole but does not make it possible to ascertain as to whether this species exists as an independent unit (molecule) or is part of a larger aggregate.

Molecules are spatially arranged according to the geometry with which the constituent atoms interact each other. Along with atoms and ions, molecules can be organized to form three-dimensional aggregates with an internal geometric order that is typical of crystalline solid materials.

When discussing aspects such as these, we are in the realm of *structural* analysis. A structural analytical technique provides information concerning the geometric shape of a molecule and/or the geometric arrangement of molecules, atoms, or ions within an ordered structure such as that of a crystalline solid. Structural analysis can be used to show how SiO_2 groups can be arranged in various ways to form crystalline solids such as quartz, cristobalite, and tridymite. The spatial disposition of the atoms present in molecules such as that of DNA (deoxyribonucleic acid) can also be ascertained by making use of structural analytical techniques.

It should now be clear that while structural techniques often provide molecular information, molecular techniques do not carry structural information in the way described.

7.2. MOLECULAR AND STRUCTURAL ANALYSIS: FURTHER COMMENTS

Readers who are not familiar with the theories and applications of analytical methods will no doubt have reached the conclusion that analytical techniques that are not elemental must necessarily be molecular or structural. Experienced readers, on the other hand, may be put off reading any further into this chapter, given the apparently loose nature of the distinction drawn so far. In reality, the boundaries separating elemental, molecular, and structural are often fogged—sometimes they do not even exist; this is especially true when referring to analytical techniques. Some analytical techniques, such as the auger electron spectroscopy (AES), are certainly elemental but can also provide a limited amount of molecular information. The wide range of techniques making use of mass spectrometer (MS) analyzers (MS is usually preceded by various acronyms according to the method used to generate the ions introduced into the analyzer; see Chapters 3, 8, and 17 for descriptions of techniques using MS analyzers) provide an illustration of how a single technique may be termed as being both elemental and molecular, there being no justifiable reason to prefer one to the other. The variety of meanings sometimes attributed to the term "structural" further complicates matters.

The remaining part of this chapter attempts to provide a brief insight into the world of molecular and structural analytical techniques, making specific reference to their use in art and archaeology. The aim is not that of providing a discussion of basic principles, for whose purposes readers are advised to consult analytical chemistry textbooks (e.g., Christian 1994; Skoog and Leary 1992), nor should it be supposed that an exhaustive list will be supplied for the variety of molecular and structural analytical techniques available today. Only certain aspects of the use of these techniques in fields of study involving art and archaeological objects will be highlighted here. An attempt has been made to put forward an across-the-board discussion that may serve to introduce nonexpert readers to the topics treated in the chapters to follow. These chapters mainly concern the principles of a variety of analytical techniques, those that may be classified as being molecular and structural along the lines outlined above, and their application to art and archaeology.

7.3. MOLECULAR AND STRUCTURAL ANALYSIS IN ART AND ARCHAEOLOGY

7.3.1. General Remarks

Art and archaeology benefit greatly from the information that molecular and structural analytical techniques can provide. This information is used both to improve our knowledge of ancient culture and usage and to find the best ways to restore and/or preserve the objects that testify to the skilful and often artistic talents of our ancestors.

Which molecules mainly constitute the yellow dye used to stain ancient Chinese documents such as the Dunhuang *Diamond Sutra*—an ancient Chinese scroll that is the earliest complete example of a dated printed book? (Dated according to the ancient Chinese calendar as the 15th of the fourth moon of the ninth year of Xiantong, corresponding to May 11, A.D. 868; Gibbs et al. 1997.) What was the composition of the dark brown paint that, on the night of the September 29, 1995, vandals poured over several of the most beautiful and important mosaics of the *Villa del Casale* (Piazza Armerina, Italy)? (The villa is the most important surviving illustration of the Roman occupation of Sicily; Ciliberto et al. 1998.) Which technique was used by Leonardo in executing the famous mural *The Last Supper*—a technique making soft modeling and the use of a wide range of transparencies possible but one that caused the immediate and continuous degradation of the mural (Matteini and Moles 1979)? These are but a few of the questions addressed in restoration and preservation studies where molecular and structural analytical techniques have helped to find answers. Archaeological studies may instead use structural analysis to date organic parts of the remains of living organisms (such as teeth, bones, or even shells) by studying specific amino acids present in them (Pollard and Heron 1996: Chapter 8; Aitken 1990: Chapter 8). Such techniques are also employed to identify food residues remaining in ancient pottery used for cooking (Oudemans and Boon 1991).

7.3.2. Which Technique for Which Sample?

All scientists involved in studies concerning works-of-art or samples of archaeological interest have experienced the importance played by the appropriate selection of the analytical method in their studies. The main questions raised usually concern how the proposed analytical procedure will affect the object to be examined in its integrity. From this point of view, only those techniques that do not alter the integrity and the appearance of artistic and/or archaeological objects are eligible as "ideal" techniques. Techniques that operate *in situ*, making sample taking unnecessary, come close to this ideal; as do those that transfer minimal amounts of energy onto the sample, thus minimizing energy-induced degradation. In this context, molecular techniques such as electronic and vibrational spectroscopy can be used with success, especially since the introduction of fiber optics to drive low-power radiation directly onto the point to be analyzed. Similar equipment is described in Chapter 12, where the potential of electronic and vibrational spectroscopy (UV-vis and IR, respectively) for *in situ* analyses of works-of-art are discussed. Similar success may be achieved with the use of Raman spectroscopy (see Chapter 10). This has further potential, in terms of spatial resolution, when a spectrometer with an adequate laser source is fitted with a microscope. In this case, the diameter of the laser beam (in the range of very few micrometers) is the limiting factor for lateral resolution (see below for a definition of this term). Raman microscopy may thus be used to analyze mostly inorganic but also organic components of works-of-art that can be placed under the microscope nondestructively (see also Clark and Gibbs 1998).

Raman spectroscopy leads back to the issue of the lines dividing molecular and structural analysis. The information provided by Raman spectroscopy can be used to obtain fingerprints for specific compounds. It provides molecular information of the kind previously discussed but may also supply a certain amount of structural information. An example of this overlap can be seen in realgar—a naturally occurring arsenic sulfide mineral used in the past as red pigment—and its yellow light-induced transformed product known today as pararealgar. Both correspond to the As_4S_4 formula but have a different structure (Bonazzi et al 1995); the term *polymorphs* is used for such cases. Raman fingerprints clearly make it possible to distinguish between the two polymorphic structures (Trentelman et al. 1996), thus providing the molecular and also structural distinction of the two compounds.

7.3.3. How May Tiny Samples Be Probed?

In the attempt to find a balance between the requirements of scientific methods and the need to maintain the integrity of the object under study, the only alternatives to *in situ* analysis require the object itself to be placed in the analyzing chamber of the analytical instrument or tiny fragments of samples to be scraped from its surface. The former approach cannot be applied in all cases, since only small objects such as coins and certain jewelry are of a size and shape that will fit chambers of common analytical tools. The destructive approach thus remains as a last resort for the extraction of analytical information from artistic and/or archaeological samples.

Modern molecular and structural analytical techniques are usually sensitive to a degree that allows samples in the range of micrograms to be analyzed. Some require, or simply sometimes prefer, the sample to be powdered; others make the analysis of tiny fragments of sample extracted from the object possible with no further treatment. We may thus speak of a microdestructive approach that often allows the fragment to be put back in its original position after analysis.

Tiny fragments may be analyzed by probing their surface with electrons, ions, or radiation. Particles or radiation are then extracted and, after detection, used for the elemental or molecular and structural analysis of the sample. AES, X-ray photoelectron spectroscopy (XPS), and secondary ion mass spectrometry (SIMS) operate on these bases. AES probes the surface using electrons, XPS makes use of X-rays, and SIMS uses ions. These techniques, however, extract information only from the first few nanometers of the samples and for this reason can be defined as surface techniques.

From a strictly chemical–physical point of view, the surface of a solid material is its outermost atomic layer. Few surface techniques extract information from the outermost atomic layer only. It is thus common procedure to distinguish between surface-specific techniques and surface-sensitive techniques that extract a small part of the overall signal from deeper regions (usually in the nanometer range). SIMS belongs to the former group when operating in appropriate conditions (static SIMS); XPS and AES belong to the latter. Techniques that are not surface techniques are defined as bulk techniques.

The application of XPS and AES to studies concerning art and archaeology that involves the surface of the materials is discussed in Chapter 13 (for the application of SIMS in studying artistic and archaeological samples, see Lechtman 1984; Schreiner et al. 1984; Stermer et al. 1996; Adiaens 1996).

The information obtained using surface techniques can be defined as vertically resolved; that is, it is possible to see how this information may vary moving down through the sample layers. Surface techniques may also play a part in the lateral resolution of the analysis, since it is common to analyze and differentiate between areas with diameters in the range of micrometers. Lateral resolution is an important point to consider when selecting the appropriate analytical technique for the study of materials coming from works-of-art or archaeological objects. When points of the sample that are often a few micrometers or less across can be analyzed, the distribution of the parameters detected for wider areas (often in the millimeters-squared range) may be mapped. A similar approach has been used in studies of ancient bronzes (Wouters et al. 1992; Allen et al. 1995; Spoto et al., in press), where the relevant mapped parameter was related to the presence of certain elements. When the parameters are specific families of electrons (i.e., electrons within specific energetic ranges), we are in the field of electron microscopy.

Transmission electron microscopy (TEM) is a specific microscopy technique that detects electrons that pass through the sample. The image reproduced results from interference between the electrons scattered by the atoms of the sample. This is also known as diffraction and is strongly dependent on the structure of the sample, which can then be seen in the image obtained. It should be pointed out that the necessity to transmit an appreciable fraction of the incident electron beam requires the sample to be prepared in the form of a very thin foil. The impact of TEM analysis on tiny fragments of samples to be studied is thus definitely destructive.

Examples illustrating the potential of TEM in terms of structural information are discussed in Chapter 17, which also provides a discussion of how TEM has been used to clarify the complex nature of the Maya Blue pigment (José-Yacamán et al. 1996).

7.3.4. How Destructive Is Science?

The destructive analytical approach should be considered as a last resort for studying objects that are often unique in their artistic or archaeological value. The necessity for such an approach stems from the inadequate and insufficient information provided by techniques that operate with an nondestructive or microdestructive approach (see Section 7.3.3). The consequences of destructive sampling have to be weighed against the advantages afforded to the subsequent analytical study. This section aims to address this point.

It should again be underlined that most modern analytical molecular and structural techniques are able to analyze samples in the range of micrograms; few require milligram weight samples. However, sample size restrictions do not always depend on analytical technique requirements. Chemical treatment may precede the analytical characterization of the sample, thus requiring greater amount of sample to be used. Similar treatment is often necessary when the molecular and structural nature of organic material is studied. In cases such as these, it should be pointed out that extremely relevant information concerning ancient organic materials is provided by modern analytical instruments and procedures.

Two of the most powerful modern analytical tools used to analyze organic materials are mass spectrometric techniques (which involve a variety of techniques using a mass spectrometry analyzer) and nuclear magnetic resonance (NMR). These techniques have been used to address a whole range of questions concerning art and archaeology. The use of appropriate mass spectrometric techniques made it possible to identify an original varnish used in an oil painting dated to 1929, thus providing fundamental information for restoration purposes (Van der Berg et al. 1998). Mass spectrometry was also used to analyze the cholesterol preserved in archaeological bones and teeth; this was a major contribution to paleodietary reconstruction studies (Stott and Evershed 1996). Another fascinating aspect of this technique is its use in the identification of volatile compounds trapped within the organic networks of archaeological plants, these being formed as a consequence of reactions established during long-term burial (Evershed et al. 1997). These illustrations are just some of the many ways in which mass spectrometry techniques can provide fundamental information in archaeochemical studies. NMR, which provides both molecular and structural information (basic principles are discussed in Lambert et al. 1998), is also widely used to study organic materials and thus makes an important contribution to the study of ancient organic materials. Since the 1970s, the possibility of analyzing solid materials in the weight range of milligrams has made it possible to study the resinous nature of amber of diverse origins around the world and to ascertain origins on the basis of NMR spectra (Lambert and Frye 1982).

7.3.5. Other Methods of Analysis

The principal aim of the above sections was to briefly inform readers unfamiliar with the use and application of molecular and structural analytical techniques of their potential in the field of art and archaeology. However, only a small selection of the wide variety of analytical tools available today has been indicated. Other techniques and methods have also been applied in studies concerning the field of interest of this book. It is necessary, for example, to acknowledge, among others, the important role played by thermal analytical techniques, especially when used in the field of study of the preservation of painting materials. Thermal analytical techniques traditionally provide chemical information but may also be useful in the study of the mechanical and electric properties of paint films. An in-depth discussion of a similar application can be found in Chapter 11.

To conclude, the great impetus seen in the study of ancient DNA molecules should also be mentioned. The invention of the polymerase chain reaction in 1983, whose inventor, Kary B. Mullis, won the Nobel Prize for Chemistry in 1993, led to tremendous improvements in the ability to replicate DNA segments of complicated genetic materials (see also Mullis 1990). This has been exploited by researchers interested in studying DNA extracted from fossil remains of living organisms, thus defining a field of research that has great potential for the reconstruction of the phylogenetic and evolutionary relationship existing between living organisms. These topics, which have also been imaginatively interpreted by some authors of science fiction, are briefly discussed in Chapter 9.

7.4. CONCLUSION

Modern instrumental analytical chemistry provides a wide range of information that can be of use for archaeologists, conservators, and all those who, for various reasons, are interested in the study of archaeological and artistic objects. In this context, the role played by molecular and structural techniques is of particular relevance, especially when the often complex chemical nature of materials constituting ancient objects is to be ascertained. This chapter has attempted to introduce the reader to the varied world of molecular and structural techniques while giving an idea of the opportunities afforded by their use. The following chapters of this book will take a closer look at these opportunities and address in depth the issues that have been outlined in very general terms in this introductory chapter.

Acknowledgments

I am indebted to Santo Di Bella and Ronald G. V. Hancock for editorial comments on an earlier draft of this paper.

REFERENCES

Adriaens, A. (1996), *Mikrochim. Acta* **124**, 89.

Aitken, M. J. (1990), *Science-Based Dating in Archaeology*, Addison-Wesley Longman, New York.

Allen, G. C., I. T. Brown, E. Ciliberto, and G. Spoto (1995), *Eur. Mass Spectrom.* **1**, 493.

Bonazzi, P., S. Menchetti, and G. Pratesi (1995), *Am. Mineral.* **80**, 400.

Christian, G. (1994), *Analytical Chemistry*, 5th ed., John Wiley and Sons, New York.

Ciliberto, E., G. Spoto, M. Matteini, and C. Puglisi (1998), *J. Chem. Ed.* **75**, 1302.

Clark, R. J. H., and P. J. Gibbs (1998), *Anal. Chem.* **70**, 99A.

Evershed, R. P., H. A. Bland, P. F. van Bergen, J. C. Carter, M. C. Horton, and P. A. Rowley-Conwy (1997), *Science* **278**, 432.

Gibbs, P. J., K. R. Seddon, N. M. Brovenko, Y. A. Petrosyan, and M. Barnard (1997), *Anal. Chem.* **69**, 1965.

José-Yacamán, M., L. Rendón, J. Arenas, and M. C. Serra Puche (1996), *Science* **273**, 223.

Lambert, J. B., and J. S. Frye (1982), *Science* **217**, 55.

Lambert, J. B., H. F. Shurvell, D. A. Lightner, and R. G. Cooks (1998), *Organic Structural Spectroscopy*, Prentice-Hall, Englewood Cliffs, NJ.

Lechtman, H. (1984), *Sci. Am.* **250**(6), 38.

Matteini, M., and A. Moles (1979), *Stud. Cons.* **24**, 125.

Mullis, K. B. (1990), *Sci. Am.* **262**(4), 36.

Oudemans, T. F. M., and J. J. Boon (1991), *J. Anal. Appl. Pyrol.* **20**, 197.

Pollard, A. M., and C. Heron (1996), *Archaeological Chemistry*, Royal Society of Chemistry, Cambridge.

Schreiner, M., G. Stingeder, and M. Grasserbauer (1984), *Fresenius J. Anal. Chem.* **319**, 600.

Skoog, D. A., and J. J. Leary (1992), *Principles of Instrumental Analysis*, 4th ed., Saunders, Fort Worth, TX.

Spoto, G., E. Ciliberto, G. C. Allen, C. M. Younes, P. Piccardo, M. R. Pinasco, E. Stagno, M. G. Ienco, and R. Maggi (in press), *Br. Corros. J.*

Stermer, E. M., S. Risnes, and P. M. Fischer (1996), *Eur. J. Oral. Sci.* **104**, 253.

Stott, A. W., and R. P. Evershed (1996), *Anal. Chem.* **68**, 4402.

Trentelman, K., L. Stodulski, and M. Pavlosky (1996), *Anal. Chem.* **68**, 1755.

Van der Berg, J. D. J., J. J. Boon, K. J. Van der Berg, I. Fiedler, and M. A. Miller (1998), *Anal. Chem.* **70**, 1823.

Wouters, H. J., L. A. Butaye, and F. C. Adams (1992), *Fresenius J. Anal. Chem.* **342**, 128.

CHAPTER

8

BIOMOLECULAR ANALYSIS BY ORGANIC MASS SPECTROMETRY

RICHARD P. EVERSHED

School of Chemistry, University of Bristol, Cantock's Close, Bristol BS8 1TS, United Kingdom

8.1. INTRODUCTION

The principle question that mass spectrometry (MS) can contribute to solving in the fields of archaeology and art is in the identification of organic raw materials. These organic materials are for the most part natural products and, hence, composed of complex mixtures of biochemical components. The activities of humans in antiquity will usually have complicated chemical composition through the intentional or unintentional mixing of different organic materials, for example, through the repeated uses of archaeological cooking vessels to prepare a range of different foodstuffs or the mixing of organic binders and pigments during the preparation of art materials. Still further increases in the complexity of composition derive from the transformation of the original natural materials either through the application of heat required to cook, extract, or otherwise mix and/or modify the properties of natural materials or through the natural degradative processes occurring through artifact or material use and environmental exposure, for example, hydrolysis, oxidation, and polymerization of oils and fats in works of art and buried archaeological artifacts. Mixtures of this nature represent a formidable analytical challenge in our efforts to define the nature and origins of such materials. While many analytical techniques can provide information relating to the chemical composition of these materials, that is, the types of functional groups present can be determined through infra red (IR) or nuclear magnetic resonance (NMR) spectroscopy, no other technique has the capability of mass spectrometry for deriving detailed compositional

Modern Analytical Methods in Art and Archaeology, Edited by Enrico Ciliberto and Giuseppe Spoto. Chemical Analysis Series, Vol. 155
ISBN 0-471-29361-X

information (i.e., structures and compound distributions) from such complex mixtures. The unmatched capacity of mass spectrometry to reveal such information derives from the ability to couple chromatographic techniques, for example, gas chromatography (GC) and high-performance liquid chromatography (HPLC), to mass spectrometers, thereby allowing separation of the individual molecular species with on-line mass spectrometric analysis of the eluting components. This capability allows structure elucidation of components of complex mixtures, with further possibilities for their detection and quantification at trace levels in complex matrices.

Many of the advances that are now being made in the use of mass spectrometry in the study of organic materials in art and archaeology are underpinned by research performed in the fields of natural product chemistry and biochemistry. A recommended starting point from newcomers to this area is Mills and Whites excellent textbook *The Organic Chemistry of Museum Objects* (Mills and White 1994), which provides a comprehensive overview and numerous bibliographic references to the chemistry of natural products relevant to art and archaeology. For more detailed accounts of the theory of mass spectrometry and ancillary techniques and reviews of the mass spectrometric analysis of natural products and biochemical materials see Waller (1972), Waller and Dermer (1980), Rose and Johnstone (1982), Lawson (1989), McCloskey (1990), Chapman (1993), De Hoffman et al. (1996), and Constantin and Schnell (1990). Specialist periodicals on mass spectrometry include *Rapid Communications in Mass Spectrometry, Journal of the American Society for Mass Spectrometry, European Journal of Mass Spectrometry, Mass Spectrometry Reviews.* Other periodicals that frequently contain articles on combined chromatography/mass spectrometry include *Analytical Chemistry, Analyst, Journal of Chromatography, Analytica Chimica Acta, Chromatographia, Journal of High Resolution Chromatography, and Chromatographic Communications.* Publications utilizing mass spectrometry in applications directly related to archaeology and art appear in a wide range of journals including *Journal of Archaeological Science, Ancient Biomolecules, Archaeometry, Science, Nature, Studies in Conservation, Antiquity*, and publications of the Materials Research Society. Access to the literature pertinent to this field can be gained through various abstracting services including *Art and Archaeology Technical Abstracts, Analytical Abstracts*, and *Chemical Abstracts.*

8.2. ANALYTICAL APPROACH

The range of raw materials that will be encountered in studies of organic materials in art and archaeology is potentially vast, ranging from various exotic plant resins, natural waxes, plant oils and animal fats, complex carbohydrates, proteins, lignins, and bituminous materials of geological age to modern synthetic polymers. Fortunately, mass spectrometry has a role to play in the analysis of all these classes of organic materials. However, when considering whether or not to use organic mass spectrometry for a particular analysis, the analyst needs to consider a number of simple questions, such as:

1. What information is required that is likely to be resolved through mass spectrometric analyses?
2. Is the sample organic in nature, or likely to contain organic constituents?
3. What class/classes of compounds are likely to be present?
4. Is sufficient material available for mass spectrometric analysis?
5. Is there an existing protocol available for the class of compound(s) being considered for mass spectrometric analysis?
6. Is specialist knowledge or are mass spectral libraries required such that on completion of analyses the data can be interpreted?
7. Is structure information required on unknown compounds or is detection required of components present at trace levels?
8. What is/are the molecular weights of the component(s) of interest?
9. Is on-line GC/MS or HPLC/MS required to resolve mixtures of compounds?
10. Will derivatization be required, for example, to improve volatility in GC, or direct MS fragmentations to provide enhanced structure information?
11. Are components of interest macromolecular in nature, for example, biopolymers, such that chemical/pyrolytic or enzymatic degradation may be required to produce subunits/building blocks that can be more easily accommodated by GC/MS or HPLC/MS.
12. Is chromatographic data available so that elution orders can be used to aid in interpretations of mass spectral data?
13. Are reference materials or compounds required for comparative purposes?
14. Are internal standards required for the quantification of analytes?
15. Is stable isotope information required at natural abundance levels?

This list is by no means exhaustive, but it does serve to illustrate the thinking required in the planning phase of an analytical program or when devising a new protocol. Mass spectrometry on its own is not a panacea, although it may ultimately be the focal point of the analytical protocol. While there is often a preferred way of undertaking a particular analysis, it pays to maintain an open mind and consider alternative approaches that may reveal new information. The examples given below have been chosen to illustrate the ways in which modern MS techniques have been employed to solve particular questions relating to the study of ancient materials, largely of archaeological interest. The general principles apply equally well to research into organic materials relating to the art world.

8.3. BASIC PRINCIPLES

Considered in its simplest form a mass spectrometer (MS) is a device for producing and mass measuring ions. In the case of organic molecules the masses and relative abundances of molecular or pseudomolecular ions and fragment ions are a direct reflection of molecular structure. Mass spectrometry is unique in its ability to yield often complete structure assignments without needing to use other physico-chemical techniques. The major advantages and attributes of modern organic mass spectrometry for the study of materials in art and archaeology are:

1. Very rapid and unambiguous structure determinations without need to use other techniques.
2. Most effective method available for analyzing complex mixtures particularly when coupled with GC or HPLC.
3. Applicable to all the major compound classes.
4. The only analytical technique that can provide exact elemental compositions of organic compounds, that is, by using high-resolution mass measurements.
5. Ultrasensitive detection method even where complex matrices are involved.
6. High precision quantitative measurements possible.
7. Reference or complementary technique to other analytical methods.

The complex nature of early mass spectrometers meant that analyses were conducted by highly trained operators. However, the increasing availability of user-friendly computer-controlled bench-top GC/MS and HPLC/MS instruments means that mass spectrometric analyses can be performed with only the minimum of expert assistance ("open access" instruments). A detailed knowledge of the workings of mass spectrometers is probably unnecessary for most analysts. Although contentious in some circles, the mass spectrometer can be regarded as a GC or HPLC detector. However, the mass spectrometer is a sophisticated device; thus an appreciation of the basic aspects of operation of the vacuum system, ion source, mass analyzer and ion detection is undoubtedly of value, ensuring safe and effective use of the instrumentation. Chapman's excellent monograph *Practical Organic Mass Spectrometry* (Chapman 1993) covers all mass spectrometric techniques commonly used in the analysis of natural products.

An overview of sample introduction and ionization methods that are commonly used (and some which may be of future use) in the mass spectrometric analyses of the materials of art and archaeology is presented below. Complementary to this is a knowledge of possible sample preparation strategies, including chemical transformations or derivatizations that can be used to enhance the structural information content of mass spectral data. Additionally, the analyst needs to be aware of the capabilities of the different operational modes of the mass spectrometer that exist. For instance, the use of full mass range scanning to provide mass spectra for structure investigations versus the use of selected ion monitoring (SIM) or selected reaction monitoring (SRM) to enhance sensitivity of detection in trace analyses. Due consideration is given to these latter points in the appropriate sections below.

8.4. INSTRUMENTATION

This section describes the components of a mass spectrometer giving due emphasis to the specific requirements of the analysis of organic materials in archaeology and art. Viewed in its simplest form a mass spectrometer consists of: (i) sample inlet system, (ii) source of ions, (iii) mass measuring system, and (iv) means of ion detection, amplification and recording (Fig. 8.1). Mass spectrometers operate at reduced pressure (typically $<10^{-6}$ Torr) to enable detection of the small amounts (sub-nanogram to microgram) of material that are analyzed.

8.4.1. Sample Inlet Systems

8.4.1.1. Direct Insertion Probe

The direct insertion probe provides the simplest means of sample introduction but is used relatively infrequently in analyses of the chemically complex materials of interest to art and archaeology since mass spectra will represent a mixture components and hence be difficult to interpret. The direct insertion probe is, however, a convenient means of obtaining electron ionization mass spectra of pure compounds for reference purposes or for compounds that

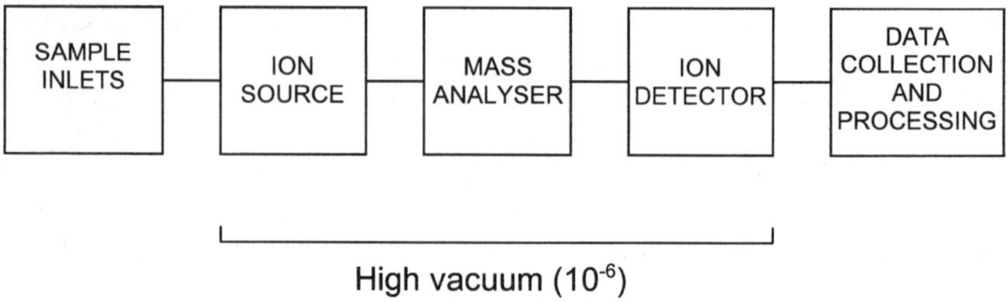

Figure 8.1. Components of a mass spectrometer.

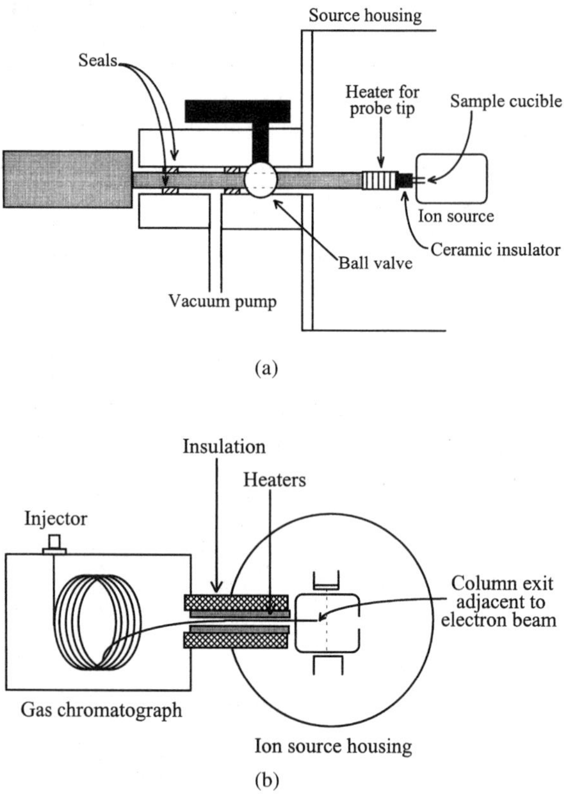

Figure 8.2. (*a*) Direct insertion probe and (*b*) GC/MS inlets.

can not be analyzed by GC/MS in laboratories that do not posses HPLC/MS instrumentation. In probe analyses the sample is added, usually as a solution in an organic solvent, to a small glass crucible mounted in the tip of a metal probe (Fig. 8.2*a*). After evaporation of the solvent the probe is introduced into the ion source of the mass spectrometer through a vacuum lock. Most low-molecular-weight apolar compounds, such as lipids, volatilize very readily at the low pressures encountered in the mass spectrometer under the influence of the heat of the ion source block. Less volatile substances may be volatilized by applying additional heat, using an integral heater in the probe tip. Some of the dedicated bench-top GC/MS instruments do not possess the facility for sample introduction by direct insertion probe. The production of gas-phase ions from an analyte is an essential part of any mass spectrometric analysis. Frequently, the heating of polar or thermally unstable materials using a direct insertion probe causes surface-catalyzed thermal decomposition on the glass crucible and derivatization, for example, acylation, alkylation, or silylation may be required to enhance volatility (Blau and Halket 1993).

8.4.1.2. Gas Chromatography/Mass Spectrometry

GC/MS is the most widely used sample introduction technique for the mass spectrometric analysis of organic materials in archaeology and art. GC and MS are especially complementary as both are gas-phase techniques, displaying optimum performance with sample sizes in the nanogram range. GC is used to separate mixtures of thermally stable volatile compounds or compounds that can be volatilized by the application of heat. Separations are achieved in GC through the partitioning of analytes between the carrier gas, typically helium, flowing through the GC column and the involatile polymeric stationary phase, commonly a crosslinked polysiloxane or polyether, coated onto the interior of the column. Flexible fused silica capillary columns are now preferred for GC/MS work as they can be fed directly from the GC oven into the ion source of the mass spectrometer (Fig. 8.2*b*). This is the optimum arrangement as it ensures efficient transfer of analytes from the GC to the mass spectrometer. The temperature of the transfer line between the GC and MS must be maintained at, or slightly above, the maximum temperature used in the GC analysis in order to avoid sample components condensing in this region. The carrier gas flows employed in capillary GC, ca. 2 ml min^{-1}, are readily accommodated by the vacuum pumping systems of modern mass spectrometers. The high flow rates employed with the increasingly popular wide bore (0.53 mm i.d.) capillary columns are incompatible with high vacuum pumping systems of mass spectrometers. In GC/MS samples are normally introduced as dilute solutions in organic solvents; specialized sample introduction devices are available that allow highly volatile/gaseous or insoluble polymeric samples to be introduced into the GC for on-line MS analysis. Examples of such applications are given below.

8.4.1.3. Liquid Chromatography/Mass Spectrometry (LC/MS)

The usefulness of LC/MS lies in its potential for the analysis of mixtures of components that are too involatile or thermally unstable to be analyzed by GC/MS. HPLC uses tubular metal columns typically 25, 15, or 10 cm in length by 4.6 or 2 mm internal diameter packed with a chromatographic adsorbant, most commonly silica gel or reversed-phase silica (octadecylsilylated silica) of 5- or 3-μm particle size. In normal phase HPLC the mobile phase is commonly a mixture of organic solvents, with more polar solvents, including methanol, acetonitrile, and water being used in reversed-phase analyses. Combining HPLC with mass spectrometry has proven much more difficult than GC. The fundamental problem that exists in any LC/MS system is the enormous volume of solvent vapor that is generated from the evaporation of the liquid eluent in the MS. The vacuum pumping systems of conventional MS instruments are simply unable to deal with the volume of vapor produced. The two common means of connecting HPLC to MS are the electrospray (ESP) and atmospheric pressure chemical ionization (APCI) interfaces. These will be discussed below in relation to ion sources since sample introduction and ionization are effectively combined into one unit.

8.4.2. Ion Sources

Electron ionization (EI) is the most widely used ionization techniques in the mass spectrometric analysis of organic materials in archaeology and art in combination with GC/MS. Other sources that are more widely used in the studies requiring HPLC/MS are electrospray and atmospheric pressure ionization. The following section describes the mechanism of operation of these three ionization techniques. Those readers requiring more detailed discussion of ionization techniques are directed to the specialist mass spectrometry literature [e.g., Chapman (1993)].

8.4.2.1. Electron Ionization

Electron ionization is produced by accelerating electrons from a hot filament through a potential difference, usually of 70 eV. Organic molecules present in this electron beam, for example, as a result of evaporation from a direct insertion probe or more commonly through elution from a GC column, will be ionized and fragmented. Extensive fragmentation occurs in the case of alkyl compounds, such as lipids. The initial product of EI in the ion source is a radical cation, resulting from the removal of an electron from the analyte molecule. If this singly charged species is stable, it will be the highest mass ion appearing in the mass spectrum for that substance. It is referred to as the molecular ion, denoted $M^{+\cdot}$, and it provides the molecular weight (measured

$$M + e \longrightarrow M^{+\cdot} + 2e \begin{cases} \nearrow A^{+} + R^{\cdot} \\ \searrow B^{+\cdot} + N \end{cases}$$

Scheme 8.1. Electron ionization (EI) induced by collision of an electron (*e*) with a molecule (M) to produce a molecular ion ($M^{+\cdot}$), which can relase internal energy through bond breaking to yield fragment ion, A^{+} (cation) or $B^{+\cdot}$ (radical cation), by loss of a radical ($R\cdot$) or a neutral molecule (N), respectively.

in atomic mass units, amu) of that substance. The electron beam usually possesses sufficient energy to induce fragmentation in organic compounds, by loss of radicals or neutral molecules (see Scheme 8.1). Although EI produces extensive fragmentation that is of use in structural investigations, the $M^{+\cdot}$ ion is frequently weak or absent from spectra. The relative abundance of $M^{+\cdot}$ may be enhanced by decreasing the ion source block temperature or electron beam potential (say to 20 eV).

8.4.2.2. Chemical Ionization

Chemical ionization (CI) is used to generate mass spectra by means of ionic reactions rather than electron bombardment. The technique is termed a "soft" ionization technique, as the spectra usually contain essentially only molecular weight information, fragmentation being either absent or much reduced compared to EI. Many variations of CI exist, but typically a reagent gas, for example, methane, ammonia, or isobutane, is introduced into the ion source at a pressure of around 1 Torr. Bombardment of the gas with electrons yields a population of ions and neutral molecules. The reagent gas ions will take part in ion–molecule reactions, with sample molecules vaporizing into the source. The most abundant reagent gas ions formed by the electron bombardment of methane are CH_5^{+} and $C_2H_5^{+}$.

As can be seen from the mass spectrum of methyl palmitate, shown in Figure 8.3, the most abundant ion arises by proton transfer and is termed the quasimolecular ion, $[M + H]^{+}$. The higher mass ions that are present in the spectrum correspond to molecular adduct ions, $[M + CH_5]^{+}$, $[M + C_2H_5]^{+}$, $[M + C_3H_5]^{+}$ and so forth, which arise by electrophilic addition of reagent gas ions to the sample molecules. The quasi-molecular ion and adduct ions are useful in inferring the molecular weights of sample molecules. The lack of fragment ions in CI spectra results from the low energy transfer involved in the CI process compared to EI. Furthermore, even-electron $[M + H]^{+}$ ions are energetically more stable than the radical $M^{+\cdot}$ ions produced by EI. Those readers requiring more detailed discussions of CI should refer to Chapman (1993) and Rose and Johnston (1982).

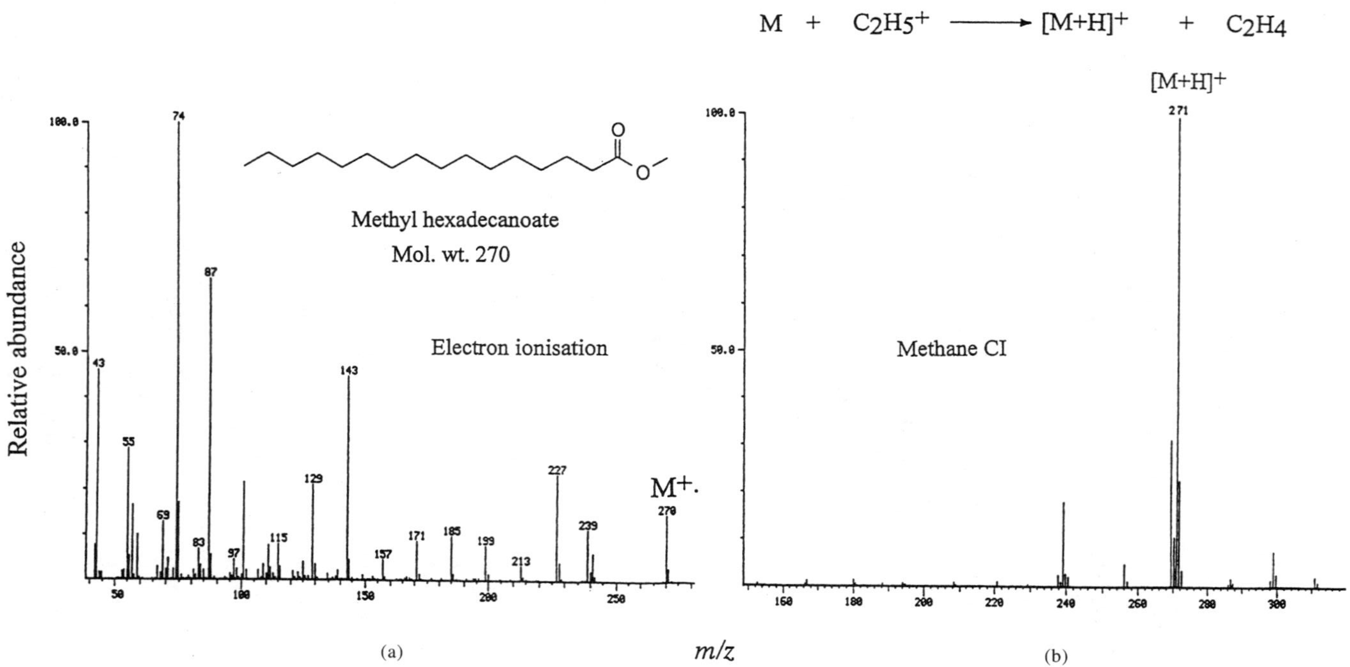

Figure 8.3. (a; 70 eV) Electron ionization and (b) methane chemical ionization mass spectra of methyl palmitate.

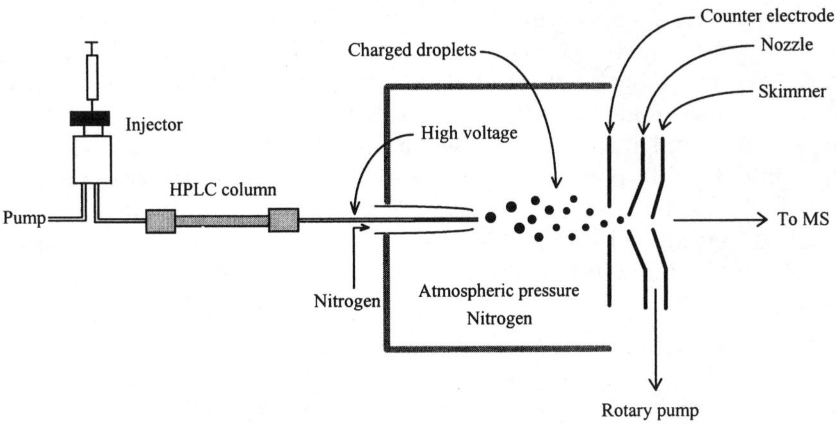

Figure 8.4. Electrospray ionization source used in HPLC/MS.

8.4.2.3. Electrospray Ionization

ESP ionization was developed by Fenn and co-workers (1989, 1990) and is characterized by its ability to generate singly and multiply charged ions from thermally labile substances. The construction of the ion source is shown in Figure 8.4. Analyses can be performed with or without the HPLC column. Samples are delivered as a dilute solution in a reasonably polar solvent such as acetonitrile/water (1:1 v/v) containing acid, for example, formic acid (1% v/v), to assist with formation of protonated pseudomolecular ions. As samples are pumped through the metal capillary, a high voltage applied to the capillary indices spraying at the tip and imparts a net charge to each droplet. As the solvent evaporates under the flow of warm nitrogen, the charge of the droplets increases and at some critical value ions are emitted from the droplet. A wide variety of compound types can be analyzed using this technique. The possibility of analyzing high-molecular-weight compounds arises through the formation of multiply charged ion, which means that their apparent mass falls within the mass range of even relatively low-performance mass spectrometers. ESP has become a very prominent technique in the field of biochemistry for the determination of molecular weights of proteins and polypeptides.

8.4.2.4. Atmospheric Pressure Chemical Ionization

The APCI ion source is rather similar in design to the ESP source. The major differences in their construction include the lack of a high voltage applied to the metal capillary and the inclusion of heaters to assist with nebulization (spraying) of the HPLC eluent and a corona discharge needle to facilitate solvent-assisted chemical ionization. As with ESP the APCI source can be used with or without an HPLC column depending on whether complex mixtures or pure substances are being studied. The APCI source has the advantage over ESP that it is readily able to accommodate the normal milliliter flow rates used in HPLC. Simple CI-type spectra are produced by a wide range of polar and apolar substances. APCI differs from ESP in that multiply charged ions are not observed. APCI is probably the closest to a universal ion source for HPLC/MS that is currently available.

8.4.3. Mass Analyzers

A number of types of analyzer are now available that will measure the mass of ions according to their mass-to-charge ratio, m/z. The instruments most commonly employed in the investigation of organic compounds in art and archaeology use quadrupole mass filters, magnetic sector, and ion trap mass analyzers. The following sections discuss the basic principles and capabilities of quadrupole, magnetic sector, and ion trap analyzers. Time-of-flight (TOF) analyzers are included for completeness although they have been used rarely in this field to date; they have considerable scope for the molecular weight profiling of synthetic polymers.

8.4.3.1. Quadrupole Analyzers

Quadrupole mass spectrometers (Fig. 8.5*a*) are constructed and function in entirely different ways to magnetic sector instruments. The quadrupole analyzer comprises four parallel rods of hyperbolic or circular cross section. A radio-frequency (RF) potential and direct current (DC) voltage is connected between opposite pairs of rods. Ions are injected into the oscillating electric fields by a small accelerating voltage (10 to 20 V) and, under the influence of the fields, begin to undergo complex oscillations. Mass separation is achieved by varying (scanning) the voltages on the quadrupole rods, while keeping the RF/DC ratio constant. At any one point in the scan only one mass can pass through the system.

Quadrupole instruments are very popular, owing to their relatively simpler operation compared to magnetic sector instruments. They are also less expensive than magnetic sector instruments and compatible with a wider range of inlet systems, including HPLC/MS. Although the mass range of quadrupole instruments is substantially less than that of many magnetic sector instruments, it is sufficient for the vast majority of analyses. A number of the properties of magnetic sector and quadrupole instruments are compared in Table 8.1.

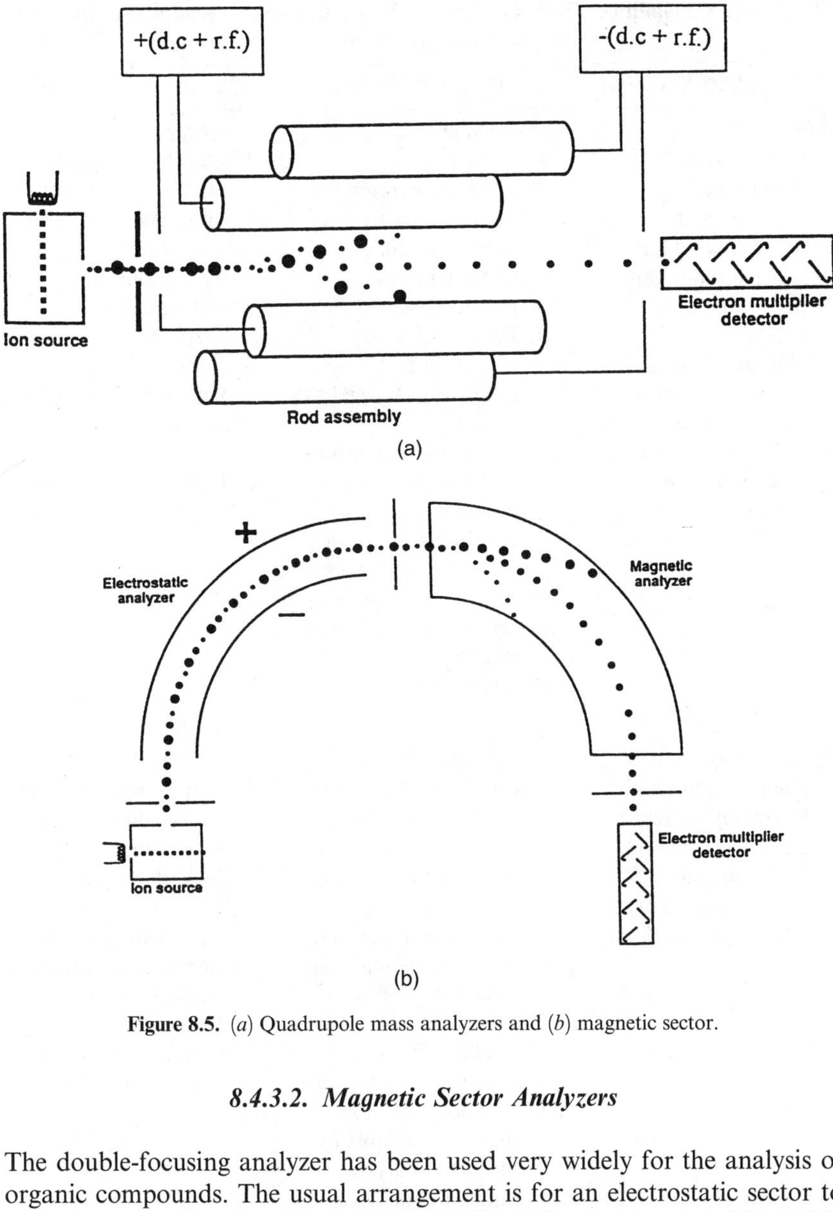

Figure 8.5. (*a*) Quadrupole mass analyzers and (*b*) magnetic sector.

8.4.3.2. Magnetic Sector Analyzers

The double-focusing analyzer has been used very widely for the analysis of organic compounds. The usual arrangement is for an electrostatic sector to precede a magnetic sector (conventional Nier–Johnson geometry; Fig. 8.5*b*). The electrostatic sector is an energy focusing device. Hence, ions produced and accelerated by the ion source are focused by the electrostatic sector according to their translational energies, irrespective of their mass-to-charge

Table 8.1. Capabilities of MS Instruments Magnetic Sector, Quadrupole, and TOF Mass Analysers

MAGNETIC SECTOR	QUADRUPOLE	TIME-OF-FLIGHT
Advantages	*Advantages*	*Advantages*
1. Greater versatility	1. Compact	1. Capable of very high-mass measurements, up to 500,000 daltons
2. Accurate mass measurement	2. Easy to operate	2. Compatible with pulsed ionization techniques, e.g., MALDI
3. More specific forms of selected ion monitoring	3. Source at earth potential makes interfacing less problematical	3. Very sensitive
4. Metastable ion analysis	4. Easy to look for positively and negatively charged ions	4. Simplest mass analyzer in common use
5. High-mass capabilities, i.e., 10,000 daltons at full sensitivity	5. Much less expensive than magnetic sector instruments	5. Very fast scanning
6. High collision energies used in MS–MS analyses	6. Readily incorporated into tandem instruments, i.e., triple quadrupoles and hybrids	6. Relatively inexpensive
	7. Relatively inexpensive	
	8. Ease of computer control	
Disadvantages	*Disadvantages*	*Disadvantages*
1. Highly sophisticated	1. Limited mass range, ca. 4000 daltons (greatly extended by use of electrospray due to formation of multply charged ions)	1. Only compatible with sources that produce pulses of ions
2. Relatively more expensive	2. No accurate mass capability; NB ca. 0.01% precision (1 in 10,000 daltons) in electrospray analyses	2. Difficult to perform combined chromatography/mass spectrometry, although GC/MS possible
3. High ion source potentials present interfacing problems	3. Only low collision energies available in MS–MS analyses	3. Accurate mass capability by use of reflectron ion mirror
	4. Limited data types compared with sector instruments	

ratio. The magnet on the other hand is a momentum analyzer, where ions are separated according to their mass-to-charge ratio. The combination of an electrostatic sector with a magnetic analyzer is highly versatile, providing the capability for operation at high resolution. The latter property enables accurate mass measurements to be performed, from which the elemental compositions of ions can be determined. Although double focusing magnetic sector instruments are well-suited to GC/MS work, they are more difficult to interface to HPLC, since the ion source operates at a high voltage (several kilovolts). The direct insertion probe technique discussed above is readily used in conjunction with magnetic sector instruments.

8.4.3.3. Ion Trap Analyzers

The three-dimensional quadrupole ion storage trap is a device comprising two end caps and a ring electrode. Ions are produced in an electron ionization source and are trapped within the quadrupole field by a combination of RF and DC potentials applied to the electrodes. Scanning of the mass range is achieved by varying the RF and DC voltages either together or singly. Each trapped ionic species (m/z) becomes unstable, that is, the ions develop trajectories that exceed the boundaries of the trapping field. The ions pass out of the field through perforations in the field-imposing ring structure and impinge on a detector such as an electron multiplier. The ion trap provides a sensitive, low-cost mass analyzer well-suited to GC/MS work. The mass range and resolution capabilities are comparable to those of a bench-top quadrupole GC/MS systems. CI can also be achieved within the ion trap using a facility for injecting ions from an external source (Chapman 1993).

8.4.3.4. Time-of-Flight Analyzers

Gas-phase ions are most commonly generated by the use of a laser to induce explosive evaporation of samples co-crystallized with a suitable matrix material; a technique termed matrix-assisted laser desorption ionization (MALDI). The ions formed are immediately accelerated then separated according to the time ions of different m/z take to reach the detector. Mass resolution is limited due to the time width of the ion packets (related to the kinetic energy spread of the ion packets) that reach the detector. Resolution is improved by use of a reflectron (ion mirror) to reduce the energy spread.

8.4.3.5. Scan Modes

No matter what type of mass spectrometer is used, when single-analyzer instruments are used two choices of mode of operation exist:

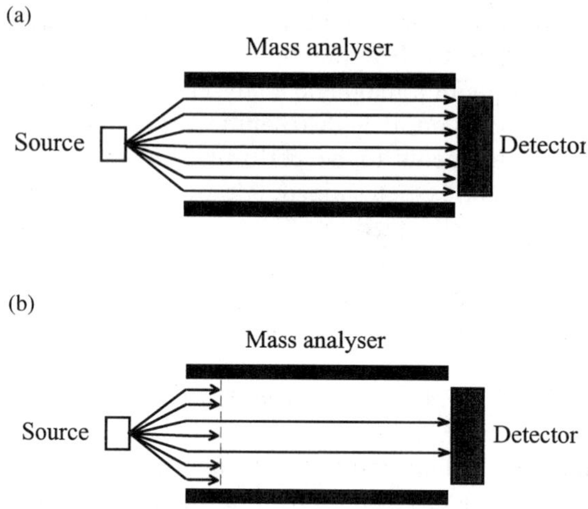

Figure 8.6. Schematic diagrams showing a mass spectrometer operating in (*a*) full scan and (*b*) selected ion monitoring modes.

Full Scan. This is the most conventional mode of operating a mass spectrometer and applies to any inlet or ionization technique and results in the familiar full-scan mass spectra (Fig. 8.6*a*). The major consideration is the m/z range to be "scanned." This will be restricted by the upper and lower mass ranges of the instrument. The required m/z range to be scanned should be chosen to encompass the highest mass ions likely to be seen in the compounds of interest. This will be dictated by the molecular weights of the compounds of interest taking into account increases in molecular weight arising through the addition of derivatizing groups. Once the appropriate parameters have been selected, the MS will be set to "scan" cyclically throughout an MS analysis such that the spectra will be recorded (and stored) at intervals, typically one second for a quadrupole scanning the m/z range 40 to 650. Full-scan mass spectra are used in structure identification and confirmation work, either by interpretation based on the known fragmentations of organic compounds or by search mass spectral databases ("libraries").

Selected Ion Monitoring (SIM). The SIM or multiple-ion detection mode ions are produced in the ion source in an exactly analogous manner as in the full-scan mode. In contrast the mass analyzer is set to pass only preselected ions characteristic of compounds of interest (Fig. 8.6*b*). This mode of operation enhances the selectivity of the analysis due to the exclusion of ions other than those of the preselected m/z values. A major advantage is the improvement in sensitivity that can be achieved since the instrument does not spend time scanning redundant regions of the spectrum. Picogram (10^{-12} g) detection limits are readily achieved. Selectivity may be further enhanced by using a magnetic sector mass spectrometer to monitor ions of specific elemental compositions. The SIM mode of operation is used in target compound analyses where interest is focused on a specific compound or class of compounds usually present as minor components of highly complex mixtures.

8.4.3.6. Tandem Mass Spectrometry (MS–MS)

Tandem MS or MS–MS is the linking of at least two sequential stages of mass analysis. Various combinations of the previously described mass analyzers are possible. The advantages of tandem MS is the ability to combine processes of separation and identification into a single instrument. The use of GC or HPLC essentially adds a further separation step with the result that very high selectivities and sensitivities can be achieved in complex mixture analysis. Triple quadrupole instruments have been most widely applied commonly to enhance structure information, confirm fragmentation pathways, and increase selectivity and sensitivity in trace analysis.

In practice precursor ions may derive from any ionization or inlet technique. Product ions generated from precursor ion selected in MS1 (*Q*1: Fig. 8.7*a*) by collisions with an inert gas, typically argon (Ar), in the collision cell (*q*; RF only quadrupole). Multiple collisions occur due to RF-induced oscillations of ions. The resulting fragmentations, more correctly termed collision-induced decompositions (CID), of the precursor ions yield product ion spectra by scanning the second mass analyzer (termed *Q*2 or MS2) in the conventional manner. The range of fragment ions may be limited by the relatively low collision energies (10- to 100-eV range) available in triple quadrupoles (or ion traps).

The major modes of operation of tandem mass spectrometers include (i) the recording of full product ion spectra of selected precursor ions (Fig. 8.7*a*), (ii) precursor ion scans reveal precursor ions of different m/z that fragment to yield a common product, (iii) neutral loss scans to reveal the loss of specific neutral fragments from precursor ions, and (iv) selected reaction monitoring (SRM) to enhance selectivity and sensitivity in trace analyses (Fig. 8.7*b*).

8.4.4. Ion Detection

The small current (typically in the range 10^{-9} to 10^{-17}A) generated by ions passing through the mass analyzer must be amplified and converted into a voltage that can be digitized or displayed on some form of recording device.

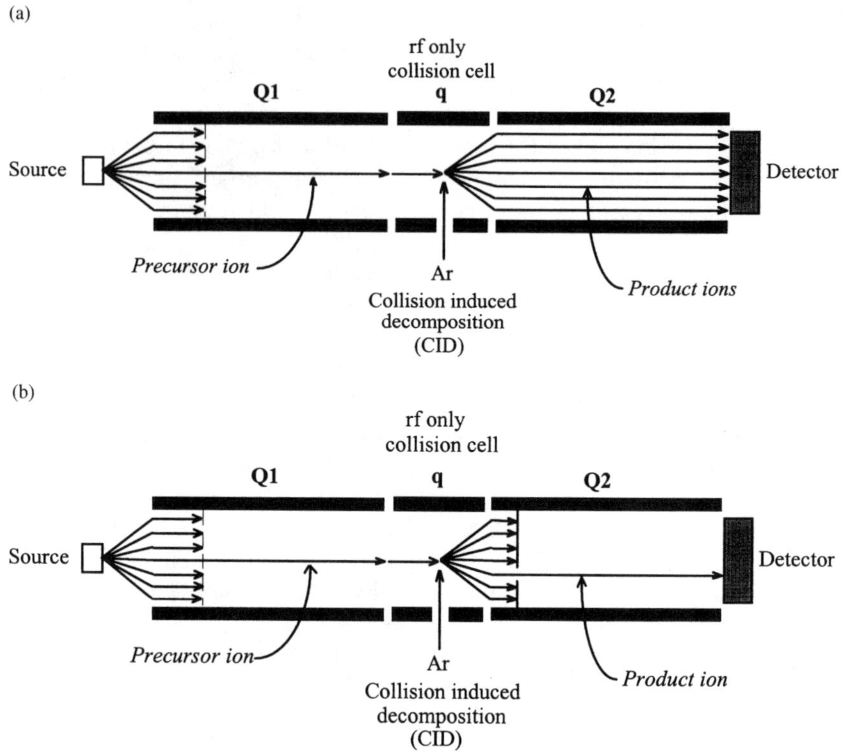

Figure 8.7. Schematic diagrams showing tandem mass spectrometers operating in (*a*) the product ion and (*b*) selected reaction monitoring modes.

Electron multipliers are widely used for this purpose. These devices comprise copper–beryllium dynodes that emit electrons when impacted by highly energetic ions, focused from the mass analyzer. Cascade emission of electrons through a series of dynodes produces gains of the order of 10^6. The final dynode of the electron multiplier is connected to a preamplifier to convert the output current to a voltage suitable for recording.

8.4.5. Mass Spectra

Figure 8.8 shows a mass spectrum in diagrammatic form. Although positive ions are most commonly studied, negative ion spectra are also used. In electron ionization (EI) the highest mass ions seen in the mass spectrum will usually be the *molecular ion*, $M^{+\cdot}$, which corresponds to the intact molecule that has lost an electron to form a cation-radical species. The ionization pro-

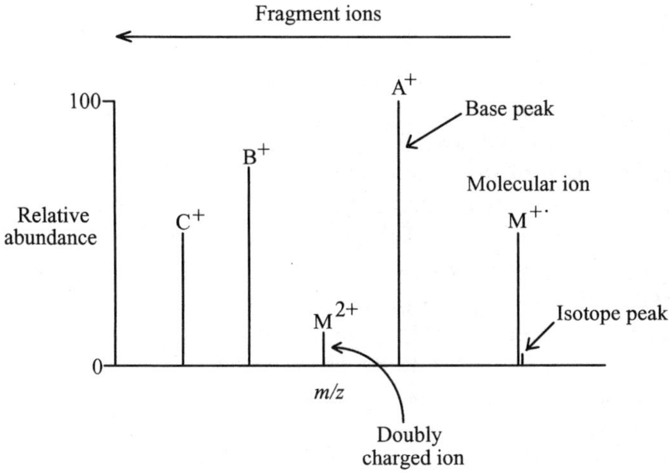

Figure 8.8. Schematic diagram of a mass spectrum.

cess will almost always impart sufficient vibrational energy into the molecule that dissociation occurs spontaneously to give lower mass ions, *fragment ions*. Neutral fragments (Scheme 8.1) are also produced, but these are not seen in the spectrum. The degree of fragmentation will depend upon: (i) the *structure* of the molecule under investigation and (ii) the *ionization technique* employed. Fragment ions may decompose further to yield still lower mass fragment ions. Hence, the mass spectrum represents the result of a series of competing and consecutive *unimolecular reactions*. The mass spectrum is a plot of the *relative abundance* of each ion (y axis) against mass-to-charge ratio (m/z; x axis). In the vast majority of cases only singly charged ions are produced, that is, $z = 1$, and since e is constant (the charge on an electron) m/z gives the mass of the ion in mass units. The production of multiply charged ions has become important in some of the newer ionization techniques, such as electrospray. For multiply charged ions z will have integer values $= 2, 3, 4, 5, 6, 7$, and so forth, and ions will appear at apparently lower masses on the mass scale, for example, for an M^{2+} ion of mass 1000, $z = 2$ and hence the M^{2+} ion will be seen at m/z 500. The most abundant ion in the mass spectrum is called the *base peak* and is set to 100% relative abundance.

8.4.6. Data Collection and Interpretation

Computers are essential to the efficient operation of modern mass spectrometers. The computer is usually fully integrated into modern instruments, controlling the various mass spectrometer, inlet, ionization, and scan functions. In addition, the computer provides an essential data collection and storage facility. The large volume of mass spectral data produced in GC/MS and HPLC/MS analyses of complex biological mixtures requires substantial postrun data processing. The interpretation of this large volume of GC/MS data is aided by automated searching of mass spectral databases held in the permanent memory of the mass spectrometer computer. A number of database libraries of EI mass spectra are available from proprietary sources. Most of these libraries contain substantial numbers of mass spectra of many of the compound commonly encountered in archaeology and art and their commonly used derivatives. Figure 8.9 shows part of the output from a commercially available GC/MS data processing package. In addition to showing the closest matching library mass spectrum, the library search also shows a range of statistical evaluations of the search. Although potentially very useful, the results of library searches are in no way definitive. The search results must be carefully assessed on the basis of the statistical evaluation of the match and consideration of GC elution orders, where mass spectra have been obtained by combined GC/MS analyses.

8.5. APPLICATIONS

The following sections provide examples of the application of modern mass spectrometry for the study of archaeological materials. The various examples have been selected to provide coverage in terms of the range of materials, compound classes, and mass spectrometric and ancillary techniques.

8.5.1. Natural Waxes

8.5.1.1. Plant Leaf Waxes

Abundant long-chain alkyl compounds known to derive from plant epicuticular waxes have been observed during GC/MS analyses of solvent extracts of archaeological pottery (Evershed et al. 1990a, 1991). One such example is shown in Figure 8.10 for the total lipid extract of a potsherd of medieval date. The major compounds in the total ion chromatogram (TIC) are acyl lipids derived from animal fats, which will be discussed further below. Three compounds seen eluting in the middle of the TIC were immediately recognizable from their EI mass spectra (shown in Fig. 8.11) as being of plant origin. The EI spectrum of the earliest eluting of these three components corresponds to *n*-nonacosane (C_{29} *n*-alkane) and is typical of the *n*-alkanes that occur widely in plant and animal waxes. The molecular weight is established from the molecular ion, $M^{+\cdot}$, at m/z 408. The remain-

```
25409 SPECTRA IN LIBRARYNB SEARCHED FOR MAXIMUM PURITY
  128 MATCHED AT LEAST 6 OF THE 16 LARGEST PEAKS IN THE UNKNOWN

RANK  IN  NAME
1   1380 HEXADECANOICACID, METHYLESTER
2   8272 PENTADECANOICACID, 14-METHYL-, METHYLESTER
3   1588 TETRADECANOICACID, METHYLESTER
4   8829 HENEICOSANOICACID, METHYLESTER
5   5497 UNDECANOICACID, METHYLESTER
```

RANK	FORMULA	M. WT	B. PK	PURITY	FIT	RFIT
1	C17. H34. O2	270	74	839	991	839
2	C17. H34. O2	270	74	816	947	854
3	C15. H30. O2	242	74	713	905	766
4	C22. H44. O2	340	74	701	942	737
5	C12. H24. O2	200	74	670	897	724

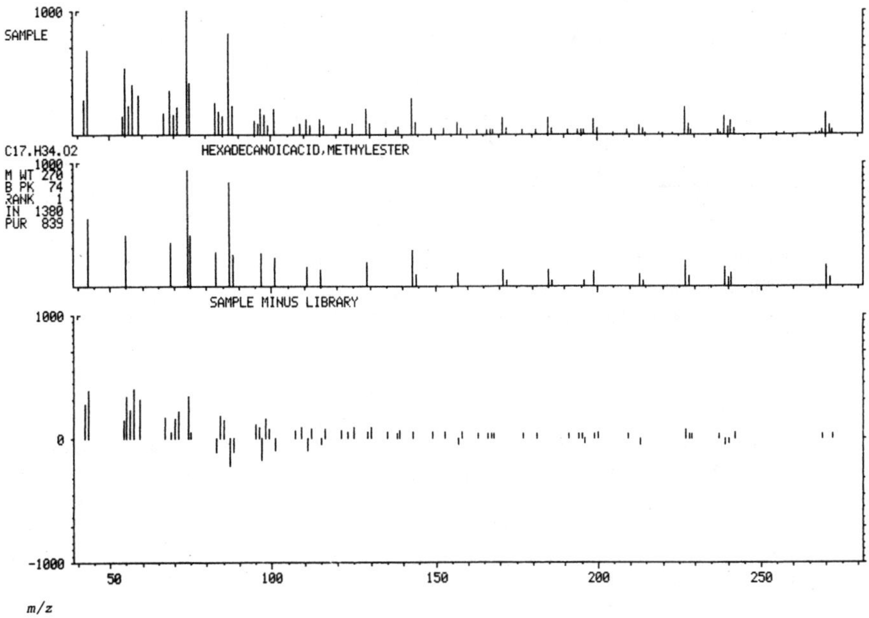

Figure 8.9. Part of the output from a commercially available library search package showing the best match of the analyte spectrum with the library database spectrum for methyl palmitate. (Reproduced with permission of Finnigan MAT, San Jose, CA).

der of the *n*-alkane spectrum is characterized by a series of ions of empirical formula C_nH_{2n+1}, hence, these are odd mass ions, separated by 14 mass unit increments that decrease in relative abundance with increasing mass. A 29-mass unit increment is always seen between the $M^{+\cdot}$ and the first C_nH_{2n+1} alkyl fragment. More complex alkane structures occur in some waxes and deducing the position of specific structural features in alkanes from EI mass spectra is difficult. The positions of double bonds are most easily deduced by reactions at the sites of unsaturation, as in the case of fatty acids (see below).

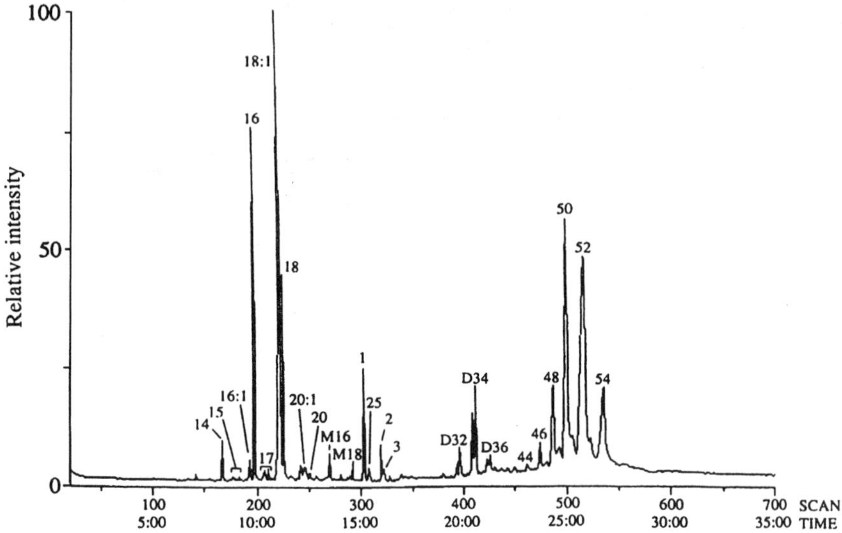

Figure 8.10. GC/MS total ion chromatogram of the total lipid extract of a potsherd recovered from the fill of a medieval ditch. Peak identities are: 14–18, 20, and 25 are saturated fatty acids (as TMS ester derivatives); 16:1, 18:1, and 20:1 are monounsaturated fatty acids (TMS esters); M16 and M18 are monacylglycerols (bis-TMS ethers); D32, D34, and D36 are diacylglycerols (TMS ethers); 44, 46, 48, 50, 52, and 54 are triacylglycerols; 1 is *n*-nonacosane, 2 is nonacosan-15-one, and 3 is nonacosan-15-ol (TMS ether). All the numbers for the acyl lipids refer to the number of acyl carbon atoms. Analyses were performed using a 12 m × 0.22 mm i.d. BP-1 coated (immobilized dimethyl polysiloxane, 0.1-μm film thickness) fused-silica capillary column. After on-column injection and a 2-min delay, the GC oven temerature was programmed to 5 to 350°C at 10°C min^{-1}. The GC/MS transfer line was maintained at 320°C. The GC/MS comprised a Pye 204 GC linked to a VG7070H double-focusing magnetic sector mass spectrometer. Spectra were recorded at 70 eV at an ion source temperature of 300°C. Data acquisition and processing employed a Finnigan INCOS 2300 data system (Evershed et al. 1990a, 1991).

The occurrence of chain branching, such as *iso*- or *anteiso*-, is best deduced from elution orders in GC/MS analyses. Prominent peaks at even mass in EI spectrum provide further indications of chain branching.

The second eluting long-chain alkyl component in Figure 8.10 is a secondary ketone. The mass spectrum shown in Figure 8.11 is typical of long-chain ketones and displays a weak molecular ion, m/z 422, that provides an indication of carbon number. The more abundant ions, appearing at m/z 225 and 241, arise by cleavage of carbon–carbon bonds α and β to the carbonyl group, and serve to identify unambiguously the position of the functional group (Budzikiewicz Djerassi et al. 1967). The ion at m/z 240 arises via the six-center McLafferty rearrangement, as shown in Scheme 8.2. Until recently mid-chain secondary ketones present in lipid extracts of arch-

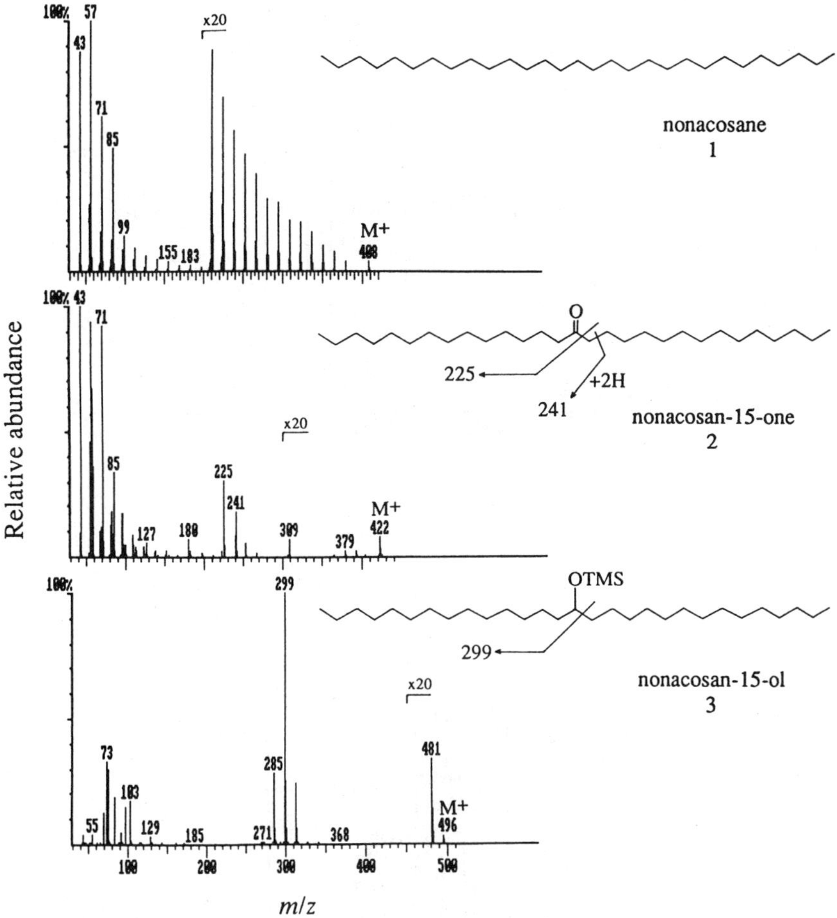

Figure 8.11. EI (70 eV) mass spectra of the components labeled 1, 2, and 3 in the TIC shown in Figure 8.10.

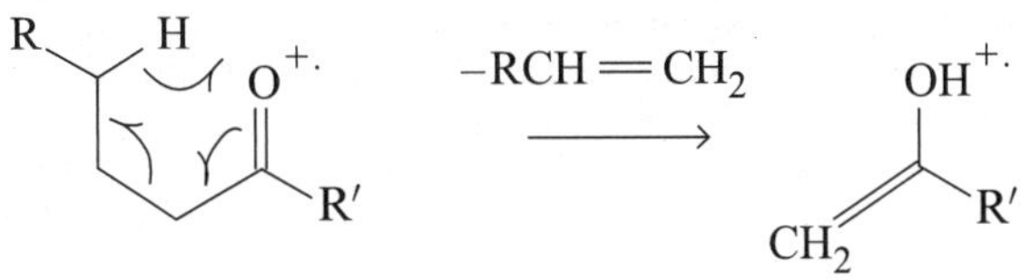

Scheme 8.2. Six-center hydrogen transfer reaction commonly referred to as the McLafferty rearrangement.

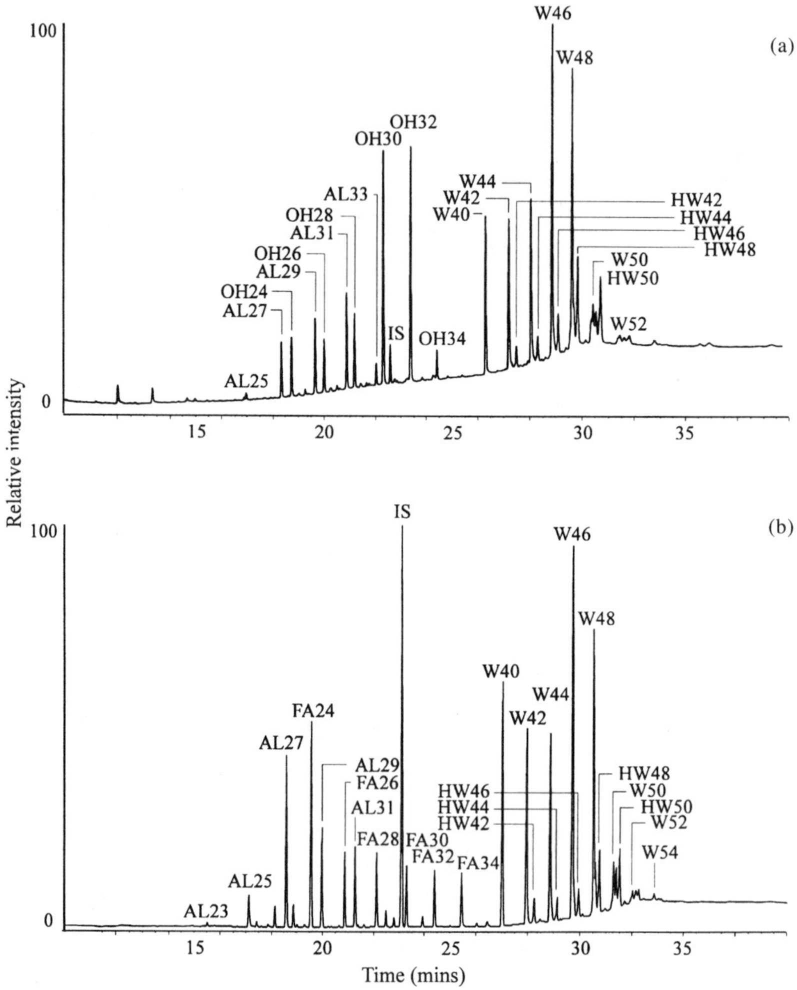

Figure 8.12. Partial GC profiles of the total lipid extracts of a Late Minoan conical cup from excavation in Crete (Evershed et al. 1997a) used as (*a*) a lamp and (*b*) 100-year-old beeswax from Crete. Components identified by high-temperature-GC/MS employing electron ionization (70 eV). Peak identities are: AL23–AL33, *n*-alkanes containing 23–33 carbons, respectively, AL23 = *n*-triacosane (C_{23}); AL25 = *n*-pentacosane (C_{25}); AL27 = *n*-heptacosane (C_{27}); AL29 = *n*-nonacosane (C_{29}); AL31 = *n*-hentriacontane (C_{31}); AL33 = *n*-tritriacontane (C_{33}); OH24–OH34, primary long-chain alcohols containing 24–34 carbons, respectively, OH24 = *n*-tetracosanol (C_{24}); OH26 = *n*-hexacosanol (C_{26}); OH28 = *n*-octacosanol (C_{28}); OH30 = *n*-triacontanol (C_{30}); OH32 = *n*-dotriacontanol (C_{32}); OH34 = *n*-tetratriacontanol (C_{34}); IS, internal standard, *n*-tetratriacontane (C_{34}); W40–W54, wax esters containing 40–54 carbons, respectively, W40 = tetracosanyl palmitate (C_{40}); W42 = hexacosanyl palmitate (C_{42}); W44 = octacosanyl palmitate (C_{44}); W46 = triacontanyl palmitate (C_{46}); W48 = dotriacontanyl palmitate (C_{48}); W50 =

aoelogical pottery were thought to derive solely from plants. However, it has recently been established that mixtures of plant-like ketones can arise through the pyrolysis of fatty acids or triacylglycerols at temperatures >300°C in the presence of fired clay. Structure analysis by GC/MS was pivotal in establishing the origins of these mixtures of ketones (Evershed et al. 1995; Raven et al. 1997).

Eluting just after the nonacosan-15-one component is the TMS ether of the corresponding secondary alcohol (nonacosan-15-ol). The EI mass spectrum shown in Figure 8.11 establishes the structure unambiguously. The molecular weight is confirmed by the weak $M^{+\cdot}$ ion at m/z 496 accompanied by a somewhat more intense $[M - 15]^{+}$ at m/z 481. The prominent fragment ion at m/z 299 is due to α-cleavage in this symmetrical secondary alcohol, thus allowing precise assignment of the position of hydroxyl group in the alkyl chain (see Fig. 8.11). Occurrence of these three long-chain alkyl compounds with an exactly analogous "fingerprint" distribution to that of the leaf wax of modern *Brassica olearacea* (cabbage) established the use of the vessel in the processing of this commodity and provided a reliable means of determining the exploitation of this vegetable in antiquity (Evershed et al. 1991, 1992, 1994; Charters et al. 1997).

8.5.1.2. Beeswax

Beeswax is frequently encountered in investigations of materials in art and archaeology, for example, organic residues in potsherds, mummies, and the like. The GC profile of fresh beeswax is shown in Figure 8.12 compared with the lipid extract of a ceramic lamp of Minoan date (Evershed et al. 1997a). The identification of beeswax is based on the presence of a characteristic mixture of long-chain alkyl compounds including wax esters, long-chain alcohols, and *n*-alkanes. GC/MS has provided mass spectra that allow confirmation of the diagnostic components. The EI mass spectrum of the major

Figure 8.12 (cont.)
tetratriacontanyl palmitate (C_{50}); W52 = hexatriacontanyl palmitate (C_{52}); W54 = octacontanyl palmitate; HW42–HW50, hydroxy–fatty acid–wax esters, containing 42–50 carbons, respectively, HW42 = hexacosanyl hydroxy-palmitate (C_{42}); HW44 = octacosanyl hydroxypalmitate (C_{44}); HW46 = triacontanyl hydroxy-palmitate (C_{46}); HW48 = dotriacontanyl hydroxy-palmitate (C_{48}); HW50 = tetratriacontanyl hydroxy-palmitate (C_{50}), and HW52 = hexatriacontanyl hydroxy-palmitate (C_{52}); FA16–FA34 correspond to saturated fatty acids bearing 16–34 carbon atoms, respectively. The peak labeled AL31 in (*b*) is a mixture of the C_{31} *n*-alkane, and the corresponding monounsaturated alkene. The close similarity between the distribution of components in (*a*) and (*b*) provide clear evidence for the presence of beeswax in the ancient conical cup.

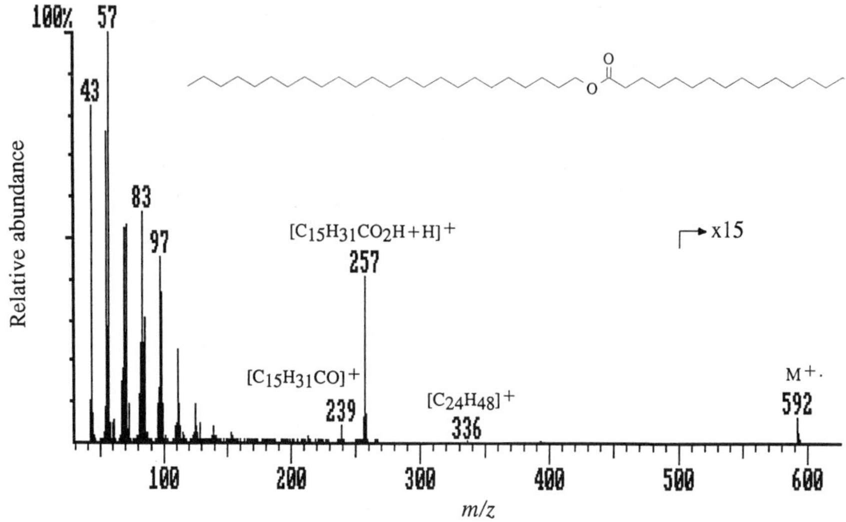

Figure 8.13. Electron ionization (70 eV) mass spectrum of tetracosanyl palmitate.

wax ester is shown in Figure 8.13. Although a molecular ion is seen (*m/z* 592) in this example, they are not always detectable in the mass spectra of wax esters obtained by GC/MS. Where an $M^{+\cdot}$ ion is seen, it can be used to determine the carbon number of the wax esters. The molecular weight of the wax ester whose mass spectrum is shown in Figure 8.13 is 592 mass units, hence, giving a molecular formula of $C_{40}H_{80}O_2$. The peak at *m/z* 257 corresponds to the $[C_{15}H_{31}CO_2H + H]^+$ ion, produced by cleavage of the alkyl oxygen bond, and provides information concerning the carbon number of the fatty acyl moiety, which is almost exclusively palmitic acid ($C_{16:0}$). This is confirmed through the presence of an intense fragment ion accompanied by an acylium ion at *m/z* 239 of empirical formula $[C_nH_{2n+1}C{=}O]^+$, which further corroborates the nature of the fatty acyl moiety. The only fragment ion derived directly from the fatty alcohol moiety is a very weak, even mass alkyl fragment ion, $[C_nH_{2n}]^+$, at *m/z* 336, which arises by fission of the alkyl oxygen bond with concomitant hydrogen transfer. Further indication of the nature of the fatty alcohol moiety can be deduced from the mass difference between the $M^{+\cdot}$ and $[RCO_2H + H]^+$ ions. In practice, however, the low abundance of the $M^{+\cdot}$ ion limits the reliability of this approach. The minor ω-hydroxy palmitic acid wax esters are usually determined by GC or GC/MS as their TMS derivatives. Weak $M^{+\cdot}$ ions are seen with fragmentations being analogous to the corresponding wax esters but with additional fragmentations being related to the terminal TMSO group, for example,

$[M - CH_3]^+$, $[M - TMSOH]^+$, $[TMSO(CH_2)_nCO_2H - TMSOH - H]^+$, and so forth.

When diagnostic ions concerning the nature of the alcohol moiety are very weak or undetectable, then the carbon number can be determined by comparing the GC retention times of wax ester peaks in a total ion chromatogram with those of compounds of known structure. The carbon numbers of the homologous wax esters found in beeswax are well established and can be used for this purpose. Alternatively, CI can be used to enhance the molecular weight information content of mass spectra through the production of diagnostic quasi-molecular ions (Wakeham and Frew 1982). The methane CI spectra of wax esters contain a variety of quasi-molecular ion species, for example, $[M + H]^+$, $[M + C_2H_5]^+$, $[M + C_3H_5]^+$, and $[M - H]^+$, the latter arising by hydride abstraction; the latter $[M - 1]^+$ ion predominates in saturated wax esters, whereas monounsaturated esters display $[M - 1]^+$ and $[M + 1]^+$ ions in approximately equal abundance. CI spectra are affected by variations in a wide range of source parameters. Hence, as a general rule, interpretations are best made by comparing the mass spectra of authentic and unknown compounds recorded under the same experimental conditions.

While free long-chain primary alcohols are insignificant in fresh beeswax, they can become major constituents of ancient beeswax as a result of diagenetic hydrolysis of the wax esters. The mass spectrum of their TMS derivative shows a weak $M^{+\cdot}$ ion. The very abundant $[M - 15]^+$ ion can be used to assign the carbon number and degree of unsaturation of the alcohol. The ions at lower mass, for example, m/z 103, are useful in distinguishing between the TMS ethers of long-chain alcohols and TMS esters of fatty acids. Note that the $[M - 15]^+$ ion of the TMS ester of a fatty acid (e.g., palmitic acid) has the same nominal mass (isobaric) as the TMS ether of the next highest carbon number fatty alcohol (*n*-heptadecanol). The mass spectra of long-chain alcohols give little other additional structural information.

Odd carbon numbered *n*-alkanes are seen in fresh and aged beeswax. The mass spectra of *n*-alkanes have already been discussed above in relation to their occurrence in plant waxes. Free fatty acids are also seen in reasonable abundance in fresh beeswax and ancient beeswax with their identities being confirmed by GC/MS as either their TMS or methyl esters derivatives, which are discussed in detail below in relation studies of animal fats and plant oils.

8.5.2. Fats and Oils

Degraded animal fats and plant oils are encountered very widely in art and archaeology as a result of the potential range of uses in antiquity. Their detection is relatively straightforward. Early work simply saponified total lipid extracts, methylated the resulting fatty acids to give fatty acid methyl esters (FAMEs), and assessed their origin by GC profiling. More recent work has shown that this approach is oversimplistic potentially limiting the information that can be obtained from lipid extracts.

8.5.2.1. High-Temperature GC/MS (HT-GC/MS)

Onc approach that has proved very effective has been to undertake GC and GC/MS of total lipid extracts using capillary columns coated with thin films of high-temperature stable stationary phases such as crosslinked dimethyl polysiloxanes (Evershed et al. 1990a). An example of an HT-GC/MS analysis of the solvent extract of an ancient potsherd is shown in Figure 8.10. The mixture comprises predominantly triacylglycerols, diacylglycerols, monoacylglycerols, and free fatty acids. The advantage of the approach lies in its simplicity and the breadth of information obtained. The solvent extract, or an aliquot thereof, is treated directly with excess *N,O*-bis(trimethylsilyl)-trifluoroaceamide/trimethylchlorosilane (99:1 v/v) for 10 min at 70°C. The derivatized extract is diluted with hexane and injected directly in the GC/MS. Fatty acids are converted to TMS esters, and mono- and diacylglycerols to mono-TMS and bis-TMS ethers, respectively (Note that since triacylglycerols lack free protic groups, they are not trimethylsilylated). HT-GC/MS using EI allows detailed confirmation of the structures of all the major acyl lipids.

Fatty Acids. The major compounds present in the chromatogram shown in Figure 8.10 are the TMS esters of the fatty acids eluting at around 10 min. The EI (70 eV) spectrum of the TMS derivative of palmitic acid shows an $M^{+\cdot}$ ion (m/z 328) with a much more prominent peak due to the $[M - CH_3]^+$ ion (m/z 313). This latter fragment ion is present in the mass spectra of the TMS derivatives of all saturated and unsaturated fatty acids (although being of somewhat reduced relative abundance in the latter) and is used to assign carbon number and the degree of unsaturation in the case of unknowns. The ions at m/z 73 and 75 corresponding to $[(CH_3)_3Si]^+$ and $[(CH_3)_2SiOH]^+$, respectively, are very common in the mass spectra of TMS derivatives. The fragment ions at m/z 117, 129, 132, and 145 are also prominent in the spectra of saturated and monounsaturated fatty acid TMS esters and help to distingiush them from the TMS ether derivatives of fatty alcohols (see above). The m/z 132 and 145 ions are related to the m/z 74 and 87 McLafferty rearrangement ions seen in FAME spectra (see below). Also in common with methyl esters, the fragmentation of the TMS derivatives becomes less distinctive with increasing unsaturation of the alkyl chain. Polyunsatured fatty acids, although common in fresh plant oils and animal fats, are rarely seen in aged materials due to oxidative degradation or polymerization (Regert et al. 1998).

Triacylglycerols. The next most abundant components seen in the TIC of the total lipid extract shown in Figure 8.10 are the triacylglycerols (TAG) eluting at long retention time. The EI spectra of TAG can be used to determine the nature of the fatty acyl moieties present in the individual TAG molecular species. The mass spectra of naturally occurring TAGs are complex due to the large number of fatty acids often present. Hence, any triacylglyceride analysis will only provide partial compositional information. The crosslinked dimethyl polysiloxane coated capillary GC columns are only capable of carbon number resolution. More polar stationary phases are available that can resolve triglyceride species according to their differing degrees of unsaturation (Evershed 1996). The EI mass spectrum of tristearin is shown in Figure 8.14. High mass ions, $M^{+\cdot}$ and $[M - 18]^{+\cdot}$, are seen in the EI spectra of TAGs, but as these are of low abundance, they are often difficult to detect above column bleed in GC/MS analyses. The lower mass ions are of greater value in analytical work. An abundant acylium ion, $[RCO]^+$, is present at m/z 267 ($[CH_3(CH_2)_{16}CO]^+$) in the mass spectrum of tristearin (Fig. 8.14). Where other acyl moieties are present, the masses of their acylium ions can be used to deduce their carbon number and degree of un-

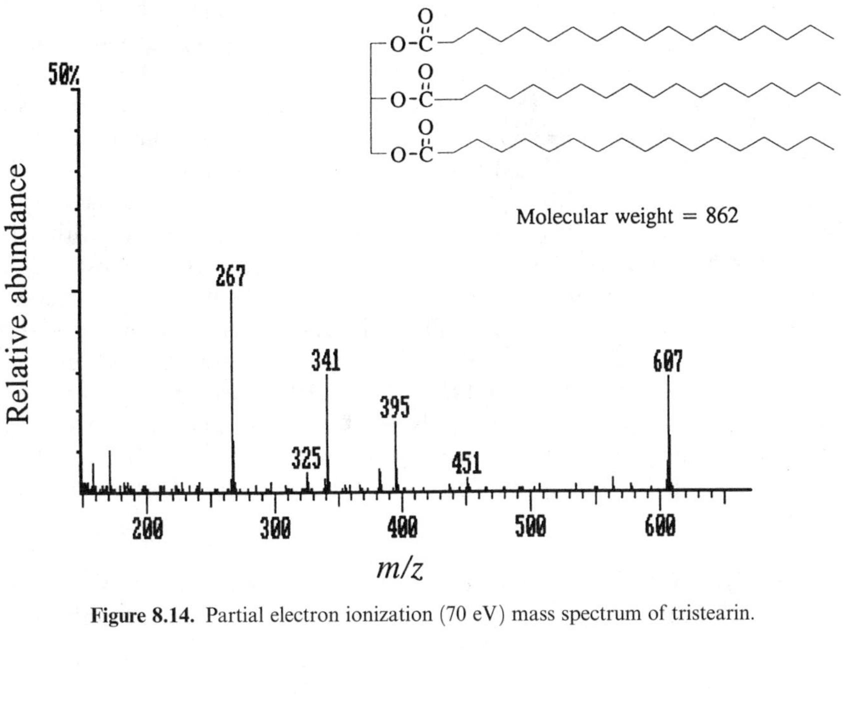

Figure 8.14. Partial electron ionization (70 eV) mass spectrum of tristearin.

saturation. It is generally accepted that the presence of unsaturation in the acyl moiety gives fragment ions of *m/z* [RCO – 1], namely a ketene structure, R – CH=C=O$^+$, of greater abundance than the acylium ions described above. However, this is not always the case as the mass spectra of triolein and trilinolein recorded by high-temperature GC/MS, under the conditions used for the analysis shown in Figure 8.14, have been found to yield spectra containing prominent acylium ions at *m/z* 265 and 263, respectively (Evershed 1996). However, as stated above, these fatty acids are uncommon in aged fats and oils. Abundant fragment ions are seen at higher mass corresponding to the loss of an acyloxy group from the molecular ion, $[M - RCO_2]^+$. Thus, the fragment ion seen at *m/z* 607, $[M - 283]^+$, is valuable in confirming the deductions made concerning the nature of the acyl moieties, based on the lower mass acylium ion appearing at *m/z* 267. A range of other fragment ions are seen in the mass spectra of triacylglycerols, which can be used to deduce the structure of unknowns present in complex biological mixtures. Where structure interpretations by EI are uncertain, positive-ion and negative-ion CI mass spectra provide useful alternatives (Murata 1977; Wakeham and Frew 1982; Evershed et al. 1990b; Evershed 1996).

Diacylglycerols. The EI mass spectra of the TMS derivatives (Curstedt 1974) of 1,2-diacylglycerols do not generally contain $M^{+\cdot}$ ions. However, high-mass ions corresponding to the loss of a methyl group, $[M - 15]^+$, and trimethylsiloxy group, $[M - 90]^+$, are useful in assigning the carbon number and degree of unsaturation of the intact diacylglycerols. As these ions are often rather weak in GC/MS analyses, carbon number is more reliably determined by retention time comparisons with authentic compounds.

The nature of the individual acyl moieties can be determined by considering the ions resulting from the loss of an acyloxy moiety, $[M - RCO_2]^+$ or $[M - RCO_2H]^+$, when the acyl moiety is unsaturated. As with triacylglycerols, the more abundant acylium ion, $[RCO]^+$, is the most useful fragment ion for determining the nature of the individual acyl moieties. Assignment of the nature of the acyl moieties can be corroborated using the $[RCO + 74]^+$ and $[RCO + 90]^+$ ions. Low-mass ions are observed at *m/z* 129, $[CH_2CHCHOTMS]^+$, and 145, $[CH(O)CHCH_2OTMS]^+$, which contain the TMS group and part of the glycerol backbone. The EI mass spectra of the TMS ethers of 1,3-diacylglycerols contain a major $[M - RCO_2CH_2]^+$ ion that does not appear in the spectra of 1,2-diacylglycerols (Barber et al. 1968; Horning et al. 1969).

Monoacylglycerols. Both acetate and TMS ether derivatives can be used in the GC/MS analysis of monoacylglycerols (Johnson and Holman 1966). The spectra of the TMS ethers are more useful, often displaying weak $M^{+}\cdot$ and $[M - CH_3]^+$ ions from which the carbon number and degree of unsaturation can be deduced. An $M^{+}\cdot$ ion is usually observed when the acyl moiety is unsaturated. Where these are not observed, this information can be gleaned from the abundant $[M - 90]^+$ (loss of CH_2OTMS) ion at lower mass. The appearance of a characteristic fragment ions at m/z 218 in the EI mass spectra of the bis-TMS of 2-monoacylglycerols and $[M - 103]^+$ in the spectra of 1-monoacylglycerols allows isomeric species to be reliably distinguished.

8.5.2.2. Structure Analysis of Fatty Acids by GC/MS

Fatty Acid Methyl Esters (FAMEs). Fatty acids are most commonly analyzed by GC/MS as their methyl ester derivatives (Blau and Halket 1993). Figure 8.3 shows the 70-eV EI spectrum of methyl palmitate. The spectrum is typical of that which would be obtained for the analysis of any fully saturated straight-chain FAME. The molecular ion ($M^{+}\cdot$) at m/z 270 in the case of methyl palmitate is generally present in FAME spectra. The mass of the $M^{+}\cdot$ ion confirms the carbon number and degree of unsaturation of the fatty acid in question. The molecular weights of other fatty acid methyl esters can be calculated by adding or subtracting multiples of 14 amu (the mass of one methylene group, CH_2) in order to establish the molecular weight of higher and lower homologues of palmitic acid (16:0). The presence of double bonds (or cyclopropane rings) are indicated by a reduction in molecular weight of 2 amu. Hence, the $M^{+}\cdot$ ions of methyl stearate (18:0), methyl oleate (18:1), methyl linoleate (18:2), and methyl linolenate (18:3) appear at m/z 298, 296, 294, and 292, respectively, although the dienoic and trienoic fatty acids are rarely seen in aged materials. Other high-mass ions, occurring at M-31 and M-43 in the spectra of saturated fatty acid methyl esters, correspond to the loss of the methoxy (CH_3O–) and propyl ($CH_3(CH_2)_2$–) groups, respectively. The loss of the propyl group involves a complex intramolecular rearrangement. The base peak, m/z 74, in the spectrum of satuarated FAMEs (e.g., methyl palmitate; Fig. 8.3) arises through the six-center McLafferty rearrangement, which results in proton transfer and cleavage of the bond linking the C-2 and C-3 carbons (see Scheme 8.2). The relative abundance of the m/z 74 ion decreases with increasing unsaturation of the alkyl chain as a result of competitive fragmentations. The mass spectra of unsaturated FAMEs differ from their saturated counterparts by exhibiting an M-32 (loss of methanol) ion rather than M-31 (loss of a methoxy radical).

Geometric and Positional Isomer Determinations. The EI mass spectra of the methyl esters and TMS esters of unsaturated fatty acids cannot be used to deduce the position of unsaturation in fatty acids, owing to migration of the

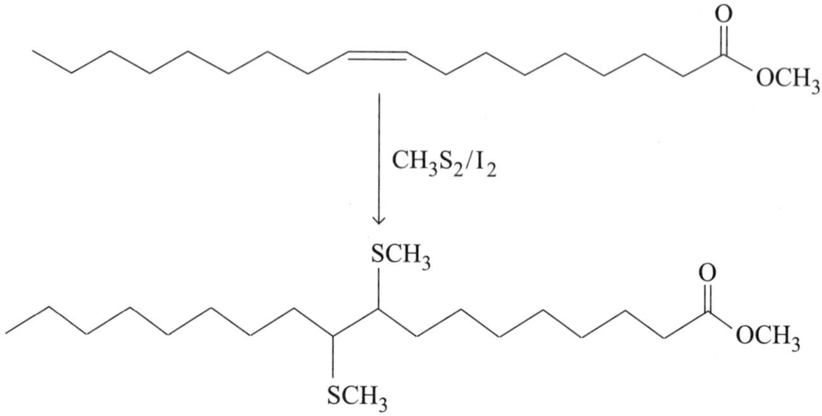

Scheme 8.3. Conversion of monounsaturated fatty acid methyl ester to DMDS derivative.

double bond(s) during the ionization process. Two approaches are commonly used to overcome this problem and locate double-bond positions. One approach is to derivatize the carboxyl moiety to stabilize the positive charge and prevent double-bond migration. The most convenient approach that has been used in studies of archaeological fats involves chemical derivatization at the double bond to produce EI mass spectra from GC/MS analyses that contain fragment ions diagnostic of the double-bond position. The preferred approach of several methods that are available involves the preparation of dimethyl disulfide (DMDS) adducts (Scheme 8.3). Although first used analytically for the location of double bonds in linear alkenes (Francis and Veland 1981), the technique is equally well applied to fatty acid methyl esters (Francis 1981).

The EI mass spectrum of the DMDS adduct of methyl oleate is shown in Figure 8.15. The molecular ion, which is visible at *m/z* 390, establishes the carbon number of the original fatty acid. The abundant ions at *m/z* 173 and 217 locate the position of the double bond unambiguously. As with the bis-TMS derivatives of the diols, the DMDS derivatives can be applied very reliably to the determination of the positions of double bonds in mixtures of positional isomers of monounsaturated fatty acids. Table 8.2 gives the diagnostic fragment ions arising from the DMDS derivatives of a range of important octadecenoic acid positional isomers. As the addition of the derivatizing agent to the double bond is stereospecific *E* and *Z* isomers of fatty acids can be discriminated in GC/MS analyses (Scheme 8.4) Although the mass spectra of the DMDS derivatives are congruent, the *erythro-* isomer (produced from the *E*-fatty acid) elutes after the *threo-* isomer (produced

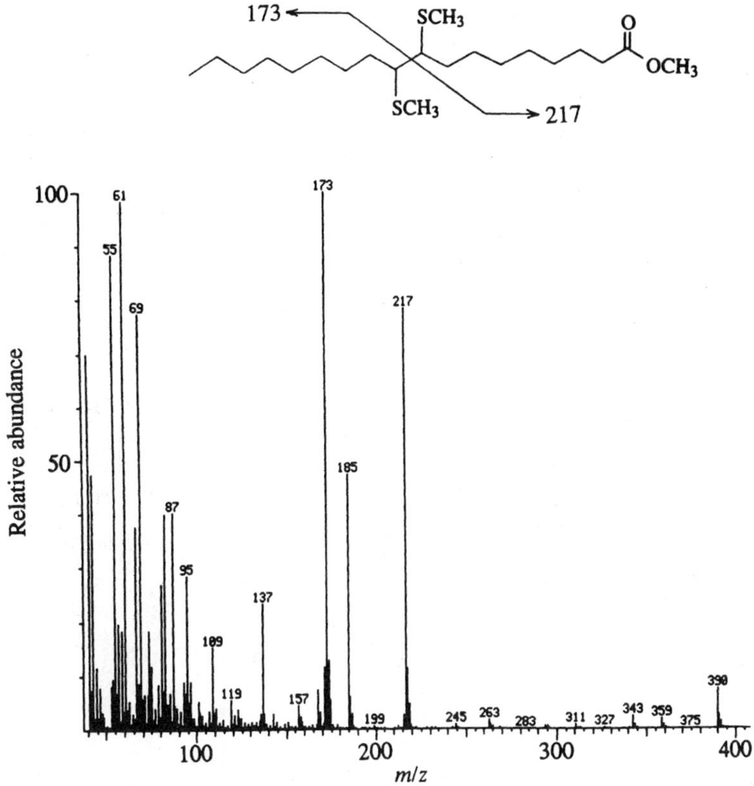

Figure 8.15. Electon ionization mass spectrum of the dimethyldisulfide adduct of methyl oleate (*Z*-9-octadecenoate).

Table 8.2. Diagnostic Fragment Ions Arising from Electron Ionization Mass Spectra of DMDS Derivatives of Octadecenoic Acid Positional Isomers

Positional Isomer	Fragment Ions (m/z)
Δ^9	173, 217
Δ^{10}	159, 231
Δ^{11}	145, 245
Δ^{12}	131, 259
Δ^{13}	117, 273
Δ^{14}	103, 287
Δ^{15}	89, 301
Δ^{16}	75, 315

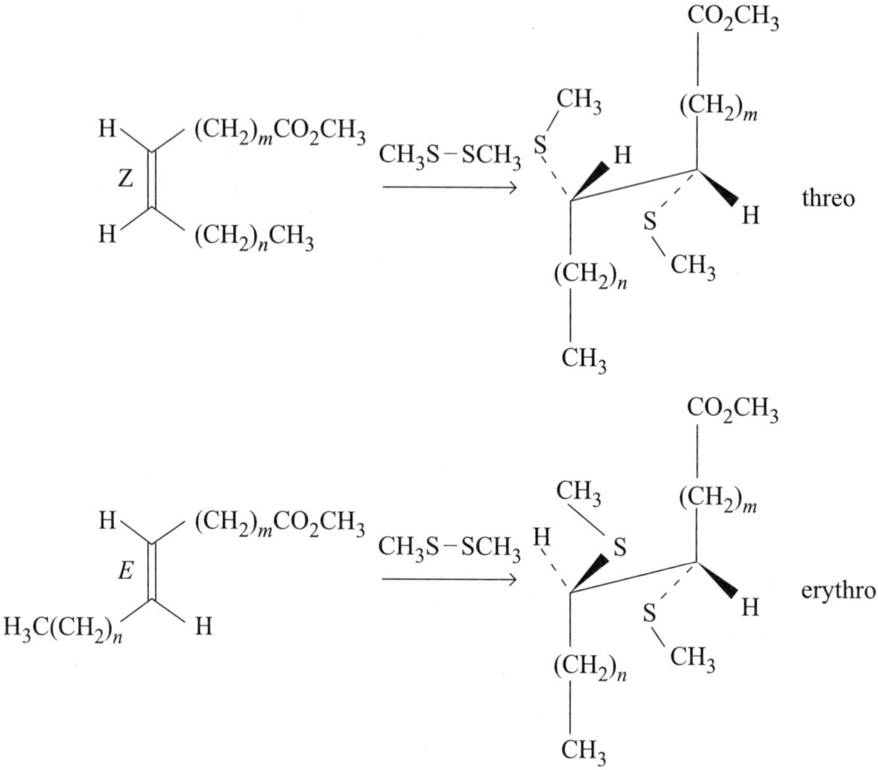

Scheme 8.4. Structures of diastereoisomeric *threo* and *erythro* forms of DMDS derivatives of monounsaturated fatty acid methyl esters.

from the *Z*-fatty acid) on an apolar dimethyl polysiloxane (OV-1 type) coated capillary column (Nichols et al. 1986; Vincenti et al. 1987).

An example of the application of this derivative in archaeology is shown in Figure 8.16 (Mottram et al. 1999). The analysis of the DMDS derivatives of the monounsaturated fatty acids present in solvent extracts of potsherds sampled from a small assemblage of medieval lamps and dripping dishes showed a complex mixture of positional isomers in the former, namely 18:1 $\Delta^{9,10,11,13,14,15,16}$. Similar mixtures of isomers are found in the fats of ruminant animals such as sheep and cattle due to biohydrogenation of unsaturated dietary fats in the rumen (Enser 1991). Such mixtures of monounsaturated fatty acids are not found in the fats of monogastric animals such as pigs, where a single isomer, 18:1 Δ^9, is observed. The presence of this isomer alone in the dripping dishes implied a monogastric or non-ruminant animal origin. The geometry of the positional isomers of the octadecenoic acid was

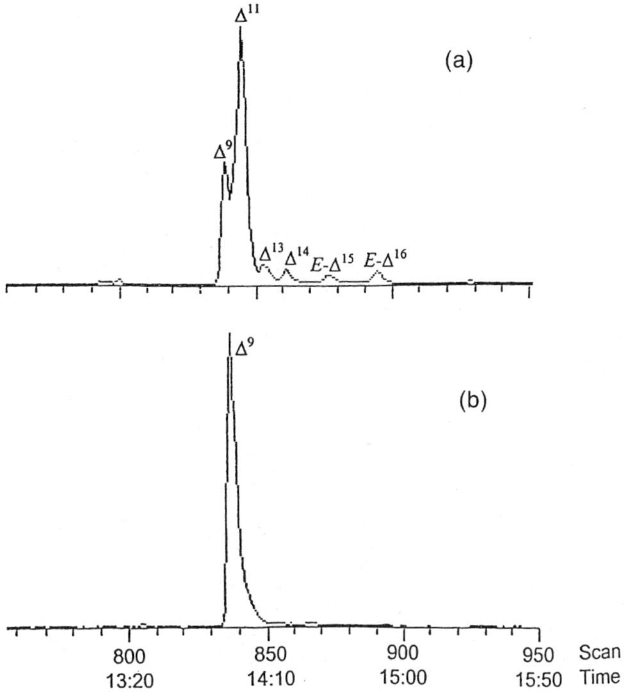

Figure 8.16. Partial ion chromatogram from the GC/MS analysis (70 eV) of the dimethyldisulfide derivatives of octadecenoic acid obtained from (*a*) a medieval lamp and (*b*) dripping dish. The results confirm the presence of ruminant tallow in the lamp based on the complex mixture of positional isomers while the dripping dish contained only nonruminant fat (from Mottram et al. 1999).

deduced from the elution orders of the *erythro-* and *threo*-forms. These results point species-specific functional interpretations for the use of the vessels (Evershed et al. 1997c, Mottram et al. 1999).

8.5.2.3. Polymerized Fats and Oils

One of the major processes of transformation of fats and oils is via the "drying" reaction to produce a semisolid polymer. While the reaction will proceed most rapidly in the case of oils and fats containing high contents of polyunsaturated fatty acids, it is anticipated that most fats will become polymerized during long-term exposure. Polymerized fats and oils are encountered in many situations in art and archaeology (Mills and White 1994). Recent work has provided an example of polymerized and oxidized fats and/or oils in archaeological pottery (Regert et al. 1998). Solvent extracts of

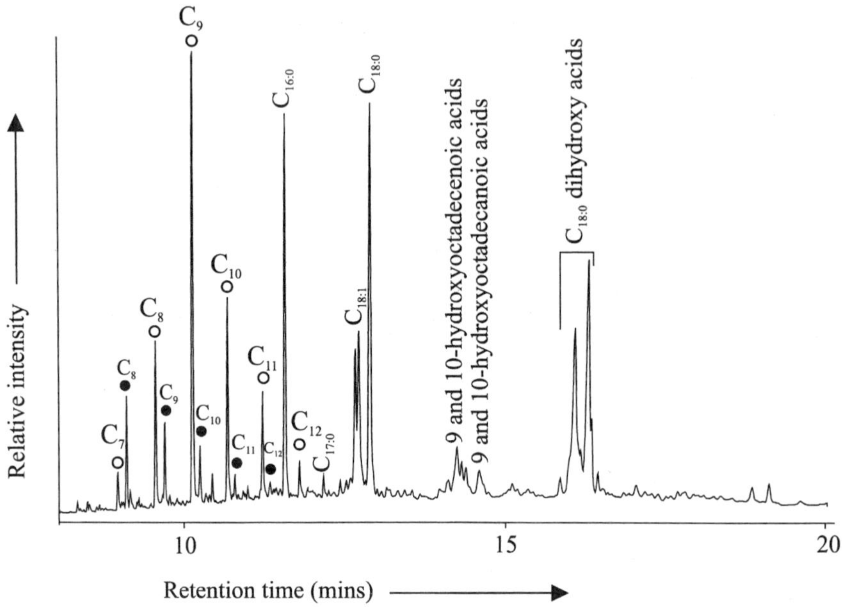

Figure 8.17. Electron ionization (70 eV) mass spectrum of methyl 9,10-*bis*(trimethylsiloxy)-octadecanoate. [Reproduced with permission from Regert *et al.* 1998].

many potsherds or carbonized residues exhibit lipid distributions of the type shown in Figure 8.10. However, elemental analyses of the extracted potsherds suggest that unextractable lipid may still be present. Alkaline treatment of the insoluble residues have yielded GC chromatograms of the type shown in Figure 8.17. GC/MS showed the mixture to comprise a range of oxygenated fatty acid derivatives. For example, α,ω-dicarboxylic acids ranging from C_7 to C_{12} were identified by GC/MS and showed azelaic (C_9) acid as the dominant component. The mass spectra of the methyl ester derivatives of the diacids are characterized by fragment ions $[M-15]^+$ and $[M-131]^+$. The latter ion only occurs in diacids with carbon chain lengths greater than C_5 and arises through scission of the C–C bond between the α- and β-carbon atoms.

A further homologous series of components present comprises C_8 to C_{12} α-hydroxy carboxylic acids. These components were identified as their methyl ester trimethylsilyl ether derivatives according to their mass spectra, which are characterized by very intense ions at $[M-15]^+$ and $[M-47]^+$ with less abundant ions at $[M-31]^+$. Other oxygenated saturated carboxylic acids found included 9- and 10-hydroxyoctadecanoic acid (cf. Evershed 1990) and two isomers of 9,10-dihydroxyoctadecanoic acid, which are characterized by

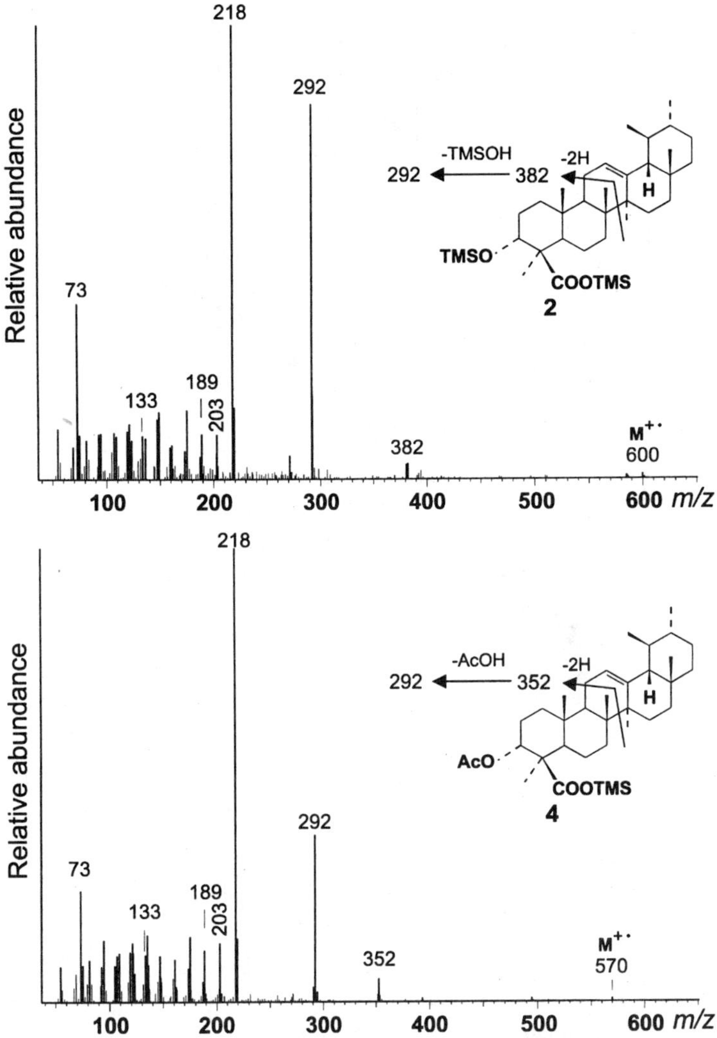

Figure 8.18. Electron ionization (70 eV) mass spectra of the trimethylsilyl derivatives of *β*-boswellic acid (upper) and *O*-acetyl-*β*-boswellic acid (lower) obtained from the GC/MS analysis of ancient frankincense. The inset structures show the origins of the fragment ions used to plot the characteristic mass chromatograms displayed over the page (Evershed et al. 1977d).

fragmentions of m/z 215 and 259 arising through cleavage of the C–C bond between the vicinal OH groups (cf. Gulaçar et al. 1990). The presence of unsaturated hydroxyoctadecenoic acids was also indicated on the basis of characteristic fragment ions seen in their mass spectra (Fig. 8.18). In addi-

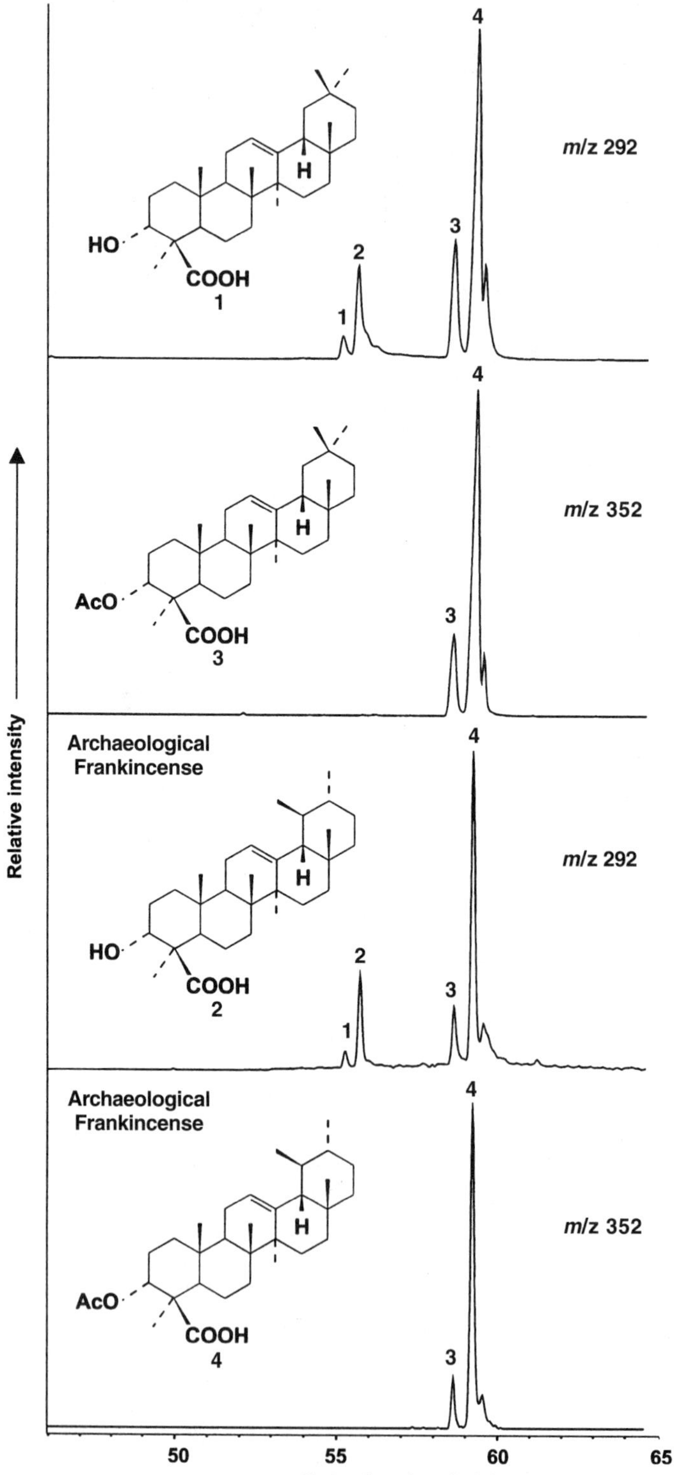

4
m/z 292
HO
COOH
1
3
2
1
4
m/z 352
AcO
COOH
3
3
Relative intensity
Archaeological
Frankincense
4
m/z 292
HO
COOH
2
2
3
1
Archaeological
Frankincense
4
m/z 352
AcO
COOH
4
3
50
55
60
65
Retention time (min)

tion to these bound oxidized compounds, saturated and unsaturated fatty acids were also liberated by base treatment. These data indicated a new mechanism for the preservation of degraded lipids in archaeological pottery possibly through chemical bonding, most likely via ester linkages, into an insoluble polymeric matrix.

8.5.3. Resins

An increasing number of reports exist of the use in antiquity of plant resins and anthropogenically altered natural materials manufactured by the pyrolysis of plant materials. These have been thoroughly reviewed (Mills and White 1994, Pollard and Heron 1996). Major sources include tars and pitches produced from various coniferous tree genera, while for angiosperms the main product derives from birch bark (see below). The complex mixtures of di- and terpenoid compounds that characterize such substances make GC/MS indespensible in their identification. The examples presented below include one of the first identifications, by means of mass chromatography, of archaeological frankincense resin recovered virtually in its natural state, with another example showing the use GC/MS to identify novel triterpenoid components formed by the heating of birch bark.

8.5.3.1. Archaeological Frankincense

Samples of amorphous material were recovered during sieving of the fill from the cellar of a house at Qasr Ibrim (Egypt) dating to around A.D. 400 to 500. The resins were characterized by GC/MS using solvent solubilization, derivatization, and pyrolysis techniques (van Bergen et al. 1997, Evershed et al. 1997d). GC/MS of the trimethylsilylated solvent soluble fraction showed the presence of triterpenoid components known to occur in modern frankincense resin (Figs. 8.19 and 8.20). Especially characteristic are the α- and β-boswellic acids and the corresponding *O*-acetyl derivatives. The mass spectra of the bis-TMS derivatives of the boswellic acids are characterized by the presence of an $M^{+\cdot}$ ion at m/z 600. The ions at m/z 382 and 218 derive

Figure 8.19. Partial GC/MS mass chromatograms of the trimethylsilylated solvent-soluble fractions of a sample of reference frankincense and one of the pieces of resin recovered from a Post Meroitic house at Qasr Ibrîm. The structures shown are: **1** = α-boswellic acid; **2** = β-boswellic acid, **3** = *O*-acetyl-α-boswellic acid, and **4** = *O*-acetyl-β-boswellic acid. The m/z 352 mass chromatograms characterize the *O*-acetyl derivatives, which are the major components of both the ancient and modern resin, while the m/z 292 chromatograms show, simultaneously, the close similarity of the distributions of the *O*-acetyl and free alcohol forms. The inset structures show the compounds in their natural (underivatized) forms (Evershed et al. 1997d).

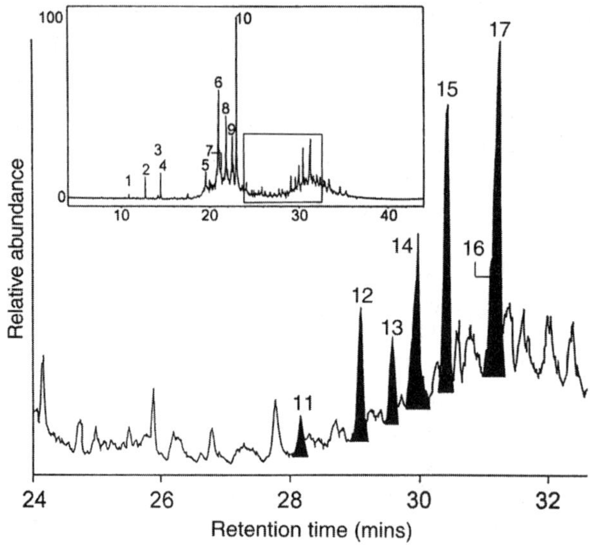

Figure 8.20. Partial high-temperature-GC profile of the trimethylsilylated total lipid extract of a Romano-British birch bark tar with the expanded section showing the distribution of high-molecular-weight esters. Peak identities: 1–4 $C_{14:0}$, $C_{16:0}$, $C_{18:1}$, and $C_{18:0}$; 5 triterpene of unknown structure possibly lupene, derived from dehydration of lupeol; 6 lupa-2,20(29)-dien-28-ol; 7 lupenone; 8 lupeol; 9 betulone; 10 betulin; 11, 12, 14 and 13, 15, 16, 17 refer to two series of high-molecular-weight ester, identified by MS (Fig. 8.21) as the triterpenols ester-linked to fatty acids seen at short retention time.

through ring scission with charge retention on either of the resulting fragment ions. The major ion at m/z 292 arises through loss of TMSOH from the A/B ring containing fragment ion. The mass spectra of the corresponding *O*-acetyl derivatives display very analogous spectra except that $M^{+\cdot}$ and A/B fragment (m/z 570 and 352) occur 30 amu lower due to the presence of the acetyl group rather than TMS at C-3. The close similarity between the compositions of the ancient and modern frankincenses is shown through the m/z 292 (*O*-acetyl derivatives and free alcohol forms) and 352 (*O*-acetyl derivatives) mass chromatograms (Figure 8.20), which reveal virtually identical "fingerprint" distributions. These data confirmed unambiguously that the archaeological resins were frankincense since the major components detected are the major constituents of the fresh aromatic gum resins from *Boswellia* trees (Khalid 1983).

8.5.3.2. Birch Bark Tar

Examination of trimethylsilylated solvent extracts of samples of hard black resinous materials recovered from Romano-British excavations by HT-GC

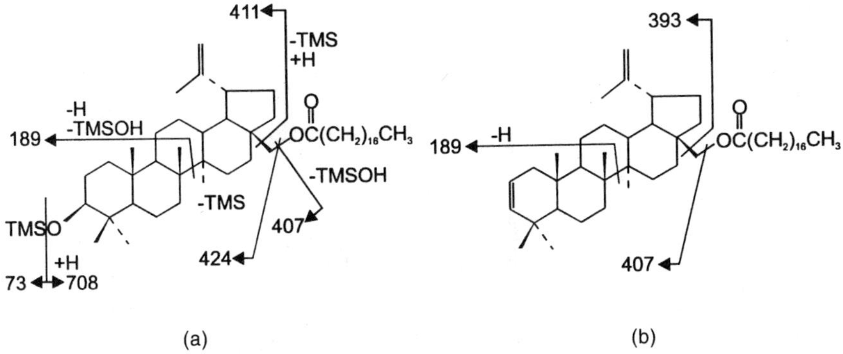

Figure 8.21. Structures and EI (70 eV) MS fragmentations for (*a*) betulin stearate and (*b*) lupa-2,20(29)-dienyl stearate.

and HT-GC/MS revealed a complex mixture of high- and low-molecular-weight components (Figure 8.21). The TMS esters of the fatty acids characteristic of animal fats were recognized through their EI mass spectra as described above. The TMS derivatives of the lupeol, lupenone, betulin, betulone, and lupa-2,20(29)-dien-28-ol were identified from previously published spectra (Dudd and Evershed 1999). Of particular interest in both these tars was the presence of two series of high-molecular-weight triterpenoid components eluting between 300 and 350°C in the temperature program. Fractionation of the solvent extracts of the tars by means of "flash" column chromatography allowed the structures of these components to be examined in more detail. EI spectra showed two series of components to be present, namely fatty acid esters of betulin and lupa-2,20(29)-dien-28-ol. $M^{+\cdot}$ ions at m/z 724, 752, 778, and 780 together with the additional fragmentations shown in Figure 8.22 confirmed the presence of $C_{14:0}$, $C_{16:0}$, $C_{18:1}$, and $C_{18:0}$ esters of betulin, with ions at m/z 634, 662, 688, and 690 confirming the presence of the corresponding ester of lupa-2,20(29)-dien-28-ol. This is the first occasion in which fatty acid esters of the triterpenoid components of birch bark tar have been observed. Their presence in two Romano-British tars suggests that animal fat was intentionally mixed with the birch bark tar to provide a material with more desirable properties.

8.5.4. Human Remains

8.5.4.1. GC/MS of Sterols

Combined GC/MS is widely used to identify the components of mixtures of sterols commonly found in lipid extracts of biological materials. EI is the most widely used ionization technique, with sterols being most frequently

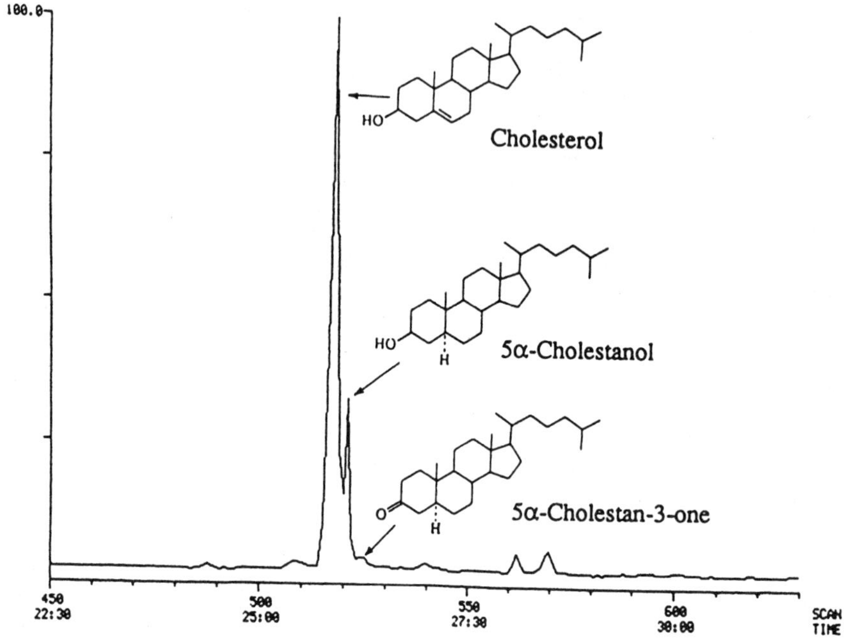

Figure 8.22. Partial GC/MS total ion chromatogram showing the steroidal components of the neutral fraction of the total lipid extract of a sample of psoas muscle tissue from Lindow Man.

analyzed as the free compounds, acetate, or TMS ether derivative. The free 3β-hydroxy compounds can yield abundant molecular ions, $M^{+\cdot}$, from which the carbon number and degree of unsaturation of the sterol nucleus can be deduced. However, excessive dehydration in the ion source can be problematical. Moreover, the GC behavior of the free compounds is more sensitive to the presence of active sites in the column and injection system than that of their derivatives. The mass spectrometry of sterols and their derivatives has been extensively reviewed (Nes 1985; Rahnier and Benveniste 1989; Goad 1991). The fragmentations of sterols are conveniently considered in terms of those arising from the ring system (including derivatizing group) and side chain.

Figure 8.22 shows the partial GC/MS total ion chromatogram of the neutral fraction of the total lipid extract of a sample of muscle from Lindow Man. The pattern of sterols is reasonably simple, being dominated by cholesterol the major sterol of fresh tissue. The EI mass spectrum of the TMS ether derivative of cholesterol is typical of TMS ethers of sterols exhibiting an abundant $M^{+\cdot}$ ion at m/z 458, and other high-mass ions corresponding to loss of angular methyl groups, $[M - 15]^+$, and 1,2-elimination of the trimethylsilanol group, $[M - 90]^+$ (m/z 368). Also present are ions at m/z 129 and 329, which correspond to the loss of the TMS group, together with a three-carbon fragment of ring A containing the C-1, C-2, and C-3 carbons, with charge retention on either fragment ion; this latter fragmentation is especially characteristic of Δ^5 sterols. The loss of a methyl group, largely from C-19, gives rise to the ions at m/z 353 ($[M - CH_3 - TMSOH]^+$) and 443 ($[M - CH_3]^+$). The relative abundances of ions resulting from the loss of the methyl group and hydroxyl moiety (including derivatizing group) are dependent on the presence and position of other structural features, such as nuclear double bonds. A fragment ion at m/z 213 arises through loss of the side chain together with carbons 15 to 17 from ring D. This fragmentation occurs commonly in sterols and is especially useful for establishing the number of double bonds in rings B and C of the sterol nucleus. Hence, the mass spectrum of the 5α-chostanol component eluting just after cholesterol in Figure 8.22 contains an abundant ion at m/z 215 reflecting the lack of the Δ^5 double bond. Fragmentations involving the sterol side chain can provide information diagnostic of the side chain itself and of the sterol nucleus. For example, the ion at m/z 255 arises by loss of the side chain and silanol group, $[M - C_8H_{17} - TMSOH]^+$. Hence, cleavage of the side chain provides evidence for the carbon number and degree of unsaturation of the side chain and the sterol nucleus. Moreover, the appearance of characteristic fragment ions will be strongly dependent upon the position of double bonds and alkyl substituents in the side chain. The reader is again directed to Goad (1991) for more detailed discussions of the specific fragmentations of the wide range of nuclei and side chain structures that occur in sterols. The 5α-cholestan-3-one, appearing as a minor component eluting just after the 5α-choestanol, is an intermediate in the reduction pathway between cholesterol and 5α-cholestanol. The mass spectrum of 5α-cholestan-3-one is characterized by an $M^{+\cdot}$ ion at m/z 386 and a major fragment ion at m/z 231 reflecting the retention of the oxo-group at C-3 on the otherwise fully saturated steroid nucleus, after the loss of the side chain. 7-Ketocholesterol is a common cholesterol degradation product seen in lipid extract of bones deposited under relatively oxic conditions. Its mass spectrum likewise reflects the presence of the C-7 oxo function with many of the corresponding ions in the cholesterol spectrum being shifted to higher mass by 14 amu.

8.5.4.2. Pyrolysis-GC/MS of Human Remains

Pyrolysis-GC/MS (py-GC/MS) is a very useful technique for the study of solid materials for which sample sizes are limited (<0.1 mg) or where they are chemically complex and/or difficult to study by other methods, for

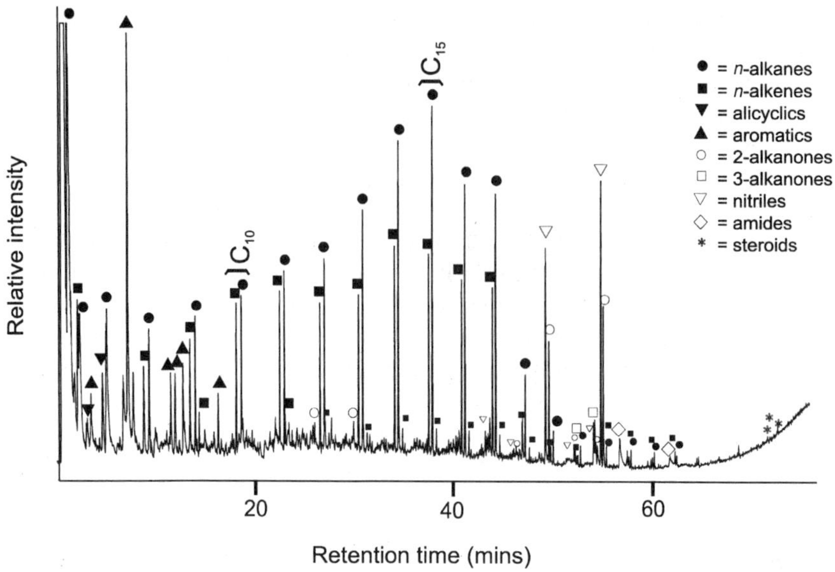

Figure 8.23. GC/MS reconstructed total ion chromatogram of the pyrolysis (610°C for 10 sec) products of bandage/tissue of an Egyptian mummy (Khnum Nakt, ca. 2000 B.C.). C_{10} and C_{15} refer to the carbon numbers of the alkenes and alkanes within the retention time windows.

example, heteropolymeric. The small sizes of samples required means the technique is ideally suited to the study of valuable museum objects, such as fossil specimens, paintings, or other art objects. The example displayed in Figure 8.23 shows the results of the py-GC/MS analysis of a microgram fragment of tissue taken from an Egyptian mummy. The analysis was aimed at determining the presence or otherwise of embalming agents. While none were detected in this particular sample, the data show homologous series of alkanes and alkenes that confirm the presence of a highly aliphatic component. The distributions of these components was established by examining individual mass spectra and use of mass chromatography using the m/z 55 (alkenes) and 57 (alkanes) ions.

A similar approach has been used to compare the relative states of survival of protein in samples of skin from human remains recovered from peat bogs, the so-called bog bodies, and Nubian mummies preserved by desiccation at arid desert sites. The study employed flash pyrolysis-GC/MS to produce "fingerprints" that were compared with samples of fresh skin. The py-GC/MS data were also correlated with more conventional amino acid data derived by hydrolysis, derivatization, and reversed-phase HPLC. The pyrolyzates of the modern and ancient skin specimens exhibited a similar mixture of products (see Fig. 8.24 and Table 8.3). All pyrolyzates were dominated by toluene, pyrrole, and a mixture of 2,5-diketopiperazines. The 2-diketopiperazines included those representing specific amino acids or dipetides, namely Pro-Gly, Pro-Ala, Pro-Val, Pro-Pro, and Hyp, which were readily identified through their mass spectra. The high abundance of diketopyrrole, a pyrolysis marker for hydroxyproline (Chiavari and Gallietti 1992), is characteristic of collagen and has not been observed in pyrolysates of other noncollagenous proteins (Stankiewicz et al. 1996). In collagen pyrrole and pyrroline derive exclusively from proline. Other important pyrolysis products of collagen include derivatives of pyrimidine. Overall, the similarity of the pyrogram of modern and ancient skins indicated remarkable preservation. This was supported by amino acid HPLC data on the same samples (Flannery et al. 1999).

8.5.5. Plant Remains

8.5.5.1. Volatile Compounds in Archaeological Seeds

It has recently been recognized that volatile compounds are preserved within ancient materials that can give clues to their biochemical nature and the degradative pathways that operated (Evershed et al. 1997b). This was recognized in studies of archaeological plant remains from Egypt. Volatile decay products were detected trapped within the internal networks of structural and storage macromolecules. Volatile compounds were released by crushing whole seeds in an all-glass headspace extraction vessel. After warming to 35°C the components were collected on a Tenax cartridge by flushing with a stream of carbon-filtered nitrogen. The volatiles were desorbed from cartridge at 250°C and cryogenically focused at the head of the GC column. GC/MS analysis allowed identification of the volatile components according to their mass spectra summarized in Table 8.4 and Figure 8.25. The major components were various alkyl pyrazines and alkyl polysulfides. The mass spectra of the former are characterized by intense $M^{+}\cdot$ or $[M-H]^{+}$ ions at high mass typical of alkylpyrazines (Budzikiewicz et al. 1964). High-mass fragment ions include those arising via loss of methyl groups or elimination of ethylene, $[M-28]^{+}$, in the case of ethyl-substituted components. While the spectra of the alkyl polysulfides exhibit intense $M^{+}\cdot$ ions including a pattern of isotope peaks reflecting the presence of increasing numbers of sulfur atoms in the respective compounds. The finding of these volatile compounds, particularly the pyrazines provides the first unambiguous evidence for the operation of the Maillard reaction during the decay of plant matter (Evershed et al. 1997b).

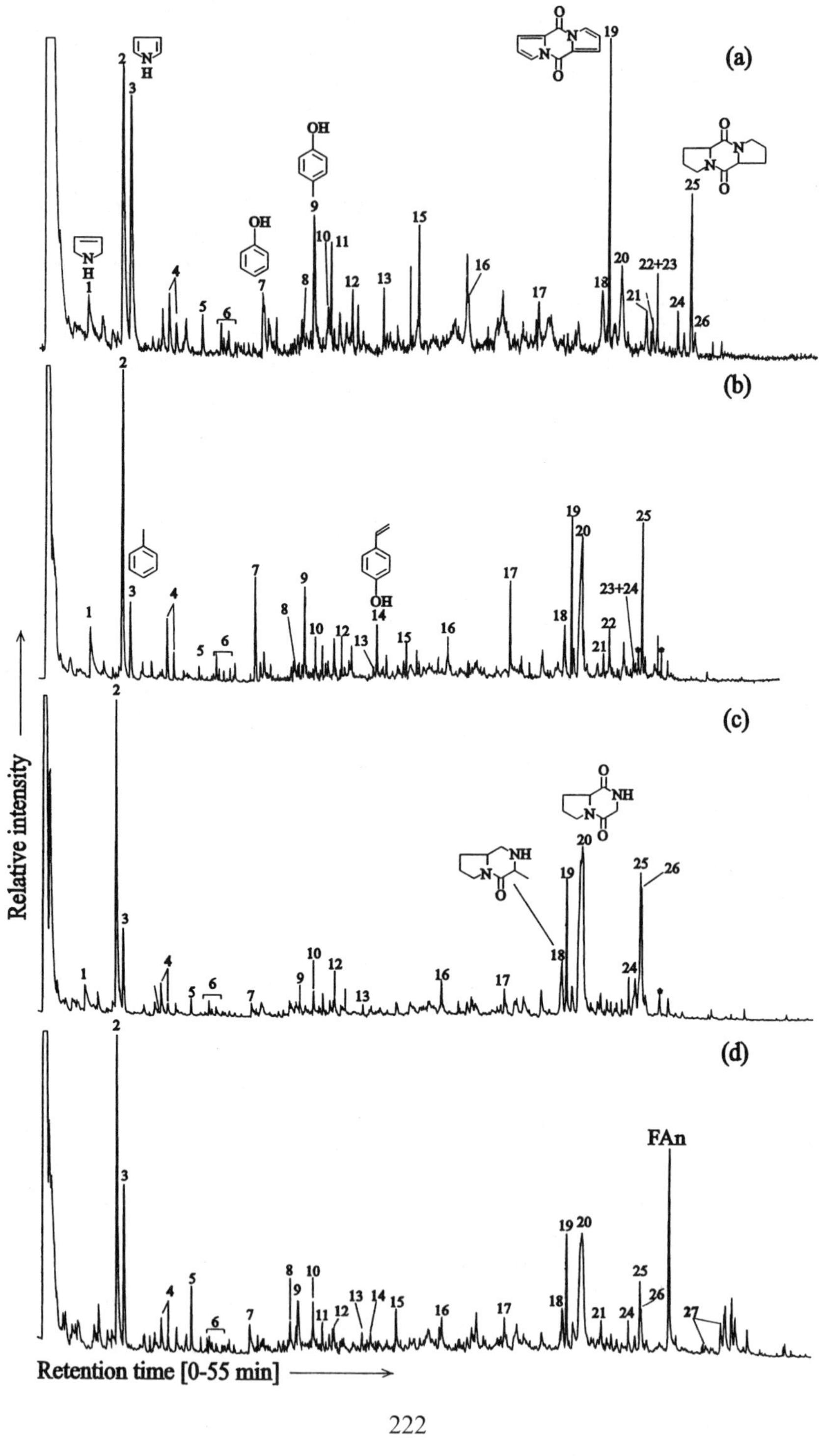
(a)
(b)
(c)
(d)
FAn
Relative intensity
Retention time [0-55 min]

8.5.6. MS Evidence for Ancient DNA

The analysis of ancient DNA (deoxyribonucleic acid) has tremendous potential in areas as diverse as archaeology, palaeontology, and conservation biology. Methods have been developed that offer the potential to chemically characterize the building blocks of ancient DNA and begin to investigate degradation processes associated with this complex biopolymer. Initial work was carried out on the individual purine and pyrimidine bases released from ancient DNA by acid hydrolysis and converted to their *tert*-butyldimethylsilyl derivatives (O'Donoghue et al. 1994). These derivatives were readily separated by GC/MS and under EI conditions yield spectra dominated by prominent $[M - 57]^+$ ions, which carry the bulk of the ion current (Fig. 8.26). The presence ions of such high intensity can be used to enhance the sensitivity and specificity of GC/MS and GC/MS/MS analyses of ancient nucleotide samples.

Figure 8.27 shows the total ion chromatograms of the *tert*-butyldimethylsilyl (TBDMS) derivatized authentic nucleotides bases. No such distinct peaks were observed when the ancient seed extract was analyzed due to the small amount of DNA preserved in the ancient seeds. However, by use of SIM, peaks were observed at the expected retention times for all five DNA bases (Fig. 8.28). While these data provide good evidence for the presence for each of the bases, more definitive assignments are obtained by SIM by monitoring of more than one ion from the mass spectrum of an analyte. The possibility then exists for confirming the presence of the analyte, by comparing the ratio of the responses of these ions determined by SIM to their abundance ratios in the full-scan mass spectrum. Figure 8.29 shows a number of possible hypothetical results that can be obtained from the search for a particular analyte using SIM for three ions. An exactly analogous approach was used in investigations of ancient nucleotide bases in desiccated seeds recovered from Qasr Ibrim, Egypt. Figure 8.26 shows the EI mass spectrum of the *bis*-TBDMS derivative of thymine. On the basis of this spectrum, ions of m/z 339, 297, and 113 were selected to screen hydrolyzates of ancient seeds for the presence of intact thymine. Figure 8.30 shows the partial TIC and mass chromatograms for three ions. The TIC shows no distinct peak at the retention time of the thymine (determined from a separate analysis using the authentic compound) due to the presence of other more abundant compounds eluting at other retention times.

Figure 8.24. Total ion chromatograms of the pyrolysates (610°C for 10 sec) of skin tissue from (*a*) modern human, (*b*) Lindow II (ca. 2000 B.P.), (*c*) Nubian mummy (241 B17; ca. 1500 B.P.), and (*d*) Nubian mummy (241 B37; ca. 1500 B.P.). Peak identifications are given in Table 8.3. Note * contaminants, FAn are *n*-fatty acids.

Table 8.3. Pyrolysis Products of Skin Samples as Observed by Py-GC/MS

Peak Number	Compound Name	*m/z* MW	*m/z* Base Peak	*m/z* Fragment Ions	Origin	Occurrence[a]
1	Pyrroline	69	41	42, 68	Pro	L, D, N, FS
2	Pyrrole	67	67	41, 40, 52	Pro	L, D, N, FS
3	Toluene	92	91	65, 51	Phe	L, D, N, FS
4	C_1-pyrroles (2 isomers)	81	80	53, 40	Pro, Hyp	L, D, N, FS
5	Styrene	104	78	51	Phe	L, D, N, FS
6	C_2-pyrroles (4 isomers)	95	94 (80)	80, 67, 53	DKP rearr.?	L, D, N, FS
7	Phenol	94	94	66, 65, 40	Tyr[b]	L, D, N, FS, S, SL
8	?	111	40	68, 83	Pro-Asp	L, D, N, FS
9	4-Methylphenol	108	107	77, 79, 66	Tyr[b]	L, D, N, FS, S, SL
10	2,5-Pyrrolidinedione	99	99	56, 40	Asp	L, D, FS
11	Ethylcyanobenzene	117	117	90, 116, 89	Phe	L, D, N, FS
12	Rearrangement product	?	54	125, 43, 68	Pro-Leu	L, D, N, FS
13	Propylcyanobenzene	131	91	66, 78, 51	Phe	L, D, N, FS
14	Vinylphenol	120	120	91, 66, 51	Tyr[b]	L, D, N, S, SL
15	Indole	117	117	90, 89, 63	Trp	L, D, N, FS
16	Trimethylpyrimidine	138	42	69, 98, 54		L, D, N, FS
17	?	136	107	53, 80, 40	Ala, Leu, Ile	L, D, N, FS
18	2,5-DKP (2 isomers)	168	70	44, 97, 125	Pro-Ala	L, D, N, FS
19	Diketodipyrrole	186	186	93, 66, 130	Hyp	L, D, N, FS
20	Pyrrolidinopiperazine derivative	154	41	83, 111, 70	Pro-Gly	L, D, N, FS
21	2,5-Diketopiperazine	196	70	154, 125, 41	Pro-Val	L, N, FS
22	2,5-Dikpzine (isomer 51)	196	70	154, 125, 41	Pro-Val	L, D, FS

23	?	190	68	189, 40, 67	Pro-Arg	L, N
24	?	192	41	68, 70, 191	Pro-Hyp	L, D, N, FS
25	2,5-Diketopiperazine	194	70	41, 96, 138	Pro	L, D, N, FS
26	2,5-Diketopiperazine	195	70	154, 41, 125	Pro-AA	L, D, N, FS
27	2,5-Diketopiperazine	210	70	41, 86, 124	Hyp, Pro-Leu, Pro-Ile	L, D, N, C

[a] L = Lindow Man, D = Drenthe body, N = Nubian mummies, FS = fresh human skin, S = *Sphagnum*-dominated peat, SL = Lindow peat. [b] Also possibly derived from plant polyphenols.

Table 8.4. Volatile Components of Ancient Radish and Barley Showing Characteristic Fragment Ions, with Their Relative Abundances, and Linear Retention Indices[a]

No.	Compound	Ancient Radish	Ancient Barley	Characteristic Fragment Ions (m/z)	LRI CPSil 19
1	Dimethyldisulphide	√	×	<u>94</u> (70), 79 (45), 64 (10), 47 (30), **45**	805
2	2-Hexanone	√	√	<u>100</u> (10), 85 (5), 70 (5), 58 (40), **43**	
3	*n*-Hexanal	√	√	<u>100</u> (tr), 82 (10), 72 (15), 57 (40), 56 (60), **44**	
4	Methylpyrazine	√	√	<u>94</u>, 67 (90), 53 (35), 42 (35), 39 (75)	905
5	2-Heptanone	√	√	<u>114</u> (5), 85 (5), 71 (15), 58 (55), **43**	
6	2,5-Dimethylpyrazine	√	√	<u>108</u>, 81 (10), 42 (85), 39 (35)	988
7	2,6-Dimethylpyrazine	√	√	<u>108</u>, 93 (5), 81 (5), 67 (5), 42 (50), 39 (35)	990
8	*n*-Heptanal	×	√	<u>114</u> (tr), 70 (5), 55 (20), **43**	
9	Dimethyltrisulphide	√	×	<u>126</u>, 111 (35), 79 (70), 64 (50), 47 (60), 45 (80)	1042
10	2-Ethyl-6-methyl-pyrazine	√	√	<u>122</u> (60), **121**, 94 (10), 56 (10), 42 (5), 39 (20)	1072
11	2-Ethyl-5-methyl-pyrazine	√	√	<u>122</u> (65), **121**, 94 (10), 56 (15), 39 (20)	1075
12	Trimethylpyrazine	√	√	<u>122</u>, 121 (35), 81 (15), 42 (70)	1078
13	2-Octanone	√	√	<u>128</u> (5), 85 (5), 71 (10), 58 (65), **43**	
14	2-Ethyl-3,6-dimethyl-pyrazine	×	√	<u>136</u> (75), **135**, 108 (10), 56 (10), 42 (25)	
15	2-Ethyl-3,5-dimethyl-pyrazine	×	√	<u>136</u> (75), **135**, 108 (10), 56 (25), 42 (15)	
16	2,5-Diethyl-pyrazine	×	√	<u>136</u> (90), 135 (95), **121**, 108 (10), 56 (20), 39 (35)	1158
17	2-Propyl-5 (or 6)-methyl-pyrazine	×	√	<u>136</u> (15), 121 (20), 108, 94 (5), 39 (25)	
18	2,5-Dimethyl-tetrahydrofuran	√	×	<u>100</u> (10), 85 (70), **56**, 41 (80)	
19	2-Nonanone	√	√	<u>142</u> (5), 71 (15), 58 (85), **43**	
20	*n*-Nonanal	√	√	<u>142</u> (tr), 127 (5), 114 (5), 98 (15), 74 (40), 57 (80), **41**	
21	C_5-Pyrazine	×	√	<u>150</u>, 149 (70), 135 (80), 121 (25), 94 (70), 56 (65), 39 (95)	
22	2,6-Diethyl-3-methyl-pyrazine	×	√	<u>150</u> (70), 149, 135 (10), 122 (15), 107 (5), 53 (35), 39 (45)	
23	2-Propyl-dimethyl-pyrazine	×	√	<u>150</u> (5), 135 (15), **122**, 107 (10), 39 (20)	
24	2-Decanone	×	√	<u>156</u> (5), 98 (5), 85 (5), 71 (20), 58 (80), **43**	
25	*n*-Decanal	×	√	<u>156</u> (tr), 128 (5), 112 (15), 82 (25), 70 (30), 57 (70), **41**	
26	Dimethyltetrasulphide	√	×	<u>158</u> (85), 111 (10), 94 (20), 79 (90), 64 (45), 47 (50), **45**	1258

[a] Ions underlined indicate molecular ion and those in bold the base peak.

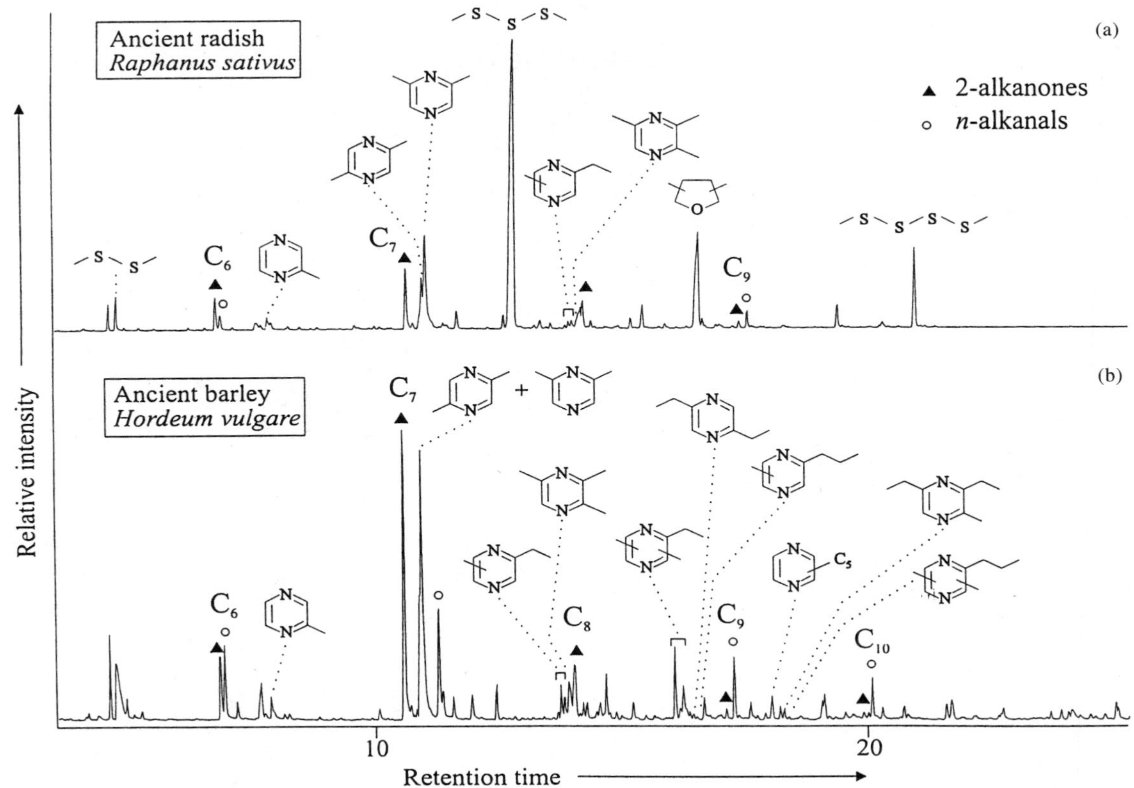

Figure 8.25. Partial desorption headspace–GC/MS total ion chromatograms for the volatile compounds trapped in the ancient propagules (*a*) radish (*Raphanus sativum*) and (*b*) barley (*Hordeum* sp.). The structures of the alkyl pyrazines and alkyl polysulfides, characteristic of the Maillard reaction and referred to in the text, are shown adjacent to the peaks on the chromatogram. The range of aliphatic carbon compounds (annotated on the chromatograms with their chain length) refer to alkan-2-ones (▲) and *n*-alkanals (○) formed through lipid oxidation.

However, by plotting the mass chromatograms for each of the characteristic ions peaks are clearly apparent at the same retention time and, most importantly, similar relative abundances to those seen in the modern specimens and authentic compounds, thus providing direct chemical evidence for the survival of thymine bases in ancient seeds (O'Donoghue et al. 1994). Mass spectral evidence for the survival of ancient DNA in these seeds has also been obtained by use of HPLC/MS following enymatic hydrolysis to yield nucleosides that were detected using electrospray ionization (O'Donoghue et al. 1996). Figure 8.31 shows the mass chromatogram for the HPLC/MS SIM analysis for 2′-deoxyadenosine, with the inset showing confirmation of the identity of the astericked peak by co-injection of the authentic compound. Interestingly, all the DNA nucleosides except 2′-deoxythymidine were detectable by HPLC/MS; however, as discussed above, the presence of the corresponding base, thymine, was confirmed by GC/MS–SIM.

8.5.7. Tandem Mass Spectrometry in Archaeology

The number of applications of tandem mass spectrometry in the study of materials in art and archaeology are very few. As stated in Section 8.4.3.6, MS–MS is most widely used to enhance structure information, confirm fragmentation pathways, and increase selectivity and sensitivity. In the study of ancient DNA discussed above (O'Donoghue et al. 1994) GC–MS–MS was used to generate product ions from the characteristic $[M - 57]^+$ ions produced through EI of TBDMS derivatives of purine and pyrimidine bases. The spectra obtained helped to confirm the presence of DNA bases in hydrolysates of ancient seeds.

GC–MS–MS has found application in archaeology in the detection of the 5β-stanols characteristic of manure in archaeological soils (Evershed and Bethell 1996). Figure 8.32 shows the product ion spectrum produced by collision-induced dissociation of the m/z 370 ion ($[M - TMSOH]^+$) of coprostanol in the collision cell of a triple-stage quadrupole mass spectrometer. The decomposition m/z 370 $\rightarrow$ 215, which arises via cleavage across the D ring of the steroid nucleus, accompanied by a proton transfer, raises possibilities for the highly selective detection of 5β-stanols by GC–MS–MS employing selected reaction monitoring, as shown schematically in Figure 8.33.

Figure 8.26. Electron ionization (EI, 35 eV) mass spectra of the TBDMS derivatives of the authentic (*a*) pyrimidines (uracil, thymine, and cytosine; spectra show the characteristic $[M - 57]^+$ and $[M - 2 \times 57]^{2+}$ ions, due to loss of one and two *tert*-butyl radicals, respectively), and (*b*) purines (adenine and guanine; spectra shows the characteristic $[M - 57]^+$ ion, due to the loss of one *tert*-butyl radical, and a weak molecular ion $[M]^{+\cdot}$). Spectra derived from the analysis shown in Figure 8.27.

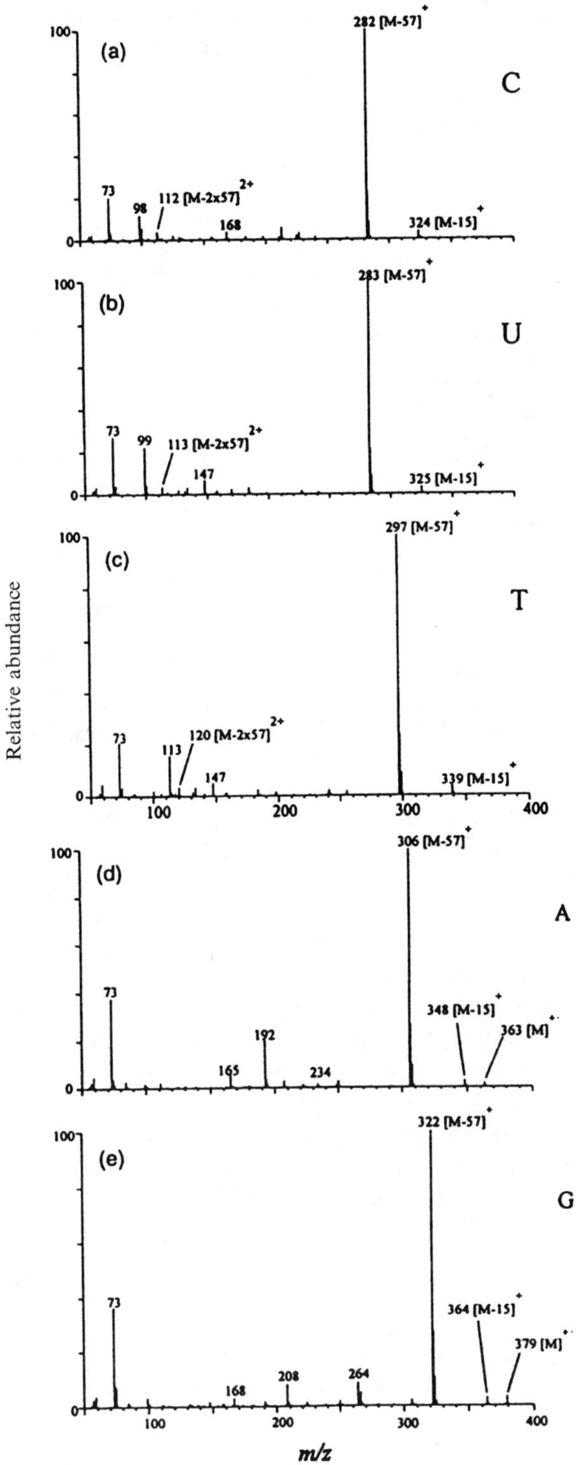

(a)
C
282 [M-57]+
73
98
112 [M-2x57]2+
168
324 [M-15]+
(b)
U
283 [M-57]+
73
99
113 [M-2x57]2+
147
325 [M-15]+
(c)
T
297 [M-57]+
73
113
120 [M-2x57]2+
147
339 [M-15]+
(d)
A
306 [M-57]+
73
192
165
234
348 [M-15]+
363 [M]+·
(e)
G
322 [M-57]+
73
168
208
264
364 [M-15]+
379 [M]+·
Relative abundance
100
0
100
200
300
400
m/z

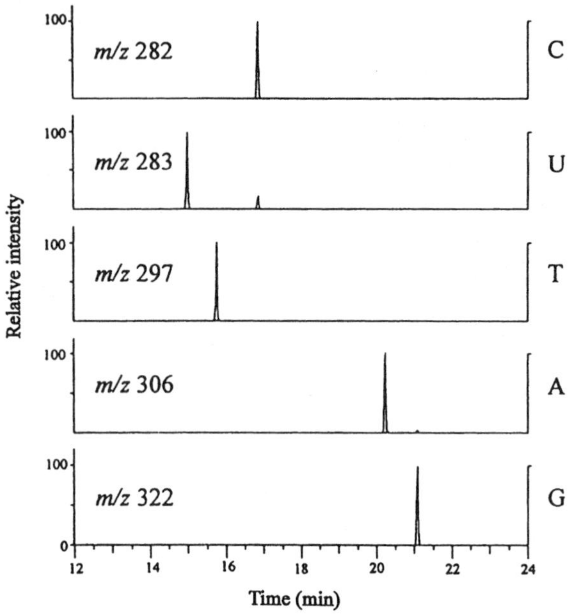

Figure 8.27. $[M - 57]^+$ mass chromatograms obtained from the GC/MS analysis of the TBDMS derivatives of authentic purine and pyrimidine bases (2 ng of each compound injected).

Indeed, in practice, the technique was found to have considerable potential for trace analysis of 5β-stanols, allowing highly selective analyses with a significantly simplified sample preparation strategy to be adopted compared to that when conventional GC–MS was employed. The hexane-soluble fraction of the total lipid extracts of soil samples from experimental ancient farm plots were trimethylsilylated and analysed directly by GC–MS–MS, using a triple-stage quadrupole instrument, without further purification. Although the hexane-soluble fraction comprises a complex mixture of cyclic and aliphatic lipid species, 5β-stanols (and 5α-stanols) were readily detected with high selectivity. The GC–MS–MS method greatly simplifies analyses, potentially increasing sample throughput by eliminating the need for time-consuming adsorption chromatographic steps, for example, thin-layer chromatography, prior to conventional GC–MS.

8.6. SUMMARY

The aim of this chapter was to provide an introduction to the use of organic mass spectrometry in the field of archaeology and art. The experiences of my

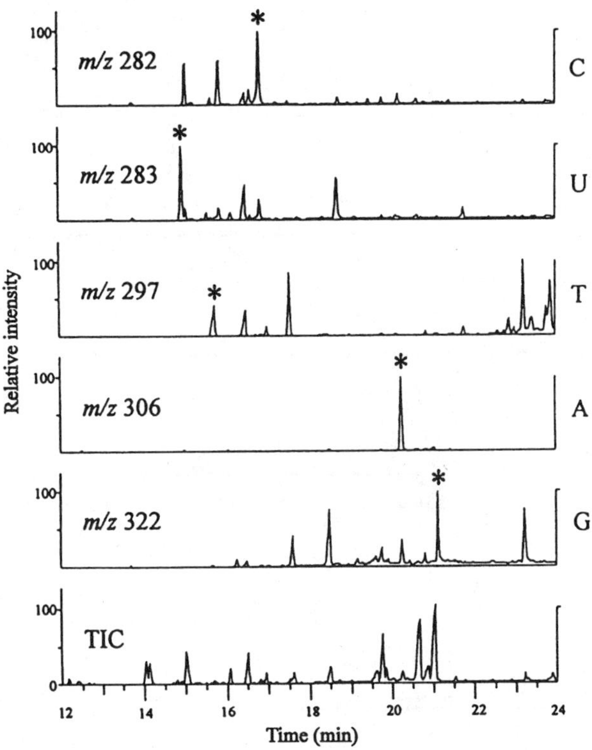

Figure 8.28. Total ion current and $[M - 57]^+$ mass chromatograms obtained by GC/MS SIM analysis of the ancient nucleic acid extract hydrolysates following *tert*-butyldimethylsilylation. Peaks assigned to each of the five major bases are indicated by asterisks. The analysis represents the equivalent of 5 mg of ancient radish seeds.

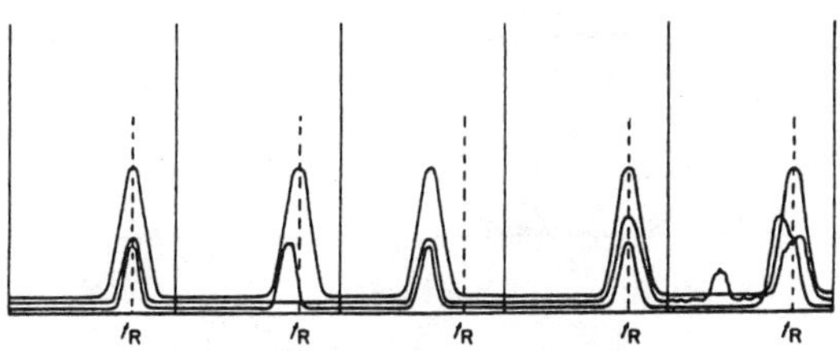

Relative intensities:	Correct	Incorrect	Correct	Distorted	Distorted
Retention times:	Correct	Incorrect	Incorrect	Correct	Distorted
Compound present?	Yes	No	No	Probably	Possibly

Figure 8.29. Hypothetical results from the GC (or LC/MS) selected ion monitoring analysis for a targeted compound. (Reproduced with permission from Chapman 1993.)

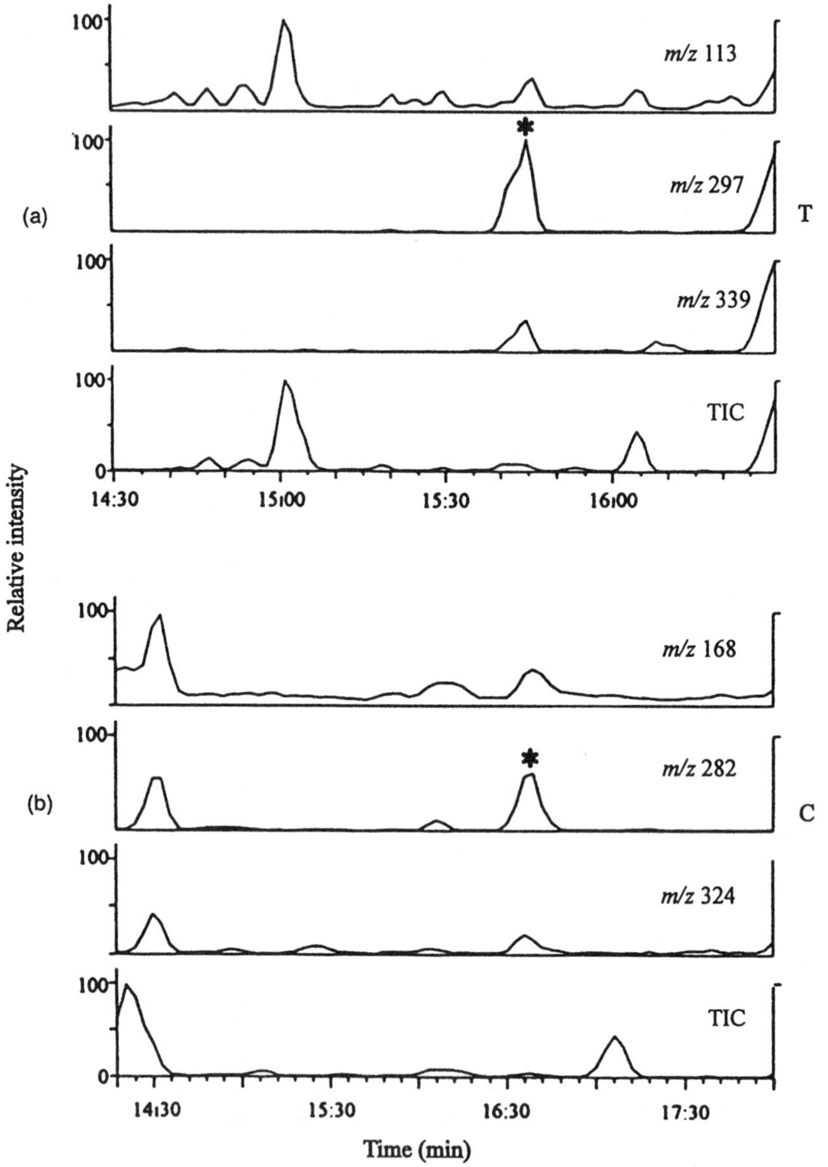

Figure 8.30. SIM mass chromatograms of TBDMS-derivatized ancient nucleic acid extract hydrolysates indicating the presence of pyrimidines; thymine.

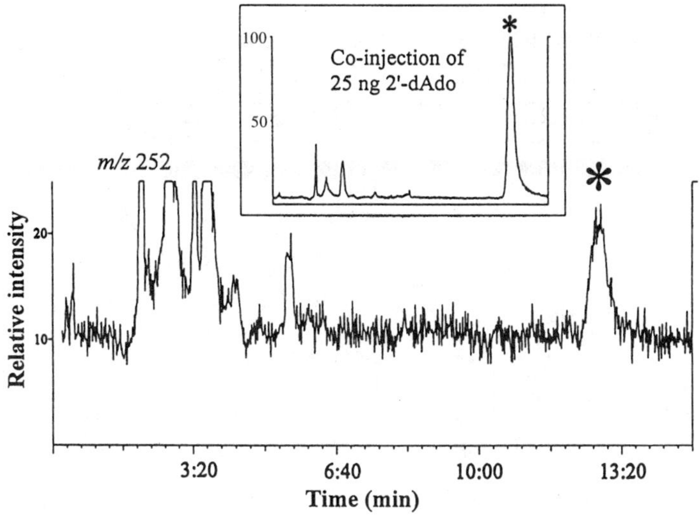

Figure 8.31. Mass chromatogram (*m/z* 252) of the $[M + H]^+$ ion of 2′-deoxyadenosine in the ancient radish seed extract. The asterisk marks peak assigned to this deoxyribonucleoside. The sample represents the total extract of 168 mg of seeds. The inset mass chromatogram shows the results of co-injecting an extract equivalent to 151 mg of seeds together with 25 ng of authentic 2-deoxyadenosine.

own laboratory have meant that the examples I have drawn on come mainly from the field of archaeology. For specific examples of the use of the technique in the field of art the reader is recommended to consult Mills and White's (1994) excellent text *The Organic Chemistry of Museum Objects*.

There are a number of different types of analytical information that can be obtained by organic mass spectrometry, and the introductory sections discussed in general terms the possibilities that exist in relation to the different classes of organic molecule that might be encountered in this field and the type of questions posed by art and archaeology. The subsequent section on instrumentation is by no means exhaustive and focused on those aspects deemed most relevant to this field. Mass spectrometry instrumentation has been a rapidly developing field over the past decade such that readers who may be involved in any way in making decisions regarding the purchase of such instrumentation would be strongly recommended to consult a range of instrument manufacturers in order to make themselves aware of the various state-of-the-art options available. For example, the brief discussion of TOF–MS given in the section on instrumentation centered very much on its use in conjunction with the matrix-assisted laser desorption ionization (MALDI) technique for the analysis of high-molecular-weight natural (and synthetic)

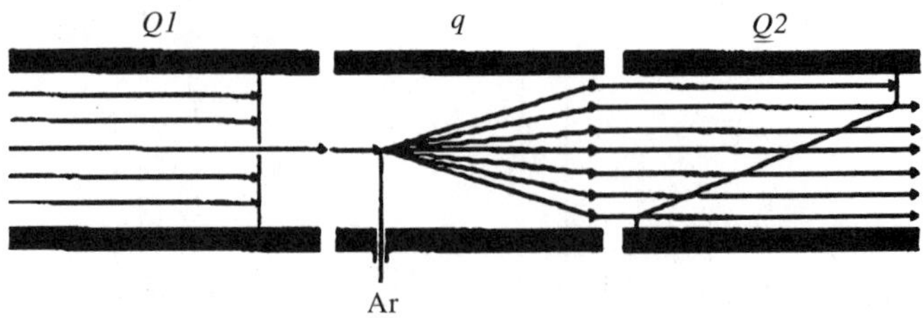

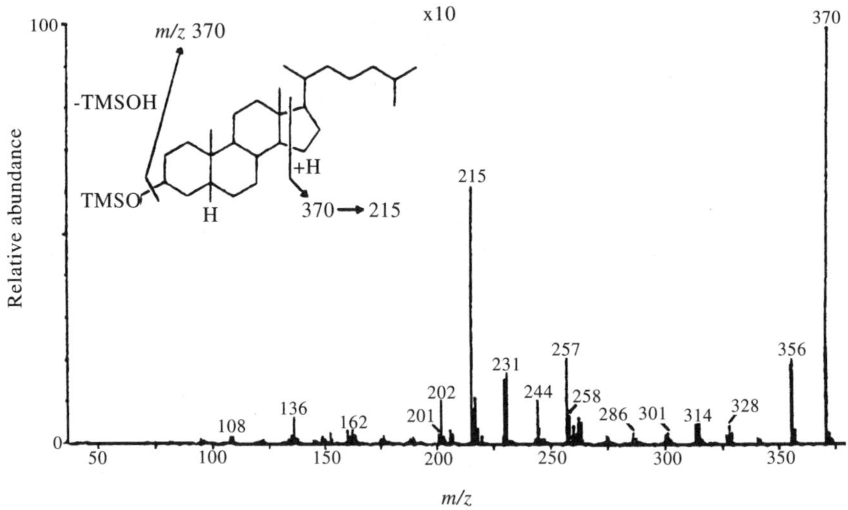

Figure 8.32. Unimolecular decomposition product ion spectrum obtained from the $[M - TMSOH]^+$ ion of coprostanol (from Evershed and Bethell 1996).

polymers. However, recent reports have described the use of TOF–MS in conjunction with GC. The fast scan rates attainable with the TOF analyzer are of special value in GC–MS analyses employing narrow bore capillary columns coated with thin films of stationary phase that yield very narrow GC peaks requiring high MS sampling rates to maintain chromatographic integrity. Such systems are now commercially available and offer considerable time savings in terms of overall analysis times. Likewise, TOF–MS instruments incorporating the latest reflectron ion mirrors are able to attain mass resolutions sufficient for making elemental determinations on ions, thus potentially removing the need for the technically more complex and more expensive magnetic sector analysers traditionally used for this type of application. Finally, in the field of tandem mass spectrometry a seemingly bewildering range analyzer combinations now exist. The major development in

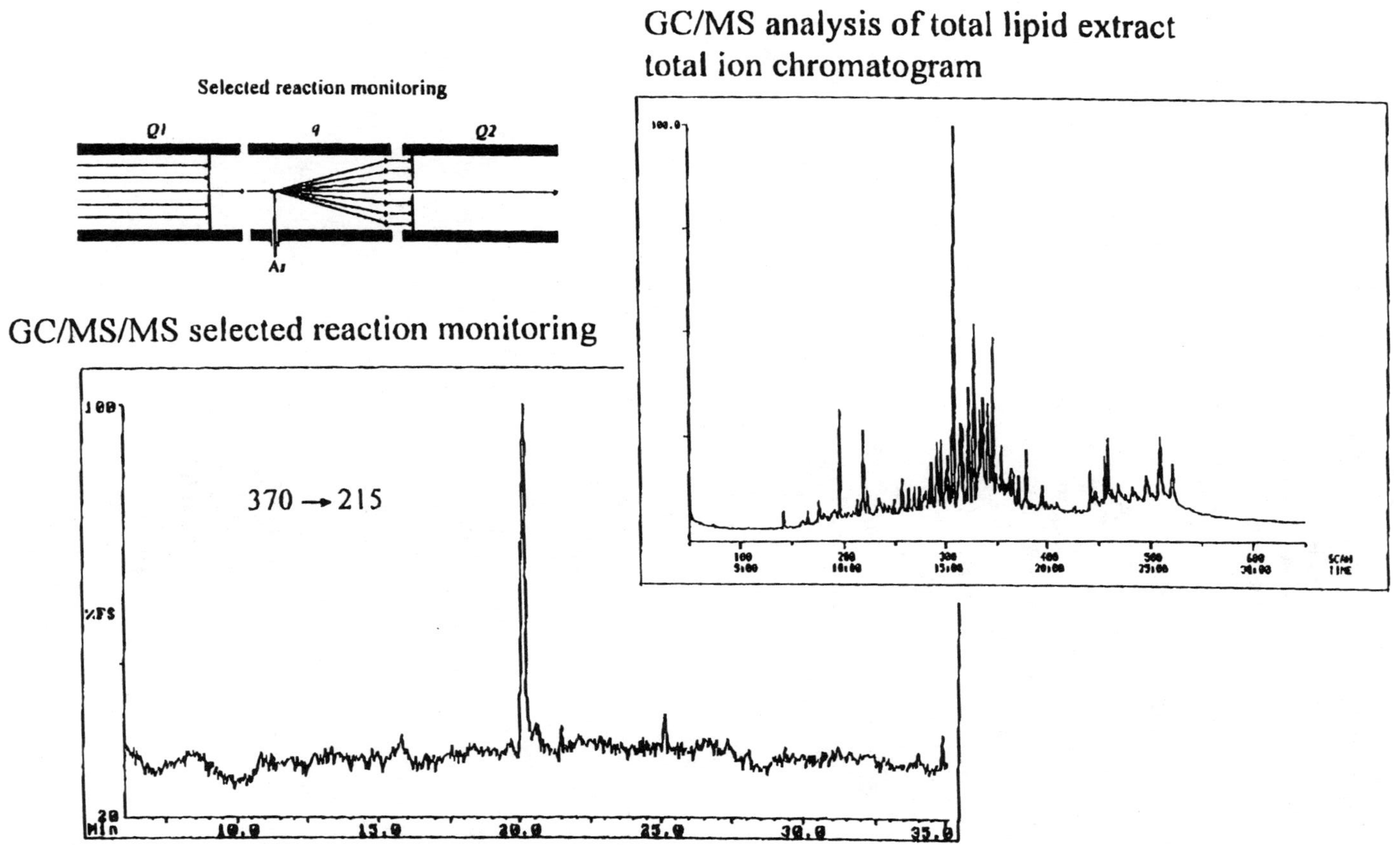

Figure 8.33. Chromatograms showing the high selectivity (lower left) that can be achieved in the analysis of 5β-stanols in complex lipid extracts (upper right) by means of tandem mass spectrometry. The inset shows schematically the mode of operation of the GC-MS-MS SRM technique (from Evershed and Bethell 1996).

this field has been in the versatility of the relatively inexpensive ion traps, which are able to perform multiple product ion experiments, that is, MS^n, at high sensitivities, thereby opening the way for in-depth studies of compound structures via collision-induced decomposition experiments.

The applications presented in the second half of this contribution were again not intended to be exhaustive. There is again a bias toward the work of our own laboratory that seems inevitable in any contribution of this nature, and the choice of examples in no way implies that these are best results that can be, or have been, achieved. However, the coverage is very comprehensive in terms of the range of compound classes covered and the MS and ancillary chromatographic and sampling techniques used to tackle particular analytical problems. Organic mass spectrometry especially when combined with chromatographic techniques (GC and HPLC) is unrivaled in its ability to yield compositional information on complex mixtures. I hope that through the range of examples presented I have achieved my aims of demonstrating the scope of applications in the field of archaeology and art and inspiring the reader to consider introducing organic mass spectrometry technique to their laboratory protocols.

Acknowledgments

In contributing this chapters I have drawn on the results, expertise, and artwork of a number of postdoctoral research assistants and Ph.D. students who have worked or are currently still working with me. Particular thanks go to Stephanie Dudd, Stephanie Charters, Helen Bland, Kerry O'Donoghue, Hazel Mottram, Martine Regert, Pim van Bergen, Carl Heron, Stephen Buckley, and Philip Bethell. I should also like to acknowledge the efforts of Jim Carter, Mark Prescott, and Andy Gledhill, who have given technical support to all the mass spectrometers used by my students, postdoctoral researchers, and myself over the years; without their painstaking maintenance of the instruments we would never have acquired the high-quality mass spectral data contained in this contribution. The SERC and NERC Science Based Archaeology Committee are thanked for their continual support for our research through grants and studentships over the past 10 years that has lead to the majority of the data presented in this chapter.

REFERENCES

Barber, M., J. R. Chapman, and W. A. Wolstenholme (1968), *J. Spectrom. Ion Phys.* **1**, 98.

Blau, K., and J. M. Halket (1993), *Handbook of Derivatives for Chromatography*, John Wiley & Sons, New York.

Budzikiewicz Djerassi, C., and D. H. Williams (1964), *Interpretation of Mass Spectra of Organic Compounds*, Holden-Day, San Francisco, pp. 258–260.

Budzikiewicz Djerassi, C., and D. H. Williams (1967), *Mass Spectrometry of Organic Compounds*, Holden-Day, San Francisco.

Chapman, J. R. (1993), *Practical Organic Mass Spectrometry*, 2nd ed., John Wiley & Sons, Chichester.

Charters, S., R. P. Evershed, A. Quye, P. W. Blinkhorn, and V. Denham (1997), *J. Archaeol. Sci.* **24**, 1.

Chiavari, G., and G. C. Galletti (1992), *J. Anal. Appl. Pyrol.* **24**, 123.

Constantin, E., and A. Schnell (1990), *Mass Spectrometry*, Ellis Horwood, New York.

Curstedt, T. (1974), *Biochim. Biophys. Acta* **360**, 12.

De Hoffman, E., J. Charette, and V. Stroobant (1996), *Mass Spectrometry*, John Wiley & Sons, Chichester.

Dudd, S. N., and R. P. Evershed (1999), *Tetrahedr. Lett.* **40**, 359.

Enser, M. (1991), in *Analysis of Oilseeds, Fats and Fatty Foods*, J. B. Rossell and J. L. R. Pritchard, Eds., Elsevier, London, p. 329.

Evershed, R. P. (1990), *Archaeometry* **32**, 139.

Evershed, R. P. (1996), *J. Am. Soc. Mass Spectrom.* **7**, 350.

Evershed, R. P., and P. H. Bethell (1996), in *Archaeological Chemistry Symposium Volume*, ACS Symposium Series, M. V. Orna, Ed., American Chemical Society, Washington, DC, pp. 158–172.

Evershed, R. P., C. Heron, and L. J. Goad (1990a), *Analyst* **115**, 1339.

Evershed, R. P., M. C. Prescott, and L. J. Goad (1990b), *Rapid Commun. Mass Spectrom.* **4**, 345.

Evershed, R. P., C. Heron, and L. J. Goad (1991), *Antiquity* **65**, 540.

Evershed, R. P., C. Heron, S. Charters, and L. J. Goad (1992), *Proc. Br. Acad.* **77**, 187.

Evershed, R. P., K. I. Arnot, J. Collister, G. Eglinton, and S. Charters (1994), *Analyst* **119**, 909.

Evershed, R. P., A. W. Stott, A. Raven, S. N. Dudd, S. Charters, and A. Leyden (1995), *Tetrahedr. Lett.* **36**, 8875.

Evershed, R. P., S. J. Vaughan, S. N. Dudd, and J. S. Soles (1997a), *Antiquity* **71**, 979.

Evershed, R. P., H. A. Bland, P. F. van Bergen, J. F. Carter, M. C. Horton, and P. A. Rowley-Conwy (1997b), *Science* **278**, 432.

Evershed, R. P., H. R. Mottram, S. N. Dudd, S. Charters, A. W. Stott, G. J. Lawrence, A. M. Gibson, A. Conner, P. W. Blinkhorn, and V. Reeves (1997c), *Naturwissenschaften* **84**, 402.

Evershed, R. P., P. F. van Bergen, T. M. Peakman, M. C. Leigh-Firbank Horton, D. Edwards, M. Biddle, B. Kjolby-Biddle, and P. A. Rowley-Conwy (1997d), *Nature* **390**, 667.

Fenn, J. B., M. Mann, C. K. Meng, S. F. Wong, and C. M. Whitehouse (1989), *Science* **246**, 64.

Fenn, J. B., M. Mann, C. K. Meng, S. F. Wong, and C. M. Whitehouse (1990), *Mass Spectrom. Rev.* **9**, 37.

Flannery, M. B., B. A. Stankiewicz, J. C. Hutchins, C. W. White, and R. P. Evershed (1999), *Ancient Biomolecules.* **3**, 37.

Francis, G. W. (1981), *Chem. Phys. Lipids* **29**, 369.

Francis, G. W., and K. Veland (1981), *J. Chromatogr.* **219**, 3799.

Goad, L. J. (1991), in *Methods in Plant Biochemistry*, Vol. 7, B. V. Charlwood and D. V. Barnthorpe, Eds., Academic, New York, pp. 369–434.

Gulaçar, F. O., A. Susini, and M. Kolhn (1990), *J. Archaeol. Sci.* **17**, 691.

Horning, M. G., G. Casparrini, and E. C. Horning (1969), *J. Chromatogr. Sci.* **7**, 267.

Johnson, C. B., and R. T. Holman (1966), *Lipids* **1**, 371.

Khalid, S. A. (1983), in *Chemistry and Taxonomy of the Rutales*, G. Waterman and M. F. Erundan, Eds., Academic, New York, Chapter 10.

Lawson, A. M., Ed. (1989), *Mass Spectrometry*, Walter deGruyter, Berlin.

McCloskey, J. A. (1990), *Methods Enzymol.* 193.

Mills, J., and R. White (1994), *The Organic Chemistry of Museum Objects*, 2nd ed., Butterworths, London.

Mottram, H. R., S. N. Dudd, G. J. Lawrence, A. W. Stott, and R. P. Evershed (1999), *J. Chromatogr. A* **833**, 209.

Murata, T. (1977), *Anal. Chem.* **49**, 2209.

Nes, W. R. (1985), *Methods Enzymol.* **111**, 3.

Nichols, P. D., J. B. Guckert, and D. C. White (1986), *J. Microbiol. Met.* **5**, 49.

O'Donoghue, K., T. A. Brown, J. Carter, and R. P. Evershed (1994), *Rapid Commun. Mass Spectrom.* **8**, 503.

O'Donoghue, K., T. A. Brown, J. F. Carter, and R. P. Evershed (1996), *Rapid Commun. Mass Spectrom.* **10**, 495.

Pollard, A. M., and C. Heron (1996), *Archaeological Chemistry*, Royal Society of Chemistry, Cambridge.

Rahnier, A., and P. Benveniste (1989), in *Analysis of Sterols and Other Biologically Significant Steroids*, W. D. Nes and E. J. Parish, Eds., Academic, New York, pp. 223–249.

Raven, A. M., P. F. van Bergen, A. W. Stott, S. N. Dudd, and R. P. Evershed (1997), *J. Anal. Appl. Pyrol.* **40/41**, 267.

Regert, M., H. A. Bland, S. N. Dudd, P. F. van Bergen, and R. P. Evershed (1998), *Proc. R. Soc. Lond. B.* **265**, 2027.

Rose, M. E., and R. A. W. Johnstone (1982), *Mass Spectrometry for Chemists and Biochemists*, Cambridge University Press, Cambridge.

Stankiewicz, B. A., P. F. van Bergen, I. J. Duncan, J. F. Carter, D. E. G. Briggs, and R. P. Evershed (1996), *Rapid Commun. Mass Spectrom.* **10**, 174.

van Bergen, P. F., T. M. Peakman, E. C. Leigh-Firbank, and R. P. Evershed (1997), *Tetrahedr. Lett.* **38**, 8409.

Vincenti, M., G. Guglielmetti, G. Cassani, and C. Tonini (1987), *Anal. Chem.* **59**, 694.

Wakeham, S. G., and N. M. Frew (1982), *Lipids* **17**, 831.

Waller, G. R., Ed. (1972), *Biochemical Applications of Mass Spectrometry*, John Wiley & Sons, New York.

Waller, G. R., and O. C. Dermer, Eds. (1980), *Biochemical Applications of Mass Spectrometry*, first supplementary volume, John Wiley & Sons, New York.

CHAPTER

9

BIOMOLECULAR METHODS

RAÚL J. CANO

Environmental Biotechnology Institute and Biological Sciences Department, California Polytechnic State University, San Luis Obispo, California

9.1. INTRODUCTION

DNA, or deoxyribonucleic acid, is the molecule that carries all of the information—or the molecular blueprint—needed to specify an organism, from bacteria to human. Height, gender, and hair color, as well as the size of red blood cells, metabolic pathways, and the ability to utilize sugar and other nutrients, is delineated in the specific order of nitrogenous bases, aligned on a sugar-phosphate "backbone." The base sequences make up genes, which interact with other genes or act alone to establish all the traits of any organism. Genes are contained on chromosomes, which (in eukaryotes) are found within the nucleus of cells. This genetic information is in every cell that is part of that organism.

Much can be learned from bits and pieces of DNA that may be preserved in "fossil" organic remains in living animals. Comparisons of selected DNA sequences allows for the determination of closeness of relationships, proximity of shared common ancestors, and the ability to estimate, at least tentatively, the time of divergence of lineages. In most cases, the molecular data thus obtained verifies and substantiates estimates based on comparisons of shared derived characters.

By targeting specific gene sequences, it may be possible to infer phylogenetic or genealogical relationships among the taxa studies. Also, through comparative studies of sequences from various extant taxa, it is possible to estimate the extent of evolutionary divergence between them (Kocher et al. 1989; Naito et al. 1991). Base-pair substitutions accumulate over time within a species genome. By comparing the amount and type of these changes, one

Modern Analytical Methods in Art and Archaeology, Edited by Enrico Ciliberto and Giuseppe Spoto. Chemical Analysis Series, Vol. 155
ISBN 0-471-29361-X

could estimate how quickly some DNA "evolves" relative to other segments or which bases or genes have the most flexibility or are more conserved over time.

Recent improvements in methods of extraction and amplification of DNA have made it possible to obtain genetic sequence information from varied sources of ancient organic matter. One of the first sources of ancient DNA was a museum skin of the extinct Quagga, (Higuchi et al. 1984; Higuchi and Wilson 1984). Their work showed that it was possible to study the DNA of extinct animals and represented the birth of molecular paleobiology. Their discovery also showed that it is possible that other taxodermic specimens as well as tanned leather and parchment might be suitable sources for ancient DNA.

Animal skins, tanned in various ways, were used as very early sources of clothing, footware, armor, and writing material. Woodward et al. (1996) showed that DNA can be extracted from parchments of the Dead Sea Scrolls, which are nearly 2000 years old. They used the cytochrome-*b* gene to identify ruminant species used to produce the parchments of the scrolls. This was done by comparing unknown cytochrome-*b* sequences with those of all the ruminant families in a database. A second region of the mtDNA, the *D* loop, was used to identify closely related animals. This study made it possible to validate matches of scroll fragments that were made previously, based solely on skin morphology. Consequently, by looking at individual and herd relationships, it is also possible that hides from known areas may serve as reference points to texts of the same provenance and may give evidence for exchange of manuscripts and trade between different geographical areas.

It can be anticipated that other archaeological and literary texts may benefit in a similar way from these observations. The European Commission has been concentrating great efforts on preserving vast archives of leather bound books, parchments and papyrus kept for hundreds of years, but facing deterioration from pollution and acid rain (STEP 1994).

De Reyer et al. (1996) in France and German scholars at Gottingen University (Herrmann, personal communication) are presently applying molecular biological tools to the conservation and study of materials of cultural heritage. De Reyer et al. (1996) have examined skin and parchment as well as the gut used as stitching, using sequences of the 18S rRNA gene for species identification.

Verification of the authenticity and context of historical records and art can be possible through DNA extraction and analysis, which would be of particular use to museums and other historical institutions. It is also possible that international customs seeking to identify goods made from endangered animals and institutions wishing to genetically tag important works made of organic material could use the techniques.

Forensic science already uses DNA extraction from a number of biological and organic sources in criminal cases. Most work has concentrated on the ability to extract human fluids from material, but it is feasible that identification of the clothing (leather) using genetic means would be of assistance in police investigations.

The isolation and characterization of fossil DNA, until recently (Higuchi and Wilson 1984), was considered unattainable as the methodologies for extracting minute quantities of partially degraded DNA and their subsequent enzymatic amplification were not available. With the advent of the polymerase chain reaction (PCR) (Mullis et al. 1986), a new analytical tool became available for the molecular study of fossils. It is now possible to conduct molecular studies of extinct organisms or fossils utilizing their DNA to unravel biological and evolutionary questions.

The value of fossil evidence is that it may demonstrate the condition of taxa before evolutionary divergence obscured phylogenetic relationships. Because they are older, ancient fossil DNA sequences should be less divergent than extant sequences and should, therefore, have value for relating more derived extant taxa. When compared with extant DNA, ancient DNA sequences may also provide an insight into the pattern of molecular evolutionary change through time.

There is already a body of scientific evidence built that supports the use of DNA from extinct animals and plants for phylogenetic studies. Higuchi et al. (1984) demonstrated that remains of a mammoth and the extinct species, the quagga, contained fragments of the original DNA. Pääbo (1989) reported the extraction of clonable DNA from a 2400-year-old mummy of a child. Subsequent DNA analysis revealed fragments measuring approximately 3.4 kilobase pairs (kbp). Thomas et al. (1989) isolated DNA from hair found in a century-old untanned hide and a piece of dried muscle collected from an extinct marsupial wolf. This DNA was later enzymatically amplified by PCR and phylogenetic studies were made. More recently, Handt et al. (1994), Rollo et al. (1994), and Cano et al. (1997) isolated and characterized ancient DNA from the 5000-year-old Tyrolean iceman.

9.2. STRATEGIES OF ANALYSIS

In terms of the analysis of ancient DNA, the single most important technology is the polymerase chain reaction. This exponential amplification of DNA targets produces enough copies of the target strand of DNA to be manipulated and analyzed through standard molecular techniques, such as cloning and enzymatically directed sequencing.

As important as PCR is, however, the success or failure of such an assay depends upon the quality and care of the sample preparation. New and refined techniques for extraction of biomolecules have been developed, but these have largely been optimized for the extraction of modern DNA. Such methods must be optimized and the solutions modified for the extraction of small fragments of DNA that may have undergone some diagenesis and for the removal of PCR inhibitors.

New techniques of DNA detection such as the fluorogenic 5′ exonuclease chemistry (Holland et al. 1991) and oligonucleotide probe array strategy (Lipshutz et al. 1997) could revolutionize the field of paleoarcheology. Such strategies have the potential of detecting a large number of genes at once and potentially arrive at a nucleic acid sequence of the extracted DNA without extended amplification protocols.

Analytical software is available, for example, CLUSTAL (Higgins et al. 1992) and FASTA (Miller et al. 1991), that allows the sequences obtained by the above methods to be matched against homologous sequences from other species that have been entered into a data bank. Statistical analyses can then be performed and estimates of relatedness and genetic distance can be obtained. Phylogenetic trees based on sequence data can be constructed, using software packages such as MEGA (Kumar et al. 1993), PAUP (Swofford 1990), and PHYLIP (Felsenstein 1989). This allows for the objective placement of an organism within the framework of known taxa (Fig. 9.1). It also allows any modern DNA that may be contaminating ancient tissues to be characterized and possibly recognized.

9.3. SELECTIONS OF GENE SEQUENCES FOR ANALYSIS OF ANCIENT DNA

When working with DNA obtained from archeological materials, the selection of gene sequences for amplification and/or analysis is a crucial step. The chances for successful amplification are increased if the target gene sequence is present in multiple copies within each cell. Nuclear DNA sequences of ribosomal constituents, such as the small and large subunits of the rRNA, as well as the glutamate synthase gene are often used in such studies. Mitochondrial DNA sequences, including the cytochrome-*b* and -*c* genes and the *D*-loop region, are also good candidates not only because they are present in multiple copies in each cell, but also because the complete mitochondrial genome has been sequenced for many taxa and has been entered into data banks, and are thus available for comparative studies. The human genome project, as well as other genome projects underway will provide additional information regarding genes or DNA segments that could be used as targets for archeological or phylogenetic analysis of ancient remains.

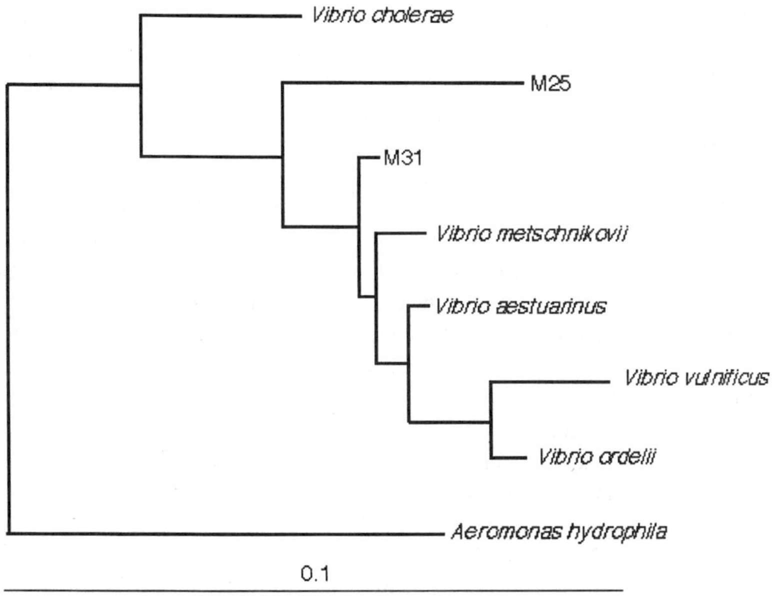

Figure 9.1. Maximum-likelihood tree showing the relationship of sequences from clones of DNA from the iceman's colon to those of *Vibrio* spp. Accession number of the sequences utilized: *Aeromonas hydrophila*, M59148; *Vibrio aestuarianus*, X74689; *Vibrio cholerae*, X74695; *Vibrio metschnikovi*, X74712; *Vibrio ordalli*, X74718; *Vibrio vulnificus*, X74727; *Aeromonas hydrophila*, X87271 was the outgroup.

For phylogenetic significance, it is desirable that the selected homologous primer sequences for known taxa flank regions of relatively high variability. This allows for better definition of phylogenetic placement than if there are relatively few changes across a broad range of taxa. Also, it is easier to tell if there is contamination with modern DNA. If DNA from ancient samples can be obtained and amplified, then genetic analysis can reveal if any base-pair changes in the sequences from ancient materials are intermediate between the modern taxa being used for comparison.

9.4. PRESERVATION POTENTIAL OF BIOMOLECULES

The DNA in living cells is continuously impacted by diagenetic processes such as hydrolysis of bonds between base and sugar backbone of the DNA and oxygen radical-induced attacks on bases and sugar residues. Living organisms are able to survive despite these diagenetic processes, however, due to efficacious DNA repair systems contained in every cell. After death, however, DNA repair ceases but many of the destructive processes may continue. The extent of DNA damage depends on the taphonomy of the specimen and the degree of hydration of the specimen. In many cases, ancient DNA molecules are reduced to a size of measuring several hundred base pairs in length and may be affected by oxidative changes. Additionally, because of bacterial or fungal contamination of some specimens, only a small fraction of a bulk of the extracted DNA may be of ancient origin.

It is a commonly held belief, based on experimental evidence as well as from extrapolated predictions based on studies of DNA in aqueous solution, that nucleic acids do not survive in fossil remains on a geological time scale. The taphonomy of the ancient DNA sample, however, is more important that the geological age of the sample itself. For example, amino acid racemization studies suggest that certain tissues of insects entombed in amber curtail amino acid racemization while 5000-year-old bone amino acids are almost fully racemized (Poinar et al. 1996).

The double-stranded, helical structure of DNA is more resistant to damage than single-stranded RNA, but its structure and chemistry make it susceptible to certain types of diagenetic damage over time. Conversion of bases through hydrolytic deamination (guanine changes to xanthine, cytosine to uracil or its derivatives) and depurination (removal of the bases guanine and adenine from the sugar–phosphate backbone) affect the informational content of the molecule.

Exposure to oxygen free radicals or ultraviolet (UV) radiation also damages DNA strands (Eglington et al. 1991). Mechanisms have evolved in living organisms for repairing such DNA damage as it occurs, maintaining genetic information, and preventing accumulation or error (Grossman 1991). With the death of the organism, this self-repair process stops, while enzymatic attack and exposure to water, oxygen, and UV radiation continue with advancing decay. Exposure to water is probably the single most destructive force acting on the DNA molecule. Water has been shown to initiate strand breaks by attacking the base–sugar bonds. Where the base is lost, the chain is weakened, and eventually cleaved (Eglington et al. 1991; Lindahl 1993).

Oxidation is another source of DNA damage, and removing DNA from water as in amber or bone does not protect the molecule from oxidative attack. Oxygen, in its molecular state, does not attack DNA, but rather it is the formation of oxygen free radicals that attack the nitrogenous bases. Oxidative attack would be rapid at first but then would level off (Pääbo 1989). It is proposed that chelation of copper or other metal ions (Eglington et al. 1991) enhances the preservation of this molecule by contributing to a reducing environment, and compensating for the production of oxygen free radicals. Exposure to UV light also causes extensive damage and degradation of DNA. Rapid burial or encapsulation of organisms in protective matrices is important to minimize the consequences of UV damage to DNA.

9.5. PROBLEMS OF WORKING WITH "FOSSIL" BIOMOLECULES

The extreme sensitivity of PCR, which opens the door to the direct analysis of DNA obtained from ancient materials, also poses the most complications. The fact that PCR technology can amplify as little as one molecule of DNA means that minute amounts of contaminating DNA from modern sources, such as bacteria, soil fungi, or human skin cells can also be amplified. Indeed, any such modern contaminant would probably be amplified preferentially over ancient target molecules, owing to the probable state of degradation of the latter.

New technologies that are not based on the enzymatic amplification of ancient DNA may be necessary to reduce the impact of contamination of ancient sequence analysis. For example, if DNA hybridizations using sensitive methods could be carried out in the absence of an amplification step, this could reduce errors due to contamination of archeological samples with modern DNA. One such method is the newly developed DNA array technology. This technology will be discussed below.

Ultimately, however, the proof of the authenticity of any DNA presumably obtained from ancient materials comes from careful analysis of sequence data. If phylogenetic analysis of the sequences does not agree with predicted relationships based on morphology, then the DNA data must be carefully reevaluated. Also the analysis of at least two different genes or gene regions should be done, and the results of both should show similar or identical phylogenies, before any claims can be made regarding the sources of the DNA.

9.6. POLYMERASE CHAIN REACTION ASSAY

Once the DNA from fossils has been successfully extracted, it is now ready for enzymatic amplification. Needless to say that gene selection and primer design are of primary importance and will depend upon the goals of the amplification assay. As each target DNA and its corresponding primer pair(s) are unique, the reaction and conditions and thermal cycling protocol will vary with each sequence, and therefore the assay must be optimized each time a new primer set is used. The Stoffel fragment of *Taq* polymerase is sometimes used for initial studies of fossil DNA as this enzyme is more tolerant to fluctuation in Mg^{2+} concentrations, and therefore, would increase the chances for initial success.

Many fossil samples have tannins, porphyrins, hematin, humic acid, and other inhibitors of the PCR reaction. For this reason, the inclusion of bovine serum albumin (BSA fraction V, Sigma) in the reagent mixture at concentrations of 2 μg/ml is sometimes added to the reaction mixture to palliate the inhibitory activity of fossil DNA contaminants.

Also, to reduce spurious hybridization of primers to nonhomologous target DNA sequences, some modification of a "hot start" PCR should be used. A method has been largely successful in our laboratory that does not require the separate addition of polymerase to each tube or the use of wax beads. It involves placing the reaction mixture and all the reagents in an ice bath throughout the preparation and dispensing of the mixture into the tubes and the addition of the template to the mixture. While this is done, a soak cycle of 80°C for 5 min is programmed into the thermocycler. When the heat block of the thermocycler reaches the 80°C, the tubes are removed from the ice and placed immediately on the heat block. From then on, the temperature cycling protocol proceeds normally.

Enzymatic amplification of DNA by PCR can be performed in as little as a single DNA molecule. Therefore, contaminating DNA, if it shares homologies with primer sequences, could be amplified as well as the desired target DNA. It is imperative that proper laboratory practices and protocols be implemented to validate all DNA isolations and amplifications. The following validation procedures should be implemented: (a) thorough training of the personnel involved in the collection and processing of specimens; (b) separate extinct and extant samples facilities; (c) separate pre- and post-PCR work areas to reduce product carry-over contamination; and (d) stringent aseptic techniques and work space DNA and biological decontamination. This latter procedure should include weak positive controls to reduce the risk of carry-over contamination with positive-control amplicons, routine monitoring of environmental contamination with DNA by performing DNA-free PCRs using a panel of commonly used primers, multiple replicates of sample preparation and DNA amplifications, and statistical evaluation of the data by sequence alignment and phylogenetic inference analyses.

The basic PCR laboratory layout should consist of three areas: (a) reagent preparation area (area 1); (b) sample preparation area (area 2); and (c) PCR analysis area (area 3). Additionally, separate laboratories should be used for the processing of extinct and extant taxa.

DNA extractions should be carried out in areas designated solely for the extraction of extinct DNA. Similarly, separate areas (in two different rooms) should be used to process extinct and extant specimens. Dedicated thermocyclers and pipetting devices (with aerosol-resistant tips) must also be used to reduce the risk of cross contamination. The preparation of reagent mixes for the polymerase chain reaction should take place within dead air boxes in designated areas away from sample preparation and PCR evaluation areas. Sample carry over must also be avoided by the use of filter-sterilized reagents and decontamination of utensils and work areas with 10% bleach followed by UV irradiation. Electrophoreses should also be carried out in a dedicated area, away from that of DNA extraction and PCR. Appropriate controls, including reagents, water, DNA extraction reagents, and bacterial DNA (which should not be amplified with the selected primers), need to be performed weekly (and concurrently with tissue amplifications). Additionally, sequences should be aligned with those from a variety of taxa in order to eliminate those samples that clearly do not represent amplification of the desired target DNA.

9.7. SEQUENCING OF AMPLIFICATION PRODUCTS

There are many suitable protocols available for determining nucleotide sequencing of PCR products, both from clones or directly from PCR reactions. These include single- and double-stranded template sequencing with Sequenase (USB, Cleveland, OH), cycle sequencing, and other techniques utilizing thermoresistant DNA polymerases. Each has its advantages and disadvantages, which must be evaluated by the investigator as best suited for the intended goal of the project.

It should be noted, however, that direct sequencing of PCR products normally yield a "consensus" sequence as the PCR product represents a "pool" of individual amplicons reflecting both template variation, template integrity, and polymerase fidelity. Sequences proceeding from cloned amplicons represent the sequence of that single amplicon ligated to the vector. It is not a consensus sequence and might reflect both template variations and/or polymerase errors. When sequencing cloned amplicons, it is recommended that a minimum of six different clones be used to generate a "consensus" sequence. Alternatively, purified plasmid DNA from 10 to 20 clones may be pooled, and a single sequencing reaction conducted as this represents a "consensus" sequence of the 10 to 20 clones pooled.

Automation of DNA sequencing, including the incorporation of fluorescent dye chemistry, has greatly improved both the quality and output capabilities of sequencing DNA from all sources. In dye terminator chemistry, fluorescent tags are attached to the chain terminating nucleotides with each of the four dideoxynucleotides carrying a spectrally different fluorophore. During cycle sequencing both dye-labeled dideoxynucleotides and deoxynucleotides are present, resulting in random chain termination during nucleotide incorporation and labeled molecules of most every possible base length. Unincorporated dye terminators are then removed from the reaction by using a spin column or ethanol precipitation step. Each reaction is subsequently electrophoresed on a polyacrilamide sequencing gel, utilizing only one lane on the gel for each primed reaction. The labeled DNA fragments are detected by their fluorescence as they migrate past the detector, which scans horizontally across the gel.

Automated fluorescent DNA sequencing systems in general offer many advantages over manual sequencing in accuracy, reproducibility, and ease of use. Both the software and basic chemistry used in automated fluorescent sequencing have drastically improved sequence quality and output. The most significant advantage of this system is the ability of computer software to perform base calling and sequence analysis; eliminating the possibility of errors arising when DNA sequences are read and processed manually. Automation also permits one to easily and qualitatively compare multiple runs of the same sequence for determining consensus sequences or heterozygous positioning. Analysis software allows the review of the run conditions (voltage, wattage, amperage, and temperature) and errors and command logs providing validation and trouble shooting of each run. Furthermore, sequence assembly software can "clean up" sequences, identifying and removing ambiguous stretches and primer or plasmid sequences.

Several modifications of fluorescent cycle sequencing chemistry have assisted in improving automated DNA sequencing and base calling. Improvements specifically in dye terminator chemistry include the incorporation of dITP in place of dGTP in the reaction mix, which aids in minimizing band compressions for more accurate base calling. Five percent DMSO (dimethyl sufoxide) included in reaction mixes decreases peak height variability, contributing to base-calling precision. Development of *Taq* polymerase FS (Perkin-Elmer, Foster City, CA), a variant of *Taq* DNA polymerase, has further refined fluorescent dye cycle sequencing processes. Mutations in the active site and the amino terminal domain result in less discrimination against the labeled terminating dideoxynucleotides and decreased $5' \Rightarrow 3'$ nuclease activity, respectively. The end result of these mutations is a much more even peak intensity pattern lending to more accurate base calling by the software.

Aside from the general advantages described above, automated fluorescent sequencing has several benefits when specifically working with ancient DNA. Accurate sequence identification is especially crucial when sequencing ancient DNA for both classification purposes and phylogenetic analysis. Manual base calling cannot always be completely eliminated with automated fluorescent systems, but the degree to which base calling relies on perception is greatly diminished. Assignment of IUB codes and viewing electropherograms of aligned forward and reverse strands can also effectively identify potential polymorphisms important in phylogenetic studies. Heter-

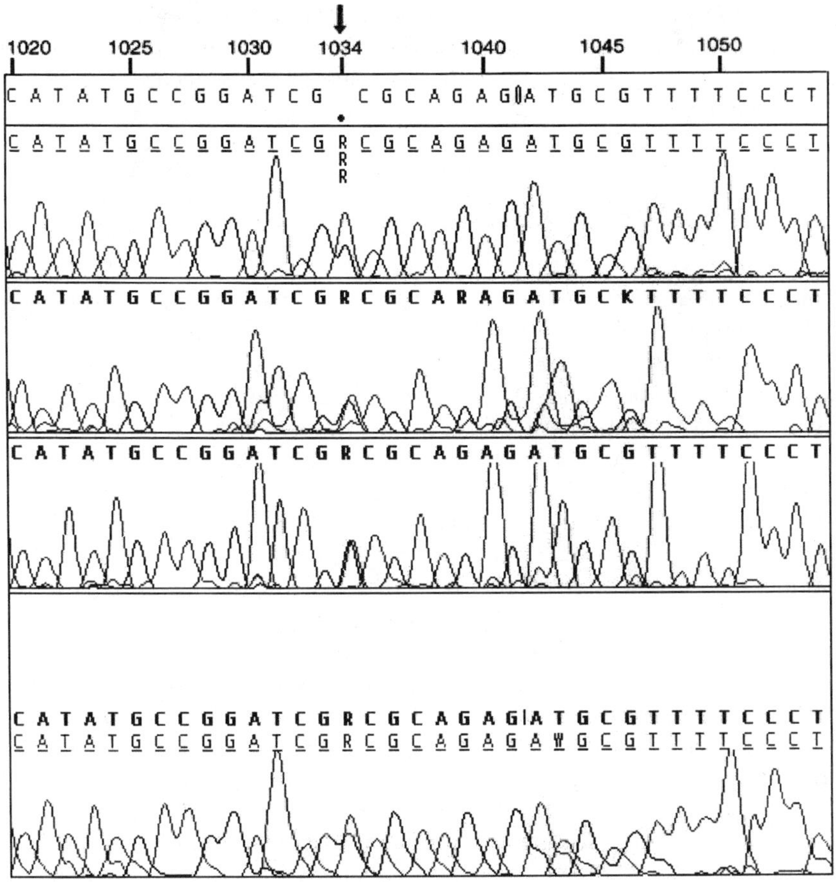

Figure 9.2. Factura output from a dye dideoxy terminator cycle sequencing reaction indicating the presence of a heterozygous position (see arrow) in the 16S rRNA of ancient *Bacillus* spp. DNA.

ozygous base positions are more easily observed by viewing an electropherogram created by automated fluorescent system software. Computer software programs such as Factura (PE/ABI) assign IUB codes to mixed base positions by using a ratio set by the user to compare the highest peak to each of the other three peaks in the same location. If the ratio between any of the three lower peaks and the higher one is above the set threshold percentage, an IUB code is assigned. This aids in the detection of heterozygous positions (Fig. 9.2), which may be present at certain positions in multicopy genes (e.g., 16S rRNA gene) and undetectable by autoradiographic methods.

Regardless of the method used and the approach to sequencing, sequence reproducibility and comparison to those of known, related taxa should be performed to increase the degree of reliability on the sequence data generated from ancient DNA

9.8. OLIGONUCLEOTIDE ARRAYS

There are several new technologies for studying the simultaneous presence of large numbers of genes. These technologies may be roughly divided into serial and parallel methods. The serial methods involve direct, large-scale sequencing of DNA molecules while the parallel approaches are based upon hybridization to DNA immobilized on glass (termed "microarrays") or to synthetic oligonucleotides immobilized on silica wafers or "chips" (termed "probe arrays"). In both parallel methods, detection of hybridized probes is by means of incorporated fluorescent or radioactive nucleotide homologs. These methods are the conceptual descendents of filter- or plastic-immobilized targets detected by radioactive or enzymatically labeled probes. Arrays of hybridization targets, constructed at high density in small areas (e.g., 100,000 different "probes" on a 2 × 2 cm filter or glass slide) are now commonly referred to as microarrays. Microarray technology now allows us to look at many genes at once and determine which are present or expressed in a particular cell or sample type. DNA molecules representing many genes are placed in discrete spots on a microscope slide. This is called a microarray. Thousands of individual genes can be spotted on a single square inch slide. Next, DNA extracted from samples (e.g., archeological remains, papyrus, parchment, etc.) is purified to remove hybridization inhibitors (or PCR inhibitors if necessary). The purified DNA molecules are then "labeled" with fluorescent dye(s) in order to be detectable, either by microscopy or by confocal laser technology, and then hybridized to the microarray. After hybridization and washing away all of the unbound DNA, the microarray can be examined under a microscope or with the aid of a laser beam and assess the hybridization pattern in the microarray. Since it is known which gene each spot represents, and the DNA only hybridizes to its complementary DNA, it can be determined which genes are present in the samples (Fig. 9.3).

DNA microarrays can be used to test for particular sets of genes, by using probes specific to those genes and their alleles. Alternatively, microarrays can be used to carry out raw sequencing of phylogenetically important genes. It is also possible to use such chips to probe archeological specimens for the presence of important human genes.

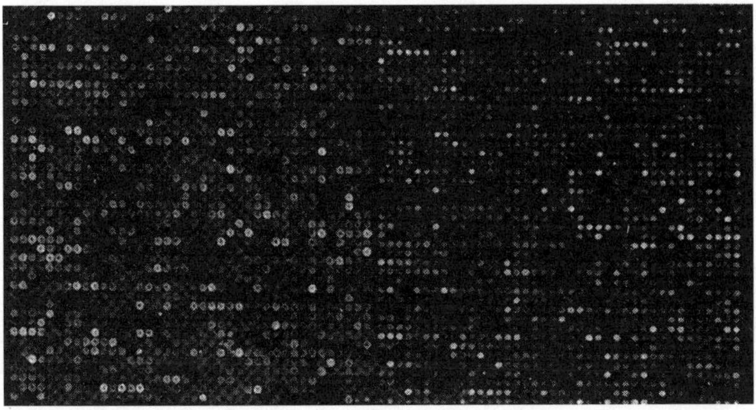

Figure 9.3. Typical results of DNA probed with an oligonucleotide array. The intensity of each "spot" indicates the degree of homology between the target DNA and the probe. Spots not illuminated by the fluorescent dye indicate not hybridization.

9.9. CONCLUSIONS

Molecular methods, when used judiciously and performed by trained technical personnel, with a thorough understanding of the procedures and their limitations, are valuable assets in the study and evaluation of archeological specimens and art.

REFERENCES

Cano, R., T. Cox, C. del Cueto, J. Tuttle, P. Orkand, F. Kunzel, and F. Tieffenbrunner (1997), *Abstract N-45*, American Society for Microbiology, Miami Beach, FL.

De Reyer, D., S. Pilbou, N. Dennebouy, and M. Monnerot (1996), *ICOM Commitee for Conservation*, Vol II, pp. 759–766

Eglinton, G., and G. A. Logan (1991), *Philos. Trans. R. Soc. London B* **333**, 315.

Felsenstein, J. (1989), *Cladistics* **5**, 164.

Grossman, L. (1991), Repair of damaged DNA, in *Encyclopedia of Human Biology*, Vol. 6, pp. 547–553.

Handt, O., M. Richards, and M. Trommsdorff (1994), *Science* **264**, 1775.

Higgins, D. G., A. J. Bleasby, and R. Fuchs. (1992), *Comput. Appl. Biosci.* **8**, 189.

Higuchi, P., and A. C. Wilson (1984), *Fed. Proc.* **43**, 1557.

Higuchi, R., B. Bowman, M. Freiberger, O. A. Ryder, and A. C. Wilson (1984), *Nature* **312**, 282.

Holland, P., R. Abramson, R. Watson, and D. Gelfand (1991), *Proc. Natl. Acad. Sci. USA* **88**, 7276.

Kocher, T. D., W. K. Thomas, A. Meyer, S. V. Edwards, S. Pääbo, F. X. Villablanca, and A. C. Wilson (1989), *Proc. Natl. Acad. Sci. USA* **86**, 6196.

Kumar, S., K. Tamura, and M. Nei (1993), *MEGA, Molecular Evolutionary Genetic Analysis*, Version 1.0, Pennsylvania State University, University Park, PA.

Lindahl, T. (1993), *Nature* **362**, 709.

Lipshutz, R., D. Morris, and M. Chee (1997), *Biotechniques* **19**, 442.

Miller, P. L., P. M. Nadkarni, and N. M. Carriero (1991), *Comp. Appl. Biol. Sci.* **7**, 71.

Mullis, K., F. Faloona, S. Scharf, R. Saiki, G. Horn, and H. Erlich (1986), *Cold Spring Harbor Symposium Quant. Biol.*, Part 1, Cold Spring Harbor Laboratory, Cold Spring Harbor, NY, pp. 263–273.

Naito, E., K. Dewa, H. Ymanouchi, and R. Kominami (1991), *J. Forens. Sci.* **37**, 396.

Pääbo, S. (1989), *Proc. Natl. Acad. Sci. USA* **86**, 1939.

Poinar, H. N., M. Höss, J. L. Bada, and S. Pääbo (1996), *Science* **272**, 864.

Rollo, F., W. Asci, S. Antonini, I. Marota, and M. Ubaldi (1994), *Experientia* **50**, 576.

European Commission (1994), STEP Leather Project, R. Larsen, Ed., Bjarnholt Repro, Copenhagen.

Swofford, D. L. (1990), *Phylogenetic Analysis Using Parsimony*, Version 3.0, Computer program, Illinois Natural History Survey, Champaign, IL.

Thomas, R., H. W. Schaffner, A. C. Wilson, and S. Pääbo (1989), *Nature* **340**, 465.

Woodward, S. R., G. Kahila Bar-Gal, P. Smith, C. Greenblatt, J. Zias, and M. Broshi (1996), in *Current Research and Technological Developments on the Dead Sea Scrolls*, W. Parry and R. S. D. Brill, Eds.

CHAPTER

10

RAMAN SPECTROSCOPY

FRANCO CARIATI and SILVIA BRUNI

Dipartimento di Chimica Inorganica, Metallorganica e Analitica, Università degli Studi di Milano, Milano, Italy

10.1. INTRODUCTION TO THE RAMAN EFFECT

The electromagnetic radiation, colliding with a molecule, can be transmitted, absorbed, or scattered. When the collision is elastic, the incident and the scattered radiation have the same frequency, producing the effect known as Rayleigh scattering. On the contrary, in an inelastic collision, the scattered radiation has either higher or lower frequency with respect to the incident one, giving the so-called Raman scattering or Raman effect. In the case of the Raman effect, the differences in energy between the incident and the scattered photons are quantized, corresponding to the energy differences within the vibrational or the rotational levels of the molecule. In Figure 10.1 the fundamental and the first excited vibrational energy levels of the molecule are reported.

When a photon having energy $h\nu_0$ collides with a molecule, two different events can take place. In one case the frequency of the radiation is suitable to induce the transition of the molecule toward one stationary excited state (electronic absorption). On the other hand, when the frequency of the radiation is not appropriate to induce this effect, the molecule itself will occupy a nonstationary state, indicated by a broken line in Figure 10.1. Consequently, the molecule is not able to maintain this state for a long time and diffuses energy. If the scattered photon has the same energy as the exciting photon, the Rayleigh scattering can be observed.

However, in a different situation, the molecule does lose energy, reaching the state with $v = 1$ and not returning to the vibrational state $v = 0$. In this case the scattered photon energy is $h\nu_0 - h\nu_1 = h(\nu_0 - \nu_1)$, where ν_1 is the

Modern Analytical Methods in Art and Archaeology, Edited by Enrico Ciliberto and Giuseppe Spoto. Chemical Analysis Series, Vol. 155
ISBN 0-471-29361-X

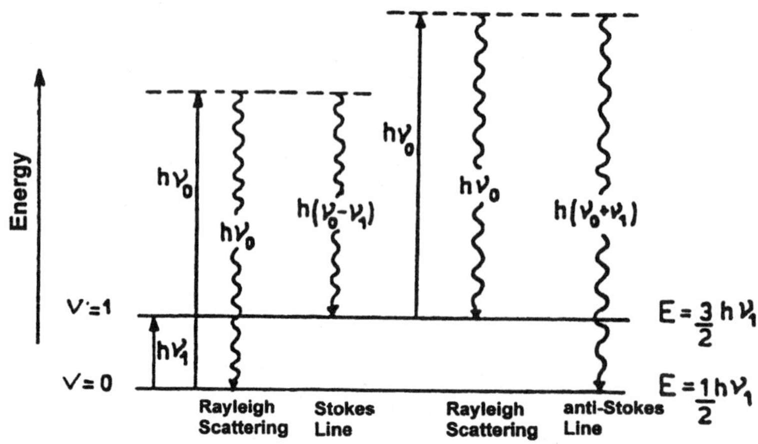

Figure 10.1. Origin of vibrational Raman effect.

frequency of the vibration ($v = 0 \rightarrow v = 1$). In addition to the Rayleigh line, the spectrum of the scattered radiation shows, at smaller frequency, the Raman line called the Stokes line. In the last possible scenario the molecule, originally in the $v = 1$ state, is first promoted to the state with energy $h(\nu_0 + \nu_1)$ and successively returns to the ground vibrational state. The energy of the scattered photon is $h(\nu_0 + \nu_1)$ and the spectral line presents a smaller wavelength and a bigger frequency than the incident radiation. This is the Raman scattering which gives rise to the anti-Stokes line. However, being the majority of the molecules in the ground vibrational state at room temperature, the Stokes lines are stronger compared to the anti-Stokes ones.

If we arbitrarily set the value of the monochromatic incident radiation frequency as zero (the Rayleigh frequency), we can obtain the Raman frequencies directly. It is worth noting that the frequencies of the scattered Raman radiation are independent from the exciting radiation wavelength. In fact, the same Raman shifts $\Delta\nu$ can be obtained, for instance, using the green line of an argon ion laser (514.5 nm, 19436 cm^{-1}) as exciting radiation or the blue line of the same laser (488 nm, 20492 cm^{-1}). It is important to highlight that the Raman effect interpretation is based on the effect of the electric dipole moment **P** induced in the molecule under the influence of an electric field **E**. This induced dipole moment generally depends on the charge distribution orientation with respect to the direction of the electric field.

When a molecule is subjected to an oscillating electric field,

$$E = E_0 \,\text{sen}\, 2\pi\nu_0 t$$

a dipole moment is induced, which assumes three frequencies: ν_0 (Rayleigh scattering), $\nu_0 - \nu_1$ (Raman scattering of Stokes type), and $\nu_0 + \nu_1$ (Raman scattering of anti-Stokes type).

It is important to note that the allowed Raman vibrational modes are those causing a variation of polarizability.

10.2. RAMAN INSTRUMENTATION AND TECHNIQUES

Up to about 1970 a Raman spectrophotometer used, as exciting source, a mercury discharge tube emitting a blue line at 435.8 nm. Before this date Raman spectroscopy, contrary to IR, was not used for routine analysis because the scattered radiation was so weak in intensity that it usually took several hours to record a spectrum. However, the potential of Raman spectroscopy greatly widened with the use of lasers as exciting sources.

The advantages of using lasers are (a) a single and intense frequency source that replaces the multiple-lined mercury lamp; (b) a better resolution; (c) a highly coherent light, which enables much easier focusing and collimation of the radiation; and (d) a large availability of possible exciting frequencies and therefore the possibility to study colored samples without causing any electronic transitions.

The scheme of a typical Raman spectrophotometer is shown in Figure 10.2. The laser light hits the sample and the scattered radiation is collected and directed to the monochromator (up to a total of three), and finally the light reaches the detector, which is usually a photomultiplier. This detector can be replaced by a multichannel photodiode array (514 or 1024 diodes) in order to register the spectrum more rapidly. In this case it is possible to accumulate a large number of spectra, thus improving the signal-to-noise ratio,

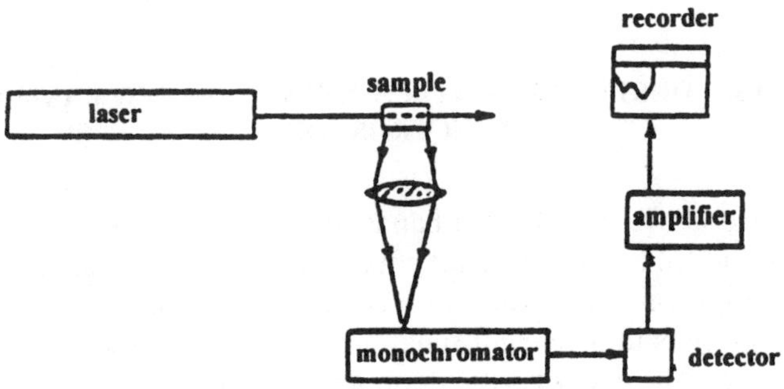

Figure 10.2. Schematic representation of Raman spectrophotometer.

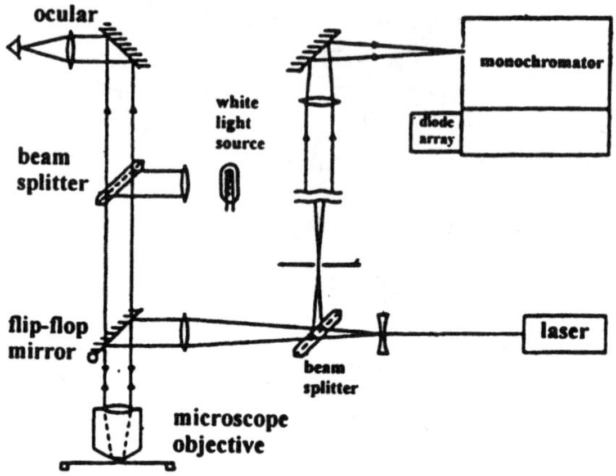

Figure 10.3. Scheme of Raman microspectrophotometer with diode array detector.

and to record the weak Raman signals. A Raman spectrophotometer can be coupled with an optical microscope, as shown in Figure 10.3.

Raman microscopy can be applied either in transmission or in reflection, depending on the type of problem to be solved or the type of sample. In Figure 10.4 the micro–Raman spectrum of a red stain on the Carrara marble of the facade of Certosa of Pavia (1473–1550) (Bruni et al. 1995) is shown. This spectrum clearly matches that of a powdered minium sample (see Fig. 10.4).

In a more expensive version, the Raman microspectrophotometer is equipped with a confocal microscope that presents two pinholes (see Fig. 10.5). With the confocal microscope, only the radiation scattered by the focused area goes through the second pinhole and reaches the detector, eliminating the radiation scattered by a not perfectly focused zone.

10.3. FOURIER TRANSFORM NEAR-INFRARED RAMAN SPECTROSCOPY

As mentioned above, the Raman bands are usually rather weak. Therefore, they can be hidden by any resonant effect such as fluorescence, either coming from the sample or from some extraneous compounds. Usually the fluorescence yields results four to eight times greater in intensity than Raman scattered radiation. The intensity of a fluorescence signal can be suitably reduced using an exciting source in the near infrared (NIR) and an interferometer

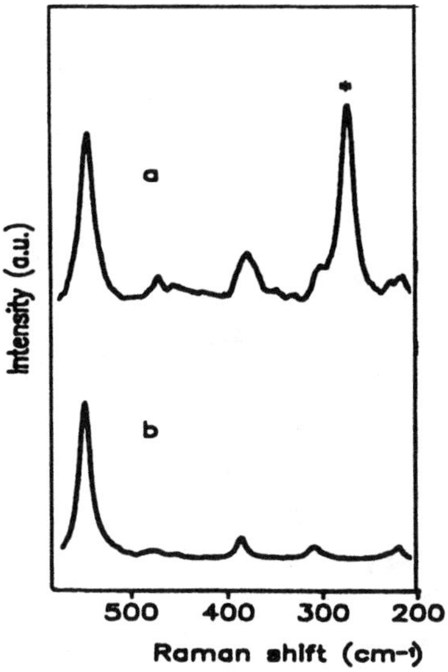

Figure 10.4. Micro–Raman spectrum of (*a*) red stain on Certosa marble; (*b*) minium ($\lambda_{ecc} = 647.1$ nm). The asterisk indicates a calcite band.

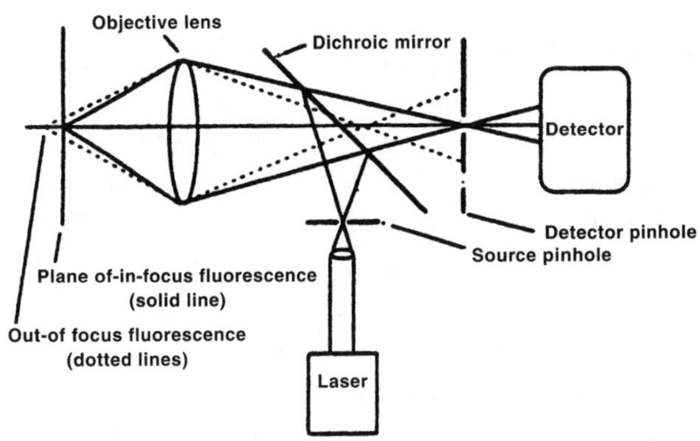

Figure 10.5. Scheme of confocal microscope with laser source.

to analyze the scattered radiation. In fact, the fluorescence yield is reduced when the photon exciting energy has value lower than that required to induce the electronic excitation.

All these drawbacks can be practically solved using a Nd–YAG laser, which allows us to eliminate the fluorescence and the sample thermal decomposition. This laser emits a radiation with wavelength of 1064 nm, which is in the NIR region of the electromagnetic spectrum. The scheme of a Fourier transform (FT) NIR Raman spectrophotometer is shown in Figure 10.6.

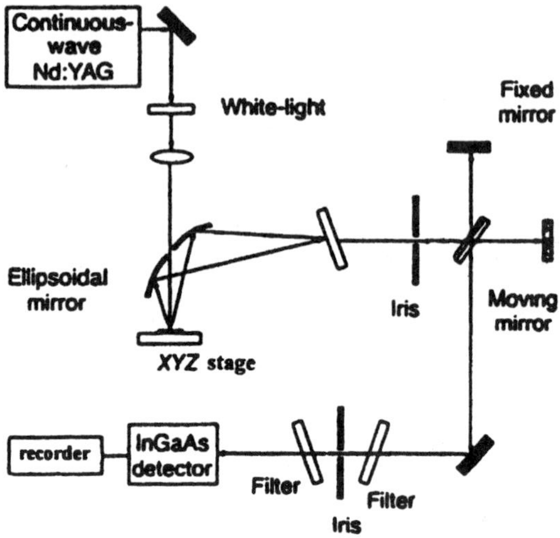

Figure 10.6. Schematic drawing of FT–NIR Raman spectrophotometer. Presence of He–Ne laser is indispensable to control optical alignment of apparatus.

10.4. RESONANCE RAMAN SPECTROSCOPY

The resonance Raman (RR) effect, or RR scattering, consists in the excitation of a molecule by using a radiation characterized by a frequency within those belonging to one of its absorption bands. As a consequence, the bands related to the vibrations of the chromophore group, responsible for the electronic absorption, are enhanced.

The conditions required to obtain strong Raman bands vary from compound to compound. In some cases the same experimental conditions as for normal Raman spectra are utilized. In other cases it is necessary to operate (i) in backscattering or (ii) at low temperatures and by using low laser power. The main advantages of RR are a strong enhancement of some Raman bands and the effect selectivity.

The record of RR spectra usually requires a normal Raman instrumentation. The more used sources are argon or krypton ion lasers, which allow use of up to 20 different wavelengths. Obviously the availability of dye lasers greatly increases the number of available exciting lines.

10.5. RAMAN SPECTROSCOPY AS SUPPORT FOR RESTORATION AND CONSERVATION

Raman microspectroscopy is a precious technique for the chemical characterization of materials used in works of art or, eventually, of their degradation products. Both small areas of relatively large artifacts or microsamples removed from a work of art can be examined, so that the integrity of the artistic product is in all cases preserved. Moreover, for a complete analysis different analytical methods must be used and, as Raman microspectroscopy is a nondestructive technique, the same sample can be subsequently employed for other measurements.

Depending on the particular purpose, a simple comparison between spectra in order to ascertain the similarity between two compounds may suffice or a detailed assignment of bands may be required. In any case, being a vibrational spectrum, the Raman spectrum provides an unique fingerprint of the examined compound.

In colored artifacts, pigments are the most significant components. A detailed knowledge of the pigments in an artwork is important in revealing information useful for characterization, restoration, conservation, dating, and authentication of the works of art. In fact, this can assist in verifying attribution and identifying forgeries as there are known dates for the introduction of certain pigments. For example, Prussian blue should not appear on a pre-eighteenth-century painting, and titanium white, introduced in the twentieth century, would be suspicious on a water color purporting to be from the nineteenth century. Consequently, the identification of the pigments used on a work of art may give an indication as to the date of the work and hence to its authenticity.

In the field of art conservation, if pigments with known susceptibility are identified, then specific recommendations for storage, display, and treatments can be made. For instance, a water color that had been identified as containing the light-sensitive pigments indigo and carmine would require strict exhibition conditions of low lux levels and controlled time on display. For works on paper the identification of pigments with certain sensitivities aids in establishing the appropriate conservation treatment.

When aqueous or deacidifying treatments are being considered to stabilize the cellulose in paper that has been degraded by acid hydrolysis or oxidation mechanisms, it is important to identify pigments that could be affected by these treatments, such as Prussian blue, which could turn brown, or chrome yellow, which converts to orange.

Raman microscopy is the ideal analytical method for pigment analysis (Clark 1995) of artifacts because this technique combines the required attributes of reliability and sensitivity and because it is nondestructive, largely immune from inteferences, and adaptable to *in situ* application. As stated above, Raman microscopy can work on nonmodified samples of very small size, usually adopting a backscattering geometry. It is also possible to analyze different layers of the cross section of the sample, eventually embedded in a polyester resin. Furthermore, since the probe laser wavelengths used are in the visible region, usually 0.5 to 0.7 μm, the spatial resolution of the experiment is of this order, and so individual pigment grains of size ≥ 0.7 μm may be identified. These advantages are unique to Raman spectroscopy and are of enormous importance when, for example, the work of art has undergone conservation treatments at different stages of its history or when a particular color is made up from a mix of pigments.

It is worth noting that the Raman spectrum is sensitive to both composition and crystal form, as is demonstrated, for example, by the results obtained for titanium dioxide, which is known to exist in three crystal modifications (rutile, anatase, and brookite), all occurring naturally (Best et al. 1992). A considerable part of the following treatment will be devoted, as a consequence of what was stated above, to use of Raman spectroscopy in the definition of the so-called palette of a work of art.

Nevertheless, the capability of this technique of yielding a fingerprint of inorganic and organic compounds, together with the opportunity of applying it to manufacts or samples of different shape and size without particular pretreatments, makes it important in the investigation of a lot of materials of either artistic or archaeological interest. Therefore, a subsequent part of this chapter will be devoted to the use of Raman spectroscopy in the characterization of various materials, ranging, for example, from minerals to resins or glasses or organic residues in burial sites. Finally, the technique has found application in the study of deterioration processes of different substrates, mainly frescoes and limestone, and examples of such application will also be reported.

10.6. APPLICATION OF RAMAN SPECTROSCOPY TO THE STUDY OF PIGMENTS IN ARTISTIC HAND-WORKS

10.6.1. Introduction

The Raman spectra of a large number of pigments, both inorganic and organic, have been reordered and now form part of a data bank. Some years ago, Guineau (1987) and recently Bell et al. (1997) published two interesting spectroscopic libraries. Of particular relevance is the work by Bell et al., which presents the Raman spectra of over 60 pigments, both natural and synthetic, known to have been in use before about A.D. 1850.

It is worth noting that the choice of the exciting laser line is extremely important to obtain a good result because absorption of the scattered light by the sample may reduce the quality of its spectrum. In such cases the exciting line is chosen to fall outside the contour of the electronic absorption bands of the pigment. For example, vermilion fails to give a significant Raman spectrum using green excitation but gives a strong Raman spectrum when excited with red radiation.

As mentioned above, when a molecule is excited with a laser line whose frequency corresponds to the maximum of an allowed electronic transition, a resonance Raman spectrum may be obtained. A classic example is provided by lapis lazuli, a complex rock mixture containing the valuable, deep-blue mineral lazurite of approximate formula $Na_8[Al_6Si_6O_{24}]S_n$, which is the essential constituent of the pigment. The aluminosilicate cage of this compound describes a truncated octahedron and contains, in the holes, trapped sulfur radical anions (notably S_3^-, but also some S_2^-) (Tarling et al. 1988; Hassan et al. 1985). The internal electronic transitions of these radicals are responsible for the color. Although accounting for less than 1% of the mass of the pigment, the S_3^- anion gives such an extremely intense Raman spectrum with 500 to 600 nm excitation that no bands due to the host lattice are seen. The Raman spectrum of lazurite, obtained by using the 647.1-nm exciting line of a kripton ion laser, shows a very intense band at 548 cm^{-1} that can be attributed to the totally symmetric stretching of the S_3^- anion (Clark and Franks 1975; Clark and Cobbold 1978). The great intensity of this band is due to the fact that lazurite shows a very broad band with a maximum at 610 nm as determined by diffuse reflectance measurements (Hofmann et al. 1969) that has been assigned to the $^2A_1 \leftarrow {}^2B_1$ transition of the S_3^- anion (Morton 1968).

As mentioned before, a possible problem when collecting Raman spectra of a pigment is that of fluorescence occasionally due to supports and binders.

10.6.2. *In Situ* Analysis of Inorganic Pigments from Illuminated Books and Manuscripts

The illuminator of books and manuscripts usually preferred to use inorganic pigments since these colors were known to be less fugitive and more stable than organic ones. Many were available as minerals or by way of simple syntheses. The more common inorganic minerals and synthetic compounds or glasses that have been used as pigments at different periods of time are listed by Clark (1995).

The pigment analysis of manuscripts was not usually performed in the past owing to the inappropriateness of the established analytical techniques, most of which required removal of pigment samples directly from the manuscript. Moreover, it may be possible to remove the pigment from offsets transferred onto the opposite page of a manuscript.

Raman microscopy is the only technique that allows an *in situ* analysis of pigments from illuminated books and manuscripts. In fact, in these cases the object to be analyzed can be stretched on a sufficiently wide translation stage of the optical microscope and excited by a laser source. As already described, the use of an aperture at the secondary focus of the scattered radiation has been found to reduce significantly any interference from this source, with the exception of the fluorescence arising from the pigment itself. In the last case, the most effective solution is to change the excitation wavelength, usually to a longer value.

The presence of different pigments of essentially the same color on the same manuscript indicates how important the concept of hierarchy was. For example, the illuminators often used azurite for the capital letters and preferred to use the more precious and expensive lapis lazuli for the garments of the Virgin Mary or the most important saints. An illustration of the close association that was made in the Middle Ages between aesthetic value and intrinsic worth of materials has been suggested by the fifteenth-century Italian painter Cennino Cennini. In other cases, to create an interesting effect and minimize the use of the more expensive lapis lazuli, the technique of layering lapis lazuli over azurite was adopted (Clark 1995).

The use of Raman microscopy in the field of illuminated manuscript analysis has been dominated by the aim of characterizing the artist's palette, providing critical information for conservation, and dating the manuscript. Furthermore, Raman microscopy can be useful to describe the story of a pigment. A good example of this comes from the studies of Coupry (Hughes 1990), who used Raman microscopy to identify lapis lazuli in an illuminated manuscript from the Auxerre Abbey. The mineral lapis lazuli originates from Afghanistan and, as the manuscript dates from ca. the year 1000, it was established that the pigment was introduced in Europe from its country of origin almost two centuries before first being mentioned in written source.

Many manuscripts have been studied by Raman spectroscopy at the University College of London (Clark 1995; Clark et al. 1995; Best et al. 1992, 1993). These include a thirteenth-century north Italian antiphonal displaying lapis lazuli and malachite; a thirteenth-century north Italian choir book displaying the technique of layering lapis lazuli over azurite; a sixteenth-century German choir book displaying white lead, carbon, azurite, vermilion, red lead, massicot, and lead tin yellow; a Paris bible, ca. 1275, displaying white lead, vermilion, red lead, lapis lazuli, azurite, orpiment, realgar, and malachite; the Skard copy of the Iceland Book of Law, ca. 1360, displaying vermilion, orpiment, realgar, red ochre, azurite, bone white, and verdigris or basic verdigris mixed with green earth (Best et al. 1995); medieval German manuscripts displaying vermilion, iron oxide, brown Siena, azurite, malachite, amorphous carbon, white lead, lead tin, and azurite in admixture with lead tin yellow (Burgio et al. 1997a); a thirteenth-century Byzantine/Syrian Gospel lectionary displaying orpiment, realgar, and pararealgar (Clark and Gibbs 1997a); a large Qur'an section, an illuminated Arabic manuscript of the thirteenth century, displaying white lead, lapis lazuli, and vermilion (Clark and Huxley 1996); and three Latin manuscripts displaying vermilion, azurite, malachite, ivory black, white lead, and tin lead yellow (Burgio et al. 1997b).

Similar studies have been carried out by other research groups. For example, J. Corset et al. (1989) studied a commentary on St Paul's epistles by Corbie Abbey, twelfth century, displaying lapis lazuli; Bussotti et al. (1997) studied the pigments of a fourteenth-century miniature displaying lead tin yellow, minium, vermilion, azurite, and lead white; and Bruni et al. (1999) studied an illuminated parchment of the fifteenth century displaying lapis lazuli, vermilion, and minium. Finally it is worth noting that blue pigments used by a nineteenth-century French artist were analyzed by Raman microscopy (Coupry et al. 1994) and that cinnabar was identified by using FT–Raman spectroscopy (Clark et al. 1997a)

10.6.3. *In Situ* Analysis of Inorganic Pigments from Pottery, Faiences, and Glasses

Due to strong fluorescence, no useful Raman spectra could be obtained for pottery or faience fragments by directing the argon ion or krypton ion lasers through the glaze. Therefore, because the pigmentation can be analyzed by Raman spectroscopy only by focusing the laser on the pigment grains exposed at the edges of the cracked fragments, it was necessary to work on the cross section of the samples, where the pigment can be exposed.

The excitation wavelengths of 488.0 and 514.5 nm, particularly suited for the Raman spectra of blue and red pigments, respectively, did not generally allow to obtain any relevant signal. In fact, despite repeated attempts to focus on colored pigment grains on the faience fragments, the same unsatisfying spectrum was always obtained (Clark and Gibbs 1997b). This spectrum showed the signal of the silica in the glaze which obscured any signal from the pigments. When on the same samples the exciting line at 647.1 nm was used, silica scattering was greatly reduced. Although it was not yet possible to obtain adequate spectra of the pigments on the green, blue, and white faience fragments, examination of the red and yellow fragments resulted in excellent spectra. Nevertheless, it is worth noting that lapis lazuli was identified as the blue pigment in glazed pottery (dating to the thirteenth and fourteenth centuries) fragments from Castel Fiorentino (Foggia, Italy) (Clark et al. 1997b) by using an argon ion laser.

The presence of the expensive mineral lapis lazuli in the blue glaze of ceramic shards, albeit in low concentration, has been established by Raman microscopy also on medieval pottery fragments from the south of Italy (Clark et al. 1997c). Probably such pottery was destined for a very important person or for religious purposes. It is well known that various pigments have been used in the past in order to obtain blue decoration. These include

Egyptian blue, cobalt oxide, all its mixtures with iron, manganese and zinc oxides, a combination of oxides of cobalt, copper, and chromium, and copper compounds.

A study of eight red and eight yellow ancient Egyptian faience fragments dated to the eighteenth dynasty uncovered at El-Amarna revealed that all the red fragments were colored with red ochre or red earth [Fe(III) oxide plus clay and silica] and the color of the yellow fragments was due to lead antimonate yellow [Pb(II) antimonate] (Clark and Gibbs 1997a). The pigments used in red on painted ceramic ware, both glazed and unglazed, from different medieval archaeological sites in the south of Italy have been identified.

The Raman spectra of various iron-based commercial pigment powders have been used for reference purpose. The research led to the identification of red pigments as Fe(III) oxide (e.g., Indian red, red ochre, or Venetian red) and yellow pigments as hydrated Fe(III) oxyhydroxide (e.g., yellow ochre and Mars yellow) (Clark and Curri 1998).

Two samples of decorated stained glass (fourteenth and nineteenth centuries) have been examined with the use of FT–NIR Raman microscopy. The mediaeval sample consisted of a pale yellow glass substratum with an applied deep red-brown cloverleaf motif. The Victorian glass sample was a bright apple-green color with a deep red-brown applied geometric pattern. The pigment used in both cases has been identified as red ochre (Edwards and Tait 1998).

10.6.4. Analysis of Inorganic Pigments for Wall Paintings and Easel Paintings

In recording Raman spectra of pigments from wall paintings done *a fresco* or *a tempera*, two different types of samples are used. The first is a powdered sample, usually available in limited amount only, that has been scraped from the mural painting. The second is a fragment of the painting that has been previously embedded in a polyester resin and subsequently cut in order to evidence its cross section. In this case, by using Raman microscopy, it is possible to obtain, in a selective manner, the spectra of pigment grains or of different painted layers.

Publications regarding the Raman spectra of pigments from wall paintings are not numerous. Some of these have been reported by B. Guineau (1987) relative to the identification of hematite in a fragment of a mural painting from Pompei (Italy) and of red ochre in a fragment of a mural painting from Karnak. In a study by Ferrer (Turrell and Corset 1996 and references therein) red-colored samples were taken from decorations of the ninth century: the pigments hematite and vermilion were present. Besides, Raman microscopy of the mural paintings decorating the church of S. Andrea

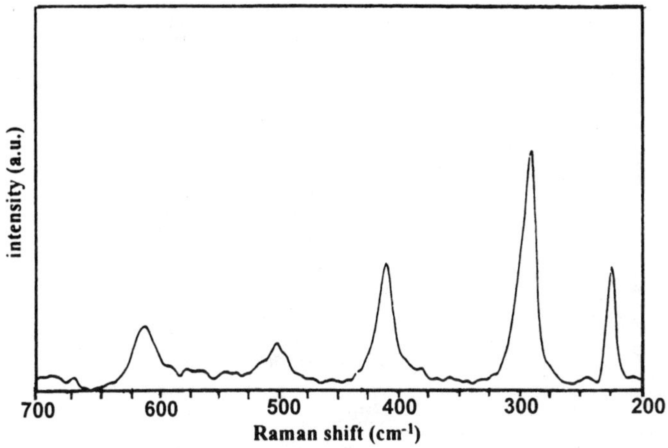

Figure 10.7. FT–NIR Raman spectrum of red pigment from frescoes of palazzo Farnese. Bands of red ochre found: 226, 296, 412, 501, and 620 cm^{-1}.

(fifteenth century) situated in Melzo near Milan (Italy) has shown the presence of azurite, vermilion, and hematite (Bruni et al., submitted).

The FT–NIR Raman microscopy was used to characterize biodeteriorated nineteenth-century Italian Renaissance frescos by Zuccari at the palazzo Farnese (Capriola). Whereas yellow, red, and violet/maroon pigments have been attributed to red and yellow ochre (see Figs. 10.7 and 10.8), the identification of the white pigments used on the frescoes remained questionable (Edwards et al. 1997a).

The FT–NIR Raman microscopy has been used for the identification of red pigments, vermilion and red iron oxide on two fragments of wall paintings representing two of the major mediaeval decorative schemes, one (late twelfth/early thirteenth centuries) from the Sherborne Abbey, Dorset (United Kingdom), and the other (twelfth century) from the Winchester Cathedral, Hampshire (United Kingdom) (Edwards et al. 1997b).

It is interesting to note that Raman microscopy was used to characterize some gilded stuccoes of Italian baroque age (Toniolo et al. 1998) from the church of S. Ambrogio (Cantù, Italy). Examination of gilded fragments by Raman microscopy established that under the gold leaf a layer composed of lead oxides (PbO, Pb_3O_4) and lead sulfate was present.

Raman microscopy has been used to study ease paintings. The examination of contemporary paintings in order to propose an authentication test has shown the capability of Raman microscopy in this field (Turrell and Corset 1996 and references therein).

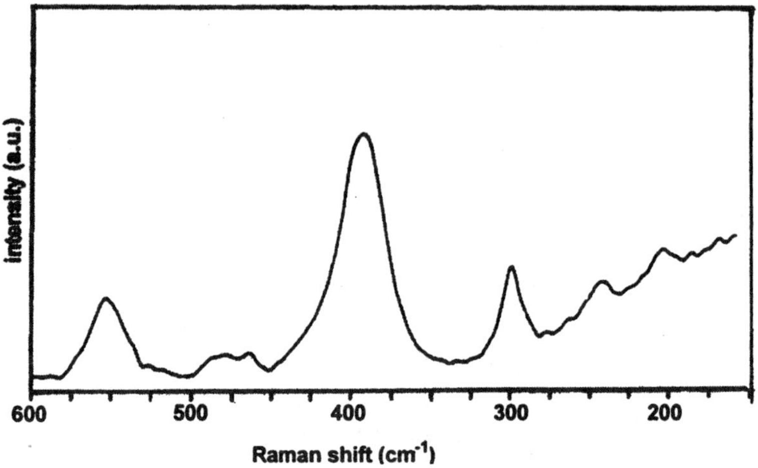

Figure 10.8. FT–NIR Raman spectrum of yellow pigment from frescoes of palazzo Farnese. Bands of yellow ochre found: 203, 298, 393, and 560 cm^{-1}.

Raman microscopy has also been used for identifying pigments taken from water colors and oil paintings. Samples of pigments were taken using a stereo microscope and a dissection needle. Where necessary, the oil medium was dissolved by application of dichloromethane. By this technique, fluorescence was minimized.

The Raman spectrum of the yellow area from *Children by a River*, a water color/gouache by H. C. D. Chorlton, 1910, showed that the pigment was cobalt yellow, aureolin (Davey et al. 1994). The Raman analysis of a sample of white pigment taken from an oil painting, ca. 1750, showed the presence of basic lead carbonate (Davey et al. 1994).

10.6.5. Organic Pigments

The first synthetic organic dye, mauveine, was obtained in 1856: since then, many hundreds of pigments and dyes, principally organic, have been synthetized. Before ca. A.D. 1850, only organic pigments of vegetal or animal origin were used, as listed by Montagna (1993). It is worth remembering the distinction between pigments and dyes (Clark 1995). Pigments are defined as finely divided coloring materials suspended as discrete particles in the painting medium, in opposition to dyes that are soluble in the medium. It should be emphasized that many different names have been used during the centuries both for organic pigments and dyes.

The use of Raman spectroscopy to identify organic pigments is often prone to some limitations, namely fluorescence and/or photochemical degradation. Moreover, such pigments often scatter very weakly, because of the fine grain size or the dilution of the pigment itself in a mordant to obtain a lake.

An example of how the background fluorescence can hide in part or completely the Raman peaks is given by the spectrum of the Kermes lac, or carmine, obtained by Clark and Huxley (1996) on a thirteenth-century illuminated Arabic manuscript. Photochemical degradation has been observed, for example, by Guineau (1989) when focusing the laser beam on violet inclusions in a chalk particle from the surface of a James Ensor pastel drawing. It has been noted by the same author that modern organic pigments are generally more lightfast than nineteenth-century dyes and can endure Raman investigation without special precautions. An example is given by the acetoacetarylide pigment known as Hansa yellow, recognized by Davey et al. (1994) in a print by V. Pasmore dated 1948. Among animal and vegetable pigments, according to Bell et al. (1997), using normal Raman spectroscopy, only a few spectra can be obtained. Among red pigments, the authors report the spectrum of purpurin, which is, with alizarin, a chromophore of the madder dye, while the spectrum of alizarin was obtained by Guineau (1987) from alizarin as such and also on an ancient bookbinding. Among yellow pigments, Bell et al. (1997) list berberine, gamboge Indian yellow (all spectra were recorded with excitation at 632.8 nm), and saffron (spectrum recorded with excitation at 514.5 nm). Finally, two black pigments, ivory black and lamp black, yielded spectra with excitation at 514.5 nm: an example regarding a real artwork, namely a mediaeval Latin manuscript, is presented by Burgio et al. (1997b).

According to Bell et al. (1997), it was impossible to obtain normal Raman spectra (i.e., with excitation at 514.5 or 632.8 nm) for the blue pigment indigo, the so-called Van Dyck brown and Tyrian purple and the yellow pigment quercitron. As regards indigo, however, it should be noted that its spectrum has been reported by Guineau (1987), both for the dye as such and fixed in the structure of a clay mineral (palygorskite) in the so-called Maya blue from a fragment of a mural painting (Fig. 10.9): the excitation wavelength is however not indicated by the author.

Using different Raman techniques, better results were obtained in some cases. One is resonance Raman spectroscopy, which allowed, for example, Bussotti et al. (1996), to obtain very good spectra of alizarin from two red lakes in a fifteenth-century painting. In order to observe the resonance Raman effect, the line at 457.9 nm of an argon laser was used: due to the signal enhancement thus obtained, relatively low laser intensities could be employed, thus avoiding the degradation of the sample.

Another important technique in the identification of organic pigments

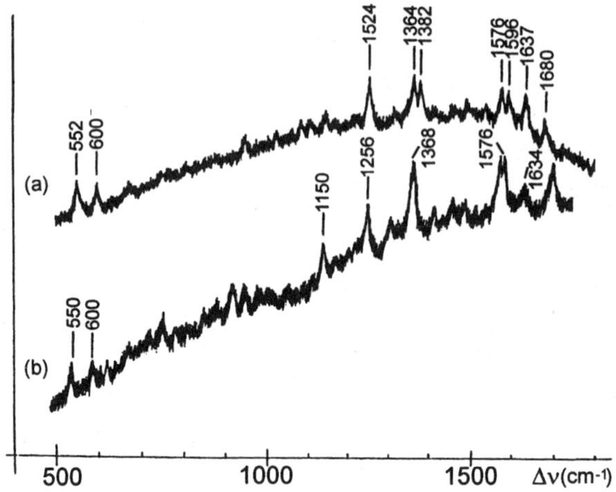

Figure 10.9. Identification by Raman spectroscopy in a fragment of a Maya blue mural painting of the dye indigo fixed in a clay mineral (*a*); for comparison, Raman spectrum of a sample of indigo is (*b*) also shown. (Reprinted, by permission, from Guineau, 1987. Copyright © 1987 by PACT.)

and dyes, even if at present not yet fully exploited, is FT–Raman spectroscopy. An example is offered by a Raman study (Edwards et al. 1997c) of the resin known as "dragon's blood," used for its fine red color in pottery, enamels, and illuminations. The use of FT–Raman spectroscopy should avoid many of the above-cited problems connected with the fluorescence background, due both to the organic molecules themselves and to their dispersing mean.

10.7. PECULIAR APPLICATIONS OF RAMAN SPECTROSCOPY IN ART AND ARCHAEOLOGY

Besides the considerable amount of work concerning the use of Raman spectroscopy in the identification of pigments employed in the field of figurative arts, two other main kinds of studies exploit this technique to characterize samples of artistic or archaeological importance.

The first is based on the use of Raman microprobe to identify the mineral phases in archaeological objects without having to destroy or prepare them. For example, it was possible to identify as chalcedony the main constituent of three early Roman Empire intaglios found in Paris (Smith and Robin 1997); to to make a hypothesis concerning the temperature of fusion of the walls of a Celtic so-called vitrified fort, a hypothesis allowed by the finding of α-cristobalite as the main component besides glass itself (Smith and Ver-

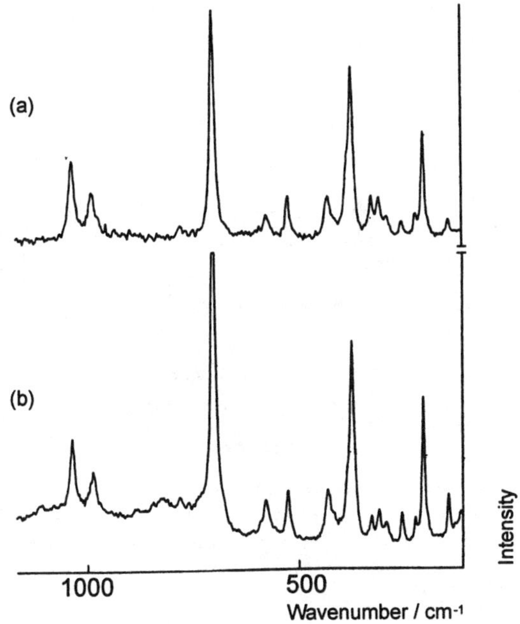

Figure 10.10. Raman spectra obtained from (*a*) pre-Colombian ceremonial "greenstone" axe-head and (*b*) jadeite jade from Burma. (Reprinted, by permission, from Smith and Gendron, 1997. Copyright © 1997 by John Wiley & Sons, Inc.)

nioles 1997); and to recognize the different rock types used in the fabrication of two ceremonial axe-heads from Mesoamerica, previously indicated generally as "greenstone" (Smith and Gendron 1997) (Fig. 10.10), performing also a semiquantitative analysis of the mineral components on the basis of their Raman wavenumber shifts.

The other main application of Raman spectroscopy in the field of archaeology concerns the use of FT–Raman instrumentation to obtain information about organic substances, including under this definition both archaeological biomaterials and organic materials employed in the manufacturing of artistic and handcrafted objects. Of course, FT–Raman spectroscopy is particularly useful for this investigation, as, being based on excitation in the near-IR, it avoids the rise of fluorescence often hindering the Raman spectra of organic compounds.

As regards human remains, which are of obvious archaeological importance, a rather extensive study was made concerning teeth: at present, it has been supplied a complete vibrational assignment both for the organic proteinaceous and inorganic hydroxyapatite components (Kirchner et al. 1997);

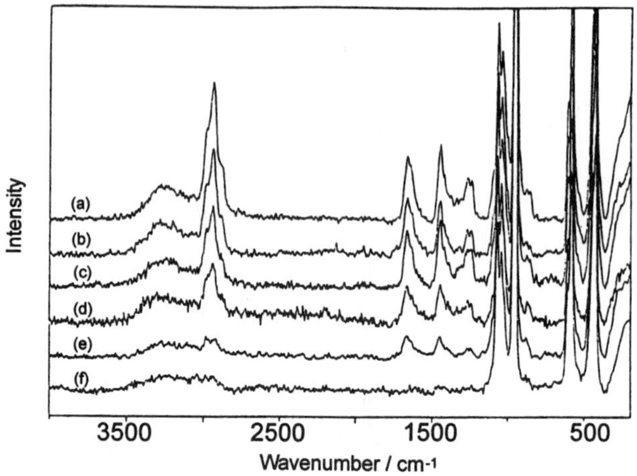

Figure 10.11. FT–Raman external enamel normalized spectra of modern deciduous canine of an infant (Bologna, Italy, 0 B.P.) (*a*) and five ancient molar teeth: (*b*) Sassari, Italy, 150 B.P.; (*c*) S. Agata, Ravenna, Italy, 650 B.P.; (*d*) Bagnacavallo, Italy, 1700 B.P.; (*e*) Monte Casasia, Italy, 2700 B.P.; (*f*) Passo di Corvo, Foggia, Italy, 5800–6300 B.P. (Reprinted, by permission, from Bertoluzza et al., 1997. Copyright © 1997 by John Wiley & Sons, Inc.)

at the same time, the intensity ratio between the band at 2941 cm^{-1}, $\nu_s(CH_2)$, of the organic component and the band at 960 cm^{-1}, $\nu_1(PO_4{}^{3-})$, as observed for enamel (i.e., the external surface of tooth) was suggested as a spectroscopic parameter to be correlated with the burial period of the teeth (Bertoluzza et al. 1997), as it decreases with the increase of such duration (Fig. 10.11).

In a similar context, a preliminary investigation was made on the changes undergone by skin in mummification processes on fifteenth-century mummies from Greenland (Gniadecka et al. 1997), thus observing a weakening of signals in the region of amide I and amide III bands, in comparison with both fresh and freeze-dried contemporary skin. The corresponding loss of protein and/or change in the secondary protein structure appears to take place a relatively short time after mummification.

Among specimen coming from burial sites, natural fibers are also of interest. Infrared spectroscopy seems to give only a poor signal-to-noise ratio when applied to such samples, probably because of the presence of quite significant amounts of free water and hydroxyl groups in the cellulose component. In contrast, good-quality FT–Raman spectra have been obtained on a series of untreated natural plant fibers and also on a 4000-year-old Egyptian linen (Edwards et al. 1997d): it has been suggested that degradation of cellulose should produce a decrease of intensity for the band at 1096 cm^{-1} due to the ν(COC) of glycosidic linkages and an increase of intensity for the 2900 cm^{-1} feature associated with the new ν(CH) modes of the decomposition products. An attempt was also made to interpret the yellow-brown coloration formed in the degradation of ancient linens from the Middle East (Edwards and Falk 1997a): a microbiological or fungal attack has been hypothesized to explain such alteration, with an associated production of carotenoids, that gives rise to the characteristic spectral feature, due to ν(C=C), at 1520 cm^{-1}.

The advantages of FT–Raman spectroscopy and microscopy have also been appreciated in the characterization of ancient resins from 3000-year-old American Indian artifacts (Edwards and Falk 1997b) and 2300-year-old Vietnamese burial jars (Edwards et al. 1997e), objects in which the resinous materials were used as preservatives, sealants, or adhesives: the vibrational assignment, particularly of bands in the 1500 to 1750 cm^{-1} range, proved to be useful for a more precise identification of different sorts of resins and for a study of their degradation under interment conditions. A systematic study of FT–Raman spectra of natural waxes and resins was also undertaken by Edwards et al. (1996) in order to provide reference data for these materials, to be used in their characterization and conservation, especially if combined with gas chromatographic analyses.

A more widespread use of resins was made as coloring agents (see, e.g., the example of dragon's blood, already discussed in the previous section) or as adornments. This last application refers especially to amber, of which FT–Raman spectra allow to make hypotheses about the geological age and the geographical provenance and to distinguish between genuine and false specimen (Edwards and Farwell 1996) (Fig. 10.12).

A similar study, aimed at both the individuation of the geographical origin and the detection of forgeries, was devoted to ivory, another biological material that has exercised great appeal as a medium for artistic carving and tool manufacture. The possibility to distinguish ivory from African or Asian elephant or even from the extinct mammoth on the basis of the Raman bands of collagen and hydroxyapatite, the main constituents of this material, has been shown by Edwards et al. (1998). The same authors (Edwards and Farwell 1995) were also able to detect various simulated ivory artifacts. In both studies on amber and ivory objects, it was possible to obtain Raman spectra from objects of different shapes and degrees of polishing, for which other identification techniques could not be applicable.

While FT–Raman microspectroscopy has been applied to the study of pigments used in decorated stained glass (Edwards and Tait 1998), the fluorescence exhibited by ancient glass objects when normal Raman spectroscopy is used appears to be, in this case, a welcome marker of the age of the samples. Bertoluzza et al. (1995) presented evidence that the logarithm of the

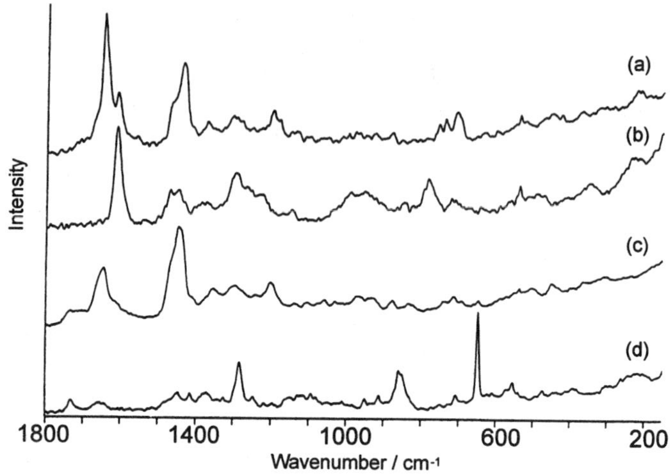

Figure 10.12. FT–Raman spectra of (*a*) contemporary pine resin, (*c*) yellow amber bead, and (*b,d*) two bangles, identified as false ambers respectively of polyurethane and acrylate origin. (Reprinted, by permission, from Edwards and Farwell, 1996. Copyright © 1996 by Elesevier Science.)

intensity ratio between the 1080-cm^{-1} band (due to the O–Si–O symmetric stretching mode) and the fluorescence band decreases linearly with increasing age of the glass samples. Such fluorescence emission is probably associated with adsorption and transformation processes of organic compounds on the vitreous surface. It is worth noting that Raman spectroscopy represents an almost unique tool of investigation for these objects, because their brittleness prevents the removal of microsamples required by other techniques.

Finally, a rather curious use has been made of Raman spectroscopy in settling a controversy concerning the archaeological meaning of a rock shelter situated in the Alpes-Maritimes in southern France. Some colored traces on the rock that were considered by a first researcher to be "miniature paintings," pointing to a sacred role of the place, were subsequently shown to be the result of the accumulation of carotenoids, produced by a green alga (de Lumley et al. 1996). The corresponding Raman spectrum exhibits, in fact, the pattern typically observed for such pigments, with bands at about 1525, 1190, 1155, and 1005 cm^{-1}.

10.8. STUDIES OF THE DETERIORATION OF ARTWORKS BY MEANS OF RAMAN SPECTROSCOPY

Perhaps the form of deterioration in artworks most extensively studied by means of Raman spectroscopy is the biological attack of different substrata

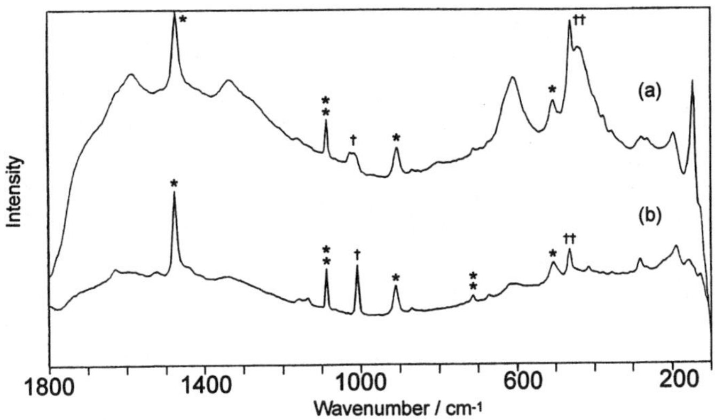

Figure 10.13. FT–Raman spectra of *Dirina massiliensis* forma *sorediata* lichen encrustations on the walls of Fiskerton church, Lincolnshire, United Kingdom: (*a*) middle, white interfacial layer; (*b*) upper surface. Bands of calcium oxalate dihydrate (∗) are observed both for upper surface and interfacial region. Substratal incorporation is shown by presence of features due to calcite stone (∗∗), gypsum mortar (†), and silica (††). (Reprinted, by permission, from Seaward and Edwards, 1997. Copyright © 1997 by John Wiley & Sons, Inc.)

perpetrated by various lichen species. This phenomenon was studied both on frescoes (Edwards et al. 1991, 1992; Edwards and Seaward 1993) and on stone surfaces (Edwards et al. 1995; Seaward and Edwards 1997) using at first normal Raman spectroscopy with excitation at 488 nm and later, with advantage, FT–Raman spectroscopy, also in conjunction with a microscope (Fig. 10.13). In lichen encrustations on frescoes and limestone, there was evidence of calcium oxalate mono- and/or dihydrate formation, the hydration state seeming to depend on the different environmental conditions in which the lichen species grew (Edwards et al. 1997f). Direct incorporation of the substrate material in the encrustation appears to take place or not depending on the particular lichen species responsible for the deterioration process (Edwards et al. 1997f). Finally, an attempt was made to individuate, on the basis of low-frequency features in the Raman spectrum, a chelation role of oxalate toward metal ions in biodeteriorated frescoes: this mechanism may help the lichen to deal with toxic salts used as pigments (Edwards et al. 1997g).

Sulfation of marble due to atmospheric pollution was studied quantitatively by means of Raman spectroscopy, and the results were compared with those obtained by means of X-ray diffraction (XRD) (Kontoyanis et al. 1997). Raman calibration curves were constructed for calcite–aragonite, calcite–gypsum, and aragonite–gypsum mixtures using the relative intensities of different bands: detection limits lower than those allowed by XRD were obtained, namely 0.1, 0.1, and 0.05 mol %, respectively, for calcite, aragonite, and gypsum. A real sulfated marble sample, removed from Athens National

Garden, was also examined in order to evidence the potential of using Raman spectroscopy for a point-by-point (or mapping) analysis. Micro–Raman spectroscopy was also suggested as a means of verifying that the cleaning of limestone in a historic building by means of a Nd–YAG pulsed laser does not alter the mineralogical composition of the stone underlying the contaminant layer (Gobernado-Mitre et al. 1996).

In conclusion, Raman and micro–Raman spectroscopy appear to be useful tools to examine the surfaces of artworks, aiming to identify at the same time the original constituents and eventually subsequently forming degradation products.

REFERENCES

Bell, I. M., R. J. H. Clark, and P. J. Gibbs (1997), *Spectrochim. Acta Part A* **53**, 2159.

Bertoluzza, A., S. Cacciari, G. Cristini, C. Fagnano, and A. Tinti (1995), *J. Raman Spectrosc.* **26**, 751.

Bertoluzza, A., P. Brasili, L. Castri, F. Facchini, C. Fagnano, and A. Tinti (1997), *J. Raman Spectrosc.* **28**, 185.

Best, S. P., R. J. H. Clark, and R. Withnall (1992), *Endeavour* **16**, 66.

Best, S., R. Clark, M. Daniels, and R. Withnall (1993), *Chem. Br.*, 118.

Best, S. P., R. J. H. Clark, M. Daniels, C. A. Porter, and R. Withnall (1995), *Stud. Cons.* **40**, 31.

Bruni, S., F. Cariati, C. L. Bianchi, E. Zanardini, and C. Sorlini (1995), *Archaeometry* **37**, 249.

Bruni, S., F. Casadio, F. Cariati, and L. Toniolo (1999), *Spectrochim. Acta Part A* **55**, 1371.

Bruni, S., F. Cariati, S. Polesello, A. Pozzi, G. Alessandrini, and L. Toniolo (submitted).

Burgio, L., D. A. Ciomartan, and R. J. H. Clark (1997a), *J. Mol. Spectrosc.* **405**, 1.

Burgio, L., D. A. Ciomartan, and R. J. H. Clark (1997b), *J. Raman Spectrosc.* **28**, 79.

Bussotti, L., E. Castellucci, and M. Matteini (1996), *Science Tech. Cult. Herit.* **5**, 13.

Bussotti, L., M. P. Carboncini, E. Castellucci, L. Giuntini, and P. A. Mandò (1997), *Stud. Cons.* **42**, 83.

Clark R. J. H. (1995), *Chem. Soc. Rev.* **24**, 187.

Clark, R. J. H., and D. G. Cobbold (1978), *Inorg. Chem.* **17**, 3169.

Clark, R. J. H., and M. L. Curri (1998), *J. Mol. Struct.* **440**, 105.

Clark, R. J. H., and M. L. Franks (1975), *Chem. Phys. Letters* **34**, 69.

Clark, R. J. H., and P. J. Gibbs (1997a), *Chem. Comm.*, 1003.

Clark, R. J. H., and P. J. Gibbs (1997b), *J. Raman Spectrosc.* **28**, 99.

Clark, R. J. H., and K. Huxley (1996), *Science Tech. Cult. Herit.* **5**, 95.

Clark, R. J. H., L. Cridland, B. M. Kariuki, K. D. M. Harris, and R. Withnall (1995), *J. Chem. Soc. Dalton*, 2577.

Clark, R. J. H., P. J. Gibbs, K. R. Seldom, N. M. Brovenko, and Y. A. Petrosyan (1997a), *J. Raman Spectrosc.* **28**, 91.

Clark, R. J. H., L. Curri, and G. S. Henshaw (1997b), *J. Raman Spetrosc.* **28**, 105.

Clark, R. J. H., L. Curri, and C. Laganara (1997c), *Spectrochim. Acta* **A25**, 597.

Corset, J., P. Dhamelincourt, and J. Babillat (1989), *Chem. Br.* 612.

Coupry, C., A. Lautié, M. Revault, and J. Dufilho (1994), *J. Raman Spectrosc.* **25**, 89.

Davey, R., D. J. Gardiner, B. W. Singer, and M. Spokes (1994), *J. Raman Spectrosc.* **25**, 53.

de Lumley, H., J. Begin, J. Blanc, A. Couté, B. Dennetière, A. Echassoux, F. Fröhlich, N. Giusto-Magnardi, S. Khatib, P. Machu, L. Mano, O. Romain, T. Serres, D. C. Smith, and F. Villain-Rinieri (1996), *C. R. Acad. Sci. Paris Série II a* **323**, 1073.

Edwards, H. G. M., and M. J. Falk (1997a), *Appl. Spectrosc.* **51**, 1134.

Edwards, H. G. M., and M. J. Falk (1997b), *J. Raman Spectrosc.* **28**, 211.

Edwards, H. G. M., and D. W. Farwell (1995), *Spectrochim. Acta Part A* **51**, 2073.

Edwards, H. G. M., and D. W. Farwell (1996), *Spectrochim. Acta Part A* **52**, 1119.

Edwards, H. G. M., and J. F. K. Tait (1998), *Appl. Spectrosc.* **52**, 679.

Edwards, H. G. M., and M. R. D. Seaward (1993), *Spectrosc. Eur.* **5**, 16.

Edwards, H. G. M., D. W. Farewell, and M. R. D. Seaward (1991), *Spectrochim. Acta Part A* **11**, 1531.

Edwards, H. G. M., D. W. Farwell, R. Jenkins, and M. R. D. Seaward (1992), *J. Raman Spectrosc.* **23**, 185.

Edwards, H. G. M., N. C. Russel, M. R. D. Seaward, and D. Slarke (1995), *Spectrochim. Acta Part A* **51**, 2091.

Edwards, H. G. M., D. W. Farwell, and L. Daffner (1996), *Spectrochim. Acta Part A* **52**, 1639.

Edwards, H. G. M., C. J. Brooke, and J. K. F. Tait (1997a), *J. Raman Spectrosc.* **28**, 95.

Edwards, H. G. M., E. R. Gwyer, and J. K. F. Tait (1997b), *J. Raman Spetrosc.* **28**, 677.

Edwards, H. G. M., D. W. Farwell, and A. Quye (1997c), *J. Raman Spectrosc.* **28**, 243.

Edwards, H. G. M., D. W. Farwell, and D. Webster (1997d), *Spectrochim. Acta Part A* **53**, 2383.

Edwards, H. G. M., M. G. Sibley, and C. Heron (1997e), *Spectrochim. Acta Part A* **53**, 2373.

Edwards, H. G. M., N. C. Russell, and M. R. D. Seaward (1997f), *Spectrochim. Acta Part A* **53**, 99.

Edwards, H. G. M., E. R. Gwyer, and J. K. F. Tait (1997g), *J. Raman Spectrosc.* **28**, 677.

Edwards, H. G. M., D. W. Farwell, J. M. Holder, and E. E. Lawson (1998), *Stud. Cons.* **43**, 9.

Gniadecka, M., H. C. Wulf, O. Faurskov Nielsen, D. H. Christensen, and J. P. Hart Hansen (1997), *J. Raman Spectrosc.* **28**, 179.

Gobernado-Mitre, I., J. Medina, B. Calvo, A. C. Prieto, L. A. Leal, B. Pérez, F. Marcos, and A. M. de Frutos (1996), *Appl. Surf. Sci.* **96–98**, 474.

Guineau, B. (1987), *PACT* **17**, 259.

Guineau, B. (1989), *Stud. Cons.* **34**, 38.

Hassan, I., R. C. Peterson, and H. D. Grundy (1985), *Acta Cryst.* **C41**, 827.

Hofmann, U., E. Herzenstiel, E. Schönemann, and K. H. Shwarz (1969), *Anorg. Allg. Chem.* **367**, 119.

Hughes, S. (1990), *New Scient.* **22/29**, 21.

Kirchner, M. T., H. G. M. Edwards, D. Lucy, and A. M. Pollard (1997), *J. Raman Spectrosc.* **28**, 171.

Kontoyanis, C. G., M. G. Orkoula, and P. G. Koutsoukos (1997), *Analyst* **122**, 33.

Montagna, G. (1993), *I Pigmenti*, Nardini Editore, Florence, pp. 226–227.

Morton, J. R. (1968), *Collogue Ampère* **15**, 299.

Seaward, M. R. D., and H. G. M. Edwards (1997), *J. Raman Spectrosc.* **28**, 691.

Smith, D. C., and F. Gendron (1997), *J. Raman Spectrosc.* **28**, 731.

Smith, D. C., and S. Robin (1997), *J. Raman Spectrosc.* **28**, 189.

Smith, D. C., and J. D. Vernioles (1997), *J. Raman Spectrosc.* **28**, 195.

Tarling, S. E., P. Barnes, and J. Klinowski (1988), *Acta Cryst.* **B44**, 128.

Toniolo, L., C. Colombo, S. Bruni, P. Fermo, A. Casoli, G. Palla, and C. L. Bianchi (1998), *Stud. Cons.*, **43**, 201.

Turrell, G., and J. Corset (1996), *Raman Microscopy Development and Application*, Academic, London, p. 420.

CHAPTER

11

THERMAL ANALYSIS

MARIANNE ODLYHA

Birkbeck College, School of Biological and Chemical Sciences, University of London, London, United Kingdom

11.1. INTRODUCTION

The topic of this chapter is the application of thermoanalytical techniques to the analysis of traditional artists' painting materials and in some cases actual samples from paintings. In polymer science extensive studies have been made on the effect of aging and of environmental fluctuations by measuring the thermal stability of polymeric materials, both pigmented and unpigmented, and parameters such as the glass transition temperature (T_g), which is the temperature at which amorphous polymeric materials change from a glassy to a rubbery state. In paintings conservation the immediate interest is in the characterization of the materials used in the painting. This information can be obtained by mass spectrometric techniques, as there is a demand for information at the molecular level. The role of thermoanalytical techniques, and these involve a range of techniques that are briefly discussed below, is to characterize materials in terms of their physicochemical properties at the macromolecular level.

These properties are described in terms of their thermal stability and their mechanical and electrical properties, which are monitored as a function of temperature or time and at a selected frequency. The measurement of changes in these properties provides information on the effects of aging, fluctuations in environmental factors such as light, relative humidity, temperature and indoor air quality (e.g., levels of pollutants), and conservation treatment. Standard paint samples have been prepared and are being evaluated and a database is being compiled on their response to the effects described above, in addition to consideration of the effects of differences in

Modern Analytical Methods in Art and Archaeology, Edited by Enrico Ciliberto and Giuseppe Spoto. Chemical Analysis Series, Vol. 155
ISBN 0-471-29361-X

pigmentation and the proportions of pigment present in the paint. It will be shown that thermoanalytical techniques, and in particular thermomechanical analysis, allow monitoring of changes due to processes used in conservation practice in the structural treatment of paintings that involve increased levels of humidification and moisture (Hedley 1993a) and new preventive conservation techniques involving deacidification of canvases (Hackney et al. 1996).

The potential of thermoanalytical techniques in the area of conservation science and in particular the area of paintings is still to be realized, and it is the aim of this chapter to provide an insight into this new area of their application.

Initially a description of these techniques will be given followed by examples where they have been applied to traditional artists' materials. The term thermal analysis is used to describe a number of techniques, which include differential scanning calorimetry (DSC), thermogravimetry (TGA), thermomechanical analysis (TMA), and dielectric analysis. Particular emphasis will be given to the description of the dynamic mechanical thermal analysis (DMTA) and its use for studying the effects of aging and conservation treatment since this work is novel (Odlyha 1998) and further developments are in progress (Foster 1997, Odlyha et al., 1999a).

11.2. DIFFERENTIAL SCANNING CALORIMETRY

Differential scanning calorimetry measures the heat flow (in milliwatts) to a sample and reference (usually an empty crucible). The basic principles, instrumentation, and applications can be found in a large number of texts (Wendlandt 1986; Widmann and Riesen 1987; Charsley and Warrington 1992; Haines 1995; Turi 1997). In this chapter measurements will be described that have been made on two different types of instruments. The rationale for this is that the methodology relating to very small sample size and suitability for samples from paintings was specially developed for the Perkin-Elmer (PE) DSC 7 (Odlyha 1990). Reproducibility of data was found to be very good provided the blank line (obtained with the two empty crucibles) was adjusted before the measurement to give a variation of the order of $\pm 30\,\mu W$ over the measured temperature range (30 to 650°C). The PE DSC 7 analyzer is based on the power compensation principle where sample and reference crucibles are located in separate furnaces maintained at the same temperature as they are linearly heated (Fig. 11.1). The measured response is derived from the differential heat supplied to the sample and reference crucibles to maintain the same temperature (Haines 1995). This type of measurement was suited to small sample size as is necessary with analysis

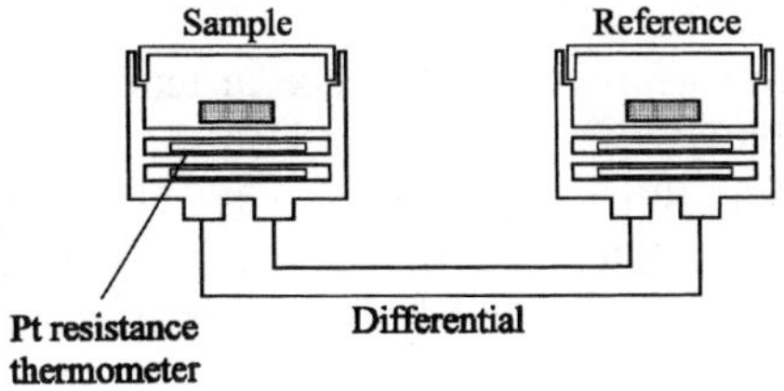

Figure 11.1. Diagram of Perkin-Elmer DSC sample cell. Sample and reference crucibles are located in separate furnaces maintained at the same temperature as they are linearly heated.

involving samples from paintings. Platinum microcrucibles (3 mm internal diameter) were specially designed for this work. Sample size was of the order of 50 μg (although in some cases as low as 20 μg) and oxygen was used as the sample purge gas (flow rate 60 cm^3/min). Optimum signal response was obtained from heating rates of 40°C/min. Baseline stability over the wide temperature range used (30 to 650°C) was made possible with additional installation of more sensitive electronic control units.

The Shimadzu DSC-50 analyzer operates using the heat flux principle. The arrangement differs in that sample and reference crucibles are located in the same furnace. The instrument signal is derived from the temperature difference between the sample and reference, which is then converted and expressed as a difference in heat flow so that the experimental measurement is similar to that obtained from the PE DSC 7. Special microcrucibles were designed. In this case the furnace type limited the high end of the temperature range to 500°C, in oxygen and so aluminum crucibles were used. For optimum measurements the sample size was larger (~1 to 2 mg) and so this restricted the technique to the measurement of standard samples such as the prepared paint tempera samples described later in this chapter. In the case of standard samples several measurements could be made, and data presented (e.g., later in Fig. 11.7) were obtained from the average of three measurements made from three different DSC curves.

In both cases the area enclosed by the DSC curve is directly proportional to the energy change occurring in the sample. Measurements made in oxygen provide information on the thermooxidative properties of materials. Samples are weighed before and after heating, and the recorded weight changes, as well as the color of the residue, provide additional information for the characterization of both binder and pigment. The shape of the resulting DSC curve (for the same experimental conditions and sample mass) is sensitive to the chemical composition of the sample. For a given composition this will alter and will depend on the state of degradation of the sample. The shape of the DSC curve therefore depends on sample type and the effect of aging. For this reason differential thermal analysis (DTA), historically an earlier technique and one that measures the difference in temperature between sample and reference crucibles, was originally used on samples from paintings (Preusser 1979; Odlyha and Burmester 1988). It was found that the resulting DTA curves were sensitive to differences in age of oil-based paint and to the type of medium used in the paintings as this varied in its chemical composition.

11.2.1. Differential Scanning Calorimetry: Applications

DSC measurements on samples from early Italian paintings demonstrated that it was possible to distinguish between oil- and non-oil-based media (Odlyha 1990). Differences were obtained on the basis of the shape of DSC curves for tempera and oil-based paint. Figure 11.2*a* shows the DSC curves obtained from samples analyzed from a sixteenth-century Italian painting by Francesco di Giorgio di Martini, *Nativity of Christ* (ca. 1500), during its restoration in the laboratories of the Opificio delle Pietre Dure in Florence, Italy. The superimposed curves originate from samples (1) from the architectural background of the painting and (2) from a figure in the lower foreground of a dog in the painting. The observed peaks correspond to exothermic reactions; that is, heat is released during the degradation reaction. Curve 1 (Fig. 11.2*a*) shows a dominant feature in the temperature range 200 to 300°C, which is a typical response of a sample based on egg tempera paint. (Curves have been inverted to maintain consistency with other DSC data shown later in this chapter, i.e. exothermic in upward direction) A similar DSC response has also been observed in a sample from another early Italian painting (thirteenth century, *Madonna del Popolo*, by Coppo di Marcovaldo, in the Capella Brancacci, Chiesa S. Maria del Carmine, Florence, Italy; Fig. 11.2*b*). By contrast, curve 2 (Fig. 11.2*a*) exhibits a dominant peak in the higher temperature range 350 to 450°C, characteristic of oil-based paint. It can thus be concluded that in the *Nativity of Christ* the background was painted using tempera, whereas the lower foreground was painted using oils.

The thermal degradation behavior of egg tempera alone (also of its components, such as cholesterol) and then egg tempera mixed with various pigments has been recently studied (Odlyha et al., 1999b). Figure 11.3 shows a series of differently pigmented paint tempera samples that have been prepared using known proportions. The curve for the unpigmented tempera is also included. In the case of the lead white (or basic lead carbonate tempera) the paint is so lean in medium content that the pigment decomposition characteristics in the temperature region of 280 to 340°C are clearly seen. This is also seen but to a lesser extent in the case of azurite (basic copper carbonate) paint where the endothermic decomposition of the pigment occurs at about 300°C. Other pigments that do not degrade in the displayed tem-

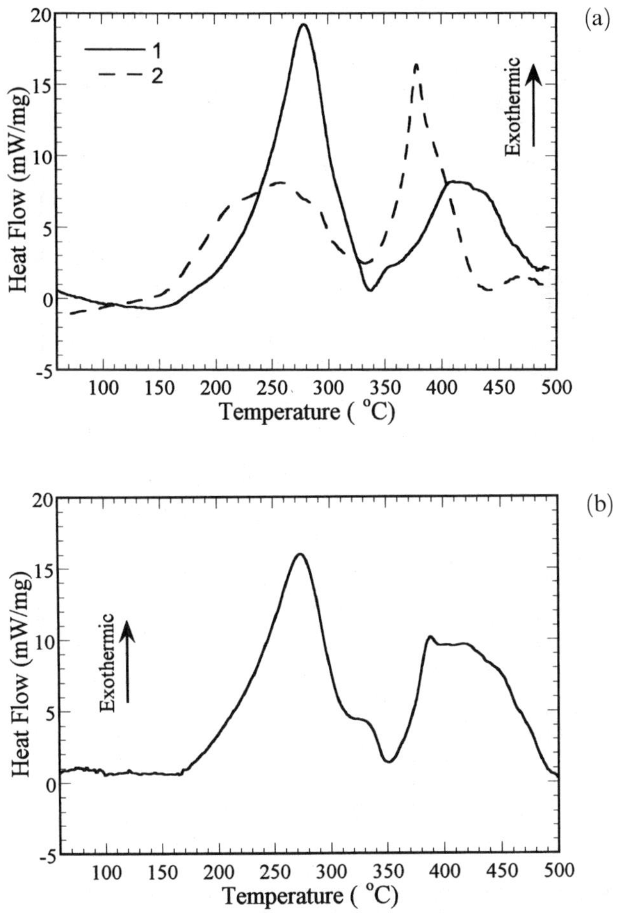

Figure 11.2. (*a*) DSC curves obtained from samples from a sixteenth-century Italian painting by Francesco di Giorgio di Martini, *Nativity of Christ* (ca. 1500), that had been heated in O_2 at 40°C/min. 1: Background 2: Foreground (*b*) DSC curve of sample from an early Italian painting that had been heated in O_2 at 40°C/min: thirteenth century, *Madonna del Popolo*, by Coppo di Marcovaldo, in the Capella Brancacci, Chiesa S. Maria del Carmine, Florence, Italy.

perature range (e.g., smalt, a cobalt glass; Naples yellow, a mixed oxide of antimony and lead) show a degradation curve similar to that of the unpigmented sample. The presence of the metallic ion in each case affects the onset temperature of degradation and the shape of the curve. This is discussed in detail elsewhere (Odlyha et al., 1999b). Each curve in Figure 11.3 has been given its own scale on the vertical axis, to show more clearly both the differences and the similarities in the curves, since peak intensities vary according to the proportion of binder present.

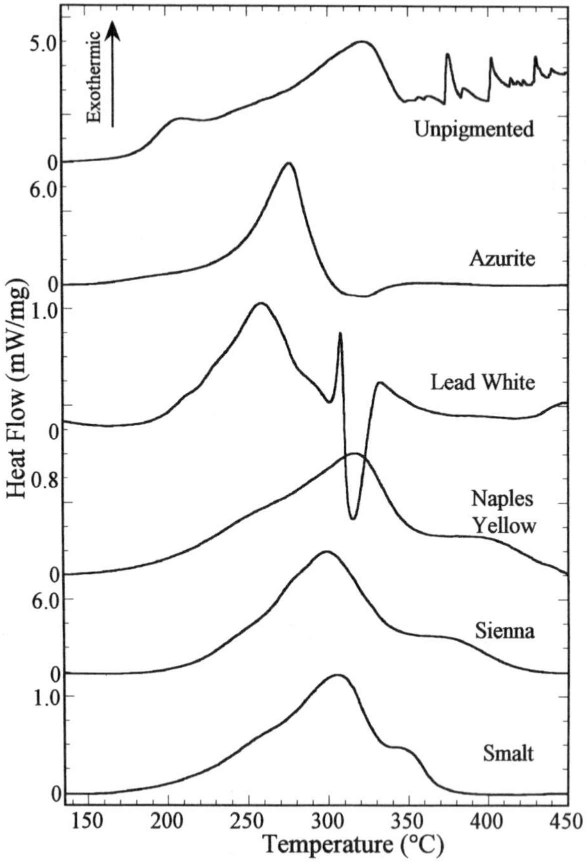

Figure 11.3. DSC curves of differently pigmented paint tempera samples heated in O_2 at 10°C/min. Samples were prepared according to traditional recipes.

Figure 11.4*a* shows the DSC curve for a sample from a ground or preparation layer of a thirteenth-century painting (Meliore, *Madonna in Trono fra i SS. Pietro e Paolo*). It shows the typical features for gypsum dehydration (endothermic reaction for loss of water molecules from $CaSO_4 \cdot 2H_2O$) and then the exothermic degradation peak for proteinaceous material (rabbit skin glue or size was generally used in ground layers of paintings). The superimposed curve in Figure 11.4*b* was obtained from a paint sample from the surface (area of white pigmentation) of Raphael's *Madonna del Baldacchino* (during restoration treatment). It shows a typical two-peak exothermic curve for oil-based paint (similar to curve 2, Fig. 11.2*a*). The medium content of the paint is leaner since some pigment features are also evident. The

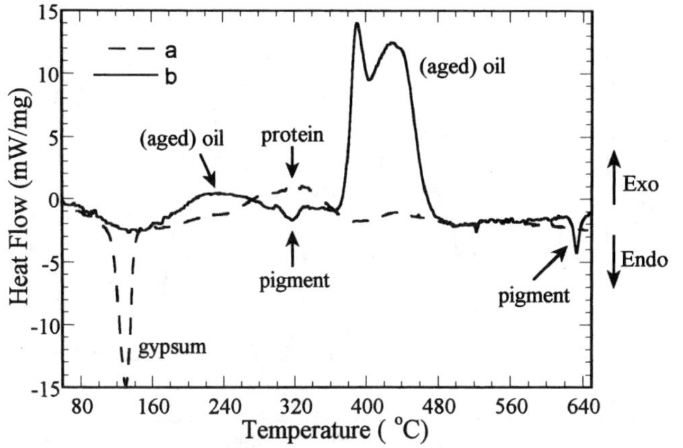

Figure 11.4. DSC curve for (*a*) sample from ground or preparation layer of thirteenth-century painting, Meliore, *Madonna in Trono fra i SS. Pietro e Paolo*, and (*b*) from the surface (area of white pigmentation) of Raphael's *Madonna del Baldacchino*. Both samples were heated in O_2 at 40°C/min.

white pigmented sample (containing basic lead carbonate) shows thermal effects around 340°C, where basic lead carbonate converts to litharge, with possible additional contributions from the decomposition of lead salts that may have formed in the paint film with age (e.g., lead stearate) and also occur in this temperature range. The second pigment feature in the temperature range 620 to 640°C corresponds to the endothermic reaction where lead tetraoxide (Pb_3O_4) converts to the yellow lead oxide form (massicot). This result demonstrates that the DSC curve over this temperature range can provide information both on the organic and inorganic components present in the paint sample. Samples are also weighed before and after heating. The recorded weight loss over the complete temperature range and the color of the residue provide additional information for the characterization of binder in terms of pigment content and the pigment itself.

Studies of nineteenth-century paint media show that in some cases the typical two-peak curve for oil-based paint was made more complex through the presence of additives in the predominantly oil-based paint. Figure 11.5 shows the DSC curve obtained from a sample taken from J. M. W. Turner's *The Opening of the Wallhalla*, 1842 (Tate Gallery N00533). Analysis of the complex curve combined with information on the thermal degradation behavior of standard materials indicated the presence of beeswax, oil, and protein (Odlyha 1995). The different materials have their onset of degradation at different temperatures so that the different components are resolved by temperature. The additional presence of resin is indicated by the temper-

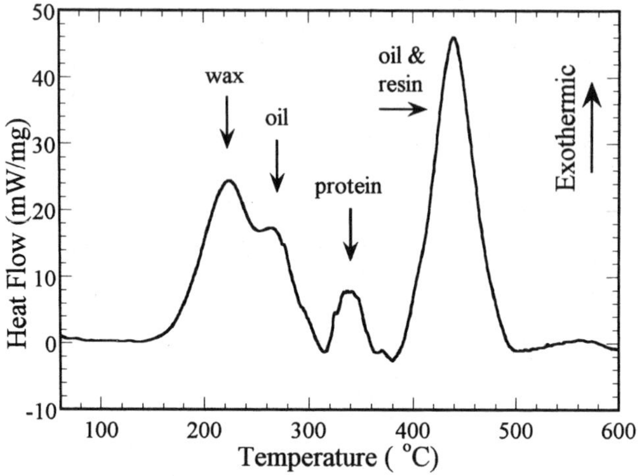

Figure 11.5. DSC curve obtained from a sample from J. M. W. Turner's *The Opening of the Wallhalla*, 1842. Sample was heated in O_2 at 40°C/min.

ature and enhanced intensity of the peak in the range 400 to 500°C. The presence of these components in the paint was confirmed by mass spectrometry (Boon et al. 1995).

More recently DSC has been used to study the effect of different metallic ions on pigmented tempera samples (Fig. 11.3). Samples were prepared and then aged under controlled conditions of light levels, relative humidity (RH), and temperature (Odlyha et al., 1999b). Figure 11.6 shows the effect of various forms of aging on the shape of the DSC curve of smalt tempera. The data are described in detail elsewhere (Odlyha et al., 1999b). For the purposes of this chapter it is important to draw attention to the sensitivity of the DSC curve to changes in the chemical composition of the sample caused by aging. The effect of light aging, for example, caused an increase in the low-temperature shoulder of the DSC curve, as indicated. This increase could be correlated with mass spectrometric data that gave evidence for the increase in levels of long-chain saturated fatty acids (e.g., palmitic and stearic acids that form and degrade in this temperature range). In the case of NO_x/SO_2 exposed samples an additional peak was observed. These characteristic changes indicate that the paint tempera samples can be used as environmental monitors (Odlyha et al., 1999b).

The effect of light aging can be quantified. The result of this calculation can be seen in Figure 11.7. The vertical axis is designated in terms of partial area, which refers to an area calculated over a selected temperature range and expressed as a percentage of the total area under the peak. Changes in the value of partial areas over a selected temperature range are directly re-

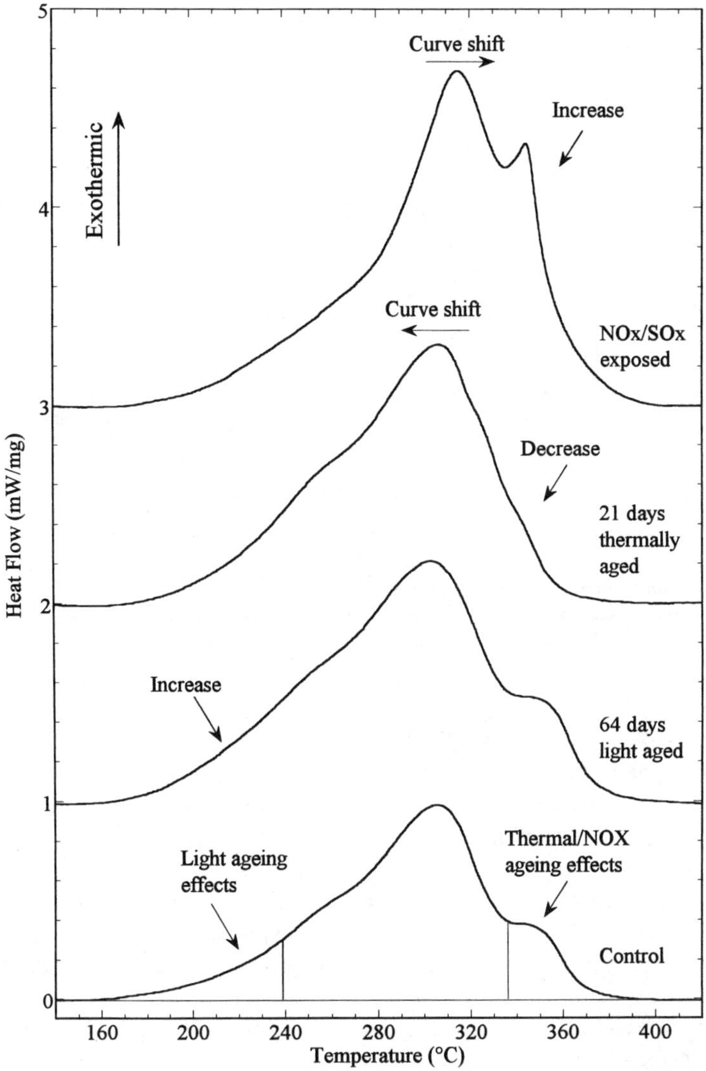

Figure 11.6. Effect of various forms of aging on shape of DSC curve of smalt tempera. Samples were heated in O_2 at 10°C/min.

lated to changes in chemical composition that have occurred. Its steady increase with light aging days (i.e., days during which samples have been exposed to set light levels) reflects the steady increase in the evolution of degradation products, as mentioned in the previous paragraph.

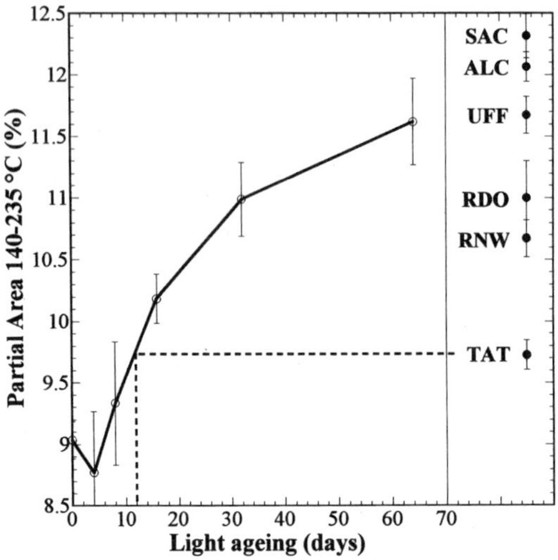

Figure 11.7. Smalt tempera light aging calibration graph. Partial areas of sensors have been plotted vertically, at side of figure: TAT, Tate gallery, U.K.; RNW, Rijksmuseum, Amsterdam; RDO, Rijksmuseum Store, Amsterdam; UFF, Uffizi Gallery, Florence, Italy; ALC, El Alcázar, Segovia, Spain; SAC, Sandham Chapel, U.K. Level of light exposure of each sensor can be read from adjoining graph as shown, with Tate sensor being used as an example.

The results obtained from the light-aged samples can then compared with those obtained with similar samples that have undergone natural aging at selected sites (art galleries and museums) in different localities where there were known differences in the light levels, environmental control, and efficiency of filtering air to minimize pollutant levels. In Figure 11.7 the partial areas of the sensors have been plotted vertically, at the side of the figure.

In determining levels of light aging and the amount of damage suffered by an object on exposure to museum conditions, it is usual to determine the total light exposure received (calculated from the product of time and light level). This relationship between time and light level is known as the reciprocity principle (Saunders and Kirby 1996). Light aging can then be related to years of museum exposure and is given by the following calculation: 200 lux $\times$ 8 hr/day $\times$ 365 days/year $= 1.6 \times 10^3 \times 365$ lux hr/year $= 5 \times 10^5$ lux hr/year $= 0.5$ Mlux hr/year. This represents typical light exposure for an oil painting under gallery conditions for one year.

Although the data discussed have focused on the binder in paint samples, in principle, DSC can be used to analyze both inorganic and organic components in cultural materials provided that the materials present undergo thermal changes when heated. These thermal changes involve, for example, thermal degradation, dehydration, glass transition, fusion, and crystalline transition (Wendlandt 1986).

11.3. THERMOGRAVIMETRY

Thermogravimetry (TGA) measures the change in weight of a sample that is linearly heated in a controlled atmosphere. Basic principles and different types of instrumentation are described elsewhere (Charsley and Warrington 1992; Haines 1995). The equipment, shown in Figure 11.8*a*, consists of an electrobalance from which a sample holder is suspended by a quartz rod. The furnace is then raised and surrounds the sample crucible. Measurements reported in this chapter were made on the Shimadzu TGA and the Perkin-Elmer TGA analyzers. The instrument configation in both cases is the same. The first derivative (DTGA) of the response (calculated from the TGA curve) provides the rate of change of mass with time. The microbalance used for this work is a null point balance. It is equipped with a sensor to detect the deviation of the beam from its null position. Deviation from the null position in this case is detected by optical methods. This set-up provides the sensitivity for measuring changes in weight of the order of micrograms. Measurements reported in this chapter were made using commercially available aluminum and platinum crucibles.

For thermal degradation studies the required sample size (about 500 μg to several milligrams) is usually greater than that for DSC. TGA measurements complement DSC data and relate observed weight changes with observed endothermic (e.g., moisture loss) and exothermic degradation peaks. The requirement of larger sample size meant that studies were limited to cases where ample sample size was available. The temperature range of the TGA analyzer generally extends from ambient to 1000°C. This means that it can be used for the analysis of carbonate-containing samples such as the ground layers of paintings. It can also be used to determine whether gypsum-containing grounds are fully hydrated. If the chemical composition of a material is known, then the measured weight loss can be compared with the expected or theoretical weight loss. In this way the observed weight changes can be related to loss of known molecules from the compound being analyzed.

In addition to thermal degradation studies, the moisture content of materials can be determined. Samples are sealed in aluminum crucibles, and a pin-hole is then made in the lid of the crucible. In this case samples are heated to just above 100°C and left for a few hours at this temperature. Recorded weight losses give a measure of the physically adsorbed moisture content of the sample.

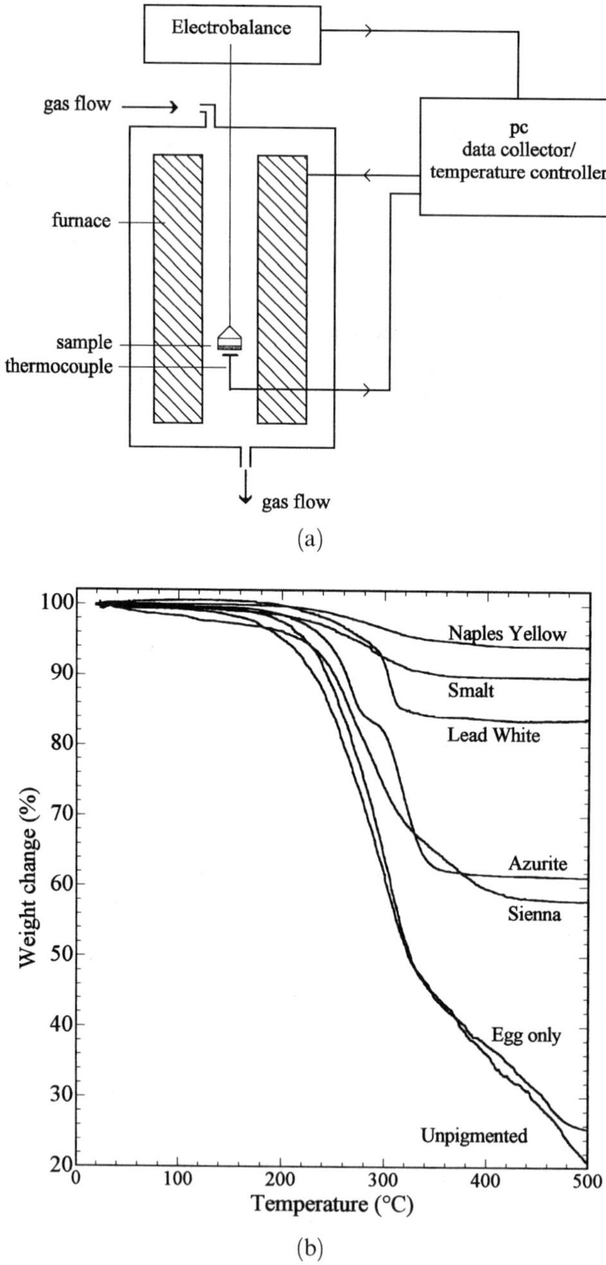

Figure 11.8. (*a*) Schematic outline of thermogravimetric analyzer. (*b*) TGA curves of pigmented tempera samples. Samples were heated in O_2 at 10°C/min.

11.3.1. Thermogravimetry: Applications

Samples from layers of priming of nineteenth-century paintings have been studied. Resulting curves from samples of priming from Sir Edwin Henry Landseer's painting *Study of a Lion* (ca. 1862, Tate Gallery, N01350) showed that it contained linseed oil, basic lead carbonate, and calcium carbonate (Odlyha et al. 1993). An extensive study was made of the thermal properties of the pigment of basic lead carbonate (Ciomartin et al. 1996). The variation in its thermal degradation behavior with the nature of the sample purge gas was examined. Some preliminary studies using high-resolution thermogravimetry (Hawkins 1996) have been made to deconvolute the complex decomposition behavior of basic lead carbonate and characterize the intermediate products.

More recent studies have been made on prepared samples of artists' materials (Odlyha et al., 1999b). Measurements were made on the differently pigmented temperas (shown in Fig. 11.8*b*). Measurement of the weight loss enables an estimation to be made of the relative medium-to-pigment concentration in the sample. For this to be calculated, it is first necessary to measure the weight loss in the pure unpigmented tempera and, where appropriate, in the pure pigment. Most of the pigmented temperas shown in Figure 11.8*b* are medium lean and contain less than 10% binder by weight. Sienna tempera, on the other hand, in medium rich and contains 42% binder by weight.

TGA was also used to study different oil–resin mixtures prepared according to recipes used in the nineteenth century by artists such as Turner. This was part of a collaborative project with the Conservation Department of the Tate Gallery. Figure 11.9*a* shows the TGA curve (and the calculated DTGA curve) for a sample with high oil content (3:1). In both cases the samples were measured three times to obtain representative TGA curves. Figure 11.9*b* shows the TGA curve (and the calculated DTGA curve) for a sample with high resin content (1:3) (Townsend et al. 1998). The increase in the amount of resin present was observed to change the shape of the TGA curve, and this can be seen more clearly in the difference in the shape of the first derivative or DTGA curve. The difference between the two samples was established by both thermogravimetry and mass spectrometry (Townsend et al. 1998).

11.4. THERMOMECHANICAL ANALYSIS

Thermomechanical analysis (TMA) measures the deformation of a sample under a static load. In the area of polymer and composite materials science

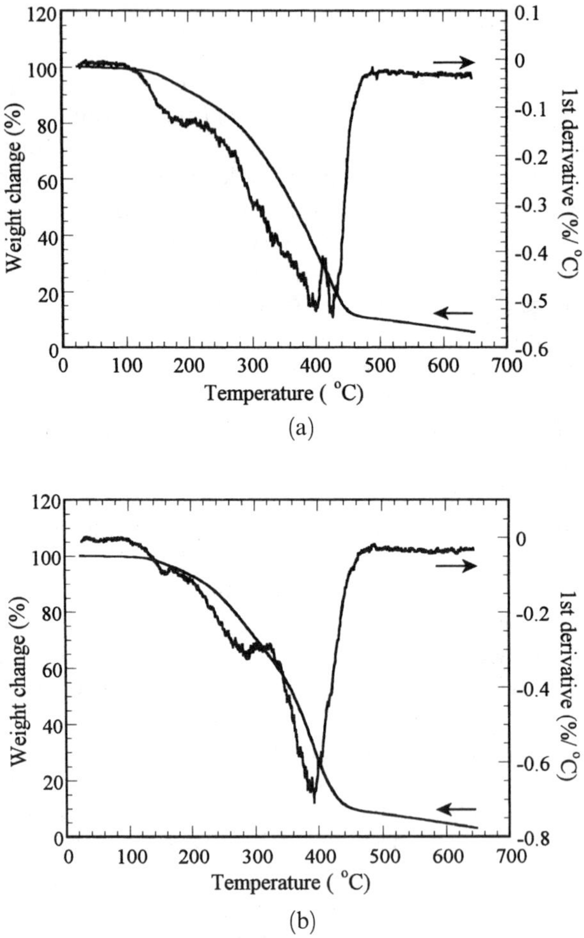

Figure 11.9. (*a*) TGA curve (and DTGA curve) for sample with high oil content (3:1). Sample was heated in N_2 at 2°C/min. (*b*) TGA curve (and DTGA curve) for sample with high resin content (1:3). Sample was heated in N_2 at 2°C/min.

thermomechanical methods, particularly Dynamic Mechanical Thermal Analysis (DMTA), have developed to the extent where there is virtually no aspect where they do not play an important role (Reading 1992). Their applications extend from measuring basic quantities such as coefficients of thermal expansion and glass transition temperatures (T_g) to assisting in defining the morphology of complex, phase-separated polymer systems and characterizing crosslinking behavior. In principle, any material that exhibits viscoelastic behavior and glass transitions can be studied using these techniques.

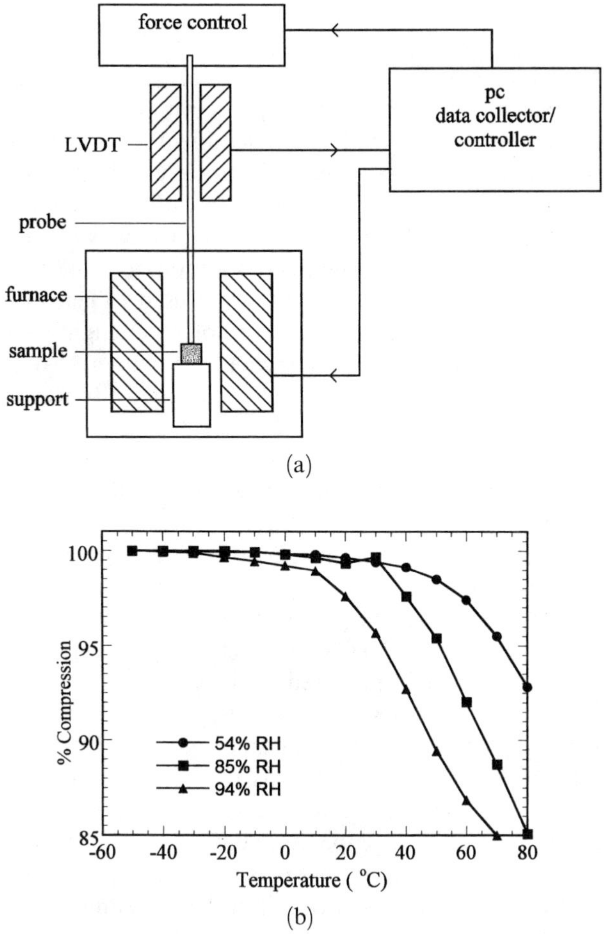

Figure 11.10. (*a*) Schematic outline of thermomechanical analyzer. (*b*) TMA curves (percent compression vs. temperature) for samples of nineteenth-century primed canvas at differing values of RH. Sample were heated in air at 5°C/min.

TMA equipment (shown in Fig. 11.10*a*) consists of a probe that rests on the sample and measures any change in sample displacement with temperature under the action of a controlled load. Measurement of displacement is achieved using a linear variable differential transducer (LVDT) attached to the sample probe. This is enclosed in the programmer-controlled furnace. Cooling of the equipment is achieved by placing a dewar flask of liquid nitrogen around the furnace.

The Stanton Redcroft TMA 690 analyzer was used to obtain the data reported in this chapter. Measurements were made in compression mode. A quartz probe with a small tip (1 mm^2) was used and a load of 100 mN was applied to the sample. For the measurement of glass transition (T_g) and softening temperatures of polymeric or paint samples, it is customary to operate in the temperature range from subambient −50°C to 120°C with a relatively slow heating rate of 5°C/min. In principle, this means that the test is nondestructive. For the evaluation of the effect of differences in RH of the environment on the T_g of samples of paint or primed canvas, small humidity chambers containing either saturated salt solutions or sulfuric acid–water solutions of known weight concentrations were used to generate the necessary conditions. Prior to the measurement samples were lightly coated with a thin film of a silicone oil to retain their moisture content. To improve this methodology, a humidity chamber is being designed for use with the dynamic mechanical analyzer.

11.4.1. Thermomechanical Analysis: Applications

In conservation science TMA was first used to study the effect of different pigment concentrations on the mechanical properties of prepared paint films. This was extended to include the effect of various solvents (Hedley et al. 1990b) and various values of RH (Odlyha et al. 1989) on the softening or glass transition temperature of prepared paint samples. The rationale for the former was to determine whether the solvents, which are generally used for cleaning the paintings, affect the mechanical properties of the paint films due to the possible leaching of the paint films that may occur. The study was conducted on 12-year-old paint films, and in these relatively fresh films changes in the T_g or softening temperature did occur. This was attributed to leaching effects and the loss of internal plasticizers (e.g., palmitic and stearic acids) from the paint. Further work would involve a similar study on artificially aged paint samples that could then simulate the behavior of historic samples. The rationale for the RH study, and additional studies where moisture itself was used, was to provide information on the response of paint films to assist in optimizing conditions where variable-RH conditions, or sometimes moisture itself, are used to remove surface deformations in paintings.

In addition to paint samples, nineteenth-century primed canvas samples (MacBeth et al. 1993) were also studied. Figure 11.10*b* shows the TMA curves (percent compression vs. temperature) for samples of nineteenth-century primed canvas at differing values of RH. Samples were provided from the original auxiliary canvas (loose lining) of Sir Edwin Henry Landseer's painting (*Study of a Lion*, ca. 1862, Tate Gallery N01350). Such samples are of extreme value for scientific investigation since they provide a source of

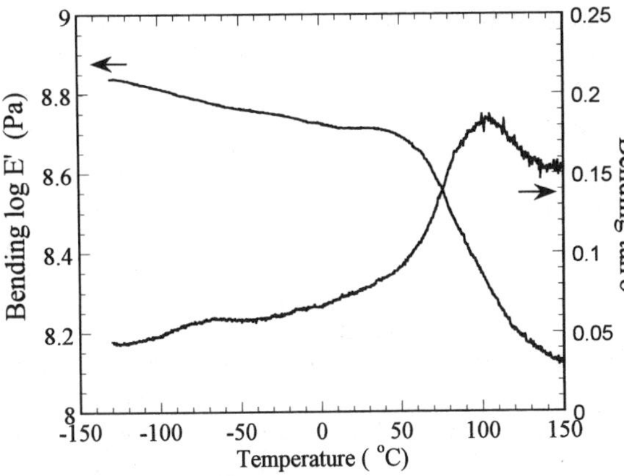

Figure 11.11. DMTA curve for sample nineteenth-century primed canvas. Sample was heated at 3°C/min.

naturally aged samples for study. The softening temperature is seen to decrease in its value as the sample is exposed to higher values of RH. The rationale for this study was to determine the response of primed canvas to humidification processes as used in the structural treatment of paintings using suction table techniques (Hacke 1963–1964; Mehra 1975). Research into the interaction of moisture with oil paintings was pioneered at the Courtauld Institute of Art, London and questions such as the quantities of moisture used and the role of treatment time and effect of moisture-based treatments were investigated (Hedley et al. 1990a; MacBeth et al. 1993; Hedley 1993a; Hedley and Odlyha 1993).

The same sample was studied by DMTA, a technique that will be discussed in the next section. The DMTA curve for the same sample (conditioned at 50% RH) is shown in Figure 11.11. In this case the decrease in the modulus (E') value or the peak in $\tan\delta$ gives the value for the glass transition temperature T_g. The significance of the parameters E' and $\tan\delta$ will be explained later in this chapter.

In the previous TGA section examples of some studies on oil–resin films were mentioned. TMA was also used to determine the glass transition temperature of the oil–resin films and how this value is affected by differing proportions of oil and resin. These films were prepared according to traditional nineteenth-century recipes. Different proportions of oil and resin (known as megilps) were used by nineteenth-century artists such as J. M. W. Turner (Townsend 1995). Results showed that in unaged samples where high proportions of oil were used two glass transition temperatures were observed: the low-temperature one corresponded to that for oil and the one at higher temperature to that for resin. This behavior is known to occur in immiscible polymers. With increasing resin concentrations the glass transition temperature moves to higher values and at a particular resin concentration only one T_g is measured (Pages et al. 1996). Further work is in progress to study the changes involved using thermally stimulated current techniques (Pages et al. 1997).

11.5. DIELECTRIC ANALYSIS

Dielectric analysis measures the electrical properties of a material when it is subjected to an alternating field. The basic prerequisite is that the material must contain dipoles. When an alternating field is applied, these dipoles will attempt to reorient with changes in the field direction. Due to the inertia of the dipoles and restrictions imposed by the surrounding environment, a phase difference occurs between the field and the response. The parameters involved are the relative permittivity, which relates to the ability of the existing dipoles in the sample to orient themselves under the action of the alternating field, and the dielectric loss, which refers to the physical inertia of the charge system to respond to the field. The principles of all dielectric techniques are on the whole similar, the differences being in the type of dipole relaxation processes observed over the various ranges of frequency and temperature. The important phenomenon contributing to the frequency dependence of the dielectric properties is the polarization arising from orientation with the imposed electric field of molecules which have permanent dipole moments. Hence the dielectric properties are no longer constant but vary as a function of frequency.

11.5.1. Dielectric Analysis: Applications

In recent years, as mentioned previously, research work at the Courtauld Institute of Art in London has concerned itself with developing new approaches to moisture treatment and studying effects of moisture treatments and cleaning agents on paintings. Two studies were carried out as final-year student projects in collaboration with the author at Birkbeck College (Husband 1990; Tillinghast 1990). They focused on the effect of solvents and moisture on the mechanical and surface properties of paint films (Hedley et al. 1990a). In the course of these studies a number of techniques (e.g., TMA, as mentioned above) were used for the first time to evaluate the effect of conservation procedures on prepared paint films. Among these was also the technique of dielectric analysis (kilohertz frequency region) over the temperature range −20 to 100°C. It was found that there were differences in the measured parameters when paint films were heated over a selected temperature range. Untreated lead white paint films and those that had been immersed in propanone and then dried showed differences in the measurements, which were ascribed to loss of polar components from paint films. This was confirmed by gas chromatography–mass spectrometry (GC–MS) (Erhardt and Tsang 1990). The sensitivity of the dielectric technique to changes in the chemical composition of the paint film led to further dielectric studies on paint and primed canvas samples (Odlyha et al. 1993, 1995; Chan and Odlyha 1994).

Low-frequency (10^{-4} to 10^{4} Hz) dielectric measurements at a selected temperature (30°C) and at stepped intervals over the temperature range of interest have shown that the response of pigmented paint films and primed canvas samples varies with the relative humidity of their microenvironment (and their moisture content) and that it depends on pigment type and medium content (Chan and Odlyha 1994). The rationale for the use of low-frequency dielectric measurements is to obtain information on the mechanism and rate of moisture transport in materials, for example, the nineteenth-century paint composite *Study of a Lion* by Landseer. A complete description of the application of this technique to conservation science is given elsewhere (Odlyha 1998).

The application of the dielectric analysis technique in the microwave frequency range has also been investigated. In this frequency range only the response of small polar molecules can be observed, and so it can respond directly to the presence or absence of water in the material. The measuring system also differs from the one used at lower frequencies where the sample is placed between two parallel plates. A coaxial line is used to convey the signal at microwave frequencies (originating from a network analyzer) to the sample, and the reflection characteristics of the signal are analyzed. A novel microprobe was specially designed for the purpose of making noninvasive measurements on paintings during treatment (Chan and Odlyha 1997) to measure directly the changes in its moisture content.

Humidification systems were set up at laboratory temperatures and at controlled temperatures between 30 and 60°C to simulate procedures used to treat deformations in canvas paintings. These humidification systems are described in detail elsewhere (Odlyha et al. 1996). The rate of humidification and the corresponding changes in the moisture content of primed canvas samples (naturally and artificially aged) were monitored with the microprobe connected to the network analyzer (Odlyha et al. 1996; Odlyha 1998). The measurement system was further improved to include the dynamic mechanical analyzer, which enabled improved control of the pressure applied to make contact between the probe and the surface (Foster 1997; Odlyha 1998). This system and the preliminary studies involving the drying of leather and groundwood paper, where the change in mechanical properties and moisture content are simultaneously monitored, are described in detail elsewhere (Odlyha 1998; Odlyha et al. 1999a).

11.6. DYNAMIC MECHANICAL THERMAL ANALYSIS

Dynamic mechanical thermal analysis (DMTA) is a thermomechanical technique that measures the mechanical or viscoelastic properties of polymeric materials and relates these, where the structure is known, to the internal motions of molecules of their segments. A stress or strain is varied in a periodic way as a function of time, usually sinusoidally (Fig. 11.12*a*). This differs from the TMA technique (described in the previous section) in which a static load is applied. Polymeric materials, when tested by DMTA, exhibit a combination of the ability to recover elastically when deformed and the ability to flow, that is, viscoelastic behavior. This is expressed in terms of a dynamic storage modulus E' for bending deformation, which quantifies the ability of the material to store energy (elastic behavior), and a loss modulus E'', which measures the ability of the material to dissipate energy by flow (viscous behavior). Here, E' is measured by quantifying that part of the sample's resistance to deformation that is in phase with the applied stress, while E'' is determined from the component that is out of phase. This is usually represented as an Argand diagram, as shown in Figure 11.12*b*. This shows that the response of the sample is split into real (in-phase) and imaginary (out-of-phase) components. For bending measurements, $E^* = E' + iE''$, where E^* is the dynamic bending modulus, E' the storage modulus, and E'' the loss modulus. Also, $\tan\delta = E''/E'$, where δ is the phase lag.

Hence the parameters measured by the instrument are the storage modulus E' (in pascals), which is a measure of the stiffness of a material; the loss modulus E'' (in pascals), which is associated with the dissipation of energy as heat when the material is deformed; and $\tan\delta$, which gives a measure of the degree of elasticity of the material (e.g., a zero value for $\tan\delta$ means that the material is perfectly elastic). Also, changes in the displacement (in terms of micrometers) under a given load (at a particular frequency) can be measured.

Figure 11.13*a* shows a schematic outline of the apparatus. There are several types of deformation: bending, tensile, and shear. The choice of an appropriate type of deformation depends on the type of material; for example, the bending mode was found to be useful for primed canvas and paint samples on a melinex substrate, whereas the tensile mode was found to be useful for unprimed canvas and unsupported paint samples. In bending the sample is held in a single cantilever clamp, as shown in Figure 11.13*b* (i), and in tensile it is clamped as shown in Figure 11.13*b* (ii). The dynamic load is applied to the sample as shown [perpendicular to the long sample axis

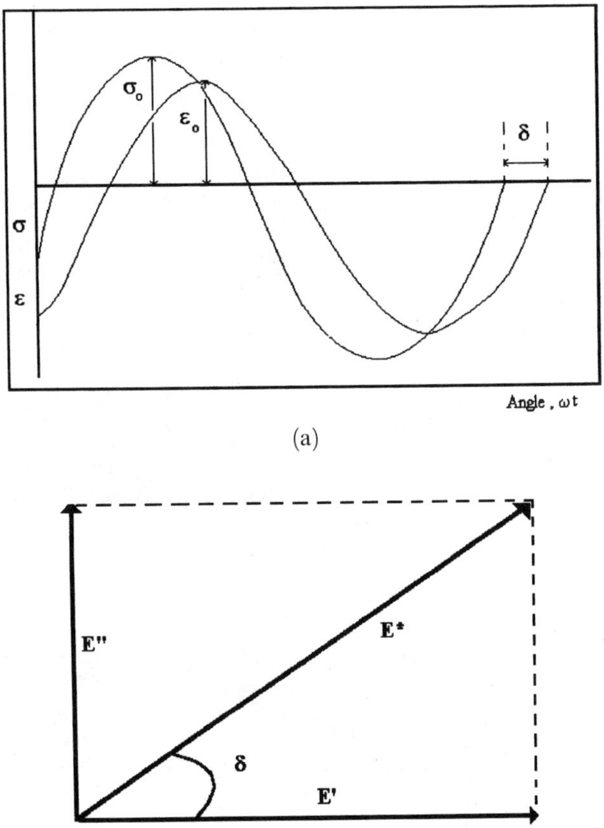

(a)

(b)

Figure 11.12. (*a*) Sinusoidal stress (σ) and stress response (ε) that lags by a phase angle (δ) because of viscoelastic behavior of sample (after Haines 1995). (*b*) Relationship between complex modulus (E^*) and real and imaginary components (E' and E'') is summarized in Argand diagram (after Haines 1995).

(bending) and parallel to the long sample axis (tensile)]. In bending the clamp can accommodate a sample from 1 mm up to 16 mm free length with a sample width of up to 12 mm and a maximum thickness of 3.5 mm. In the tensile mode the clamp holds a sample from 1 mm up to 30 mm free length.

The Rheometric Scientific MkIII (Epsom, United Kingdom) instrument consists of a dynamic mechanical spectrometer head surrounded by a temperature enclosure. This is interfaced with both a dynamic control system/analyzer and a temperature controller with its cryogenic unit. The instrument is interfaced to a Dell 486 PC and is controlled by DMTA Plus 5 operating

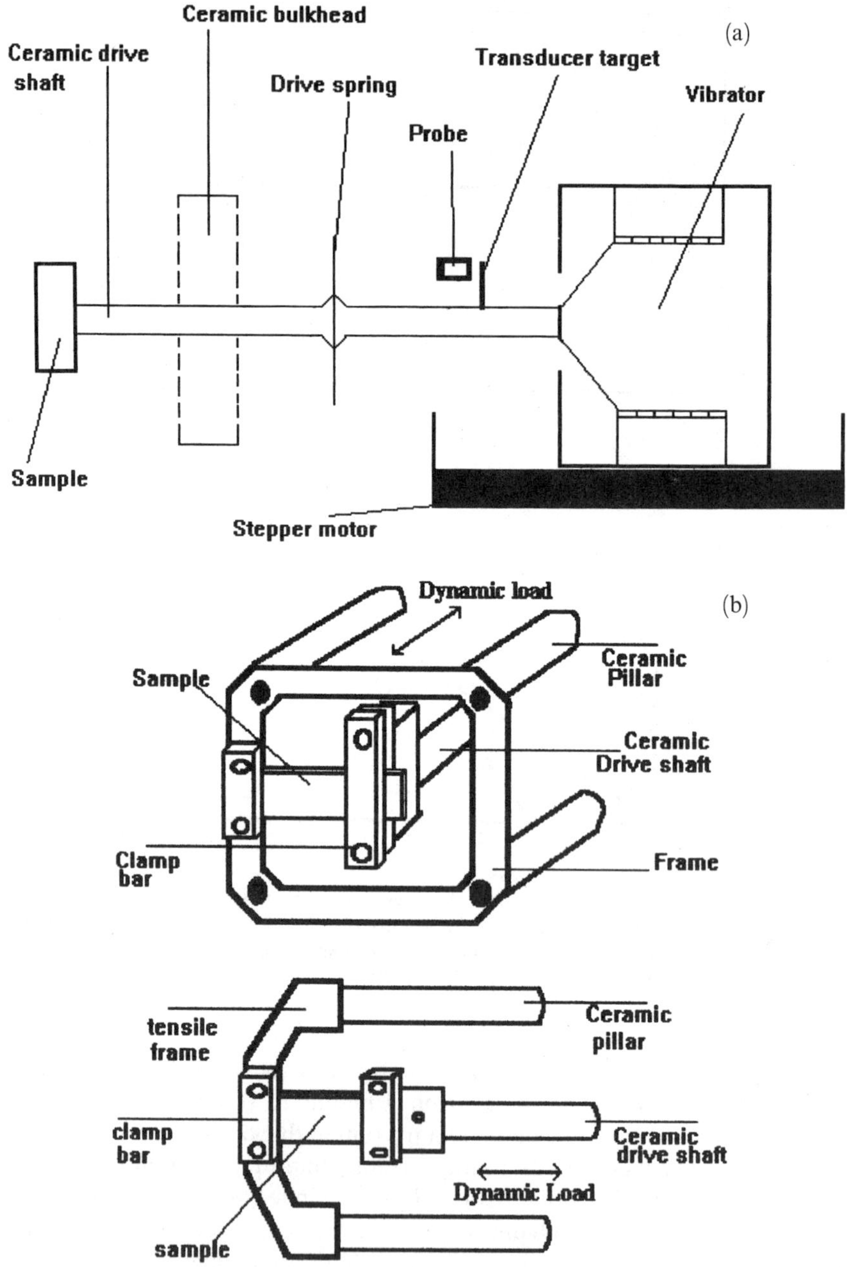

Figure 11.13. (*a*) Schematic drawing of DMTA analyzer (after Hill and Gilbert 1993). (*b*) (i) single cantilever clamp; (ii) tensile clamp.

software (V5.42). The drive shaft is controlled by an electromagnetic vibrator. The resulting displacement is monitored by means of a transducer whose probe is attached to the shaft behind the baffle where the temperature is maintained at ambient (Fig. 11.13*a*) (Hill and Gilbert 1993). This part of the apparatus is isolated from the temperature-controlled furnace by means of a large ceramic baffle. The heating apparatus consists of an insulated heater winding surrounded by a closed cooling chamber and insulation. The heater coils are connected to the temperature programmer, which controls the temperature of the furnace using a platinum resistance temperature sensor.

The dynamic behavior is monitored by means of an analyzer that measures the displacement resulting from the application of a known sinusoidal force. The frequency of oscillation can be selected from 0.01 to 200 Hz and strain can be varied over a 20 times range in 10 steps. Strain is proportional to the displacement of the drive clamp and is monitored by a noncontacting eddy current transducer. The strain signal is resolved into its in-phase and out-of-phase components to the applied stress and the values of $\log E'$ and $\tan\delta$ are computed (Wetton and Marsh 1989).

11.6.1. Dynamic Mechanical Thermal Analysis: Applications

DMTA has been used to evaluate the effect of nonaqueous deacidification procedures applied to canvas (Odlyha et al. 1997) as a preventive measure to assist in retarding the deterioration of painting canvases, as suggested by the Conservation Department of the Tate Gallery (Hackney and Torben 1994).

DMTA measurements are described on test samples of canvas. The approach taken was to record the mechanical response of samples under conditions of selected load or strain by (1) heating over a defined temperature range, (2) holding the samples at a set temperature and varying the RH immediately surrounding the sample, and (3) monitoring the displacement during drying of unprimed canvases at a selected temperature.

Deacidification procedures used on the test canvas samples are described in detail elsewhere (Odlyha 1998). For the purposes of this chapter it is sufficient to say that the material used is sold under the trade names Wei T'o and pHizz and is described according to the manufacturer as containing methoxy magnesium methyl carbonate (MMC) and ethyl magnesium ethoxy carbonate (MEC). Some samples were also treated by the Battelle process (Wittekind 1994). This uses a mixture of alkoxides of magnesium and titanium.

The rationale for using DMTA is that it may reveal changes that may occur to the essentially rigid cellulosic chain (canvas or fresh linen has as its main constituent cellulose) where the arrangement is stabilized by hydrogen bonding between the hydroxyl groups on the adjacent chains of glucopyranose rings. A more detailed discussion of the chemical composition and structure of linen is given elsewhere (Odlyha 1998).

In addition, the effects of artificial aging studies on the viscoelastic parameters of primed canvas samples are also presented. These have been previously reported elsewhere (Foster et al. 1997; Odlyha 1998). For this purpose, unaged and artificially aged canvas samples were prepared at the Tate Gallery Conservation Department. This was done by applying Winsor and Newton Cremnitz white tube paint (oil) on sized canvas. Samples were then aged: 17 days at a temperature of 70°C and RH 50% and 34 days of intermittent periods of heat and light aging (light exposure was 18,000 lux). The artificial aging was performed in the Conservation Department of the Tate Gallery. Light aging was carried out using filtered fluorescent tubes where wavelengths below 400 nm have been removed.

Light aging can then be related to years of museum exposure, as described in the previous DSC section, and is given by the following calculation: 200 lux $\times$ 8 hr/day $\times$ 365 days/year $= 1.6 \times 10^3 \times 365$ lux hr/year $= 5 \times 10^5$ lux hr/year $=$ 0.5 Mlux hr/year. This represents typical light exposure for an oil painting under gallery conditions for one year. The amount of light that was used for the aging test can then be equated to years of aging: 18,000 lux $\times$ 24 hr/day $\times$ 34 days $= 50 \times 10^4 \times 34$ lux $=$ 20 Mlux hr. Hence 34 days of light aging are equivalent to 40 years of gallery or museum exposure at 200 lux. In the case of the samples studied, thermal aging, in addition to light aging, was used, and it is not possible to relate the aging produced to an equivalent natural exposure, as is done with light aging. With complex materials, the relationship between natural aging and thermal conditions cannot readily be simplified (Howells et al. 1993) to a single activation energy in the Arrhenius equation. Six variables have been proposed (Gray 1977) for aging of paper, and this negates any attempt to apply the Arrhenius equation. Thermal aging includes the possibility of a number of reactions proceeding simultaneously at different rates and of reactions being interdependent. These considerations highlight the difficulties in preparing aged material that will be similar to naturally aged materials several hundred years old.

11.6.1.1. Experimental Procedure

Effect of Change in Temperature. Samples (8 mm long $\times$ 4 mm wide) were cut in the weft direction from pieces of untreated and treated (sprayed and brushed) canvas and placed in the tensile clamp under a defined load of 2 N. They were then heated at 3°C/min over the temperature range −130 to 275°C and a frequency of 1 Hz. Measurements were also made at a selected temperature (30°C) as a function of time.

Effect of RH Variation at 30°C. Samples of untreated and treated (sprayed and brushed) canvases (5 to 10 mm long × 4 mm wide) were cut in the weft direction. They were mounted in the tensile clamp and tested using a load of 1.5 N at a frequency of 1 Hz and a temperature of 30°C. The sample environment was controlled by a simple humidification system that was set up alongside the instrument. Air (from a gas cylinder) was bubbled through a Dreschel bottle containing a known (w/w) sulfuric acid–water mixture and then introduced into the sample chamber of the DMTA measuring head. Three values of RH were used for the test: 50, 75, and 100% RH. Each sample was initially exposed to 50% RH and allowed to equilibrate. When steady values of the storage modulus were obtained, conditions were changed to 75% RH and then 100% RH. The test was designed to monitor the change of modulus of the samples with RH at a given temperature and to determine whether the response to changes in RH was affected by the deacidification treatment. The RH values were checked by measuring with a digital thermohygrometer (HT-40, A.T.P instrumentation) connected on line to the measuring system. The test was repeated on primed canvas (Superfine Linen, Universal primed).

Monitoring Displacement during Drying of Samples at 30°C. Samples (10 mm long × 4 mm wide) were cut in the weft direction from the selected canvas sample and soaked in water (24 hr). Prior to measurement the excess moisture was removed using a paper tissue. Sample were then mounted in the tensile clamp using a static load of 0.05 N. The amount of creep or displacement (in micrometers) was measured by monitoring the drying profiles of untreated and treated (brushed and sprayed) canvas samples at 30°C. The drying profile of nineteenth-century unprimed canvas was measured *in situ*. It was achieved by adapting the DMTA so that the measuring clamps could be immersed in water. This is described elsewhere (Foster et al. 1998). The whole process (i.e., wetting and drying) could then be monitored without any disturbance to the sample.

11.6.1.2. Results

Thermal Scans. Figure 11.14 shows the superimposed thermal scans ($\log E'/\tan\delta$) for the unprimed canvas samples (untreated and MMC/MEC treated) from −130 to 275°C. The starting value for $\log E'$ for the untreated sample is higher than for the sprayed sample and remains higher over the complete thermal scan. In both cases there is a decrease in modulus or stiffness of the samples until the temperature approaches −30°C. This is followed by either (1) steady $\log E'$ values (untreated sample) or (2) a slight increase in $\log E'$ values (sprayed sample) until 20 to 30°C, at which tem-

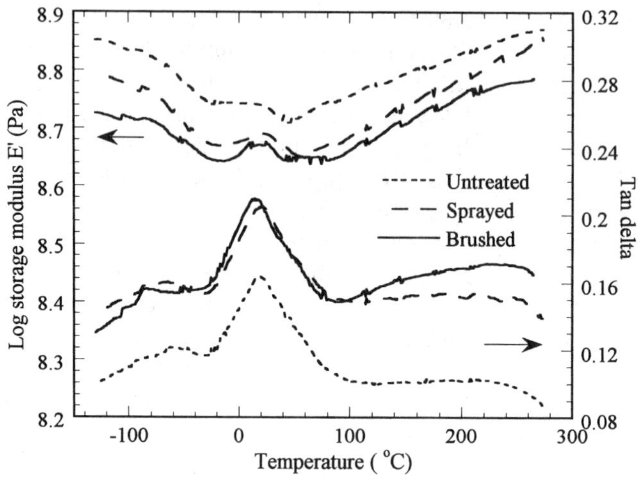

Figure 11.14. log E' (in pascals) and tan δ curves for unprimed canvas samples (untreated and MMC/MEC treated). Samples were heated at 3°C/min.

perature the samples soften. After 50°C they progressively increase in stiffness as moisture is lost from the samples. There is a single peak in tan δ in both cases with a peak maximum at 20°C. The intensity of the tan δ peak is higher for the sprayed than for the untreated sample. The untreated sample is stiffer (higher log E' value at 20°C), and this gives rise to a lower value for tan δ. Figure 11.15 shows the displacement traces. As the temperature increases from −130°C samples contract until they reach the vicinity of 20°C. This is followed by brief expansion of the samples to about 100°C and contraction until the onset of degradation in the region of 250°C.

There is a difference in the measured parameters between the untreated and treated samples but an almost indistinguishable difference between the sprayed and brushed samples. Table 11.1 shows the actual values for the parameters for the unprimed canvas samples that have been measured at selected temperatures.

Previous thermomechanical data on cellulose has been reported and demonstrated that several transitions occur during heating: −83 to −73°C, 7 to 27°C, 117 to 127°C, and 217 to 227°C (Toth et al. 1984), where the last transition has been identified as the glass transition temperature. In the present studies on unprimed canvas (Odlyha 1998) the dominant peak observed in the unprimed canvases was in the region of 20°C [moisture content (MC) of the canvas was about 6%]. Thermogravimetry was used to record the moisture content of the samples. Samples were initially heated at 20°C/min to 105°C and then held at that temperature for 1 hr. Small differences in the

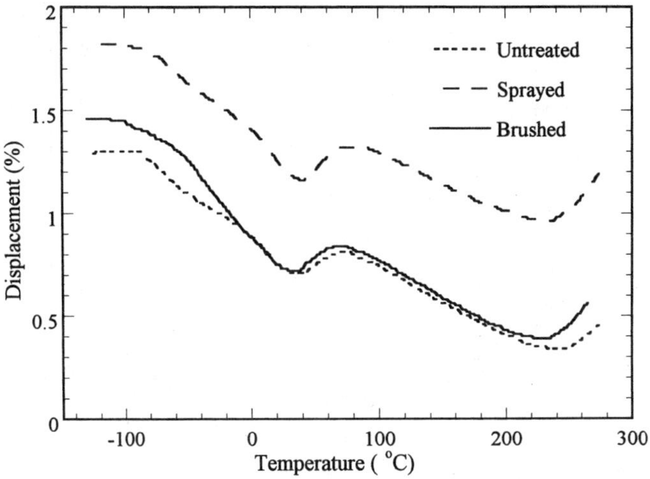

Figure 11.15. Displacement (%) curves for unprimed canvas samples (untreated and MMC/MEC treated). Samples were heated at 3°C/min.

Table 11.1. DMTA Parameters of Unprimed Canvas Samples[a]

	Pretransition (−20°C)		At Transition (20°C)		Posttransition (50°C)	
Treatment	log E' (Pa)	E' (MPa)	log E' (Pa)	E' (MPa)	log E' (Pa)	E' (MPa)
Untreated	8.74	550	8.74	550	8.72	525
Sprayed	8.67	468	8.69	490	8.66	457
Brushed	8.64	437	8.67	468	8.64	437

Treatment	tan δ, Pretransition (−20°C)	tan δ, At Transition (20°C)	tan δ, Posttransition (50°C)
Untreated	0.12	0.16	0.13
Sprayed	0.16	0.21	0.17
Brushed	0.16	0.21	0.18

Treatment	Displacement (%) Pretransition (−20°C)	Displacement (%) at Transition (20°C)	Displacement (%) Posttransition (50°C)
Untreated	0.98	0.75	0.75
Sprayed	1.50	1.26	1.21
Brushed	1.03	0.78	0.76

[a] Measured from Figures 11.14, 11.15, and 11.16.

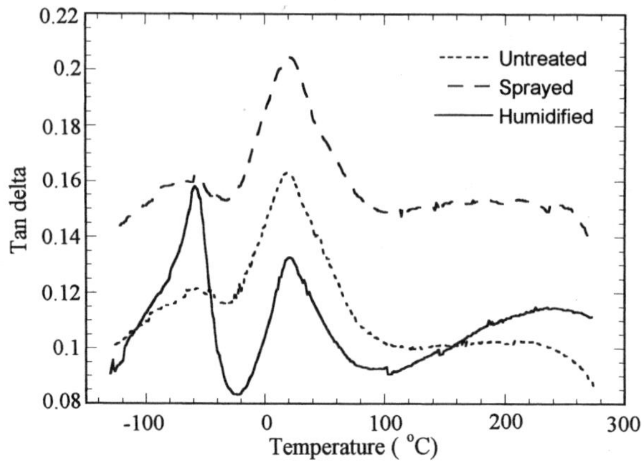

Figure 11.16. Tan δ curves for untreated, sprayed, and humidified unprimed canvases. Samples were heated at 3°C/min.

moisture content of untreated, sprayed, and brushed MMC/MEC samples were observed. The moisture content for the untreated unprimed canvas is 6%, and it coincides with the expected value for the moisture content of paper at about 50% RH (Browning 1977).

The effect of moisture content on the shape of the DMTA curves of unprimed canvas was tested. A sample that had been humidified at 80% RH until it had reached equilibrium moisture content (10% moisture content as determined by TGA) was measured by DMTA. Figure 11.16 shows the presence of an intense peak in the low-temperature region (−50°C). This peak has been observed in the relaxation behavior of the amorphous components of wood (Kelly et al. 1987) and has been attributed to the motion of a diluent (i.e., site exchange) rather than a specific side group of a wood component, and in regenerated cellulose solids (Manabe and Fujioka 1996) it was attributed to the local twisting motion of the main chain in the amorphous region. In the canvas it was only seen in the humidified sample (10% moisture content).

Figure 11.17 relates $\log E'$ (as measured at 30°C for treated and untreated samples) to the displacement (also measured at 30°C): as the samples reduce in stiffness, that is, reduction in the value of ($\log E'$), the amount of displacement increases: the untreated sample is stiffer and extends less under an applied load than the sprayed and the brushed (MMC/MEC) samples, and the Battelle sample shows the highest value for modulus and the lowest value of displacement. Values shown represent the average of two measurements made on the same day. Isothermal measurements were made to reduce errors arising from differences in the moisture content of the samples. The next

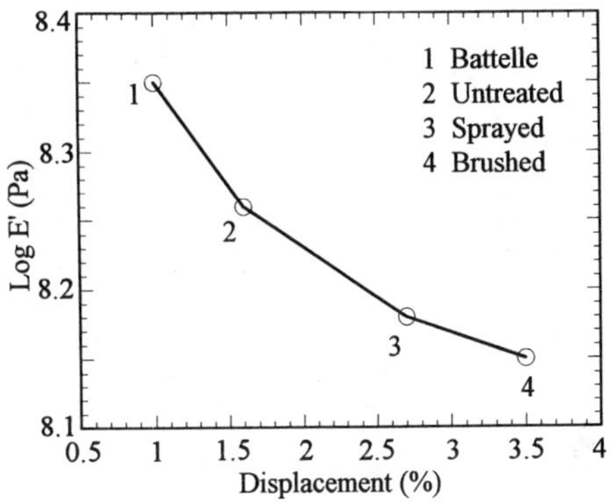

Figure 11.17. Log storage modulus vs. displacement of unprimed canvas (warp direction) untreated, brushed, sprayed (MMC/MEC), and Battelle.

section measures the response of treated and untreated samples to variations in relative humidity.

Effect of RH Variation at 30°C. Figure 11.18*a* shows the variation in storage modulus E' (in megapascals) with time and changing RH conditions. Samples were initially equilibrated (i.e., until a steady modulus value was obtained) at 50% RH. The starting E' values show that the untreated sample has a higher modulus and is stiffer than the brushed and sprayed samples. This confirms the previous results from the thermal scans. With increasing RH, differences between untreated and treated samples were accentuated and could be clearly seen from the change in modulus (or stiffness) (untreated, 245 MPa; treated, 180 and 190 MPa) accompanying changes in sample exposure from RH 75% to 100%. In the case of the untreated canvas both the magnitude of the change and the rate at which it occurs is greater than for either the brushed or the sprayed (MMC/MEC) samples.

The stiffening of the canvas can be explained in terms of what is already known about the behavior of canvas, that is, the shrinkage that occurs above 80% RH in unrestrained samples (Hedley 1988). Canvas is a woven structure and Hedley (1993b) has suggested that as a woven fabric absorbs moisture the yarns expand; in a tightly woven stressed material they exert a mutual force upon each other where they cross. In unrestrained specimens this force leads to shrinkage; in restrained specimens it causes fabric tension. In these measurements the sample is restrained, and so the tension that develops in the sample at elevated RH as it attempts to shrink is measured as an increase in stiffness (or E').

The fact that the extent of change in E' is different for the treated samples can be explained by considering that the degree of wettability or response of fibers to moisture has changed due to a change in the nature of the fiber

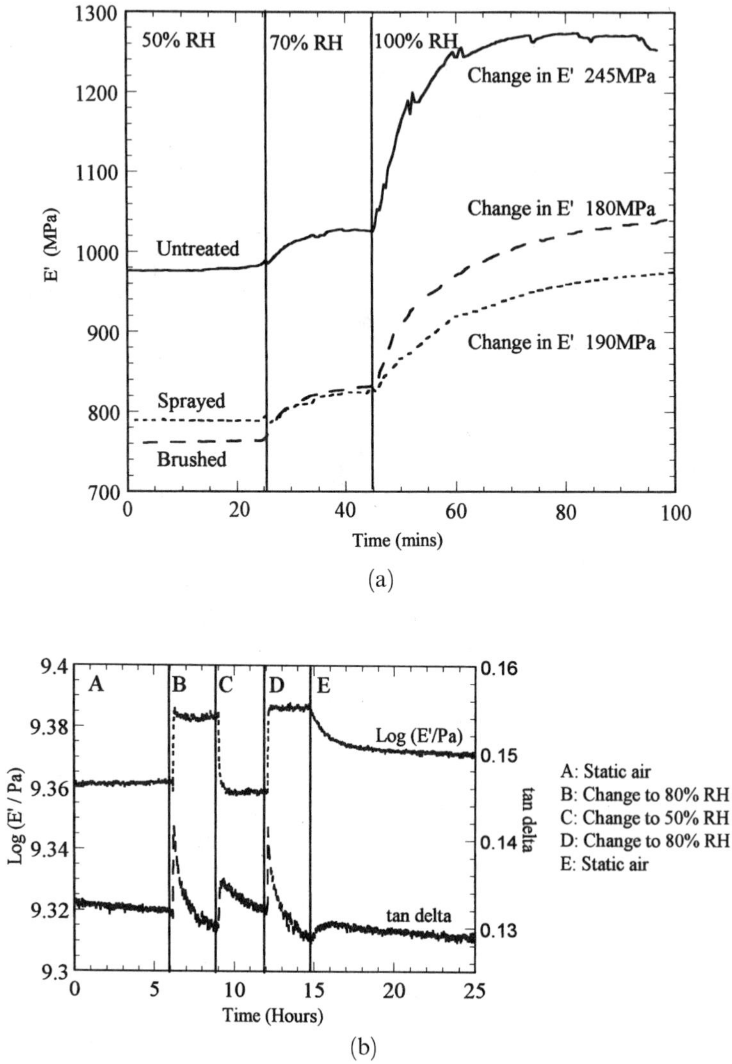

Figure 11.18. (*a*) Effect of change in RH on unprimed canvas (untreated and MMC/MEC treated) at 30°C. (*b*) Effect of change in RH on primed canvas at room temperature.

surfaces. In other studies of paper documents the extent of surface treatments on papers with sizing agents, fillers, and additives has been determined by the degree of wettability of paper fibers with water (Hon 1981). The reduction in moisture sorption may also be due to the smaller amount of available hydroxyl groups on the surface. Cellulose acetate, for example, is less

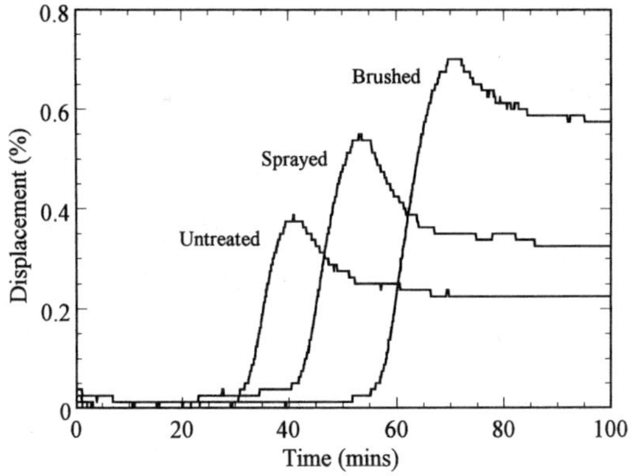

Figure 11.19. Displacement (%) (expansion followed by contraction) that occurs during drying of sample under action of static load (0.05 N) as function of time at 30°C.

water absorbing than cellulose due to the smaller number of hydroxyl groups in the molecules (Tsuge and Wada 1962).

Figure 11.18*b* shows the effect of change in RH on Universal primed canvas. The increase in stiffness with increasing RH is observed in the primed canvas (Universal, Russell and Chapple) in a manner similar to that which occurs in the unprimed canvas.

Monitoring of Displacement during Drying at 30°C. Figure 11.19 shows the displacement (%) (expansion followed by contraction) that occurs during the drying of the wet sample under the action of a static load (0.05 N) as a function of time. Results show that the rate of loss of moisture from all samples is approximately equal. There is a difference, however, in the time (minutes) of onset of expansion that occurs in the samples during drying and the amount of displacement that occurs. The time of onset of expansion increases in the following order: untreated < sprayed < brushed; the extent of initial expansion also increases in the same order: untreated < sprayed < brushed. The contraction that follows the expansion is much smaller in the brushed sample.

The observed expansion can be explained by considering previous observations where the tension development in restrained samples was measured during RH cycles (Hedley 1988). High-stress development was found to occur in dry and moist conditions with minimum-stress values at around 70 to 85% RH. When the wet canvas sample is placed in the tensile clamp as it loses moisture, there comes a point (when its MC reaches a value equivalent to that of a sample at 75 to 80% RH) of minimum or reduced stress levels in the sample, and this is observed as an expansion before it contracts with

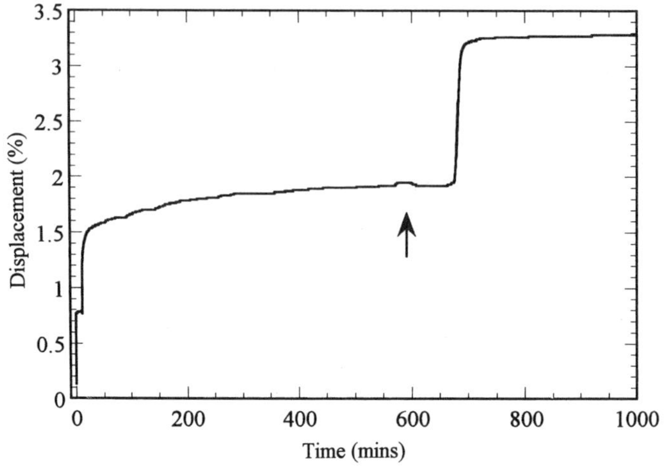

Figure 11.20. Displacement (%) of nineteenth-century unprimed canvas during drying under load (0.1 N). Arrow denotes removal from water at laboratory temperature.

increasing stress levels (and decreasing MC) under the constraint of the imposed load.

Water is considered nearly exclusively to be absorbed by the amorphous phase of cellulose (Salmén and Back 1980). The effect of the moisture is to soften or plasticize the amorphous regions and to make them swell. Foreign molecules, derived from carbonate ions, which may be present as a result of the deacidification treatment, may impede the loss of moisture from the structure. The longer time in the case of the brushed sample implies the presence of more foreign molecules.

Figure 11.20 shows the displacement of nineteenth-century unprimed canvas during drying under a load (0.1 N). Drying occurs with a sudden expansion of the canvas and there is no contraction, as in Figure 11.19. This measurement was performed by initially immersing the measuring head of the DMTA analyzer in water to monitor the displacement on wetting the fabric and then removing it and monitoring the drying process (Foster et al. 1998). This is an improvement on the measurement in Figure 11.19, where the sample had to be clamped in its wet state.

11.6.2. Effects of Artificial Aging on the Viscoelastic Parameters of Primed Canvas Samples

Figures 11.21*a* and *b* show DMTA curves obtained from the aged and unaged samples. These samples had been prepared and aged as described

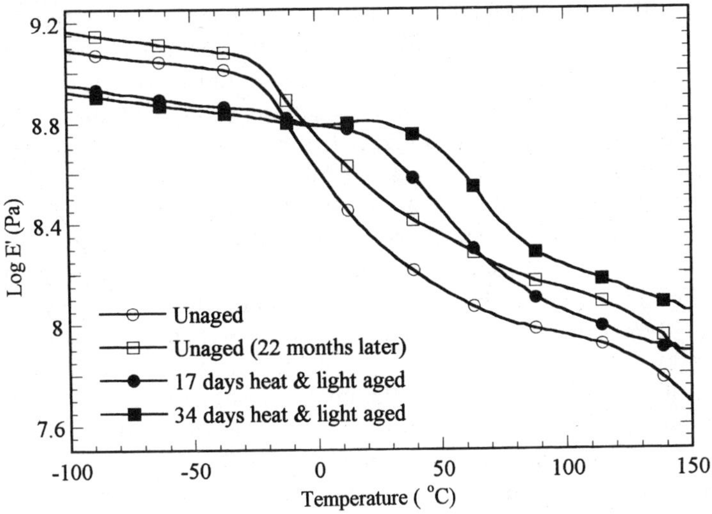

(a)

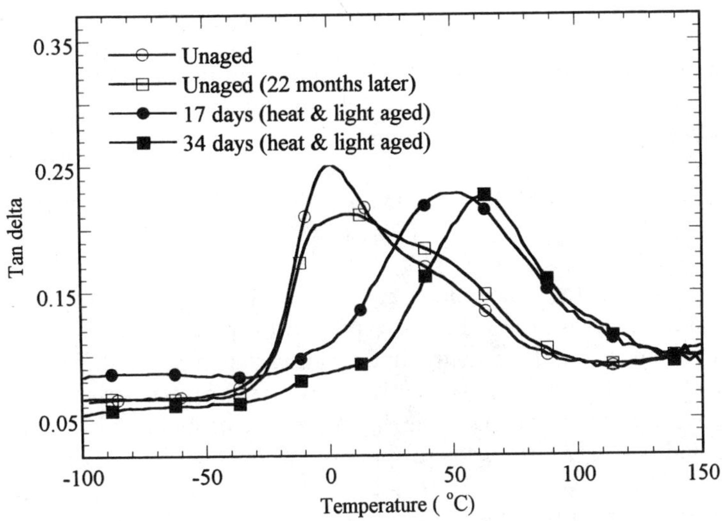

(b)

Figure 11.21. (*a*) Log storage modulus (E') for unaged and artificially aged primed canvas samples heated at 3°C/min. (*b*) Variation in $\tan\delta$ with temperature for unaged and artificially aged primed canvas samples heated at 3°C/min.

previously. From the $\tan\delta$ information it is possible to see two transitions: a peak at approximately 0°C (1 Hz), which is the main peak in the unaged sample and reduces its magnitude after aging, and a second transition at approximately 50°C (1 Hz), which appears as a shoulder in the case of the unaged sample and shifts to higher temperature after artificial aging. In Table 11.2 the modulus data (e.g., at 30°C, 1 Hz) show the increase in modulus and stiffness of the material as the sample is aged by heat and light.

Results show that for an unaged canvas values of modulus and $\tan\delta$ change significantly with small variations in room temperature (Figs. 11.21*a* and *b*). For aged samples, however, small variations in room temperature do not cause these values to change significantly; changes now occur at a higher temperature in the region of 60°C. The main transition ($\tan\delta$ peak), which starts in the subambient region in the unaged sample, is seen to decrease in intensity after 22 months of natural aging (stored in the dark under laboratory conditions). In the artificially aged samples this is radically reduced in intensity. The general shape of the peak in the unaged sample also differs from that of the aged samples. The modulus values below −50°C are significantly higher for the unaged sample. This can be explained by considering that the higher moisture content of the unaged samples makes them stiffer at those temperature. Above 0°C the modulus of the unaged sample then falls below that of the aged samples. Basically with heat and light aging the sample becomes stiffer, more stable, and more brittle at room temperature. Table 11.2 shows how the values for modulus and $\tan\delta$ of the samples change with artificial and natural aging.

11.6.3. Summary of DMTA Results on Effect of Preventive Conservation Treatment

Results show that the deacidification (MMC/MEC) and Battelle treatments have had an effect on the viscoelastic properties (modulus and $\tan\delta$) of unprimed canvases. The Battelle treatment causes a significant stiffening of the unprimed sample whereas the MMC/MEC treatment causes a softening (sprayed is stiffer than brushed; the differences are probably due primarily to the methods of application, where spraying would produce a more even distribution of the treatment). With the Battelle-treated samples the increase in stiffness can be attributed to the deposition of fine particles of TiO_2 and eventually $MgCO_3$ in the cellulosic structure. Some caution is necessary before making a final judgment on the result of the Battelle treatment as factors such as interfiber adhesion may cause a sample to appear stiffer immediately after treatment. Then with movement of the fibers the effects of interfiber adhesion will disappear and the sample will become softer.

In the MMC/MEC treatment the sample's response to moisture sorption

Table 11.2. DMTA Parameters of Unaged and Artificially Aged Primed Canvas Samples[a]

Temperature (°C)	Unaged		Unaged (22 months later)		17 days (Heat Aged)		34 days (Heat and Light Aged)	
	E' (MPa)	$\log E'$ (Pa)	E' (MPa)	$\log E'$ (Pa)	E' (MPa)	$\log E'$ (Pa)	E' (MPa)	$\log E'$ (Pa)
10	301	8.48	444	8.65	598	8.78	625	8.79
20	238	8.38	348	8.54	550	8.74	640	8.81
30	190	8.28	295	8.47	464	8.67	617	8.79
40	163	8.21	250	8.39	360	8.56	560	8.75

Treatment	Main Transition ($\tan\delta$)		Shoulder Peak	
	Position (°C)	Height	Position (°C)	Height
Unaged	1.4	0.25	41.4 } (R.H.)	0.17
Unaged (22 months later)	6.0	0.21	43.0 } (R.H.)	0.18
17 days (heat)	49.3	0.23	−6.7 } (L.H.)	0.10
34 days (heat and light aged)	62.3	0.23	−4.7 } (L.H.)	0.09

[a] Measured from Figures 11.21*a* and *b*.

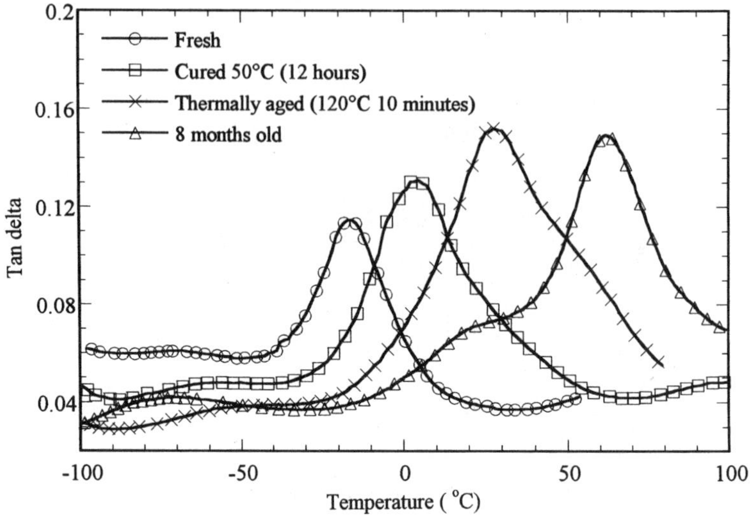

Figure 11.22. Tanδ curves for resin mastic samples and variation with aging. Samples were heated at 3°C/min.

and drying after treatment has been measured. The measurements reveal differences in the response of unprimed canvas samples that have been treated either by brushing or spraying. These are expressed in terms of the change in the modulus values and response times for the sprayed and brushed samples. The loss of moisture for the brushed sample occurs over a longer time scale than it does for the sprayed or untreated samples.

The application of DMTA to the measurement of unprimed canvases has demonstrated that it is possible to determine some of the effects of the preventive conservation treatment. The changes in the unprimed canvases indicate that surface changes on the fibers have occurred that have altered their response to moisture. The decrease in stiffness implies that the rigid straight chain where the arrangement is stabilized by hydrogen bonding between the hydroxyl groups on the chains of glucopyranose rings may have been weakened. It has also been shown that DMTA can detect the effect of natural and artificial aging of materials. In paint varnishes DMTA has been used to evaluate changes that occur on aging. This can be measured as a shift in the tanδ peak (which is also the glass transition temperature T_g of resin mastic varnish) to higher temperatures (Fig. 11.22).

A major advantage of DMTA is that it enables bulk properties to be related to molecular processes that involve changes in crosslinking and chemical structure. Where structures are known, this information is used effectively. With the type of samples available using traditional artists' ma-

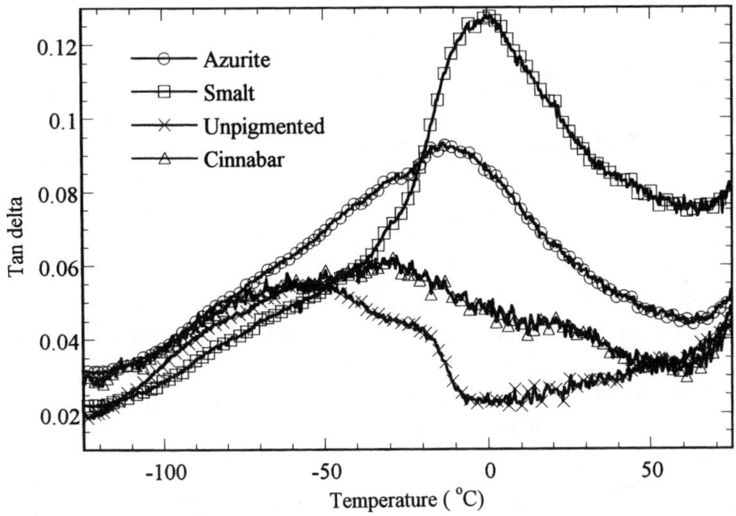

Figure 11.23. Tanδ curves for differently pigmented tempera samples heated at 3°C/min.

terials, the observed changes are complex, as seen, for example, in Figure 11.23, where the peak (tanδ) is shown for the differently pigmented paint tempera samples (Odlyha et al., 1999b). (Results on some of these pigmented temperas were presented earlier in the DSC and TGA sections.) Since the value of this peak provides a measure of the viscoelastic properties of the samples, its appearance and magnitude indicate the presence of a polymeric network and the extent to which it has developed in the paint tempera samples.

11.7. CONCLUSIONS

The chapter has demonstrated that the application of thermoanalytical techniques to measure the physical and chemical properties of samples from artists' materials and paintings can provide information on the nature of materials, the effects of aging, effects of conservation treatment, and environmental fluctuations. In particular, it has highlighted the value of dynamic mechanical techniques for the measurement of the effect of preventive conservation treatment and the effect of humidification processes on canvases. Further developments are in progress with the DMTA analyzer where the sample head with extended clamps can be immersed in water or solvents or an atmosphere of defined RH to measure *in situ* the sample response. The potential of dielectric techniques has also been mentioned. The novel microprobe connected to the network analyzer and measuring in the microwave region provides a tool for monitoring changes directly in the moisture content of the sample (Odlyha et al. 1997), and it has been used in conjunction with the DMTA analyzer to monitor the drying behavior of materials (Foster 1997; Odlyha 1998; Odlyha et al. 2000).

Acknowledgments

Special thanks to Dr. Neil S. Cohen (Birkbeck College, Chemistry Department) for the preparation of all the figures used in this chapter and his critical reading and contributions to the chapter. Thanks also to Gary Foster for the DMTA measurements and processing the DMTA figures. Thanks also go to Dr. Joyce H. Townsend, Stephen Hackney of the Tate Gallery (London), Caroline Villers, Courtauld Institute of Art, (London), Professor Dr. J. J. Boon and Drs. O. van den Brink (FOM Institute, The Netherlands), and Dr. Mauro Matteini, Opificio delle Pietre Dure, Florence, Italy. Part of this work acknowledges the support of the European Commission (EV5V-CT94-0548).

REFERENCES

Boon, J. J., J. Pureveen, D. Rainford, and J. H. Townsend (1995), in *Turner's Painting Techniques in Context*, J. H. Townsend, Ed., United Kingdom for Conservation of Historic and Artistic Works, London, pp. 35–45.

Browning, B. L. (1977), *Analysis of Paper*, 2nd ed., Marcel Dekker, New York, p. 42.

Chan, T. Y. A., and M. Odlyha (1994), *Life Chem. Repts.* **10**, 269.

Chan, T. Y. A., and M. Odlyha (n.d.), U.K. patent application no. 9503975.6, "Dielectric Probe."

Chan, T. Y. A., and M. Odlyha (n.d.), U.S. patent application no. 08/524759, "Dielectric Probe" (allowed Sept. 1997).

Charsley, E. L., and S. B. Warrington (1992), *Thermal Analysis—Techniques and Applications*, Royal Society of Chemistry, Cambridge.

Ciomartin, D. A., R. J. H. Clark, L. J. Mc Donald, and M. Odlyha (1996), *J. Chem. Soc. Dalton Trans.* **18**, 3639.

Erhart, D., and J. S. Tsang (1990), in *Cleaning, Retouching and Coatings Preprints IIC (International Institute for the Conservation of Historic and Artistic Works)*, J. S. Mills and P. Smith, Eds., London, pp. 93–98.

Foster, G. M. (1997), *J. Thermal Anal.* **49**, 1069.

Foster, G. M., M. Odlyha, and S. J. Hackney (1997), *Thermochim. Acta* **294**, 81.

Foster, G. M., R. Ibbett and M. Odlyha (1998), Birkbeck College, London, and Courtaulds Textiles, Coventry, U.K., Internal Report.

Gray, G. G. (1977), *Adv. Chem. Ser.* **164**, 286.

Hacke, B. (1963/1964), *Meddelser om Konservering*, Denmark.

Hackney, S. J., and E. Torben (1994), in *Preventive Conservation Practice, Theory and Research*, Preprints IIC, International Institute for the Conservation of Historic and Artistic Works, London, pp. 223–227.

Hackney, S. J., J. H. Townsend, and N. Wyploscz (1996), in *ICOM Committee for Conservation Preprints (11th Triennial Meeting, Edinburgh)*, Vol. 1, James and James Science Publishers, London, pp. 271–275.

Haines, P. J (1995), *Thermal Methods of Analysis, Principles, Applications and Problems*, Chapman and Hall, London.

Hawkins, N. (1996), TA Instruments, Leatherhead, U.K., personal communication, June.

Hedley, G. (1993a), in *Measured Opinions*, C. Villers, Ed., United Kingdom Institute for Conservation of Historic and Artistic Works, London, pp. 123–128.

Hedley, G. (1993b), in *Measured Opinions*, C. Villers, Ed., United Kingdom Institute for Conservation of Historic and Artistic Works, London, pp. 112–123.

Hedley, G., and M. Odlyha, (1993), in *Measured Opinions*, C. Villers, Ed., United Kingdom Institute for Conservation of Historic and Artistic Works, London, pp. 99–102.

Hedley, G., *Studies in Conservation* **33** (1988), 133–148.

Hedley, G., C. Villers, R. Bruce-Gardner, and R. MacBeth (1990a), in *ICOM Committee for Conservation Preprints (9th Triennial Meeting, Dresden)*, Vol. 1, K. Grimstad, Ed., Getty Conservation Institute, Marina del Rey, CA, pp. 119–124.

Hedley, G., M. Odlyha, A. Burnstock, J. Tillinghast, and C. Husband, (1990b), in *Cleaning, Retouching and Coatings*, Preprints J. S. Mills and P. Smith, Eds., International Institute for the Conservation of Historic and Artistic Works, (IIC) London, pp. 98–105.

Hill, R., and P. Gilbert (1993), *J. Am. Ceram. Soc.* **76**, 417.

Hon, D. N.-S., in *Preservation of Paper and Textiles of Historic and Artistic Value 11*, Advances in Chemistry Series, 193 (1981), ACS, Los Angeles, 109–141.

Howells, R., A. Burnstock, G. Hedley, and S. Hackney (1993), in *Measured Opinions*, C. Villers, Ed. United Kingdom Institute for Conservation of Historic and Artistic Works, London, pp. 27–34.

Husband, C. (1990), *The Effect of Relative Humidity on the Mechanical Properties of Oil Paint Films*, Final Year Project, Courtauld Institute of Art, London.

Kelly, S. S., T. G. Rials, and W. G. Glasser (1987), *J. Mater. Sci.* **22**, 617.

MacBeth, R., M. Odlyha, A. Burnstock, C. Villers, and R. Bruce-Gardner (1993), in *ICOM Committee for Conservation Preprints* (10th Triennial Meeting, Washington D.C.), Vol. 1, J. Bridgland, Ed., James and James Science Publishers, London, pp. 150–156.

Manabe, S., and R. Fujioka (1996), *Polym. J.* **28**, 860.

Mehra, V. R. (1975), in *ICOM Committee for Conservation Preprints* (4th Triennial Meeting), Venice.

Odlyha, M. (1990), in *ICOM Committee for Conservation Preprints* (9th Triennial Meeting, Dresden), Vol. 1, K. Grimstad, Ed., Getty Conservation Institute, Marina del Rey, CA, pp. 57–62.

Odlyha, M. (1995), in *Turner's Painting Techniques in Context*, J. H. Townsend, Ed., United Kingdom for Conservation of Historic and Artistic Works, pp. 29–34.

Odlyha, M. (1998), Characterisation of Cultural Materials by Measurement of Their Physicochemical Properties, Ph.D. thesis, University of London, London.

Odlyha, M., G. Hedley, C. D. Flint, and C. F. Simpson (1989), *Anal. Proc. R. Soc. Chem.* **26**, 399.

Odlyha, M., and A. Burmester (1988), *J. Thermal Anal.* **33**, 1041.

Odlyha, M., D. Q. M. Craig, and R. M. Hill (1993), *J. Thermal Anal.* **39**, 1181.

Odlyha, M., T. Y. A. Chan, and O. Pages (1995), *Thermochim. Acta* **263**, 7.

Odlyha, M., G. Foster, and M. Scharff (1996), in *ICOM Committee for Conservation Preprints (11th Triennial Meeting, Edinburgh)*, Vol. 1, James and James Science Publishers, London, pp. 297–303.

Odlyha, M., N. S. Cohen, R. Ibbett, and G. M. Foster (2000), Birkbeck College, London, ACORDIS, Sponden, U.K., and University of Exeter, Exeter, U.K., Internal Report.

Odlyha, M., G. M. Foster, N. S. Cohen, and R. Larsen (1999a), *International Council of Museums (ICOM) Committee for Conservation Preprints (12th Triennial Meeting, Lyon)*, Vol. II, ed. J. Bridgland, James and James Science Publishers, London, pp. 702–707.

Odlyha, M., N. S. Cohen, R. Campana, and G. M. Foster (1999b), *J. Therm. Anal. Cal.* **56** 1219–1232.

Pages, O., B. Legendre, M. Odlyha, and D. Craig (1996), *Thermochim. Acta* **287**, 53.

Pages, O., A. Lamure, C. Lacabanne, M. Odlyha, and D. Craig (1997), *J. Mater. Res.* **12**, 2784.

Preusser, F. (1979), *Maltechnik* **85**, 54.

Reading, M. (1992), in *Thermal Analysis—Techniques and Applications*, E. L. Charsley and S. B. Warrington, Eds., Royal Society of Chemistry, Cambridge, p. 108.

Salmén, L. N., and E. L. Back (1980), *Tappi* **63**, 117.

Saunders, D., and J. Kirby (1996), in *ICOM Committee for Conservation Preprints (11th Triennial Meeting, Edinburgh)*, Vol. 1, J. Bridgland, Ed., James and James Science Publishers, London, pp. 87–90.

Tillinghast, J. E. (1990), *The Effect of Solvent Exposure on the Mechanical Properties of Oil Paint Films*, Final Year Project, Courtauld Institute of Art, London.

Toth, F. H., Pokol, G., Gyore, J., and Gal, S., *Thermochimica Acta*, **80** (1984), 281–286.

Townsend, J. H. (1995), in *Turner's Painting Techniques in Context*, J. H. Townsend, Ed., United Kingdom for Conservation of Historic and Artistic Works, pp. 5–11.

Townsend, J. H., L. Carlyle, A. Burnstock, M. Odlyha, and J. J. Boon, "Nineteenth-

Century Paint Media Part 1: The Formulation and Properties of Megilps," Preprints IIC meeting Dublin, September 1998.

Turi, E. A. (1997), *Thermal Characterisation of Polymeric Materials*, 2nd ed., Academic, San Diego.

Tsuge, K., and Y. Wada (1962), *J. Phys. Soc. Jpn.* **17**, 156.

Wendlandt Wesley, W. M. (1986), *Thermal Analysis*, 3rd ed., John Wiley & Sons, New York.

Wetton, R., and R. D. L. Marsh (1989), in *Rheology of Good, Pharmaceutical and Biological Materials with General Rheology*, R. E. Carter, Ed., Elsevier, pp. 231–247.

Widmann, G., and R. Riesen (1987), *Thermal Analysis Terms, Methods and Applications*, Hüttig Verlag, Germany.

Wittekind, J. (1994), *Restaurator* **15**, 189.

CHAPTER

12

UV–VIS–NIR, FT–IR, AND FORS SPECTROSCOPIES

MAURO BACCI

Istituto di Ricerca sulle Onde Elettromagnetiche, Firenze, Italy

12.1. INTRODUCTION

The birth of spectroscopy goes back to when Isaac Newton performed his well-known experiments with sunlight and a triangular glass prism. The resulting group of colors ranging from violet to red, which could be observed on the wall of the room, was given the name *spectrum* by Newton himself, who arbitrarily assumed the existence of only seven fundamental colors, probably by analogy with the seven musical notes.

Much later, at the very beginning of the nineteenth century, it was demonstrated that the spectrum in fact extended beyond the visible range. In 1800, Herschel (Herschel 1800), who was studying the different distribution of heat throughout the spectrum with the aid of a thermometer, pointed out the existence of a warmer area beyond the red, thus discovering infrared (IR) radiation. Just one year later, Ritter (Ritter 1801) observed that AgCl, which was decomposed by the blue-violet light of the spectrum, reacted more easily when it was placed beyond the violet area, thus discovering ultraviolet (UV) radiation. However, more than 50 years elapsed before the first analytical applications of spectroscopy, which are due to G. Kirchoff and R. Bunsen (Born and Wolf 1965), were made. To these scientists can be attributed the first chemical analysis of the sun's atmosphere. This was done by comparing the Fraunhofer lines of the solar spectrum with the lines emitted by pure elements when excited by a flame or a spark. To carry out this assignment, the general statement made by Kirchoff 2 years earlier (1859) was necessary, that is, each element is characterized by its own unique spectrum and can absorb exactly the light of the same frequency that it can reemit (Kirchoff 1859). In the meantime, some molecular spectra, mainly in the gas phase,

Modern Analytical Methods in Art and Archaeology, Edited by Enrico Ciliberto and Giuseppe Spoto. Chemical Analysis Series, Vol. 155
ISBN 0-471-29361-X

were also recorded, while the first application of IR spectroscopy to chemical analysis is probably due to Abney and Resting (Abney and Resting 1881; Jones 1987a, 1987b, 1987c), who in 1881 reported the absorption spectra of 46 organic liquids in the near infrared (NIR). Many more spectra of organic compounds were recorded by Coblentz a few years later (Coblentz 1905; Jones 1987a, 1987b, 1987c).

It is evident that all the experimental data that were accumulating in the field of optics and spectroscopy during the nineteenth century were a challenge to theoretical interpretation. However, first the electromagnetic theory of light, which was formulated by Maxwell (for an exhaustive discussion of the theory see Born and Wolf 1965), and subsequently the quantum theory (Hollas 1992) opened up the way to a full understanding of atomic and molecular spectra over the whole range of the electromagnetic spectrum. Moreover, it is quite obvious that, for an extensive application of spectroscopic methods to chemical analysis, not only a theoretical interpretation of the spectra but also instrumentation developments were necessary. Indeed, Coblentz himself built some IR spectrometers; but it was only during the period between World War I and II that the analytical potentialities of molecular spectroscopy were appreciated, although it was necessary to wait until the end of the World War II for a wide diffusion of instrumentation, when industries realized the big market that could be opened in the field. As regards the different spectral regions, near infrared (NIR) spectroscopy (700 to 2500 nm), which was born together with UV and visible (VIS) spectroscopy, was not used much by either the theoretical/interpretative or the analytical points of view. This can be attributed to the greater difficulty in interpreting all the spectral features and to the weaker signal in comparison to the mid-IR region. Therefore, for many years, the NIR range was not very appealing for analytical purposes. However, the development of new detectors and of fast data-recording systems, together with the setting up of chemometric procedures, produced an astonishing renewed interest in NIR spectroscopy beginning in the 1980s, particularly in the fields of agriculture and food chemistry (McClure 1992).

It is not easy to establish when spectroscopy was first applied to works of art. In fact, the materials that can be investigated using UV, VIS, or IR spectroscopy (pigments, dyes, ligands, varnishes, ceramics, glasses, stones, marbles, bricks, wood, fabrics, paper, inks, and so on) are of interest also for industry. It could thus happen that a given material was investigated in a different context but that the resulting data were utilized for the knowledge and/or conservation of works of art. To the best of our knowledge the first spectrophotometric studies in the VIS region that were intentionally devoted to works of art are due to N. F. Barnes (Barnes 1938–1939; Barnes 1939), who reported both spectral and colorimetric data of many natural and artificial artists' pigments. Around the same time (1936 to 1942) F. I. G. Rawlins published a set of studies (Rawlins 1936, 1937, 1941, 1942) in which color change of paintings due to aging or cleaning processes was analyzed. It has also to be recalled that, as early as in 1917, H. E. Merwin (Merwin 1917) published a study concerning the optical properties (mainly refractive index and size of grains) and the color of pigments, without, however, reporting any reflectance spectrum. It is more difficult to establish the beginning of IR studies: A report on the IR absorption properties of pigments is found in the literature as early as in the 1930s (Farnsworth 1938–1939). However, the first applications of "real" IR spectroscopy to works of art are found in the literature mainly since the 1960s (Feller 1954; Abrahams and Edelstein 1964; Gettens and Stout 1966; Afremow and Vandeberg 1966; de Barros et al. 1967a, 1967b), even if reports concerning IR spectra of materials of potential interest for the fine arts can also be found in the 1950s or before (Hunt et al. 1950).

12.2. THEORETICAL PRINCIPLES

Electromagnetic radiation is absorbed selectively by a material when the electromagnetic field of the radiation induces a transition between two different energy levels of the molecules or ions constituting the material. The absorbed radiation has a frequency corresponding to the energy gap between the two levels as expressed by the Bohr condition:

$$h\nu = E_f - E_i$$

where ν is the frequency and f and i stand for final and initial states, respectively. The main contribution to the transition is given by the electric dipole moment μ, and therefore hereinafter we refer only to electric dipole transitions. From a quantum mechanical point of view, the absorption intensity I is proportional to the square of the transition moment

$$I \propto \left| \int \Psi_f^* \mu \Psi_i \, d\tau \right|^2 \tag{12.1}$$

where Ψ_f and Ψ_i are the total wave functions of the final and initial states, respectively, and $d\tau$ is the volume element in the coordinates of all the nuclei and electrons. On the basis of symmetry rules, it is possible to establish whether the integral in Eq. (12.1) vanishes or not. In fact, let Ψ_f, μ, and Ψ_i transform as the irreducible representations Γ_f, Γ_μ, and Γ_i, respectively; the transition moment may exist if the direct product $\Gamma_f \times \Gamma_\mu \times \Gamma_i$ contains the totally symmetric representation (Hollas 1992). Accordingly, some selection rules can be derived. The vibrational transitions, which occur within a unique electronic level, are allowed only if the variation of the vibrational quantum number Δv ($= v_f - v_i$) is ± 1, which reduces to $\Delta v = 1$, if absorption, and not emission, is considered. Usually, at room temperature (or below), only the so-called *fundamental* transitions ($v = 0 \rightarrow v = 1$) are observed in the mid- or far-IR region, with the bands corresponding to the bond length deformation (stretching) modes expected at higher frequencies than those corresponding to the bond angle deformation (bending) modes. However, the previous rule is valid when the harmonic oscillator wave functions are considered. Indeed, in actual cases, anharmonicity is always present so that it is often possible to observe, though much less intense, vibrational transitions for which $\Delta v \neq 1$. These transitions, which occur in the near- or mid-IR region, are named *overtones* and *combination* transitions. The former originate from excitations to upper levels ($\Delta v = +2, +3, \ldots$) within the same normal vibration, while the latter involve transitions to levels in which more than one normal vibration is excited. Owing to anharmonicity, the corresponding bands are expected at frequencies lower than the multiple (overtones) or the linear combinations (combinations) of the fundamental frequencies. Usually, the most intense bands to which the stretching vibrational modes contribute are expected in the NIR range. Several electronic transitions that are transitions between different electronic levels can be also observed in this spectral range. Depending on the nature of the electronic levels involved, these transitions can occur throughout the entire UV–VIS–NIR interval. Keeping in mind the materials constituting cultural properties, the following types of electronic transitions are of main interest.

Transitions between Delocalized Molecular Orbitals. These are typical of organic pigments and dyes (e.g., indigo, alizarin, curcumin). The bands are very intense and occur usually in the VIS or very near UV. It can in fact be demonstrated that, for an organic molecule to absorb visible radiation, that is, at not too high energy, it must have an electronic structure that permits electron delocalization over a wide space interval. That can be realized when conjugate double bonds are present in the molecule, better if accompanied by electron-donor and electron-acceptor groups far from each other. In this way, the energy levels involved in the transition (as a rule, the highest occupied and the lowest unoccupied ones) are separated by energies corresponding to the frequencies in the VIS range.

Charge-Transfer Transitions. Like the above-mentioned transitions, these are transitions between molecular orbitals. However, in the present case, the molecular orbitals are localized in different sites of the molecule or crystal so that a real electron transfer occurs during the transition. Some examples concerning materials that may be found in works of art can better clarify this point. Blue sapphire consists of corundum crystals Al_2O_3, where two close Al^{3+} ions are substituted by Fe^{2+} and Ti^{4+} ions; during the transition, one electron is transferred from iron to titanium. The overall process can be sketched as:

$$Fe^{2+} + Ti^{4+} \xrightarrow{h\nu} Fe^{3+} + Ti^{3+}$$

A completely analogous mechanism is found in Prussian blue, $Fe_4[Fe(CN)_6]_3.xH_2O$, but the charge transfer occurs between two non-equivalent iron ions:

$$Fe_A{}^{2+} + Fe_B{}^{3+} \xrightarrow{h\nu} Fe_A{}^{3+} + Fe_B{}^{2+}$$

Other charge-transfer transitions are the so-called ligand-to-metal charge transfers, which often occur in inorganic pigments such as chrome yellow, $PbCrO_4$, where the absorption of the radiation causes an electron transfer from oxygen to chromium into the chromate ion.

Ligand Field Transitions. These are characteristic of many inorganic pigments (azurite, malachite, verdigris, smalt, cobalt violet, Egyptian blue) and gemstones (ruby, emerald, turquoise). The transitions occur between levels that are localized mainly on the metal ion. In general, the latter is a transition-metal ion, which may be one of the main constituting elements of the material, as in azurite, or an impurity substituting at a little extent and randomly other metal ions such as Cr^{3+}, which, in ruby, replaces Al^{3+} of corundum. The metal ion is surrounded by ligand ions (or atoms or molecules) that produce a field acting upon the central ion orbitals, so as to resolve the degeneracy of the five *d* orbitals of the transition-metal ion (Lever 1984). Accordingly, the energy of the different *d* orbitals is a function of the ligand type, number, and geometrical arrangement. A typical example is given by the Co^{2+} ion, which is coordinated to 4 oxygen atoms in a near tetrahedral geometry in smalt and to 6 oxygen atoms in a near octahedral geometry in cobalt violet. Consequently, in the former case only one transition is observed in the VIS range, while in the latter case two transitions, also with a different line shape, occur in the same interval, with a resulting color shift from blue to violet (Bacci and Picollo 1996).

Energy Band Transitions. These occur in metals and semiconductors and are, therefore, observed in every metal work of art and in those inorganic pigments that are constituted by semiconductors (cinnabar, minium, cadmium yellow, litharge, orpiment). In these systems, the molecular orbitals arising from the interaction of the atomic orbitals of each atom are so close to one another that they yield a continuum, that is, an energy band. The so-called conduction band is partially filled in metals so that radiation of any sort of energy can be absorbed and, subsequently, re-emitted, thus causing the typical luster of the metals. On the contrary, in semiconductors, two bands, one filled (valence band) and the other empty (conduction band), are separated by an energy gap so that the radiation begins to be absorbed continuously starting from the frequency that corresponds with an energy quantum equal to the band gap. Accordingly, the absorption band is characterized by a "S" shape, instead of the bell shape encountered for transitions between isolated molecular orbitals.

When a beam of electromagnetic radiation impinges upon an object, a part of it is reflected by the surface and a part enters into the object, where it is absorbed and also scattered, depending upon the physical nature of the material. For sufficiently transparent objects, a part of the radiation can pass through the object itself. In such a case, a transmittance measurement can be made to obtain qualitative and quantitative information about the materials constituting the object under examination. However, except for a few exceptions such as glass objects or stained glass windows, works of art are opaque; therefore, spectrophotometric data can be obtained from reflectance measurements only. Reflectance spectroscopy (Kortüm 1969) measures the spectral composition of the radiation specularly or diffusely reflected by the object surface, to which radiation emerging from the inside of the object (scattered, multireflected, and so on) is added. The balance between specular and diffuse reflection depends on how smooth or mat the surface is, respectively. Apart from attenuated total reflection (ATR) techniques, which are used in IR spectroscopy, reflectance spectroscopy usually is concerned with diffuse radiation, so that experimental equipment is planned so as to avoid the recording of specular radiation. In applied spectroscopy, the Kubelka–Munk theory is widely accepted (Kubelka 1948, 1954), which relates diffuse reflectance to the absorption and scattering coefficients of a light-diffusing material. Usually, the Kubelka–Munk function:

$$F(R_\infty) = \frac{(1 - R_\infty)^2}{2R_\infty} = \frac{K}{S}$$

is utilized, where R_∞ is the diffuse reflectance of an infinitely thick material that is such that the reflectance does not increase for any sort of small increase in the thickness; $K(\lambda)$ and $S(\lambda)$ are the absorption and scattering coefficients, respectively. The Kubelka–Munk function is very useful from a practical point of view because it provides data proportional to the absorption of the material. However, the basic theory starts from several simplifying assumptions, so that only a few applications to works of art were fairly successful, as with paper or textiles (Van Asperen de Boer 1970).

Lastly, since color is of utmost importance for art objects, few concepts about colorimetry are in order. Indeed, it is not easy to give a univocal definition of color because this depends on both physical and psychological factors. In fact, it must be taken into account that the concept of color is intrinsically linked to the potential observer and, therefore, even if the physical origin of colors is well understood, problems arise when one aims to quantify what an observer sees. It has been ascertained that humans have a trichromatic vision, which is based on the different spectral responses (centered in the blue, green, and red) of three kinds of photopigments in the retina. This fact reflects on the way in which colors and color differences are perceived. Besides the observer's eye, two other factors contribute to the color vision, that is, the lighting source and the chemical and physical nature of the object. According to the Commission Internationale de l'Eclairage (CIE) convention (Judd and Wyszecki 1975; Wyszecki and Stiles 1982; Hunt 1991), colors can be expressed quantitatively by the so-called tristimulus values:

$$X = k \int \beta(\lambda) S(\lambda) \bar{x}(\lambda)\, d\lambda \tag{12.2}$$

$$Y = k \int \beta(\lambda) S(\lambda) \bar{y}(\lambda)\, d\lambda \tag{12.3}$$

$$Z = k \int \beta(\lambda) S(\lambda) \bar{z}(\lambda)\, d\lambda \tag{12.4}$$

where $\beta(\lambda)$ is the object reflectance, which can be obtained by recording the reflectance spectrum of the object; $S(\lambda)$ is the relative spectral power of the source, assuming 100 to be the spectral power at 560 nm [$S(560) = 100$]; and $\bar{x}(\lambda)$, $\bar{y}(\lambda)$, and $\bar{z}(\lambda)$ are the color-matching functions that take into account the eye response of a normal vision observer. The normalizing factor k is given by:

$$k = \frac{100}{\int S(\lambda) \bar{y}(\lambda)\, d\lambda}$$

As the color-matching functions are tabulated for both 1931 and 1964 CIE conventions, a simple measurement of the object reflectance and source emission spectra makes it possible to obtain the X, Y, and Z values. Commonly, reference is made to standard illuminants for which the normalized spectral power curves are known (Judd and Wyszecki 1975; Wyszecki and Stiles 1982; Hunt, 1991). At present, the most common illuminant is D_{65}, where the symbol stands for a daylight at a temperature of 6500 K.

Starting from X, Y, and Z, it is possible to obtain the chromaticity coordinates defined as

$$x = \frac{X}{X + Y + Z}$$

$$y = \frac{Y}{X + Y + Z}$$

which single out univocally every color. Since $x + y + z = 1$, the third coordinate z is automatically known, when x and y are known. Thus, each color can be represented in a two-dimensional chromaticity diagram having a triangle shape (Fig. 12.1) that depends on the adopted CIE convention. Inside the triangle are situated the points corresponding to white lighting sources,

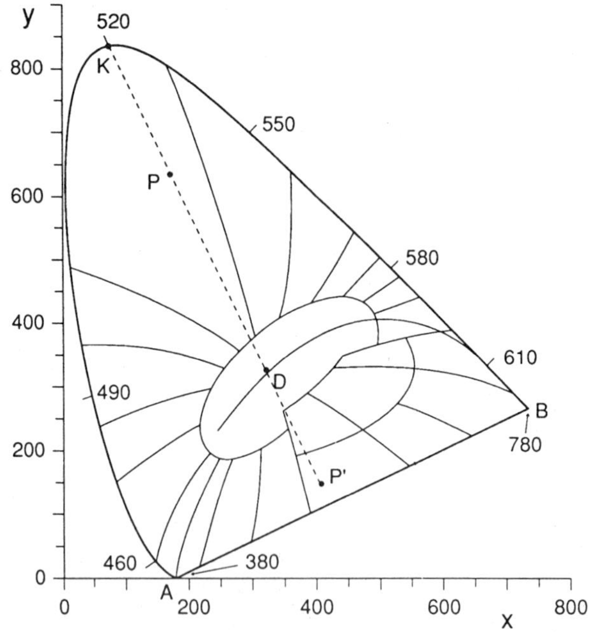

Figure 12.1. Chromaticity diagram. Dominant wavelengths are displayed on the boundary line, D is the representative point of the illuminant, while P and P' stand for two generic colors.

for example D_{65}, while the points situated on the outline correspond to pure colors, that is, without any white contribution. A generic point $P(x_P, y_P)$ in the graph can be characterized by two other quantities, widely used in colorimetry: the dominant wavelength λ_D and the purity p. The dominant wavelength corresponds to the wavelength reflected by a sample of the same color, but the reflectance of which is limited to λ_D only. In practice, the dominant wavelength is obtained by the chromaticity diagram by taking into account the straight line connecting point P and point D, which is representative of the illuminant, and by determining the crossing point K with the outline. Purity p can be evaluated in an analogous manner. It is defined as the ratio between the two segments PD and KD:

$$p = \frac{\overline{PD}}{\overline{KD}} \%$$

Accordingly, purity can vary between 0 and 100%, ranging from a pure white to a pure color with dominant wavelength λ_D. The ABD triangle, having as its vertices the extreme points of $\lambda_D = 380$ and 780 nm and the point representing the illuminant, is named the region of the purple colors, for which no dominant wavelength exists. In fact, using the procedure described above, segment AB, with which real wavelengths do not correspond, is crossed. Accordingly, if a color is represented by a point P' inside this ABD triangle, it is characterized conventionally by a complementary wavelength $\lambda_c = -\lambda_D$, which can be obtained by extending the segment up to cross the chromaticity curve on the side of the real wavelengths.

When works of art are considered, it is also very important to quantify color changes. In fact, environmental factors can heavily affect color by means of different processes: (i) soot or dust deposit upon the painted surface; (ii) chemical modification of the pigment or dye induced by pollutants, humidity, and the like; (iii) fading of organic dyes induced by light or redox reactions with ozone, sulfur, or nitrogen oxides; (iv) photo-oxidation of oils and varnishes and consequent yellowing of the painting. A long debate has developed over the years regarding the most suitable system for defining color changes. At present, the CIE definition of 1976, CIELAB 1976, is the most widely accepted one (Judd and Wyszecki 1975; Wyszecki and Stiles 1982; Hunt 1991). This definition starts from the tristimulus values defined in Eqs. (12.2) to (12.4) and defines three other quantities, L^*, a^*, and b^*:

$$L^* = 25\left(100\frac{Y}{Y_0}\right)^{1/3} - 16 \qquad (12.5)$$

$$a^* = 500\left[\left(\frac{X}{X_0}\right)^{1/3} - \left(\frac{Y}{Y_0}\right)^{1/3}\right] \quad (12.6)$$

$$b^* = 200\left[\left(\frac{Y}{Y_0}\right)^{1/3} - \left(\frac{Z}{Z_0}\right)^{1/3}\right] \quad (12.7)$$

where X_0, Y_0, and Z_0 define a nominally white color (for the illuminant $D_{65}: X_0 = 95.017;\ Y_0 = 100;\ Z_0 = 108.842$). From Eqs. (12.5) to (12.7) it is clear that L^* is closely related to the lightness of the object, and a^* and b^* correspond to two antagonistic chromatic processes (red-green and yellow-blue, respectively). Accordingly, the color of sample 1 can be represented in the space $L^*a^*b^*$ by a vector $\mathbf{E}_1$ with components $L_1{}^*$, $a_1{}^*$, $b_1{}^*$ and, analogously, a second sample 2 is represented by a vector $\mathbf{E}_2$. Therefore, color difference is provided by the modulus of the difference vector:

$$\Delta E = \mathbf{E}_2 - \mathbf{E}_1 = [(\Delta a^*)^2 + (\Delta b^*)^2 + (\Delta L^*)^2]^{1/2}$$

As a rule, the human eye is able to detect color changes such as $\Delta E \geq 1$.

12.3. INSTRUMENTATION

Owing to the uniqueness of works of art, analytical techniques, able to provide reliable data from the smallest possible sampling, have been developed. However, at times even these techniques may be unsatisfactory if no alteration at all of the object is desired. Therefore, completely nondestructive and noninvasive techniques are needed. Indeed, in recent years several spectroscopic methods have been set up that match these requirements. So, here, only these types of instrumentation will be treated, while the reader is referred elsewhere (Schrader 1995) in regard to the traditional ones.

12.3.1. Image Spectroscopy

The term *image spectroscopy* (IS) usually is taken to mean the recording of a two-dimensional image at different wavelengths or, better, at different ranges of wavelengths. From a historical point of view, the first attempt at applying image spectroscopy to paintings can be considered the series of studies that Rawlins (1936) performed in the 1930s at the National Gallery in London. Indeed, his interest was in detecting color changes of paintings, and for this purpose he used a visual colorimeter (Lovibond Tintometer) fitted with a series of accessories, which made it suitable for the investigation of hung pictures. The working principle was very simple: The values of the three primary colors were measured by the instrument when the color of a portion of painting matched with the color of a magnesium carbonate standard, lighted by a lamp whose light was filtered by a set of colored glasses.

Still remaining in the field of a trichromatic response, a rough broad-band image spectroscopy can be performed by photographing the work of art after inserting RGB (Red, Green, Blue) filters in front of the lens of the camera. A further improvement has been attained by the use of IR false color film (Matteini et al. 1980; Aldovrandi et al.1988), which is sensitive up to 900 nm. In this case, the emulsion of the film consists of three layers that are sensitive to the green, red, and IR portions of the electromagnetic spectrum, respectively; a suitable filter gets rid of blue light, which affects all three layers. After processing, the above three layers yield blue, green, and red colors, respectively, thus giving conventional trichromatic false color images of the painting. Owing to the different NIR absorption/reflection properties of some pigments, which otherwise are quite similar in the visible region, it is possible to distinguish these in a painting by means of the different colors under which they appear. A typical example is given by the pigments azurite and ultramarine, which have been widely used as blue pigments ever since ancient times. Although pure pigments can easily be distinguished with the naked eye and, even better, by using a microscope, their differentiation is more difficult in an actual painting when aging processes have washed out the original visual differences. However, ultramarine, differently from azurite, strongly reflects NIR radiation (see Fig. 12.2) so that it appears cherry red, while azurite appears bluish (Matteini et al. 1980; Hoeniger 1991). Two other pigments that are easily identifiable by using this technique are cobalt green and verdigris, which are rendered as magenta and blue, respectively (Aldovrandi et al. 1988).

Owing to the intrinsic drawbacks of the photographic method, since the 1980s the latter has been replaced more and more frequently by solid-state TV cameras connected with computers for the storing and retrieving of images. A first attempt was made at the National Gallery in London (Saunders 1988), where a prototype system was set up that worked in the VIS range using three standard color separation filters. Indeed, in the same study, Saunders (1988) suggested the use of a set of narrow-band interference filters and the extension of the investigations to the NIR region, although he limited himself to trichromatic images in the VIS range for practical reasons. Subsequently, this idea was developed and applied to actual cases (Burmester et al. 1992; Casini et al. 1992; Burmester and Bayerer 1993, 1994; Bacci et al. 1996), following two alternative approaches for the selection of the wavelengths. The former (Burmester et al. 1992; Burmester and Bayerer 1993, 1994) illuminates the painting by light from a monochromator with a band-

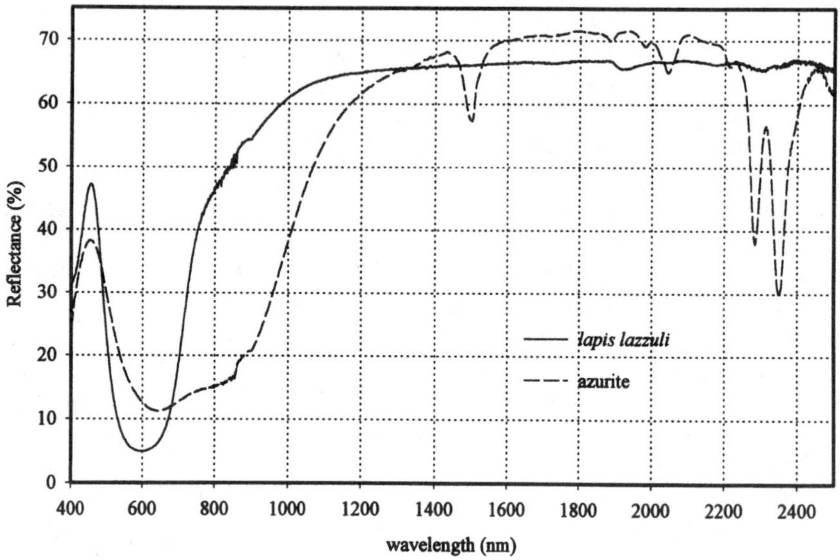

Figure 12.2. VIS and NIR reflectance spectrum of lapis lazuli and azurite (as pure powder).

width of 40 nm for UV–VIS and 80 nm for NIR, while the latter (Casini et al. 1992; Bacci et al. 1996) uses white light and narrow-band interferential filters (at present 10 nm) in front of the video-camera lens. A schematic diagram of the measurement setup for the latter case is displayed in Figure 12.3. Two tungsten–halogen lamps (L) illuminate the sample in a uniform way with incidence angles near 45° to avoid specular reflections. The shutters (S) are driven by the PC so that the painting is illuminated only during the measurement as to reduce possible damage from light. The radiometric calibration of the images can be obtained by placing calibration standards in the scene and recording their diffuse reflectance. In spite of its conceptual plainness, the method makes possible the recording and storage, for later comparisons, of reflectance spectra of each pixel, in good agreement with the spectra obtained using fiber optics reflectance spectroscopy (FORS) (see later). In Figure 12.4, the spectra of the same area of a painting obtained by IS and FORS are compared. A good correspondence is quite evident, at least when the bandwidth is sufficiently narrow. However, both approaches have some drawbacks that have to be taken into account, particularly in view of possible improvements. In particular (i) much attention has to be paid so that the illumination is homogeneous; (ii) when the monochromator is used, higher order overtone radiation has to be filtered; (iii) when interferential filters are used, an image refocusing is needed for each wavelength, mainly in

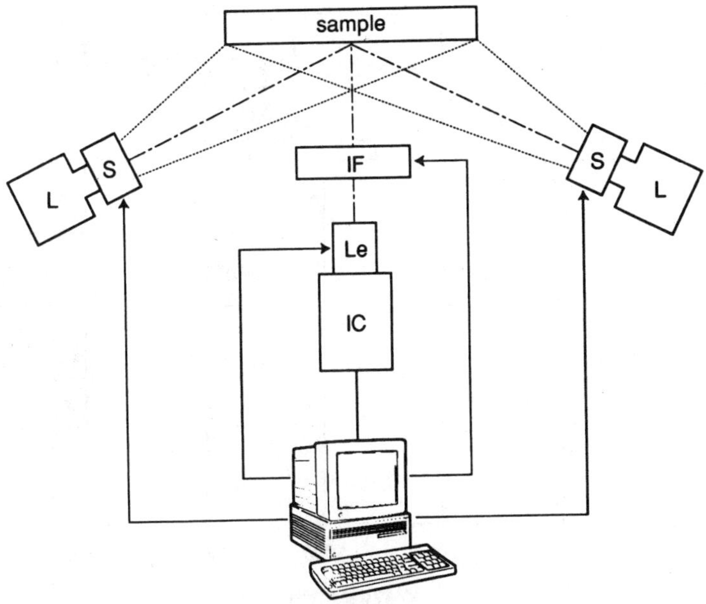

Figure 12.3. Sketch of the experimental setup for image spectroscopy. L: lamp; S: shutter; IF: narrow-band interferential filter; Le: lens; IC: IR camera.

the NIR; however, in doing so, magnification is altered so that the recorded images can be compared only after software correction, which make the images geometrically coincident (Baronti et al. 1998); and (iv) there is the problem of repositioning the object. This problem was tackled within the European Commission VASARI project (see, e.g., Burmester et al. 1992, and references quoted therein) by means of a very sophisticated instrumentation, which allowed the three-dimensional movement of the recording system using stepper motors with a step width as small as 2.5 μm. This device also makes possible studies on large (1.50 × 1.50 m) paintings, as the only limitation is the acquisition time and the computer memory. However, if small paintings or a portion of a painting is under investigation, a simpler, though rough, approach can be used (Bacci et al. 1996). In such a case, a card frame printed in a black and white chessboard pattern is inserted into the scene and related to some reference points of the painting so as to yield an easy realignment of the images.

12.3.2. Fiber Optics Reflectance Spectroscopy (FORS)

Fiber optics permit the propagation of light over long distances, even along a winding path. Their operation is based on the principle of total internal

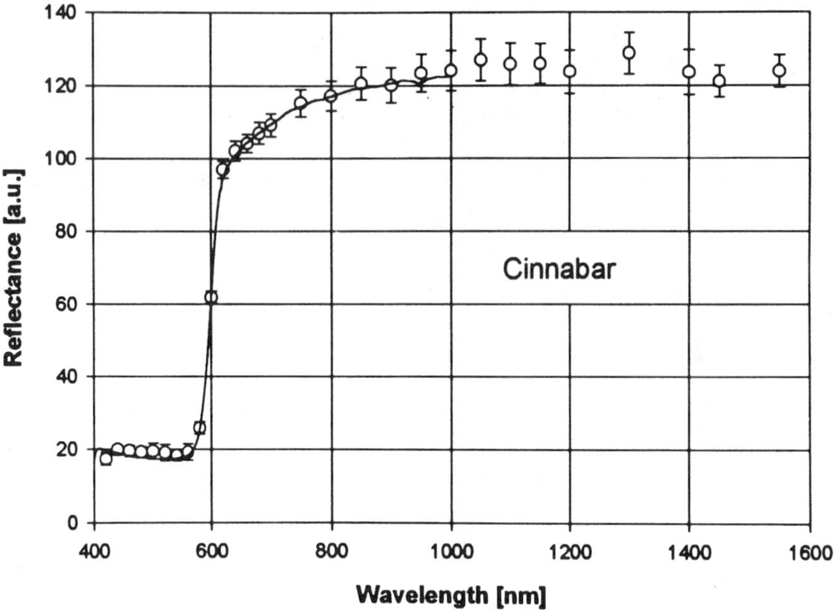

Figure 12.4. Comparison between reflectance spectra recorded by means of a fiber optic spectrophotometer (solid curve) and image spectroscopy (circles). [Reprinted, by permission, from S. Baronti et al., *Applied Optics* **37**, 1299 (1998), copyright © 1998 by Optical Society of America.]

reflection, according to which light incident on the interface between two dielectric media of refractive indices n_1 and n_2 ($n_1 > n_2$), respectively, does not enter medium 2 and is totally reflected into medium 1, if the incident angle exceeds the so-called critical angle (Born and Wolf 1965). Accordingly, fiber optics consist of two extended coaxial cylinders, the inner one of which is called *core* and has high refractive index, while the exterior one is called *cladding* and has a lower refractive index. The whole system is usually protected by a jacket made of nonoptical materials, such as metal or plastic. When light is launched at a sufficiently high angle into the core, it is guided by the fiber and can travel for long distances, depending upon the attenuation of the material constituting the core. The latter is selected on the basis of applications and cost. In fact, glass and plastic fibers are the cheapest ones, but their use is practically limited to only the VIS region; fluorinated glass fibers can cover the NIR range, while the whole range UV–VIS–NIR (200 to 2400 nm) is covered by fused silica fibers. As regards the mid-IR range, very expensive calchogenide fibers, suitable from about 1.8 up to 11.0 μm, are now on the market. Fiber optics are currently well known for their use in telecommunications; however, owing to their flexibility, inertness, and capa-

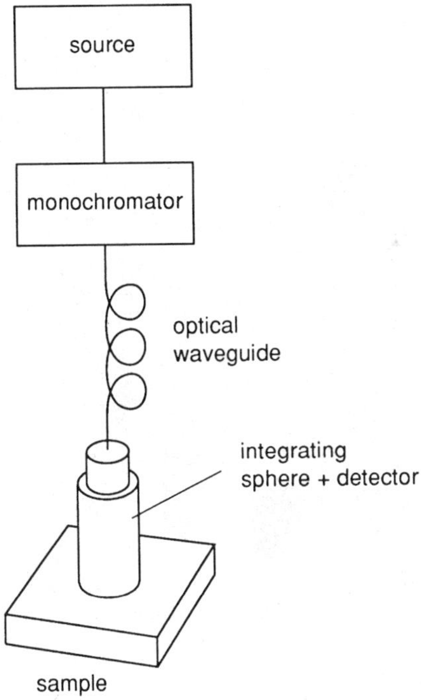

Figure 12.5. Sketch of experimental setup for FORS using a spectrophotometer equipped with an external integrating sphere.

bility of conveying light in hostile environments or upon large objects, they are very suitable for remote spectroscopy and, in particular, for noninvasive and nondestructive analyses of works of art.

FORS can be accomplished by two different types of instrumentation: spectrophotometers and spectroanalyzers, whose configuration is sketched in Figures 12.5 and 12.6, respectively. From a historical point of view, FORS by means of spectrophotometers originates from the Wright–Wassall spectrophotometer, which was designed and constructed during the 1970s on commission for the National Gallery and the Courtauld Institute in London (Bullock 1978; Saunders 1986). In this apparatus, light from a tungsten lamp enters a double monochromator and then, through an optical fiber, is conveyed on the painting. Reflected light is collected at 45° and detected by a photomultiplier. Moreover, a beam splitter is inserted just after the optical fiber, with a twofold purpose: (i) to send light directly to the photomultiplier, to obtain an internal standard, with which to compare the amount of light reflected by the painting and (ii) to reflect an image of the area under investigation, thus allowing a photographic record for later comparison.

Successive commercial dual-beam spectrophotometers were equipped

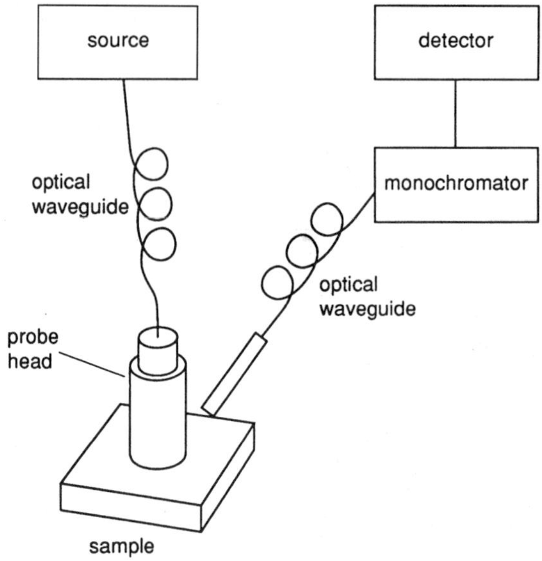

Figure 12.6. Sketch of experimental setup for FORS using a spectroanalyzer.

with an external integrating sphere connected to the spectrophotometer via a fiber optics bundle. In such an instrumentation, the monochromatic light reflected by the surface under examination is detected and compared with that of the reference beam reflected by the white coating of the integrating sphere (usually $BaSO_4$ or Spectralon). A photomultiplier, which is placed just upon the sphere, reveals the light directly so as to avoid further attenuation due to the backward path; the electric signal is then sent to the instrument for processing. This method is very accurate, in particular when color measurements are required; however, the instruments so far available are suitable when they are located in laboratories inside the museums where paintings can be transferred under safety conditions. Actually in some very interesting cases, the instrumentation was transported outdoors also on scaffolding (Bacci et al. 1991, 1993a), but this cannot be the rule owing to the heaviness and poor handiness of the apparatus.

Therefore, if a light and easily transportable instrument is required, spectroanalyzers have to be used. As shown in Figure 12.6, in the latter case white light from a source (usually a tungsten–halogen lamp) is sent to the sample by a fiber optics bundle, and then the backscattered light is collected by other fiber optics, selected by a monochromator and revealed by a suitable detector [photomultiplier, photocell, charge-coupled device (CCD) array]. Since only one light beam is used, a calibration is needed by measuring the backscattered light from a $BaSO_4$ or Spectralon tablet and assuming it as a 100% reflectance for all the wavelengths.

Although spectroanalyzers are easier to handle and are often preferable for practical reasons, several factors have to be properly considered in order to avoid possible sources of error. For example, $BaSO_4$ is a good white reference for most part of NIR, but it can absorb water from the air so that particular care must be given when comparing measurements done with different calibrations. Moreover, coaxial optical fibers used for sending and receiving light are more practical; but they can also collect part of the white light totally reflected by the surface of the sample, so that measurement of the color is affected. In fact, the purity is underestimated because the color is less saturated (less pure) than the real one, and, consequently, the chromaticity coordinates are affected by error, which is quite negligible for mat, but not for glossy, surfaces. Instead, the dominant wavelength and the qualitative features of the spectrum generally remain unchanged. To overcome this problem, a different geometrical arrangement of the fiber optics can be adopted: Specularly reflected light is avoided if the sending light fibers are situated at 45° with respect to the collecting ones.

As for IS, the problem of repositioning is an important one also for FORS. In the Wright–Wassall spectrophotometer, the problem was resolved by taking a photograph of the area under examination during the first experiment and then by placing the negative in the focal plane of the camera and adjusting the position of the measuring head so as to match the negative and the new image of the painting exactly (Bullock, 1978; Saunders 1986). Another approach (Bacci 1995) particularly suitable for easy-to-handle spectroanalyzers, utilizes a transparent film of Mylar, which is fixed on the back of the painting by means of a self-adhesive tape; some reference points and areas to be investigated are then marked on the film by a pencil. Subsequently, the film is removed and holes are made slightly larger than the light spot of the fiber optics probe (usually a few millimeters). The film is then repositioned by matching the reference points, and the reflectance spectrum of each selected area is recorded by fitting the head probe to the holes. The Mylar film can be stored in a controlled environment and subsequently used for further measurements and controls of the painting. When large paintings or frescoes have to be examined, it is easier to find sufficiently wide, homogeneously painted areas, so that a reliable repositioning can be obtained by simply referring with a coordinate system to photogrammetric marks placed on the work (Bacci et al. 1991).

It is quite obvious that repositioning is one of the main causes of error in the measurement, and this fact has to be taken mainly into account when controls are made at different times to monitor possible alterations in the object. Therefore, it has to be ascertained whether the measured change is really significant or is within the measurement error. Accordingly, it is advisable to record a set of five spectra for the measurement of each spot, to calculate the average spectrum and the corresponding standard error for each set, and then to compare the average spectra. The difference can be considered significant if it is larger than the sum of the standard errors of the two sets of measurements.

If reflectance spectra are used for qualitative analysis, looking only at the materials constituting the work of art, the problem of repositioning is less compelling. However, in such a case, particular care must be paid, particularly in the mid-IR, because the shape and intensity of the spectrum can be seriously distorted by specular reflectance and by the balance between absorbed and scattered radiation. In other words, they also depend on the matrix where the material is "diluted" and, as a consequence, on the penetration depth of the radiation (Hawranek et al. 1976; Fraser and Griffiths 1990). For example, the NIR and mid-IR reflectance spectrum of pure calcite is shown in Figure 12.7. It is quite evident that the antisymmetric C–O stretching usually seen as a strong band at about 1430 cm^{-1} in the transmission mode has nearly disappeared, while the five overtone and combination bands between 1.80 and 2.55 μm (Hunt and Salisbury 1971) can easily be detected.

12.3.3. Mid-IR Analysis on Bulky Samples

As was shown in the previous section, the use of fiber optics is particularly suitable when a UV–VIS–NIR analysis of objects that are too big to be placed in the sample compartment is required. Instead, as regards the mid-IR region, the introduction on the market of mid-IR optical fibers is too recent to have reliable data at present on the actual applicability of the method. Therefore, several other methods have been proposed for recording IR spectra without any damage to the artwork. A necessary condition is to have available a Fourier transform (FT) IR spectrophotometer because of its higher signal-to-noise ratio and larger energy throughput with respect to the traditional double-beam dispersive spectrophotometers. Since a full description of the technique would require a great deal of space, while we are mainly interested in the applications of FT–IR spectroscopy to works of art and the related results, we refer the interested reader to excellent introductory publications, from both a general (Bell 1972; Perkins 1986, 1987a, 1987b) and specialist (Low and Baer 1977) point of view.

The simplest idea is to draw the radiation beam out of the spectrophotometer by means of a mirror set. For example, quite recently (Poslusny and Daniel 1995) this method was successfully used to check the authenticity of an old newspaper without damaging it: three gold flat mirrors drew the

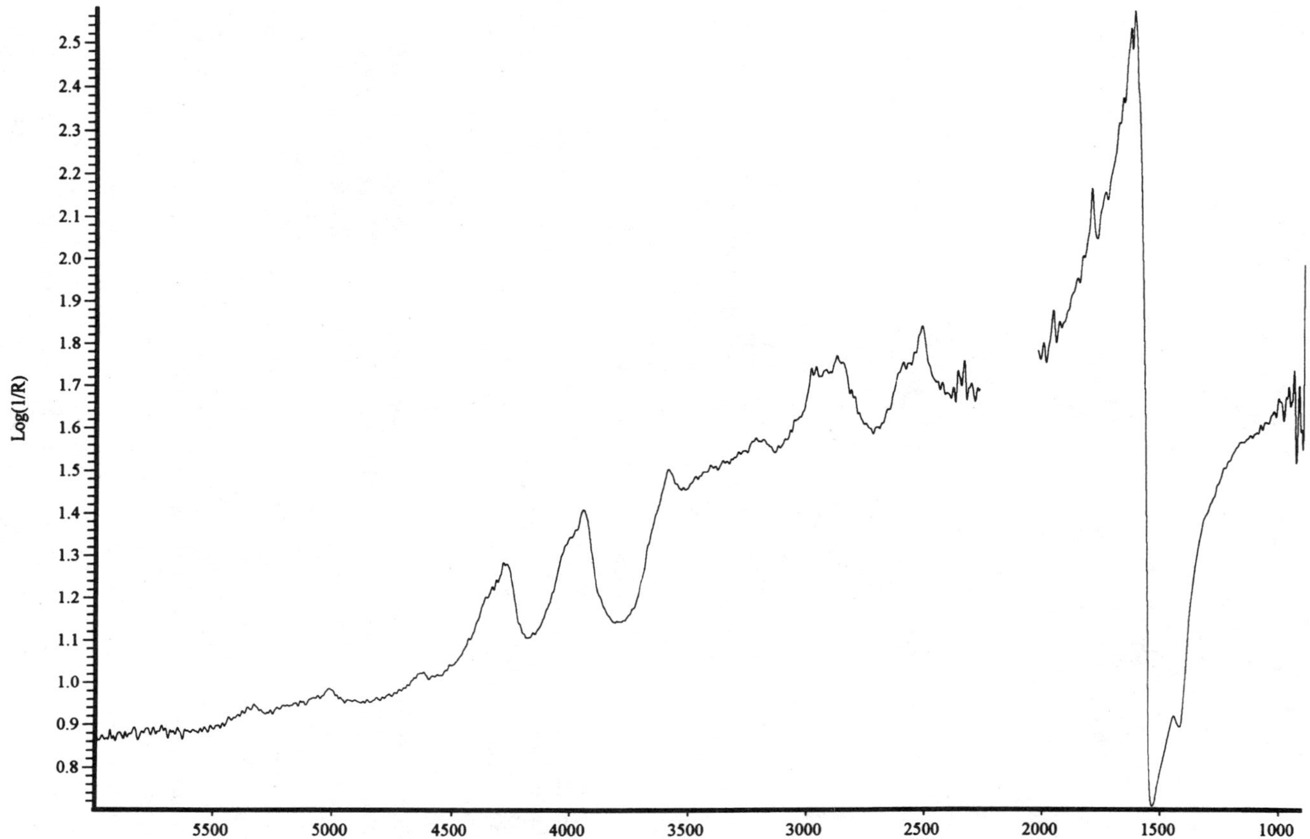

Figure 12.7. NIR and mid-IR reflectance spectrum of calcite.

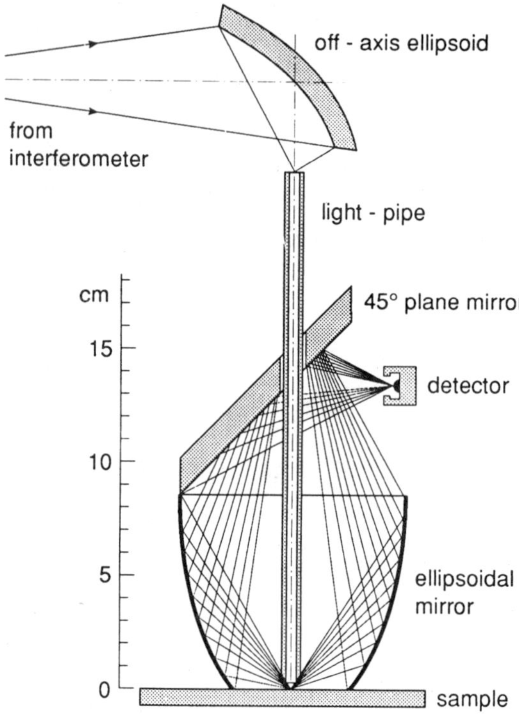

Figure 12.8. Optical layout of the infrared diffuse reflectance accessory for local analysis on bulky samples. [Reprinted, by permission, from E. H. Korte and A. Otto, *Applied Spectroscopy* **42**, 38 (1988), copyright © 1988 by Society for Applied Spectroscopy.]

radiation beam from the sample compartment of the spectrophotometer and directed it onto the detector, to record the background spectrum; the second mirror was then replaced by the sample. Of course, such an arrangement cannot avoid the presence of water and carbon dioxide peaks and also specular reflectance. This latter aspect does not present too many problems in a case like the previous one, but it is particularly important when smooth surfaces have to be considered. Therefore, an interesting accessory was proposed to reject the specularly reflected radiation so as to record diffuse reflectance spectra of varnished objects like violins (Korte and Otto 1988). The optical layout of the accessory is displayed in Figure 12.8. The radiation beam from the spectrophotometer is focused inside the bore of the light pipe by means of an off-axis ellipsoidal mirror, and impinges the sample, which is located in one focus of the gold-plated ellipsoidal mirror, at an angle $\alpha = 0°$. The radiation diffused by the sample is deflected and focused onto the detector by an additional flat mirror at 45°. The geometry of the accessory is such that the optical paths for diffusely and specularly reflected radiation are different, and only the former is properly focused on the detector. The measurements can be performed under nitrogen flux, so as to avoid interferences from carbon dioxide and water.

Lastly, FT–IR microspectroscopy, which by now is one of the most widespread analytical techniques, has to be mentioned also with regard to nondestructive analyses. Indeed, at the beginning, the rapid spread of this technique among investigators of works of art was mainly due to the very small quantity of material (few tenths of a micrometer and sometimes even less) necessary in order to obtain a reasonably good IR spectrum from which to infer the composition of the sample (Van't Hul-Ehrnreich 1970, Orna et al. 1989). However, the energy throughput made possible only transmission measurements; it was not sufficient for recording reflectance spectra, which is the necessary condition for examining the entire work of art without sampling. Recent improvements in the instrumentation have made available FT–IR microscope systems that are particularly suitable for investigations of paintings because, in addition to permitting recording of reflectance spectra, these make it possible to examine upright objects and to map the selected area.

12.4. APPLICATIONS TO ART AND ARCHAEOLOGY

Electronic and vibrational spectra of the different materials that can be encountered during investigations of works of art will be reported here together with a possible interpretation. As was said in Section 12.1, many of these materials have an interest that goes beyond the restricted field of studies on cultural heritage. However, we have attempted to focus mainly on the literature concerning straightforward applications to works of art.

12.4.1. Electronic Spectroscopy

Electronic spectroscopy spans the whole range of the UV, VIS, and NIR regions of the electromagnetic spectrum. In fact, while organic compounds absorb mainly in the VIS and UV region, inorganic compounds, particularly those containing transition metals, can present *d-d* absorption bands also at low energies that correspond to the NIR frequencies. In Table 12.1 a short summary of the electronic spectra of some common pigments is reported, while much more information concerning absorbance/reflectance spectra of pigments can be found in the literature (Barnes 1939; Feller 1986a; Roy 1993; West FitzHugh 1997a). An identification of the constituting material can be attempted on the basis of the typical features of the spectra by making

Table 12.1. Summary of Electronic Spectra of Some Common Pigments

Color	Pigment	Main Chemical Composition	Reflectance (Absorption)[a,b] Peaks (nm)	Origin of Electronic Transition[c]	Ref.
Violet	Cobalt violet	Cobalt(II) phosphate	(531, 594, 860)	*d-d*	Bacci and Picollo (1996)
Blue	Azurite	$2CuCO_3 \cdot Cu(OH)_2$	454	*d-d*	Bacci et al. (1991)
	Ultramarine (lapis lazuli)	Aluminosilicate doped with S_3^-	460; $\lambda > 800$	CT	Bacci et al. (1991)
	Smalt	Potash glass doped with Co^{2+}	(544, 593, 642) (1314, 1534, 1750)	*d-d*	Bacci and Picollo (1996)
	Indigo	$C_{16}H_{10}N_2O_2$	411, $\lambda > 800$ (599)	$\pi \rightarrow \pi^*$	This work
	Prussian blue	$Fe_4[Fe(CN)_6]_3.xH_2O$	440, $\lambda > 1000$ (568)	CT	This work
Green	Malachite	$CuCO_3 \cdot Cu(OH)_2$	518	*d-d*	Bacci et al. (1991)
	Green earth	Aluminosilicate doped with Fe^{2+} and Fe^{3+}	556, 680, $\lambda > 800$	*d-d*	Bacci et al. (1991)
	Verdigris	$Cu(CH_3COO)_2 \cdot H_2O$	498	*d-d*	Kühn (1993b)
	Viridian	$Cr_2O_3 \cdot 2H_2O$	410, 510 (460, 600)	*d-d*	Newman (1997)
Yellow	Yellow ochre	Aluminosilicate with hydrated ferric oxide	450, 600, 760	*d-d*	Bacci et al. (1991)
Yellow	Orpiment	As_2S_3	$\lambda > 600$	Band-to-band	Barnes (1939)
	Chrome yellow	$PbCrO_4$	(470)	CT	This work
Red	Haematite	Fe_2O_3	650 sh, 745 (430, 550, 640, 900)	*d-d*	Bacci et al. (1991) Sherman and Waite (1985)
	Cinnabar	HgS	$\lambda > 630$	Band-to-band	Bacci et al. (1991)
	Red Lead	Pb_3O_4	$\lambda > 620$	Band-to-band	Bacci et al. (1991)

[a] Absorption peaks in parentheses.

[b] sh = shoulder.

[c] CT = charge-transfer transition.

suitable correlation to reference compounds. However, some points have to be taken well into account when operating with real cases.

The presence of powder or soot, as well as the yellowing of varnishes covering a painting, not only affects the intensity of the reflected light but also shifts the absorbance/reflectance bands (Bacci et al. 1991). Therefore, the matching between the sample and reference spectra must be done after removal of the upper foreign deposit and/or of the aged varnish, by means of a suitable cleaning procedure.

The presence of a pure or diluted pigment, as well as the particle size, can strongly affect the intensity of the reflectance bands. As a rule, the whole reflectance spectrum appears flattened, with differences between maxima and minima reduced in comparison with the corresponding transmittance spectrum (Kortüm 1969).

Lastly, for strongly absorbing materials and/or large particle size, the balance between absorption and scattering coefficients can simulate S-shaped absorption bands even for nonband-to-band transitions: for example, lead(II) chromate (Chrome yellow), whose absorption band is due to a ligand-to-metal charge-transfer transition, shows a typical S-shaped reflectance band with an inflection point around 500 nm (Kühn and Curran 1986).

By taking into account the above-mentioned warnings, sometimes the simple reflectance spectrum in the visible region can be sufficient for an identification of the material, in particular for pure pigments and when the author's palette is known. Usually, pigments characterized by *d-d* transition bands present the most favorable conditions for identification because the small absorption coefficient produces a compression of the spectrum only to a small extent and maintains the typical spectral features that can be observed in the transmission mode. Azurite is easily detected and distinguished from other blue pigments, owing to its reflectance peak around 450 nm and the low reflectance in the red and NIR region, which corresponds to a broad and structured *d-d* absorption band, while indigo, Prussian blue, lapis lazuli (or artificial ultramarine) show a strong reflectance in the same region. Indigo and lapis lazuli are characterized by not too different visible spectra, and only a good practice can allow their differentiation in some favorable cases (Orlando et al. 1995); instead, vibrational spectroscopy in the NIR and mid-IR (see later) can give a definite answer. Among blue pigments, smalt is characterized by a typical electronic spectrum consisting of two *d-d* bands, one in the VIS and the other in the NIR region, which makes possible a sure identification (Fig. 12.9). In fact, each band, centered at about 600 and 1520 nm, respectively, is structured into three subbands, which originate from the Jahn-Teller split transitions $^4A_2 \rightarrow {}^4T_1(P)$ and $^4A_2 \rightarrow {}^4T_1(F)$ in a tetrahedral coordination, respectively (Bacci and Picollo 1996). Moreover, it has been suggested (Bacci and Picollo 1996) that useful information about the

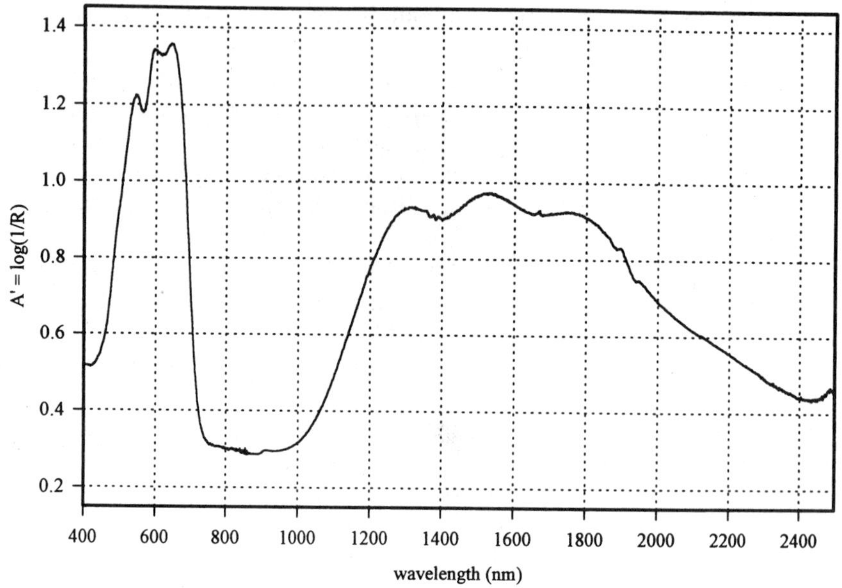

Figure 12.9. VIS and NIR absorption spectrum of smalt (as pure powder).

origin and/or manufacturing of ancient glasses can be obtained by utilizing the same bands as a marker for the sodic or potassic nature of the glass: in fact, the crystal lattice is slightly modified on passing from soda to potash glass, so that in the latter case the cobalt(II) ion is affected by a ligand field strength that is smaller than in the former one. As a consequence, in potash glasses, the center of the two bands is shifted toward the lower frequencies (higher wavelength), and the overall band splitting is reduced. The presence of tetrahedrally coordinated cobalt(II) ion is always accompanied by these typical two structured absorption bands, the position of which depends, of course, on the host crystal lattice. For example, in cobalt green, that is, ZnO containing Co(II), the absorption bands are slightly red-shifted, so that the resulting color is green. However, if the appearance of the above-mentioned structured bands is always symptomatic of the presence of cobalt(II), the absence of this structure does not imply the absence of pigments containing cobalt. For example, cobalt violet, that is, a cobalt(II) phosphate, shows a strong doublet centered at about 560 nm, a less intense absorption band at 860 nm, and no significant absorption in the remaining NIR region. In this case, the cobalt (II) ion is six-coordinated in a pseudo-octahedral environment, so that the most intense *d*-*d* transitions are ${}^4T_{1g}(F) \rightarrow {}^4T_{2g}$ and ${}^4T_{1g}(F) \rightarrow {}^4T_{1g}(P)$, in order of increasing energy (Lever 1984).

Another typical *d-d* spectrum is observed in Cr_2O_3, where the two transitions $^4A_{2g} \rightarrow {}^4T_{2g}$ and $^4A_{2g} \rightarrow {}^4T_{1g}(F)$ due to the Cr(III) ion in an octahedral environment are observed with similar intensity at 600 and 460 nm, respectively. Another characteristic feature of Cr_2O_3 is the high (about 60%) and constant reflectance throughout the entire NIR range, starting from 900 nm. Instead, other common green pigments containing copper(II) ions, such as malachite, $CuCO_3.Cu(OH)_2$, have a broad and structured absorption in the NIR, which is due to the copper(II) *d-d* transitions in an approximately square planar geometry and corresponds to the one observed in azurite, but red-shifted owing to the smaller ligand field.

In spite of their use since ancient times, the electronic spectra of yellow and red ochres have been studied little and an interpretation has been proposed only recently (Sherman and Waite 1985). These pigments are natural earths consisting of clay and silica and owe their color to the presence of goethite (yellow), an iron(III) oxide hydroxide, α-FeOOH, and haematite (red), iron(III) oxide, α-Fe_2O_3. The visible reflectance spectra are dominated by a peak at 740 to 760 nm and a more or less pronounced shoulder at 640 to 660 nm; very often in yellow ochres the shoulder transforms into a real band and is shifted to 600 nm. Moreover, it is also accompanied by a weak reflectance peak at 450 nm. These features are not due to ligand-to-metal charge-transfer transitions, as was first supposed, but to Fe(III)–Fe(III) pair excitations and *d-d* Fe(III) transitions, which are intensified by the magnetic coupling of adjacent Fe(III) ions, while charge-transfer transitions are responsible for the absorption in the UV. Similar spectra are observed also in other earths, such as brown ochre or green earth. In this latter case, a further reflectance peak around 550 nm, responsible for the green color, is observed, which is due to the presence of Fe(II) in the lattice.

Also for lapis lazuli and the corresponding artificial pigment ultramarine, the interpretation of the absorption spectrum was a mystery for a long time. Indeed, the anion S_3^-, which is present in the lattice of a very complex aluminosilicate, is responsible for the strong absorption at about 600 nm (Clark and Cobbold 1978; Clark et al. 1983). The simultaneous presence of S_2^- anions, which absorb at 380 to 400 nm, can produce a green-bluish or green color.

Many pigments, such as cinnabar (or vermilion), minium, cadmium sulfide or selenide, orpiment, Naples yellow, and lead–tin yellow, owe their color to band-to-band transitions, and the corresponding absorbance/reflectance spectrum has a typical S-shape (Fig. 12.4). Moreover, as previously said, pigments or dyes that strongly absorb radiation in the blue–violet region, such as some organic compounds, show similar S-shaped bands, when examined in actual solid samples, thus making identification only on the basis of the visible spectrum very difficult. However, if the artist's palette is known

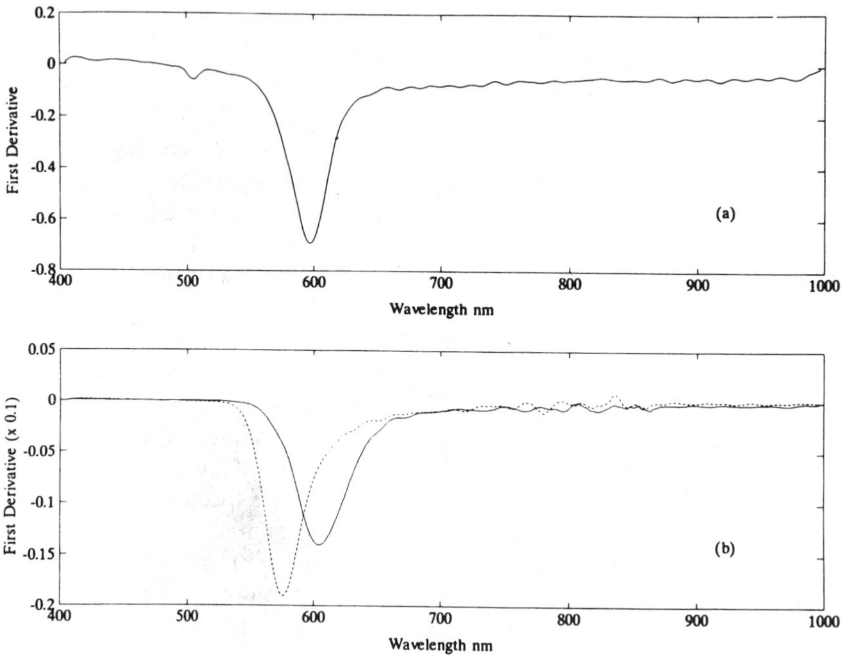

Figure 12.10. VIS and very NIR first-derivative spectra of (*a*) a red spot in the *Madonna d'Ognissanti* by Giotto and (*b*) cinnabar (solid line) and minium (dashed line). [Reprinted, by permission, from M. Bacci, Sensors and Actuators B **29**, 190 (1995), copyright © 1995 by Elsevier Science.]

approximately, an attempt to identify pigments can be made. In this context, derivative spectroscopy can be of help for a correct assignment. An example that shows the application of the first derivative of the absorbance spectrum is reported in Figure 12.10. The absorbance/reflectance spectrum obtained by using FORS on some red areas of the *Madonna d'Ognissanti* by Giotto could be interpreted as being due to minium or cinnabar. The comparison between the first derivative spectrum of the sample (Fig. 12.10*a*) with the first derivative spectra of the two pure pigments (Fig. 12.10*b*) has made possible the correct identification of cinnabar in Giotto's painting (Bacci 1995).

12.4.2. Vibrational Spectroscopy in the NIR and Mid-IR Regions

The region of the mid-IR is the most fruitful for identification purposes because the resulting spectra are typical for each compound, and it can be said that there are not two identical spectra for different compounds. Indeed, for this reason, the mid-IR region is also called a "fingerprint" region. However, when raw materials and/or materials deriving from the aging of an original one are considered, a very complex spectrum with overlapping bands is recorded. Therefore, the interpretation is difficult and often only the class of the main constituting materials can be identified. As regards inorganic compounds, as well as pigments, good references can be found in the literature (Hunt et al. 1950; Afremow and Vandeberg 1966; Nyquist and Kagel 1971; Feller 1986a; Roy 1993; West FitzHugh 1997a), while, to the best of our knowledge, the information about IR spectra of organic material of interest for art are scattered throughout a large quantity of published articles. Here, a survey of the main results reported in the literature for different classes of materials will be given (Table 12.2). Particular attention will be paid also to NIR data whenever possible because of the opportunity offered by fiber optics for noninvasive spectroscopy in this range of the electromagnetic spectrum.

12.4.2.1. Inorganic Pigments

All the pure pigments that are constituted by heavy metal sulfides or oxides do not present any absorption band in the whole NIR and mid-IR range. Only the presence of impurities in the lattice or some kind of mixture with other materials can produce absorption bands that could have been observed in the above range. In fact, for the pure compounds, lattice vibrations can occur only below 400 cm^{-1}, that is, in the so-called far-IR region and can also be used for identification purposes. In cinnabar and synthetic red mercuric sulfide (HgS), a doublet is observed at 347 and 285 cm^{-1} and one more intense band around 130 cm^{-1} (Nyquist and Kagel 1971). The presence of hydrated ferric ions as impurities in natural cinnabar produces small absorption bands at 850 nm and near 1.4, 1.9, and 2.2 to 2.4 μm superimposed on the high reflectance observed in the NIR range. The first band is a *d-d* transition of the iron(III) ion, while the other three bands are overtones and combination bands of water (Hunt et al. 1971b). Cadmium sulfide (CdS), which is often mixed with HgS in modern red pigments, has a strong and broad band around 250 cm^{-1} (Flieder and Bayar 1986), while zinc sulfide (ZnS), which can be also found associated with CdS in yellow and orange pigments, has a similar strong band at 290 cm^{-1} (Flieder and Bayar 1986). Barium sulfate ($BaSO_4$) is another compound that can be found together with sulfides in the so-called lithopones. The real lithopone is a mixture of ZnS and $BaSO_4$; however, the name is extended to the mixtures with other sulfides and even selenides (Afremow and Vandeberg 1966). Accordingly, in recent paintings, the bands of $BaSO_4$ can be found superimposed on those of the sulfides. In this case, strong bands are observable in the mid-IR (Feller

Table 12.2. Main Infrared Spectral Data of Pigments and Other Inorganic Materials Commonly Found in Works of Art

Compound	Main Chemical Composition	NIR Absorption Peaks (μm)	Mid- and Far-IR Absorption Peaks (cm^{-1})	Ref.
Cinnabar	HgS		347, 285, 130	Nyquist and Kagel (1971)
Cadmium sulfide	CdS		250	Nyquist and Kagel (1971)
Zinc sulfide	ZnS		290	Flieder and Bayar (1986)
Orpiment	As_2S_3		305, 183, 139	Forneris (1969); WestFitzhugh (1997b)
Realgar	As_4S_4		373, 367, 343, 225, 170	Forneris (1969); WestFitzhugh (1997b)
Zinc white	ZnO		400–500	Afremow and Vandeberg (1966); Nyquist and Kagel (1971)
Litharge	PbO		375, 295	WestFitzhugh (1986)
Red lead	Pb_3O_4		530, 455, 320, 152, 132	WestFitzhugh (1986)
Naples yellow	$Pb_3(SbO_4)_2$		666, 408	Wainwright et al. (1986)
Chromium oxide	Cr_2O_3		632, 566	Newman (1997)
Viridian	$Cr_2O_3 \cdot 2H_2O$		3500–3000, 1600, 555, 481	Newman (1997)
Haematite	Fe_2O_3		544, 472, 344	Afremow and Vandeberg (1966)
Azurite	$2\ CuCO_3 \cdot Cu(OH)_2$	1.498, 2.044, 2.286, 2.350	3425, 1490, 1415, 1090, 952, 837	Goldsmith and Ross (1968), this work
Malachite	$CuCO_3 \cdot Cu(OH)_2$	1.914, 2.268	3400, 3320, 1500, 1400, 1095, 1045	Goldsmith and Ross (1968), this work
Lead white	$2\ PbCO_3 \cdot Pb(OH)_2$	1.445, 2.006, 2.323	3535–3530, 1400, 1047–1045, 693–683	Bessière-Morandat et al. (1970)
Lead chromate	$PbCrO_4$		905, 860, 830	Kühn and Curran (1986)

Prussian blue	$Fe_4[Fe(CN)_6]_3.xH_2O$	2.370, 2.404	3500–3000, 2083	Berrie (1997), this work
Barium sulfate	$BaSO_4$		1185, 1128–1120, 1082, 639, 614, 200	Feller (1986b)
Gypsum	$CaSO_4{\cdot}2H_2O$	1.446, 1.490, 1.538, 1.748, 1.90	3500–3400, 1700–1600, 1150–1100, 700–600	Hass and Sutherland (1956); Afremow and Vandeberg (1966); Nakamoto (1978); Hunt et al. (1971b); this work
Calcite	$CaCO_3$	1.88, 2.00, 2.16, 2.35, 2.55	1492–1429, 879, 706	Nakamoto (1978); Hunt and Salisbury (1971)
Whewellite	$CaC_2O_4{\cdot}H_2O$	1.518, 1.960, 2.001	1650–1620, 1330–1315, 790–780	Schmelz et al. (1957); Bacci et al. (1993b)
Weddelite	$CaC_2O_4{\cdot}2H_2O$	1.440, 1.939	1650–1620, 1330–1315, 790–780	Schmelz et al. (1957); Bacci et al. (1993b)

1986b), which are due to the site symmetry split ν_3 and ν_4 transitions in the tetrahedral SO_4^{2-} ion (Nakamoto 1978). Additional weak bands occur at 983 cm^{-1} and around 375 cm^{-1}, while a strong doublet appears in the far-IR around 200 cm^{-1}, which can be used to ascertain the presence of barium sulfate together with zinc sulfide in lithopone (Feller 1986b).Yellow orpiment (As_2S_3) and red realgar (As_4S_4) are two other sulfides that were commonly used as pigments in ancient times, but now they are no longer in use owing to their toxicity. They have many absorption bands below 400 cm^{-1}, which can be used for their discrimination and also to distinguish them from other pigments (Forneris 1969; West FitzHugh 1997b).

Zinc oxide or zinc white (ZnO) shows an intense, broad, and structured band in the 400 to 500 cm^{-1} range (Afremow and Vandeberg 1966; Nyquist and Kagel 1971); a similar band, but shifted toward 300 cm^{-1}, is observed in litharge (PbO). The latter pigment may be present also together with lead(II) lead(IV) oxides, or minium or red lead (Pb_3O_4), which shows more bands at 530, 455, and 320 cm^{-1} and a further doublet at 152 and 132 cm^{-1} (West FitzHugh 1986). Lead antimoniate yellow or Naples yellow, likely $Pb_3(SbO_4)_2$, shows a characteristic spectrum with two intense bands around 666 and 408 cm^{-1}. These permit the pigment to be distinguished from other yellow pigments such as the two forms I and II of lead–tin yellow (Wainwright et al. 1986; Kühn 1993a). However, it must be noted that identification of the aforesaid pigments on the basis of IR spectroscopy alone could be a difficult task, owing to the uncertainty about the composition of the real pigments. Because of the minor mass of chromium, in chromium oxide (Cr_2O_3) and hydrated chromium oxide or viridian ($Cr_2O_3.2H_2O$), the lattice vibrations are shifted toward shorter wavelengths (Newman 1997). In the latter, moreover, the characteristic bands of water are, of course, present in the 3000 to 3500 cm^{-1} range (O–H stretching) and around 1600 cm^{-1} (O–H bending). All the pigments containing red iron(III) oxide or haematite (Fe_2O_3), such as red ochres, Indian red and so on, have three strong absorption bands in the range 600 to 300 cm^{-1}. The hydrated yellow forms of iron(III) oxides have a strong doublet at 800 and 900 cm^{-1} (Afremow and Vandeberg 1966) that add to the O–H stretching and to a series of structured bands below 600 cm^{-1}. Moreover, these latter pigments show typical water bands in the NIR near 1.4 and 1.9 μm (Hunt et al. 1971a).

Many important pigments contain the carbonate ion CO_3^{2-}, which is characterized by three main absorptions in the mid-IR around 1400 to 1500 cm^{-1} (asymmetric C–O stretching), 800 to 900 cm^{-1} (out-of-plane bending), and 700 cm^{-1} (in-plane bending) (Nakamoto 1978). Indeed, the blue basic copper(II) carbonate or azurite, $2CuCO_3.Cu(OH)_2$, shows two very strong absorption bands at 1490 and 1415 cm^{-1}, due to asymmetric stretching mode split by the site symmetry, plus a strong band at 837 cm^{-1} and a medium band at 1090 cm^{-1}. The latter can be attributed to the symmetric C–O stretching, allowed by the low symmetry. Moreover, the presence of $O–H^-$ ions in the lattice produces new strong absorption bands at 952 cm^{-1} (O–H bending) and 3425 cm^{-1} (O–H stretching) (Goldsmith and Ross 1968). Similar features are also present in the other natural basic copper(II) carbonate, that is, malachite, $CuCO_3.Cu(OH)_2$. In particular, the two bands corresponding to the asymmetric stretching mode are found at 1500 and 1400 cm^{-1}, while the symmetric stretching has a medium intensity at 1095 cm^{-1}. Moreover, the strongest O–H out-of-plane bending is shifted to 1045 cm^{-1}, and a strong doublet corresponding to the O–H stretching is observed at 3400 and 3320 cm^{-1} (Goldsmith and Ross 1968). The discrimination between malachite and other green copper pigments, such as verdigris [i.e., a mixture of basic copper(II) acetates], and copper resinate [i.e., a mixture of copper(II) salts of resin acids such as abietic or pimaric acid] is based on the presence in the latter of some bands around 1600 cm^{-1} due to the carboxylic group COO^-, whereas it is not easy to distinguish between verdigris and copper resinate on the basis of the IR spectrum only (Kühn 1993b). Another well-known basic carbonate is lead white, $2PbCO_3.Pb(OH)_2$, which shows analogous spectral features as the previous basic carbonates: strong absorptions at 1400 and 693 to 683 cm^{-1} for the asymmetric stretching and the in-plane bending modes, respectively. A set of weak bands in the range 770 to 850 cm^{-1} is probably attributable to the out-of-plane bending mode of the $CO_3{}^{2-}$ ion, while the sharp band at 1045 to 1047 cm^{-1} is more probably attributable to the O–H bending (Bessière-Morandat et al. 1970) than to the carbonate ion, as suggested by some authors (Gettens et al. 1993). Another sharp band is observed in the same pigment at 3530 to 3535 cm^{-1} and is clearly due to the O–H stretching mode. The presence of the $O–H^-$ ions produces typical strong and sharp absorption bands in the NIR range, at 1.498 and 1.445 μm for azurite and lead white, respectively, which are due to the first overtone of the O–H stretching. Instead, in malachite, this band is masked by the broad absorption band due to the low energy *d-d* transitions (see above). The presence of the $CO_3{}^{2-}$ ion is also detectable in the NIR region where, due to combination and overtone transitions, up to five bands are expected (Hunt and Salisbury 1971).

Lead(II) chromate or chrome yellow, $PbCrO_4$, as well as other pigments containing the tetrahedral ion $CrO_4{}^{2-}$ usually show three strong absorption bands in the 800 to 1000 cm^{-1} range, which are due to the symmetric and asymmetric Cr–O stretching modes (Nakamoto 1978); the addition of silica to some commercial products in order to improve light fastness is revealed by an extra band around 1100 cm^{-1} (Kühn and Curran 1986). No characteristic band is observed in the whole 0.8 to 2.5 μm range, at least for pure chromates.

Prussian blue, $Fe_4[Fe(CN)_6]_3.xH_2O$, is one of the first synthetic pigments, and was widely used just after its discovery at the beginning of the eighteenth century. Its presence is revealed by a single, strong band at 2083 cm^{-1} (Berrie 1997), that is, in a range usually free of bands in other blue pigments. Variable intensity bands due to bound water can be observed in the 3500 to 3000 cm^{-1} range. Prussian blue also shows a characteristic NIR spectrum, with a strong band at 2.370 μm and a shoulder at 2.404 μm, which are likely due to the combination of the aforesaid IR active C–N stretching of T_{1u} symmetry with the Raman active C–N stretchings of A_{1g} and E_g symmetry, which are found at 2155 and 2095 cm^{-1}, respectively. Another four bands of medium intensity are present around 2 μm, in which a contribution from bound water cannot be excluded.

Smalt, a potassium glass doped with Co(II), and ultramarine blue or the corresponding natural pigment, lapis lazuli, which are complex aluminosilicates containing sulfur, cannot be easily distinguished by using mid-IR spectroscopy only. In fact, both pigments show a strong band due to the Si–O stretching around 1000 cm^{-1}, with a slight shift toward the higher frequencies in smalt. But this cannot be relied upon for identification. Instead, Raman spectroscopy is able to discriminate the two pigments on the basis of the strong band at 448 to 551 cm^{-1}, due to the ion S_3^-, which is observed in lapis lazuli and ultramarine blue (Guineau 1984; Clark and Cobbold 1978; Clark et al. 1997). However, NIR spectroscopy can also easily discriminate the two pigments because ultramarine blue does not show any absorption, while the spectrum of smalt is dominated by a strong and typical *d-d* absorption band (see above).

12.4.2.2. Other Inorganic Materials

Calcium carbonate, $CaCO_3$, mainly in its most common natural form (i.e., calcite) frequently is found in works of art such as statues, monuments, frescoes, and also in paintings as white pigment or in the preparatory layers. As discussed in Section 12.4.2.1 calcite, like other carbonates, shows three main peaks in the mid-IR (Nakamoto 1978) and five bands in the NIR (Hunt and Salisbury 1971). Owing to atmospheric pollutants such as SO_2, calcium carbonate is often transformed into gypsum or calcium sulfate, $CaSO_4.2H_2O$, so that it is of interest to follow the extent of the attack. The presence of gypsum is revealed by the appearance of two doublets due to the asymmetric stretching ν_3 and the bending ν_4 of the ion SO_4^{2-} in the 1100 to 1150 cm^{-1} and 600 to 700 cm^{-1} ranges, respectively (Hass and Sutherland 1956; Afremow and Vandeberg 1966; Nakamoto 1978). Moreover, structured bands due to water are observed in the O–H stretching and bending regions. However, if a noninvasive monitoring is required to follow possible

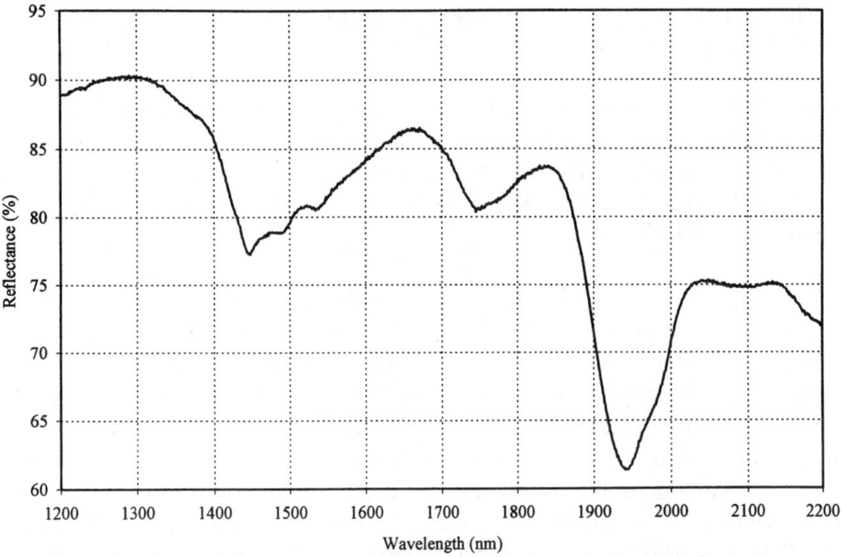

Figure 12.11. NIR reflectance spectrum of gypsum, $CaSO_4{\cdot}2H_2O$ (as pure powder).

alterations, time after time, in exactly the same portions of the object, FORS in the NIR has to be preferred (Bacci 1995; Bacci et al. 1997). In fact, gypsum shows a typical absorption constituted by three subbands between 1.45 and 1.55 μm, due to the first overtone of the O–H stretching of the water, plus other water combination bands at higher wavelengths (Hunt et al. 1971b (Fig. 12.11)), while calcite, as reported above, shows no bands up to 1.88 μm. When FORS in the NIR is used on paintings, it is important to recall that IR radiation can penetrate the paint layer and impinge on the preparatory layer that often is made by gypsum and animal glue, so that the spectrum of gypsum can appear superimposed on the spectrum of the pictorial layer.

Another compound frequently found as an alteration product of calcite is calcium oxalate, regarding the origin of which the debate is still open (Sabbioni and Zappia 1991; Bacci et al. 1993b). Calcium oxalate can occur as whewellite, $CaC_2O_4.H_2O$, or weddellite, $CaC_2O_4.2H_2O$, which show strong absorption bands due to the oxalate ion in the mid-IR (Schmelz et al. 1957). It is not easy discriminating between them because in real cases the presence of other compounds can complicate the spectrum. Instead, the NIR range makes possible an easy and also noninvasive discrimination between the two species. This is because whewellite shows a typical doublet at 1.960 and 2.001 μm due to combination bands of water and, probably, oxalate ion, whereas in the same region weddellite has only a single strong band at 1.939 μm (Bacci et al. 1993b).

The nitrate ion, NO_3^-, is also frequently encountered in monuments, statues, or frescoes as a result of the action of atmospheric nitrogen oxides upon the materials constituting the artifact or the diffusion of the soil salts caused by humidity. Its presence is characterized by a strong band at 1385 cm^{-1}, which corresponds with the asymmetric N–O stretching mode (Nakamoto 1978).

12.4.2.3. Organic Materials

Owing to the large number of bands usually found in organic compounds, even more problems are encountered in the latter case, considering also the possibility of modifications of the original materials, due to aging processes. In general, it can be said that it is not reasonable to perform a blind search on an unknown material because, at most, some information about the presence of a few functional groups can be attained. In fact, the usefulness of an IR investigation is evident when a particular problem has to be resolved, for example, ascertaining the presence or absence of a given material. Here, as follows, only a few particular cases will be treated as examples.

Many IR spectra of organic pigments have been reported in the literature (Flieder 1968; Masschelein-Kleiner and Heylen 1968). Indigo is a widely used blue dye, which is present, as glycoside, in some plants but is now produced by synthesis. It is also present in the so-called Maya blue, in which indigo is absorbed on a clay, and its molecule is closely related to those of other dyes, such as Tyrian purple (Mills and White 1987; see also Chapter 14 this volume). The mid-IR spectrum is dominated by strong bands due to the N–H and C=O stretchings, which are found at 3269 cm^{-1} and as a doublet at 1626 and 1613 cm^{-1}, respectively. The first overtone of the N–H stretching appears as a medium band in the NIR at 1.547 μm, while three most intense bands occur at 2.058, 2.154, and 2.273 μm, where a contribution from C=O and C=C stretchings is likely. A sharp and medium intensity band at 1.674 μm could be tentatively attributed to the first overtone of aromatic C–H stretching. Many natural organic dyes, such as carmine (Schweppe and Roosen-Runge 1986) and madder (Schweppe and Winter 1997), contain molecules related to anthraquinone, for example, kermesic acid, alizarin, and purpurin. The position of the C=O stretching bands is in the 1600 to 1680 cm^{-1} range, and is situated below 1600 cm^{-1} only in highly substituted anthraquinones. In the latter case, it is indistinguishable from the aromatic C–C stretchings (Bloom et al. 1959). The presence of hydroxyl groups on the aromatic ring is revealed by strong absorption between 3200 and 3500 cm^{-1}. The number and the position of the O–H and C=O stretching bands are very sensitive to the number and position of the substitutents on the anthraquinone ring and can be used for discriminating the different molecules (Bloom et al. 1959). For example, purpurin, 1,3,4-trihydroxyanthraquinone, which is an oxidation product of alizarin, 1,2-dihydroxyanthraquinone, both present in madder, shows only two bands around 1600 cm^{-1}, namely at 1621 and 1580 cm^{-1} instead of the three bands that occur in alizarin (Bloom et al. 1959). The two dyes show very similar spectra in the NIR region, as the first overtone of the O–H stretching found at 1.471 μm in alizarin and at 1.492 μm in purpurin is the only remarkable difference. Yellow turmeric contains the curcumin molecule, a β-diketone, 1,7-bis(4-hydroxy-3-methoxyphenyl)-1,6-heptadiene-3,5-dione, which displays features similar to the above dyes around 1600 cm^{-1} (Bellamy et al. 1952), while the occurrence of keto–enol tautomerism produces a broad O–H stretching band around 3400 cm^{-1}, in addition to the sharp and strong band at 3514 cm^{-1} due to the phenolic O–H. In the NIR region, curcumin displays a very rich spectrum in which the sharp band of the first overtone of the phenolic O–H stretching at 1.457 μm stands out and four weak bands in the range 1.0 to 1.3 μm, which arise from a constant and high reflectance plateau. An accurate investigation devoted to the identification of carminic acid and anthraquinone lakes in real cases can be found in the literature (Lang et al. 1992). Here, the presence of bands typical of proteic groups in the red-painted areas only is used as a proof of the presence of an insect-derived dye. Indeed, the presence of protein can be inferred from the occurrence of the N–H stretching mode between 3280 and 3300 cm^{-1} and the so-called amide I and amide II bands at 1630 to 1660 and 1520 to 1550 cm^{-1}, respectively (Campbell and Dwek 1984). Proteins are frequently encountered when paintings are examined because they are the main constituents of the binding media such as egg, milk, and glue. Other common binding media are drying oils, such as linseed oil, which are characterized by the bands typical of the triglyceride ester around 1750 cm^{-1} and of the olefinic group at 3000 to 3100 cm^{-1}, 1600 to 1700 cm^{-1}, and 700 to 1000 cm^{-1}, for the C–H stretching, the C=C stretching and the out-of-plane C–H bending, respectively (Dolphin and Wick 1977). It is interesting to note that aging can appreciably modify the spectral features of the binding media, and this fact has to be taken in due account when the spectra of fresh materials are compared with aged samples from paintings (Meilunas et al. 1990). For example, in fresh egg yolk some spectral features attributed to the triglyceride esters and to the unsaturation are present, which are, on the contrary, lacking in the naturally aged samples. Moreover, in the latter new carbonyl bands appear between 1700 and 1800 cm^{-1}, probably due to oxidized triglycerides (Meilunas et al. 1990). Very often, the identification of the constituting materials in actual cases is also complicated by the presence of resins, which were and are widely used, not only as a protective coating, but also as a paint medium (Gettens and Stout 1966). From a chemical point of view, resins belong to different series of chemicals, but the most common of them (like dammar, mastic, turpentine, and sandarac) are terpenoids (Mills and White 1987). Indeed, IR spectroscopy can be of help in identifying pure terpenoids (Derrick 1989); but the simultaneous presence of other materials and the modifications induced on the component molecules by aging make the investigation using only IR spectra very difficult. However, in the literature attempts can be found at identifying resins together with other materials in several painted works of art (Doménech Carbó et al. 1996, 1997).

12.5. FUTURE TRENDS

It is quite clear that, owing to their characteristics and value, works of art require the development of noninvasive techniques to as large an extent as possible. The use of optical fibers in the VIS and NIR region for several years now, and now their extension to the mid-IR (Williams 1997), appears to be the most promising technique, not only for its nondestructiveness but also because it makes it possible to set up transportable instrumentation for *in situ* investigations. Advances in IR detectors and cameras suggest that image spectroscopy could efficiently stand by FORS so as to obtain general spectral information on the whole work of art. Of course, it has to be kept well in mind that noninvasivity means having to deal with a generally complex system, which cannot be simplified as is done when a sample is selected from the work. Undoubtedly, the possibility of a safe recording of multiple spectra can be of help in making possible the application of statistical methods of data processing; however, on the other hand, this requires the elaboration of suitable comparison models. It can thus be affirmed that, in the near future, the challenge will concern the development and improvement of the noninvasive instrumentation and procedures that conservators and restorers rightfully are asking for from the scientists.

Acknowledgments

The author wishes to thank his co-workers, A. Casini, P. Castagna, F. Lotti, M. Picollo, S. Porcinai, and B. Radicati, for their invaluable help in gathering part of the material quoted in the bibliography and in checking several spectra in our laboratory.

REFERENCES

Abney, W. de W., and E. R. Resting (1881), *Proc. R. Soc. (London)* **172A**, 887.

Abrahams, D. H., and S. M. Edelstein (1964), *Am. Dyestuff Reporter* **55**, 19.

Afremow, L. C., and J. T. Vandeberg (1966), *J. Paint Technol.* **38**, 169.

Aldovrandi, A., D. Bertani, M. Cetica, M. Matteini, A. Moles, P. Poggi, and P. Tiano (1988), *Stud. Cons.* **33**, 154.

Bacci, M. (1995), *Sensor Actuat. B* **29**, 190.

Bacci, M., and M. Picollo (1996), *Stud. Cons.* **41**, 136.

Bacci, M., F. Baldini, R. Carlá, and R. Linari (1991), *Appl. Spectrosc.* **45**, 26.

Bacci, M., F. Baldini, R. Carlá, R. Linari, M. Picollo, and B. Radicati (1993a), *Appl. Spectrosc.* **47**, 399.

Bacci, M., R. Linari, A. Orlando, and B. Radicati (1993b), *Science Tech. Cult. Herit.* **2**, 21.

Bacci, M., A. Casini, F. Lotti, and M. Picollo (1996), Fiber optics reflectance spectroscopy and spectral imaging for conservation, in *Proceedings of the International Symposium on the Conservation and Restoration of Cultural Property—Spectrometric Examination in Conservation*, Tokyo, Oct. 31–Nov. 2, 1994, Tokyo National Research Institute of Cultural Properties, Tokyo, p. 97.

Bacci, M., S. Porcinai, and B. Radicati (1997), *Appl. Spectrosc.* **51**, 700.

Baronti, S., A. Casini, F. Lotti, and S. Porcinai (1998), *Appl. Opt.* **37**, 1299.

Barnes, N. F. (1938/1939), *Technical Studies in the Field of Fine Arts* **7**, 120.

Barnes, N. F. (1939), *J. Opt. Soc. Am.* **29**, 208.

Bell, R. J. (1972), *Introductory Fourier Transform Spectroscopy*, Academic, New York.

Bellamy, L. J., G. S. Spicer, and J. D. H. Strickland (1952), *J. Chem. Soc.* 4653.

Berrie, B. H. (1997), in *Artists' Pigments. A Handbook of Their History and Characteristics*, Vol. 3, Oxford University Press, Oxford, pp. 191–217.

Bessière-Morandat, J., V. Lorenzelli, and J. Lecomte (1970), *J. Phys.* **31**, 309.

Bloom, H., L. H. Briggs, and B. Cleverley (1959), *J. Chem. Soc.* 178.

Born, M., and E. Wolf (1965), *Principle of Optics*, Pergamon, New York.

Bullock, L. (1978), *Natl. Gallery Tech. Bull.* **2**, 49.

Burmester, A., and F. Bayerer (1993), *Stud. Cons.* **38**, 145.

Burmester, A., and F. Bayerer (1994), Remote imaging spectroscopy of drawings, in *Proceedings of the 4th International Conference on Non—Destructive Testing of Works of Art*, Berlin, Oct. 3–8, 1994, Deutsche Gesellschaft für Zerstörungsfreie Prüfung e. V., Berlin, p. 183.

Burmester, A., J. Cupitt, H. Derrien, N. Dessipris, A. Hamber, K. Martinez, M. Müller, and D. Saunders (1992), The examination of paintings by digital image analysis, in *Proceedings of the 3rd International Conference on Non-Destructive Testing, Microanalytical Methods and Environment Evaluation for Study and Conservation of Works of Art*, Viterbo, Oct. 4–8, 1992, ICR-AIPnD-BetaGamma, Rome, p. 201.

Campbell, I. D., and R. A. Dwek (1984), *Biological Spectroscopy*, Benjamin/Cummings, Menlo Park, CA, p. 54.

Casini, A., F. Lotti, S. Baronti, P. Castagna, and L. Stefani (1992), *Sci. Tech. Cult. Herit.* **1**, 33.

Clark, R. J. H., and D. G. Cobbold (1978), *Inorg. Chem.* **17**, 3169.

Clark, R. J. H., T. J. Dines, and M. Kurmoo (1983), *Inorg. Chem.* **22**, 2766.

Clark, R. J. H., L. Curri, G. S. Henshaw, and C. Laganara (1997), *J. Raman Spectrosc.* **28**, 105.

Coblentz, W. W. (1905), Investigations of Infrared Spectra. Part I. Infrared Absorption Spectra. Part II. Infrared Emission Spectra, Publication No. 35, Carnegie Institution, Washington.

de Barros e Vasconcelos, M., A. Cailliez, and A. Toussaint (1967a), *Chim. Peint.* **8/9**, 295.

de Barros e Vasconcelos, M., A. Cailliez, and A. Toussaint (1967b), *Chim. Peint.* **30**, 357.

Derrick, M. (1989), *J. Am. Inst. Cons.* **28**, 43.

Dolphin, D., and A. Wick (1977), *Tabulation of Infrared Spectral Data*, John Wiley & Sons, New York.

Doménech Carbó, M. T., F. Bosch Reigh, J. V. Gimeno Adelantado, and V. Peris Martinez (1996), *Anal. Chim. Acta* **330**, 207.

Doménech Carbó, M. T., V. Peris Martinez, J. V. Gimeno Adelantado, F. Bosch Reigh, and M. C. M. Moya Moreno (1997), *J. Mol. Struct.* **410/411**, 559.

Farnsworth, M. (1938/1939), *Technical Studies in the Field of the Fine Arts* **7**, 88.

Feller, R. L. (1954), *Science* **120**, 1069.

Feller, R. L., Ed. (1986a), *Artists' Pigments. A Handbook of Their History and Characteristics*, Vol. 1, Cambridge University Press, Cambridge.

Feller, R. L. (1986b), in *Artists' Pigments. A Handbook of Their History and Characteristics*, Vol. 1, R. L. Feller, Ed., Cambridge University Press, Cambridge, p. 47.

Flieder, F. (1968), *Stud. Cons.* **13**, 49.

Flieder, I., and M. A. Bayard (1986), in *Artists' Pigments. A Handbook of Their History and Characteristics*, Vol. 1, R. L. Feller, Ed., Cambridge University Press, Cambridge, p. 65.

Forneris, R. (1969), *Am. Mineral.* **54**, 1062.

Fraser, D. J. J., and P. R. Griffiths (1990), *Appl. Spectrosc.* **44**, 193.

Gettens, R. J., and G. L. Stout (1966), *Painting Materials. A Short Encyclopedia*, Dover, New York.

Gettens, R. J., H. Kühn, and W. T. Chase (1993), in *Artists' Pigments. A Handbook of Their History and Characteristics*, Vol. 2, Oxford University Press, Oxford, pp. 67–81.

Goldsmith, J. A., and S. D. Ross (1968), *Spectrochim. Acta* **24A**, 2131.

Guineau, B. (1984), *J. Forens. Sci.* **29**, 472.

Hass, M., and G. B. B. M. Sutherland (1956), *Proc. Roy. Soc. A*, **236**, 427.

Hawranek, J. P., P. Neelakantan, R. P. Young, and R. N. Jones (1976), *Spectrochim. Acta* **32A**, 85.

Herschel, W. (1800), *Philos. Trans. R. Soc. (London)* **90**, 255.

Hoeniger, C. (1991), *J. Am. Inst. Cons.* **30**, 115.

Hollas, J. M. (1992), *Modern Spectroscopy*, 2nd ed., John Wiley & Sons, Chicester.

Hunt, R. W. G. (1991), *Measuring Colour*, 2nd ed., Ellis Horwood, Chichester.

Hunt, G. R., and J. W. Salisbury (1971), *Modern Geol.* **2**, 23.

Hunt, J. M., M. P. Wisherd, and L. C. Bonham (1950), *Anal. Chem.* **22**, 1478.

Hunt, G. R., J. W. Salisbury, and C. J. Lenhoff (1971a), *Modern Geol.* **2**, 195.

Hunt, G. R., J. W. Salisbury, and C. J. Lenhoff (1971b), *Modern Geol.* **3**, 1.

Jones, R. N. (1987a), *Eur. Spectrosc. News* **70**, 10.

Jones, R. N. (1987b), *Eur. Spectrosc. News* **72**, 10.

Jones, R. N. (1987c), *Eur. Spectrosc. News* **74**, 20.

Judd, D. B., and G. Wyszecki (1975), *Color in Business, Science and Industry*, John Wiley & Sons, New York.

Kirchoff, G. (1859), *Monatsber. Akad. Wissensch. Berlin*, 662.

Korte, E. H., and A. Otto (1988), *Appl. Spectrosc.* **42**, 38.

Kortüm, G. (1969), *Reflectance Spectroscopy*, Springer-Verlag, Berlin.

Kubelka, P. (1948), *J. Opt. Soc. Am.* **38**, 448.

Kubelka, P. (1954), *J. Opt. Soc. Am.* **44**, 330.

Kühn, H. (1993a), in *Artists' Pigments. A Handbook of Their History and Characteristics*, Vol. 2, Oxford University Press, Oxford, pp. 83–112.

Kühn, H. (1993b), in *Artists' Pigments. A Handbook of Their History and Characteristics*, Vol. 2, Oxford University Press, Oxford, pp. 131–158.

Kühn, H., and M. Curran (1986), in *Artists' Pigments. A Handbook of Their History and Characteristics*, Vol. 1, R. L. Feller, Ed., Cambridge University Press, Cambridge, pp. 187–217.

Lang, P. L., M. V. Orna, L. J. Richwine, T. F. Mathews, and R. S. Nelson (1992), *Microchem. J.* **46**, 234.

Lever, A. B. P. (1984), *Inorganic Electronic Spectroscopy*, Elsevier, Amsterdam.

Low, M. J. D., and N. S. Baer (1977), *Stud. Cons.* **22**, 116.

Masschelein-Kleiner, L., and J. B. Heylen (1968), *Stud. Cons.* **13**, 87.

Matteini, M., A. Moles, and P. Tiano (1980), *Le Scienze* **142**, 40.

McClure, W. F. (1992), in *Making Light Work: Advances in Near Infrared Spectroscopy*, I. Murray and I. A. Cowe, Eds., Ian Michael, Chichester, pp. 1–13.

Meilunas, R. J., J. G. Bentsen, and A. Steinberg (1990), *Stud. Cons.* **35**, 33.

Merwin, H. E. (1917), *Proc. Am. Soc. Testing Materials* **17**, 494.

Mills, J. S., and R. White (1987), *The Organic Chemistry of Museum Objects*, Butterworths, London.

Nakamoto, K. (1978), *Infrared and Raman Spectra of Inorganic and Coordination Compounds*, Wiley-Interscience, New York.

Newman, R. (1997), in *Artists' Pigments. A Handbook of Their History and Characteristics*, Vol. 3, Oxford University Press, Oxford, pp. 273–293.

Nyquist, R. A., and R. O. Kagel (1971), *Infrared Spectra of Inorganic Compounds*, Academic, New York.

Orlando, A., M. Picollo, B. Radicati, S. Baronti, and A. Casini (1995), *Appl. Spectrosc.* **49**, 459.

Orna M. V., P. L. Lang, J. E. Katon, T. F. Mathews, and R. S. Nelson (1989), in *Archaeological Chemistry*, Vol. 4, R. O. Allen, Ed., American Chemical Society, Washington DC, p. 265.

Perkins, W. D. (1986), *J. Chem. Educ.* **63**, A5.

Perkins, W. D. (1987a), *J. Chem. Educ.* **64**, A269.

Perkins, W. D. (1987b), *J. Chem. Educ.* **64**, A296.

Poslusny, M., and M. Daniel (1995), *Appl. Spectrosc.* **49**, 395.

Rawlins, F. I. G. (1936), *Technical Studies in the Field of Fine Arts* **4**, 179.

Rawlins, F. I. G. (1937), *Technical Studies in the Field of Fine Arts* **5**, 150.

Rawlins, F. I. G. (1941), *Technical Studies in the Field of Fine Arts* **9**, 207.

Rawlins, F. I. G. (1942), *Technical Studies in the Field of Fine Arts* **10**, 230.

Ritter J. W. (1801), *Gilbert's Ann. Physik* **7**, 527.

Roy, A., Ed. (1993), *Artists' Pigments. A Handbook of Their History and Characteristics*, Vol. 2, Oxford University Press, Oxford.

Sabbioni, C., and G. Zappia (1991), *Aerobiologia* **7**, 31.

Saunders, D. (1986), *Eur. Spectrosc. News* **67**, 10.

Saunders, D. (1988), *Natl. Gallery Tech. Bull.* **12**, 66.

Schmelz, M. J., T. Miyazawa, S. Mizushima, T. J. Lane, and J. V. Quagliano (1957), *Spectrochim. Acta* **9**, 51.

Schrader, B. (1995), in *Infrared and Raman Spectroscopy*, B. Schrader, Ed., VCH, Weinheim, pp. 63–188.

Schweppe, H., and H. Roosen-Runge (1986), in *Artists' Pigments. A Handbook of Their History and Characteristics*, Vol. 1, R. L. Feller, Ed., Cambridge, University Press, Cambridge, pp. 255–283.

Schweppe, H., and J. Winter (1997), in *Artists' Pigments. A Handbook of Their History and Characteristics*, Vol. 3, Oxford University Press, Oxford, pp. 109–142.

Sherman, D. M., and T. D. Waite (1985), *Am. Mineral.* **70**, 1262.

Van Asperen de Boer, J. R. J. (1970), Infrared Reflectography. A Contribution to the Examination of Earlier European Paintings, Ph.D. thesis, University of Amsterdam.

van't Hul-Ehrnreich, E. H. (1970), *Stud. Cons.* **15**, 175.

Wainwright, I. N. M., J. M. Taylor, and R. D. Harley (1986), in *Artists' Pigments.*

A Handbook of Their History and Characteristics, Vol. 1, R. L. Feller, Ed., Cambridge University Press, Cambridge, pp. 219–254.

West FitzHugh, E. (1986), in *Artists' Pigments. A Handbook of Their History and Characteristics*, Vol. 1, R. L. Feller, Ed., Cambridge University Press, Cambridge, pp. 109–139.

West FitzHugh, E., Ed. (1997a), *Artists' Pigments. A Handbook of Their History and Characteristics*, Vol. 3, Oxford University Press, Oxford.

West FitzHugh, E. (1997b), in *Artists' Pigments. A Handbook of Their History and Characteristics*, Vol. 3, Oxford University Press, Oxford, pp. 47–79.

Williams, R. S. (1997), *Mat. Res. Soc. Symp. Proc.* **462**, 25.

Wyszecki, G., and W. S. Stiles (1982), *Color Science: Concepts and Methods, Quantitative Data and Formulae*, 2nd ed., John Wiley & Sons, New York.

CHAPTER

13

X-RAY PHOTOELECTRON SPECTROSCOPY AND AUGER ELECTRON SPECTROSCOPY IN ART AND ARCHAEOLOGY

GIUSEPPE SPOTO and ENRICO CILIBERTO

Dipartimento di Scienze Chimiche, Università di Catania, Catania, Italy

13.1. INTRODUCTION

In 1981 Kai M. Siegbahn won the Nobel prize for physics for his contribution to the development of high-resolution electron spectroscopy. His pioneering work began in the 1940s with β-emission high-resolution spectroscopy that paved the way for the development in the late 1960s of one of the most powerful analytical tools able to extract chemical information from the surface of materials (compare Section 7.3.3). To identify this new technique, he coined the acronym ESCA (electron spectroscopy for chemical analysis), also known as XPS (X-ray photoelectron spectroscopy), the term that will be used in this chapter. In today's scientific world XPS and the closely connected Auger electron spectroscopy (AES) are important analytical techniques used to investigate the surface properties of materials and a variety of applications of these techniques in studies of materials constituting works of art and archaeological artifacts has been documented.

The aim of this chapter is to provide a brief introduction to applications of XPS and AES techniques in the study of materials constituting works of art and archaeological artifacts. The basic principles of XPS and AES will be outlined briefly in the following section. For an in-depth evaluation of the principles, instruments, and applications of these techniques, the reader is advised to consult specific books (Briggs and Seah 1990; Hüfner 1995) and the Fundamental and Application Reviews regularly published in *Analytical Chemistry* (see Turner and Schreifels 1998). Specialized journals such as the *Journal of Electron Spectroscopy and Related Phenomena* and *Surface and*

Modern Analytical Methods in Art and Archaeology, Edited by Enrico Ciliberto and Giuseppe Spoto. Chemical Analysis Series, Vol. 155
ISBN 0-471-29361-X

Interface Analysis regularly publish research articles concerning both XPS and AES. An introductory overview of the use of XPS in archaeology have been recently provided by Lambert et al. (1999).

13.2. BASIC PRINCIPLES

13.2.1. Background Information

In the last 30 years numerous surface-sensitive and surface-specific techniques (cf. Section 7.3.3) have been developed for the study of properties such as surface structure, surface chemical composition, the oxidation states and chemical environment of elements, and electronic and mechanical features. The most commonly used techniques involve scattering, absorption, or emission of photons, electrons, atoms, and ions.

Figure 13.1 outlines the mechanisms involved in the emission of photoelectrons and Auger electrons. In the first case, a photon of sufficiently high energy interacts with the electron shell of an atom and causes ionization as a result of the photoelectric effect (Einstein 1905). Here, the kinetic energy of the emitted electron, that is, the photoelectron, can be represented as follows:

$$\tfrac{1}{2}mv^2 = h\nu - \mathrm{BE} - \phi$$

where $h\nu$ is the energy of the incident photon; BE is known as the binding energy, that is, the difference in energy between the ground state and the ionic state corresponding to a specific ionization; and ϕ is the so-called work function, an energy shift that depends on both the sample and the instrument in use. BE, which is the most important physical observation in XPS, can be regarded as the ionization energy of the atom for the particular shell involved. Since each type of atom can give a variety of possible ion states, there is a corresponding variety of kinetic energies for the emitted electrons. Moreover, if p, d, or f orbitals are ionized, their energy levels split owing to spin-orbit coupling and $p_{3/2}$ $p_{1/2}$, $d_{5/2}$ $d_{3/2}$, and $f_{7/2}$ $f_{5/2}$ ionization, with corresponding probability ratios of 2:1, 3:2, and 4:3, is observed.

An ion with an electron vacancy at the core level can lose energy as a result of any of three processes: the emission of an X-ray photon (fluorescence), the emission of an electron in an Auger process (discovered by Pierre Auger 1925), and the emission of an electron in a Coster-Kroning process (this latter process will not be treated in this chapter. For details see Briggs and Seah 1990). The first two occurrences are set out in Figure 13.1. The probability of an X-ray photon being produced or an electron being emitted

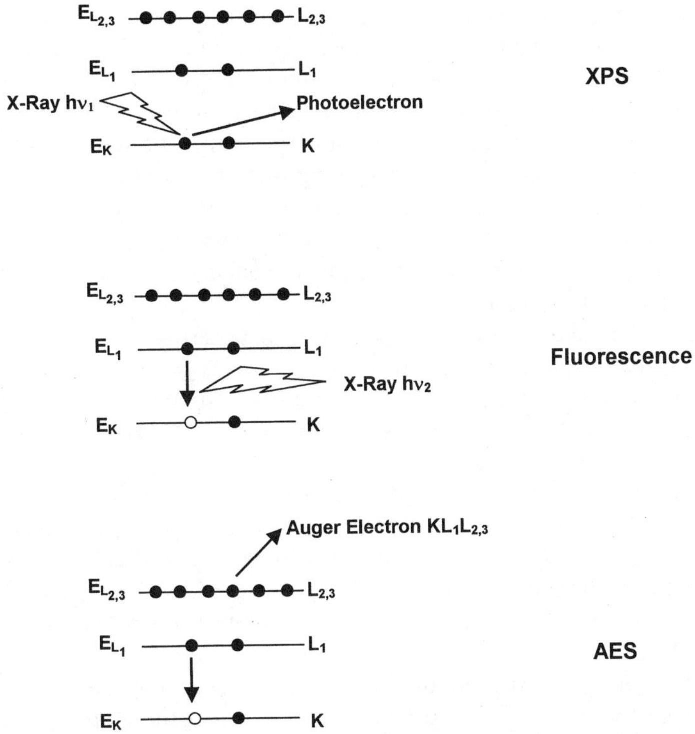

Figure 13.1. Schematic diagrams of the processes of photoelectron emission, X-ray fluorescence emission, and Auger emission. Incident X-ray or electrons cause an electron to be emitted from an electronic core level (photoelectron). The vacancy generated in a core level (K in the figure) can be filled by the relaxing of an electron belonging to successive energetic levels (L_1 in the figure). A similar transition may cause the emission of energy in the form of an X-ray photon of energy $(E_K - E_{L_1})$ or the emission of another electron (Auger electron). In the case shown, the Auger electron is emitted from the $L_{2,3}$ level and possesses an energy of $(E_K - E_{L_1} - E^*{}_{L_{2,3}}) E^*{}_{L_{2,3}}$ is asterisked as it indicates the energy of the $L_{2,3}$ level in the presence of a vacancy in the L_1 level.

is complementary. Both depend on the atomic number of the element in question and on the particular emission mechanism involved.

The kinetic energy of the Auger electrons will, of course, depend on the energy of the three levels involved in the process, that is,

$$E_{KL_1L_{2,3}} = E_K - E_{L_1} - E^*{}_{L_{2,3}}$$

$E^*{}_{L_{2,3}}$ is asterisked since it represents the energy of the $L_{2,3}$ level in the presence of an electron vacancy at level L_1 and is different from $E_{L_{2,3}}$.

It is known that electron energy levels vary according to the particular type of atom concerned and its chemical environment. This being so, an analysis of the kinetic energies of these groups of electrons can provide reliable information not only concerning the element itself, but also as regards the particular chemical state of the atom in question. Since the probability of interaction with atoms or ions constituting matter is very high, the electron path length is of the order of tens of Ångstroms, and only the electrons that originate within a few nanometers below the solid surface may leave the surface without any loss of energy. Applied to solids, XPS and AES are thus surface-sensitive techniques.

Even if photoelectron and Auger electron emission occur in the same context, a distinction must be made as regards the type of spectroscopy these phenomena imply. When spectroscopic information derives essentially from the analysis of the kinetic energies of photoemitted electrons, we speak of photoelectron spectroscopy (Siegbahn et al. 1967). When Auger electrons are chosen for analysis, we are in the realm of Auger electron spectroscopy (Harris 1968a, 1968b). In the former type of spectroscopy, ionization is induced by a suitable, constant, and distinctive energy radiation. In the XPS we may be dealing with X-rays such as Al K_α (1486.6 eV) and Mg K_α (1253.6 eV) or properly monochromatized synchrotron radiation as is the case for XPS.

In AES, ionization is generally induced by bombarding the material with primary electrons of an appropriate energy, and it is therefore necessary to use an electron gun. Both XPS and AES require the use of an electron kinetic energy analyzer. This is usually either a hemispherical electrostatic system, such as the concentric hemispherical analyzer (CHA) or a cylindrical electrostatic system, such as the cylindrical mirror analyzer (CMA). Further information regarding these devices may be found in Riviere (1990).

Figure 13.2 shows a general diagram of a combined XPS/AES apparatus. The equipment works in an ultrahigh vacuum to ensure maximum electron path length within the instrumental chamber and to avoid surface contamination during analysis. For the majority of experiments it is adequate to operate at a pressure of the order of 10^{-10} Torr.

13.2.2. Sample Preparation

Sample preparation and mounting are rarely of critical importance when working with XPS or AES. A standard guide for sample handling has been provided by American Society for Testing and Materials (ASTM) (see ASTM 1991b). The sample may be mechanically attached to the holder and analyzed in an "as-received" condition, and in many cases the samples can

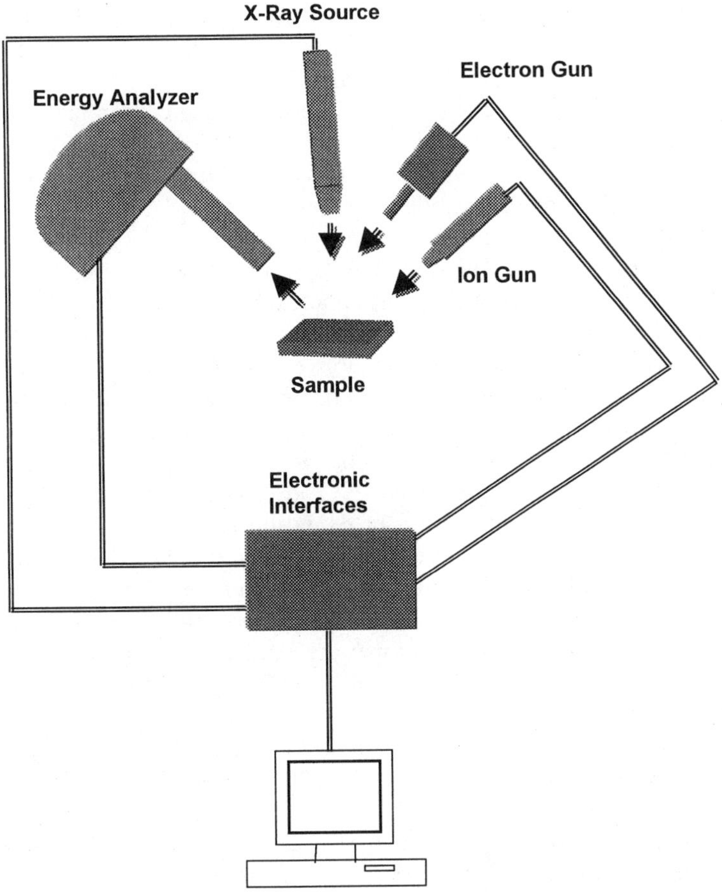

Figure 13.2. Schematic diagram of an integrated XPS and AES spectrometer.

simply be replaced in their original location after analysis. This is of considerable importance when studying samples from works of art or from valuable archaeological objects.

In XPS and AES, spectra featuring bulk composition are frequently obtained by grinding the sample to a powder in an alumina mortar. This powder is then analyzed after mounting on adhesive copper tape or indium foil or after compressing so as to obtain a smooth and homogeneous pellet.

Any volatile species must be removed from the material because degassing processes can cause vacuum conditions to deteriorate, especially when porous materials such as those of stone or ceramics are analyzed. This can be done by setting up a separate long-term vacuum pump in a chamber properly interfaced with the main analysis chamber.

Ion etching can also be used to dispose of surface contamination. Contamination usually involves the appearance of a ubiquitous "hydrocarbon" $C1s$ peak in the spectrum. It can be caused by atmospheric exposure or simply from lack of careful handling.

Ion sputtering is sometimes employed for the removal of contamination layers. It is also frequently used for the depth profiling of the sample, which provides information concerning the concentration of elements orthogonal to the material surface. For this purpose an ionic gun is usually used to accelerate ions (often Ar^+) to an energy above 100 eV (typically 0.5 to 5 keV). The ionic beam bombards the material, and a fraction of the energy is transferred to surface atoms and causes them to leave the sample; they are sputtered away. It should be noted that ion-etching procedures generally have the effect of changing the chemical and/or the structural nature of the surface and that the identification of oxidation states may not necessarily reflect the surface's original condition (Naguib and Kelly 1975; Lam 1988; Mitchell et al. 1990). Sputtering induced by ion bombardment has been extensively studied both theoretically and experimentally, and those readers wishing to learn more about this topic are advised to refer to introductory studies in this field (Hofman 1990; Taglauer 1994) and to standard guides provided by ASTM (1991a).

Another feature particular to XPS and AES applied to solids is the sample charging effect. Insulating and semiconducting samples acquire a surface charge when bathed in high-intensity ionization radiation or high-energy ionizing particles. This effect is an equilibrium phenomenon determined by (i) the rate at which electrons are ejected from the surface and (ii) the rate at which stray electrons in the vacuum and the electrons conducted through the bulk of the sample arrive at the surface and neutralize the positive charges. These conditions normally induce a positive charge equilibrium on the surface of the sample during XPS or AES experiments. Negative charging can result from handling procedures. The surface potential for insulating systems is commonly of the order of a few volts, but it may sometimes exceed +25 V in the case of highly insulating samples. As this phenomenon generally results in an energetic shift of the electron bands in the spectrum, an accurate calibration of the spectrum must be accomplished. A more serious problem arises if the charging of the sample surface is not homogeneous. This can lead to extensive broadening or even a multiplication of individual signals. To inhibit charging effects, a source of thermal electrons produced by a filament positioned near the sample is sometimes used.

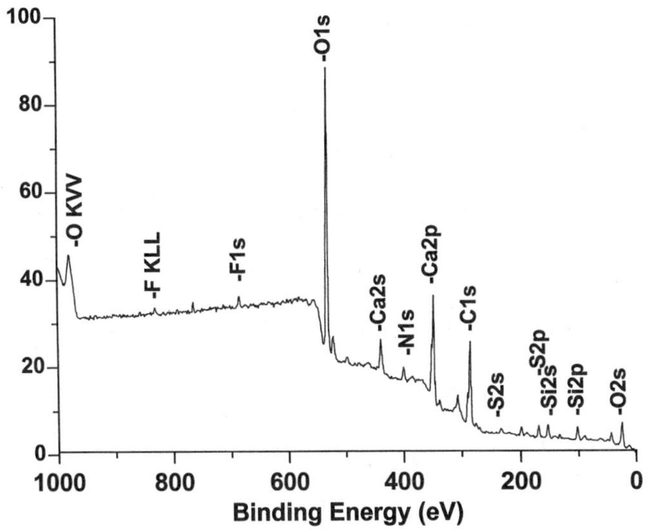

Figure 13.3. XPS wide spectrum obtained from a fragment of a Carrara marble sample taken from the Donatello pulpit in Prato Cathedral (Italy). The sample had been exposed to the atmosphere for more than 500 years and was consolidated during the 1940s by treatment with $ZnSiF_6$ water solution.

13.2.3. X-Ray Photoelectron Spectra

An X-ray photoelectron spectrum is displayed as a plot of the electron binding energy versus the number of electrons detected. A series of peaks can be observed. The corresponding binding energies reflect the electronic state of the material and provide information on its elemental chemical composition. This chemical specificity is the principal strength of XPS as an analytical technique. However, it has been found that nonequivalent atoms of the same elements in a solid give rise to core level peaks with considerably different binding energies, a discovery that has enormously stimulated research in this field. The critical evaluation of an X-ray photoelectron spectrum can thus provide much more chemical-physical data that is not merely concerned with the element itself. As an example, Figure 13.3 shows the spectrum of a Carrara marble sample from the Donatello pulpit in Prato Cathedral, which had been exposed to the atmosphere for more than 500 years and was consolidated during the 1940s by treatment with $ZnSiF_6$ water solution. A number of ionization events and Auger transitions are clearly shown. Looking at the *C*1*s* band envelop (Fig. 13.4), the signals centered at 285 and 289.6 eV indicate more than one chemical environment for the carbon

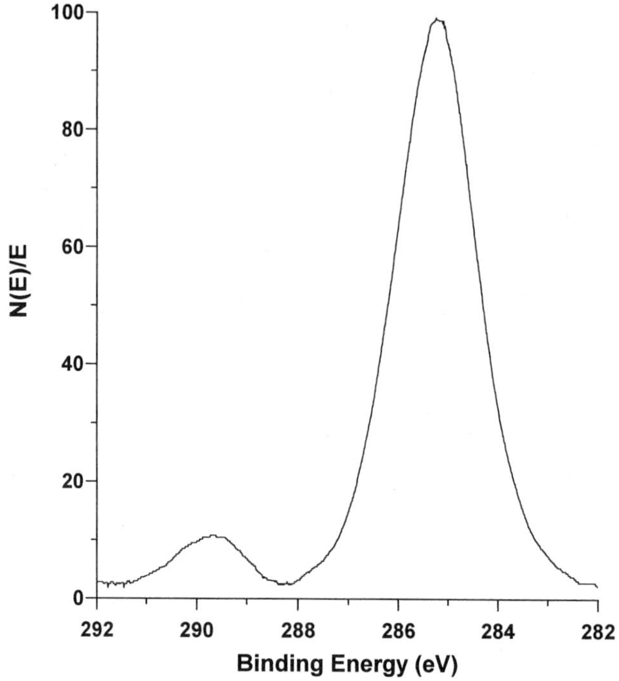

Figure 13.4. The C 1*s* band envelop obtained from the XPS analysis of a Carrara marble sample taken from the Donatello pulpit in Prato Cathedral (Italy). The signals centered at 285 and 289.6 eV indicate the presence of different chemical environments for the carbon atoms. These refer to the carbons coming from surface pollution (285 eV) and carbonate carbon atoms (289.6 eV), which are typical of the calcite phase.

atoms. These refer to the carbons coming from surface pollution (285 eV), such as hydrocarbons or hydrocarbon moieties of organic compounds partially coating the surface, and carbonate carbon atoms (289.6 eV), which are typical of the calcite phase. The large 4.6-eV difference between the two signals, originating from the extremely different formal oxidation states of the carbon atoms, leads us to predict that the *C*1*s* binding energies of most functional groups will fall within this range.

Binding energy shifts can be observed in many kinds of atoms, but, for insulator or semiconductor samples the measured energy values are often dependent on the electrostatic charging of the specimen during the experiment. An accurate evaluation of the chemical environment of an atom may be often accomplished by using an Auger parameter. This concept, introduced by Wagner (1975), consists in making use of the energy shift in both the Auger and photoelectron peaks in the same spectrum by recording the separation of the two lines. The Auger parameter (α) is numerically defined as:

$$\alpha = \mathrm{BE} + E_K - h\nu$$

where BE is the binding energy of the photoelectron peak, E_K is the kinetic energy of the Auger transition, and $h\nu$ is the energy of the X-ray source. The above equation can sometimes produce negative values of α, and a modified Auger parameter α^* was therefore subsequently introduced (Wagner et al. 1979a) as follows:

$$\alpha^* = \alpha + h\nu$$

Both α and α^* are empirical parameters with unique values for each chemical state, and they have been extensively used as fingerprints for characterizing such states.

For further information on the preliminary identification of XPS lines and the relative chemical shifts, readers are advised to consult one of the very useful introductory handbooks available that deal with this topic (Briggs and Seah 1990; Moulder et al. 1992).

13.2.4. Auger Electron Spectra

Even though Auger lines are detectable in XPS, AES more commonly refers to a technique using an electron beam as the primary source of ionizing energy. An electron gun containing an electron source and lenses that focus and scan the beam is therefore necessary in order to generate an electron flux that remains stable over relatively long periods and can be used to bombard the sample surface. An Auger electron spectrum is usually displayed, in a differentiated mode, as a plot of dI/dE versus E, where I is the electron intensity measured and E is the electron kinetic energy. The Auger energies should thus correspond to the points where the first derivatives are zero, but the position of an Auger peak is conventionally taken as that of the minimum in the high-energy negative excursion. The identification of the signals in the Auger spectrum is easily accomplished by using the chart of principal Auger electron energy and by comparison with standard spectra (Childs et al. 1995). The elemental characterization of the sample can thus promptly be performed. On the contrary, the chemical or molecular specificity of AES is not so straightforward as in XPS. This is because the Auger transition involves three electrons, two of which are often in the band structure of the materials. This leads to very broad and poorly defined peaks for most elements and the assignment of the chemical state tends to rely on the shape of the peak rather than on its position.

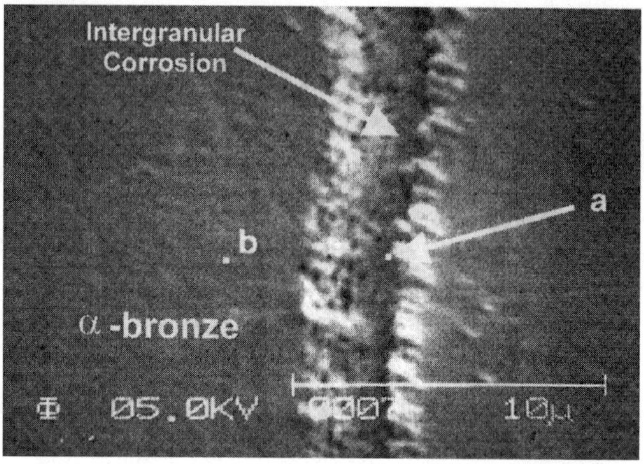

Figure 13.5. Secondary electron image of the cross section of a bronze-filled bracelet or *armilla* dated to the eighth century B.C. (Chiavari, Italy). The image shows advanced oxidation products into the metallic matrix alongside grain joints showing the typical feature of intergranular corrosion.

The use of a finely focused electron beam (diameter around 100 Å in modern instruments) makes analysis at high spatial resolution possible. Figure 13.5 shows an SEM image of a polished cross section of an archaeological corroded bronze. Two Auger electron spectra (Figs. 13.6*a* and 13.6*b*) have been recorded at the point indicated. They clearly show the Auger signals coming from the corroded and uncorroded regions. The different composition of these two regions, discussed in the Section 13.6.2.2, is evident. When an electron spectrometer is used together with a scanning electron microprobe, it is also possible to carry out scanning Auger microscopy (SAM). In this way, scanning electron images (SEM) and chemical mapping procedures can be employed in the context of the same analytical system.

13.2.5. Quantitative Analysis

The XPS and AES techniques also make the quantitative analysis of the studied surfaces possible through the accurate evaluation of the spectroscopic peak areas if the preliminary calibration of the equipment in use is carried out. The quantification of X-ray photoelectron spectra is rather more straightforward and potentially more accurate than that of Auger electron spectra. The intensity of a photoelectron peak is given by:

$$I_x = J n_x \sigma K \lambda$$

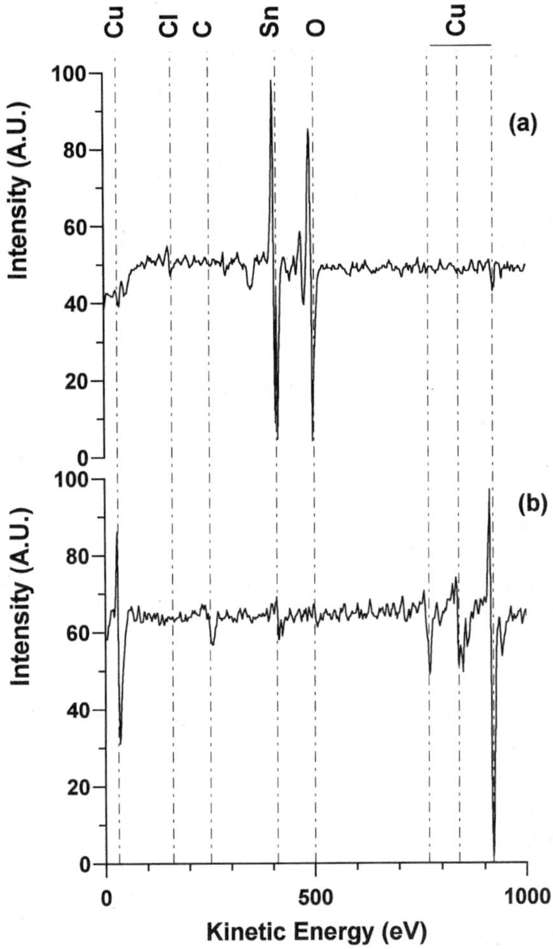

Figure 13.6. Auger spectra taken from the (*a*) intergranular corroded region and (*b*) uncorroded α-bronze grain. The analysis shows that the intergranular regions have a higher tin and oxygen content in comparison with the uncorroded areas.

where J is the X-ray flux, n_x is the number of atoms per cubic centimeters of the atomic species x, σ is the photoelectron cross section, K is an instrumental factor, and λ is the electron inelastic mean-free path. The equation can be used for direct quantification, the so-called first principle, but experimentally determined parameters are more commonly employed. Generally, $\sigma K \lambda$ are included in an experimental parameter F called the sensitivity factor. By considering the ratio between two strong bands from each of the two elements, x and y, which constitute a homogeneous solid solution, the following equation may be formulated:

$$\frac{n_x}{n_y} = \frac{I_x/F_x}{I_y/F_y}$$

The term Fx/Fy becomes almost matrix independent if instrumental conditions are maintained constant. After developing a set of relative values of F (relative sensitivity factors), atomic concentrations of the atomic species x in a matrix of different atomic species i may thus easily be determined as follows:

$$c_x = \frac{I_x/F_x}{\sum_i I_i/F_i} \times 100$$

Similar procedures make it possible to determine element concentrations within an accuracy that can sometimes reach 5%.

The quantitative interpretation of Auger spectra is not as straightforward, even though similar equations are often used. These include elemental sensitivity factors obtained from pure elements and compounds. However, if highly accurate AES quantitative analysis is desired, standards of known concentrations close to those of the specimen analyzed must be used. In this way, electron escape depths, backscattering factors, and chemical effects are similar in both samples. It should be emphasized that the quantitative data obtained using these techniques refer to the surfaces of materials, and that in principle they cannot be generalized to provide the quantification of bulk elemental compositions.

13.3. XPS AND AES IN ART AND ARCHAEOLOGY

The earliest applications of XPS to the study of archaeological artifacts date back to the beginning of the 1970s, while AES began to be used at the beginning of 1980s. Lambert and McLaughlin (1976) and Leventhal and Thompson (1978) analyzed a variety of archaeological materials with XPS, ranging from glass, pottery, bronze, stone, and mummy cartonagge. Their studies clearly show the potential of this technique as far as elemental oxidation state determination and the variety of elements detected are concerned. The earliest application of AES to the study of archaeological materials should be attributed to Cline Love et al. (1980) and Polak et al. (1983). Since then, a variety of materials and processes of relevance for archaeological and/or artistic objects have been investigated using XPS, and

AES and the potential and limitations of these techniques are now comparatively clear.

In the following sections the use of XPS and AES in studies involving a variety of materials is discussed, focusing specifically on materials concerning archeological and/or artistic artifacts. Particular attention has been paid to highlighting the potential and limitations of these two techniques in solving problems arising from the analysis of specific kinds of materials. Each section is prefaced by a brief overview of the history and the evolution of the materials discussed. These brief introductions aim to provide the reader with the basic information necessary to understand the archaeological interests that lie behind the applications discussed. An exhaustive and authoritative discussion of the historical and technological aspects of the evolution of materials can be found in Goffer (1980), Hodges (1989), and more recently in the work of Lambert (1997).

13.4. POTTERY

13.4.1. Introduction

The discovery of fire allowed humans to process natural materials to improve or simply change their characteristics (ca. 1,600,000 years ago; see Rowlett 1999). One of the earliest uses of fire concerned cooking. Food became safer and tasted better after cooking. Later, stones were heated to improve their hardness (ca. 80,000 B.C.). The complex technology required for the making of pottery was not to develop until thousands of years after fire had been discovered. Paleolithic objects from Dolni Vestonice in the Czech Republic are probably pottery's earliest ancestors and can be dated to 24,000 B.C. (Vandiver et al. 1989). The development of pottery is still, however, a subject for debate, and its origins have been placed between 12,000 and 10,000 B.C. (see Lambert 1997; Copper 1972). Pottery technology evolved with time, and more and more complex procedures allowed the production of finely finished ceramic artifacts that combined aesthetics with usefulness. The introduction of glaze certainly improved both the appearance and utility purposes of pottery, especially when the use of proper fluxes provided suitable glazes for curved surfaces such as those of pots (see Tite et al. 1998). Flat surfaces were in fact already being glazed in Egypt in ca. 5000 B.C., but the glazed pots found in Syria have been dated as being from a later period, ca. 1700 B.C.

Different methods for finishing pottery required control of the oxidizing/reducing atmosphere during firing. The most famous and striking examples of this procedure were provided by the Greeks after the invention of the so-called red-figures technique (Boardman 1976). It involved producing red and black paints on pottery as a result of switching from an oxydizing firing atmosphere to one that was reducing so as to form Fe_2O_3 (red) or Fe_3O_4 (black) on the surface of the pottery previously covered with suitable clay materials. The complex decorations developed during the Middle Ages and the Renaissance bear witness to the tremendous evolution of pottery decoration techniques through successive ages.

13.4.2. XPS in Pottery Analysis

The potential of XPS for pottery analysis has been highlighted since the earliest applications of XPS to the study of archaeological materials (Lambert and McLaughlin 1976; Lambert et al. 1978). The use of XPS as a merely quantitative analytical technique for pottery analysis has been discussed by Lambert and McLaughlin (1976) and more recently by Torrisi et al. (1996). With its potential for light element detection, the ease of sample preparation and the low destructiveness of samples, XPS offers numerous advantages when compared to other elemental analytical techniques largely used for the quantitative analysis of pottery such as XRF (X-ray fluorescence; for details see Chapter 4); NAA (neutron activation analysis; for details see Chapter 5), and atomic spectroscopic and atomic spectrometric techniques (see Chapter 3). However, XPS quantitative analysis does have some drawbacks that should be correctly addressed before inhomogeneous matrixes may be analyzed. XPS provides quantitative data that express the surface composition of the material. It should thus be considered as not exactly reproducing the bulk composition of the pottery under analysis. Roughness, the irregular shape of the sample, and inhomogeneity within the matrix may also introduce random uncertainties, and these must be added to others normally associated with spectrum acquisition (Harrison and Hazell 1992), peak intensity determination (Cumpson and Seah 1992a), and the route used for quantification (Cumpson and Seah 1992b). In general, accuracy in the range of $\pm 5\%$ is attainable by XPS quantification.

XPS's potential for chemical state determination remains, however, the main point in favor of its use in pottery analysis. The technique provides valuable information for the identification of silicate compounds (Seyama and Soma 1988; Wagner 1990) as well as for the study of glaze and pottery surfaces.

In 1983, Gillies and Urch studied a variety of black-surfaced Early Iron Age potteryware from Little Walthman in Essex and from Servia, and Iron Age Northern Black Polished Ware (NBPW) from India. XPS made it possible to see that carbon was responsible for the black coloration of the surface of pottery from Little Walthman and Servia, whereas the black surfaces of the NBPW were seen to be due to iron oxides.

In 1990, Lambert et al. provided a significant demonstration of the potential of XPS in chemical state determination by distinguishing the different oxidation states of iron present in 23 pottery sherds coming from Bahamas and Puerto Rico. The samples, coming from sites dated from A.D. 600 to A.D. 1360 showed colors that well correlate to the iron oxidation state revealed by XPS. Other applications of XPS in the study of pottery concern Neolithic Chinese ceramics from Liangzhu, China (Dajnowski et al. 1992), and Medieval sherds from Southern Italy (Bruno et al. 1994, 1997). Hematite and chromium-containing compounds (chromium was present in both +3 and +6 oxidation states) were found in the red finished ceramics dated to 400 to 300 B.C. and coming from Canosa, Italy (Sabbatini et al. 1994).

13.5. GLASS

13.5.1. Introduction

The development of the technology necessary to obtain glass could be linked to the smelting of metal ores or to the manufacturing of glazed pottery. The earliest known glass material have been supposed to be linked to smelting technology and have been dated to ca. 2000 B.C. (Wertime 1973; Brill 1963). Glazing technology may, however, have anticipated the origin of glass (Lambert 1997: Chapter 5). Mesopotamia was probably the region in which glass production was first established but was in Egypt after 1500 B.C. under the XVIII dynasty that glass production found its first landmark.

13.5.2. Surface Compositional Analysis of Glass by XPS and AES

Surface-sensitive analytical techniques have been widely used in the study of glass due to the importance that surface composition plays in the determination of its performance (Fox 1981). Corrosion processes in archaeological glasses have recently been reviewed by Pollard and Heron (1996: Chapter 5). Few studies have tested the potential of XPS for identifying colorant material present in archaeological glasses. In this context, Lambert and McLaughlin (1978) showed how XPS can identify the different oxidation states of copper-containing compounds to ascertain their presence in red, blue, and green colored glasses.

The effects of the electron beam on the surface composition of glass has been consistently pointed out since 1969 (Vassamillet and Caldwell 1969). The majority of studies that discuss the influence of the electron beam on glass surfaces, with specific reference to AES analytical information, do not specifically refer to ancient glasses. The conclusions drawn are, however, of general interest and should thus also be considered for archaeological glass AES analysis.

The comparison of X-ray photoelectron spectra carried out before and after electron irradiation of a glass surface with the usual irradiation conditions for AES analysis (3 keV, 30 μA over a 3 × 3 mm area) shows a depletion in the Na surface composition while Ca, Si, K, and B increase in concentration. The effect described was initially attributed to beam-assisted segregation or decomposition of the glass surface (Coad et al. 1975). Desorption of surface contamination was excluded. Na depletion is dependent on the electron beam current and thus depends on the irradiation time (Pantano et al. 1976). Moreover, an induction period before decay in the Na concentration is established upon sample cooling. Dawson et al. (1978) concentrated their attention on the role played by carbon surface contaminants and saw the electromigration caused by beam-induced surface charging as possibly being responsible for alkali metal depletion. The incidence angle of the electron beam on the surface also affects surface composition (Chappell and Stoddart 1977). The increase in the amount of Si detected after electron bombardment has been attributed to the chemical reduction of silicon present in SiO_2 species (Pantano et al. 1975). The oxygen outgassing has been also proposed to cause the formation of direct Si–Si bonds (Puglisi et al. 1983)

Bearing all this in mind, it is clear that AES analysis of ancient glass requires the accurate control of the parameters reported as affecting surface composition. Important limitations are, however, associated with all analytical techniques that require the sample to be irradiated with an electron beam taking into account that similar modification have been reported to be induced by ion beam (Torrisi et al. 1988). Surface compositional information for glass should thus be obtained using a variety of analytical techniques that minimize surface composition alteration. XPS has been widely used to study modifications induced by electron beam on the surface of irradiated glasses. This case acknowledges the Na near-surface segregation induced by the X-ray flux (Puglisi et al. 1986).

13.6. METALS AND ALLOYS

13.6.1. Introduction

Metals and alloys are commonly involved in XPS and AES analysis. They are highly electric conductive materials, can easily be handled so as to obtain clean, flat, and smooth analyzable surfaces, and, finally, they are usually ultra-high-vacuum friendly. Archaeological and/or artistic metallic objects do not always possess all these properties because of degradation processes that often chemically modify the external surface and deep regions of the samples.

A great variety of metallic objects of archaeological and artistic interest have been studied making use of XPS and AES. These include bronze artifacts, lead pipes, slags, and ironworks. Uneven applications involving coins (Gillies et al. 1981; Cline Love et al. 1980) and pre-Columbian pectorals (Cline Love et al. 1980) have also been documented, their main aim being that of showing the potential of XPS and AES for chemical state analysis and spatial resolution. Due to the high affinity of XPS and AES for metallic samples studies, the techniques have often been used together in the study of the same selected object(s). For this reason, differences as regards the material studied rather than the technique used are referred to in the following subsections.

13.6.2. Copper and Copper Alloys

The production of copper and copper alloy objects developed alongside the evolution of the technology linked to the extraction of the metal from ores—smelting, the improved ability to work the metal, and, finally, the ability to form alloys (Tylecote 1992). Much of this evolution was a consequence of the ability of metalworkers to find and control processes that allowed them to heat the metal, in its native state or in ores containing copper compounds, at increasingly high temperatures so as to reach the temperature necessary to melt copper (1083°C). The shaping of native copper was a well-established custom in Southwest Asia from ca.10,000 B.C. onward. This area was by far the most advanced in copper-work technology. In fact, evidence for the smelting of copper-based ores, which leads us to suppose that casting skills were already established, has been dated to 7000 to 6000 B.C. (Çatal Hüyük, Anatolia).

The addition of elements other than copper to form alloys with better properties in terms of castability, hardness and, appearance may originally have been accidental. Tin was the most important alloying element in the Old World from 4000 B.C. onward, the resulting alloy being *bronze*. Lead and zinc were two other important elements in forming or modifying the characteristics of copper alloys. The former was used especially by the Greeks who intentionally added lead to the tin–bronzes used to make statues both to improve the fluidity of the molten alloy and to enhance appearance. The intentional alloying of zinc—thus forming *brass*—was certainly successive to the use of the mentioned elements in copper alloys preparation. The earliest brass objects were probably manufactured in eastern Turkey during the first millennium B.C. The *calamine process* having been established and perfected, brass production increased rapidly thanks to its use in minting coins during the Roman Empire (Pollard and Heron 1996: p. 196).

13.6.2.1. General Remarks on Copper-Compound Analysis by XPS

The XPS analysis of copper and copper compounds requires various aspects to be taken into account to interpret spectra correctly. Various features allow XPS to distinguish between Cu^{2+} and Cu^{+} compounds. Shift in peak position is the most common feature for distinguishing oxidation states by XPS. However, copper spectrum profile provide extra structures for distinguishing between different oxidation states, such as *shake-up* satellites that are evident for Cu^{2+} compounds but absent for Cu^{+} and Cu. Shake-up satellites appear in the X-ray photoelectron spectra of a large number of elements in both organic and inorganic compounds (Brisk and Baker 1975). They have been associated with the valence electron relaxation that occurs when the nuclear charge is raised due to the photoemission of a core electron. The resulting extra energy allows the excitation of a valence electron, which is raised to a higher level (transitions of types $3d^n \rightarrow 3d^{n-1}4s$ or $3d^n \rightarrow 3d^{n-1}4p$). Another proposed mechanism for the shake-up process involves a charge transfer from ligand orbitals to metal orbitals ($3d$ orbitals for copper). Both processes result in the lowering of the energy available for the primary outgoing electron. A structure on the low kinetic energy (low KE) side of the spectrum (high BE) results (i.e., the shake-up satellite). This ligand–metal charge-transfer mechanism may explain why shake-up satellites are observed in Cu^{2+} ($3d^9$) compounds and not in Cu^{+} ($3d^{10}$) compounds. In fact, only unfilled d orbitals will allow a ligand-to-metal charge transfer. Factors that may result in the alteration of the oxidation state of copper should be considered before analysis. Ion bombardment constitutes an important source of alteration (Naguib and Kelly 1975; Mitchell et al. 1990). The reduction of CuO after ion bombardment should thus be taken into account when performing depth profiles. Loss of oxygen and lowering of the shake-up structure are two signs of reduction during profiling. Other reasons for the alteration of the shape and position of relevant peaks for copper oxidation state identification could come from the X-ray exposure involved in XPS analysis. A partial X-ray-induced reduction of Cu^{2+} has been attributed to the different excitonic probabilities for Cu–O bonds in different crystalline sites (Marletta et al. 1982), *excitons* being electron–hole pairs that can migrate through the crystal lattice. This reduction has also been seen as being caused by photoreactions or thermal and electronic reactions (Frost et al. 1972). More recently, sample biasing has been observed to minimize X-ray-induced CuO reduction (Iijima et al. 1996). An accurate evaluation of possible X-ray-induced alteration should therefore also be carried out when a variety of elements other than copper are taken into account (Copperthwaite 1980).

13.6.2.2. Corrosion of Copper and Copper Alloys

The corrosion of copper and copper alloys has been studied since the 1800s (Davy 1826). Numerous studies have been made of the formation of various corrosion products, the influence of different environmental and burial conditions, and the role of various copper-containing compounds resulting from the corrosion process. This section focuses on how XPS and AES can be used to study the corrosion of copper and copper alloys. For in-depth information on the corrosion processes established in these materials, the reader is advised to consult specific books (Uhlig 1971), papers (Strandberg and Johansson 1997; Scott 1990; Kruger 1992; Pollard et al. 1992; Walker 1980), and reports (Edney n.d.).

When exposed to the atmosphere, copper gradually develops a dull red patina that with time turns a different color. For example, a green patina with malachite [$Cu_2(OH)_2CO_3$] as its main component is formed in most burial environments. The dark layer is mainly made up of cuprite (Cu_2O) but even traces of copper hydroxy-sulfate and copper hydroxy-chloride may develop. The development of this patina on pure copper and copper alloys was first studied in the early 1930s (Vernon 1932) when the influence of atmosphere composition was established. Since then, the contribution of various pollutants in the development of the copper patina has been studied, and various experiments have been performed in an attempt to understand the effect of typical pollutants such as SO_2 and chlorides. Both XPS and AES have been used extensively for the identification of compounds formed after the exposure of copper and copper alloys to controlled atmospheres.

Exploiting the fact that XPS is sensitive to formal oxidation states, Strandberg and Jonasson (1997) used this technique to detect the presence of Cu^{2+}-containing species on the surface of a copper sample exposed to a flow of humidified SO_2 gas [69 ppb SO_2 at 90% relative humidity (RH)]. A shake-up satellite was evident in the copper 2*p* profile. Moreover, a different S(IV)/S(VI) ratio was calculated for the black and shiny areas of the exposed sample. This information led the authors to propose that a passivation effect, attributed to the formation of a chemisorbed sulfite layer, is established when a high SO_2 relative concentration (higher than 69 ppb) or relative humidity lower than 75% is maintained during exposure. XPS and AES have also helped researchers to understand the importance of relative humidity in the sulfidation of copper by atmospheric H_2S and OCS (carbonyl sulfide) (Graedel et al. 1985).

The compositional profile distribution of Cu, O, S, and Cl in four copper samples exposed to ambient atmosphere for 1, 12, 44, and 100 years, respectively, has been studied by Opila (1987) making use of AES coupled with Ar^+ sputtering. Opila found that, beneath a superficial layer of carbon, sulfur was primarily present near the surface of the patina while chlorine was distributed throughout. Mass balance evaluations coupled with thermodynamic considerations allowed Opila to make hypotheses about the patina profile composition. However, it must be emphasized that such a procedure does not account for the complex distribution of compounds within the copper and copper alloy patina, especially when buried samples are taken into consideration (Robbiola et al. 1988; Robbiola and Hurtel 1997). The usual bombarding ion flux and the typical surface roughness of corroded samples together with the alteration of the near-surface composition induced by the ion sputtering itself (Lam 1988) mean that similar procedures may not be employed for the study of thick and irregular patinas such as those typically found on ancient copper and copper alloy objects. As the most important and widespread copper alloy of the past, bronze suffers corrosion processes similar to those seen for copper. Corrosion resistance may even be reduced. A polished cross section of an archaeological bronze typically shows that corrosion extends into the metallic matrix alongside grain joints (see Fig. 13.5). Auger analysis shows that intergranular regions have a higher tin and oxygen content (Fig. 13.6*a*) in comparison with the uncorroded areas in α-bronze grains where tin concentration is lower than 9.1 at% (Fig. 13.6*b*). SAM may be used to detect the segregation of other elements such as chlorine in the infiltrated corrosion products (Spoto et al., in press). The preferential corrosive attack along grain joints has been supposed to be a consequence of tin segregation at the boundaries of α-bronze grains.

AES can be used to detect submonolayers of elements and makes it possible to study the segregation phenomena that occur at grain boundaries. A typical procedure used to expose grain boundaries for analysis is to fracture the sample under high-vacuum conditions, thus minimizing reactions at the solid–gas interface. Due to its ductile nature, the fracturing of bronze materials in the high-vacuum chamber of an AES instrument requires the samples be cooled with liquid nitrogen to achieve intergranular fractures.

Auger analysis performed on different areas of the fractured surfaces of a bronze-filled bracelet or *armilla* (dated to the eighth century B.C. and found in a tomb of the Necropolis of Chiavari, Italy) shows no detectable amount of oxygen or chlorine. This evidence accounts for the assertion that no corrosion processes were present in the regions from which the fractured sample were taken, that is, a few millimeters beneath the corrosion crust.

When quantifying tin segregation at the grain boundary, it is assumed that the segregant forms an overlayer (expressed as a *fractional monolayer*) on the substrate. In the case discussed this substrate is constituted by the copper–tin alloy. The fractional monolayer of tin on the substrate Φ_{Sn} should be calculated using the following equation (Seah 1990):

$$\Phi_{Sn} = mK_{SnCu}\frac{I_{Sn}}{I_{Cu}}$$

where $m = 2$, I_{Sn} and I_{Cu} are the intensities of the Auger electron signals for Sn (KE = 430 eV) and Cu (KE = 920 eV), respectively, and K_{SnCu} is

$$K_{SnCu} = \left[\frac{\lambda_{Sn}(430\text{ eV})\cos\theta}{a_{Sn}}\right]\left[\frac{1 + r_{Sn}(430\text{ eV})}{1 + r_{Cu}(430\text{ eV})}\right]\frac{I^{\infty}_{Cu}}{I^{\infty}_{Sn}}$$

where $\lambda_{Sn}(430\text{ eV})$ represents the attenuation length (Seah and Dench 1979); $r_{Sn}(430\text{ eV})$ and $r_{Cu}(430\text{ eV})$ are the backscattering terms (Ichimura and Shimizu 1981); a_{Sn} the atom size; θ the angle emission of the Auger electrons from the surface normal; and I^{∞}_{Sn} and I^{∞}_{Cu} are the relative sensitivity factors (Childs et al. 1995).

A level of Φ_{Sn} in the range of 50 to 60% was calculated for the intergranular fractured surfaces of the bronze-made *armilla* (Spoto et al., in press). Such enrichment of tin at the bronze grain boundary must certainly have contributed to the preferential corrosive attack observed at the grain joints. In fact, the higher reactivity of tin to oxygen compared to that of copper renders the enriched surface more susceptible to oxidation, thus enhancing the corrosive attack localized at the grain boundaries.

13.6.3. Lead

The smelting of lead has been assumed to precede that of copper. The lower temperature requested for lead smelting (800°C) is the most obvious reason for this assumption. The use of lead was at a height when the Greeks used it with copper–tin alloys to make statues. However, during the Roman Empire lead production increased tremendously and reached levels of production that were to be reproduced only after the Industrial Revolution (see Hong et al. 1994 and references cited therein). Under the Romans, lead was used to produce a great variety of objects ranging from coins to pipes for the distribution of water.

13.6.3.1. Roman Lead Pipes

Lead pipes—*fistulae*—were among the most common metallic objects made during the Roman Empire. Writers of the period described the technology used to produce fistulae and the standards used for their classification (see references cited in Paparazzo 1994). Fistulae were cast in lengths longer than 10 ft and joined with a soldering technique that made use of a lead–tin alloy —*tertiarium*—that consisted of two parts lead and one part tin.

Paparazzo's XPS and AES analyses confirmed the different compositions of the joint and the pipe itself (Paparazzo 1994; Paparazzo and Moretto 1995, 1997; Paparazzo et al. 1995). Detectable amounts of tin (up to 3 at%) were found only at the joint of a fistula classified as an *octonaria*—internal diameter 3.7 cm—and dated between A.D. 222 and A.D. 235 (Paparazzo 1994). A detailed study of the $3d_{5/2}$ Sn peak position—Sn $3d_{5/2} = 485.0$ eV (Moulder et al. 1992), SnO_2 $3d_{5/2} = 486.4$ eV (Wagner et al. 1979b)—coupled with a study of the shape of the Sn MNN Auger peak led to the identification of the SnO_2 species. An evaluation of the vertical distribution of tin in the soldered region of the fistula was carried out using Ar^+ sputtering followed by XPS and AES analysis—vertical range in the order of few hundred Ångstroms. This confirmed the thorough oxidation of the joint region. As has already been discussed for copper compounds (see Section 13.6.2.1), here the ion etching is seen to have altered the stoichiometry of the etched surface (Larciprete et al. 1993; Mitchell et al. 1990), thus inducing the appearance of features attributed to the presence of a substoichiometric tin oxide species. The comparison of the shape and position of the Pb $4f$ peaks obtained after analyzing regions corresponding to the joint and the pipe before and after Ar^+ sputtering showed a higher content of oxidized lead in regions far from the soldered joint, that is, Pb $4f_{7/2} = 136.9$ eV (Moulder et al. 1992), PbO $4f_{7/2} = 138.9$ eV, and PbO_2 $4f_{7/2} = 137.4$ eV (Bertrand and Fleischauer 1980). Lead oxides were also seen to suffer reduction caused by Ar^+ sputtering. Components due to the metal thus appeared in the spectra (Kim et al. 1974; Mitchel et al. 1990). XPS sensitivity for carbon and oxygen allowed Paparazzo to suppose a lower thickness of the poisoning corrosion product *White Lead* [$2PbCO_3 \cdot Pb(OH)_2$] in the areas corresponding to the joint where *tertiarium* alloy was used (Paparazzo et al. 1995). The higher percentages of hydrocarbon detected in the joint areas were attributed to the custom of wetting the rims to be soldered with oil so as to minimize oxidation during soldering.

The soldering technique employed in the Roman era is thus seen to have involved the use of the *tertiarium* after wetting the surfaces to be joined with oil. The presence of *tertiarium* lowers the oxidation of lead in the soldered areas, thereby minimizing the formation of poisoning corrosion products.

13.6.4. Iron and Steel

The beginning of the Early Iron Age (Maddin et al. 1977)—fixed as being 1200 B.C.—coincides with the ability of people from west Asia to smelt iron and alloy carbon so as to obtain steel. Up to 1500 B.C. the Hittites had the most developed technology for the working of iron. The advantages of iron over copper and its alloys had, however, been known since the Bronze Age. Evidence of uneven uses of iron, sometimes in its native state, have been dated to 2500 B.C. Iron-smelting technology presumably dates back to the Bronze Age as the temperature required for smelting iron (1100 to 1150°C) is similar to that required for copper. However, iron smelting necessitated a more accurate control of the carbon and oxygen present in the furnace so as to maximize the percentage of iron present in the spongy mass, or *bloom*, obtained after the furnace was cooled (Hodges 1989). Glasslike materials, or *slag*, were formed during smelting due to a reaction between the silica impurities present in the ore and fluxes. The formation of slag, inevitably containing certain amounts of iron, was sometimes due to the intentional addition of fluxes to the furnace to help rid the ore of silica impurities.

13.6.4.1. XPS and AES in the Study of Early Iron Age Slag and Roman Ironworks

The development of iron-smelting technology in the northwest region of Syria has been assumed as having begun in the years 1200 to 1000 B.C. Little slag and few iron artifacts have, however, been found in this area to aid in our understanding of the level of technology reached in iron smelting at this time in northwestern Syria. Slag from Tell Afis and dated to ca. 800 B.C. has been studied by Ingo and Scoppio (1992) using XPS and XAES (X-ray induced Auger electron spectroscopy) to give an idea of the level of knowledge reached in the northwest region during the Early Iron Age.

Slag is assumed to form as a result of iron smelting when traces of other elements, which would indicate copper or tin smelting (i.e., copper, tin, nickel, cobalt, and zinc), are not present. Detailed XPS studies of the position of the Fe $2p_{3/2}$ peak, shake-up satellites, and the general shape of the spectrum in the region of interest for Fe $2p_{3/2-1/2}$ photoemissions (see Moulder et al. 1992; Sherwood 1990; Welsh and Sherwood 1989; Allen et al. 1982; Harvey and Linton 1981) led to the identification of iron oxides (FeO_{1-x}, wustite, and Fe_3O_4, magnetite) and hydrous ferric oxide (γ-FeOOH, goethite or lepidocrocite) in different structures that were shown up after analyzing the polished cross-sectioned surfaces of the slag. The evaluation of the energies of the $2p$ photoemitted electrons and the KLL X-ray stimulated Auger electrons for aluminum and silicon suggested the contemporary presence of silicate compounds (Mg_2SiO_4, $MgSiO_3$) together with alluminosilicate compounds such as melilite [$Na_2(Ca,Mg)_{11}(Al,Fe)_4(SiO_4)_9$]. Such assignments can be performed by making reference to the two-dimensional Auger parameter plots proposed by Wagner and co-workers (Wagner 1978;

Castle and West 1980; Wagner et al. 1981; West and Castle 1982). These plots can be used as fingerprints for characterizing chemical states.* The compositional data obtained—expressed as $(CaO + MgO)–SiO_2–Al_2O_3$ diagrams—was compared with values reported in the literature for slag composition (Committee for Fundamental Metallurgy 1981) and a furnace temperature of around 1210 to 1250°C was supposed. After the collapse of the Hittite Empire (1200 B.C.), ironworking is supposed to have developed in the Mediterranean area and subsequently spread over Europe to reach Britain during the pre-Roman and Roman Iron Age (600 B.C. to A.D. 300). Phosphorous has been found to be a common impurity in many of the iron objects and artifacts produced in this period. Particularly when present at concentrations higher than 0.08 at %, phosphorus reduces the ductility of steel. Grain boundary corrosion, creep embrittlement, and stress corrosion cracking have also been reported as being caused by the grain boundary segregation of phosphorus (see Kaloyeros and Ehrenreich 1991 and references therein). Under certain conditions, phosphorus can, however, act as a hardening agent for steel. It may thus have been used by the Romans to produce harder tools and weapons. Phosphorous impurities found in artifacts for which toughness was required confirm this apparent control over the raw materials used for smelting iron (Goodway 1987). In particular, two Roman ironworks from Irchester, Northampthonshire, with a respectively high and low phosphorus impurity content (higher and lower than 0.08 at%) were studied by coupling AES depth profile with other analytical techniques such as SIMS (secondary ion mass spectrometry) and EDX (energy-dispersive X-ray fluorescence) (Kaloyeros and Ehrenreich 1991). A standard sample with an iron/phosphorus composition similar to that found in the artifacts was used to take into account the alterations introduced by Ar^+ sputtering. The study confirmed the high phosphorus content after both vertical and lateral semiquantitative AES analyses in one of the artifacts (0.25 at% of P) while no phosphorus was detected in the other sample studied.

It should be underlined that accurate studies of phosphorus grain boundary segregation requires that intergranularly fractured samples be analyzed following procedures similar to the procedure described in Section 13.6.2.2.

* Si KLL Auger energy is centered, for most inorganic compounds, at around 1609 to 1610 eV (KE). This peak appears on the usual binding energy (BE) scale used to plot X-ray photoelectron spectra at negative energetic values [−122/−123 eV (BE) when a standard Al $K_{\alpha 1,2}$ source is used]. The apparent non-sense of this can easily be explained by taking into account the background radiation emitted from the source (bremsstrahlung radiation), which contributes for emissions at kinetic energies higher than the $K_{\alpha 1,2}$ radiation energy (Al $K_{\alpha 1,2}$ = 1486.6 eV) (see also Wagner and Ashley-Taylor 1980). These emissions can thus appear as being on the negative side of the BE scale.

13.7. DYES, PIGMENTS, AND PAINTINGS

13.7.1. Introduction

Humans started to paint the wall of caves more than 30,000 years ago leaving imprints of hands that had previously been plunged into liquid coloring materials. Complex images of animals dated back to 32,000 B.C. were found to adorn the walls of an ancient cave in France (Grotte Chauvet, Ardeche) (Balter 1999; Fischman 1995). At that time pigments were essentially inorganic materials such as iron oxides for red or yellow and carbon (from soot or charcoal) for black. Manganese oxide, calcined bone, and white clays completed the palette. The technique for painting walls evolved alongside architecture. As soon as the irregular rock surfaces were replaced by almost smooth walls, techniques to paint the mud or clay layers previously laid onto the wall developed. Paintings from Catal Hüyük in Anatolia date back to 6000 B.C. and represent one of the earliest examples of this evolution (Mellaert 1962).

Coloring pottery surfaces also required the use of pigments. Pottery produced in Mesopotamia 7000 years ago shows black, brown, and red colors on the surfaces. The color palette was even more varied due to the rise of the civilization in the Mediterranean region. A wealth of examples have shown the variety of coloring materials possessed by the Egyptians and the evolution of wall paintings technology since the predynastic period (Mora et al. 1984; Lambert 1997).

A great variety of organic dyes were also obtained in the past from insects, molluscs, and plants (Schweppe and Roosen-Runge 1985; Mills and White 1994: Chapter 10; Lambert 1997). The Aztecs used to extract a red dye called *nochezli* from cochineals, while the most famous organic dye in the Mediterranean area was Tyrian purple, also known as Royal purple. This was obtained after processing the secretions of three different molluscs, namely *Murex trunculus*, *Murex Brandaris*, and *Purpura haemastoma*. More than 5000 molluscs were necessary to obtain one gram of dye. The dye was used to dye fabrics from the first millennium B.C. and successively acquired such a high sociopolitical significance that in the first century A.D. Nero issued an edict making Royal purple the prerogative of the emperor.

13.7.2. XPS in Study of Dyes, Pigments, and Paintings

For the characterization of coloring materials and paintings and the study of their alteration and preservation, XPS is a valuable technique that provides a wealth of significant information. XPS can distinguish signals coming from light elements such as carbon and oxygen, which constitute the fundamental structures of organic materials, and can separate them according to their different chemical environment.

The study of organic structures with XPS requires that X-ray-induced degradation processes be prevented by minimizing both exposure time and the power used to accelerate electrons in the source. Monochromatic sources should preferentially be used and fast acquisition should be carried out. With this background in mind, XPS has been coupled with other analytical techniques to identify the purple deposit found on the pottery sherds discovered at Sarepta (Sarafand) (McGovern and Michel 1985, 1990). The deposit was assumed to contain components that reproduced the 6,6′-dibromoindigotin structure present in the Royal purple dye whose Phoenician origin had been previously established as dating to 1000 B.C. The comparison of the binding energy of the bromine 3*d* signal (centered at 70.2 eV) detected after the unknown purple was analyzed by XPS, with the energy of the signals revealed after analyzing a 6,6′-dibromoindigotin standard compound allowed McGovern and Michel to ascertain its organic nature. However, information obtained with other analytical techniques was necessary to define the deposit found on these pottery sherds as being Royal purple and as representing the earliest evidence of the production of this dye (1300 B.C.).

13.7.2.1. Deterioration of Paintings and Pigments

The tombs at Beni Hasan (Middle Egypt) dated to 2400 to 1700 B.C. have been subjected to significant deterioration processes that have covered the wall surface with an almost opaque layer, disfiguring the painted murals in 12 of the upper-level tombs. XPS has been used with the aim of identifying the physico-chemical processes responsible for such deterioration (Wilson-Yang and Burns 1987). Wilson-Yang and Burns (1987) reviewed two mechanisms that had already been proposed for the formation of the opaque layer. The starting point for the deterioration process was either the percolation of water through rock or the condensation of CO_2-saturated water on the surface of the paintings. Samples to be analyzed by XPS were taken from (i) the painted surfaces covered by the opaque layer, (ii) the painted surfaces that had been cleaned with HCl around 10 years earlier, and (iii) the painted surfaces of walls that slanted to prevent the deposit of dust and sand by gravity. XPS analyses showed the presence of carbonate in samples belonging to the first and the latter group, while no significant amount of carbonate was detected in the sample treated with HCl. Carbonate species are easily detected by XPS because the photoelectrons emitted from the C 1*s* core level undergo a clearly visible energy shift (around + 4.5 eV BE). This is distinct from the signals coming from the carbon present in the hydrocarbon species usually detected as contaminant on the surface of most samples. Further differences between the three groups of samples were also observed. The first group showed the presence of SiO_2, Al_2O_3, MgO, NaCl, and amounts of sulfates and phosphates. The treated sample showed only SiO_2 and Al_2O_3 with a lesser quantity of chlorine on its surface. The latter group showed only traces of chlorine and phosphates present with carbonate. These findings allowed Wilson-Yang and Burns to support the hypothesis of the condensation of CO_2-saturated water on the surface of the paintings as contributing to the deterioration of the paintings even if the deposition of dust or sand still appeared to be the initial step in the process. The samples treated 10 years earlier showed only SiO_2 and Al_2O_3. Water had condensed on the surface where sand was present and seeped through the murals to the limestone wall causing $CaCO_3$ to continuously and progressively dissolve. Daily fluctuations in temperature and humidity caused Ca^{2+} to spread to the surfaces where it was fixed as carbonate. Further studies confirmed that the subsurface alteration was established after the surface layer had formed (Wilson-Yang and Burns 1988).

Such studies highlight the importance that surface-sensitive techniques have in the evaluation of processes that start at the interface between the atmosphere and the material in question. The degradation of painting often starts here and can thus be properly studied by XPS, especially if the process is in its initial stages. Degradation processes can sometimes affect the pigments present in paintings by altering their chromatic characteristics. The darkening of lead-containing pigments (Carlyle and Townsend 1990; Mora et al. 1984) is a typical example of the alterations to which inorganic pigments can be subjected. In this context, XPS can usefully contribute to the understanding of such degradation processes even when complex systems such as lake pigments are taken into account. The lake pigments of the past were usually obtained by coordinating dyestuff ligands to metallic species such as Al, Sn, or Fe. Dyestuffs were organic compounds extracted from insects such as cochineals and kermes or plants such as madder. Most of these pigments have been observed to fade under natural light (Saunders and Kirby 1994) and in the ozone-enriched atmospheres (Whitmore et al. 1986) that are typical of museum environments. Modern investigative techniques used before restoration such as radiography and *in situ* XRF further expose lake pigments to potentially degradative agents such as X-rays. XPS provides chemical evidence of the degradation induced by X-rays. The exposure of madder lake obtained after coordinating Al^{3+} ions with dyestuffs extracted from the roots of *Rubia Tinctorum* under un-monochromatized Al K_α radiation causes the loss of oxygen-containing species, as shown in Figure 13.7.

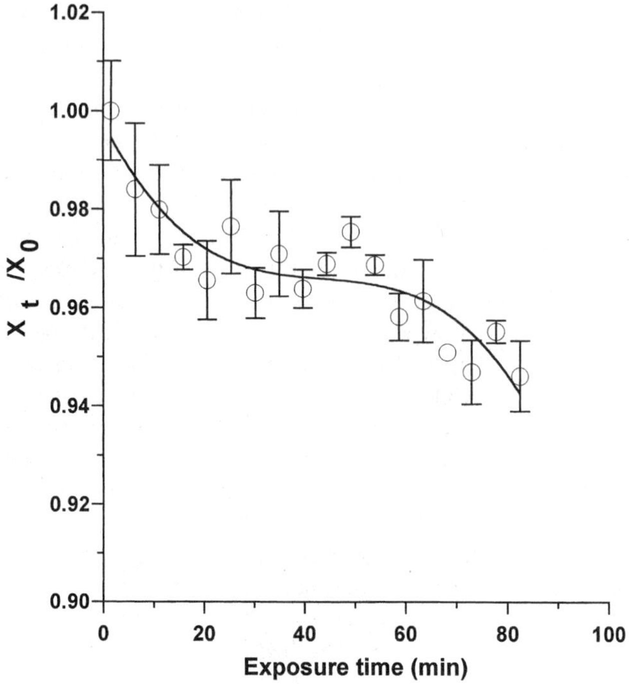

Figure 13.7. Diagram showing the variation of the X_t/X_0 quantitative ratio ($X = O/C$) with the exposure time for a madder lake. A 400-W Al K_α radiation (1486.6 eV) was used for the experiment. An average flux of 2×10^{11} photons cm^{-2} sec^{-1} was maintained during exposure.

13.8. PAPER

13.8.1. Introduction

The manufacture of paper, that is, one of the most familiar and important materials in the Modern Era, is assumed to date back to A.D. 200 in China. A member of the imperial guard first made sheets of paper by drying natural fibers—flax, rags, rice stalks—that were then macerated in water and drained on molds (Hon 1989). Paper was introduced to Europe by Arabs 1000 years later, reaching the Americas only at the end of the sixteenth century. Only after the development of a grinder for the mechanical pulping of wood in 1847 did wood become the prevalent source for fibers in paper production (Hall 1974). Today, almost 95% of the world's manufactured paper comes from wood.

From a chemical point of view, wood is mainly composed of cellulose,

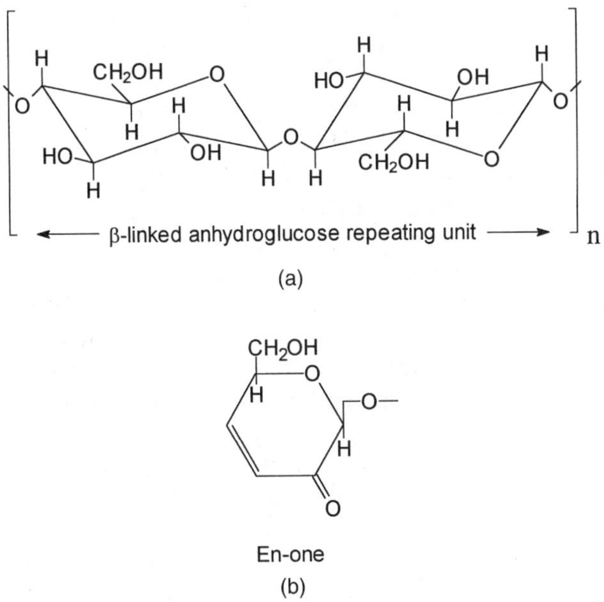

Figure 13.8. (*a*) β-Linked anhydroglucose repeating unit that constitutes the cellulose and (*b*) en-one structure generated by the exposure of the cellulose X-ray.

hemicellulose, lignin, and extractive fractions—components that can be extracted from wood with water or organic solvents. These are mainly resins and polysaccharides. The relative percentages of these components depend on the variety of tree from which the wood derives. For softwood (pine, spruce, western hemlock, etc.) they are $42 \pm 2\%$, $27 \pm 2\%$, $28 \pm 3\%$, and 1–5%, respectively (Wilson and Hamilton 1986). After pulping enrichment in cellulose and hemicellulose is obtained. The properties of the end product strongly depend upon the amount of linked carbohydrate material still present. Cellulose thus represents the main component of paper and gives it many of its distinctive properties. It is made up of linear polymers of anhydroglucose units linked through 1,4-β-glycosidic bonds crosslinked with hydrogen bonds (Fig. 13.8*a*). The manufacturing of paper requires more complex procedures aimed at modifying its final characteristics. These often result in the introduction of acidic species that cause important degenerative processes in the cellulose structure. The acid-catalyzed hydrolysis of the cellulose involves the depolymerization of the chain. This is caused by the cleavage of the β-acetal linkages between the anhydroglucose units (Bunton et al. 1955; Whistler and Richards 1958; Blazej and Kosik 1985), making the paper brittle. Various factors contribute to the overall acidity paper develops during storage. Some of these are intrinsic to the wood-extracted fibers. Lignin and hemicellulose along with other components not completely eliminated during pulping slowly degrade to form acidic species. External factors such as pollutants (sulfur dioxide, nitrogen dioxide) (Hudson et al. 1964) and photooxidative effects (Hon 1979) also cause the acidification of paper. However, the major source of acidity in paper arises during its manufacture. Sizing agents are usually added to the pulp slurry (internal size), and/or to the surface of the paper after the paper sheet is formed (surface size), with the aim of rendering the cellulose more hydrophobic. Cellulose is naturally highly hydrophilic due to the presence of many hydroxyl groups. After this process, the paper increases its resistance against penetration by aqueous liquids, thus preventing the printing ink from running. The development of wood-pulping technology has led to animal gelatin, used as an internal sizing agent until the second half of the nineteenth century, being replaced by resinous rosin precipitated by an aluminum salt—usually $Al_2(SO_4)_3 \cdot 18H_2O$ and called "alum" by papermakers. The acid hydrolysis of the complex hydrated aluminum ions generated by this salt was the main source of degradation in the vast majority of books produced until the 1950s (Abelson 1987). The need to preserve the immense heritage contained in these books of the past has led researchers to study and develop deacidification methods using a variety of analytical methods to investigate their effectiveness.

13.8.2. XPS in Study of Paper

The evaluation of the macroscopic properties of paper quickly shows that it is not an ideal material to be studied with surface analytical techniques. Paper is a nonconductive material. It is porous and often nonhomogeneous and certainly does not possess smoothness in the nanometer range. Nonetheless, combined with other analytical techniques used in the monitoring of the properties of paper for conservation purposes (Sitholé 1995), XPS has been shown to be a useful analytical technique in the study of the surface composition of paper and the degradative processes to which it is subjected (Dorris and Gray 1978a, 1978b; Istone 1995). XPS is sensitive for carbon and oxygen and makes it possible to distinguish between contributions from atoms with different chemical environments. This means that it can provide relevant information for the surface composition of paper while maintaining a microdestructive approach. The X-ray photoelectron spectrum of pure cellulose, such as that obtained from Whatman filter paper, shows only signals coming from carbon and oxygen atoms (Beamson and Briggs 1992: p. 159). A detailed study of the different components of the C 1*s* and O 1*s* peaks shows three bands for the former and two bands for the latter element.

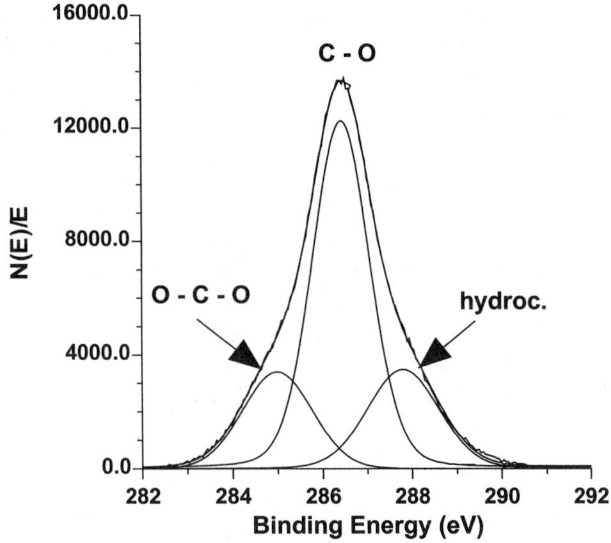

Figure 13.9. Typical C 1*s* band envelop of a Whatman paper composed almost entirely of cellulose. The bands are respectively due to (1) hydrocarbon impurities with direct C–C links positioned at 285.0 eV; (2) carbon atoms directly linked to an oxygen atom (C–O) (+1.7 eV); and (3) carbon atoms from hemiacetalic groups (O–C–O) (+3.1 eV). Carboxyl impurities contribute for bands positioned at a higher binding energy.

The three contributions for the C 1*s* peak are due, respectively, to (Fig. 13.9) (1) hydrocarbon impurities with direct C–C links where the signal is used to correct the binding energy scale for charging effects and is positioned at 285.0 eV; (2) carbon atoms directly linked to an oxygen atom (C–O), which produce a signal with a shift of around 1.7 eV; and (3) carbon atoms from hemiacetalic groups (O–C–O), which generate a signal shifted to around 3.1 eV as refers to the signal positioned at 285.0 eV. The carboxyl impurities sometimes detected contribute for bands positioned at a higher binding energy. The theoretical atomic ratio between the C–O and O–C–O carbon atoms is 83.3/16.6. A similar intensity ratio is thus to be expected for the two bands they generate. Two bands contribute to the oxygen peak. These come from the oxygen atoms belonging to the C–O–H and O–C–O groups and are positioned at 532.9 and 533.5 eV, respectively. A theoretical intensity ratio of 60.0/40.0 can be expected for these contributions. Deviations from the theoretical quantitative ratios expected for cellulose are due to the presence of impurities on the surface of the paper analyzed. In particular, the presence of lignin or extractive fraction components lowers the overall oxygen/carbon quantitative ratio detected on the surface of the cellulose.

This is due to both the prevalent aromatic structures present in the lignin and the extractives fraction components such as lignans and tannins, as well as to the organic chains making up the basic structure of other extractive components such as resin, fatty acids, and their ester (Dorris and Gray 1978b).

These features may suffer damage induced by exposure to X-rays during analysis. In fact, the energy transferred onto the analyzed surface by X-rays induces the double dehydroxylation of the anhydroglucose subunit of cellulose and generates an en-one structure (Fig. 13.8*b*) (Ahmed et al. 1987; Istone 1994). As a consequence of these modifications, a change in the relative intensity of the bands contributing to both the C 1*s* and O 1*s* peaks is observed. X-ray irradiation induces a relative increase in the hydrocarbon (C–C) and highly oxidated carbon atom (O–C–O and C=O) signals compared to the signal coming from the C–O carbon atoms. The lowering of the relative abundance of oxygen and carbon is also observed. Slower degradative processes have been observed after the cooling of the analyzed sample (Ahmed et al. 1987), thus suggesting that the source heating is an important factor to consider. In general, the study of paper requires brief exposure to X-rays. It thus requires fast XPS analyses to be carried out using instruments equipped with cold stage and/or monochromatic X-ray sources that allow the heating of the irradiated area to be kept at a minimum.

13.8.3. XPS in Study of Processes for Paper Deacidification

XPS has been shown to be a useful tool for studying the effects of deacidification, which is carried out to preserve paper from degenerative processes caused by the acidic catalyzed hydrolysis of cellulose. Various processes have been tested to deacidificate paper and to establish an alkaline buffer that can compensate for acid species that may be further generated by degradative processes (Hon 1989). Today's scientists are now devoting their interest mostly to developing processes that avoid alterations in the appearance and physical nature of books and minimize reactions which may involve inks or colors. For these reasons, organic basic solutions (for nonaqueous deacidification) have been preferred to aqueous solutions. XPS has been used to show the increased contribution of signals coming from C–O and O–C–O carbon atoms in the C 1*s* spectrum for a letter coming from the Fanshawe archive dated 1665 before and after a nonaqueous deacidification process based on an organic solution containing methyl magnesium carbonate (HCMC) (Odhlya et al. 1992). The presence of a band positioned at 290.1 eV accounts for a carbonate buffer (probably $MgCO_3$) formed on the surface of the treated sample. Aqueous deacidification processes based on treatments with solutions of $Ca(OH)_2$ or $MgCO_3$—in the latter case CO_2 being added to the solution—led, instead, to a negligible increase in signals attributed to carbon atoms linked to oxygen, while a significant increase in hydrocarbon contaminants was evident. The nonaqueous HCMC treatment thus appears to be more effective in providing a buffer of basic compound on the surface of the treated paper.

The efficacy of other nonaqueous solution based deacidification processes has been studied using XPS together with other analytical techniques. The FMC process (FMC Corporation, Bessemer City, North Carolina), based on a solution of magnesium butoxytriglycolate (MG-3) and Freon (CFC-113) has been studied using XPS coupled with EDX and SIMS. The hydrolysis of the magnesium butoxytriglycolate forms a buffer of $MgCO_3$. The combined analysis techniques made it possible to identify the compounds after hydrolysis in the paper and showed the distribution of the basic buffer formed (MacInnes and Barron 1992). Comparison between the XPS derived Mg atomic concentration before and after a-2 min 3 kV Ar^+ etching cycle showed a nonsignificant amount of magnesium on the surface. This was attributed to an insufficient reaction time before heptane rinse. Samples of uncoated paper treated with the MG-3 solution but not subjected to the post-treatment rinse or exposed to CO_2 before rinsing showed significant amount of Mg (up to 20 at%).

The immense amount of material requiring deacidification has led to the development of gas deacidification processes that allow the safer, large-scale treatment of books. In the 1970s the Library of Congress developed a process based on diethylzinc (DEZ), which forms ZnO as a basic buffer in the paper. XPS analysis of thus treated uncoated paper shows up the ZnO compound (modified Auger parameter $\alpha^* = 2010$ eV) and makes it possible to ascertain the almost homogeneous vertical distribution of the buffer in the near-surface region of the paper (MacInnes and Barron 1992).

13.8.4. Dating Manuscript Inks with SAM

Auger spectroscopy has not found specific and significant applications in the study of paper due to its lower ability compared to XPS in providing information on the chemical environment of carbon and oxygen and because of the pronounced charging effects caused by electron bombardment. However, the dating of manuscript inks is one interesting application where typical spatial resolution of SAM can be exploited (McNeil 1984, 1988).

The basic feature of this application of SAM for the dating of manuscript inks is the time-dependent migration of metal ions from the ink. This migration causes the formation of a skewed tail on the ink edge profile beyond the dried capillary drawn ink area that increases with time. Ion migration is assumed to be related to the presence of carboxyl groups in the cellulose-

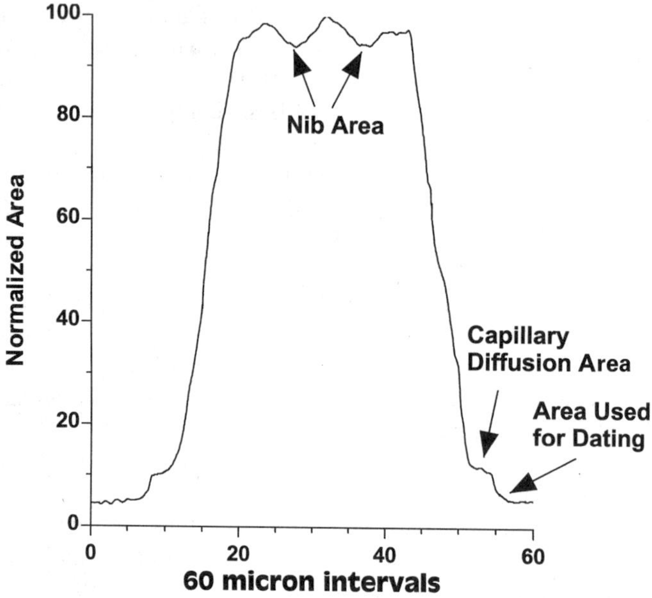

Figure 13.10. Auger lateral scan showing the distribution of iron across the ink profile. The shoulders at either side of the curve are generated by diffusion driven by capillary forces at the surface of the fiber. The area under the tail that shows up beyond the capillary diffusion-induced shoulder is used for calculating the period of time that has elapsed since the ink was deposited on the fiber. [Reprinted, by permission, from McNeil, R. J., in *Archaeological Chemistry III*, J. B. Lambert, Ed., Advances in Chemistry Series 205, p. 255. Copyright © 1984 by American Chemical Society.]

based fibers upon which the ink is deposited. It develops following a form of ion exchange with the concentration gradient (McNeil 1988).

The high lateral resolution of SAM (up to a few hundred Ångstroms) makes it possible to emphasize the distribution of a selected element along the ink profile simply by performing a line scan acquisition. Iron is the most common element used for SAM dating purposes since the wide use of ferrogallic inks in the past (Arpino 1977). A typical profile, possibly taken from fibrils that run perpendicular to the ink line so as to avoid morphological induced alterations of the elemental profile, shows two shoulders at either side of the curve that define the area on which the ink was distributed by the nib of the pen (Fig. 13.10). These shoulders are generated by diffusion driven by capillary forces at the surface of the fiber. The area under the tail that shows up beyond the capillary diffusion-induced shoulder is used for calculating the period of time that has elapsed since the ink was deposited on the fiber.

Accurate sampling methods are required to maximize reliability and reproducibility. A preliminary evaluation of the presence of elements that indicate previous treatment of the document—mainly for deacidification purposes—should be performed, since the suppressive action of such treatment may affect the ion migration process (McNeil 1988).

13.9. STONE DEGRADATION

13.9.1. Introduction

Stone is certainly one of the earliest materials used by humans. A variety of lithic objects have survived decay, while this has prevented objects made of other natural materials, for example, reeds and wood, being attested for their use by early humans. The chemical and physical "resistance" of lithic materials has provided us with the knowledge that flint was used in the Paleolithic period for cutting and pounding. Unfortunately, a variety of physical-chemical factors can render stone artifacts unstable as regards external environments (see Amoroso and Fassina 1983).

13.9.2. XPS in Study of Surface Stone Degradation

Studies concerning the first stages of reactions established between stones and reactive atmospheres have been shown to be greatly improved using XPS for surface analysis and chemical environment detection. The effects of polluted urban atmospheres on marble have been studied by Garbassi et al. (1985) by exposing samples of Carrara marble to the urban atmosphere of Rome (Piazza Colonna). After only 5 days of exposure, low amounts of sulfur (S $2p = 168.3 \pm 0.2$ eV BE) and chlorine (Cl $2p = 198.4 \pm 0.3$ eV BE) were detected (around 1 at %). Traces of lead (Pb $4f7/2 = 139.0 \pm 0.2$ eV BE) were revealed after 30 days' exposure. Due to the high solubility of calcium nitrate in water, provided by rain, no traces of nitrogen were found, even if significant amounts of NO_x pollutants were present in the atmosphere during the exposure (0.1 to 0.2 ppm). Sulfur oxides are common anthropogenic air pollutants that accelerate weathering processes in carbonaceous stones such as limestone and marble. In the presence of humidity, they form $CaSO_4$ (in the hemi-hydrate or bi-hydrate form according to relative humidity value) and CO_2. In the experiment discussed, the deposition of gypsum particles was excluded due to the absence of other common atmospheric particulates such as silica or silicate. A reactive interaction was thus taken into account. The formation of $CaSO_4$ species on the marble surface appeared to be a multistep process because of the discontinuities found in the variation with time of the S/Ca atomic ratio. This showed two different plateaus after a few days' exposure and roughly 150 days' exposure, respectively. The formation process of the crust, which develops on marble after exposure to SO_2 and NO_x pollutants in the atmosphere, has also been studied by Gauri et al (1990), who also hypothesized a three-step process. On the basis of this, Yerrapragada et al. (1994) provided a mathematical model to evaluate the growth rate of the crust. In this context, XPS depth profiling was used to evaluate the thickness of the crust grown on samples of marble exposed to the atmosphere for 3 and 6 months, respectively. They found values that were nearly half of those obtained by using an accurate ion chromatography method, thus confirming the difficulties found in using depth profile analysis to evaluate the thickness of overlayers when sputtering-induced processes are not taken into account.

13.10. FUTURE TRENDS

We have outlined the majority of studies concerning archaeological and/or artistic objects that have used XPS or AES, paying particular attention to the potential of these techniques for chemical state identification and high lateral resolution—a factor that makes a microdestructive approach possible. This potential together with the intrinsic surface-sensitive nature of these techniques is extremely useful when model systems are built to study processes that involve archaeological and artistic materials. They are especially relevant when looking at the initial stages of degradation or when processes that start at the material–atmosphere interface are taken into account. In this context, XPS and AES will provide the most significant contribution to the study of topics of archaeological and artistic interest. Studies of real samples can be properly carried out while taking into account and preventing any processes that may alter the composition of the surface to be analyzed, paying particular attention to the damage induced on the surface by the analytical techniques themselves. The diversity of areas in which XPS and AES can be of use provides ample proof of their value in the field of art and archaeology—a value that has been attested in the past and will continue to be attested in the future.

Acknowledgments

We are indebted to Giovanni Marletta, Joseph B. Lambert, and Alberto Torrisi for comments on an earlier draft of this chapter. Joseph B. Lambert is also thanked for access to as-yet unpublished material. Financial support for some of the studies cited here was provided by the Consiglio Nazionale delle Ricerche, Progetto Finalizzato Beni Culturali.

REFERENCES

Abelson, P. H. (1987), *Science* **238**, 595.

Ahmed, A., A. Adnot, J. L. Grandmaison, S. Kaliaguine, and J. Doucet (1987), *Cellulose Chem. Technol.* **21**, 483.

Allen, G. C., P. M. Tucker, and R. K. Wild (1982), *Philos. Mag. B* **46**, 411.

Amoroso, G. G., and V. Fassina (1983), *Stone Decay and Conservation*, Materials Science Monographs, Vol. 11, Elsevier Science Publisher B.V., Amsterdam.

Arpino, P., J. P. Moreau, C. Oruezabal, and F. Flieder, (1977), *J. Chromatogr.* **134**(2), 433.

ASTM (1991a), *Surf. Interf. Anal.* **17**, 961.

ASTM (1991b), *Surf. Interf. Anal.* **17**, 965.

Auger, P. (1925), *J. Phys. Radium* **6**, 205.

Balter, M. (1999), *Science* **283**, 920.

Beamson, G., and D. Briggs (1992), *High Resolution XPS of Organic Polymers. The Scienta ESCA300 Database*, John Wiley & Sons, New York.

Bertrand, P. A., and P. D. Fleischauer (1980), *J. Vac. Sci. Technol.* **17**, 1309.

Blazej, A., and M. Kosik (1985), in *Cellulose and Its Derivatives*, J. F. Kennedy, G. O. Phillips, D. J. Wedlock, and P. A. Williams, Eds., John Wiley & Sons, New York, 238.

Boardman, J. (1976), *Athenian Red-Figure Wares: The Archaic Period, a Handbook*, Thames and Hudson, London.

Briggs, D., and M. P. Seah, Eds. (1990), *Practical Surface Analysis*, Vol. 1: *Auger and X-ray Photoelectron Spectroscopy*, 2nd ed., John Wiley and Sons, New York.

Brill, R. H. (1963), *Sci. Am.* **209**, 120.

Brisk, M. A., and A. D. Baker (1975), *J. Electron. Spectrosc. Relat. Phenom.* **7**, 197.

Bruno, P., M. Caselli, M. L. Curri, P. Favia, R. Lamendola, A. Mangone, A. Traini, and C. Laganara (1994), *Fresenius J. Anal. Chem.* **350**, 168.

Bruno, P., M. Caselli, M. L. Curri, P. Favia, C. Laganara, and A. Traini (1997), *Ann. Chim. Rome* **87**, 539.

Bunton, C. A., T. A. Lewis, D. R. Lewellyn, and C. A. Vernon (1955), *J. Chem. Soc.* 4419.

Carlyle, L., and J. H. Townsend (1990), in *Dirt and Pictures*, United Kingdom Institute for Conservation, London, p. 40.

Castle, J. E., and R. H. West (1980), *J. Electron. Spectrosc. Relat. Phenom.* **18**, 355.

Chappell, R. A., and C. T. H. Stoddart (1977), *J. Mater. Sci.* **12**, 2001.

Childs, K. D., B. A. Carlson, L. A. LaVanier, J. F. Moudler, D. F. Paul, W. F. Stickle, and D. G. Watson (1995), *Handobook of Auger Electron Spectroscopy*, 3rd ed., C. L. Hedberg, Ed., Physical Electronics, Eden Praire, MN.

Cline Love, L. J., L. Soto, and B. T. Reagor (1980), *Appl. Spectrosc.* **34**(2), 131.

Coad, J. P., M. Gettings, and J. C. Rivière (1975), *Farad. Discuss. Chem. Soc.* **60**, 269.

Committee for Fundamental Metallurgy (1981), *Slag Atlas*, Verlag Stahleisen, Düsseldorf.

Copper, E. (1972), *A History of Pottery*, Longman, London.

Copperthwaite, R. G. (1980), *Surf. Interf. Anal.* **2**(1), 17.

Cumpson, P. J., and M. P. Seah (1992a), *Surf. Interf. Anal.* **18**, 345.

Cumpson, P. J., and M. P. Seah (1992b), *Surf. Interf. Anal.* **18**, 361.

Dajnowski, A., E. Farrell, and P. Vandiver (1992), in *Materials Issues in Art and Archaeology III*, Materials Research Society Symposium Proceecings, Vol. 267, P. B. Vandiver, J. R. Druzik, G. S. Wheeler, and I. C. Freestone, Eds., Materials Research Society, Pittsburgh, PA, p. 609.

Davy, J. (1826), *Philos. Trans. R. Soc. London* **116**, 55.

Dawson, P. T., O. S. Heavens, and A. M. Pollard (1978), *J. Phys. C: Solid State Phys.* **11**, 2183.

Dorris, G. M., and D. G. Gray (1978a), *Cellulose Chem. Tecnol.* **12**, 9.

Dorris, G. M., and D. G. Gray (1978b), *Cellulose Chem. Tecnol.* **12**, 721.

Edney, E. O. (n.d.), in *Acidic Deposition: State of Science and Technology*, Vol. III, Report 19, Final Assesment Report of the National Acid Precipitation Assessment Program, Washington, DC, p. 37.

Einstein, A. (1905), *Ann. Phys.* **17**, 132.

Fischman, J. (1995), *Science* **267**, 614.

Fox, P. G. (1981), *Glass Technol.* **22**(2), 67.

Frost, D. C., A. Ishitani, and C. A. McDowell (1972), *Mol. Phys.* **24**, 861.

Garbassi, F., E. Mello, and M. Laurenzi Tabasso (1985), *Durability Build. Mater.* **3**, 51.

Gauri, K. L., A. R. Punuru, and G. C. Holdren (1990), *Environ. Geol. Water Sci.* **15**, 55.

Gillies, K. J. S., and D. S. Urch (1983), *Archaeometry* **25**, 29.

Gillies, K. J. S., G. P. Moss, and D. S. Urch (1981), *Rev. d'Archéometr.* **3**, 121.

Goffer, Z. (1980), *Archaeological Chemistry: A Sourcebook on Applications of Chemistry to Archaeology*, Chemical Analysis—A Series of Monographs on Analytical Chemistry and Its Applications, Vol. 55, John Wiley & Sons, New York.

Goodway, M. (1987), *Science* **236**, 932.

Graedel, T. E., J. P. Franey, G. J. Gualtieri, G. W. Kammlott, and D. L. Malm (1985), *Corros. Sci.* **25**, 1163.

Hall, F. K. (1974), *Sci. Am.* **230**, 52.

Harris, H. A. (1968a), *J. Appl. Phys.* **39**, 1419.

Harris, H. A. (1968b), *J. Appl. Phys.* **39**, 1428.

Harrison, K., and L. B. Hazell (1992), *Surf. Interf. Anal.* **18**, 368.

Harvey, D. T., and R. W. Linton (1981), *Anal. Chem.* **53**, 1684.

Hodges, H. (1989), *Artifacts: An Introduction to Early Materials and Technology*, Gerald Duckworth, London.

Hofman, S. (1990), in *Practical Surface Analysis*, Vol. 1: *Auger and X-ray Photoelectron Spectroscopy*, 2nd ed., D. Briggs and M. P. Seah, Eds., John Wiley and Sons, New York, pp. 144–199.

Hon, D. N. S. (1979), *J. Polym. Sci. Polym. Chem. Ed.* **17**, 441.

Hon, D. N. S. (1989), in *Historic Textile and Paper Materials II: Conservation and Characterization*, ACS Symposium Series, Vol. 410, S. H. Zeronian and H. L. Needles, Eds., American Chemical Society, Washington, DC, p. 13.

Hong, S., J. P. Candelone, C. C. Patterson, and C. F. Boutron (1994), *Science* **265**, 1841.

Hudson, F. L., R. L. Grant, and J. A. Hockey (1964), *J. Appl. Chem.* **14**, 444.

Hüfner, S. (1995), *Photoelectron Spectroscopy: Principles and Applications*, Springer-Verlag, Berlin.

Ichimura, S., and R. Shimizu (1981), *Surf. Sci.* **112**, 386.

Iijima, Y., N. Niimura, and K. Hiraoka (1996), *Surf. Interf. Anal.* **24**, 193.

Ingo, G. M., and L. Scoppio (1992), *Surf. Interf. Anal.* **18**, 551.

Istone, W. K. (1994), *J. Vac. Sci. Technol. A* **12**, 2515.

Istone, W. K. (1995), in *Surface Analysis of Paper*, T. E. Conners and S. Banerjee, Eds., CRC Press, Boca Raton, FL, p. 235.

Kaloyeros, A. E., and R. M. Ehrenreich (1991), in *Materials Issues in Art and Archaeology II*, Materials Research Society Symposium Proceedings, Vol. 185, P. B. Vandiver, J. R. Druzik, and G. S. Wheeler, Eds., Materials Research Society, Pittsburgh, PA, p. 725.

Kim, K. S., W. E. Baitinger, J. W. Amy, and N. Winograd (1974), *J. Electron. Spectrosc. Relat. Phenom.* **5**, 351.

Kruger, J. (1992), in *Dialogue/89—The Conservation of Bronze Sculpture in the Outdoor Environment*, T. Drayman-Weisser, Ed., National Association of Corrosion Engineers, Houston, TX, p. 135.

Lam, N. Q. (1988), *Surf. Interf. Anal.* **12**, 65.

Lambert, J. B. (1997), *Traces of the Past: Unraveling the Secrets of Archaeology through Chemistry*, Addison-Wesley, Reading, MA.

Lambert, J. B., and C. D. McLaughlin (1976), *Archaeometry* **18**(2), 169.

Lambert, J. B., and C. D. McLaughlin (1978), in *Archaeological Chemistry II*, Advances in Chemistry Series 171, G. F. Carter, Ed., American Chemical Society, Washington, DC, p. 189.

Lambert, J. B., C. D. McLaughlin, and A. Leonard (1978), *Archaeometry* **20**(2), 107.

Lambert, J. B., L. Xue, J. M. Weydert, and J. H. Winter (1990), *Archaeometry* **32**(1), 47.

Lambert, J. B., C. D. McLaughlin, C. E. Shawl, and L. Xue (1999), *Anal. Chem.* **4**(8), 614A.

Larciprete, R., E. Borsella, P. De Padova, M. Fanfoni, M. Mangiantini, and P. Perfetti (1993), *Appl. Surf. Sci.* **72**, 313.

Leventhal, E. T., and M. Thompson (1978), *J. Intern. Inst. Cons. Can. Group* **3**(2), 16.

MacInnes, A. N., and A. R. Barron (1992), *J. Mater. Chem.* **2**(10), 1049.

Maddin, R., J. D. Muhly, and T. S. Wheeler (1977), *Sci. Am.* **237**(4), 122.

Marletta, G., O. Puglisi, S. Pignataro, G. Alberti, and U. Constantino (1982), *Chem. Phys. Lett.* **89**, 333.

McGovern, P. E., and R. H. Michel (1985), *Anal. Chem.* **57**, 1515A.

McGovern, P. E., and R. H. Michel (1990), *MASCA* **7**, 69.

McNeil, R. J. (1984), in *Archaeological Chemistry III*, Advances in Chemistry Series 205, J. B. Lambert, Ed., American Chemical Society, Washington, DC, p. 255.

McNeil, R. J. (1988), *Literary Res.* **13**, 137.

Mellaert, J. (1962), *Anatolian Studies* **12**, 41.

Mills, J. S., and R. White (1994), *The Organic Chemistry of Museum Objects*, 2nd ed., Butterworth–Heinemann, Oxford.

Mitchell, D. F., G. I. Sproule, and M. J. Graham (1990), *Surf. Interf. Anal.* **15**, 487.

Mora, P., L. Mora, and P. Philippot (1984), *Conservation of Wall Paintings*, Butterworths, London.

Moulder, J. F., W. F. Stickle, P. E. Sobol, and K. D. Bomben (1992), *Handbook of X-ray Photoelectron Spectroscopy*, J. Chastain, Ed., Perkin-Elmer, Eden Prairie, MN.

Naguib, H. M., and R. Kelly (1975), *Radiat. Eff.* **25**, 1.

Odlyha, M., R. M. Walker, and W. H. Liddell (1992), in *Conservation of the Iberian and Latin American Cultural Heritage*, preprints of the contributions to the IIC Madrid Congress, September 9–12, 1992, H. W. M. Hodges, J. S. Mills, and P. Smith, Eds., International Institute for Conservation of Historic and Artistic Works, London, p. 104.

Opila, R. L. (1987), *Corros. Sci.* **27**, 685.

Pantano, C. G., D. B. Dove, and G. Y. Onoda (1975), *J. Non-Cryst. Solids* **19**, 41.

Pantano, C. G., D. B. Dove, and G. Y. Onoda (1976), *J. Vacuum Sci. Technol.* **13**, 414.

Paparazzo, E. (1994), *Appl. Surf. Sci.* **74**, 61.

Paparazzo, E., and L. Moretto (1995), *J. Electron. Spectrosc. Relat. Phenom.* **76**, 653.

Paparazzo, E., and L. Moretto (1997), in *Materials Issues in Art and Archaeology V*, Materials Research Society Symposium Proceecings, Vol. 462, P. B. Vandiver, J. R. Druzik, J. F. Merkel, and J. Stewart, Eds., Materials Research Society, Warrendale, PA, p. 53.

Paparazzo, E., L. Moretto, C. D'Amato, and A. Palmieri (1995), *Surf. Interf. Anal.* **23**, 69.

Polak, M., J. Baram, and J. Pelleg (1983), *Archaeometry* **25**, 59.

Pollard, A. M., and C. Heron (1996), in *Archaeological Chemistry*, Royal Society of Chemistry, Cambridge.

Pollard, A. M., R. G. Thomas, and P. A. Williams (1992), in *Dialogue/89—The Conservation of Bronze Sculpture in the Outdoor Environment*, T. Drayman-Weisser, Ed., National Association of Corrosion Engineers, Houston, TX, p. 123.

Puglisi, O., G. Marletta, and A. Torrisi (1983), *J. Non-Crystalline Solids* **55**, 433.

Puglisi, O., G. Marletta, and A. Torrisi (1986), *J. Non-Crystalline Solids* **83**, 344.

Riviere, J. C. (1990), in *Practical Surface Analysis*, Vol. 1: *Auger and X-ray Photoelectron Spectroscopy*, 2nd ed., D. Briggs and M. P. Seah, Eds., John Wiley and Sons, New York, pp. 19–83.

Robbiola, L., and L. P. Hurtel (1997), in *Metal 95, Proceedings ICOMCC Conference Internationale sur la Conservation des Metaux*, Semur en Auxois, France, p. 109.

Robbiola, L., I. Queixalos, L. P. Hurtel, M. Pernot, and C. Volfovsky (1988), *Stud. Cons.* **33**, 205.

Rowlett, R. M. (1999), *Science* **284**, 741.

Sabbatini, L., R. Guascito, F. Palmisano, and P. G. Zambonin (1994), *Sci. Tech. Cult. Herit.* **3**, 9.

Saunders, D., and J. Kirby (1994), *Natl. Gallery Tech. Bull.* **15**, 79.

Schweppe, H., and H. Roosen-Runge (1985), in *Artists' Pigments. A Handbook of Their History and Characteristics*, R. L. Feller, Ed., Vol. 1, Oxford University Press, New York, pp. 255–283.

Scott, D. A. (1990), *J. Am. Inst. Conserv.* **29**, 193.

Seah, M. P. (1990), in *Practical Surface Analysis*, Vol. 1: *Auger and X-ray Photoelectron Spectroscopy*, 2nd ed., D. Briggs and M. P. Seah, Eds., John Wiley and Sons, New York, p. 311.

Seah, M. P., and W. A. Dench (1979), *Surf. Interf. Anal.* **1**, 2.

Seyama, H., and M. Soma (1988), *Application of X-ray Photoelectron Spectroscopy to the Study of Silicate Minerals*, National Institute for Environmental Studies, Onogawa, Tsukuba, Japan.

Sherwood, P. M. A. (1990), in *Practical Surface Analysis*, Vol. 1: *Auger and X-ray Photoelectron Spectroscopy*, 2nd ed., D. Briggs and M. P. Seah, Eds., John Wiley and Sons, New York, p. 577.

Siegbahn, K., C. Nordling, A. Fahlman, R. Nordberg, K. Hamrin, J. Hedman, C. Johansson, T. Bergmark, S. E. Karlson, I. Lindgren, and B. Lindberg (1967), *Nova Acta Regia Soc. Sci. Upsaliensis*, Ser. IV, Vol. 20, Kongliga Vetenskaps-Societeten, Upsala.

Sitholé, B. B. (1995), *Anal. Chem.* **67**, 87R.

Spoto, G., E. Ciliberto, G. C. Allen, C. M. Younes, P. Piccardo, M. R. Pinasco, E. Stagno, M. G. Ienco, and R. Maggi (in press), *Br. Corros. J.*

Strandberg, H., and L. Johansson (1997), *J. Electrochem. Soc.* **144**, 81.

Taglauer, E. (1994), *Surf. Sci.* **299/300**, 64.

Tite, M. S., I. Freestone, R. Mason, J. Molera, M. Vendrell-Saz, and N. Wood (1998), *Archaeometry* **40**(2), 241.

Torrisi, A., G. Marletta, A. Licciardello, and O. Puglisi (1988), *Nucl. Instrum. Methods B* **32**, 283.

Torrisi, A., G. Arena, G. Bellia, A. Contino, G. Falco, L. Grasso, and S. Ingrassia (1996), *Ann. Chim. Rome* **86**, 329.

Turner, N. H., and J. A. Schreifels (1998), *Anal. Chem.* **70**, 229R.

Tylecote, R. F. (1992), *A History of Metallurgy*, 2nd ed., Institute of Metals, London.

Uhlig, H. H. (1971), in *Corrosion and Corrosion Control*, John Wiley & Sons, New York, p. 322.

Vandiver, B. V., O. Soffer, B. Klima, and J. Svoboda (1989), *Science* **246**, 1002.

Vassamillet, L. F., and V. E. Caldwell (1969), *J. Appl. Phys.* **40**, 1637.

Vernon, W. H. J. (1932), *J. Inst. Met.* **49**, 153.

Wagner, C. D. (1975), *Farad. Discuss. Chem. Soc.* **60**, 291.

Wagner, C. D. (1978), *J. Vac. Sci. Technol.* **15**, 518.

Wagner, C. D. (1990), in *Practical Surface Analysis*, Vol. 1: *Auger and X-ray Photoelectron Spectroscopy*, 2nd ed., D. Briggs and M. P. Seah, Eds., John Wiley and Sons, New York, p. 595.

Wagner, C. D., and J. Ashley-Taylor (1980), *J. Electron. Spectrosc. Relat. Phenom.* **20**, 83.

Wagner, C. D., L. H. Gale, and R. H. Raymond (1979a), *Anal.Chem.* **51**, 466.

Wagner, C. D., W. M. Riggs, L. E. Davis, J. F. Moulder, and G. E. Muilenberg (1979b), *Handbook of X-ray Photoelectron Spectroscopy*, G. E. Muilenberg, Ed., Perkin-Elmer, Eden Prairie, MN.

Wagner, C. D., H. A. Six, W. T. Jansen, and J. A. Taylor (1981), *Appl. Surf. Sci.* **9**, 203.

Walker, R. (1980), *J. Chem. Educ.* **57**, 277.

Welsh, I. D., and P. M. A. Sherwood (1989), *Phys. Rev. B* **40**, 6386.

Wertime, T. A. (1973), *Am. Sci.* **61**, 670.

West, R. H., and J. E. Castle (1982), *Surf. Interf. Anal.* **4**, 68.

Whistler, R. L., and G. N. Richards (1958), *J. Am. Chem. Soc.* **80**, 4888.

Whitmore, P., R. G. Cass, and J. R. Druzik (1986), in *Preprints of the 14th Annual Meeting of the American Institute for Conservation of Historic and Artistic Works*, American Institute for Conservation of Historic and Artistic Works, Chicago, IL, p. 114.

Wilson, J. D., and J. K. Hamilton (1986), *J. Chem. Educ.* **63**, 49.

Wilson-Yang, K. M., and G. Burns (1987), *Can. J. Chem.* **65**, 1058.

Wilson-Yang, K. M., and G. Burns (1988), *Can. J. Chem.* **66**, 2348.

Yerrapragada, S. S., J. H. Jaynes, S. R. Chirra, and K. L. Gauri (1994), *Anal. Chem.* **66**, 655.

CHAPTER

14

ELECTRON MICROSCOPY AND ITS APPLICATION TO THE STUDY OF ARCHAEOLOGICAL MATERIALS AND ART PRESERVATION

MIGUEL JOSÉ-YACAMÁN

Instituto Nacional de Investigaciones Nucleares, México City, México Instituto de Física, Universidad Nacional Autónoma de México, México City, México

JORGE A. ASCENCIO

Instituto Nacional de Investigaciones Nucleares, México City, México

14.1. INTRODUCTION

Modern electron microscopy is now widely considered a fundamental tool in the study and characterization of archaeological materials. It has proved to be an invaluable aid for the development of conservation procedures that involve ancient objects, works of art, and buildings. This chapter provides an outline of the main electron microscopy techniques in use alongside examples of their application to the study of ancient materials. The principal areas for discussion are scanning electron microscopy (SEM), conventional transmission electron microscopy (TEM) (including sections on the X-ray microanalysis coupled with these systems), environmental scanning electron microscopy (ESEM), and high-resolution electron microscopy (HREM). We have also included a section on electron energy loss spectroscopy (EELS)—a technique used in TEM. Ample examples are provided to show how these approaches are applied on a practical level, such as reference to the study of ancient pigments used in Mesoamerica (e.g., Maya blue). There will also be an opportunity for the reader to learn more about the application of diffraction techniques associated with the use of the electron microscope and to see how these have a bearing on specific problems. The chapter concludes by

Modern Analytical Methods in Art and Archaeology, Edited by Enrico Ciliberto and Giuseppe Spoto. Chemical Analysis Series, Vol. 155
ISBN 0-471-29361-X

pointing out how important modeling is in achieving a thorough analysis and interpretation of electron microscopy images.

14.2. CHOOSING THE BEST INSTRUMENT

Before starting any work with electron microscopy it is crucial to decide what kind of microscope or which technique will be most useful to solve a specific problem. The first question to ask is what kind of resolution is required to study a given sample. A good general principle to use is that of the "Ocam razor," or in other words "Less is more." A popular Mexican saying states that "you don't kill ants with a cannon." It follows that we may be able to select a technique of limited scope as long as it can solve the problem in question.

The first key parameter to be considered is the resolution, that is, the minimum feature distance required for the images. Figure 14.1 shows the range of resolution of optical microscopy (OM), SEM, and TEM. The kind of objects observable are indicated in the figure. Of course, the techniques overlap in places and other parameters should also be considered.

A second important parameter to be considered is the depth of field. In general, the depth of field (D) of an image can be defined as the vertical distance in which two features in an image stay in focus and can be defined as (Goldstein et al. 1984):

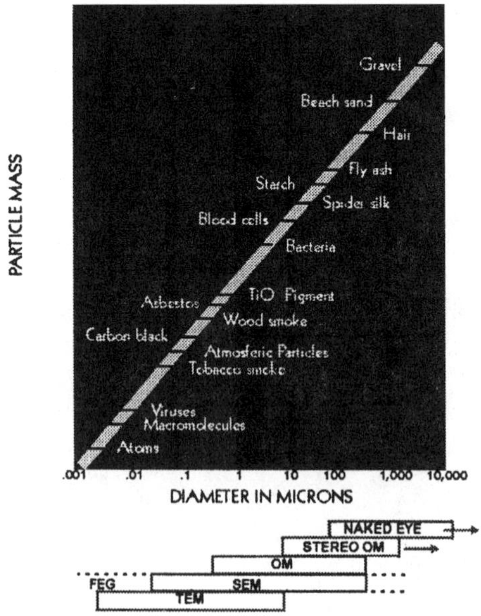

Figure 14.1. Range of different instruments including the human eye.

$$D = \frac{0.2\ \text{mm}}{\alpha M}$$

where α is the beam divergence in radians and M is the magnification.

Therefore, whereas an optical microscope at a magnification of ($\times$) 1200 has a depth of focus of size 6.08 μm, SEM can reach ~100 to 500 μm, thus providing a tremendous gain in depth of field. TEM has the highest resolution but the lowest depth of field (Gabriel 1985).

If chemical analysis information is required in addition to topography, then SEM and TEM will both be useful, although the minimum concentration and particle size to be analyzed will be the key parameter. The minimum particle size that can be analyzed in modern TEM using a field emission gun drops to ~5 Å in diameter (Otten and Coene 1993).

Other important considerations in preservation studies using TEM are sample sensitivity to radiation damage, mechanical stability, and whether the sample is wet or dry. The following sections will discuss the different techniques of electron microscopy with their benefits and limitations, with the aim of choosing the right technique for each sample under study.

14.3. SCANNING ELECTRON MICROSCOPY

When an electron beam irradiates a sample, several types of phenomena may occur. For instance, electrons can be transmitted through the sample. Some of those electrons will be scattered at a well-defined angle. This signal is studied using a transmission electron microscope and will be discussed in the following sections. In SEM images, the signals produced on the incident surface of the sample are what we look at. Corresponding signals can be picked up with an appropriate detector and then used to form an image, as shown in Figure 14.2. In this particular method the beam is scanned through the sample synchronically with the beam raster of a chatodic ray tube (CRT). The signal intensity in the CRT beam is modulated by the intensity of the sample signal. Contrast is produced by variations in the topography and composition of the sample. Each kind of signal generates a different observation technique. Table 14.1 offers a summary of these techniques and illustrates how each one should be used for a different purpose (Jeol Report 1996).

Another relevant factor that should not be overlooked is that each signal comes from a different region on the sample. Each one carries information at a different level. If we assume that we have a thick sample, a spot-sized beam will be effectively blurred by the multiple scattering of the electrons in the sample. This, of course, will change according to the acceleration voltage,

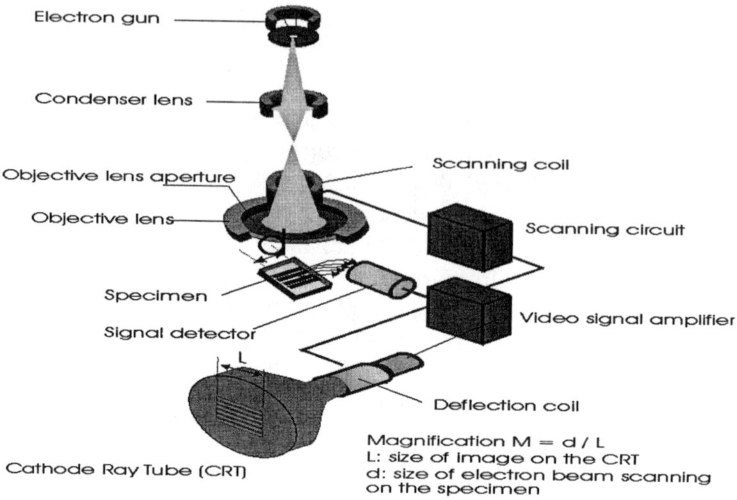

Figure 14.2. System diagram of a scanning electron microscope.

Table 14.1. Uses of Different Types of SEM Signals

Signal	Mode of Operation	Purpose Information Carried
Secondary electrons	SEI	Topographical observation of surface
Backscattered electrons	BEI	Compositional observation of surface
X-rays	X-ray	Elemental analysis of sample
Transmitted electrons	TEI	Internal structure observation
Cathodo luminiscence	CL	Internal characteristics observation
Electromotive force	EBIC	Internal characteristics observation
Secondary electrons or	ECP	
backscattered electrons	MDI	Magnetic domain observation

incidence angle, and sample kind. It is thus always advisable to calculate the beam-spreading effect, that is, the effective beam size *d*. This is especially important in the case of X-ray analysis where misleading chemical information may be picked up from regions that are not of specific interest (Shimizu and Mirata 1971). Commercial software exists for the calculation of beam spreading. One of the most useful techniques in SEM is secondary electron image (SEI). Secondary emission is strongly dependent on the incidence angle of the electrons (Fig. 14.3). These are produced at lower energy and come from areas close to the surface. The area in which these electrons are generated is so close to the electron spot diameter.

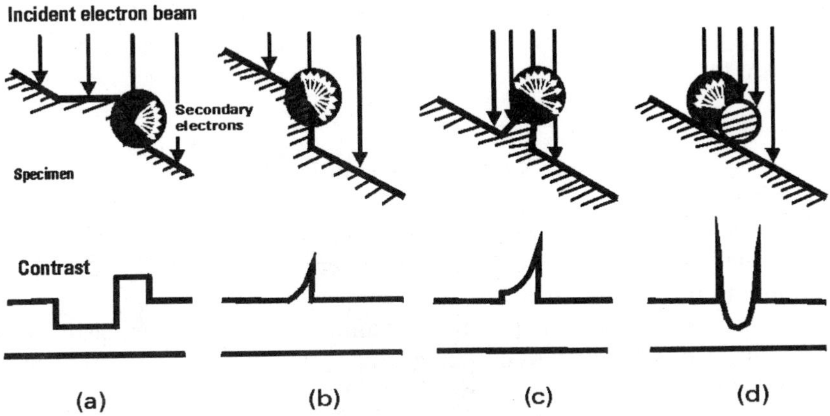

Figure 14.3. Secondary emission dependence of the incidence angle.

The resolution (δ) in SEM is related to the spot size in the ratio:

$$\delta \sim d_0$$

Maximum resolution is therefore limited by the minimum spot size that can be achieved. Lens aberrations may affect the probe. The aberrations are diffraction d_D, chromatic d_C, and the spherical aberration d_S.

Total beam degradation is thus

$$D_0 = d_S + d_C + d_D \approx 8 \text{ nm}$$

The highest resolution images will be those produced using the secondary electron signal. Sensitivity to the incidence angle also results in a greater depth of field. This technique is ideal for the study of surface feature, roughness/grain size evaluation, studies of inclusions, and the like.

Backscattered electrons, on the other hand, leave the surface with a very high energy, close to the incident beam, and are less influenced by stray fields in the SEM chamber. In addition to topographical information, backscattering can also provide important chemical information. This is because the backscattering effect is roughly proportional to the atomic scattering factor. The higher the atomic number the greater the intensity of backscattered electrons. Solid-state detectors used in modern microscopes can pick up small differences in atomic number, to give stark compositional contrast. Arrays of symmetrically located detectors can be used to select between topographic and compositional contrast. The subtraction, summation, or the derivative of the images can be used to improve the contrast. An example of this is shown in Figure 14.4, which is an image of a sea shell, showing the secondary and backscattered electron image. It is clear that in this case the topography is best shown by the backscattered electrons.

14.4. X-RAY ANALYSIS IN SEM

The general principles of X-ray analysis apply to both SEM and TEM and will therefore be discussed together. The differences between X-ray analysis in both machines will also be discussed.

One of today's most important techniques in studying archaeological materials is the X-ray analysis (see also Chapter 4). X-rays are produced when the incident electron beam interacts with the inner shell electrons of the atom. The ejected electron produces a hole that is filled by an electron of the outer shells. The transitions produce X-rays with characteristic energy, that is, specific energy for each element (cf. Fig. 4.3). We must consider that X-rays are also produced by the interaction of the electron with the Coulomb field of the atom giving rise to white or bremsstrahlung radiation. It is essential to characterize this radiation for the purpose of quantitative analysis. Two main types of detectors have been used to analyze X-ray energy: energy-dispersive system (EDS) and wavelength-dispersive system (WDS). EDS was originally developed for nuclear physics, and it is very popular in modern microscopy. The system uses a Si(Li) detector, which in many cases has a beryllium window (see also Section 4.3.2). The detected signal can be also displayed as maps of elements. Figure 14.5 shows an example of elemental mapping in a sample of atmospheric particles from Mexico City.

For a quantitative analysis many physical effects have to be taken into account. These include white radiation, double ionization peaks, overlaps of peaks from other elements, and beam spreading. Most manufacturers of EDS equipment provide reliable software for standard analysis. The detection limits of this technique are about 1% in weight for elements from C to Si and 0.1% for elements with $Z > Z_{Si}$. The other detection method for quantitative analysis is WDS. In this case, a crystal spectrometer analyses the wavelength (λ) using Bragg diffraction. The diffracted X-rays are detected by a proportional counter. The signal is amplified and converted to a standard size pulse in a single-channel analyzer and counted. Very stringent constraints are placed on the sample/spectrometer geometry by the necessity to satisfy Bragg's law. This kind of detector can only count the elements sequentially by varying the position of the crystal detector. To cut down on analysis time, it is necessary to have at least two detectors. Some instruments where chemical analysis is the principal activity incorporate up to four spectrometers and computer-controlled crystal movement. WDS is a more accurate technique for quantitative analysis (as far as flat surfaces are con-

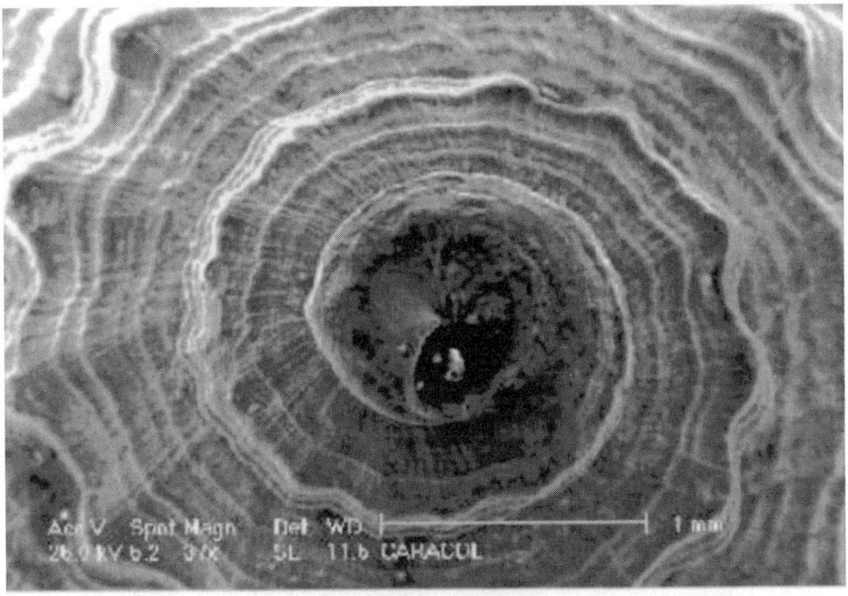

Secondary Electrons

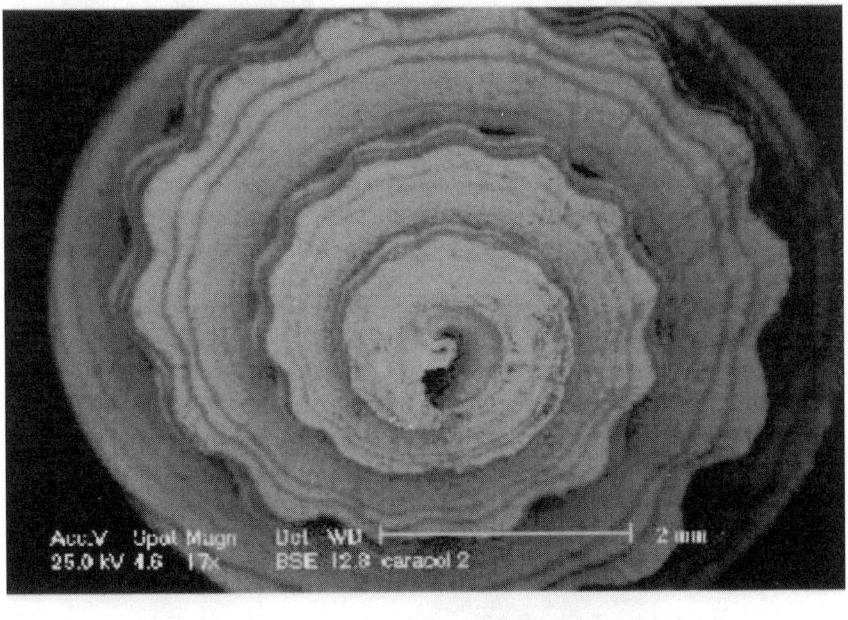

Backscattering Electrons

Figure 14.4. SEM image of a sea shell indicating the differences between secondary and backscattering electron images.

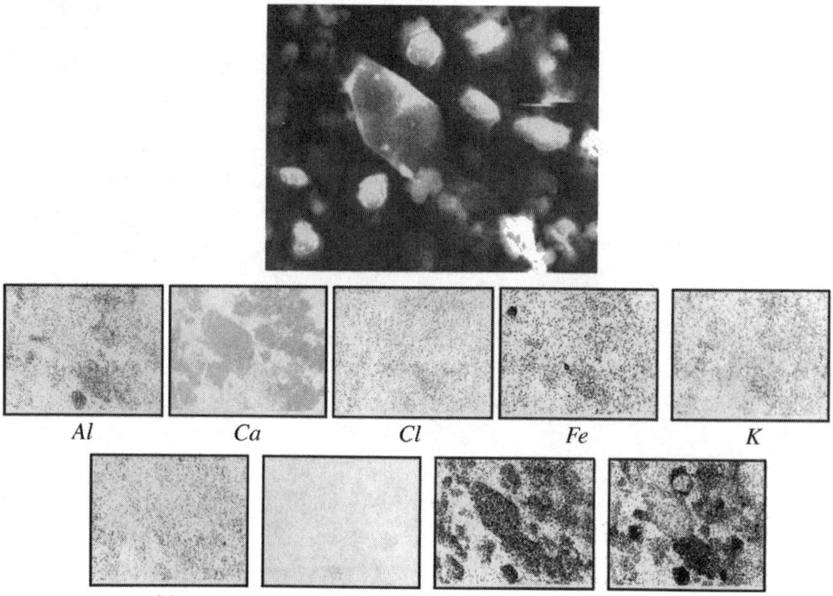

Figure 14.5. Image and elemental mapping from an air sample of an urban area that contains pollutant particles.

Table 14.2. X-Ray Fluorescence Analysis

Performance	EDS	WDS
Measurable element range	B–U	$z = 5–92$
	Be–U	$z = 5 - Z = 92$
Energy resolution	$E \sim 150$ eV	$E \sim 20$ eV
Detection limit	1500–200 ppm	5–100 ppm
Qualitative analysis	Fast	Slow
Counting rate	~2000 cts/sec	~50,000 cts/sec
Signal-to-noise ratio	Low	High
Efficiency of results and cost	Low	High

cerned) and can detect much lower concentrations than 0.01 wt%. It is also a better technique for lighter elements. There are, however, several drawbacks to WDS that are summarized in Table 14.2.

14.5. ENVIRONMENTAL AND LOW-VACUUM SEM

Two techniques that are now becoming increasingly important in today's study of archaeological and historical materials are environmental scanning electron microscopy (ESEM) (ElectroScan Corporation 1996; Danilatos 1998) and low-vacuum scanning electron microscopy (LVSEM) (Jeol Report 1994).

In many cases it is necessary to observe samples that are oily or wet or samples that are vacuum unfriendly and will contaminate the microscope. In addition to this, archaeology samples may not be sufficiently stable to be placed in a vacuum chamber. Finally, many of the samples in archaeology are poor conductors and must be coated with a metal film in order to be observed. These problems can all be overcome using LVSEM and ESEM.

Let us first consider the charging effect in SEM. Secondary electrons leaving a nonconductive sample generate an excess of positive charges on its surface. This will in turn deflect the electrons traveling to the detector resulting in a blurred image. An equally undesirable effect can be prevented by covering the sample with an evaporated metal film that renders its surface electrically conductive.

However, even if the coating is sufficiently thin, it has an internal structure that will interfere with the sample features. In the case of LVSEM the sample chamber of the microscope is maintained at a much higher pressure than the rest of the column. The sample is typically placed in an environment with a ~0.01 to 1 Torr gas pressure (or even much higher in ESEM). The rest of the column is set at 10^{-2} Torr. The gas molecules are ionized by the incident and secondary electrons and results in the neutralization of charge excess on the sample. This allows the noncoating of the sample and no preparation is required. This effect is also an advantage for X-ray detection. Due to problems with detector performance in the case of LVSEM, only the backscattering signal is clearly observed. In an ESEM machine both SEI and BSE are clear. The pressure operation range of both machines are shown in Figure 14.6. It can be seen from this figure that water and oil remain stable under observation.

In the case of water, an ~4.6-Torr pressure is required (at a temperature of 0°C) to maintain the liquid state. As far as archaeological materials are concerned, it is important that the researcher knows the thermodynamic

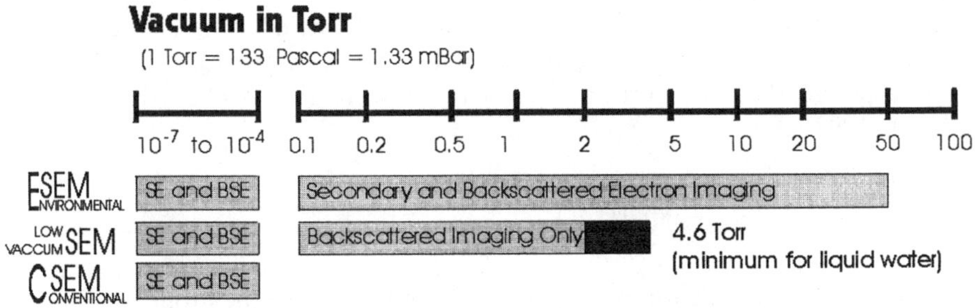

Figure 14.6. Working of different types of environmental and conventional microscopes.

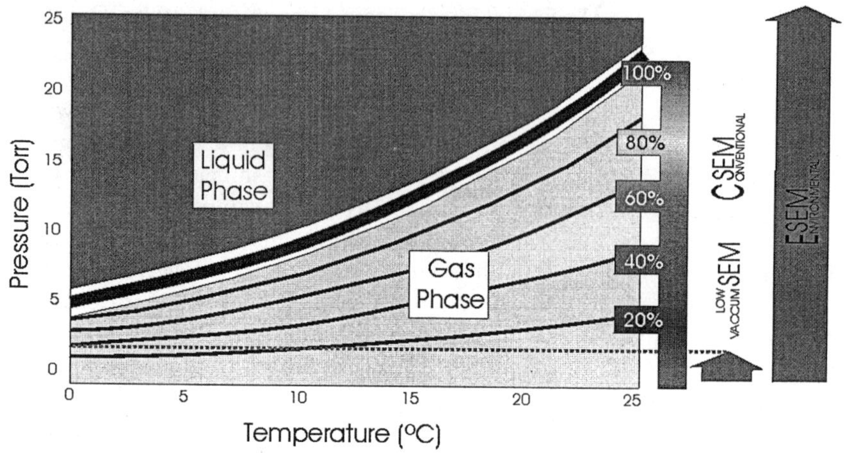

Figure 14.7. Relative humidity isobars for low-vacuum and environmental SEM. The minimum pressure that can sustain water in the liquid phase is ~4.6 Torr at 0°C.

data of the sample to be studied. The relative isobars (plots of pressure vs. temperature) for a sample containing different humidity is plotted in Figure 14.7. This kind of plot is very useful for understanding the work and observation conditions of a sample.

14.6. TRANSMISSION ELECTRON MICROSCOPY: GENERAL CONSIDERATIONS

The following section discusses the use of transmission electron microscopy (TEM) in characterizing archaeological materials. The discussion initially outlines the basic principles underlying TEM that are useful for understanding the contrast mechanism produced on the microscope. First, we assume that the incident electron beam wave $\psi_0(x, y)$ phases through the crystal and becomes an exit wave ψ_{exit}. The amplitude does not change, only the wave phase changes in a way that reflects the variations on the electrostatic potential $\varphi(x, y)$ of the sample. The exit wave can thus be written (Zou et al., 1993):

$$\psi_{exit}(x, y) = e^{-i\sigma N_Z \varphi(x,y)}$$

where σ is the interaction parameter of the electron with the atom:

$$\sigma = \frac{2\pi m e \lambda}{h^2}$$

N_Z is proportional to the thickness of the sample and $\psi(xy)$ reflects potential distribution on the sample. This is an approximation and is called the "phase object approximation." The electron wave can be expressed simply in terms of a Taylor series, as follows:

$$\psi_{\text{exit}}(x, y) = 1 - i\sigma N_z \varphi(x, y) - \frac{\sigma^2 N_z^2}{2} [\varphi(x, y)]^2 + i\frac{\sigma^3 N_z^3}{6} [\varphi(x, y)]^3 + \cdots$$

The exiting waves travel in different directions. All the waves will be focused in the back focal plane of the objective lenses. The wave function on the back focal plane $\hat{\psi}_{\text{exit}}(\mathbf{u})$ is the Fourier transform of $\psi_{\text{exit}}(x, y)$ and is given by:

$$\hat{\psi}_{\text{exit}}(\mathbf{u}) = f[\psi_{\text{exit}}(xy)]$$

where f is the Fourier Transform and becomes

$$\hat{\psi}_{\text{exit}}(\mathbf{u}) = \delta(\mathbf{u}) - i\sigma N_Z f[\psi_{\text{exit}}(xy)] - \frac{\sigma^2 N_Z^2}{2} f[\psi_{\text{exit}}(xy)]^2 + \cdots$$

Thus it may be proved that

$$\hat{\psi}_{\text{exit}}(\mathbf{u}) = \delta(\mathbf{u}) - iF^{\text{I}}(\mathbf{u}) - \frac{F^{\text{II}}(\mathbf{u})}{2} + i\frac{F^{\text{III}}(\mathbf{u})}{6} \cdots \tag{14.1}$$

where

$$F^{\text{I}}(\mathbf{u}) = \frac{\lambda\tau}{\Omega} F(\mathbf{u})$$

$$F^{\text{II}}(\mathbf{u}) = \frac{(\lambda\tau)^2}{\Omega} F(\mathbf{u}) \cdot F(\mathbf{u})$$

$$F^{\text{III}}(\mathbf{u}) = \frac{(\lambda\tau)^3}{\Omega} F(\mathbf{u}) \cdot F(\mathbf{u}) \cdot F(\mathbf{u})$$

where τ is the thickness of the sample, Ω is the volume of the unit cell, and $F(\mathbf{u})$ is the structure factor defined as:

$$F(\mathbf{u}) = \sum_{j=1}^{N} \rho_j e^{2\pi i \mathbf{u} \cdot \mathbf{r}_j}$$

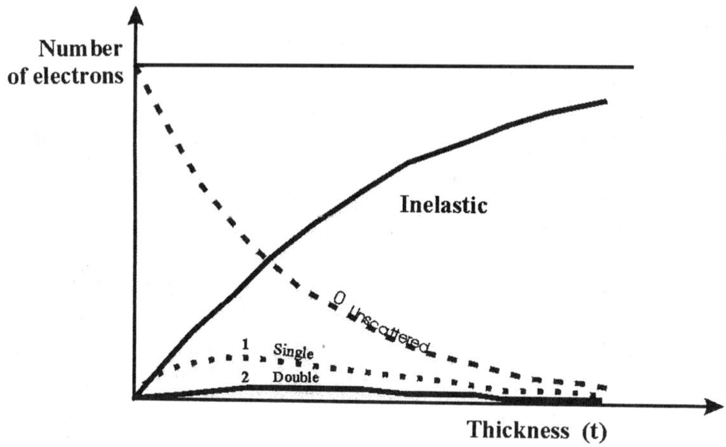

Figure 14.8. Proportions of different electrons as a function of crystal thickness showing unscattered (0), singly scattered (1), and doubly scattered (2) electrons.

where ρ_j is the atomic scattering factor. This is a quantity that has been calculated (Doyle and Turner 1968) and values can be found in the literature. In Eq. (14.1) the terms represent the single, double, and triple scattering of electrons by the sample. Figure 14.8 shows the fraction of electrons that is unscattered by a sample and the fraction with single and double diffraction as a function of thickness. The inelastic fraction grows in importance as thickness increases and is at the basis of electron energy loss spectroscopy (EELS). However, it is clear that for thin samples single scattering is the dominant factor, and the contributions from double and higher scattering terms can be ignored. Therefore, ψ_{exit} can be written as:

$$\hat{\psi}_{\text{exit}}(\mathbf{u}) = \frac{\lambda\tau}{\Omega} e^{-i\pi/2} F(\mathbf{u})$$

This approximation is basically to formulate the linear approximation and is called the weak phase object approximation (WPOA). It means that the exit wave will be shifted by $-90°$ regarding the structure factor and as follows:

$$I(\mathbf{u}) \sim |\hat{\psi}(\mathbf{u})|^2 \sim |F(\mathbf{u})|^2$$

The intensity of the diffraction pattern spots is given by the structure factor. This is also known in the literature as the kinematical approximation. However, in the microscope the lenses can introduce the additional phase. The defocusing of the objective lenses can add other phase shifts. These phase shifts can then be used to increase the contrast for a thin sample representing a thin weak phase object for lenses with a spherical aberration C_S. The change of phase due to defocus can be given by (Cowley 1981):

$$\chi(\mathbf{u}) = \pi\Delta f \lambda \mathbf{u}^2 + \frac{\pi C_s \lambda^3 \mathbf{u}^4}{2}$$

We can now consider the diffraction effect of the objective aperture as:

$$A(u) = \begin{cases} 1 & \text{If } u \text{ is inside the aperture} \\ 0 & \text{If } u \text{ is outside the aperture} \end{cases}$$

The chromatic aberration caused by lens instabilities can be modeled as a spread function:

$$E_D = e^{-(1/2)\pi^2 \Delta f \lambda^2 \mathbf{u}^4}$$

The total effect on the wave is thus

$$\psi_{\text{image}}(\mathbf{u}) = A(\mathbf{u})E_D = e^{i\chi(\mathbf{u})}\Sigma\psi_{\text{exit}}$$

Under this effect it can be shown that the intensity of the image may finally be expressed as:

$$I_{\text{Image}}(\mathbf{u}) = D(\mathbf{u})2F^{\text{I}}(\mathbf{u}) \sin\chi(\mathbf{u})$$

If we take double scattering into account, a more general expression is

$$I_{\text{Image}}(\mathbf{u}) = D(\mathbf{u})[2F^{\text{I}}(\mathbf{u}) \sin\chi(\mathbf{u}) - F^{\text{II}} \cos\chi(\mathbf{u})]$$

When defocusing Δf occurs so that $\chi(\vec{u}) = n\pi/2$ for all **u**, we obtain

$$\sin\chi(\mathbf{u}) = \pm 1 \quad \text{and} \quad \cos\chi(\mathbf{u}) = 0$$

which eliminates the double scattered signal, and the intensity becomes

$$I_{\text{Image}}(\mathbf{u}) = \pm 2D(\mathbf{u})F^{\text{I}}(\mathbf{u}) = \pm\frac{2\lambda t}{\Omega} D(\mathbf{u})F(\mathbf{u})$$

The defocus conditions for which this situation holds are approximated by the Scherzer focus given in Buseck et al. (1988):

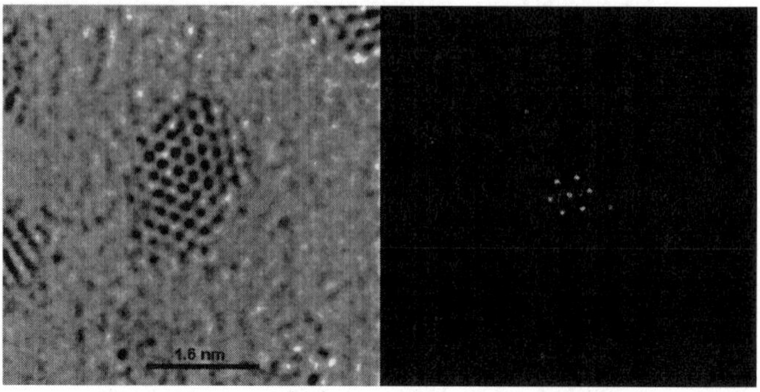

(a) (b)

Figure 14.9. (*a*) Image of a gold particle taken at Scherzer focus for a Jeol-4000EX microscope ~402 Å in which the black dots represent an atomic columns. (*b*) Fourier transform of the particle showing the fcc structure.

$$\Delta f = -\left(\frac{4C_S\lambda}{3}\right)^{1/2}$$

for a typical high-resolution microscope such as the JEOL4000EX, $\Delta f = -405$ Å. In these focusing conditions black dots on the image represent the projection of a row of atoms. In this case, the image may be interpreted in terms of the crystal structure. The Scherzer condition represents the case where the double scattered electrons do not contribute to the image. Figure 14.9*a* shows an image of a gold particle where the dark atom contrast is shown. In this kind of image the fast Fourier transform obtained from the photographic plate can be used to determine the crystal structure. In the case of Figure 14.9*b*, a face-centered cubic (fcc) crystal can be seen with a $\langle 110 \rangle$ axis parallel to the electron beam.

Recently, however, there have been theoretical advances that allow information to be obtained from pictures that were not taken at the correct focus. The reader s referred to the work of Hovmoller's group in Sweden (Zou et al. 1993) and Van Dick's group in Belgium (Van Dick and De Jong 1992) who have discussed these subjects in detail. It is now possible to obtain images that can be interpreted beyond the resolution of the microscope up to what is called its "information limit."

In modern microscopy it is very important to calculate images and compare them with experimental ones. In most practical cases the sample is mounted on a support such as in the case of metal particles in ancient pigments that are supported on clay. In this case the visibility of the particle will be limited by the support. An example of this is shown in Figure 14.10

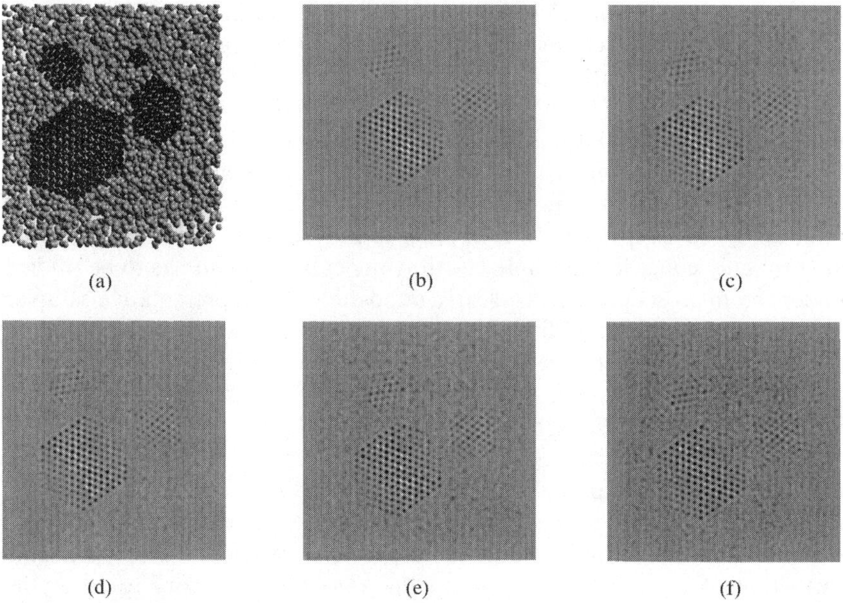

Figure 14.10. Calculation of the contrast of gold particles supported on amorphous carbon. (*a*) Shows the model, (*b*)–(*f*) show the particles for carbon thickness of 25, 75, 130, 200, and 400 Å, respectively.

in which the image of a gold particle supported on carbon is shown. Four differently sized particles on substrates with different thickness are shown. Figure 14.10*a* shows the model used in the calculation (José-Yacaman et al. 1998); Figures 14.10*b* to 14.19*f* show images of the particles on carbon substrates 2.5, 7.5, 13, 20, and 40 nm thick, respectively. It is clear that the substrate produces a number of effects, for example, for thickness >40 nm the smaller particles are not visible. On the other hand, the contrast varies with the thickness of the substrate. In fact, it is clear that the contrast of atomic lines changes from black to white. So in many cases particle contrast in the Scherzer focus will not show the typical contrast.

14.7. SAMPLE PREPARATION

Sample preparation is one of the most critical aspects of obtaining adequate results in electron microscopy. It is impossible to obtain good results with a poorly prepared sample. For the electron microscopist involved in studying archeological samples or pieces of great historical value, the task is often more difficult because the pieces may be unique and impossible to reproduce.

Unfortunately, electron microscopy is a destructive technique. Often samples cannot be obtained from historical pieces without causing some damage.

The high magnification of the microscope does, however, allow the use of very small samples. This means that observing a 1-cm sample at a regular magnification of $\times 100,000$ is like being able to observe 1 km of sample under the microscope or even 10 km when observing at a higher magnification ($\times 1,000,000$). This also illustrates the great problem of assuring the sample to be analyzed is representative. In order to fully analyze a mural of 1 m^2, an area roughly equal to the whole North American continent has to be studied under the microscope. This is clearly impossible and representative samples have to be chosen. An additional complication is that we frequently have to deal with samples that have been restored and contain materials mixed with those originally present. It is therefore necessary to have some preliminary information on the history of the piece to be analyzed.

In the case of regular SEM examination many of the samples will be nonconductive. Examples of this are ceramics and some pigments. When the electron beam strikes an insulator, the electrons that are absorbed accumulate on the surface, and this builds up a space-charge region that prevents the formation of high-quality images. In this case it is necessary to cover the sample with a metallic layer. The most common metal used is Au–Pd. This layer should be applied using a vacuum evaporator ($\sim 10^{-5}$ Torr). The sample should be rotated during evaporation to ensure homogenous coating. It is important that the grains of the metal are <10 nm in size to avoid interference with the topographic features in the sample. In the case of metals this is not required, and it is simply necessary to have good electrical contact between the sample and the specimen holder. This may be achieved using a colloidal Ag paint. In the case of coated samples the best results are obtained by first applying a carbon coating of ~ 3 nm in thickness and then a metallic layer.

An additional problem is the presence of hydrocarbons in the sample. The electron beam causes the hydrocarbons to crack, resulting in carbon deposition on the sample. The image thus deteriorates. Thorough cleaning of the sample before examination is therefore essential. Techniques used to degrease the sample may include cleaning with adequate solvents in ultrasonic bath, mechanical brushing, and, in extreme cases, chemical etching. However, in most practical cases cleaning with acetone or alcohol in an ultrasonic cleaner is often all that is necessary to produce good results.

If the main purpose of the examination is X-ray microanalysis, the sample should be as flat as possible to avoid analysis errors due to differences in local inclination, which introduce statistical incertainty. This means that the sample should be mechanically polished to produce a flat surface. This is possible when the sample is a bulk metal or insulator.

Sample preparation for TEM is a little more complicated and time consuming. However, success in TEM observations depends a great deal on good sample preparation. The most simple case is when the sample is a powder with small grain size. In this case a solution of the sample is prepared using alcohol or acetone. A drop of this solution is then placed on a copper grid covered with a very thin layer of carbon. Most of the time there will be areas that are thin enough to be observed under the microscope. It is sometimes useful to grind the powder using a ball mill. As a rule it is always possible to obtain a very small area of the sample left undamaged by the milling process.

Another case is when the samples are metals in bulk form. In this case thinning of the sample to a thickness of $\sim$100 Å should be achieved. The most common method used to achieve this is to thin the sample electrolytically using a proper acid solution until a hole opens in the center of the sample. When the thinning process is slow enough, there will be regions thin enough for observation around the hole. To ensure the mechanical stability of the sample, it is important to cut the sample to the same size of the copper grids. This is achieved by commercially available mechanical punch instruments. The use of dimpling devices is also recommended for optimum sample preparation. It is usually possible to thin the sample to about 10 μm. Final electropolishing may also enhance the quality of the sample.

In the field of art and archeology studies it is often necessary to examine ceramic materials, fibers, polymers, or particulate material that is not easy to grind. When this is the case, the best method is ion milling. In this technique the specimen is bombarded with an ion or neutral atom beam that brings out neutral and/or charged species from the specimen. An ion beam milling apparatus is an essential tool in the TEM lab. This technique usually produces high-quality samples.

However, before thinning, the sample should be embedded in an epoxy resin. Before being cured and mixed with the sample, the epoxy should be forced into a brass cylinder. After curing, the sample should be cut into thin disks using a diamond knife or an ultra microtome. The disk is now ready for ion milling. This allows the sample to be observed at high resolution. This method can also be used to obtain cross sections of the sample, which are often important in the case of analyzing paints.

14.8. IMAGE PROCESSING

Image processing is another key technique related to modern microscopy. It is now possible to apply methods that improve the information that can be obtained from an image. The steps for image processing are shown in

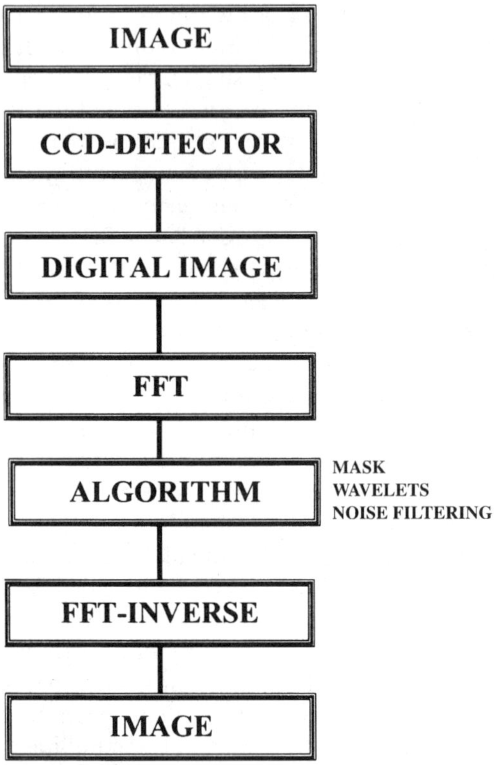

Figure 14.11. Block diagram showing the steps in image processing.

Figure 14.11. Once the fast Fourier transform (FFT) of the digital image is obtained, then a filtering process, which blocks a portion of the FFT, is applied. An FFT is now again applied. Thus the original image, now formed by a limited number of frequencies, is restored. This procedure allows the separation of different kinds of information.

A variety of filters can be used. Figure 14.12 shows an example of the image processing of a nonperiodic object. Figure 14.12*a* shows the object while Figure 14.12*b* shows the FFT of the object. Figures 14.12*c* and 14.12*d* show the reconstruction of the original image using a gradient filter and a low-pass filter. The former results in an image that shows the edges of the object, the latter produce an image with smeared-out edges.

Figure 14.13 shows an example of relevance to the study of archaeological materials regarding the application of image processing to a palygorskite clay (de Pablo 1996; Galan and Castillo 1984). The sample was obtained from an ancient paint of the Aztec culture (ca. 1400 A.D.). Figures 14.13*a*

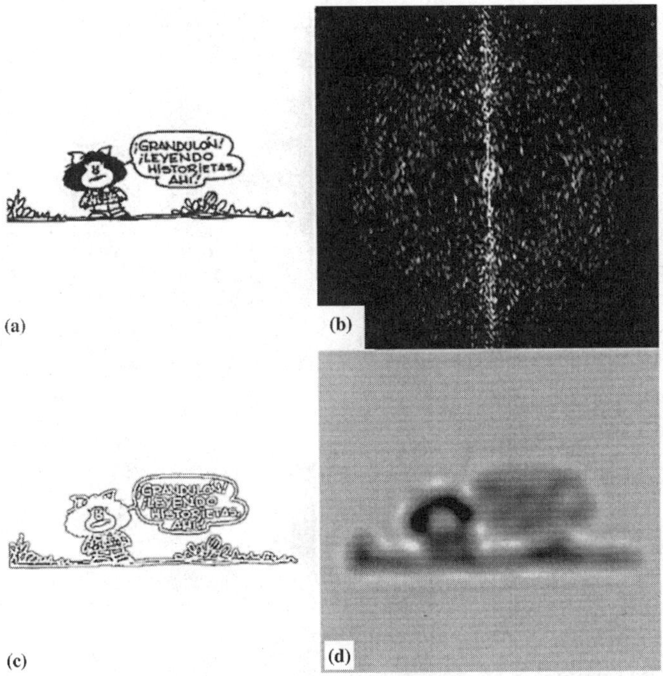

Figure 14.12. Example of the image processing of a nonperiodic object. (*a*) The object. (*b*) The Fourier transform. (*c*) Processed image using a gradient filter. (*d*) Image using a low-pass filter.

and 14.13*b* show the original image and corresponding FFT, respectively. Figure 14.13*c* shows a filter formed by an array of circular apertures with smooth borders that avoid edge effects. This is a kind of multiple Bragg filter that enhances the structure and allows the closer study of the periodicity of the channels present in the considered structure (see latter sections). The image in Figure 14.13*d* shows the resulting image produced using the filter of Figure 14.13*c*. The channel structure on the clay can be seen in this kind of image. The image of Figure 14.13*e* is obtained by blocking the central beam, which produces a kind of dark-field image in which the clay channels are visible. For comparison purposes Figure 14.13*f* shows an image in which all the information from the crystal structure is blocked. This shows the non-periodic features on the image.

14.9. ELECTRON DIFFRACTION

One of the key techniques in the study of ancient materials is electron diffraction. This technique allows the identification of compounds that are

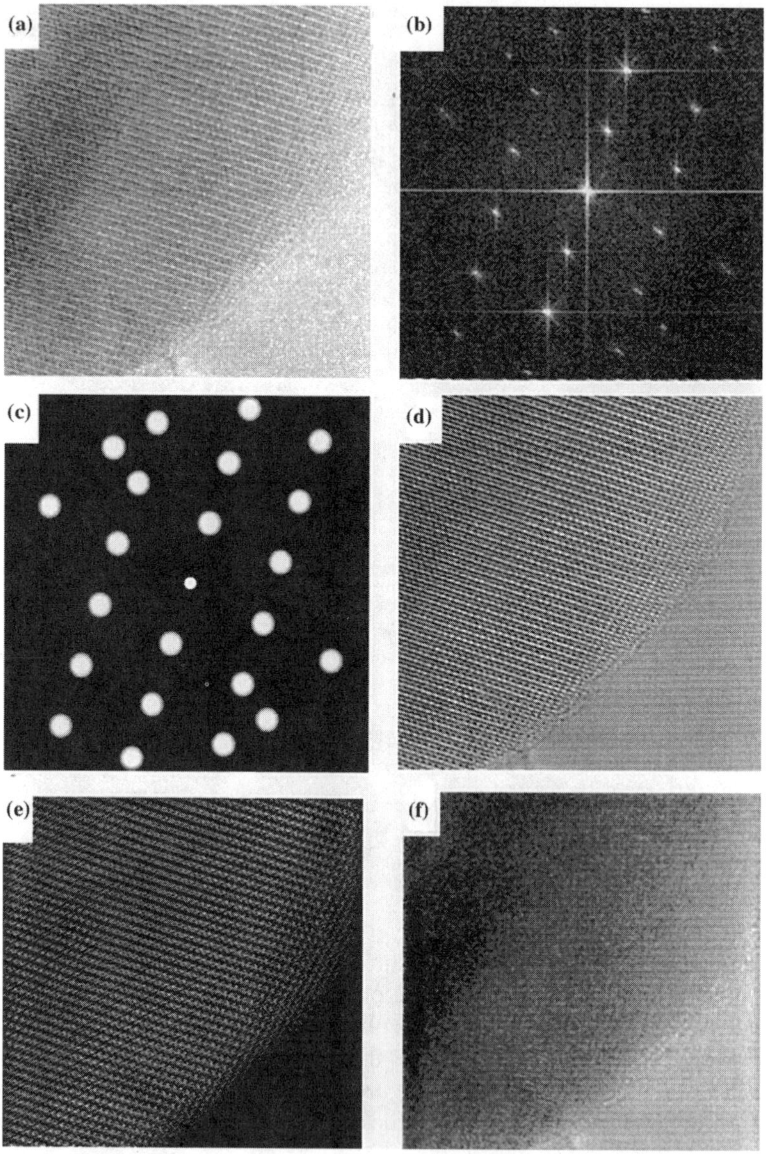

Figure 14.13. Paligorskite fibers obtained from a sample of paintings from the Aztec Templo Mayor ~A.D. 1450 (*a*) Original high-resolution image. (*b*) Fourier transform of the image. (*c*) Filter used on the image. (*d*) Resulting image, which enhances the channel structure of the clay. (*e*) Image produced by blocking the central beam, which also enhances the chanels. (*f*) Image in which the crystal information is blocked and only the nonperiodic information is present.

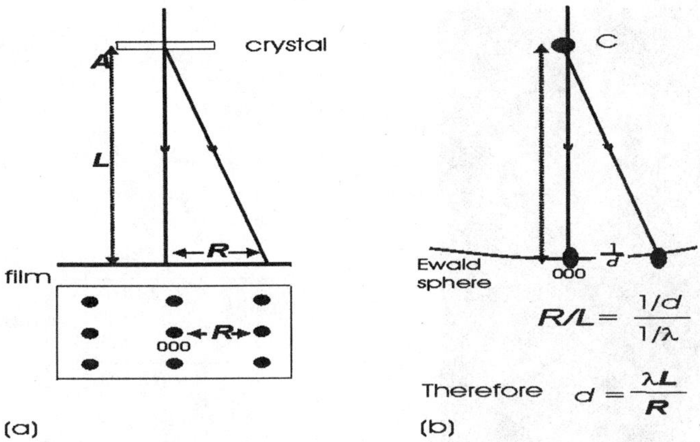

Figure 14.14. (*a*) Geometry of the electron diffraction patterns showing the Ewald sphere intersection with the reciprocal lattice. (*b*) Geometry of the camera constant λR, which allows the measurement of interplanar distances.

otherwise not easy to characterize. If the aim of this kind of study is to determine the structure and composition of an unknown material, then X-ray diffraction should be used. In this technique the scattering is kinematical, and theory can predict the observed intensities. However, in many cases, when studying archaeological or historical materials, the amount of material is very scarce. X-ray diffraction (at least using conventional sources) will not give information above the noise threshold, and, in this case, electron diffraction is the most suitable technique. Figure 14.14 shows the geometry of diffraction pattern formation in a thin crystal. Diffraction is produced by planes that are parallel to the electron beam. The lattice distance in the real crystal d is related to the measured distance as:

$$d = \frac{\lambda L}{R} = \frac{2\lambda L}{D} \tag{14.2}$$

where R is the measured distance between spots or $D = 2R$ in the magnified diffraction pattern, λ is the electron wavelength, and L is the distance traveled by the electrons (camera length). This formula can be used to determine interplanar distances. By analyzing the systematic absences in the pattern, it is possible to determine the structure. The most suitable samples for the determination of unknown structure are the polycrystalline. In principle, all the orientations and rotations on a plane are present in a polycrystalline sample. The spot thus becomes rings, as shown in Figure 14.15*a*. This figure shows an indexed gold ring pattern. Every ring is associated with a lattice plane.

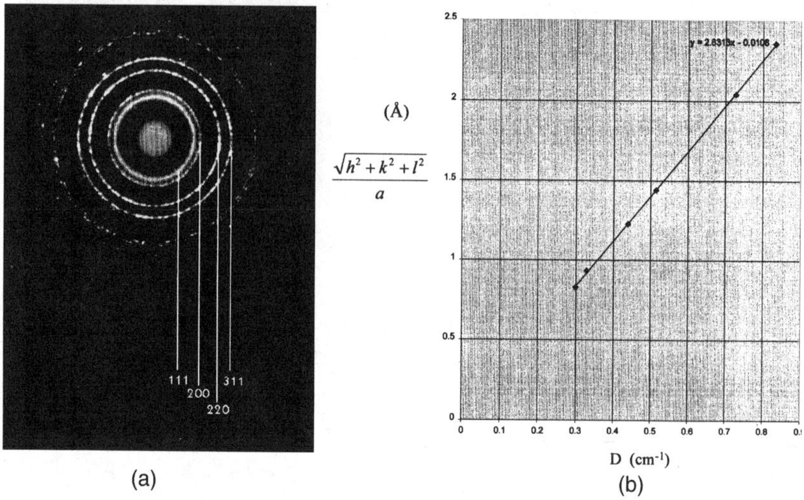

Figure 14.15. (*a*) Diffraction pattern of polycrystalline gold film. (*b*) Plot of $\sqrt{h^2+k^2+l^2}/a$ vs. D for the camera constant calibration of a Jeol-4000EX microscope.

This material is ideal to determine the parameter $2\lambda L$, which is called the *microscope constant*. This is a parameter that should be calibrated on every instrument so that Eq. (14.2) can be written as:

$$D = C\frac{\sqrt{h^2+k^2+l^2}}{a}$$

where C is the microscope constant $2\lambda L$, hkl are the Miller indexes of each reflection on the calibration pattern, and a is the lattice constant. By taking a series of measurements of D values and plotting vs. $D = (\sqrt{h^2+k^2+l^2}/a)$ the slope of the curve is seen to correspond to the microscope constant. An example of the calibration of a Jeol 4000FX microscope is shown in Figure 14.15*b*. Diffraction patterns can be obtained from very small areas from 1 μm to 1 nm (in modern field emission gun microscopes). In addition, we can also obtain chemical analysis information plus morphological information on the same area. This is a fundamental point in the application of the TEM to the study of archaeological materials. When local information about unknown substances is needed, the TEM technique should be used. The following steps should be followed to identify phases:

Obtain a diffraction pattern from the area to be analyzed. If it is a single-crystalline phase and a spot pattern is obtained, then it will be necessary to tilt the sample to obtain different zone axes.

Determine a table of interplanar distances values by indexing the pattern. Symmetry of the pattern, angles between spots, and systematic absences determine the crystal symmetry.

From the X-ray chemical analysis identify the elements that form the phase obtained at an approximate stoichiometry of the phase. This generates a short list of compounds as candidates.

From the short list of compounds compare the $d(hkl)$ values with those reported in the standard tables of crystallography (Lyman and Carr 1992; Wyckoff 1963), databases such as the JCPDS-ICDD PDF-1 and PDF-2, which is a computer-readable version of files (JCPDS 1981), the NIST crystal data file (NBS crystal data evaluated and compiled by NIST Gaithersburg, MD) or the NIST/Sandia/CCD electron diffraction database (NIST–Sandia.ICCD electron diffraction database available from J. CPDS International Center for Diffraction Data, Park Ione, Swarthmore, PA). In most practical cases the compound will be found in those databases. A new unreported compound will require more involved work.

Commercially available software can also be used to speed up the identification of unknown compounds. It is possible to digitize film through a scanner or a CCD camera, feed this data into the computer, let the software measure the distances, and then work out the structure. Common occurrence when working with historical materials is that of coming up against a naturally occurring mixture of different clays—the palygorskite clay, which was often used in antiquity. In most cases the sample is a mixture of palygorskite, sepiolite, and calcite. Figure 14.16 shows the structure of the three models and the corresponding diffraction patterns. In this case the patterns were simulated using the Cerius^2 program (Cerius^2, Version 3.5 Molecular Simulation, Inc., San Diego, CA, 1997).

14.10. ENERGY LOSS MATERIALS ANALYSIS

One of the most important techniques used in modern TEM for the study of nanostructured material is EELS. This technique can provide key information about the sample and helps to solve the problem of determining sample structure and composition. Energy loss is the result of inelastic scattering from nanomaterial. Four principal events should be considered: phonon excitation, inter- and intraband transitions, plasmon excitation, and inner-shell ionization (Egerton 1996). Table 14.3 shows the amount of energy that is typically lost in each of these events and the angle at which the electrons travel.

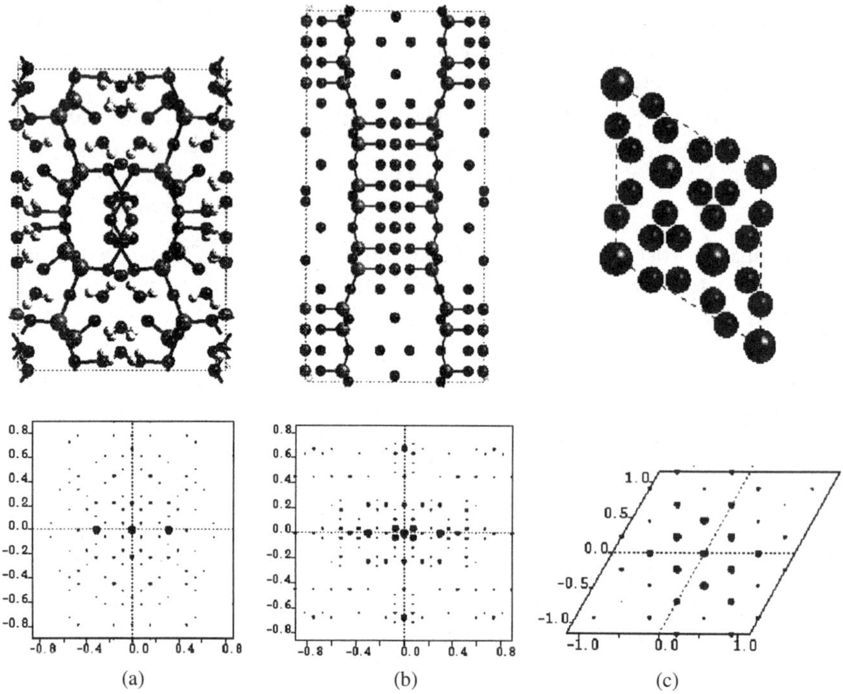

Figure 14.16. Structure of the three most commonly found compounds in natural minerals used in producing paints. (*a*) palygorskite, (*b*) sepiolite, (*c*) calcite and their corresponding electron diffraction patterns.

Table 14.3. Elemental Interaction to Be Considered in EELS Spectrum

Type of Interaction	Energy Loss (eV)	Angle θ_E (mrads)
Electron–phonon	−0.02	5–15
Electron inter- or intraband transitions	5–25	5–10
Electron–plasmon	3–25	Less than 0.1
Inner shell ionization	10–1000	1–5

The electron angle is given by

$$\theta_E \simeq \frac{E\sqrt{1 - v^2/c^2}}{m_0 v^2}$$

where v is the velocity at which the electrons travel, c is the light speed, and m_0 is the mass of the electron at rest.

It is clear from Table 14.3 that we can filter the different inelastic signals according to their angles or energy value. For instance, plasmon loss will be found at a very low angle, close to the zero loss peak or central spot of a diffraction pattern (Williams and Carter 1996). In practice, most electron spectrometers have a resolution of ~0.3 eV, which means that the phonon peak is contained in the zero loss peak. Phonon excitation is therefore not relevant in the case of TEM. The peak due to electron plasmon interaction is very intense; in fact, it is the most intense after zero loss. This peak carries interesting physical information about the electronic states in the sample.

The inter- and intraband transitions are also important if we wish to understand the physical nature of a sample. It should be said that they are of little use when attempting to extract analytical information, and this signal is difficult to interpret for the purposes of microanalysis. It is, however, possible to make the direct comparison of spectra (Ahn and Krivanek 1983).

The previous excitations—phonon excitation, inter- and intraband transitions, and plasmon excitation—correspond to the so-called low-loss part of the spectra. This helps us to learn about the interaction of electrons with free electrons, valence electrons, and the bonding in materials. The so-called high-loss portion of the spectra, which contains the information related to the interaction with the inner atomic shells, is perhaps the most interesting part of the spectrum. In this case the energy loss is characteristic of the material. EELS can then be used for the chemical analysis of nanostructured materials, providing information similar to that obtained using X-ray (EDS or WDS). However, there is considerably more information on the EELS spectrum that can be put to use to characterize nanostructured materials. The same notation used for X-ray microanalysis is used in EELS. This refers to the *K*, *L*, and *M* peaks in the spectra and the ionization edges from the *K*, *L*, *M* electron shells are present. As shown in Figure 14.17, around the critical ionization energy E_0, we have oscillations that are termed the energy loss near edge structure (ELNES). Beyond the edge, we may find oscillations due to diffraction effects from the atoms surrounding the ionized atom. These are called the energy loss fine structure (EXELFS). This structure is very similar to the technique known in X-ray diffraction as EXAFS (extended X-ray absorption fine structure). This technique is now widely used in connection with X-ray synchrotron sources. EXELFS and EXAFS basically yield local information about the structure of the sample. Those techniques are particularly useful for crystal structure determination in an unknown sample.

14.11. USE OF ELECTRON MICROSCOPY IN STUDY OF ARCHAEOLOGICAL MATERIALS

Electron microscopy continues to be a very important tool in archaeological material research. An example of this may be seen in the way Magaloni and

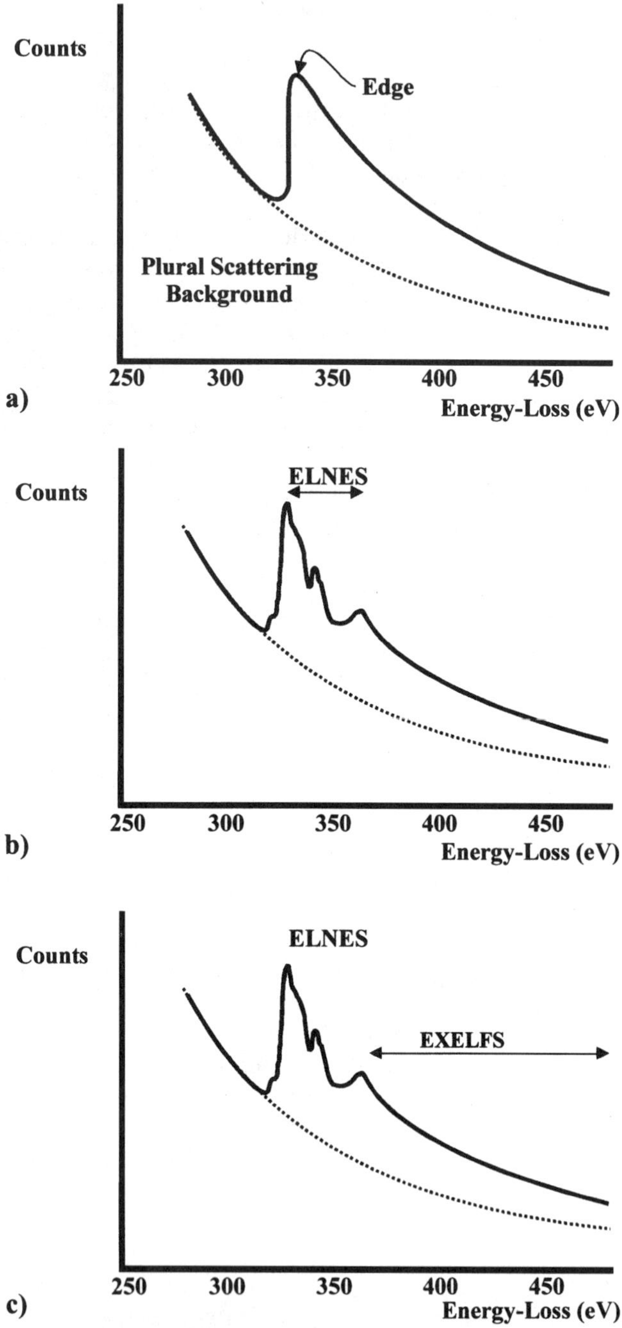

Figure 14.17. Typical energy loss spectrum of a sample. Around the critical ionization energy there are oscillations that can be used to obtain the coordination around a ionized atom.

co-workers (1991) have applied SEM and X-ray microanalysis to the study of murals of the prehispanic cultures in Mesoamerica. A further example of the application of SEM with EDS in the study of painting is the work of San Andrés Moya et al. (1995) who applied this technique to the study of *EL Bautizo de Cristo*, a work of the eighteenth century belonging to the collection of the Ermita de Santa Inés in Spain. Likewise, Ortega and Calleja (1995) have applied SEM coupled with EDS to the study of Roman mural paintings.

SEM also has been applied in numerous studies of the deterioration of historical monuments. For instance, De la Torre and co-workers (1995) have applied SEM with EDS to the study of the deterioration of the Alhambra in Granada, Spain. These examples are not exhaustive, and it is generally sufficient to note that the application of electron microscopy techniques has grown very rapidly over the last few years. Almost any major meeting on electron microscopy nowadays includes several contributions applying TEM to archeological materials (Calderón and José-Yacaman 1998). In the following sections more detailed examples of the applications of TEM to the study of archeological materials will be provided by drawing upon the work carried out by our group at the Institute of Nuclear Research of Mexico (ININ) and in the National University of México (UNAM).

14.11.1. Maya Blue Pigment

Maya blue was first reported by Merwin (1931) when studying the beautiful murals of the Maya culture in Yucatan early this century. It was, however, also widely used in the spectacular murals of Mesoamerica. Figure 14.18 shows two examples of paintings using Maya blue pigment. The first (Fig. 14.18*a*) was found in Bonampak (Chiapas) and the other (Fig. 14.18*b*) in the murals of Cacaxtla in Tlaxcala (Mexico), painted circa A.D. 700. It is evident that Maya blue was resistant to acids. The most stunning feature of this pigment is probably its resistance to corrosion (Gettens 1955). The paintings have survived some of the worst conditions, from rain forest, bacteriological corrosion, to wearing due to rains, winds, and so on. The blue pigment has a tonality and brightness that has remained through the centuries. Maya blue was characterized using electron microscopy techniques. It was found that the color contains three main ingredients: palygorskite clay, Indigo molecules, and small metallic particles of nanometer size (José-Yacamán et al. 1996).

The found clay corresponds to a mineral called *Sacalum* by the Mayas. This mineral contains palygorskite as its main component with sepiolite and calcite impurities. Figure 14.19*a* provides the SEM image of the sample that clearly shows two kinds of crystal habits: fibers and square-shaped crystal-

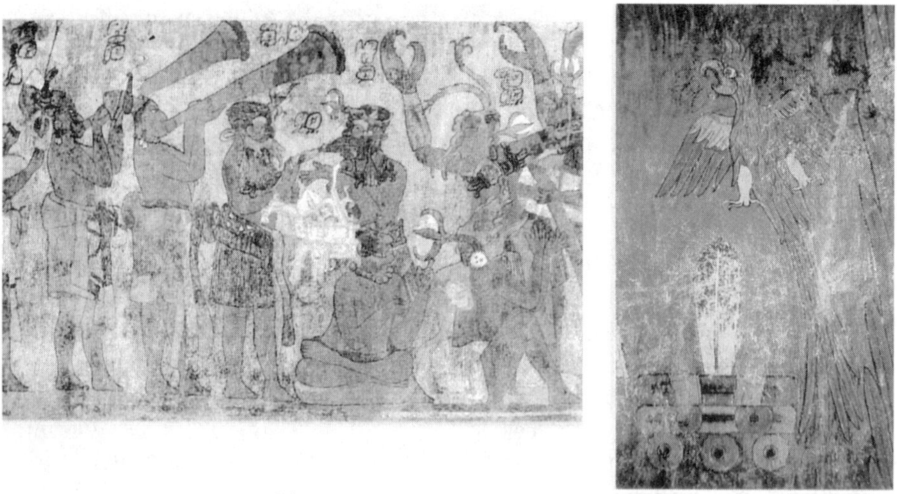

(a) (b)

Figure 14.18. Two examples of murals from Mesoamerican cultures that used Maya blue. (*a*) Mural from Bonampak and (*b*) mural from Cacaxtla. (Reproduced with permission of INAH, Mexico.)

lites. Figure 14.19*b* shows the corresponding EDS spectra, and Figure 14.19 shows also the crystal habits (assuming energy minimization) that are taken on by the calcite, sepiolite, and palygorskite. We can see right away that the fibers in the image correspond to either palygorskite or sepiolite, whereas the square crystals correspond to calcite.

Further identification of these fibers can be achieved through electron diffraction and X-ray analysis. Figure 14.20 shows a high-resolution TEM image of the fibers and reveals the layered type of material. We can also see regions in which a superlattice is produced. This seems to indicate an intercalation phenomena. We assume that in this pigment the palygorskite contains Indigo molecules intercalated in such a way that a periodic contrast is seen. However, the problem of exactly how the Indigo got into the clay is still not fully understood. The Indigo molecules tend naturally to interchange with water. Therefore, unless a mechanism that stabilizes the Indigo molecules were established, the color would wash away with rain. It is necessary to identify the mechanism capable to explain the durability of the murals throughout the centuries. An important factor to be considered is the acidity of the clay, which might play an important role in the Indigo absorption. High resolution is also useful to identify the metal particles revealed by EELS, as shown in Figure 14.21. The metal particles are supported on the clay surface. Their structure is very similar to the particles produced by colloidal chemistry in modern technology. These particles definitely play a role

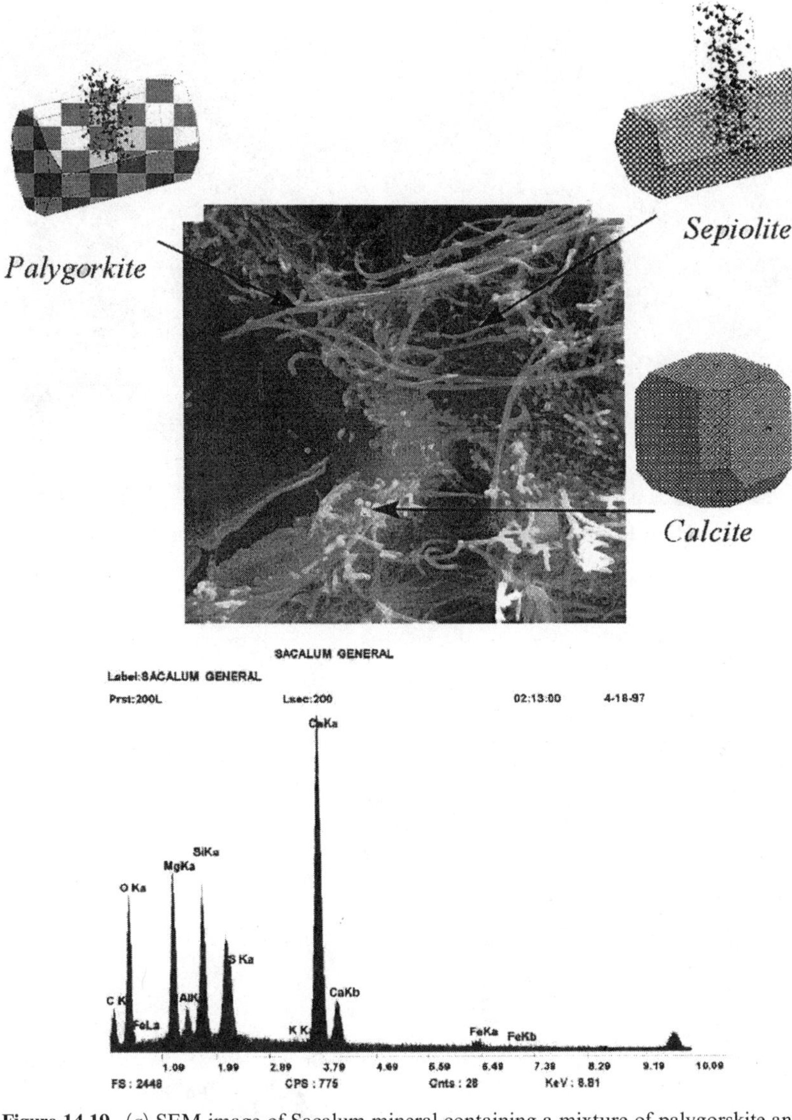

Figure 14.19. (*a*) SEM image of Sacalum mineral containing a mixture of palygorskite and calcite crystals. (*b*) EDS spectrum. Crystal habits of palygorskite, sepiolite, and calcite are, respectively, shown in the figure.

in producing the tonality of the Maya blue color (José-Yacamán et al., 1996). With the help of an electron microscopy it was possible to reproduce the pigment using the original Maya technology. These pigments were used in mural restoration.

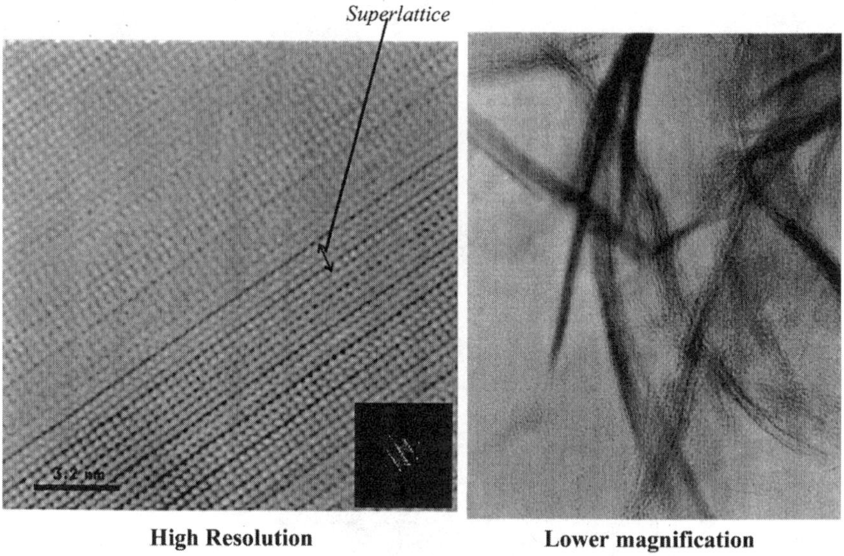

Figure 14.20. High-resolution TEM image showing the channel structure of the clay. A low-magnification image of the fibers is also shown.

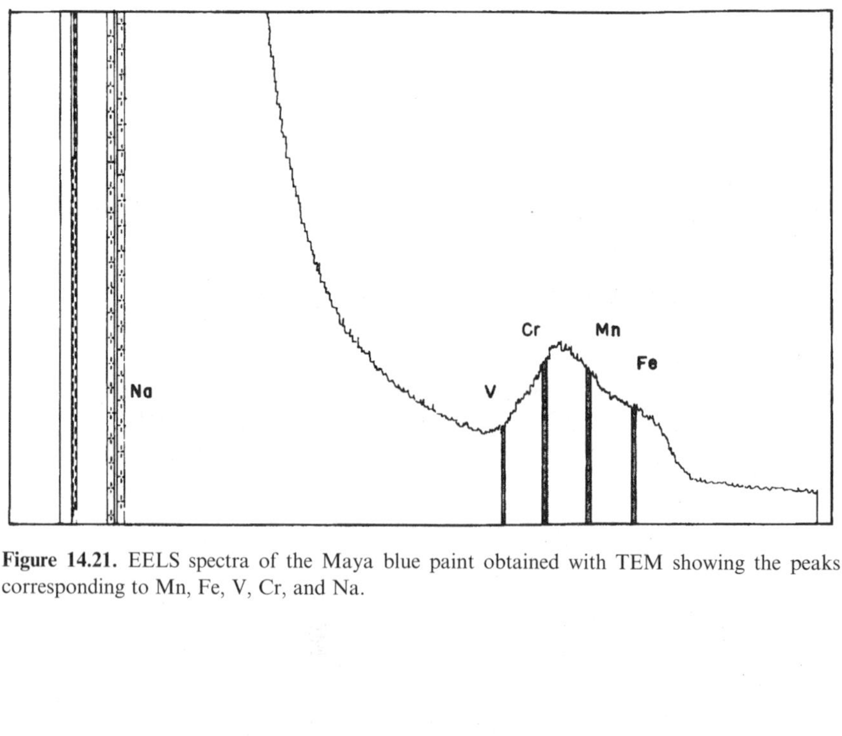

Figure 14.21. EELS spectra of the Maya blue paint obtained with TEM showing the peaks corresponding to Mn, Fe, V, Cr, and Na.

Table 14.4. Elements (wt%) Identified by EDX in Some Paintings from the Archaeological Site in Tajín

Element	S-2 wt% Pink	S-5 wt% Green	S-6 wt% Green	S-7 wt% Green	S-6 wt% Ochre	S-7 wt% Ochre	S-8 wt% Ochre	S-3 wt% Grey
C	26.75	2.18	2.84	2.62	0.75	2.77	2.46	2.19
O	25.43	54.79	47.36	56.76	62.76	64.39	56.49	54.08
Mg	0.40	1.48	0.87	0.47	0.600	0.38		0.73
Al	0.79	2.41	1.01	2.45	7.07	4.18	2.40	4.17
Si	1.71	6.96	4.24	9.06	9.26	6.73	7.02	15.19
K	0.508		1.16	0.76	0.47	0.70	1.21	
Ca	23.51	29.26	43.03	26.46	7.19	16.33	28.12	19.61
Fe	0.68	1.97	0.64	1.02	11.25	4.75	2.79	2.82
Hg	15.14							
Na		0.43						
Ti					0.36			
S	5.09							
P	0.50							

14.11.2. Paints from Other Cultures

We have made a systematic study of the pigments used in Mesoamerica by the Maya, Totonaca, Olmeca-Xicalanca, and Aztec civilizations using electron microscopy. The principle of producing pigments was very similar in all these cultures. It was a combination of clay, organic pigments, and small particles often nanometers in dimension. They used natural polymers extracted from cactus plant, very common in Mesoamerica, as a binder.

Table 14.4 shows a summary of the results obtained for the paints of the Totonaca culture found on the Tajin site. This shows the whole variety of colors used in their murals. The resistance of the pigment is due to the great stability of the metal particles, which are oxidized and form stable compounds. Table 14.5 shows the results of paintings in the Aztec culture at the Templo Mayor (great temple) of Tenochtitlan in Mexico City Valley. This culture, which met with the European conquest of the sixteenth century, clearly showed a sophisticated level of technology, as their art works clearly testify.

Generally, it was found that the paints used by the Aztec culture were similar to those from the Mayas, implying strong connections between these two cultures and at least a shared part of technology. It should be finally mentioned that the Olmeca Xicalanca culture, which developed in Cacaxtla, also had similar features in their pigments.

Table 14.5. Maya Blue Pigment Analysis from *Templo Mayor*[a]

Element	Fibers	Aggregates
C	X	X
O	X	X
Mg	X	X
Al	X	X
Si	X	X
K		X
Ca		X
Fe	X	X
Na		X
S	X	X

[a] Crystalline structures found are sepiolite, ferrite, and palygorskite.

14.11.3. Metallurgy in Ancient Cultures of Peru

Another example of the use of the high resolution in TEM comes from the metal artifacts of the Moche culture in South America (North Coastal Peru 1, ca. A.D. 100 to A.D. 800). These objects are considered to be remarkable cases in ancient metallurgy and provide a good example of how metal artifacts can be studied using TEM. Metal workers in ancient America used a number of techniques to improve the surface appearance of their metals. In particular, great effort was made in the production of gilded or silvered metals (Forty 1979; Letchmann 1979). The Moche metal objects are basically copper with a layer of gold or silver deposited on their surface. The smoothness of the layer indicates that the surface layer was most likely deposited using electrochemical techniques. Figure 14.22*a* is a high-resolution image of copper covered with a gold layer. The image indicates irregular patches of gold cover the copper surface. In Figure 14.22*b*, two regions where image processing has been used to improve the images are shown. The difference between pure copper and the area with the Au–Cu sandwich is evident. A full account of how TEM and SEM can be applied to this problem can be found in the research of Centeno and Schorsch (Letchmann et al. 1982; Centeno, Private communication).

14.12. APPLICATIONS OF ENVIRONMENTAL SEM TO STUDY OF ANCIENT MATERIALS AND ART PRESERVATION

14.12.1. SEM Studies of a Sixteenth-Century Oil Canvas

Figure 14.23 shows a picture composition containing information on a Spanish oil canvas from the sixteenth century. The sample was placed on the

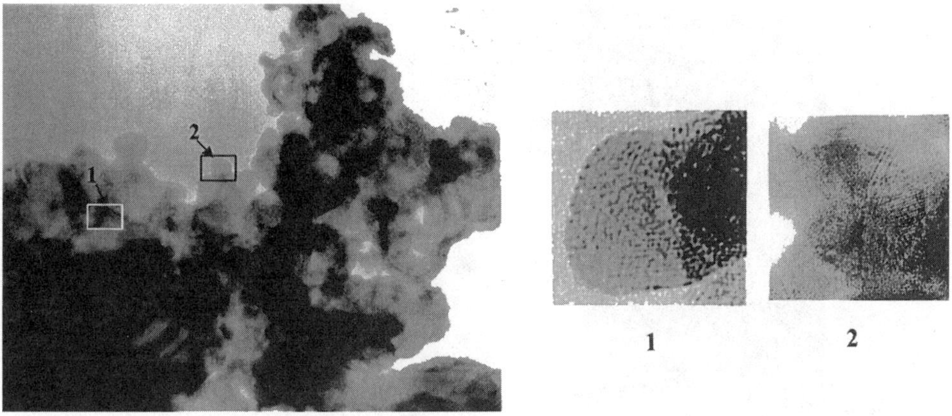

Figure 14.22. (*a*) High-resolution image of a gold–copper sample coming from the Moche culture in North Peru. (*b*) Image processing of two regions of the previous picture showing the Moire patterns produced between gold and copper and some high-resolution images.

microscope without further preparation, taking advantage of the properties of the low-vacuum microscope.

The figure shows the images and corresponding spectra of three different particles. A general view of the sample can be seen in Figure 14.23*a*, and the images and spectra of three particles can be seen in Figures 14.23*b*, 14.23*c*, and 14.23*d*, respectively. Here carbon was included in the analysis. The first particle contains mainly C and O with Si being the other conspicuous element. Fe is very scarce and S and Sr are seen as being relatively important. The second particle is rich in C and O but also in Si and Fe. The third particle is much richer in Fe but the Si concentration is lower. Several conclusions can be drawn from this analysis:

- The carbon and oxygen are mostly present in the form of carbonates and oxides.
- Silicon oxides, most likely quartz, play a very important role in the pigmenting principle.
- Metal oxides, particularly iron oxides, are present in the samples and also contribute greatly to the pigmenting principle.
- In all cases the evident mixture of coloring components is very important. The techniques were able to analyze nanometer sized particles.

14.12.2. TEM and Environmental SEM in Study of Old Paper

A common effect in paper fabrication in Europe in the sixteenth to nineteenth centuries was the appearance of yellow coloring on paper. This effect

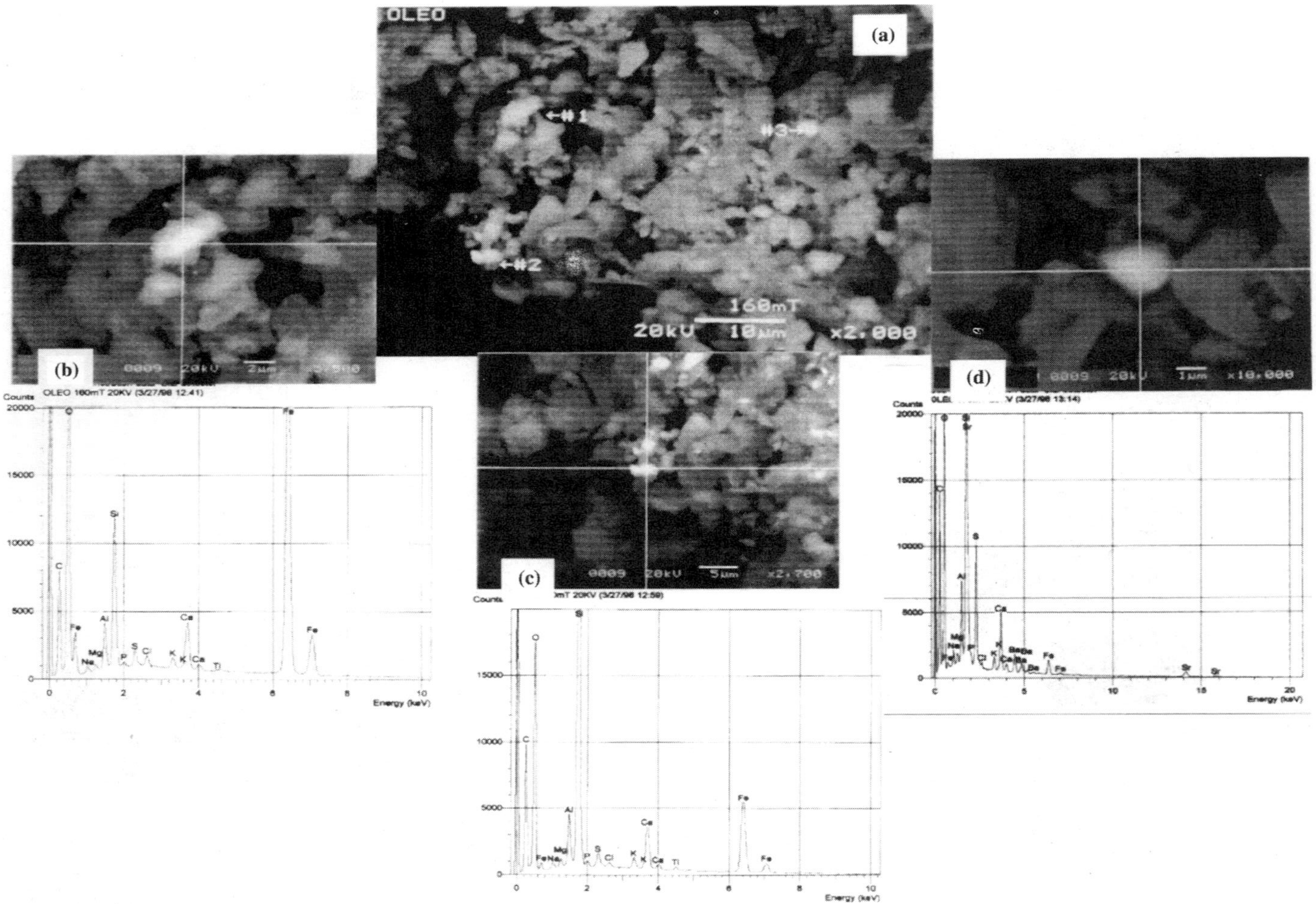

Figure 14.23. Images and analysis of three different particles in a sixteenth-century oil canvas. (*a*) general image, (*b*) image and analysis of the first particle,

causes degradation on the quality of the document. It is well known that the yellowing of the paper is due to a photoxidation process that affects the component of the paper (cellulose, lignin, hemicellulose, etc.). In particular, yellowing seems to be the result of the formation of aldehydic groups and on the oxidation of the lignin. However, additional insight in this important phenomenon can be obtained using electron microscopy. When studying this effect, environmental SEM is one of the most suitable techniques to be employed. Figure 14.24 gives a general view of a nineteenth-century paper that has not deteriorated and shows the typical fibers that form the paper. Two images of secondary electrons and backscattered electrons are shown in Figures 14.24*a* and 14.24*b*. It is clear from the comparison of these two images that the morphology of the paper is more or less uniform. However, in the yellow regions many of the particles are seen to be present on top of the fibers.

Bacteria affecting the paper were also found. These are shown in Figure 14.25. Bacteria were only found in the yellow regions, and it is likely that they play a role in the yellowing effect.

Environmental scanning electron microscopy is a technique that has only been fully developed in the last few years. It has, however, found a very vast application in the field of art preservation. Two really spectacular examples have been reported by Doehne at the Betty Conservation Institute (Doehne 1997).

The first is the cleaning of the Sistine Chapel, where the cotton swabs used during the cleaning were analyzed using ESEM. The issue to be resolved was whether pigment particles were being removed in addition to soot particles. Image and X-ray analysis indicated the presence of silicates, salt particles, sodium, magnesium, iron, and sulfur. These particles were most likely the result of air pollution.

The second example is the study of the Dead Sea scrolls. The aim of the study was to determine the mechanism of their remarkable preservation over the centuries. Doehne (1997) reports the study of samples of the Dead Sea scrolls using ESEM. The samples were undried and uncoated. The samples were subjected to dynamic studies of their response to moisture. The experiment showed the presence of soluble salts on the scrolls, in particular KCl and BrCl salts. These introduced salts, probably present as a result of the preparation method, most likely contributed to preservation by impeding the action of microorganisms. It is clear that ESEM will continue to be a fundamental tool in the field of art preservation in the future.

Acknowledgments

The authors are indebted to Luis Rendón, Cristina Zorrilla, Leticia Carapia, Samuel Tehuacanero, M. Eufemia Fernández, Tomás Santamaría, and

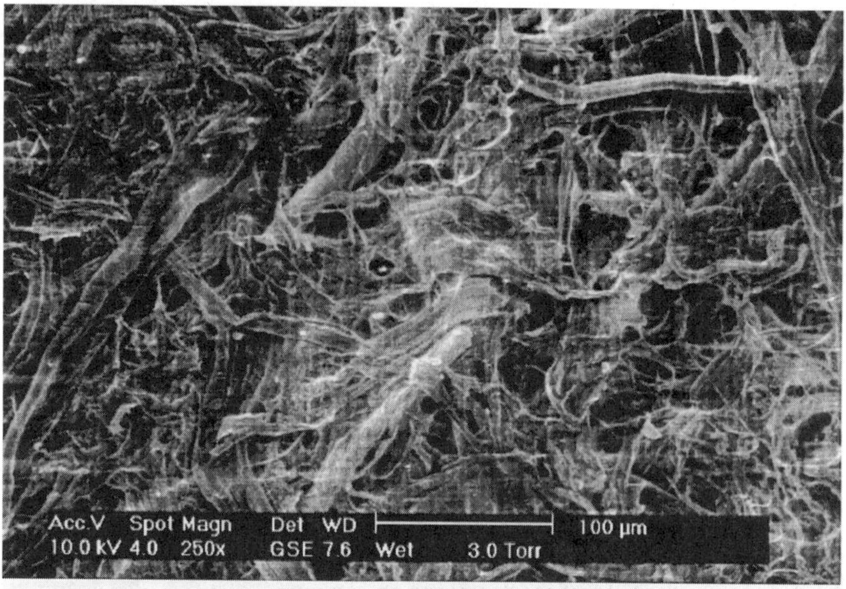

(a)

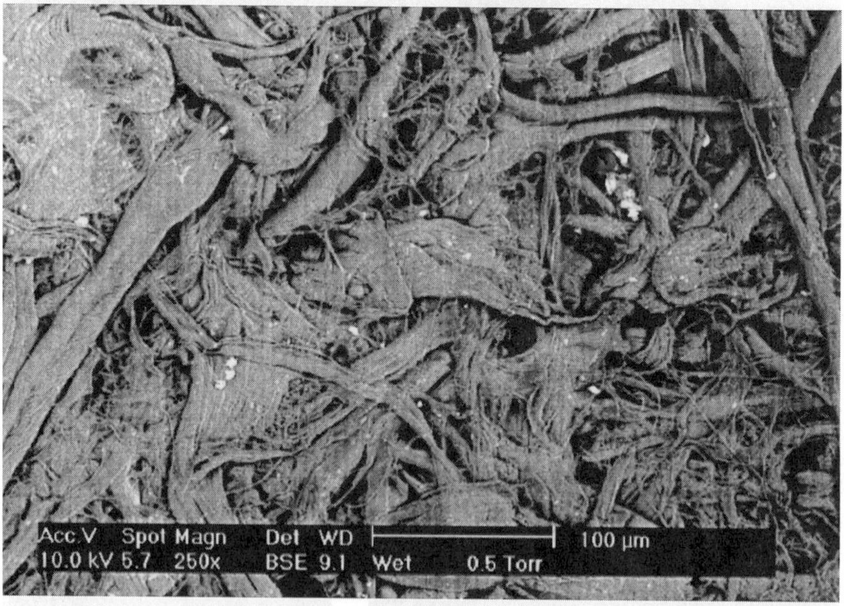

(b)

Figure 14.24. Images of paper from the nineteenth century in which the fibers can be seen: (*a*) secondary electron image and (*b*) backscattering image.

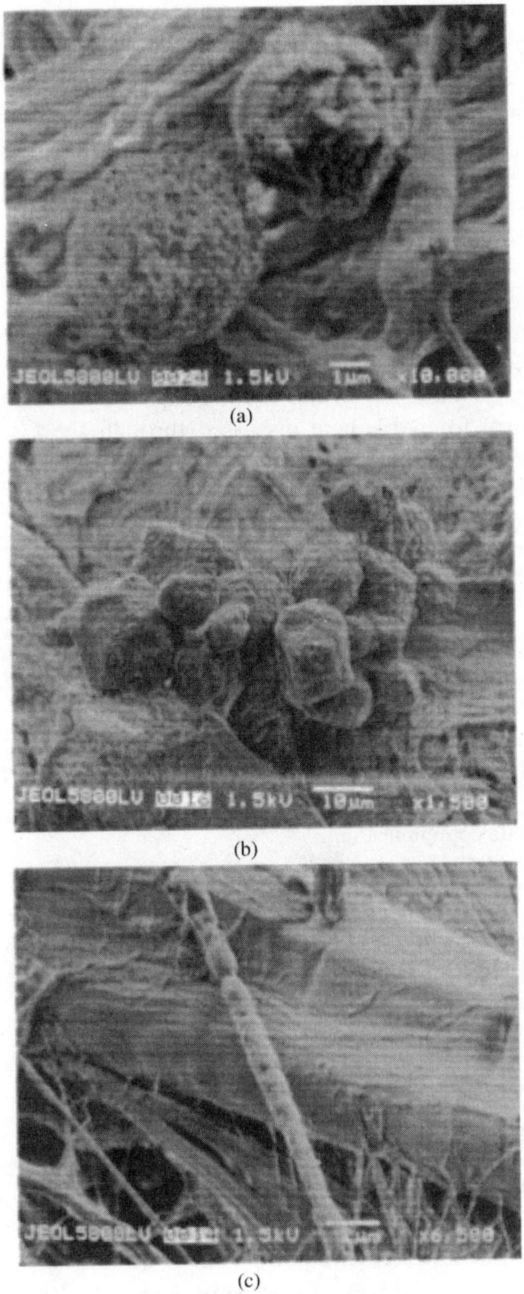

Figure 14.25. Images of bacteria in the yellow portion of the paper: (*a*) single bacteria, (*b*) cluster of bacteria, and (*c*) string of bacteria.

Ricardo Reglero for their technical assistance. We would also like to thank M. en C. Jesús Arenas for obtaining some of the ESEM and channeling images. JEOL de México and Philips de México should be acknowledged for making their low-vacuum and environmental scanning microscopes available. Gustavo Martínez, Ventura Rodríguez-Lugo, Demetrio Mendoza, Elizabeth Romero, Mayahuel Ortega, and Carmen Reza must be thanked for helping with the study of archaeological samples. We also thank to Dr. Leonardo López Lujan for providing samples of the Aztec culture. To Dra. Mari Carmen Serra for providing samples of Maya blue, to Dr. José L. Galván for providing the samples of oil canvas, to Dr. Silvia Centeno for her collaboration and the providing of the Moche samples, and to the National Archives of México for providing samples of old documents used in this study. We are also indebted to the CONACYT for providing the financial support necessary for work here discussed through the Grant «Coloides Cuánticos». We also thank INAH for permission to reproduce the Maya murals.

REFERENCES

Ahn, C. C., and O. L. Krivanek (1983), *EELS*, Atlas Gatan, Palo Alto, CA.

Buseck, P. R., J. M. Cowley, and L. Eyring (1988), in *High Resolution Electron Microscopy and Associated Techniques*, P. R. Buseck, J. M. Cowley, and L. Eyring, Eds., Oxford University Press, New York.

Calderón, H. and M. José Yacamán (1998), *Proceedings of the XIV International Congress on Electron Microscopy*, Cancún, México. Institute of Physics of Bristol. UK.

Centeno, S., private communication.

Cowley, J. M. (1981), *Diffraction Physics*, 2nd ed., North-Holland, Amsterdam.

Danilatos, G. S. (1998), *Adv. Electron. Electr. Phys.* **71**, 109.

de Pablo, L. (1996), *Revista Mexicana de Ciencias Geológicas* **13**, 94.

De la Torre López, M. J., E. Sebastián Pardo, and J. Rodríguez Gordillo (1995), In *Proceedings of XVII Congress of the Spanish Electron Microscopy Society*, **328**, University of Oviedo Publications.

Doehne, E. (1997), *In situ Microscopy in Materials Research*, P. L. Tai, Ed., Kluwer Academic Publishers, Norwell, MA, USA.

Doyle, P. A., and P. S. Turner (1968), *Acta Cryst.* **A24**, 390.

Egerton, P. F. (1996), *Electron Energy Loss Spectroscopy in the Electron Microscope*, 2nd ed., Plenum, New York.

ElectroScan Corporation (1996), *Environmental Scanning Electron Microscopy, an Introduction to ESEM*, ElectroScan Corporation Publications, Boston, MA.

Fernández, M. E., J. A. Ascencio, D. Mendoza-Anaya, V. Rodríguez-Lugo, and M. José-Yacamán (1999), *J. Mater. Science.* **34**, 1.

Forty, J. (1979), *Nature* **282**, 579.

Gabriel, B. L. (1985), *SEM, A Users Manual for Materials Science*, American Society of Metals, Metals Park, OH.

Galan, E., and A. Castillo (1984), *Develop. Sedim.* **37**, 87.

Gettens, R. J. (1955), *Carnegie Institution Washington Publication* **602**.

Goldstein, J., D. Newburg, P. Echlin, D. Joy, C. Fiori, and G. Lifshin (1984), *Scanning Electron Microscopy and X-Ray Analysis*, Plenum Press, New York.

Jeol (1994), *Invitation to the SEM World*, Jeol Publications.

Jeol (1996), *A Guide to Scanning Microscope Observation*, Jeol Publications Tokyo, Japan.

José-Yacamán, M., L. Rendón, J. Arenas, and M. C. Serra Puche (1996), *Science* **273**, 223.

José-Yacamán, M., S. Tehuacanero, C. Zorrilla, J. A. Ascencio, and A. Gómez (1998), *Nanostruct. Mater.* **10**, 1.

Letchmann, H. N. (1979), *J. Metals* **31**, 154.

Letchman, H. N., A. Erlij, and E. Barry (1982), *Am. Antiquity* **47**, 3.

Lyman, C. E., and M. J. Carr (1992), in *Electron Diffraction Techniques*, Vol. 2, J. M. Cowley, Ed., Oxford Science Publications, Oxford.

Magaloni, D., M. Aguilar, and V. Castaño (1991) *Mat. Res. Soc. Symp. Ser.*, **185**, Materials Research Society Publications.

Merwin, H. E. (1931), *Carnegie Institution of Washington Publication* **406**.

NIST-Sandia, ICCD Electron Diffraction database available from J. CPDS International Center for Diffraction Data, Park Ione Swarthmore, PA, USA.

Ortega, F., and J. Martin Calleja (1995), In *Proceedings of XVII Congress of the Spanish Electron Microscopy Society*, **328**, University of Oviedo Publications.

Otten, M. T., and W. M. J. Coene (1993), *Ultramicroscopy* **48**, 77.

San Andrés Moya, M., M. I. Baez, D. Cornejo Sastre, and J. N. Baldonedo (1995), In *Proceedings of XVII Congress of the Spanish Electron Microscopy Society*, **332**, University of Oviedo Publications.

Shimizu, R., and K. Mirata (1971), *J. Appl. Phys.* **42**, 387.

Van Dick, D., and O. A. F. De Jong (1992), *Ultramicroscopy* **47**, 266.

Williams, D., and B. Carter (1996), *Transmission Electron Microscopy*, Plenum, New York.

Wyckoff, R. W. J. (1963), *Crystal Structures*, John Wiley & Sons, New York.

Zou, X., Sukharev, Y. and S. Hovmoller (1993), *Ultramicroscopy* **52**, 436.

CHAPTER

15

ISOTOPE ANALYSIS, DATING, AND PROVENANCE METHODS

GÜNTHER A. WAGNER

Forschungsstelle Archaeometrie der Heidelberger Akademie der Wissenschaften, am Max-Planck-Institut fuer Kernphysik, Heidelberg, Germany

15.1. INTRODUCTION

Most chemical elements consist of a mixture of several isotopes. The term *isotope* was coined in 1913 by Frederick Soddy (1877 to 1956) to allow for the observation that more than one atomic species may occupy the same (Greek *isos*) place (*topos*) in the periodic table of elements. Although in nature the isotopic composition of an element is almost unvarying, there exist minor but significant deviations from the normal isotopic abundances. Such deviations bear information on the age as well as the provenance of archaeologic materials. But it took several decades before this potential was realized. In 1947 the detection of radiocarbon in nature by Willard Libby (1908 to 1980) and co-workers opened the possibility to exploit the radioactive decay of ^{14}C for dating ancient objects. Also stable isotope geochemistry began in 1947 with Harold Urey's (1893 to 1981) studies on oxygen isotope fractionation for determining paleotemperatures of ocean water. In the meantime, the knowledge of isotopic shifts in many elements became a wealthy and indispensable source of information for archaeology and art history. This chapter intends to introduce the underlying principles of isotopic variations as well as to provide a general overview of their archaeologic application, with emphasis on methods not treated in detail elsewhere in this book.

Modern Analytical Methods in Art and Archaeology, Edited by Enrico Ciliberto and Giuseppe Spoto. Chemical Analysis Series, Vol. 155
ISBN 0-471-29361-X

15.2. PRINCIPLES

15.2.1. Elements, Nuclides, and Isotopes

Natural matter is built up of chemical elements. An *element* consists of atoms with a specific number of protons, expressed as atomic number, in the nucleus, for example, all atoms of the element uranium (symbol U) have 92 protons. In its atomic shells the nucleus is surrounded by an equal number of orbiting electrons. The positively charged protons and the negatively charged electrons compensate each other electrically. If there is a deficiency or surplus of electrons in the shells, the atom becomes a positively or negatively charged *ion*, respectively. In addition to protons, the nucleus also bears neutrons that have nearly the same mass as protons but carry no charge. The sum of protons and neutrons in the nucleus defines the mass number of the atom. A *nuclide* is an atomic species with a definite number of protons and of neutrons. The mass number is indicated as a superscript to the upper left, whereas the atomic number is sometimes given as a subscript to the lower left of the elemental symbol, for example, the nuclide ${}^{238}_{92}U$ has 92 protons and 146 neutrons and, thus, the mass number is 238 (= 92 + 146). It is sufficient to write simply ${}^{238}U$ since the element uranium implicitly is already specified by the atomic number 92. *Isotopes* are two or more nuclides that belong to the same element, that is, they have the same number of protons but different numbers of neutrons. For example, both uranium isotopes ${}^{235}U$ and ${}^{238}U$ have 92 protons, but 143 or 146 neutrons, respectively. Since isotopes are atomic species of the same chemical element, they show identical chemical—but not physical—behavior.

As to their natural sources nuclides may be *primordial* (i.e., formed during elemental synthesis), *radiogenic* (i.e., the result of radioactive disintegration), and *cosmogenic* (i.e., produced by the interaction of cosmic rays with the atmosphere and Earth's surface). To give examples, ${}^{204}Pb$ is purely of primordial origin, while ${}^{206}Pb$ is both primordial as well as radiogenic, and ${}^{3}He$ may be—apart from primordial and radiogenic—also of cosmogenic origin.

15.2.2. Isotopic Homogeneity and Variation

The vast majority of the naturally occurring nuclides is primordial. Thus they were synthesized in the solar nebula and thoroughly mixed before they were incorporated into Earth. Consequently, the chemical elements that consist of more than one nuclide have the same specific isotopic compositions, regardless of where they are found on Earth (and even on the moon). In first approximation this actually holds true; for example, the isotopic abundance ratio ${}^{235}U/{}^{238}U$ is everywhere 7.2527×10^{-3}. However, there are natural processes that upset this global isotopic homogeneity and lead to isotopic variation. Such processes are the radioactivity, the cosmogenic production, and the fractionation.

15.2.2.1. Radioactive Decay

One distinguishes stable and unstable, or radioactive, nuclides. The latter disintegrate spontaneously and transmute into nuclides of a different element—a process called *radioactivity* by which a radioactive parent nuclide decays into a radiogenic daughter nuclide. Although in nature, of course, primarily stable nuclides occur, also numerable radioactive nuclides are found. Owing to their sources, the radioactive nuclides also can be primordial, radiogenic, and cosmogenic. Primordial nuclides with correspondingly long half-lives, such as ^{238}U with $t_{1/2} = 4.5 \times 10^9$ a (a is used instead of year), are still present from the time of element synthesis and, thus, are older than the formation of Earth. The shorter lived radiogenic nuclides, for example, ^{234}U with $t_{1/2} = 2.5 \times 10^5$ a, are formed intermediately in the decay chains of uranium and thorium, before they decay themselves. The cosmogenic nuclides, for example, ^{14}C with $t_{1/2} = 5.73 \times 10^3$ a, are steadily produced by cosmic rays.

Several types of radioactive decay can be used for archaeologic dating (Fig. 15.1): *alpha decay* under emission of an α-particle, which is a ^4He nucleus, *beta decay* under the emission of a β^--particle, which is an electron whereby a neutron is converted into a proton, *electron capture* under the capture of an extranuclear, orbiting electron from the innermost atomic shell whereby a nuclear proton is converted into a neutron, and *spontaneous fission* under splitting into two heavy nuclides and two to three neutrons. Most of these types of radioactive decay are accompanied by the emission of energetically discrete γ-rays—a type of electromagnetic radiation coming from excited nuclei.

Radioactivity is a statistical process. Its quantitative treatment requires a sufficient number N of atoms of the radioactive nuclide because it is impossible to predict the single disintegration event. The probability of disintegration, the decay constant λ (a^{-1}) or inversely expressed as mean life-time $\tau = 1/\lambda$ (a) or more commonly as half-life $t_{1/2} = 0.693/\lambda$ (a), is a nuclear property and is specific for the nuclide. The activity dN/dt [measured in Becquerel Bq (s^{-1})] is defined as the fraction dN of N (with initial amount N_0) that disintegrates within the time interval dt into the stable daughter atoms D ($= N_0 - N$):

$$\frac{dN}{dt} = -\lambda N = -\frac{\ln 2}{t_{1/2}} \times N \tag{15.1}$$

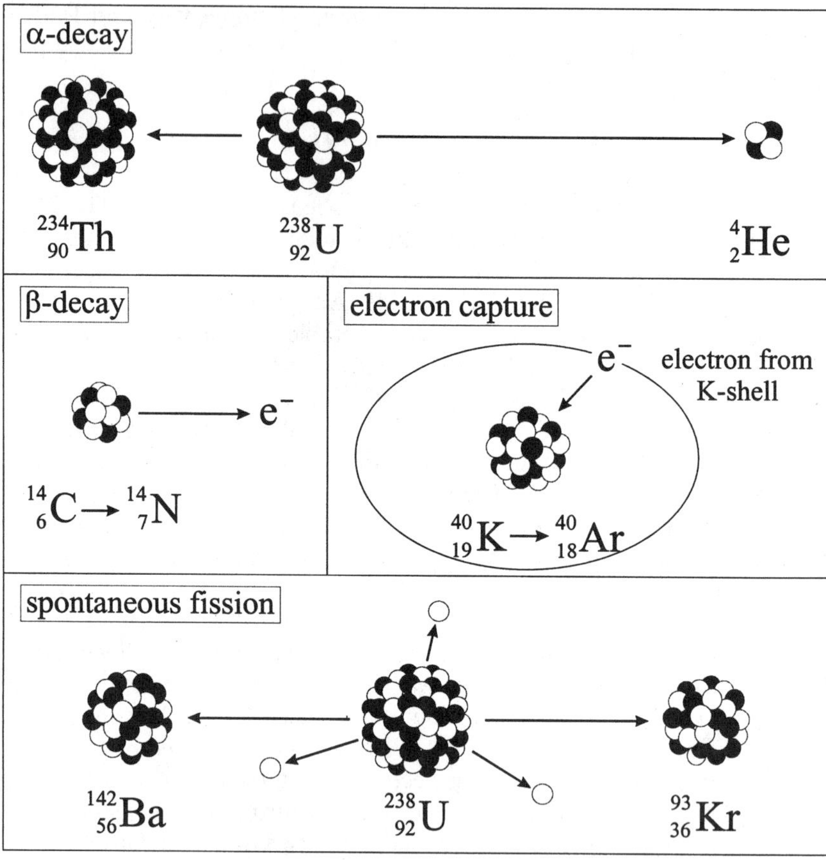

Figure 15.1. In radiometric dating four different types of natural radioactive decay are used: α-decay, β-decay, electron capture, and spontaneous fission; white: neutrons, black: protons (from Wagner 1998).

By integration one obtains time-dependent functions of nuclide pairs (Fig. 15.2) that can be used for dating. The age t (a) can be derived, depending on whether two out of three quantities N_0, N, or D are known, according to the equations

$$t = \frac{1}{\lambda} \times \ln\left(\frac{D}{N} + 1\right) \tag{15.2}$$

$$t = \frac{1}{\lambda} \times \ln\left(\frac{N_0}{N}\right) \tag{15.3}$$

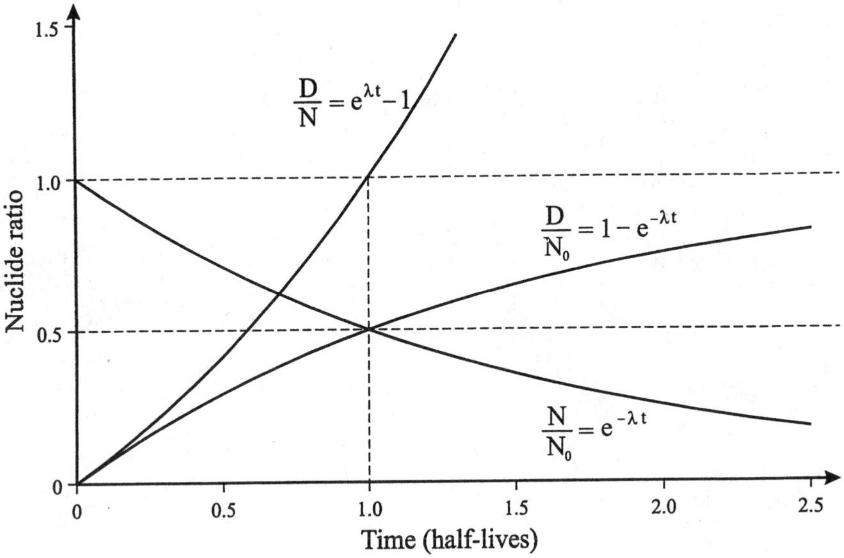

Figure 15.2. Change of nuclide ratios with time t (normalized to half-life of N) due to radioactive decay; N: amount of radioactive parent nuclei; N_0: initial amount of N at time $t = 0$; D: amount of radiogenic daughter nuclei (from Wagner 1998).

Instead of being stable, the radiogenic daughter nuclide may be radioactive and decay itself. If several such radioactive daughter nuclides follow each other, a decay chain is formed until finally a stable end member is reached. This is, for instance, the case in the decay chain starting from ^{238}U and ending at ^{206}Pb involving several steps of α- or β-decays (Fig. 15.3). In such decay chains a balance between production and decay of the intermediate members is established after a certain time (*secular equilibrium*) in which for all radioactive members N_1, N_2, N_3, and so forth the activity dN/dt becomes the same:

$$\lambda_1 \times N_1 = \lambda_2 \times N_2 = \lambda_3 \times N_3 \ldots \tag{15.4}$$

The state of equilibrium in which the activity ratio $(\lambda_D \times D/\lambda_N \times N)$ equals 1 can be disturbed by chemical, physical, and biological processes through which parent or daughter atoms are added or removed, resulting in disequilibrium, with an initial ratio $(\lambda_D \times D/\lambda_N \times N)_0 \neq 1$ between daughter and parent nuclides. After the disturbance has ended, that is, the system has closed again, gradually a radioactive equilibrium is attained (Fig. 15.4). The daughter/parent ratio changes as a function of the age t, and it takes at least

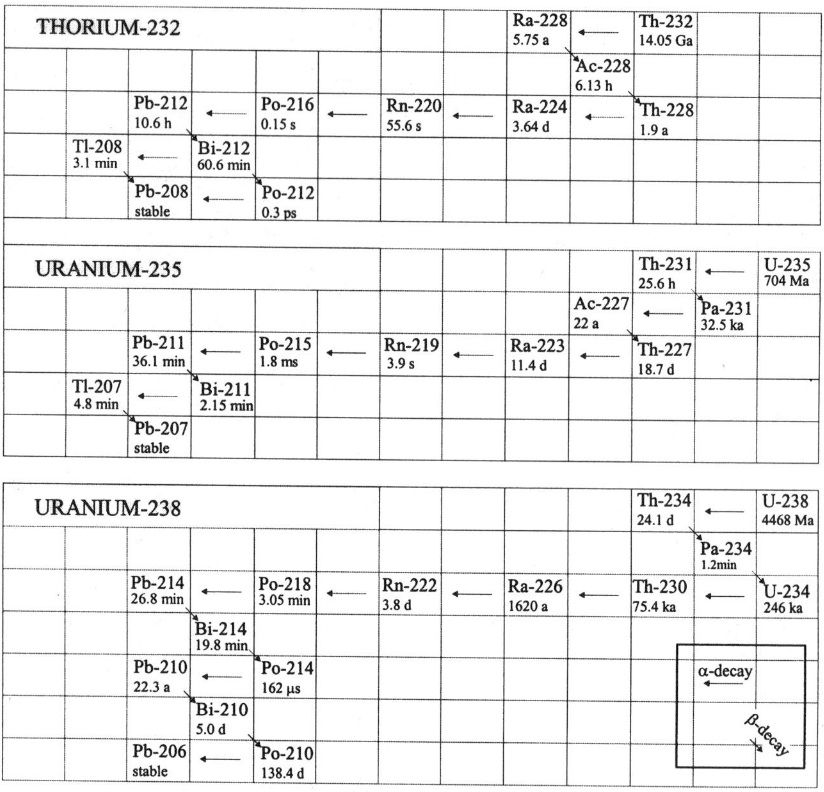

Figure 15.3. Radioactive decay series of ^{232}Th, ^{235}U, and ^{238}U; these parent nuclides decay through various unstable daughter nuclides to stable lead isotopes (from Wagner 1998).

$5 \times t_{1/2,D}$ until the daughter nuclide is practically again in equilibrium with the parent nuclide. This time-dependent process enables to date the moment of disturbance according to

$$t = \frac{1}{\lambda_D} \times \ln \frac{\left\{ \left(\frac{\lambda_D \times D}{\lambda_N \times N} \right)_0 - 1 \right\}}{\left\{ \left(\frac{\lambda_D \times D}{\lambda_N \times N} \right) - 1 \right\}} \tag{15.5}$$

15.2.2.2. Cosmogenic Production

Earth is exposed to a steady flux of cosmic rays that interact with the atoms of the atmosphere and Earth's surface. The particles of cosmic rays induce

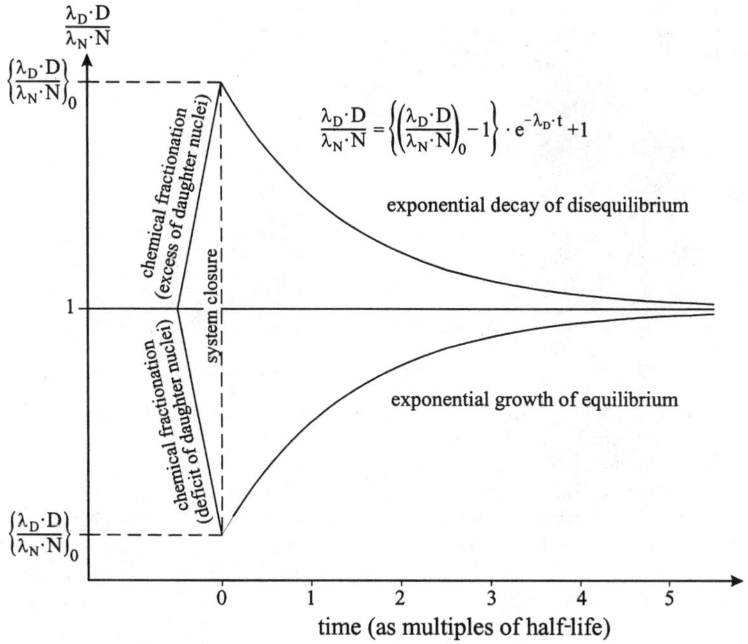

Figure 15.4. Radioactive system, which is in secular equilibrium, has an activity ratio equal to one between daughter D and parent N. After events that disturb the equilibrium the daughter/parent activity ratio approaches exponentially again the ratio equal to one. This allows the dating of such events. After about five half-lives the equilibrium is reestablished (from Wagner 1998).

nuclear reactions in which both radioactive nuclides (e.g., ^{14}C) and stable nuclides (e.g., ^{3}He) are formed with the production rate P (atoms g^{-1} a^{-1}). According to the place of formation, one distinguishes *atmospheric* (in the air produced, e.g., ^{14}C) from *in situ* (in the uppermost meters of the rock surface produced, e.g., ^{41}Ca) cosmogenic nuclides. The cosmogenic nuclide concentration N (atoms g^{-1}) grows gradually with the duration t of exposure. For stable nuclides the growth function is linear, and for radioactive nuclides it saturates exponentially (Fig. 15.5). The time-dependent accumulation of N allows to determine the exposure age t of samples that at first had been completely shielded from and later were suddenly exposed to cosmic rays. The exposure age is evaluated for stable cosmogenic nuclides according to

$$t = N/P \tag{15.6}$$

and for radioactive cosmogenic nuclides according to

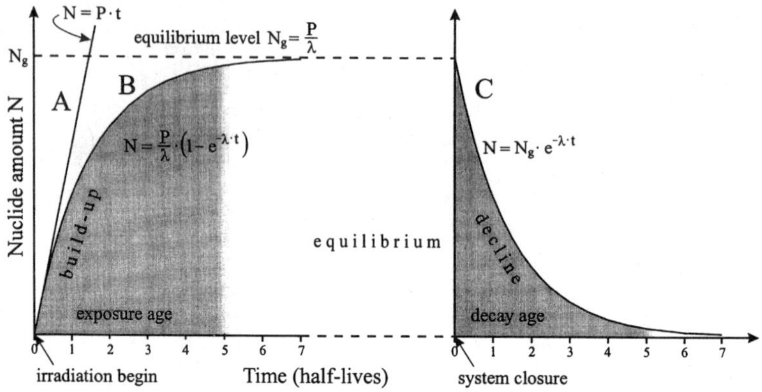

Figure 15.5. Production and decay of cosmogenic nuclides. Their amount N increases with irradiation time t (normalized to half-life of N): proportionally for stable nuclides (curve A) and exponentially saturated for radioactive nuclides (curve B); P = production rate, λ = decay constant. The radioactive nuclides approach an equilibrium level N_g, at which the production and decay rates equal each other. When the system is removed from further production or supply (e.g., by shielding from cosmic rays or by death of organism), N decreases exponentially (curve C). Surface exposure ages can be obtained for the buildup phase and decay ages for the decline phase (from Wagner 1998).

$$t = -\ln(1 - \lambda \times N/P)/\lambda \qquad (15.7)$$

The concentration of the radioactive atoms approaches gradually the equilibrium level $N_g = P/\lambda$, which is practically reached after about five half-lives. When a sample, which attained already its level N_g, is suddenly shielded from the cosmic radiation, so that the production is suspended, N declines exponentially. The cut-off from cosmogenic supply can be dated analogous to Eq. (15.3).

15.2.2.3. Fractionation

It is usually assumed that isotopes belonging to the same element show identical chemical behavior, but this is not strictly valid, in particular not for light elements. To give an example, photosynthesis enriches the light ^{12}C over the heavy ^{13}C, and the latter one in turn over the even heavier ^{14}C, so that the carbon in plants is isotopically lighter than the carbon in the atmosphere. This phenomenon, known as isotope fractionation, causes isotopic variation. There are various reasons for isotope fractionation. For quantum mechanics reasons the chemical bonds involving lighter isotopes require less energy to break than those of the heavier ones and, thus, molecules bearing the lighter isotopes react chemically more readily. For reaction kinetics reasons molecules containing lighter isotopes move more rapidly than those with heavier ones, resulting also in a higher reaction rate of the isotopically lighter molecule, that is, in irreversible chemical reactions the lighter isotope is enriched in the reaction product. Also during evaporation and condensation as well as during melting and crystallization, isotope fractionation occurs, for example, the vapor phase is isotopically lighter than the liquid one. It is important to note that isotope fractionation effects are strongly temperature dependent. They decrease with increasing temperature. The isotope fractionation of a sample is expressed as δ-notation, which is the relative deviation (in per mill) between sample x and a standard s. For example, the $\delta^{18}O$ value for the $^{18}O/^{16}O$ fractionation is calculated according to

$$\delta^{18}O = 1000 \times [(^{18}O/^{16}O)_x - (^{18}O/^{16}O)_s]/(^{18}O/^{16}O)_s \qquad (15.8)$$

A negative δ-value means that the sample is isotopically lighter than the standard and vice versa.

15.2.3. Isotope Analysis

The most important analytic method for the determination of the isotopic abundance is mass spectrometry. Depending on the sample material, the elements to be investigated and the problems to be solved, various types of mass spectrometers are used. Basically the mass spectrometer is an evacuated, bent tube consisting of three components: the ion source, the ion deflector, and the ion detector. After extraction and various purification steps the sample is introduced into the mass analyzing system of the spectrometer. At the ion source the atoms are ionized. The ions are focused and accelerated by electrostatic lenses. During the passage of the ion beam through a magnetic field, the ions are more or less deflected depending on their mass—more precisely on their charge/mass ratio. Finally, the mass-separated ions are collected at the ion detector. For special techniques, such as thermoionization, accelerator, and stable isotope mass spectrometry, see the following three chapters in this book.

Radioactive isotopes may also be analyzed by decay counting dN/dt. The type of nuclide can be identified by the discrete energies of its emitted α- or γ-rays (α- and γ-spectrometry). The number of radioactive atoms is derived from the measurement of radioactivity using the relation $dN/dt = \lambda \times N$. It becomes clear that this indirect determination of N becomes inefficient for small λ, that is, for long-lived nuclides. The techniques most commonly used are β-counting for ^{14}C dating and α-spectrometry for uranium-series dating.

Instead of analyzing the quantity of the radiogenic daughter nuclide explicitly—as is done with mass spectrometry—the amount of the daughter

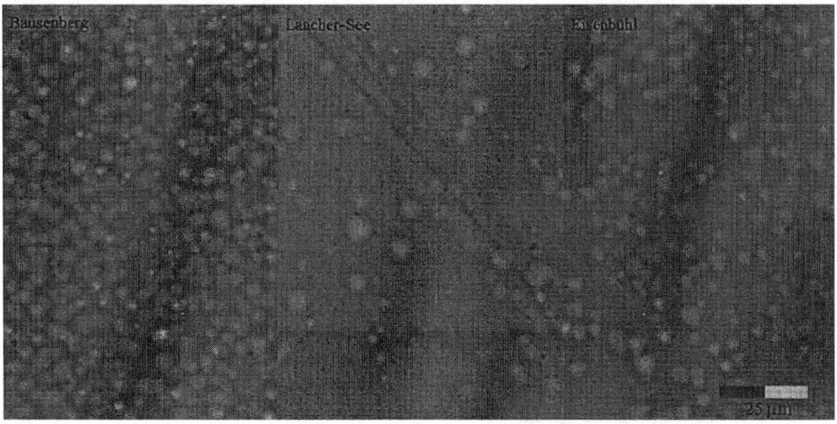

Figure 15.6. Micrograph of α-recoil tracks in dark micas from three Quaternary volcanoes: Bausenberg, Laacher See (both Germany), and Eisenbühl (Czechia). The samples were etched 10 min in 40% hydrofluoric acid at 20°C (from Wagner 1998).

may be determined indirectly by counting particle tracks. The areal density of the particle tracks is proportional to the number of daughter atoms. Particle track counting is the base for the fission track and α-recoil track dating methods. The heavy fragments of the spontaneous fission of uranium as well as the recoil nuclei of the uranium and thorium α-decays leave in their host minerals etchable tracks that are visible under the microscope (Fig. 15.6).

The energy of ionizing radiation released during radioactive decay is to some extent deposited and accumulated as radiation damage in the mineral lattice. The energy dose (= absorbed energy per unit mass), which a mineral has received from α, β, γ, and cosmic rays, can be measured by the luminescence and electron-spin-resonance signals (see Chapters 18 and 19).

15.3. APPLICATION

15.3.1. Dating

The time-dependent change of the isotopic composition, predominantly caused by radioactive decay, but also by temperature-controlled fractionation processes, is utilized for dating. The great advantage that radioactivity, as a chronometric tool, has over other time-dependent physical and chemical processes is its independence of environmental influences, such as temperature or pressure. In other words, nuclear clocks run even and accurately. The use of a radiometric system for age determination presupposes that after closure neither the parent nor the daughter nuclides are lost or gained except through the decay process itself, a condition known as "closed system behavior." In order to yield meaningful ages there must be a defined event of archaeologic relevance at which the radiometric system is reset and closed. This allows to determine the time ("age") elapsed since the system was at zero. Such events are (1) the formation of a sample (e.g., the precipitation of travertine), (2) the last heating (e.g., the burning of a flint artifact), or (3) the last bleaching (e.g., the sedimentation of dune sand). Owing to the underlying principles the various radiometric dating techniques can be divided in several groups (Table 15.1).

For the Quaternary age range there are two chronometric methods that are based on *radiogenic noble gas* nuclides: the K–Ar method uses the decay of the potassium isotope ^{40}K to the argon isotope ^{40}Ar, and the U–He method rests on the accumulation of the radiogenic helium isotope ^{4}He, which is produced during the α-disintegrations within the uranium and thorium decay series (Fig. 15.1). The ages are calculated according to Eq. (15.2). Noble gases are chemically inert and thus do not enter the compounds of minerals. Consequently, the daughter nuclide should be absent at mineral formation so that the "noble gas age" dates the formation of the sample. If sufficiently heated, any noble gas already contained in the crystal should be completely driven out so that one dates the moment of last degassing. In practice, such clear-cut cases are not always met. Minerals may already incorporate some amount of the daughter as an excess component. Furthermore, a defined thermal event may not result in complete degassing so that the daughter is partially inherited by the subsequent system. In both cases the obtained age would be an overestimate of the event of interest. Another difficulty is the possible leakage of the daughter from the mineral. Any such loss results in an underestimate of the age. The ability to retain noble gases differs among the minerals. The K–Ar method is a basic tool for establishing the chronometric framework of early hominids, while the U–He method finds only little archaeologic application. In its general conception the K–Ar method comprises several dating techniques. Apart from the conventional K–Ar technique (^{40}Ar/^{40}K), these are the argon–argon (^{40}Ar/^{39}Ar) and the argon–argon laser techniques. A great advantage of the latter one is that it enables single-grain analysis, whereby potassium-rich minerals, such as sanidine, are preferred. The available ultrasensitive analytical technology developed lately allows to determine ages down to few ka (kiloan = thousand years) on milligram-small samples.

Since the K–Ar system is reset by complete degassing, datable events are the solidification of volcanic minerals as well as the volcanic contact heating of preexisting minerals, and this means in both instances the time of volcanic eruption. An important material in this connection are the widespread tephra

Table 15.1. Isotopic Dating Methods in Archaeology and Quaternary Geology

Method	Age Range	Material
Radiogenic Noble Gases		
K–Ar	>10 ka	Basalt, tephra, obsidian, tektite, impact glass
U–He	>10 ka	Coral, mollusk, bone, basalt
Uranium-series		
$^{230}Th/^{234}U$	100 a–500 ka	Coral, calcareous cave-deposit, travertine, limnic carbonate, mollusk, volcanite
$^{231}Pa/^{235}U$	1 ka–150 ka	Coral, calcareous cave-deposit, travertine, limnic carbonate, mollusk, volcanite
$^{234}U/^{238}U$	10 ka–1 Ma	Calcareous cave-deposit, travertine, bone, tooth
Excess ^{230}Th and ^{231}Pa	1 ka–300 ka	Deep-sea sediment, volcanite
Excess ^{210}Pb	<150 a	Lead pigment and alloy
$^{226}Ra/^{230}Th$	<10 ka	Marine sediment, volcanite
$^{206,207,208}Pb/U$, Th	>1 Ma	Volcanite
Cosmogenic Nuclides		
^{3}He	250 a–1 Ma	Basalt, moraine
^{10}Be	1 ka–10 Ma	Soil, limnic and fluvial sediments, loess, deep-sea sediment, volcanite, quartz-bearing rock
^{14}C	100 a–45 ka	Wood, charcoal, grains, pollen, phytoliths, peat, paper, textile, bone, soil, limnic sediment, calcareous cave-deposit, travertine, mollusk, eggshell, coral, rock varnish, mortar, stone tool, rock painting, ceramics, metallurgical slag
^{21}Ne	1 ka–1 Ma	Basalt, moraine
^{26}Al	1 ka–1 Ma	Deep-sea sediment, quartz-bearing rock
^{36}Cl	1 ka–1 Ma	Moraine, lava flow, impactite, evaporite
^{41}Ca	20 ka–500 ka	Bone, calcareous sinter
Particle Tracks		
Fission track (FT)	>100 a	Volcanite, tephra, deep-sea volcanite, impact glass, pseudotachylite, burnt stone, artificial glass
α-Recoil track (ART)	>100 a	Volcanite, ceramics

Table 15.1. (Continued)

Method	Age Range	Material
Radiation Dosimetry		
Thermoluminescence (TL)	100 a–500 ka	Ceramics, burnt clay, burnt stone, slag, artificial glass, volcanite, impactite, loess, sand, colluvium, archaeosediment, phytoliths
Optically stimulated luminescence (OSL)	100 a–100 ka	Loess, sand, colluvium, archaeosediment, tephra, ceramics
Electron spin resonance (ESR)	1 ka–1 Ma	Calcareous cave-deposit, travertine, mollusk, tooth, bone, coral, clastic sediment, burnt stone, volcanite
Stable Isotopes		
$^{18}O/^{16}O$	10 ka–2 Ma	Deep-sea sediment

layers for which K–Ar dating yields excellent tephrochronologic time marks, in particular for intercalated early-hominid-bearing sedimentary layers. The knowledge that the earliest fossil hominids *(Australopithecus afarensis)* are at least 3.9 Ma old (Ma = million years) essentially was gained by K–Ar analysis (Hall et al. 1984). At Aramis in Ethiopia the Gaala Vitric Tuff, which occurs stratigraphically below the hominid fossils, was dated by single-grain Ar–Ar laser fusion. Despite dominant contamination with older grains an age of 4.387 ± 0.031 Ma was calculated from the juvenile feldspar fraction placing an upper age limit to these earliest hominids (WoldeGabriel et al. 1994). The oldest stone tools, so far discovered, are those from the Oldowan Industrial Complex at Gona, Ethiopia, with an age of 2.6 to 2.5 Ma based on single feldspar laser fusion Ar–Ar analyses and magnetostratigraphy on underlying and overlying tephra beds (Semaw et al. 1997). Furthermore, K–Ar dating delivers the chronometric calibration of the geomagnetic polarity time scale. Singer and Pringle (1996) found for the Matuyama/Brunhes transition an $^{40}Ar/^{39}Ar$ age of 778.7 ± 1.9 ka.

The various *uranium-series dating* techniques are based on radiometric disequilibria within the radioactive decay series arising from the two uranium isotopes ^{238}U and ^{235}U. The individual techniques are named after the respective daughter/parent nuclide pairs, and in archaeologic or art historic contexts the important ones are $^{230}Th/^{234}U$ and excess ^{210}Pb. Provided the initial daughter/parent ratio $(D/N)_0$ is known and the system behaved closed, the age is derived from Eq. (15.5). The recent shift from the α-spectrometry to the thermoionization mass spectrometric analysis, which requires less (ca. 10^{-2}) sample material and provides higher age precision,

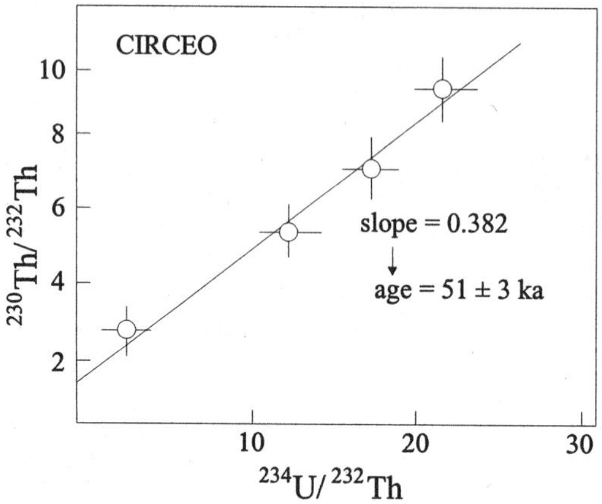

Figure 15.7. $^{230}Th/^{232}Th$ versus $^{234}U/^{232}Th$ isochron of calcareous sinter crusts attached to the cranium of the Neanderthal Man at Monte Circeo, Italy. The different fractions were gained through leaching (from Schwarcz 1989).

gave a fresh impetus to the uranium series dating methods. The various techniques as well as their applications are comprehensively treated by Ivanovich and Harmon (1992).

$^{230}Th/^{234}U$ dating—sometimes combined with $^{231}Pa/^{235}U$ or $^{234}U/^{238}U$—is applicable to marine carbonates, fresh-water calcium carbonates, mollusk shells, bones, teeth, and peat. The $CaCO_3$ of these materials incorporates during precipitation uranium, and thus ^{234}U, without thorium, and thus without ^{230}Th ($t_{1/2} = 75.4$ ka), as a result of the higher water solubility of uranium than that of thorium. For peat it is assumed that the organic matter adsorbs uranium from the groundwater during peat formation. Fossil bones, teeth, and mollusk shells take up uranium from the groundwater after their deposition, seriously violating the requirement of closed-system behavior. In order to date these materials in spite of this shortcoming, models about the timing of the uranium uptake need to be adopted. A common problem is the presence of initial ^{230}Th, which enters the carbonate deposits together with the detritus, such as wind-borne dust in travertine. A technique for the recognition and avoidance of initial ^{230}Th contamination offers the analysis of chemically or mineralogically separated fractions in combination with an isochron plot in the form of $^{230}Th/^{232}Th$ versus $^{234}U/^{232}Th$ (Fig. 15.7).

There are many dating studies on archaeologically relevant cave carbonates, such as the sinter crusts attached to the cranium of the Neanderthal

Man at the Grotta Guattari at Monte Circeo, Italy (Schwarcz 1989). In the latter example the encrustation on the cranium consisted of a bright inner and a dark outer layer. Subsamples from the inner layer revealed detrital ^{230}Th contamination. From the slope of the straight line through the data points in the isochron plot the age is calculated to 51 ± 3 ka (Fig. 15.7). Examples for successful ^{230}Th/^{234}U dating of travertine are the Middle Pleistocene profiles Weimar-Ehringsdorf and Bilzingsleben, Germany, with important finds of hominids (Blackwell and Schwarcz 1986, Schwarcz et al. 1988). By mass spectrometric ^{230}Th/^{234}U dating of mammal teeth from the Middle Pleistocene sites Tabun, Qafzeh, and Skhul, Israel, McDermott et al. (1993) were able to confirm the existence of anatomically modern humans (*Homo sapiens*) for the Levant already before 100 ± 5 ka, contemporary with the Neanderthal Man (*Homo neanderthalensis*). Important to note is the fact that the age values of both subsamples—dentine and enamel—from the same tooth are in agreement, although these materials differ strongly in the uranium content, being suggestive of an early uranium uptake soon after the deposition and of a closed system. For art historic authentication it is fortunate that young lead-bearing objects can be identified by their content of excess ^{210}Pb. Lead ores most often contain traces of uranium, being in equilibrium with its decay product ^{210}Pb. During the smelting of lead ores, ^{210}Pb together with the lead is extracted from the ore without the parent nuclides, hence, the smelted lead contains unsupported ^{210}Pb, which decays within several half-lives ($t_{1/2} = 22.3$ a). The presence of ^{210}Pb activity can be used to distinguish young (<150 a) from older, inactive lead as was demonstrated by Keisch (1968) on several questionable paintings, among them *Jesus and His Disciples in Emmaus*, which had been claimed to be a genuine Vermeer. By analyzing its lead-white pigment, it was clarified beyond doubt that the supposed "Vermeer paintings" are forgeries from the twentieth century. Similar authenticity tests can also be conducted on other lead-bearing pieces of art, including lead bronzes.

Although for dating purpose various *cosmogenic nuclides*, that is, the radioactive nuclides ^{3}H, ^{10}Be, ^{14}C, ^{26}Al, ^{32}Si, ^{36}Cl, ^{39}Ar, ^{41}Ca, and ^{81}Kr, and the stable nuclides ^{3}He and ^{21}Ne, can be used, presently only ^{14}C is important for archaeology. ^{14}C is produced and equilibrated in the atmosphere from where it is exchanged with other reservoirs, in particular the biosphere. At the moment of death the organism is barred from further ^{14}C supply so that its ^{14}C content starts to decline (Fig. 15.5), allowing to date a wide spectrum of archaeologic materials. The ^{14}C half-life of 5730 a is convenient to cover the age range from the Late Paleolithic to the Middle Ages. The principles, procedures, and the great archaeological potential of ^{14}C dating are dealt with in Chapter 16. Of the remaining cosmogenic nuclides above all ^{41}Ca is promising for archaeology. It is produced in the uppermost few meters of the Earth's surface. From there it reaches the biosphere. If calcium-bearing material is moved from this surface zone to deeper zones, the ^{41}Ca concentration declines and the radiometric age can be derived. Bones, calcareous sinter, and concretions should be datable. The ^{41}Ca half-life of ca. 103 ka enables the application in the 20- to 500-ka age range. The method has not yet developed into practical use. Major problems, such as the improved ^{41}Ca detection, the wide variation of the initial $^{41}Ca/Ca$ ratio, and the geochemical behavior of the ^{41}Ca system, have still to be overcome (Taylor et al. 1989; Fink et al. 1990). The other *in situ* produced cosmogenic nuclides, in particular ^{3}He, ^{10}Be, ^{21}Ne, ^{26}Al, and ^{36}Cl, enable surface exposure dating [Fig. 15.5, Eqs. (15.6) and (15.7)] and thus may be of geomorphologic and geoarchaeologic interest (Bierman 1994). The exact knowledge of their production rates is a prerequisite for the dating application.

Although counting *particle tracks* does not present, strictly speaking, an isotope analytic technique, the fission track and α-recoil track dating methods are treated here briefly because they are based on radioactive decay. Both methods utilize the phenomenon that particles, released during nuclear decay, leave etchable tracks in minerals and glasses. In that way single nuclear disintegration events can be detected. If all tracks were preserved since the formation of a sample, that is, the system stayed closed, the total number of accumulated tracks allows the age determination whereby the particle tracks represent the daughter product in Eq. (15.2). The age value dates the sample's formation or last heating with complete erasure of the previously present tracks. The fission track method rests on the spontaneous ^{238}U fission. It covers a wide age range from a few hundred years upwards. The principles and application of fission track dating are described by Wagner and Van den Haute (1992). In the young age range, zircon-bearing tephra that are intercalated in sedimentary sequences—among them hominid remains and Paleolithic implements—are datable. Also burnt stone artifacts, in particular those made of obsidian, can be dated, provided these objects had been sufficiently (>ca. 500°C) heated by ancient man. Uranium-rich artificial glass, such as the nineteenth century uranium glass, is also datable with fission tracks. The α-recoil track dating method (Fig. 15.6), on the other hand, is still under development. It is very promising for dating mica-containing tephra layers and thus has a great potential for Quaternary tephrochronology (Gögen and Wagner 2000).

The three *dosimetric dating* methods thermoluminescence (TL), optically stimulated luminescence (OSL), and electron spin resonance (ESR) are based on the time-dependent accumulation of radiation damage, measured as energy dose. The mineral sample is used as natural dosimeter for the omnipresent ionizing radiation, which originates from natural radioactivity and cosmic rays, and which is determined as dose rate. Hence the dosimetric methods have many features in common. The age is obtained by dividing the natural dose by the dose rate. What basically distinguishes the methods from each other is the physical phenomenon that is exploited to record the energy dose. Each of these phenomena has its specific potential for archaeology. The remarkable sensibility of the dosimetric systems to thermal or optical resetting enables the dating of ancient heating and bleaching events. The principles, procedures, and applications of the dosimetric dating methods are treated in more detail in the Chapters 18 and 19 of this book.

Climatically controlled *stable isotope* abundance variation is observed for the light elements, predominantly for hydrogen, carbon, and oxygen in deep-sea sediments, ice cores, and tree rings. These variations have different physico-chemical causes and play an important role in paleoenvironmental reconstruction. For the purpose of Quaternary chronology the $^{18}O/^{16}O$ changes are by far the most important and solely will be treated here briefly. Oxygen isotope studies on calcareous shells of fossil foraminifera from deep-sea sediments reveal a long-term variation, expressed as $\delta^{18}O$ values [Eq. (15.8)], that correlates with the climatic history of the Quaternary (Shackleton and Opdyke 1973). Observable are high values for cold and low values for warm periods. The main cause of this variation is the fractionation between the vapor phase above the ocean surface and the ocean water. The isotopically lighter moisture above the ocean surface is blown by the wind across the continents, where it precipitates. In interglacial periods most of the precipitation returns through the hydrologic cycle back to the ocean. In glacial periods, however, a significant fraction of the precipitation is fixed in the growing ice caps. Since this ice is isotopically lighter than the ocean water, the ocean water experiences a depletion of ^{16}O with increasing ice volume, that is, the $\delta^{18}O$ value of the remaining ocean water increases. The foraminifera incorporate oxygen from the ocean water into their calcitic shells so that the fossil foraminifera, found in deep-sea sediments, record the climatic fluctuation and thus enable the construction of the marine $\delta^{18}O$ curve (Fig. 15.8). The marine $\delta^{18}O$ curve exhibits a distinct signature that allows its division into stages with uneven numbers for the warm and even ones for the cold periods. The striking correlation between the marine $\delta^{18}O$ variation and the Milankovitch insolation oscillation allows the precise astronomical dating of the $\delta^{18}O$ stages (Bassinot et al 1994). The marine $\delta^{18}O$ chronology forms the backbone of the Quaternary chronology. Not confused with the marine $\delta^{18}O$ curve should be the ice core $\delta^{18}O$ curves (Lorius et al. 1985). The oxygen–isotope ratio in ice cores reflects essentially the variation in the mean air temperature during snow formation whereby the $\delta^{18}O$ value increases with temperature, that is, in the opposite sense compared to the marine curve. Both types of $\delta^{18}O$ curves can be correlated to each other.

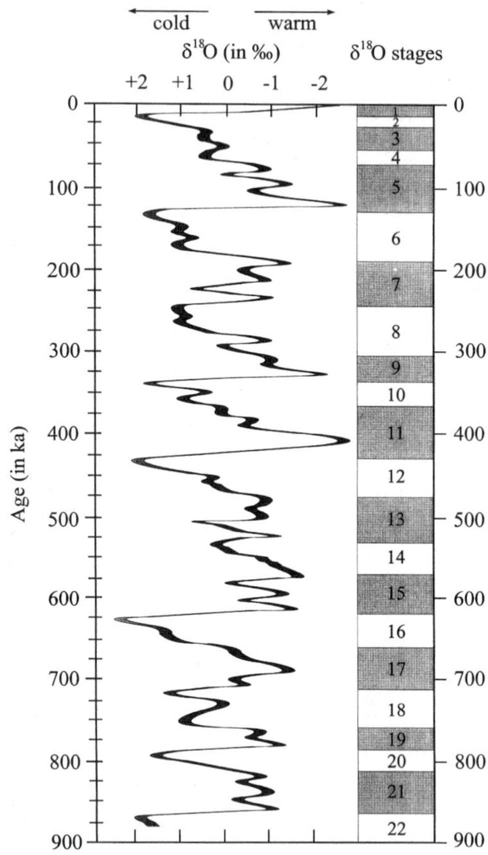

Figure 15.8. Marine $\delta^{18}O$ curve with the astronomically dated $\delta^{18}O$ stages. The even-numbered stages represent glacial and the odd-numbered ones interglacial periods (from Wagner 1998, after Bassinot et al. 1994).

15.3.2. Provenance

Provenancing of archaeological materials by means of diagnostic features ("fingerprints") presupposes that (1) the possible source regions differ significantly with respect to the features, and (2) these characteristic features pass over unchanged from the source to the archaeologic object. Diagnostic features can be of mineralogic, chemical, or isotopic nature. As to the utilization of the isotopic signatures for provenancing, one has to distinguish between heavy and light elements. The matching of features of an object with one of the potential sources does, however, not prove that the object actually derived from this source since there might be other, still unknown sources having the same features. However, discrepant features logically exclude a potential region from being a source. In other words, the correspondence of features is necessary but not sufficient for the positive assignment of an archaeological object to a certain geographic source. Despite their clarity, these principles of provenancing are often disregarded, for example, in sourcing of Late Bronze Age copper ingots by lead isotopes, as pointed out by Budd et al (1995).

Suited heavy elements are those that have radiogenic isotopes; these are essentially lead and strontium. Depending on the age and the geochemistry of the source region, the isotopic composition of these elements varies. Therefore, one observes in rocks and soils regional differences of the radiogenic/nonradiogenic isotope ratios $^{206}Pb/^{204}Pb$, $^{207}Pb/^{204}Pb$, and $^{208}Pb/^{204}Pb$ as well as $^{87}Sr/^{86}Sr$. A great advantage of the heavy elements for isotopic provenancing is that they resist isotope fractionation and thus their isotopic abundance—as opposed to the chemical composition of the material—is not changed by technologic and weathering processes. The lead isotopy finds manifold applications in archaeometallurgy, as described in Chapter 17. In skeletal materials, such as bones and teeth, the lead isotopic composition reflects via the food chain that of the rocky underground and thus is helpful in migration studies of men and animals. Strontium is used for tracing stone implements, potter's clays, as well as skeletal materials.

REFERENCES

Bassinot, F. C., L. D. Labeyrie, E. Vincent, X. Quidelleur, N. J. Shackelton, and Y. Lancelot (1994), *Earth Planet. Sci. Lett.* **126**, 191.

Bierman, P. R. (1994), *J. Geophys. Res.* **99**(*B7*), 13885.

Blackwell, B., and H. P. Schwarcz (1986), *Quaternary Res.* **25**, 215.

Budd, P., A. M. Pollard, B. Scaife, and R. G. Thomas (1995), *J. Mediterranean Archaeol.* **8**, 1.

Fink, D., J. Klein, and R. Middleton (1990), *Phys. Res. B* **52**, 572.

Gögen, K., and G. A. Wagner (2000), *Chem. Geology.* **166**, 127.

Hall, C. M., R. C. Walter, J. A. Westgate, and D. York (1984), *Nature* **308**, 26.

Ivanovich, M., and S. Harmon (1992), *Uranium-Series Disequilibrium: Applications to Earth, Marine, and Environmental Sciences*, Clarendon, Oxford.

Keisch, B. (1968), *Science* **160**, 413.

Lorius, C., J. Jouzel, C. Ritz, L. Merlivat, N. I. Barkov, Y. S. Korotkevich, and V. M. Kotlyakov (1985), *Nature* **316**, 591.

McDermott, F., R. Grün, C. B. Stringer, and C. J. Hawkesworth (1993), *Nature* **363**, 252.

Schwarcz, H. P. (1989), *Quaternary Int.* **1**, 7.

Schwarcz, H. P., R. Grün, A. G. Latham, D. Mania, and K. Brunnacker (1988), *Archaeometry* **30**, 5.

Semaw, S., P. Renne, J. W. K. Harris, C. S. Feibel, R. L. Bernor, F. Nesseha, and K. Mowbray (1997), *Nature* **385**, 333.

Shackleton, N. J., and N. D. Opdyke (1973), *Quaternary Res.* **3**, 39.

Singer, B. S., and M. S. Pringle (1996), *Earth Planet. Sci. Lett.* **139**, 47.

Taylor, R. E., P. J. Slota, W. Henning, W. Kutschera, and M. Paul (1989), in *Archaeological Chemistry IV*, R. O. Allen, Ed., American Chemical Society, Advances in Chemistry Series 220, Washington, DC, pp 321–335.

Wagner, G. A. (1998), *Age Determination of Young Rocks and Artifacts—Physical and Chemical Clocks in Quaternary Geology and Archaeology*, Springer-Verlag, Berlin, p. 466.

Wagner, G. A., and P. Van den Haute (1992), *Fission-Track Dating*, Ferdinand Enke, Stuttgart; Kluwer Academic, Dordrecht, p. 285.

WoldeGabriel, G., T. D. White, G. Suwa, P. Renne, J. de Heinzelin, W. K. Hart, and G. Heiken (1994), *Nature* **371**, 330.

CHAPTER

16

RADIOCARBON DATING

ROBERT E. M. HEDGES

Oxford Radiocarbon Accelerator Unit, Research Laboratory for Archaeology and the History of Art, Oxford University, Oxford, United Kingdom

16.1. INTRODUCTION

The calendrical date of an archaeological event is arguably the single most important datum that can be established in the study of archaeology. The method of radiocarbon dating, described here, is overwhelmingly the most frequently used technique to do this on an absolute basis. The reasons for this are that, in the first place, radiocarbon dating covers the range of time in which most archaeological events, certainly most human lives, elapse; in the second place, the method is applicable in a wide range of circumstances and to a wide range of materials; and in the third place, the method is sufficiently simple that usefully accurate dates can usually be determined. However, although the method is simple in principle, the complexities of chemical change on Earth mean that in practice there are many issues that can influence the outcome. These are discussed in detail in this chapter.

Radiocarbon dating was developed about 50 years ago (Arnold and Libby 1951). Since then, tens of thousands of archaeological dates have been produced, and the main chronological framework for most archaeology of the last 50,000 years set up. As a rule it is not possible to verify any given date without extra information, but sets of dates can be critically assessed for their internal consistency, their stratigraphic coherence, and their relationship to the overall dating framework. This is not easy to do objectively and invites circular argument. In any case, there is a general perception that in many periods and regions of archaeology radiocarbon dates determined many years ago are not necessarily very reliable. This is an issue that will be taken up further. It underlines the observation that methods in radiocarbon

Modern Analytical Methods in Art and Archaeology, Edited by Enrico Ciliberto and Giuseppe Spoto. Chemical Analysis Series, Vol. 155
ISBN 0-471-29361-X

dating are still improving: in the methodology of measurement, in our understanding of how best to apply the method, and in our knowledge of the global context (such as the calibration curve, see Section 16.4.5).

16.2. BASIC PROCESS OF RADIOCARBON DATING

16.2.1. The Principle

Any dating method relies on tracking a time-dependent process. Here the process is the radioactive decay of an isotope of carbon, ^{14}C (also called radiocarbon or carbon-14). ^{14}C has additional neutrons in its nucleus, two more than the common isotope ^{12}C, and one more than the stable isotope ^{13}C (which on our planet is present at about 1% of the abundance of ^{12}C). The excess of neutrons is sufficient to cause an average ^{14}C atomic nucleus to disintegrate in a few thousand years. A collection of ^{14}C atoms will diminish in time according to the well-known law of radioactive decay, first enunciated by Rutherford:

$$N(t) = N(0)e^{-\lambda t} \tag{16.1}$$

where λ is the characteristic decay probability (closely related, but not equal to, the half-life). The important point is that λ is not influenced by any environmental factor and so is a truly independent "clock." The value of $1/\lambda$ is set by the structure of the nucleus and happens, fortunately for archaeology, to be in the region of 8000 years—a timescale comparable to the rise of human dominance on the planet. Unlike much longer-lived radioactive nuclides, such as uranium or potassium-40, whose presence is a leftover from the nuclear explosions that formed our solar system, carbon-14 must be very recently formed to be present at all. The formation process was shown, 50 years ago, to be the impact of cosmic rays on nitrogen atoms in the stratosphere. This is more or less continuous, so the ^{14}C content of our planet is being constantly replenished, and the abundance of ^{14}C is at a level where the rate of loss through radioactive decay (which depends on the amount present) is balanced by the rate of formation (which depends on the cosmic ray flux). This abundance is usually expressed not as so many atoms of ^{14}C on the planet but as a ratio to the essentially constant ^{12}C level, that is, as $^{14}C/^{12}C$. The value of this ratio is about 10^{-12}, that is to say, on the planet—or more correctly in the biosphere—for every carbon-14 atom there are a trillion carbon-12 atoms. It is a most fortunate aspect of radiocarbon (and one that is not, unfortunately, shared by other cosmogenic nuclides, which might also have been the basis of simple dating methods) that it is first formed in the stratosphere, so that it is rapidly converted to the stable gas, carbon dioxide, which is then able to mix uniformly and rapidly about the atmosphere, and from there into the oceans and biosphere. Mixing into the oceans produces a complication that will be dealt with later—it takes a long time (see Section 16.2.1.1). Mixing into the biosphere is accomplished through photosynthesis (incorporation of CO_2 into plant biomolecules) and then via food chains into animals. As a first approximation, all living creatures have the same $^{14}C/^{12}C$ ratio as in the atmosphere, namely 10^{-12}.

Having a clocklike rate is not enough for a dating process; events are to be measured in time, and these must somehow start or stop the clock. Radiocarbon dating is really the dating of the moment of death of living things, because it is at this point that the flow of carbon, from biosphere to the dying organism, stops. A living plant maintains the level of 10^{-12}, but on death no more carbon enters or leaves; however, as nothing prevents the radioactive decay of ^{14}C, the ratio of $^{14}C/^{12}C$ decreases over time. Indeed Eq. (16.1) can be rewritten to produce the time since death:

$$t = 18500 \log_{10}\{[^{14}C/^{12}C(t)]/[^{14}C/^{12}C(0)]\} \qquad (16.2)$$

where the value of 18,500 corresponds to a half-life for ^{14}C of 5568 years.

It is clear from this equation that a measurement of $^{14}C/^{12}C$ at time t, together with a knowledge of $^{14}C/^{12}C$ at time 0 (identified with the "present" as in "before present," and conventionally taken to be A.D. 1950) is sufficient to calculate the date t. In this sense, radiocarbon dating is simple since only one measurement need be made on the sample; the other [$^{14}C/^{12}C$ for $t = (0)$] need be made only once.

However, several assumptions are involved if this equation is to be applied to archaeological (or indeed geochemical) material. These are:

1. The $^{14}C/^{12}C$ value used for the "present" has remained the same throughout the dating timescale.
2. The distribution of ^{14}C is uniform at any point in time throughout the biosphere.
3. The point in time at which ^{14}C atoms in a sample become isolated from the biosphere, that is, the actual "radiocarbon event," corresponds to the archaeological event whose date is sought.
4. All the carbon atoms in the sample being measured were present when the radiocarbon event took place.

Assumptions 1 and 2 are not strictly true; they depend on global processes that are quite well-understood in principle, though not in full detail. Assumptions 3 and 4 can be true but depend on the specific sample chosen and its context; so they require consideration of the particulars of the dating project.

16.2.1.1. Global Assumptions

Radiocarbon dating of samples whose age at death is known (e.g., tree rings) enables the assumptions, and measurements, to be checked; careful work has identified the very significant degree to which the level of $^{14}C/^{12}C$ in the atmosphere has fluctuated over the last few millennia. Because this fluctuation is global, a single correction applies to all samples for that particular time. Therefore a single curve showing the offset between measured radiocarbon age and the known age (essentially the dendrochronologically estimated age, which is taken to be the calendrical age) throughout the period suffices to "calibrate" the radiocarbon date. This therefore allows for violations in assumption 1, although at a price. The *calibration curve* will be discussed in Section 16.4.5.

Although atmospheric exchange with plants is fast (and can be tracked through the analysis of tree rings during the rapid increase in atmospheric $^{14}C/^{12}C$ during the 1960s when nuclear bombs were detonated in the atmosphere), the global turnover of carbon has many slower components. While most terrestrial situations are fairly direct, the roughly 1000-year turnover time in the deep ocean, where >90% of the planet's circulating CO_2 is located, means that marine systems can be depleted in ^{14}C. This applies especially to the deep ocean, but also to surface waters, to an extent depending on upwelling currents. In general, food chains based on marine carbon are measured to be 300 to 600 years "older." There are numerous other "reservoirs" for carbon participating in living processes that are of an older age. These include calcite rocks releasing bicarbonate ions, volcanic vents releasing CO_2, soil organic matter that may become reoxidized, biological recycling, as with bacteria, and the different turnover times associated with biological processes that can be comparable to the errors of the measurement for long-lived species, such as humans. Most of these effects are well recognized, and generally can be included in a consideration of the dating.

16.2.1.2. Sample and Context Assumptions

The event dated by radiocarbon is the time elapsed since the sampled atoms were in full metabolic exchange with the atmosphere (or its equivalent). This generally does not exactly correspond to the archaeological event purportedly being dated, and it is always essential to determine the relationship between these events. To take but one example, radiocarbon dating on a sample of wood or charcoal containing many annual growth rings will return a weighted average of the date for when the cellulose of the wood was formed—a quite different date from that of felling the tree, and even more different than the date of a dwelling, say, in which the timber was reused from an earlier phase.

A different, though related, problem is where the sample combines carbon from various sources. Bone collagen from animals eating both marine and terrestrial foods will incorporate carbon from reservoirs of two different initial radiocarbon compositions. A more complex example is that of pottery, which can contain residual carbon from the clay (geological age), carbon from unburnt organic temper (age of firing), from food remains (age of use), and from mobile organic molecules in the burial environment (generally a postdepositional age). Sometimes it is possible to extract one source (e.g., characteristic lipids from food remains). A difficulty common to many samples is the uptake by organic molecules from the burial environment. Various "pretreatment" methods have been devised to deal with this, but it is difficult to prove that these are always completely successful since neither is the burial environment well characterized nor do the original organic molecules usually remain in a chemically pristine form.

Therefore much of the effort in radiocarbon dating aims to understand better the various ways in which carbon atoms are replaced or otherwise altered during burial, as well as critically assessing the possible sources of carbon in the sample in the first place. In addition, dialog between the archaeology and the physics and chemistry of dating is needed for any dating, other than for standard situations, to have a useful archaeological interpretation.

16.3. DATING IN SEQUENCE

From the above, it is clear the dating process involves linking many elements together. In brief, these are, in operational order:

- Formulating the archaeological events to be dated
- Selecting the most appropriate samples for dating
- Ensuring that the samples can, after pretreatment, register an unambiguous radiocarbon event
- Understanding the relationship between the archaeological event and the radiocarbon event
- Selecting and carrying out suitable chemical pretreatment on the sample
- Measuring the $^{14}C/^{12}C$ ratio of the sample and estimating the errors in the measurement

- When possible, calibrating the radiocarbon date to produce a calendrical date
- Interpreting the calendrical date as pertaining to the archaeological event in the context of the events of the site as a whole

16.3.1. Measurement Methods

Measuring the sample $^{14}C/^{12}C$ is at the core of radiocarbon dating, and technical developments still form a large part of international conferences on the subject. The principles are reviewed here and detailed in Section 16.4.

Because $^{14}C/^{12}C$ is very small (i.e., 10^{-12}), ^{14}C has to be detected by very sensitive methods. Until recently, the only method available used the signal obtained from the radioactive decay of the ^{14}C nuclei. This is called the *beta decay* or *beta counting method* (Arnold 1954; Vries et al. 1959) (often, the "conventional" method), since the beta particle (high-energy electron) emitted from a ^{14}C nucleus can be detected as a single event. Two drawbacks exist: (1) The long half-life of ^{14}C (i.e., about 5730 years) implies that in any sample, only a very small fraction actually decays within a reasonable measurement time (about 1 in 400,000 in a week); it follows that comparatively large samples are needed to produce an adequate estimate of the ^{14}C content. (2) The signal of the beta decay is comparatively weak and not easily distinguished from general background radioactive signals. This makes for considerable difficulty when measuring old samples, where the ^{14}C will have decayed to the point where $^{14}C/^{12}C$ is reduced to 10^{-14} or less. Since the 1980s, an alternative method, accelerator mass spectrometry or AMS, has been developed. In this, ^{14}C is detected mass spectrometrically, that is to say, ^{14}C atoms are separated from all other atoms and isotopes by methods that recognize their unique combination of nuclear charge and mass (Hedges 1987; Litherland 1987). The AMS system therefore analyses all the ^{14}C in a sample (in principle) but in practice analyses about 1%. This is some 4000 times more than the beta method and so enables samples of at least 1/1000 less in size to be measured, thereby opening up a much wider range of possibilities. AMS methods also have difficulty in measuring samples where $^{14}C/^{12}C$ is less than about 5×10^{-15}, since the small sample size makes it extremely vulnerable to any contamination by modern carbon. AMS methods can now, after a decade's development, be said to equal the accuracy and precision of beta methods. Another important comparison is cost; AMS methods require much more capital investment, but one laboratory can make far more measurements per year. At the time of writing, the costs have become roughly comparable; although it should be borne in mind that the cost of chemical preparation accounts for about half the total cost.

16.4. DATING OPERATION IN THE LABORATORY

This section treats all the laboratory issues of radiocarbon dating, from the choice of a particular sample to the production of a radiocarbon date complete with laboratory reference number.

16.4.1. Sample Pretreatment Chemistry

As explained above, pretreatment aims to remove carbon from sources that are irrelevant to the radiocarbon event. These are of two types: those that are somehow incorporated into the sample as an archaeological entity and those that become added as a result of the burial environment or conservation practice. In any case, they depend on the type of sample being studied, and it is simpler to consider various classes of material in turn. In all cases, a preliminary removal of extraneous material (e.g., plant rootlets) and a cleaning of dirt (as by shot-blasting) from any exposed surfaces (e.g., of shell and bone) is to be recommended.

16.4.1.1. Plant Remains

The dominant chemical material in dead plant material is cellulose. This is degraded in most environments where microbiological action is possible, but cellulose in somewhat degraded form often survives in waterlogged or very dry or frozen conditions for millennia. Examples are wood, straw, or woven textile. *Macrofossils* such as twigs or seeds are often recovered from lake sediment cores. Pollen spores are noted for their resistance to decay, although here the cellulose component is minor. Identification of such material, for example, as to species, is important, at least in order to assess for any contribution of dissolved bicarbonate to photosynthesis, as in aquatic plants. In hard-water regions such plants can incorporate carbon from carbonate derived from bedrock, which will be radiocarbon depleted. There is no evidence, despite careful search, that plants incorporate CO_2 from the soil. Plants growing near volcanic fumaroles have been found to be severely depleted in ^{14}C, but the effect is negligible beyond a hundred meters or so. Apart from what is often called the "hard-water effect," plants do not as a rule present complex sources of carbon with different radiocarbon ages, and the main question of pretreatment is how best to remove contaminants acquired during burial.

A very general approach is to "wash" with dilute hydrochloric acid (mainly to dissolve particles of carbonate) and to extract soluble "humic" components with a mild alkali, generally dilute sodium hydroxide. Groundwater percolating through soils contains a number of poorly defined, mobile complex organic molecules, varying in molecular weight and structure, but characterized by carboxyl groups, a degree of aromaticity, and a degree of chemical stability. Such material can be operationally divided into fulvic and humic acids and humin, depending on the solubilities at a given pH. The relevant point here is that there is potential for buried radiocarbon samples to absorb weakly reactive mobile organic material. Well-preserved material can usually be further purified by "bleaching" (oxidizing) with an acidic solution of sodium hypochlorite. This has the effect of oxidizing lignins and possibly humic contaminants (though the evidence for this is weak). It is found that wood, so treated, gives more consistent radiocarbon dates than when simply treated with acid/alkaline washes (Hoper et al., 1998).

Generally, however, the best chance for plant material to survive is if it has been charred, that is, heated to pyrolyse the carbohydrate content but not oxidize reduced compounds or residual carbon. Charcoal, especially where this occurs in fires, is not susceptible to microbiological attack and decays through comminution and slow oxidation (accelerated in hot climates). Most charcoal seems to be satisfactorily freed from contamination by a simple acid wash and alkaline humic extraction. If charcoal is reduced to a very small grain size, any sample is liable to be contaminated by other fine-grained organic deposits, which may be of different ages. To some extent, these can be differentially removed by oxidation with acid dichromate or permanganate (Gillespie 1990). However, they are not samples of choice, simultaneously raising difficult chemical and taphonomic problems. Charcoal fragments of a few millimeters in diameter are very suitable for dating, however (provided they do not represent long-term growth). Many plant materials, and especially charred grains, may only be partially carbonized; humic extraction with alkali frequently totally dissolves such material; more gentle methods can be applied, such as ultrasonication in hot distilled water or ammonia solution and sodium pyrophosphate.

Charcoal, in general, probably constitutes more than 50% of all archaeological dating. This reflects its ubiquity and quantity since it is not necessarily the material of choice.

16.4.1.2. Animal Remains

Animals are mainly composed of fats and proteins. As with plant remains, these become food for many other creatures and, apart from bone, quite special circumstances that inhibit macro- and microbiological action are needed for their survival. Chemical deterioration of protein is fairly rapid, and many proteins only survive in very cold conditions. The more stable crosslinked proteins, such as keratin (hair, wool), elastin (skin), and collagen (skin, bone) do survive better, and much recent research has been expended on understanding the general survival of protein (Collins et al. 1998). So hair, horn, skin, and bone can be found in dry desert, dry tundric, and also temperate waterlogged peat-bog environments. In the latter, it is probable that humic acids in peat help preservation through a crosslinking reaction very similar to that of leather tanning. The preservation of animal fat is less obvious, but most lipids are in fact chemically rather stable in the burial environment. A special case is the exoskeleton of insects (and arthropods), which is composed of chitin, a polymer of an amino sugar and therefore chemically in between cellulose and protein. Chitin also preserves quite well in waterlogged environments.

Bone. In most animal remains, bone is present and is the material of choice. Bone contains 20% by weight of protein, of which 90% is the structural protein collagen. It also contains small amounts of lipid (fat) and of carbonate, in addition to many complex molecules whose presence is, so far, irrelevant to radiocarbon dating. As with plants, the main aim of pretreatment is to remove potential contaminants since there is only one source of carbon, that of the food. However, the food chain may well contain components originally from different sources (such as, but not only, marine and terrestrial foods), although these cannot be chemically disentangled once in the animal bone.* The question is one that must be addressed through considering the paleoecological context. Another issue is that of the turnover time of the bone collagen. All the evidence points to this being long for humans—typically up to half the age of the individual. Under some conditions this could influence the date and needs to be borne in mind.

Provided at least 1%, and preferably more than 5%, of the collagen remains, it can be extracted and dated. Collagen extraction methods first dissolve the inorganic matrix, either with dilute HCl or EDTA, to leave insoluble protein, and then exploit the ability of collagen to be made soluble in hot water. The resulting solution, known as gelatin, can be further purified, if thought necessary, by selecting out a molecular weight range (Brown et al. 1988), by elution from an ion-exchange column (Law and Hedges 1989), by hydrolysis to amino acids (Gillespie et al. 1983), and by purification of individual amino acids (Stafford et al. 1991). The technical effort and expense involved usually does not justify extreme methods, although a problem of collagen dating is that it is difficult to be certain how well purified the colla-

* This may not be true in the future; for example, an animal that eats fish protein and terrestrial starch should have essential amino acids in its collagen derived only from fish, and an high proportion of non-essential amino acids derived from starch; separate dating of the amino acids should establish this; the experiment is not an easy one to do; the true date would be closer to the nonessential amino acid date.

gen actually is. Amino acid analysis is helpful, as is the measurement of the carbon:nitrogen ratio in the preparation. Indeed, obtaining a C/N (atom ratio) value within the limits of 2.9 to 3.5 is considered an essential prerequisite for dating or stable isotopic measurement. However, an admixture of only 5% extraneous carbon may severely alter the date, and it is very difficult to devise analytical tests that can be applied to the often badly degraded collagen that remains in bone yet be sensitive enough to detect 5% admixture of noncollagenous material. It is possible that contamination is more widespread and less removable beyond the 1 to 5% level than is usually thought, but that the radiocarbon date is usually unaffected because the contamination has a similar radiocarbon age. There is no doubt that the advent of the AMS method, for which samples can be very much smaller and which can receive more complete and elaborate treatment, has enabled the reliability of dates on bone to be very greatly increased. Nevertheless, we do not have a good estimate for exactly what percentage of bone dates, in what circumstances, can be said to be reliably within their error limits. The best guess is that less than 10%, but possibly only 5%, of bones may give dates that are compromised by contamination effects.

One way to achieve much greater certainty that contamination has been removed is to extract specific amino acids or peptides. Hydroxyproline has been isolated and measured (Stafford et al. 1982), but the process is very labor intensive. In many ways a more suitable technique is to select tripeptides obtained by the specific digestion of collagen by the enzyme collagenase (Van Klinken et al. 1994). These are then separated from other contamination and degraded material by high-pressure liquid chromatography (HPLC), and tests show that a more reliable date is produced in conditions where contamination is known to be difficult to remove. However, the method requires 10 to 20 times the quantity of bone and so is not applicable to bone that is very low in collagen.

Low-Collagen Bone. Low-collagen bone is particularly difficult to date satisfactorily but is frequently submitted especially from countries with warmer climates. Unfortunately, there are no good alternatives at present. Methods have not been devised that satisfactorily isolate the carbonate content of bone for dating; too much exchange with environmental carbonate can take place during burial. Some recent attempts have been made with tooth enamel (Hedges et al. 1995) (which is less porous and more crystalline), and perhaps a better understanding of how bone interacts with its environment will lead to methods that better remove or reverse the effects of burial. Again, the important issue is to be able to rely on the date obtained. Too little lipid survives in bone to be useful in dating, although recent work on measuring the $\delta^{13}C$ of cholesterol from bone suggests that, on occasion, heroic efforts could be made to extract sufficient material in this way (Stott and Evershed 1996).

Burnt Bone. Bone that appears charred often contains no amino acids or protein but substantial quantities of reduced carbon. This carbon probably comes from the collagen, rather than fat from flesh, as it has the same stable isotopic value. Strongly burnt bone, as results from cremations, usually has a white, gray, or blue appearance and generally contains rather little carbon of any kind. The reduced carbon may be treated similarly to charcoal or charred seeds and often gives correct dates, where these may be externally checked. However, it does not necessarily do so, perhaps because it is virtually impossible to ensure removing humic acid contamination from material where the carbon is so finely distributed. Some burnt bone dates are certainly wrong. Therefore, although not recommended as a suitable material, it is possible that research aiming to give a better understanding of the environment and of pretreatment approaches would enable burnt bone dates to be reasonably reliably dated; this could be important in sites where collagen does not seem to survive in bone and where there is little else to date.

Shell. Shell commonly occurs as the mineral calcite reinforced with structural protein. Some animals secrete calcium carbonate shells as aragonite; this enables recrystallization (which is always to calcite) to be easily recognized by X-ray diffraction analysis. Pretreatment usually consists of a superficial etch of the carbonate by acid and oxidation of extraneous organic matter by acidified hypochlorite solution. The main problems with carbonates are replacement or addition with diagenetic carbonate, and it is often impossible to detect this. It is also very difficult to remove such carbonate chemically without removing the original material. It is more worthwhile to put effort into careful examination of the material, using scanning electron microscopy and X-ray diffraction, selecting only those shells that appear free from diagenetic alteration. Carbonate shells generally contain substantial protein, although this is not easily characterized chemically. Some success has been obtained in dating the protein component; this has worked well for ostrich egg shell, which seems to act as a closed system (Johnson et al. 1997). Again, the biological sources of carbon can be problematic; shells in marine environments naturally represent the marine reservoir (for both carbonate and protein). Terrestrial shell carbonate often contains carbon derived from both country rock carbonate as well as atmospheric carbon, while the protein content is much more liable to be derived entirely from plants; the degree depends on both species and environment.

Chitin. This material only preserves well in waterlogged or very dry or cold conditions and is rarely used in dating. One application is the dating of insect cuticles, for example from lake sediments. It should be possible to split chitin with enzymes to *N*-acetyl glucosamine, the monomer from which it is built, and to isolate this by HPLC. In general, the few chitin dates that have so far been made are on the insoluble polymer after cleaning with acid and base extractions.

16.4.1.3. Anthropogenic Materials

Many human artifacts with organic components survive quite well; they often have complex carbon sources, however, which makes their dating especially problematic. Furthermore, the carbon is often rather diluted, and the artifacts are often valuable so that samples have to be small and the likelihood of contamination by the environment is large. The advent of AMS methods has meant many more such samples can be submitted for dating, but the golden rule, that one should not date material that is not chemically well understood or characterized, is very often completely compromised for these materials.

The main classes of material include textiles, pottery, adhesive residues, and paintings (usually on rock), and remarks relevant to pretreatment chemistry are made here. Archaeological dating is also briefly mentioned in Section 16.5.3.

Textiles. Textiles are straightforward, being cellulosic (linen, cotton, bark, etc.) or proteinaceous (silk and wool). The main problem is if they have been conserved or otherwise exposed to unknown cleaning agents.

Pottery. Pottery, as already mentioned, can contain carbon from multiple sources, and its dating is difficult unless carbonized macrofossils can be picked out or obvious macroscopic food remains can be sampled. However, these are not usually able to be chemically characterized in bulk. Methods are being developed to extract specific lipids, such as palmitic acid, from the interior of cooking pots. This requires separation of extracted and saponified lipids by derivitization and gas chromatography, and since analytical separations only yield about 1 μg per compound, the separation process must be repeated about 100 times to accumulate sufficient yield for dating (Eglinton et al. 1995). This is a good example, however, of the advantages of exactly chemically characterizing what can be dated, although at the cost of considerable time and effort. In this case, the source of the palmitic acid can be reliably identified with the foodstuff and distinguished from older or younger carbon sources in the pottery.

Organic Residues. Residues of gums and glues are occasionally found, for example, on flint implements for hafting. It is difficult to separate out any one compound, but the homogeneity and high concentration of material precludes contamination except that due to mobile humics. Dating so far has been based on pretreatment using the standard acid–base–acid extraction, although it would be interesting and preferable to date a soluble lipid extraction where sufficient material exists. In any case, such residues should at least be separately analyzed (Pollard and Heron 1996).

Painting. Rock paintings constitute one of the greatest challenges for radiocarbon dating, and success has been very patchy. Where charcoal can be positively identified as the black pigment, and exists in sufficient quantity, dating based on the standard pretreatment for charcoal, using gently humic extraction with phosphate, appears to be very satisfactory (Clottes et al. 1995). Where more complex materials have been suspected or identified, generally introduced as binder, results have been more controversial. Natural deposits may also be involved, as with calcium oxalate formation (Watchman 1990) and can form a stratigraphy whose dating depends on understanding the formation process. The same material, this time sourced from plants, may also be used as a pigment (Hedges et al., 1998) Dating paintings based on large quantities of beeswax has been more straightforward (Nelson et al. 1995) because of the greater molecular simplicity, the better preservation, and the easier extractive methods of lipids.

16.4.1.4. Paleosols, Peats, and Sediments

These materials are chemically very complex and heterogeneous. Furthermore, the radiocarbon date of any one chemical component is not easily related to any particular archaeological or depositional event. Insofar as the material is derived from what was a contemporary soil, it should be noted that soil components vary in their ^{14}C date; as a rule, the greater the chemical disruption required to extract a given fraction, the older the radiocarbon date. This is in accordance with the turnover time of different components of the soil organic matter (Harkness and Becker-Heidman 1996). It is best to select specifically identifiable macrofossils, where possible, although the danger of such material having been redeposited must be recognized. In terms of pretreatment, a cellulose fraction (i.e., as for wood) can often be obtained from peat, and sometimes from sediments; otherwise fulvic and/or humic acids as well as humin can be extracted using standard methods. These frequently give dates in accord with each other, but any particular sample must be considered on its own terms. Lipids can also be extracted, the most commonly used solvent being a methanol–chloroform mixture, and bulk lipid dates have been reported (Fowler et al. 1986). It has also recently become possible to date single specific lipids (Eglinton et al. 1995) in a way similar to that for lipids extracted from pottery. However, the cost of obtaining such detailed information should be justified by the value of the result.

16.4.2. Chemical Conversion of Sample

Pretreatment removes contaminating and other misleading sources of carbon. It is next necessary to convert the purified material into a form that can be assayed for ^{14}C content. In all cases, the first stage is to oxidize the carbon in the sample to CO_2. Subsequent processing depends on the method of measurement.

Oxidation to CO_2 is generally accomplished by combustion in a stream of O_2 or by heating with CuO in a sealed quartz tube. Using a stream of O_2 has become popular recently because of the development of continuous-flow systems that can be easily interfaced to mass spectrometers and typically operate on sample sizes suitable for AMS (Hedges et al. 1992b). It is important that the conversion, to CO_2 and for any subsequent reactions, is close to 100% in order to minimize any change in isotopic composition. This topic is relevant to the discussion on systematic errors (see discussion below), but here it should be said that it is usual to measure the $^{13}C/^{12}C$ ratio of the CO_2 (referred to as $\delta^{13}C$), by injecting an aliquot into a stable isotope mass spectrometer. The value obtained is assumed to apply to the pretreated material and also to the sample as measured for ^{14}C (whether it be CO_2, graphite, or benzene). (In fact, small systematic errors in this assumption are canceled since they will apply equally to standards treated in the same way.) The main point of the $\delta^{13}C$ measurement is to determine how much isotopic fractionation, or discrimination, has taken place in the transfer from the atmosphere (where the value of $\delta^{13}C$ is known) to the sample being measured. Such fractionation can occur through many processes in the environment—photosynthesis being the most important—but potentially occurs in laboratory processes too. If laboratory isotopic fractionation can be kept small (e.g., by maintaining the yields for all processes close to 100%), the measured $\delta^{13}C$ value can be interpreted as resulting from wholly environmental processes and thereby provides additional information about how the sample was formed (Deines 1980). However, from the point of view of radiocarbon dating, isotopic discrimination against ^{14}C, from atmosphere to measured sample, will give an error, unless corrected for. The correction is evaluated from a measurement of $\delta^{13}C$ since ^{14}C is discriminated by twice the magnitude of the discrimination of ^{13}C.

In AMS systems, CO_2 is usually converted to graphite [although one laboratory—Oxford (Bronk Ramsey and Hedges 1994)—also can measure directly from CO_2]. The process entails reduction of CO_2 usually on a catalytic surface of Fe or Co, in the presence of H_2 or Zn at a temperature of a few hundred degrees Celsius. The graphite forms over a period of some hours on the Fe powder, which is then available to be pressed into a pellet of about 1 mm diameter for loading into the ion source of the AMS system. In beta systems (Kromer and Munnich 1992; Polach 1992), either CO_2 can be directly used in gas counting or, it is converted to benzene for scintillation counting. The conversion to benzene (which is done on a gram scale) is effected by reduction of CO_2 with hot Li to Li_2C_2, followed by hydrolysis to C_2H_2, which is then catalytically converted to C_6H_6. Although seemingly daunting, experience has developed this process to give very high yields, and in fact more laboratories use scintillation counters than gas counters.

16.4.3. Measurement of Radiocarbon Content

The main points of interest in the measurement process are how techniques in physics are applied and to consider the sources of error, which will determine performance. It should be borne in mind that the two main methods, AMS and beta, differ somewhat in both these features.

16.4.3.1. Technicalities

This section is deliberately brief. Fuller details can be found from citations in the Bibliography.

Scintillation Counting. Scintillation counting has been the main "workhorse" for radiocarbon dating, but recent work has shown that, with care, results can reach the highest accuracy and precision. To the benzene of the sample is added a compound that scintillates (i.e., flashes) for every beta emission, as well as a filler to achieve constant volume. The scintillations are counted photoelectrically, and the counting of samples scheduled such that calibration, using standards in the same form as samples, is able to compensate for drifts in the counting efficiency. Thus one counter will typically have 30 to 50 samples measured serially, with many cycles before a sufficient number of counts (scintillations identified as ^{14}C beta decays) has been obtained for each sample. This may take several weeks. To achieve the highest accuracy (of 0.2% in the measurement), a very large number of contributing sources of error must be identified, minimized, and controlled.

Gas Counting. Counting using gas (almost always CO_2 nowadays) has tended to be regarded as a more difficult method, requiring the gas to be of exceptional purity (especially free from O_2) and the counting electronics of great stability. But for many years the highest precision results were obtained from a few gas counting laboratories. In gas counting, a specially designed proportional counter is filled with CO_2, and each beta decay is registered as an electrical signal. The counts are accumulated over a period of about a week. One feature for both beta systems is the need to avoid registering spurious signals from radioactive and cosmic ray events, and much effort has gone into the active and passive designs for shielding the equipment.

AMS Systems. These tend to be large installations, built around tandem electrostatic accelerators. The smallest of these operate at about 2 MV, and the total beamline length is several meters. All AMS systems operate a tandem type of ion beam mass spectrometry; the ion source generates C^- ions of which a mass 14 ion beam (containing such ions as $^{13}CH^-$ as well as $^{14}C^-$) is injected into an accelerator; this converts the beam to various positive charged ions at greatly increased energy, from which the mass 14 ions of $^{14}C^{3+}$ are able to be separated and positively identified by their characteristic signal in a suitable detector. Usually 30 min of beam time is sufficient to accumulate sufficient ^{14}C counts in the detector for the sample to be measured. Most AMS laboratories run samples serially, interspersed with calibrating standards, in much the same way as for scintillation counters, usually on a 24-h schedule. However, the dating output from an AMS installation is much higher, and some laboratories achieve 3000 to 4000 measurements per year. AMS systems are not subject to spurious signal detection due to radioactive backgrounds; rather their backgrounds, which tend to be lower, are from either misidentified non-^{14}C ions or ^{14}C ions originating from sources other than the sample.

16.4.3.2. Performance

In those aspects actually relevant to the characteristics of the measurement, AMS and beta methods differ rather little. Overall accuracy and age range are similar. The treatment of errors tends to be rather different, and, of course, the methods differ greatly in the size of sample and in measurement time and cost. It is worth pointing out here that the smaller sample size requirement of AMS often permits multiple dating, for example, if it is desirable to repeat a particular result, and this can have an important effect on reliability.

16.4.4. Calculation of Result

In all cases, ^{14}C content is not measured absolutely but by comparison with a standard that is measured in the same batch. The standard is related to a primary standard originally prepared by the U.S. National Bureau of Standards (NBS) and to which a radiocarbon date is referred, by the definition:

$$\text{Conventional radiocarbon age} = t = -8033 \ln[1 + D^{14}C/1000] \quad (16.3)$$

where

$$D^{14}C = 1000[(1 + \delta^{14}C/1000) \times 0.975^2/(1 + \delta^{13}C/1000) - 1]$$

where $\delta^{14}C$ is the relative difference between the sample activity [i.e., $^{14}C/(^{12}C + {}^{13}C)$] and 95% of the primary standard (NBS oxalic acid), and $\delta^{13}C$ is the relative difference between the sample $^{13}C/^{12}C$ ratio and that of the PDB standard.

Note that this definition relates the date to a nominal $\delta^{13}C$ value of −25‰, which is close in value to most radiocarbon samples (being appropriate for wood). Thus the definition automatically incorporates the correction for isotopic fractionation (where this is measured; otherwise a value must be estimated). To an extent, the definition of the radiocarbon date is arbitrary in several ways (e.g., the half-life is known not be exactly correct), although the parameters have been chosen to be close to the calendar date; this arbitrariness is canceled on calibration. The date is called a conventional radiocarbon age and always referred to as so many years BP (i.e., before present). [This can now give confusion when compared with dates by other methods, such as thermoluminescence, where dates are also often quoted as BP, but where the authors consider the date scale to be absolute (calendrical).]

In practice, beta methods actually measure $^{14}C/(^{12}C + {}^{13}C)$, while AMS methods usually measure $^{14}C/^{13}C$, or sometimes $^{14}C/^{12}C$. However, these can be sufficiently accurately related to the proper definition, but care must be taken to consider the consequences when there has been substantial isotopic fractionation. As already mentioned, the sample $\delta^{13}C$ must be separately measured, and a correction applied. The correction can be quite significant—up to 1% or so in $^{14}C/^{12}C$, or about 100 years. The radiocarbon result is generally quoted along with the $\delta^{13}C$ measurement and can only be regarded as a properly constituted radiocarbon date if it has also been given a Reference Number that identifies the particular Laboratory responsible for the measurement.

16.4.4.1. Consideration of Errors

The mathematics of the radiocarbon decay curve imply that an error of, say, 1% in the measurement (of $^{14}C/^{12}C$) is equivalent to a fixed error in years (i.e., 83 years)—no matter whether the date is at 1000 BP or 20,000 BP. However, it is much more difficult to measure old samples to a precision of

1% because the ^{14}C content is so very small. At 37,000 years the $^{14}C/^{12}C$ ratio has fallen to 1% of the modern value, and it becomes increasingly difficult to detect ^{14}C with any precision. Eventually, generally between 40,000 and 50,000 years, depending on circumstances, the estimated errors outweigh the value, and the limit of dating has been reached.

Two types of error are distinguished; the uncertainty in the result, which shows as a spread when the measurement is replicated many times, and the bias in the result, which can only be properly known if the "true" result is available for comparison. The first type, called random error, or (lack of) precision, can be estimated fairly well since although few measurements are replicated frequently, the behavior of a group of measurements can substitute quite well. The second, systematic error, can only be corrected for by measuring known material, with the assumption that the same correction applies generally. Since so few dates can be checked, radiocarbon laboratories have to devote a large amount of effort both into understanding and reducing the sources of both random and systematic error, and in performing checks of replication and on known age material to ensure that the errors behave as expected.

Random Error. Since $^{14}C/^{12}C$ is measured by counting ^{14}C detections, the ratio for the archaeological sample is estimated from what is only a very small subsample, with a statistical error that is proportional to $1/N^{0.5}$, where N is the number of detections. This tends to be the dominant random error in radiocarbon dating and ranges from less than 0.5% for recent samples to about 3% at 25,000 years. It can be reduced by longer counting times, but the square root relationship makes for diminishing returns. Other random errors may result from replication tests—such as testing if repeated measurements of the same sample incorporate more scatter than expected from counting statistics, and also, equally importantly, from the uncertainty of applied corrections. For example, beta laboratories have to subtract a background, which itself has a statistical error. AMS laboratories usually subtract a background due to extra ^{14}C being introduced during sample processing (e.g., combustion); the average value and variability must be estimated by processing blanks. For older samples, the error in the background subtraction becomes the dominant random error. For samples so old that the magnitude of $^{14}C/^{12}C$ is reduced to less than twice the standard deviation of the random error, the date should be reported as being "older than" the date corresponding to the measured $^{14}C/^{12}C$ plus two sigma. That is, for a date where $^{14}C/^{12}C = R \pm r$ (where r is the random error quoted as one sigma), if R is smaller than $2r$, the date is reported as $^{14}C/^{12}C < (R + 2r)$. (Of course, the $^{14}C/^{12}C$ values are to be translated into dates in radiocarbon years.) Thus the ultimate age for radiocarbon dating depends on the estimate of r, which would include counting statistics, variability in the background subtraction, and so on.

Systematic Errors. Essentially, if these can be estimated, they can be corrected for; the uncertainty in the estimate will be added to the quoted random error (see above). If such errors cannot be estimated, then nothing can be done about them (except to make vague warnings). There are many causes of bias in measurements, and it is the art of the laboratory measurement to aim to detect all that are significant (mainly through measuring known age material). If, for example, laboratory processing systematically introduces additional carbon into the sample with a $^{14}C/^{12}C$ value equivalent to, say, 5000 years, this will make younger samples slightly too old, older samples too young, and have no effect on samples of 5000 years. Therefore such systematic effects need careful monitoring. The main biases are due to this type of effect (mainly for AMS samples), to isotopic fractionation (discussed above), and to possible biases in the measurement equipment (such as in the ion beam optics or in the counting electronics). One very important systematic effect is due to the fact that very old samples are much more likely to be contaminated by younger carbon, to an extent that may not be entirely removed by pretreatment. Since this will be sample and context dependent, there is no way to estimate the error for any given sample; although accumulated knowledge of the efficacy of the pretreatment method, for samples of the type being processed, and from environments similar to the sample context, may provide a general guide. It is therefore quite possible that some older dates are in fact older than their radiocarbon measurement, due to this bias, and this must be borne in mind when discussing the reliability of dates (Hedges and Pettitt, in press).

16.4.4.2. Quality Assurance

Interlaboratory comparisons now play an increasingly important part in maintaining the quality of radiocarbon laboratory measurements. A series of three International comparisons has been made in the last decade (Scott et al. 1990, in press; Rozanski et al. 1992, Scott et al. 1998) in which over 30 laboratories participated. The comparisons were made on materials such as bone and wood, which also require pretreatment, so that the complete process can be compared. The value of these exercises is demonstrated by the decline, over the decade, in the proportion of laboratory measurements that were well outside the expected error band of agreement. The overall picture shows that the top half of the laboratories are operating more or less according to specification, but that this is far less true for the bottom half. At one time it seemed as if the divergence of measurements from the consensus value was about twice as great as the estimated errors (i.e., suggesting, on average, that the laboratory error is a serious underestimate), but that no longer is the case, mainly due to improvement in the performance of the less accurate laboratories. In fact, the errors quoted by laboratories claiming high precision results, and by the great majority of laboratories with modern facilities, including the AMS labs, are being correctly estimated or only slightly underestimated. Such international exercises are a valuable addition to essential "in-house" quality control, and most laboratories nowadays operate a formal procedure for their own checking for bias and uncertainty estimates. There is in any case a move toward consolidation, with fewer laboratories producing more dates.

16.4.4.3. Reporting of Radiocarbon Dates

Radiocarbon dates are reported according to a well-defined protocol. As already stated, the designation BP stands for Before Present (defined as before A.D. 1950), and in the context of radiocarbon dating, is not necessarily equivalent to a date in calendrical years. Where measurements appear more recent than this, the value is expressed as so many "percent modern carbon" or pmc. The protocol includes the laboratory reference, the conventional radiocarbon age as calculated according to Eq. (16.3), together with the estimated error (quoted at one standard deviation). The error is assumed to cover all the total random errors known by the laboratory (though this is not always very clear). Correction for isotopic fractionation is included, but any other correction, e.g., for reservoir effects, is not; such a corrected value would be designated a "reservoir-corrected radiocarbon age" (and so on). The standard form of reporting is to be found in the date lists of the journal *Radiocarbon*, although rather fewer dates are published in this way than formerly. Nevertheless, inclusion of comments by the submitter, and by the laboratory, for each date, together with such basic information as the exact sample location and context and material does enable the information to remain indefinitely relevant. This has its own danger, for once it has entered the literature, it is almost impossible to expunge a "wrong" date. Given that the practice of radiocarbon dating has improved enormously over each decade, both in terms of the techniques of chemical pretreatment and isotope measurement and in the archaeological interpretation, dates in the literature require a degree of individual scrutiny they rarely receive.

16.4.5. Calibration from a Radiocarbon Date to a Calendrical Date

It is important that the calculation of the radiocarbon date does not change over time, despite improvements in method, and therefore the calculation algorithm has some fixed conventions. For example, the half-life is taken to be 5568 years, rather than the measured and accepted value of 5730 years. In any case, radiocarbon years are not the same as calendar years, and the relationship can only be found by measuring material whose calendrical age is known. The relationship is known as the calibration curve, and has been determined, with the maximum accuracy, for about the last 10,000 years (Stuiver and Kra 1986; Stuiver et al. 1993; Stuiver et al. 1998). A portion is shown as Figure 16.1*a*. To go further back in time it is necessary to rely on material other than tree rings for a known calendrical age. An extended curve, based on dating by uranium series disequilibrium of corals and on lake sediments dated by counting annual layers, is included in the "Intcal 98: Calibration Issue" of the journal Radiocarbon (Stuiver and van der Plicht, 1998). This also includes further work on the Holocene calibration.

There is a substantial divergence (increasing to 1000 to 2000 years) between calendrical age and radiocarbon age, and, what is worse, this divergence has a lot of high-frequency structure (the curve "wiggles"). The wiggles are not yet discernible on the far less detailed calibration beyond 10,000 years, but no doubt are present since they seem to be caused by changes in solar activity (probably through changes in magnetic shielding of cosmic rays in the solar system). The major divergence is believed to be due to slow changes in Earth's magnetic field, having a similar effect. Other effects may also operate; for example, the major changes in reservoirs during climatic changes (such as the change of atmospheric CO_2 abundance during the last Ice Age) may well cause release or uptake of older ocean CO_2. Curves measured for trees growing in different parts of the earth are in very good agreement (although trees in the Southern Hemisphere are about 20 years "older" due to the stronger effect of the oceanic reservoir), and the need to obtain accurate calibration curves has helped drive the methodology for high-precision measurements and their accurate interlaboratory comparison.

To calibrate a date, one checks, from the calibration curve, which calendrically dated material would have a similar radiocarbon age to that of the measured material. The probability of any given age is based on a Gaussian distribution for the radiocarbon measurement, defined by the error at one standard deviation. This, unfortunately, produces a calendrical date with a probability distribution that is very far from Gaussian (it will often have multiple peaks); also, as a rule the distribution in the calibrated (calendrical) date is wider than for the radiocarbon date (though occasionally the calibration curve is such that the probability is actually sharpened after calibration). A notorious example is for radiocarbon dates of around 2400 to 2500 BP, which correspond, whatever the precision, to the unfortunately wide range of calendrical dates from 800 to 400 Cal B.C. There are several other similar periods in the Holocene, and those involved in archaeological radiocarbon dating need to be familiar with them since they impose impor-

(a)

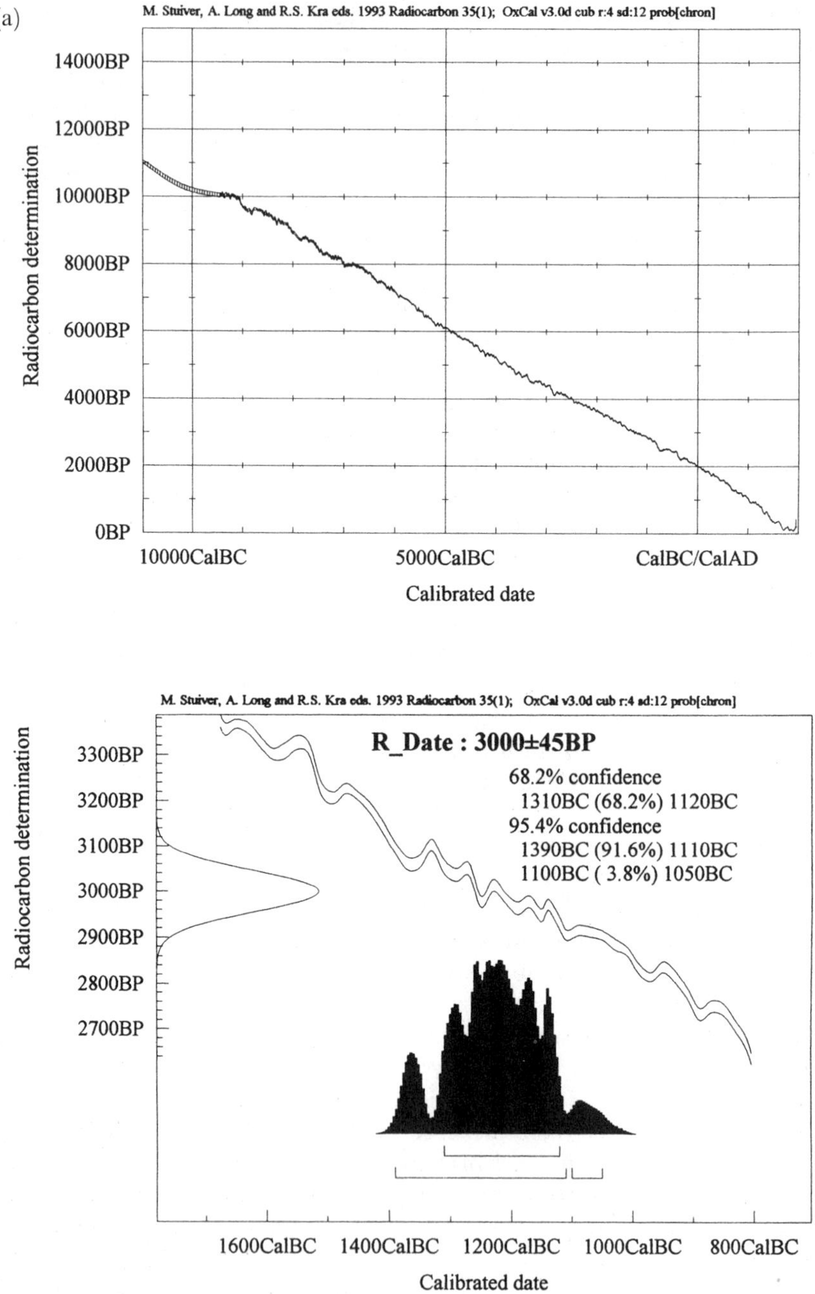

Figure 16.1. (*a*) Calibration curve for the last 10,000 years. (*b*) calibrating a radiocarbon date.

tant limitations on the information that radiocarbon dates provide. Calibrated dates often do not have well-peaked probabilities and are normally expressed as a range, or series of ranges, usually defined to be that within which it is estimated that the true date lies with a probability of 68% (equivalent to one standard deviation) or 95% (2 standard deviations). The result is expressed in so many years "Cal BP". The Cal standing both for calibration and calendrical. If so desired, 1950 years may be subtracted, and the resulting range expressed as years "Cal B.C." Expressions such as "bc" (which has been favored in some quarters in the United Kingdom) or indeed any form other than the above, is to be strongly discouraged as it violates the agreed convention and is apt to mislead. Computer programs are now well developed to calibrate dates and produce suitable outputs for the range, and the form, of the probability distribution of the calendrical date, in a user-friendly way (Bronk Ramsey 1995; Stuiver and Reimer 1993; Van der Plicht 1993). See Figure 16.1.

16.4.5.1. Statistics of Groups of Dates

Before Calibration. Multiple dates from portions of the same sample may be averaged, and the combined *random* error will be reduced by the square root of the number dated. This obviously improves precision, but since undetected small systematic errors will not be reduced, the practice cannot be taken too far. For example, AMS labs can produce single measurements to about 0.3 to 0.4% but that may well contain systematic errors at around 0.1%. High-precision counting laboratories, using sufficiently large samples, produce a measurement to 0.2% quoted error, implying that any systematic error is much smaller than this. If the dates are on different fractions or different samples suspected of being of the same radiocarbon date, the hypothesis can be statistically tested (as by a χ^2 test). It is then reasonable to treat the group as pertaining to one date and combine them as above.

After Calibration. Since the probability distribution of a calibrated date is non-Gaussian, the relationship between dates on a calendrical basis requires more sophisticated methods for analysis. For a single date, external information may reduce the probability distribution. For example, the radiocarbon date for the Shroud of Turin (Damon et al. 1989) was 750 $\pm$ 30 BP which calibrates to two peaks at 1280 and 1370 Cal A.D.; since the shroud was historically attested by A.D. 1352, the second peak can be rejected. The use of external information can be combined with probability estimates using Bayes's theorem, and the application of Bayesian statistics to groups of calibrated dates, although relatively new, has become very fruitful (Buck et al. 1991).

"Wiggle matching". A large piece of wood contains rings with a clear internal chronological relationship; several dates on single rings can be calibrated, and the rather wide individual probabilities are collectively constrained by Bayesian statistics using the chronological information. In favorable cases (i.e., favorable wiggles) the resulting probability distribution is so sharp as to amount to fixing the date to within a few calendar years. A good example is to be found in Kuniholm et al. (1996) in which an important but dendrochronologically "floating" tree ring section was matched rather exactly by radiocarbon dating.

Site Stratigraphy. Similar, though less precise, relationships can be recognized on archaeological sites. These usually involve stratigraphic succession; groups of dates can be collectively treated as phases, and different phases can be regarded as successive, with or without overlap. Different hypotheses of such kinds, often with quite complex logical structure, can be set up and tested for their consistency with the results, and the calendrical probabilities can be evaluated based on such hypotheses. As with wiggle matching, this often leads to a useful reduction in the range of probable calendrical date for each sample. In fact, it has been reported that, on average, the incorporation of stratigraphically based chronological information into the probability distributions for groups of dates brings about a roughly 50% increase of precision in the calibrated date. A good example of such an exercise is to be found in (Bayliss et al. 1997) in which the calibrated dates of stratigraphic features of Stonehenge could be made more precise by taking the relationships between phases into account. Dating Stonehenge is described in more detail below, but the original study should be consulted for explanation of the application of Bayesian methods. Programs (Bronk Ramsey 1995) have been constructed that allow users to frame their own specific hypotheses concerning external relationships between dates and to evaluate the calibrated probabilities from their particular set of radiocarbon dates. (The calculation of probability is best done by simulation rather than analytically, Gibbs sampling being the method most commonly employed.) This development is a particularly good example of how cooperation between field archaeology and laboratory scientist can improve results far beyond what each can achieve separately.

16.5. DATING PROCESS IN ARCHAEOLOGY

This section addresses the type and quality of information required for archaeological dating and follows with an overview of the contribution of radiocarbon dating to archaeology, with some selected examples.

16.5.1. Relationship Between Archaeological Event and Radiocarbon Event

However routine the dating, the relationship between archaeological and radiocarbon event should never be taken for granted. Usually, the archaeological event is either intrinsic to the sample or is determined by the context of the sample. In any case, the archaeological event may involve assumptions, often implicitly, that can complicate or invalidate the connection with the radiocarbon date.

16.5.1.1. Samples Dated for Their Intrinsic Information Concerning Archaeological Events

These have recently increased in importance because AMS has allowed such samples, which tend to be intrinsically valuable, not to be destroyed during dating. The greater the information content of the sample, the greater the implications of the date. So a date on cut-marked, species or genus-identified bone would be especially worthwhile, for the date shows what animal species was living in the particular environment of the time and also dates the human behavior. Even with this example, one must be satisfied that fresh rather than curated bone was being cut for, for example, decoration, or, for instance, that mammoth tusks and skulls, used in building huts in Gravettian central Europe, have not been gathered from previous ancient deposits. Documenting the occurrence, and disappearance, of particular species, such as horse or mammoth (Stuart 1991) over time has provided important data for the paleoecological setting of past humans. A second class of example is of domesticated grain. Certainty in dating the early appearance of cultivated species with morphologies diagnostic has been vital in showing, for example, how past climate has influenced the start of cultivation (Hillman et al. 1989) Also, such studies have been important in demonstrating that many apparently (contextually) early occurrences of domesticates are due to taphonomic irregularities. (For example, two dates of grape-pips from archaeological sites in the United Kingdom have been carried out at Oxford. One—from Neolithic Hambledon Hill—is at present the earliest record of grapes being grown in Britain; the other was shown by dating to be modern and testifies to the effect of international trade on twentieth-century luncheon diets.) Similar taphonomic problems arise with many human burials, especially in caves; many have been found in apparently Paleolithic levels, but, on radiocarbon dating the bones themselves, frequently turn out to be Neolithic or later in date, it being difficult, especially in old excavations, to detect the intrusive nature of the burial. Anthropogenic materials also come into this category. Since most organic artifacts are often chemically very complex, and present in only trace amounts, it is generally difficult to date a chemically defined component (see above). An interesting case is the dating of unidentified adhesive from a microlith from the Mesolithic site of Rekem in Belgium (Gowlett et al. 1987). This site also provided several small samples of charcoal, from a generally stratigraphically poorly controlled context. While the charcoal samples were very variable in date, and much younger than the lithic assemblages would imply, the adhesive was found to be of the expected age. Rock paintings often present problems of stratigraphy and sampling as well as those to do with chemical identification. The question, and difficulties, of pottery dating has already been mentioned (see 16.4.1.3).

16.5.1.2. Samples Dated Because Their Context Relates to Archaeological Events

This is probably the case for the majority of samples dated and certainly has been for most of the history of radiocarbon dating, so that field archaeologists tend to regard radiocarbon dating as the dating of contexts rather than as the dating of events. There are then two issues—the taphonomy of the selected samples and the relationship of the date of the context to that of the chronological framework for the site. In other words, has the context been well chosen? Has the sample been well chosen? Two questions about context one should be clear about are: What event does it represent (e.g., the development of a soil, the silting of a ditch, the filling of a post hole, the firing of a hearth, the creation of an incendiary destruction layer)? How well defined is this event in time? The questions concerning the selection of the sample within the context are essentially taphonomic and amount to being assured that the processes by which the sample was, and remains, deposited are known. Many difficulties arise at this point; perhaps the greatest is residuality—stable material, such as bone or charcoal, being redeposited from some other context. Articulated bone almost certainly has not been redeposited (over any length of time) nor large fragile charcoal pieces; one has to consider each specimen and its context on its merits and make the best possible choice. Finely disseminated charcoal can often be reworked and collected into lenses and is generally unsuitable, but may easily be the only kind of organic material on a site. Subsequent mobility of material should not be underestimated, and this is often not identifiable, especially in sandy soils. The larger the specimen, the better, from this point of view, and bone does seem to be, in general, stratigraphically a more reliable kind of sample. Even here, large debris flows, as in Kent's Cavern (United Kingdom), can juxtapose bones of very different age.

Radiocarbon dating can help contribute to sample taphonomy, partly by testing the various models one might construct for a given site, but also through measuring different fractions. A comparison has been made (Batten et al. 1986) between the humic acid fraction in charred material and the chemically immobile carbonaceous residue, for a range of materials (seeds, burnt bone, charcoal) from different sites. In general, there was agreement between the dates of the two fractions where the sample had not been stratigraphically misplaced (by whatever agent), but where the date of the sample and of its context were seriously different, this was reflected in the humic acid fraction, which tended to follow the date of the context. But this is only one study, and far more could be learnt when comparing radiocarbon measurement with taphonomic information.

16.5.1.3. Dating Samples Whose Archaeological Events Involve Complex Relationships with Radiocarbon

The classic example is the "old wood effect." The large quantities of charcoal needed for beta measurement inevitably are often derived from massive pieces of wood. The radiocarbon event is the time the particular tree rings formed; the archaeological event is likely (but not necessarily) to be the time when the wood became charred. This is later, at least by the time between growth and cutting down the tree or branch (which may be from, say, ten to hundreds of years). But it may be very much later still, if the wood was part of massive timbers, which themselves may be recycled from one construction to another; an example of this is the use of driftwood in northern marine sites such as Norway (where wood can be scarce), which may have grown centuries before being transported to the site. Many radiocarbon dates appear to be too old and have had to be corrected by redating more suitable, short-growth material [e.g., dating charred seeds from early Linien Band Keramik (LBK) sites in northwestern Europe (Whittle 1990)].

A second area of difficulty involves radiocarbon reservoirs (see mainly Section 16.2.1.1). Where, say, marine shells are dated, it is obvious that a marine correction is in order, and this is applied by calibrating according to the marine calibration curve (Stuiver and Braziunas 1993). (Additional local adjustment is also necessary since the surface marine correction is not uniform over the globe.) However, animal bone from marine sites may also contain a portion of carbon from the marine food chain—this is detectable in humans from many coastal Mesolithic sites (Tauber 1983)—but the magnitude of correction is not obvious and indeed is not normally made. Therefore humans are likely to appear 100 to 400 years older than they really are. Where this is detected by the radiocarbon laboratory, it should be commented upon in publication, but this has not always been the case. The difficulty is compounded when freshwater food chains are involved. These are not readily detectable in the analysis of archaeological bone (whereas marine food chains leave a characteristic $\delta^{13}C$ signature in the bone collagen, though this can be overlaid by signatures from C4 and CAM plants). Both geological calcite, in hard waters, and "old" carbon from soils can deplete the ^{14}C content in food chains, though the evidence for this happening in practice is scarce and circumstantial. This is another area where more research is required. However, it is only really serious for dating human bone since freshwater fish was potentially an important source of protein. At present the likelihood must simply be borne in mind that humans, who could possible have consumed substantial freshwater food, might date to perhaps some 100 years or so too old.

16.5.2. Selection and Handling of Archaeological Samples

Selection should be based on the issues considered above, together with the most efficient deployment of limited financial resources for dating. This will involve the level of precision required (and so of multiple dating), the use of stratigraphic information when calibrating, and the taphonomy of samples from targeted contexts. In addition the importance of both the intrinsic information carried by the sample and the chemical issues relating to sample type should be investigated and understood. The widespread use of AMS dating, at an increasingly competitive cost, has enabled selection to be practiced far more rigorously, with a consequent improvement in the quality and relevance of the archaeological information. Small samples are, naturally, more susceptible to contamination. The ideal is to avoid contact with human skin and the like and deposit samples in clean glass containers with no other material. Less breakable plastic is often preferred, and the sample can be kept away from abrading the plastic when wrapped in aluminum foil. While both container materials can contain volatile organic plasticizers, the effect is too small to matter, except in extreme cases (such as a tiny sample left in a plastic bag in the hot sun). In one case, small bones were stored in medical capsules made out of gelatin; the dates obtained showed a truly impressive degree of contamination.

16.5.2.1. Conservation of Material

Often organic material, especially bone, is consolidated with glue or other organic polymer, either during or after excavation. It was not uncommon for animal glue (essentially collagen) to be used in museums until 50 years ago. This is almost impossible to remove completely and adds a recent signal to the extracted indigenous bone collagen; its presence is also very difficult to detect. The use of poly vinyl acetate (PVA) is easier to contend with, as it is generally extractable with a solvent such as acetone and detectable, at least at the 10% level, by infrared spectrometry. This same approach can be used for wood preservation. However, all materials that have been, or are suspected of being, treated with consolidants or substantial organic preservatives are compromised and may give unreliable dates. Where suspected, one can compare the dates obtained from a full, consolidant-extracted, pretreatment, and ones with partial and no consolidant extraction, and this will give a measure of the effectiveness of the pretreatment—though at a cost. Fortunately any consolidant treatment carried out at recent excavations is usually well documented.

16.5.3. Examples of Applications of Radiocarbon Dating in Archaeology

These are chosen to give a sense of the scope of radiocarbon dating in archaeology and the different ways in which the dating information relates to other kinds of information. They give a picture of what is happening now, rather than attempt to describe how the whole edifice of the radiocarbon dating framework has built up over 50 years.

16.5.3.1. Paleolithic Archaeology

Many Paleolithic sites are unsuitable for dating by beta counting, and the greater selectivity in sample brought about by AMS methods has had a particularly beneficial influence in the older part of radiocarbon's dating range. In Europe the transition from Middle to Upper Paleolithic remains an active topic, but has been well pinned down to between 40 and 45 ka BP at various points from Bulgaria to northern Spain (ka = kiloan is used instead of thousand years). This is of course on the limit of the technique, but the weight of evidence that these are finite dates (rather than just defining the younger edge of a date range) is now very strong. Recent work in Siberia shows Upper Paleolithic industries there as early as 40 ka BP, while the Ahmarian, at Kebara Cave in Israel, is also dated to this period. The poor precision, coupled with uncertainty in calibration, leaves room for future work to establish precisely the timing and movement of the transition. However, in several limited areas, the Middle Paleolithic apparently continued for some millennia, notably in the Dordogne in southwestern France, in the Crimea, and perhaps in southern Spain. This has become clear, despite the ease with which contaminated material can produce younger dates, and allowing for comparisons that involve different material (e.g., nearly all the dates from the Dordogne are on bone) (Hedges and Pettitt, in press). Hominid remains are too scarce to support these dates in detail; the earliest anatomically Modern Humans to be directly dated by radiocarbon, so far, are no older than 32 ka BP (Hedges et al. 1989). However, Modern Human behavior at its most impressive display of artistry, has been dated to at least as old as 32 ka BP directly from the rock art of the Chauvet Cave (Ardeche, France) (Clottes et al. 1995). Much of the pigment used in this cave is charcoal, and it is still possible to see cellular structure in the 10- to 50-mg samples that could be obtained for radiocarbon dating. It is possible to be confident that the actual pigment material is being dated, and not reworked organic "dust." While most cave art dates to the period of the height of the Ice Age or later, here we are lucky to see such impressive artistic productions at what, with these dates, must be close to the earliest made by Modern Humans.

Migrations and Colonizations. The transition to Upper Paleolithic industries may not necessarily involve wholesale migration, but the peopling of Australia definitely did. Radiocarbon dates from archaeological sites older than about 30 ka BP are hard to find on the continent, and, in comparison with dates of purely environmental events have suggested that human presence can be documented to be absent at the beginning of the radiocarbon timescale. However, it appears that taphonomic issues do bias the result; and the survival of organic material in a predominantly hot dry country is in any case poor. The argument has been blunted by increasing evidence for earlier occupation (to 60 to 70 ka BP—well before the limit of radiocarbon dating) in North Australia using luminescent dating of sediments formed on sites with lithic assemblages. In the Americas the main question is whether there was an earlier migration than one (apparently after the Last Glacial Maximum, LGM) mainly documented as the Clovis culture in temperate North America at 12 ka BP. Evidence remains controversial; the earliest well-dated site (reliable dating of unquestionably human artifacts) is Monte Verde, southern Chile, at 12 to 13 ka BP (Dillahay 1997) uncomfortably early for the consensus opinion of a post-LGM migration. At other sites, with dates well before the LGM, either the radiocarbon dating has been questioned (as at Meadowcroft, USA) or the association with human activity (as at Pedra Furada in Brazil), while many other candidates for "earliest man in America" have had both aspects doubted (Haynes 1992). Well-dated sites in eastern Siberia dating to before 25 ka BP are now well established (Goebel and Aksenov 1995), and it seems only a matter of time before unambiguous evidence for at least some earlier human presence is unearthed. Radiocarbon evidence for the colonization of the Pacific Islands has been comparatively free from problems, but one interesting twist is the dating of the Pacific rat or Kiore (*Rattus exulans*), which, on New Zealand, apparently precedes the first obviously human presence by some 1000 years, despite humans being the only apparent vector. This has raised the question whether the rats were apparently "too old" through eating a mainly marine diet, but it is now clear that the dates are not greatly affected by a marine reservoir effect. Therefore it has been concluded that humans (and rats) reached these distant islands much earlier; but that humans, at least, did not last (Holdaway 1996). However, the issue (and the dates) remains controversial.

The best documented case for Paleolithic movements, at present, is of the repeopling of northern Europe following the Last Glacial Maximum (Housley et al. 1997). This shows refugia during the LGM, determines the order of recolonization, discovers some of the complexity in that process (probably at least two stages are involved), and determines an overall rate as being comparable with the later movement of Neolithicization into Europe.

16.5.3.2. Technological and Demographic Change

Neolithicization is associated with two particular technological inventions: how to make pottery and how to cultivate plants and domesticate animals. Radiocarbon dating has shown that pottery seems to have been invented in many places at different times, and often earlier than expected. Early dates for Jomon pottery (ca. 10 ka BP) based on radiocarbon dating of shell middens (Kendrick 1995) and supported by thermoluminescence dating of the pottery have now been supplemented by early pottery (8 to 9 ka BP) in South China (Hedges et al. 1992), in Siberia north of Manchuria (by 8 ka BP), and also by at least 6 ka BP in northern South America (Roosevelt et al. 1991). Most of these dates are from less than ideal fractions taken from pottery rich in low fired organic temper; a precise chronology will emerge once purified lipid fractions are able to be extracted and dated. As for domestication and cultivation, the site of Abu Hureyra (Syria) has perhaps been foremost in providing a comprehensively dated record, throughout a long stratigraphy, of the change, at the end of the Pleistocene, from wild-type seeds and hunted animals (such as gazelle) to fully cultivated emmer and domesticated sheep (Moore et al. 1999). Issues such as the relationship between rapid climate changes following the end of the glacial period and the adoption of agriculture, or the rate at which cultivars are able to be selected, are now open to examination through radiocarbon dating. The whole question of the adoption of farming practices in Europe, in its relationship to demographic growth, the movement of ideas, and the long-distance migration of human groups, remains unresolved. The interesting "wave of advance" model of demographic growth, based on radiocarbon measurements, is seen to be an oversimplification (Zvelebil 1989). Some movements, for example, of the LBK culture from the Danube basin to northwestern Europe, seems to have taken place very rapidly in radiocarbon terms; while the movement along the Mediterranean coast and the Atlantic coast seems to have been a quite separate, if simultaneous, process. Therefore a far more detailed dataset of dated cultural activity is required, ideally in combination with other information such as genetic affinity, in order to provide and constrain a model consistent with the facts.

16.5.3.3. Specific Projects

The accounts above represent rather sweeping syntheses of many individual studies. Radiocarbon is more often employed in providing information more limited in scope but important for specific sites.

Stonehenge. Many sites can be complex and enduring and require carefully planned strategies to be elucidated through radiocarbon dating. One such is Stonehenge (Bayliss et al. 1997). Phases 1 and 2 comprise a ditch and bank, with secondary silting and various alterations (e.g., backfilling episodes), together with the circle of Aubrey holes. Twenty-four samples were selected for dating from the ditch sequence, including nine antler picks from phase 1 and a bone chisel from phase 2, as well as many animal bones, including two that were articulated. The results, after synthesizing the stratigraphic information with the radiocarbon dates, show the main ditch to be cut between 3015 and 2935 Cal B.C. (at 95% confidence) and that therein was immediately deliberately placed some bones that themselves were between 70 and 420 years old at the time. Secondary silting took place in less than 70 years, being complete within 460 to 740 years, by which time a later burial had been inserted. Its date was 2400 to 2140 Cal B.C. Phase 3 was the period of constructing the stone settings and obtaining representative samples was more difficult. Similar analysis to the earlier phases shows phase 3 to occupy 850 to 1000 years before the final form was reached. The sarsen circle was in place by 2620 to 2480 Cal B.C., with the trilithons by 2440 to 2100 Cal B.C. Remodeling of the bluestones continued into the circle and then into the horseshoe at 2280 to 2030 Cal B.C. and 2270 to 1930 Cal B.C., and the last major modifications were made by 1640 to 1520 Cal B.C.

Volcanic Eruption of Thera. When radiocarbon dating is applied to more recent history, results must be related to temporal connections established by such archaeological methods as cultural contacts. These become increasingly evident as societies and technologies increase in sophistication, and, with the advent of writing and king lists, provide alternative sources of calendrical dating. The dating of the eruption of Thera (the present day island of Santorini in the Aegean Sea) in Bronze Age times is an interesting example of the problems and issues involved for a single point. The cataclysmic eruption of the island, leaving a submerged caldera several kilometers across, destroyed the town of Akrotiri on the island, and marks the end of the cultural phase characterized as Late Minoan 1A. The dating of this phase, recognized in other parts of the Aegean, especially Crete, has special interest. At one time it was suggested that the eruption, accompanied by earthquakes and tidal waves, may have been directly responsible for the destruction of the palaces on Crete and the decline of Minoan civilization. However, tephra from the eruption is found between LM 1A and LM 1B at Knossos, showing a degree of continuity there. Dating the LM 1A/1B transition has relied on three main lines of evidence: artifacts of Egyptian origin (mainly Tuthmosis III, ca. 1450 B.C.) found in various relevant contexts, dating the start of LM 1A on similar connections as not being earlier than about 1600 B.C., and dating LM IIIA to Amenophis III and estimating the time from LM1A. The consensus on these data has been for the eruption to be 1450 to 1500 B.C., and at the very earliest at 1550 B.C. This has seemed at odds with the radiocarbon dating, first undertaken in 1976 (beta dating, at Philadelphia), which included some long-term growth material, and, once accurate calibration was available, suggested 1600 to 1650 Cal B.C. Such an early result started various arguments about the possible effect of volcanic CO_2 on making dates too old, although such an effect has only been observed in quite extreme cases (dates from Iceland have been much studied in this way). Careful redating was planned for a major conference (Hardy 1990) in 1989, using charred seeds from storage jars at Akrotiri. The range of ideal material for dating is quite restricted, since other material is not well related to the event being dated. One laboratory, averaging several dates, reported an average of 3338 ± 30 BP (Oxford; AMS), while another (Copenhagen; beta counting) reported 3355 ± 32 BP. Questions were raised about pretreatment methods and, in particular, whether circulating humic acids, contaminating the seeds, might contain "old" volcanic carbon from CO_2. These were addressed and shown not to cause significant error. The combined date of 3346 ± 25 could then be calibrated to give a range of 1680 to 1527 Cal B.C. at 95% confidence, with the main probability lying between 1610 and 1675 Cal B.C. Thus unfortunately the calibration curve has unhelpfully wiggled extra uncertainty into what was a very precise measurement, and while suggesting a date that is acceptable only to a minority of Aegean Bronze Age specialists, is consistent with a date of, say, 1550 B.C. However, the story is by no means finished. A major volcanic event, marked by leaving a layer of extra acid ice in the ice deposits of Greenland has been dated, by annual ice layer counting, to be close to 1640 B.C.; while a global climatic blip, of the kind caused by the massive injection of ash into the stratosphere, has been registered in tree rings in Ireland and the United States dated to 1628 B.C. There is no direct evidence that either (or both) of these were caused by Thera, but one is impressed by the coincidence. The same event is apparently registered very clearly in tree rings from Anatolia, although the dating of the rings depends on wiggle-matching with the radiocarbon curve (so is not yet absolute).

Dating Bronze Age Metalwork. Another example of the matching of a radiocarbon chronology to that established by "conventional" archaeological means is provided by the dating of bronze artifacts. The change in style of such high status objects as axe heads and spears, together with an analysis of how assemblages of such artifacts occur, has enabled typochronologies to be constructed, including long-distance links, for instance, between Britain and the Continent. This has been of necessity since bronze, in itself, cannot be dated. However, often fragments of the wood hafting are preserved, through the biocidal properties of copper leaching from the bronze or through the predilection of Bronze Age folk to deposit their more valuable possessions in waterlogged places. Only fairly recently has the AMS method had the precision to make a useful contribution to such studies. The results (Needham et al., 1997) have been to sharpen precision and to establish the chronological reliability, with various recognized phases being confirmed in their sequence but generally changed in duration (usually toward longer times; in one case dramatically so).

16.5.3.4. Dating in the Historic Period

Radiocarbon dating tends to be less important during the historic period, being more useful in confirming the date of unusual specimens—for example, showing that a macaque (Barbary ape) found in excavations in Roman Dublin is indeed of Roman date. However, in many parts of the world, dating in the last 2 or 3 millennia is still essential to providing the basic chronological framework—such as establishing the early dates for iron working in Africa. As already mentioned, it is not possible to achieve good calendrical precision in much of the first millennium B.C. due to the calibration curve.

Venice. A final example of an application of dating is in the establishment of settlements in the region of Venice during the Dark Ages of the later first millennium A.D. (Ammerman et al. 1995). Recent excavations have disclosed remains of wooden poles forming a quay, together with a boat; these have been radiocarbon dated to the same period as when the lagoon dwellers were addressed by Cassiodorus (A.D. 537). Radiocarbon dating of samples of wood or reeds, from cores taken below the present pavement of the Piazza San Marco, first provide a chronology for the environmental setting of the lagoon sedimentation, but also provide evidence for occupation, in Venice itself, in the late seventh to early eighth century A.D. Nearby cores provide the first evidence for a canal used in the eighth and ninth centuries and then allowed to silt up. Good organic preservation allows a quite detailed history of the origins of Venice to be built, dependent only on the scale at which the site can be examined.

These examples cannot give a balanced picture of radiocarbon dating of human activity around the whole planet. The bias is to European archaeology, although this reflects where most radiocarbon dates, now numbered at well over 100,000, and perhaps over 200,000, have been concentrated. But they do show, at both a strategic and a detailed level, laboratory techniques and environmental science collaborating in conjunction with archaeological questions and knowledge to produce an efficient and trustworthy dating framework.

Acknowledgments

I would like to acknowledge the support and expert work of the whole of the ORAU team at Oxford and specifically stimulating discussions with Christopher Ramsey and Paul Pettitt.

BIBLIOGRAPHY

References to specific points are given separately. Radiocarbon dating is treated in Mook and Waterbolk (1985) and, in relationship to other dating techniques, in Aitken (1990). A useful edited book on the techniques of radiocarbon dating is Taylor et al. (1992). The Triennial International Radiocarbon Conference Proceedings have, for the most part, been published in *Radiocarbon*, although papers specifically dealing with archaeological dating, as opposed to dating methodology, are rare. The journal *Radiocarbon* also publishes date lists and research articles. There have been three Conferences on Archaeology and Radiocarbon, the proceedings being published as Radiocarbon and Archaeology, PACT 8 1983 and PACT 29 1989 (Hacken 1983; Hacken and Mook 1989). The most recent, held in Lyon, 1998, is to be published as a special issue in *Revue d'Archéométrie* (expected 1999).

REFERENCES

Aitken, M. (1990), *Science-Based Dating in Archaeology*, Longmans, London, 1990.

Ammermann, A. J. M. de Min, R. Housley, and C. E. McClennen (1995), *Antiquity* **69**, 501.

Arnold, J. R. (1954), *Science* **119**, 155.

Arnold, J., and W. Libby (1951), *Science* **113**, 111.

Batten, R. J., R. Gillespie, J. A. J. Gowlett, and R. E. M. Hedges (1986), "The AMS dating of separate fractions in archaeology," *Proc. Radiocarbon* **28**(2A), 698.

Bayliss, A., C. Bronk Ramsey, and F. G. McCormac (1997), *Proc. Bri. Acad.* **92**, 39.

Bronk Ramsey, C. (1995), *Radiocarbon* **37**(2), 425.

Bronk Ramsey, C., and R. E. M. Hedges (1994), *Nucl. Instrum. Methods B* **92**, 100.

Brown, T. A., D. E. Nelson, and J. R. Southon (1988), *Radiocarbon* **30**, 171.

Buck, C. E., J. B. Kenworthy, C. D. Litton, and A. F. M. Smith (1991), *Antiquity* **65**, 808.

Collins, M. J., D. M. J. Walton, and A. King (1998), in *The Geochemical Fate of Proteins in Nitrogen-Containing Macromolecules in the Biosphere and Geosphere*, American Chemical Society Symposium Series 707, B. A. Stankiewicz and P. F. van Bergen, Eds., American Chemical Society, Washington, DC 74–86.

Clottes, J., J-M. Chauvet, E. Brunel-Deschamps, C. Hillaire, J-P. Daugas, M. Arnold, H. Cachier, J. Evin, P. Fortin, C. Oberlin, N. Tisnerat, and H. Valladas (1995), *C. R. Acad. Sci. Paris Ser. 11a* 320, 1133.

Damon, P. E., D. J. Donahue, B. H. Gore, A. L. Hatheway, A. J. T. Jull, T. W. Linick, P. J. Sercel, L. J. Toolin, C. R. Bronk., E. T. Hall., R. E. M. Hedges, R. Housley, I. A. Law., C. Perry, G. T. Bonani, S. Trumbore, W. Woelfli, J. C. Ambers, S. G. E. Bowman, M. N. Leese, and M. S. Tite (1989), *Nature* **337**, 611.

Deines, P. (1980), in *Handbook of Environmental Isotope Geochemistry*, P. Fritz and J. Fontes, Eds., Elsevier, Amsterdam, pp. 329–406.

Dillehay, T. D. (1997), *Monte Verde: A Late Pleistocene Settlement in Chile. 2: The Archaeological Context*, Smithsonian Institution Press, Washington, DC.

Eglinton, T. I., J. Bauer, A. P. McNichol, and E. R. M. Druffel (1995), in *Organic Geochemistry: Developments and Applications to Energy, Climate, Environment and Human History*, J. O. Grimalt and C. Dorronsoro, Eds., AIGOA, Conostia-San Sebastian, pp. 697–698.

Fowler, A., R. Gillespie, and R. E. M. Hedges (1986), *Radiocarbon* **28**, 441.

Gillespie, R. (1990), *Nucl. Instrum. Methods B* **52**, 345.

Gillespie, R., R. E. M. Hedges, and N. R. White (1983), *Radiocarbon* **25**(2), 771.

Goebel, T., and M. Aksenov (1995), *Antiquity* **69**, 349.

Gowlett, J. A. J., R. E. M. Hedges, I. A. Law, and C. Perry (1987), *Archaeometry*, Datelist 5, **29**(1), 125.

Hacken, T., Ed. (1983), *PACT (J. Eur. Study Group Techn. Appl. Archaeol.)* **8**.

Hacken, T., and W. Mook, Eds. (1989), *PACT (J. Eur. Study Group Techn. Appl. Archaeol.)* **29**.

Hardy, D. A. (1990), Thera and the Aegean world III, in *Proceedings of the Third International Congress*, Santorini, Greece, 1989, The Thera Foundation, London.

Harkness D. D., and P. Becker-Heidman (1996), *Radiocarbon* **38**(2), 175.

Haynes, C. V. (1992), in *^{14}C Dating and the Peopling of the New World*, R. E. Taylor, A. Long, and R. Kra, Eds., Springer-Verlag, New York, pp. 355–374.

Hedges, R. E. M. (1987), *Philos. Trans. R. Soc. London A* **323**, 57.

Hedges, R. E. M., and P. B. Pettitt (1999), On the validity of archaeological radiocarbon dates beyond 30,000 years BP, in *Proceedings of the Third International Symposium "¹⁴C and Archaeology," Lyon 1998, Revue d'Archéométrie, (Supplement)*, G.M.C.P.A., pp. 137–141.

Hedges, R. E. M., R. A. Housley, I. A. Law, and C. R. Bronk (1989), *Archaeometry*, Datelist 9, **31**(2), 207.

Hedges, R. E. M., C. Tiemei, and R. A. Housley (1992a), *Radiocarbon* **34**(3), 906.

Hedges, R. E. M., M. J. Humm, J. Foreman, G. J. Van Klinken, and C. R. Bronk (1992b), *Radiocarbon* **34**(3), 306.

Hedges, R. E. M., J. A. Lee-Thorp, and N. C. Tuross (1995), *Radiocarbon* **37**(2), 285.

Hedges, R. E. M., C. Bronk Ramsey, G. J. van Klinken, P. B. Pettitt, C. Nielsen-Marsh, A. Etchegoyen, J. Niello, M. T. Boschin, and A. M. Llamazares (1998), *Radiocarbon* **40**(1), 35–44.

Hillman, G., S. Colledge, and D. Harris (1989), in *Foraging and Farming*, D. Harris and G. Hillman, Eds., Unwin Hyman, London, pp. 240–266.

Holdaway, R. N. (1996), *Nature* **384**, 225.

Hoper, S., F. McCormac, A. Hogg, T. Higham, and M. Head (1998), Evaluation of wood pretreatments on oak and cedar, in *Proceedings of the Sixteenth International Radiocarbon Conference, Groningen*, Netherlands, 1997, *Radiocarbon* **40**(1), 45–50.

Housley, R., C. Gamble, M. Street, and P. Pettitt (1997), *Proc. Prehist. Soc.* **63**, 25.

Johnson, B. J., G. H. Miller, M. L. Fogel, and P. B. Beaumont (1997), *Palaeogeogr. Palaeoclimatol. Palaeoecol.* **136**(1–4), 121.

Kendrick, D. M. (1995), *Jomon of Japan: The World's Oldest Pottery*, Kegan Paul International, London.

Kromer, B., and K. Munnich (1992), CO_2 gas proportional counting in radiocarbon dating—review and perspective, in *Radiocarbon after Four Decades*, R. Taylor, A. Long, and R. Kra, Eds., Springer-Verlag, New York.

Kuniholm, P. I., B. Kromer, S. W. Manning, M. Newton, C. E. Latini, and M. J. Bruce (1996), *Nature* **381**(6585), 780.

Law, I. A., and R. E. M. Hedges (1989), *Radiocarbon* **31**(3), 247.

Litherland, A. (1987), *Philos. Trans. R. Soc. London A* **323**, 5.

Mook, W. G., and H. T. Waterbolk (1985), *Radiocarbon Dating: Handbooks for Archaeologists No. 3*, European Science Foundation, Strasbourg, 1985.

Moore, A., T. Legge, and G. Hillman (1999), *Abu Hureyra: A Village on the Euphrates*, Oxford University Press, New York.

Needham, S. P., C. B. Ramsey, D. Coombs, C. Cartwright, and P. Pettitt (1999), *Archaeol. J.* **154**, 55–107.

Nelson, D. E., G. Chaloupka, C. Chippindale, M. S. Alderson, and J. R. Southon, (1995), *Archaeometry* **37**(1), 151.

Polach, H. (1992), Four decades of progress in ^{14}C dating by liquid scintillation counting and spectrometry, in *Radiocarbon after Four Decades*, R. Taylor, A. Long, and R. Kra, Eds., Springer-Verlag, New York.

Pollard, A. M., and C. Heron (1996), in *Archaeological Chemistry*, Royal Society of Chemistry, London, pp. 239–270.

Roosevelt, A. C., R. A. Housley, M. Imazio da Silveira, S. Maranca, and R. Johnson (1991), *Science* **254**, 1621.

Rozanski, K., S. W., Stichler, R. Gonfiantini, E. M. Scott, R. P. Beukens, B. Kromer, and J. Van der Plicht, (1992), *Radiocarbon* **34**, 506.

Scott, E. M., A. Long, and R. Kra, Eds. (1990), *Radiocarbon* **32**(3), 253.

Scott, E. M., D. D. Harkness, G. T. Cook (1998), Inter-laboratory comparisons, lessons learned, in *Proceedings of the XVI International Radiocarbon Conference*, Groningen, Netherlands, 1997, *Radiocarbon* **40**(1), 331–340.

Stafford, T., R. Duhamel, V. Haynes, and K. Brendel (1982), *Life Sci.* **31**, 931.

Stafford, T., P. Hare, L. Currie, A. Jull, and D. Donahue (1991), *J. Archaeol. Sci.* **18**, 35.

Stott, A. W., and R. P. Evershed (1996), *Anal. Chem.* **68**, 4402.

Stuart, A. J. (1991), *Biol. Rev.* **66**, 453.

Stuiver, M., and T. F. Braziunas (1993), *Radiocarbon* **35**(1), 137.

Stuiver, M., and R. S. Kra Eds. (1986), Calibration issue, *Radiocarbon* **28**(2B), 805–1030.

Stuiver, M., and P. J. Reimer (1993), *Radiocarbon* **35**(1), 215.

Stuiver, M., A. Long, and R. S. Kra (1993), *Radiocarbon* **35**(1).

Stuiver, M., P. J. Reimer, E. Bard, J. Warren Beck, G. S. Burr, K. A. Hughen, B. Kromer, G. McCormac, J. van der Plicht, and M. Spurk, *Calibration issue Radiocarbon* **40**(3), 1041–1083.

Stuiver, M., and J. van der Plicht (2000), *Intcal 98: Calibration Issue, Radiocarbon* **40**(3).

Tauber, H. (1983), *PACT (J. Eur. Study Group Techn. Appl. Archaeol.)* **8**, 365.

Taylor, R., A. Long, and R. Kra, Eds. (1992), *Radiocarbon after Four Decades*, Springer-Verlag New York.

Van der Plicht, J. (1993), *Radiocarbon* **35**(1), 231.

Van Klinken, G. J., A. D. Bowles, and R. E. M. Hedges (1994), *Geochim. Cosmochim. Acta* **58**(11), 2543.

Vries, H., M. de Stuiver, and I. Olsson (1959), *Nucl. Instrum. Methods B* **5**, 111.

Watchman, A. (1990), *Rock Art Res.* **7**, 44.

Whittle, A. (1990), *Antiquity* **64**, 297.

Zvelebil, M. (1989), *Antiquity* **63**, 379.

CHAPTER

17

LEAD ISOTOPE ANALYSES APPLIED TO PROVENANCE STUDIES

NOEL H. GALE and ZOFIA STOS-GALE

Isotrace Laboratory, Oxford University, Oxford, United Kingdom

17.1. INTRODUCTION

The development of the lead isotope technique has provided a unique scientific tool for provenance studies, which has had a significant impact on archaeology. It is applicable to artifacts made of many materials, but it is perhaps especially in metal provenance studies that its full impact has been felt, if only in contrast with the complete failure of other attempts to provenance metals. For example, recent work using this technique has resulted in discoveries that have radically changed the archaeological perceptions of the past 40 years about the sources of copper used by the principal peoples of the Mediterranean Bronze Age. There have been previous short reviews of particular aspects of this subject that, though now out of date in some respects, readers may wish also to consult (e.g., Farquhar and Vitali 1987; Gale 1989; Gale and Stos-Gale 1982, 1992a; Stos-Gale 1993d; Renfrew and Bahn 1996: 344–352; Lambert 1997).

17.2. RADIOACTIVITY AND RADIOGENIC ISOTOPES

Lead is an ubiquitous constituent of ore deposits, rocks, and minerals. Lead occurs in major amounts in lead–zinc and lead–silver ore deposits and in lesser amounts in copper and iron ore deposits; it occurs in at least parts-per-million levels in all rocks and nonmetallic minerals. This would be of little significance for geochemistry or for archaeology but for the fact that the isotopic composition of naturally occurring lead is variable. Nowadays we

Modern Analytical Methods in Art and Archaeology, Edited by Enrico Ciliberto and Giuseppe Spoto. Chemical Analysis Series, Vol. 155
ISBN 0-471-29361-X

know that this variability is due to the fact that three of the four natural isotopes of lead are the end products of the radioactive decay of isotopes of uranium and thorium (Evans 1955: 470–523). That the elements uranium and thorium are radioactive was discovered in 1895 to 1898 by Becquerel (1896) and Maria Sklodowska-Curie and Pierre Curie (Curie and Curie 1898). Subsequent work on the radioactive decay series of uranium and thorium showed that there are several kinds of radiothorium that decay at different rates. In 1913 T. W. Richards and M. E. Lambert (1914) found that lead produced by the decay of uranium had an atomic weight different from that of ordinary lead; he had already shown that the atomic weights of natural elements are not whole numbers. These and similar problems led Frederick Soddy (1911, 1914) to suggest that chemical elements are in general composed of isotopes, that is, atoms having the same chemical properties but differing mass. J. J. Thomson directly proved that this was so for neon in 1913. Aston, also working in the Cavendish Laboratory, immediately set out to build ever more sophisticated mass spectrographs to extend this finding to other elements, but it was not until 1927 that he directly proved the existence of the isotopes of lead (Aston 1942).

17.2.1. Geological Applications

In 1905, even before the discovery of isotopes, Rutherford and Boltwood had independently recognized that radioactivity could provide a method of measuring the ages of rocks and minerals. By 1907 Boltwood had made the first crude estimates of the ages of three uraninite samples based on their U–Pb ratios. By 1929 Rutherford was able, on the basis of the uranium–lead decay scheme, to estimate the age of the earth and to get it right within a factor of 2. In the late 1930s Nier perfected the first mass spectrometers capable of relatively high precision and ushered in modern lead isotope studies (Nier 1939, 1940). Subsequent to 1945 geologists and geochemists have been ever more active in using the uranium–lead and thorium–lead radioactive decay schemes (Köppel and Grunenfelder 1979). They not only have used them for dating rock formations and measuring geological time but also have used lead isotope compositional variations as a tracer and provenance indicator for ores and rocks in studies of ore genesis and petrogenesis (Doe and Zartman 1979) and to aid the exploration for ore bodies (Gulson 1986).

17.2.2. Archaeological Applications

The successful applications of lead isotope studies in the earth sciences led Brill and his collaborators to speculate that such analyses might have an application in provenance studies of archaeological materials. By 1962 Brill and Wampler already had unpublished data showing that they could distinguish between a few lead objects excavated in Greece and some lead pigs from Roman Britain (R. H. Brill, personal communication, 1976). In December 1964 they publicly presented their first studies of ancient lead metal before the 66th General Meeting of the Archaeological Institute of America, followed by publications in 1965 (Brill and Wampler 1965a, 1965b) and 1967 (Brill and Wampler 1967). In 1966 Grögler, Geiss, Grunenfelder, and Houtermans independently published a comparative lead isotope study of Roman lead ingots and lead ores (Grögler et al. 1966). The subject of lead isotope archaeometry had been born. In the years 1968 to 1981 there followed applications, principally by Brill and his colleagues, of lead isotope analyses to glass, glazes, leaded bronzes, kohls, silver and gold coins, and artists' pigments (Brill 1968, 1969, 1970a, 1970b, 1972, 1976, 1978; Brill and Shields 1972; Brill et al. 1970, 1974, 1979).

Unfortunately these early studies were vitiated, in their utility to archaeology or numismatics, by the low accuracy of measurement of isotope composition then attainable and the lack of a sufficient database for ore samples and a coherent archaeological strategy. The method of comparison of isotopic data between artifacts and ores was also defective, as discussed by Gale and Stos-Gale (1992a: 70–71, Fig. 3); it is disturbing that this old and discredited approach has still been used quite recently (Lilyquist and Brill 1993). The birth of the subject of lead isotope archaeology had to await new developments, but the deficiencies of the early work in no way detract from the remarkable prescience shown by Robert Brill. He must be acknowledged as the founder of this approach to provenance studies in a sense similar to that in which Demokritos foreshadowed modern atomic theory, an approach that has eventually become of importance to archaeology. Currently it is possible to apply this technique to the range of materials shown in Table 17.1, which shows also other provenance applications of thermal ionization mass spectrometry.

17.3. BASIC PRINCIPLES

17.3.1. Variations in the Isotopic Composition of Natural Lead

Natural lead consists of the four isotopes ^{208}Pb, ^{207}Pb, ^{206}Pb, and ^{204}Pb; all have essentially identical chemical properties. The first three of these isotopes derive, in terrestrial lead, partly from the radioactive decay of long-lived naturally radioactive isotopes of uranium and thorium, but the isotope of ^{204}Pb is nonradiogenic in origin. This is the basic reason for the variability of the isotopic composition of natural lead. The earth was created about

Table 17.1. Applications of TIMS to Provenance Studies

1. Lead isotopes
   - Linking mineral pigments from ancient sites (galena, kohl, copper and iron minerals, etc.) with specific ore deposits
   - Identifying ore sources for lead and silver artifacts, including certain types of silver coinage and leaded alloys (pewter, leaded bronze, etc.)
   - Identifying copper ore deposits used for production of unleaded copper-based alloys or the lead ore deposits used for leaded bronzes
   - Provenance of iron and zinc artifacts
   - Finding the source of mineral pigments used in production of glass and glaze
   - Provenance of artists' pigments used on canvases and wood
   - Identification of sources of lead in human remains (lead poisoning)
2. Strontium isotopes
   - Provenance of obsidian
   - Provenance of gypsum (within limitations)
   - Additional information for oxygen and carbon isotope provenance studies of marble
   - Possibilities in studies of provenance of pottery, glass, faience, and glaze
   - Possibilities in paleodiet and ancient migration studies
3. Isotope dilution analyses (very accurate quantitative elemental analyses)
   - Provenance of obsidian (Rb, Sr, U, Th)
   - Provenance of glass and glaze (Rb, Sr, U, Th, Pb)

4.5 thousand million years ago, so although they are long lived, appreciable fractions of the earth's primaeval budget of these radioactive parent isotopes have decayed and have substantially increased the earth's complement of ^{208}Pb, ^{207}Pb, and ^{206}Pb. As a result the lead isotopic compositions of rocks in the external parts of the earth have evolved through geological time. Metallic ore deposits owe their origin to the processing of large volumes of rock at various times in the past, usually by hydrothermal processes in which the metals were transported to the site of ore deposition by hot (50 to 450°C) aqueous brines. Where, as in all lead deposits and in most copper deposits, the processes of ore formation have resulted in very low concentrations of uranium and thorium relative to lead, the isotopic composition of the lead is "frozen" and records the evolution of the isotopic composition of lead up to that point in time when the metals in the ore deposit were segregated. As a result, the isotopic composition of lead in an ore deposit varies from deposit to deposit in a remarkably systematic way, depending to first order on the geological ages of the deposits.

The isotopes of ^{208}Pb, ^{207}Pb, and ^{206}Pb (but not ^{204}Pb) in any present-day orebody may conveniently be thought of as composed of a mixture of two parts:

Present-day lead = primaeval lead + radiogenic lead

The isotopic composition of lead existing at the time when the earth was formed is termed primaeval lead; this composition has to be inferred from the isotopic composition of lead in the uranium-free troilite (FeS) phase of the Canyon Diablo iron meteorite (Göpel et al. 1985). Long radioactive decay chains are involved in the production of radiogenic $^{206,207,208}Pb$ [see Evans (1955: 511–523) or Faure (1986: 282–287)], but they may conveniently be summarized as

$$^{238}U \xrightarrow[\text{6 beta particles}]{\text{8 alpha particles}} {}^{206}Pb + \text{energy} \qquad \text{half-life} = 4468 \text{ Ma}$$

$$^{235}U \xrightarrow[\text{4 beta particles}]{\text{7 alpha particles}} {}^{207}Pb + \text{energy} \qquad \text{half-life} = 704 \text{ Ma}$$

$$^{232}Th \xrightarrow[\text{4 beta particles}]{\text{6 alpha particles}} {}^{208}Pb + \text{energy} \qquad \text{half-life} = 14010 \text{ Ma}$$

where Ma means million years. The isotope composition of a given sample of lead is usually specified by one of two sets of atomic ratios:

$$\frac{208\text{ Pb}}{206\text{ Pb}}, \frac{207\text{ Pb}}{206\text{ Pb}}, \frac{206\text{ Pb}}{204\text{ Pb}} \quad \text{(archaeological)} \rightarrow [A]$$

$$\frac{208\text{ Pb}}{204\text{ Pb}}, \frac{207\text{ Pb}}{204\text{ Pb}}, \frac{206\text{ Pb}}{204\text{ Pb}} \quad \text{(geological)} \rightarrow [B]$$

However, other sets of ratios have been used, even in geological applications (see Gulson 1986: 25–27).

For a particular ore sample the present-day isotope ratios will be determined by the integrated effects of all associations with uranium and thorium between some initial time and a later time, when the ore was formed, after which the lead is no longer associated with uranium and thorium. The three isotope ratios are conventionally plotted in two diagrams. For archaeological applications these diagrams usually take the form of plots of $^{208}Pb/^{206}Pb$ versus $^{207}Pb/^{206}Pb$ and of $^{206}Pb/^{204}Pb$ versus $^{207}Pb/^{206}Pb$. As a provenance technique lead isotope analysis depends on the variability of the isotopic composition of lead (present either as the chief or subsidiary element) in metal ore sources and on comparative analyses of the isotopic composition of lead present in artifacts and ore sources. It is free from most of the limitations of provenance methods based on chemical analysis, in that the isotopic composition of lead usually varies only within narrow limits in the majority of ore bodies and (for metal provenance studies) the isotopic composition of lead in the ore passes unchanged through the smelting, refining, cupellation, and casting processes into the metal. In contrast, all these metallurgical processes are strongly selective chemically (e.g., Tylecote et al. 1977), while minor and trace element concentrations tend to vary through various parts of an ore body both in absolute amounts and relative to each other (e.g., Constantinou 1982), so that provenancing of metals based solely on minor or trace element analysis can never be expected to have more than limited success.

17.3.2. Lead Isotope Evolution for Ore Bodies

The simplest possible history for, say, a galena is that the parent material for the galena suffered no chemical fractionation of U and Th from Pb from the time t_p for formation of the earth until production of the galena at time t: such a history is referred to as single-stage growth of lead in a closed chemical system. It can be characterized by single-stage growth curves with associated μ value [($^{238}U/^{204}Pb$), as measured today)] for the parent U–Pb system: and W value [$^{232}Th/^{204}Pb$] for the parent Th–Pb system (see Köppel and Grunenfelder 1979). The growth curves originate at time t_p from an initial primaeval lead isotope composition defined in practice by lead of the troilite phase of the Canyon Diablo meteorite (Tatsumoto et al. 1973; Göpel et al. 1985). This model is the Holmes–Houtermans model (Houtermans 1946; Holmes 1946, 1947, 1949). From this model it is possible to compute, from the measured lead isotope compositions of ore deposits, so-called model ages which should give the geological age of the ore deposit if the model assumptions are satisfied (see Köppel and Grunenfelder 1979). In this model the growth of the lead isotope ratios is described by the following equations which give the isotope ratios at geological time t measured from t_p, the time of formation of the earth:

$$\left(\frac{^{206}\mathrm{Pb}}{^{204}\mathrm{Pb}}\right)_t = \left(\frac{^{206}\mathrm{Pb}}{^{204}\mathrm{Pb}}\right)_{t_p} + \mu(\exp\lambda_8 t_p - \exp\lambda_8 t)$$

$$\left(\frac{^{207}\mathrm{Pb}}{^{204}\mathrm{Pb}}\right)_t = \left(\frac{^{207}\mathrm{Pb}}{^{204}\mathrm{Pb}}\right)_{t_p} + \frac{\mu}{137.8}(\exp\lambda_5 t_p - \exp\lambda_5 t)$$

$$\left(\frac{^{208}\mathrm{Pb}}{^{204}\mathrm{Pb}}\right)_t = \left(\frac{^{208}\mathrm{Pb}}{^{204}\mathrm{Pb}}\right)_{t_p} + W(\exp\lambda_2 t_p - \exp\lambda_2 t)$$

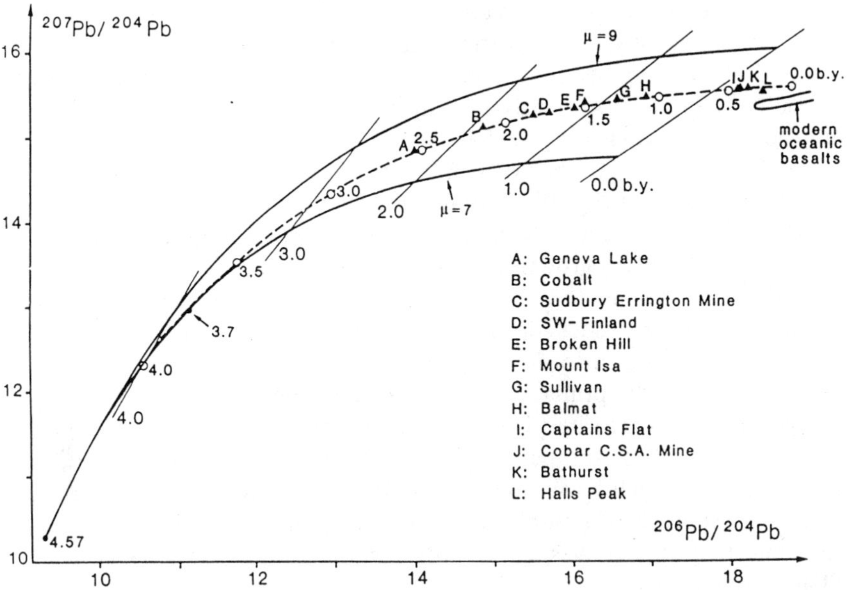

Figure 17.1. Lead isotope compositions for conformable ore deposits and modern oceanic basalts. Solid curves are single-stage model growth curves for μ values of 8 and 9. Hatched curve is Cumming and Richards (1975) model fit to conformable data, forced through data for Captains Flat and troilite from Canyon Diablo meteorite.

where $(^{206}Pb/^{204}Pb)_{t_p}$, $(^{207}Pb/^{204}Pb)_{t_p}$, $(^{208}Pb/^{204}Pb)_{t_p}$ are the primordial lead isotope ratios and λ_8, λ_5, λ_2 are the radioactive decay constants for ^{238}U, ^{235}U, and ^{232}Th, respectively. These equations describe growth curves; examples are plotted in Figure 17.1. The lead isotope composition for the galena at the present day is obtained by setting $t = 0$. If the galena throughout a lead ore body has followed this simple, single-stage, Holmes–Houtermans evolution, then the lead isotope composition will be strictly constant for all galena samples no matter from where they originate in the ore deposit.

Many ore deposits, especially the so-called strata-bound (Faure 1986: 309–335) or conformable ore deposits, have a lead isotope composition that varies very little indeed throughout the whole deposit. In some cases the variation is as little as 0.1%; an example is the Bleiberg deposit in Austria (Köppel and Kostelka 1976). Quite often the variation in lead isotope composition is less than 0.3%; an example is from the Amelia mine in the Upper Mississippi Valley ore district (Heyl et al. 1966). Such ore deposits commonly have lead isotope compositions that lie close to a single-stage model (Faure 1977: 227) evolution curve, as was first noticed by Stanton and Russell (1959: 288–607). Many of these, possibly syngenetic, lead ore deposits are apparently associated in a conformable way with rocks of marine origin and of island-arc-volcanic affiliation. Since near-surface concentrations of U, Th, and Pb are very heterogeneous, the conformable ores were suggested to be the product of a lower crustal or upper mantle environment, which were emplaced in an environment that virtually precluded mixing with crustal rocks (Russell and Farquhar 1960: 53). The single-stage growth curve (shown as a dashed curve in Fig. 17.1) passing through these lead ore deposits, each of which exhibits a very uniform lead isotopic composition, has come to be known as the primary growth curve for terrestrial lead.

Today it is accepted that the concept of a single-stage lead isotope history cannot be entirely correct for any real ore deposit. Improvements in the accuracy of measurement of lead isotope compositions brought about the redetermination of the primaeval lead isotope composition as recorded in the Canyon Diablo meteorite (Tatsumoto et al. 1973; Göpel et al. 1985). This new primaeval lead isotope composition shifted the Holmes–Houtermans growth curve in such a way that it was no longer possible to fit the ore lead of strata-bound (conformable) deposits on a single-stage evolution curve such that their model ages agreed with the stratigraphic ages, especially for Phanerozoic deposits.

17.3.3. Recent Models for Lead Isotope Evolution

Three new models have been proposed to accommodate the conformable ore deposits on revised growth curves. Stacey and Kramers (1975) introduced a two-stage model that involved a major first differentiation of the earth into a crust and mantle at about 3.7×10^9 years ago, whereby the crust was enriched in U and Th with respect to the mantle. The effects of later differentiation within the crust or mantle were assumed to be largely obliterated by mixing processes. The nature of the fundamental event required by the model is however not clear, through it may represent an approximation for a prolonged period of crustal formation that ended approximately 2.5×10^9 years ago.

Cumming and Richards (1975) proposed another model in which μ and W increase linearly with time, forcing the growth curves through the data points of Canyon Diablo lead and of the Captains Flat (Australia) Pb–Zn–Cu deposit of Silurian age (430 my; my = million years). The continuous linear increase of the μ and W values can be related to continuous differentiation processes, but the implied continuous linear increase of U and Th concentrations in the crust cannot occur forever. A rather elaborate multistage model of discontinuous evolution, the "Plumbotectonics" model of Doe and Zartman (1979), attempts to involve lead isotope evolution in different zones of the crust and mantle in connection with their dynamic development and interaction. More recently Amov (1983) has constructed what is currently perhaps the most realistic model of conformable ore lead evolution using a dynamic model of the continuous isotopic evolution or uranogenic and thorogenic lead. His model depends essentially on the assumption that the rate of change of $\mu = {}^{238}U/{}^{204}Pb$ and $W = {}^{232}Th/{}^{204}Pb$ (present-day values) increased initially and, after attaining a maximum, decreased due to surface cooling of the earth. An important parameter in Amov's model is the time t_m when the rates of change of μ and W attained their maxima; it depends on the age of formation of the different layers of the crust or, in mixing processes, on the participation of materials from the older or younger crust and mantle. The value of t_m (or a related parameter T) indicates the time of enrichment in U/Pb, which is an extremely important index for identifying the lead source-type.

17.3.4. Lead Isotope Model Ages

The importance of these models is that, from the observed lead isotope composition of a particular ore body, they allow so-called model ages to be calculated that should give the geological age of the time of emplacement of the ore body. The value of this for isotope archaeology is that, once the lead isotope composition of say an archaeological metal artifact has been measured, a model age can be computed that will give a guide to the rough geological age of the ore deposit that supplied its metal. This can be useful as a guide, though not an infallible one, to which ore deposits should be investigated and included in further provenance studies. Needless to say, attempts to provenance metal artifacts by chemical analysis do not have this advantage.

Both of the 1975 models yield model ages for strata-bound sulfide deposits that agree quite well with geological ages and the ages obtained by other geochronological methods. Amov's model yields model ages which are in even better agreement with independently derived ages of such deposits, especially for the younger deposits.

17.3.5. Ore Bodies with Inhomogeneous Lead Isotope Compositions

Conformable (strata-bound) ore deposits, together with many for which the geology is not yet completely understood, each have a relatively constant isotopic composition within the particular ore body. In contrast, some vein-type ore deposits have revealed the presence of linear arrays of data points in one or both lead isotope diagrams. They may be mixing lines or they may have the significance of isochrons; their interpretation is discussed by Köppel and Grunenfelder (1979). Such anomalous, or multistage, ore deposits can sometimes present difficulties for archaeological provenance studies. Thus far they have not been found among Mediterranean ore deposits, but they have figured in studies of galena sources for Archaic/Woodland sites in North America (Farquhar and Fletcher 1980, 1984). Gale and Mussett (1973) have discussed the isotopic systematics of multistage ore deposits.

The causes of the various types of isotopic variability within a given ore deposit are not in every case known with certainty but may often be the result of selective extraction of the metals from their source rocks prior to their transport and ultimate deposition. Again, lead that developed in several stages in the crust and was incompletely mixed prior to deposition in an ore body will have variable isotopic composition and variable lead isotope model ages that do not correspond with the geologically established age of deposition.

Two geological situations that have the potential for causing isotopic inhomogeneities are as follows:

1. Metamorphic events subsequent to the time of emplacement of an ore body may cause some mixing with lead from the surrounding country rocks.
2. Geochemists concur that over geological time uranium has migrated toward the upper crust faster than thorium and that both have moved in preference to lead. It follows that rocks in the lower crust and upper mantle contain slightly larger proportions of lead relative to uranium and thorium than rocks in the outer continental crust; hence the isotopic evolution of lead in these reservoirs has proceeded at different rates. Some ore deposits appear to contain mixtures in varying proportions from these reservoirs.

The elegant work of Farquhar and Fletcher (1980, 1984) have introduced archaeometrists to the ores of the Upper Mississippi Valley lead–zinc district, whose lead isotope compositions vary over the range of about 3%. It must be emphasized that this variation is for an ore district of about 10,000 km^2 in area that comprises many individual ore deposits. Surprisingly little detailed work has been done on individual deposits within this district, but the Amelia mine shows only about 0.25% variation in lead isotope composition throughout its extent (see Fig. 17.2). Isotopic inhomogeneity on the scale of an ore district does not necessarily preclude lead isotope compositions from pinpointing the source of metal to an individual ore deposit within that district. On the other hand, some of the individual mines within the Mississippi Valley ore district have a large range of variation in isotopic composition. For instance, the Buick Mine (Heyl et al. 1966) exhibits a range of about 2% on average in lead isotope composition. Moreover, a total variation of lead isotope ratios of about 3% has been demonstrated by ion microprobe anal-

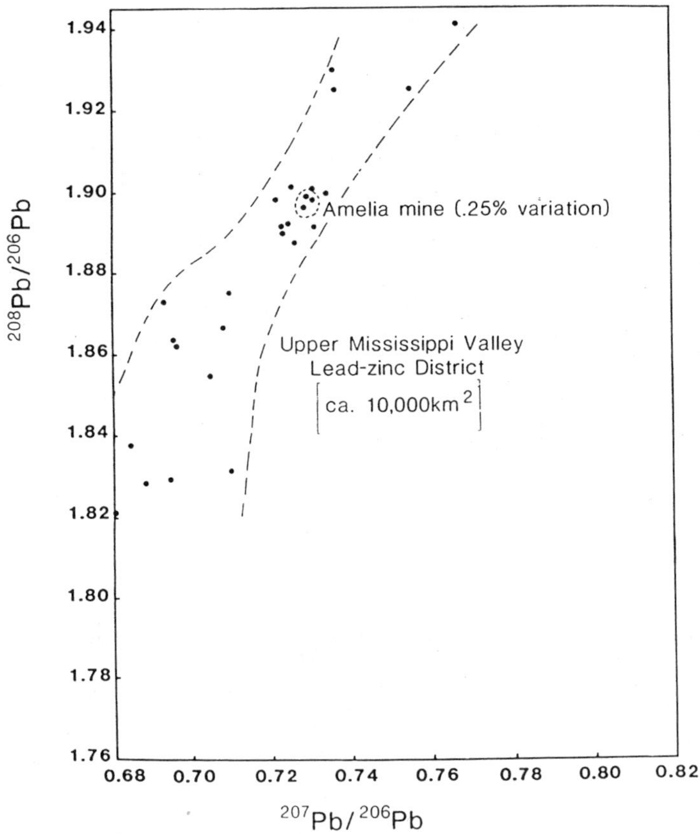

Figure 17.2. Lead isotope compositions for ores from individual ore deposits comprising Mississippi Valley lead–zinc ore district.

ysis in a single crystal of galena from this same Buick Mine (Hart et al. 1981). It should again be stressed that this is an unusual situation and one that does not significantly affect the ability to make provenance studies using lead isotope analyses. It should of course be remembered that the lead isotope compositions of individual galena crystals will be mixed and averaged in a smelting process.

Figure 17.3 shows the chief evolutionary path of lead isotope composition in terrestrial ores. The breadth of the hatched area represents the extent to which some ore deposits have departed from the relatively simple evolution assumed in the Cumming–Richards average growth curve for crustal ore deposits. The dashed curve is the Cumming–Richards average growth curve for crustal ore deposits.

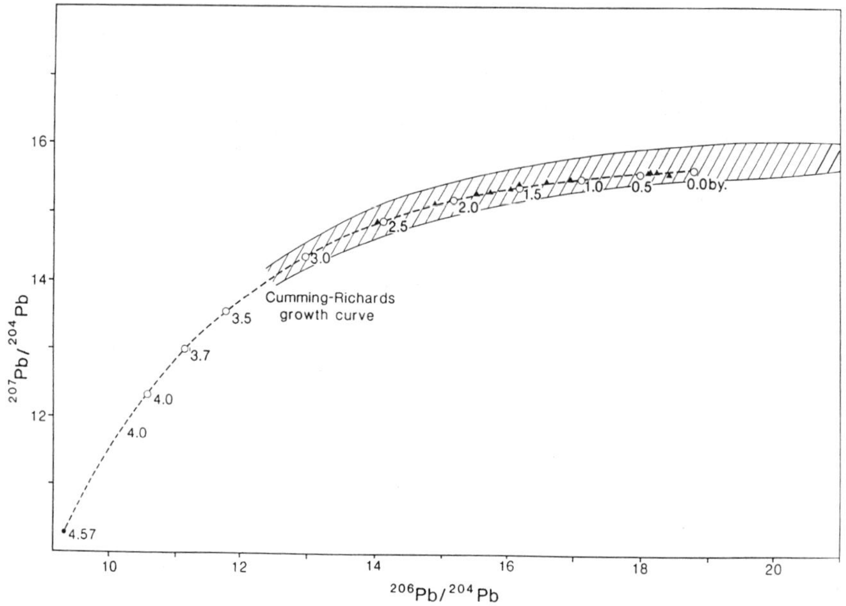

Figure 17.3. Lead isotope data for ore deposits of various ages and geographical locations lie roughly in hatched area of graph, though most modern data lie rather closer to growth curve (Cumming and Richards 1975). Data used to construct hatched region taken from Russell and Farquhar (1960) and Doe and Rohrbough (1977). Times marked on growth curve are in units of 10^9 years.

Figure 17.4 shows on an expanded scale the lead isotopic compositions, in relation to the Cumming–Richards growth curve, for some individual groups of deposits typical of the many ore deposits that have now been investigated. Some of these deposits have isotopic compositions well away from the growth curve for conformable leads; some deposits, such as the Indonesian island ore deposits, have isotopic "fields" that cross the growth curve. Yet other deposits, such as deposit WA, have isotopic compositions that plot to the right of the zero time point on the main evolutionary growth curve. Figure 17.5 shows that some ore deposits have even more extreme lead isotope compositions that lie well beyond the zero time point of the Cumming–Richards growth curve. It should be noted that the large spread of lead isotope compositions for, say, the Mississippi Valley ore deposits or the New Mexico deposits are not for individual ore deposits but for extensive ore districts; in Figure 17.5 the Washington Metalline ore deposit is typical of the usual small spread of lead isotope compositions for an individual deposit.

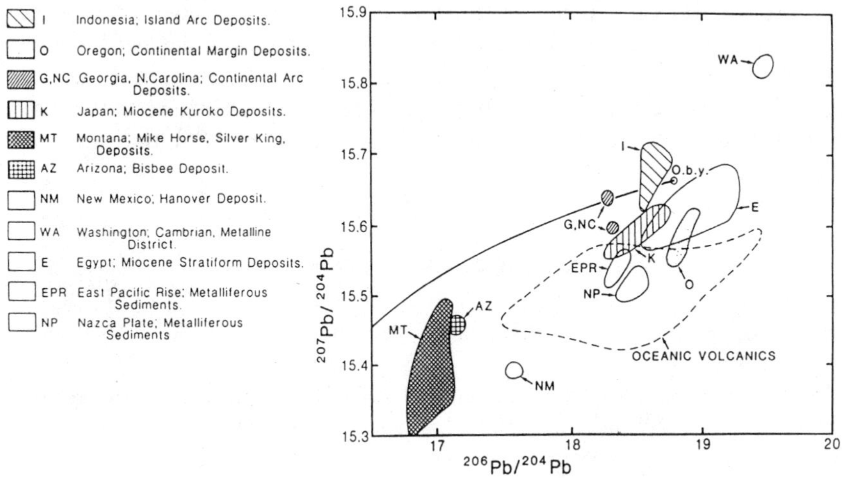

Figure 17.4. Lead isotope data for ore deposits of various ages and localities plotted on an expanded scale and shown in relation to the Cumming–Richard growth curve. References for sources of data are given by Doe and Zartman (1979).

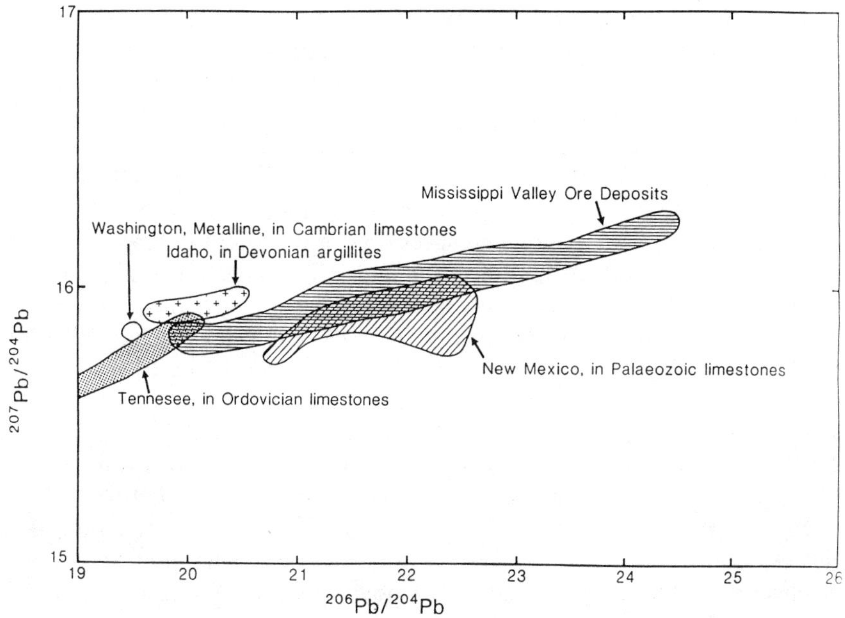

Figure 17.5. Extremely radiogenic lead isotope data for certain ore districts in the U.S. All data lie well to the right of the zero age point of Cumming–Richards growth curve and has negative model ages. Original sources data given by Doe and Zartman (1979). Radiogenic nature of data may be due to major contributions of lead from crustal regions of the earth to these ore deposits.

Here we are content merely to draw attention to the general trends and extent of the lead isotope compositional variations in ore deposits. For detailed discussions of the underlying causes of these variations and their implication in particular cases for hypotheses of ore genesis we refer the interested reader to Doe and Zartman (1979) and Gulson (1986) and references given therein.

17.4. MEASUREMENT TECHNIQUES

17.4.1. Mass Spectrometry

Until recently the only technique capable of sufficient accuracy both for isotope geochemistry and for archaeological applications is solid source thermal ionization mass spectrometry using mass spectrometers of the magnetic sector type [but see later for very recent developments of multiple collector–inductively coupled plasma mass spectrometry (MC–ICPMS)]. The thermal ionization technique used by Dempster (1918) in his original mass spectrograph depends on the fact that when an atom or molecule evaporates from a heated surface, there is a probability that it will evaporate as a positive ion. For lead isotope mass spectrometry, by far the most common arrangement used for production of positive ions from a solid sample employs a single heated ribbon filament of rhenium on which the sample is placed, though where large enough samples are available, a more accurate triple filament technique is used (Cantanzaro 1968; Elliott 1963). For mass analysis a spectrometer using a sector magnetic analyzer (introduced by Nier in 1940) is used, with the source region and spectrometer tube both pumped to about 10^{-8} Torr using ultrahigh-vacuum techniques. Measurement of ion currents down to 10^{-12} A usually makes use of a Faraday cup and a vibrating reed or other type of electrometer amplifier. Use of a Daly detector (Daly 1960) or an electron multiplier affords a potential increased sensitivity by a factor of at least 10^4, though for precise measurements such detectors are usually used to provide a gain of 20 to 100 relative to a Faraday cup system. Further details of mass spectrometric techniques are available in McDowell (1963), Milne (1971), Mayne (1960), Webster (1960), and Faure (1977, 1986).

Modern mass spectrometers at first employed automatic peak-jumping between the isotopes of interest, but state-of-the art thermal ionization mass spectrometry (TIMS) spectrometers are now equipped with up to nine movable Faraday ion beam collectors arranged side by side across the focal plane. This allows simultaneous collection of the ion beams for the four isotopes of lead, which averages out ion beam fluctuations and thus greatly improves the precision of isotope ratio measurements, at the same time reducing measurement time.

17.4.2. Mass Fractionation and Accuracy

The accuracy of determination of lead isotope abundance ratios using the thermal ionization method is limited by time-dependent variable mass fractionation (Tuttas and Habfast 1982; Habfast 1998) in the mass spectrometer. With the lead sulfide or lead nitrate technique in general use in the 1960s, the minimum sample size was about 1 μg of lead and the absolute 2σ errors in the isotope ratios were about 1% or larger. Unfortunately this source of error greatly limits the utility of all archaeometric lead isotope work reported up to about 1972, including the pioneering papers by Brill and his colleagues.

For some elements corrections can be applied for variable mass fractionation during mass spectrometry in terms of an inherent isotope ratio that is naturally constant (Faure 1986; Habfast 1998). For example, natural strontium is composed of four isotopes of atomic mass 88, 87, 86, and 84. Only ^{87}Sr varies in abundance in nature, due to augmentation by the decay of naturally radioactive ^{87}Rb; the other isotopes of strontium are not fractionated by geological processes. Very accurate measurements of the ratios ^{87}Sr/^{86}Sr and ^{84}Sr/^{88}Sr can be made by normalizing to the constant ratio ^{86}Sr/^{88}Sr $= 0.1194$; in practice, ^{87}Sr/^{86}Sr ratios can routinely be measured to an accuracy of better than 50 ppm (2σ level) (Ludwig 1997). This approach is not available for determination of the three isotope ratios for lead, since all three vary in nature due to augmentation from the naturally radioactive isotopes ^{238}U, ^{235}U, and ^{232}Th. The first real advance in lead isotope mass spectrometry was made by the introduction of a triple-filament thermal ionization technique (Catanzaro 1968; Catanzaro et al. 1968) that allowed the isotopic ratios of lead to be measured to about $\pm 0.05\%$, but at the expense of requiring a very large sample of about 200 μg to 1 mg per analysis. Such large samples are not commonly available, either in geological or archaeological applications.

The next advance was the independent investigation of the use of emitters, which not only greatly increased the sensitivity of the thermal ionization method for lead isotope analysis but also improved the precision by an order of magnitude. (Akishin et al. 1957; Amov 1968; Cameron et al. 1969; Arden and Gale 1974). Both isotope geochemists and archaeometrists have almost universally adopted the silica-gel emitter technique to produce lead ions for mass analysis (Cameron et al. 1969; Arden and Gale 1974). A maximum of 200 to 500 ng of lead in solution, usually as the nitrate, is placed on a thin substrate of silica gel with a small amount of phosphoric acid on a rhenium metal ribbon filament. Lead ions are produced by heating this filament to 1100 to 1200°C. This technique routinely produces ^{208}Pb ion currents of the order of 10^{-11} A and within-run precision of the order of 0.005% for the ratios ^{208}Pb/^{206}Pb and ^{207}Pb/^{206}Pb and better than 0.02% for the ^{206}Pb/^{204}Pb ratio (at the 2σ level). However, according to the exact details of the technique used, most laboratories experience interrun mass-dependent isotopic fractionation of the order of 0.05 to 0.1% per atomic mass unit difference between the two isotopes that comprise a measured ratio. Interlaboratory bias is avoided by correcting the raw data according to the ratios measured for a lead isotopic standard, usually one of the U.S. NBS/NIST Standard Reference Materials SRM 981 or SRM 982 (Todt et al. 1996). Such an approach leads to routine interlaboratory precision of 1 part in 10^3, which is sufficient for most purposes. An improved silica gel emitter technique has recently been published that greatly improves the lead ion yield and the constancy of the ion beam and allows as little as 10 ng of lead to be isotopically analyzed by TIMS in routine fashion with very high precision (Gerstenberger and Haase 1997).

A considerable advance was made some 15 years ago in the direction of allowing accurate isotopic composition measurements on samples usually of 200 ng size, but down to as little as 1 to 10 ng of lead, by the perfection of commercial multicollector TIMS mass spectrometers. McKinney et al. (1950) used a double-collector mass spectrometer more than 47 years ago and achieved a precision of 0.01% by measuring the ion beams from two isotopes simultaneously using twin collectors. This technique compensates for fluctuations of thermal emission, eliminates problems due to the decay of emission, and, by eliminating peak jumping, utilizes all the ions emitted. It is a technique that raises sample throughput rate and enhances the internal reproducibility of isotope ratio data for small samples. There are now excellent commercial automated mass spectrometers that are fitted with up to seven adjustable and two fixed Faraday cups plus a secondary emission electron multiplier or Daly detector to detect very small ion beams. It is the combined use of such multicollector TIMS spectrometers with silica gel emitters that has allowed the precisions and accuracies in lead isotope ratio measurements mentioned above. Roddick and Loveridge (1986), using such a system with special attention to detail for measuring very radiogenic lead from zircon samples, demonstrated precision and accuracy of better than 0.1% for samples of lead down to about 0.5 ng.

17.4.3. Chemical Separation and Purification of Lead for TIMS

Thermal ionization mass spectrometry is sensitively dependent on the chemical purity of the lead loaded onto the filament. The presence of other elements can reduce both the accuracy and the sensitivity (ion yield) of the measurements, so that careful and time-consuming chemical separation of lead from the ore or artifact is necessary prior to isotopic analysis (Fig. 17.6). If ores or artifacts with trace lead contents are to be analyzed, then the blank level has to be controlled (to well below 1 ng) by the special preparation and

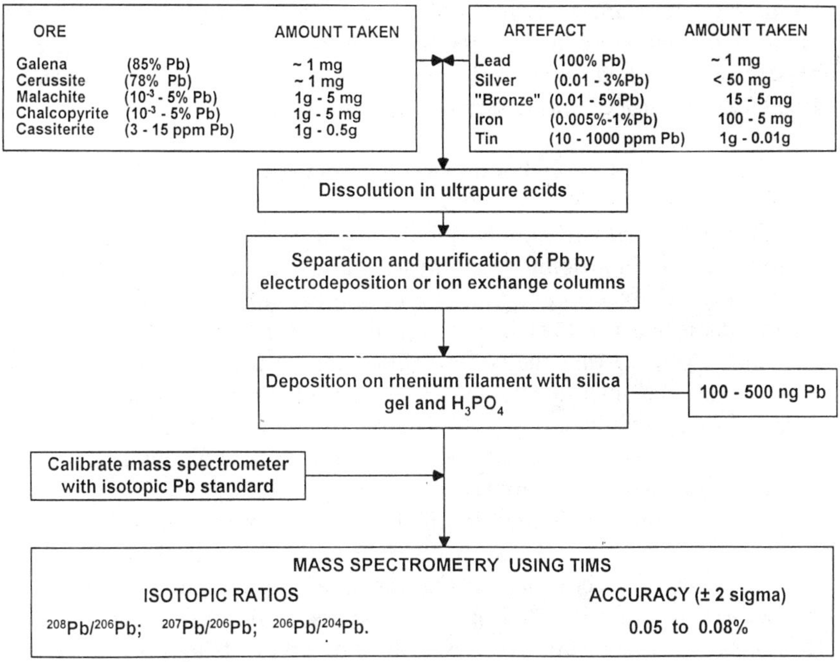

Figure 17.6. Block diagram illustrating typical steps in measuring lead isotope composition of sample using TIMS.

use of ultrapure reagents (Mattinson 1972) and by conducting the chemical separation in an overpressured ultraclean laboratory fed with charcoal and Hepa filtered air to the standards of the semiconductor industry. Initial dissolution of ancient metal samples is easily achieved using mineral acids. However, some ores and all slags need a more aggressive attack to achieve complete solution, using hydrofluoric (HF) with other acids and microwave oven assisted solution (Kingston and Jassie 1988) or a high-pressure "bomb" technique (based on Krogh 1973). The required low blank levels can be achieved by high-yield lead separation methods either based on electrodeposition (Arden and Gale 1974) or by using anion ion exchange columns (e.g., Göpel et al. 1985; Gulson 1986: 195–197) or a combination of both. The necessity of separating lead from iron-rich slags or ores, often with large amounts of copper, renders useless techniques devised by geochemists for the Pb isotope geochemistry of ordinary rocks and rock-forming minerals. New low blank separative schemes have recently been devised for these difficult materials, based on ion exchange procedures using a new generation of extraction chromatographic materials developed at the Argonne National

Laboratory (Dietz and Horwitz 1993). These special resins are commercially produced by Eichrom; a Pb separative technique based on one termed PbSpec was published by Philip-Horwitz et al. (1994), while a superior technique based on SrSpec was described by Gale (1996).

17.4.4. Interlaboratory Comparisons

As examples of what can now be achieved in any good laboratory, one can mention data for an absolute lead isotope standard published by the Oxford laboratory (Chamberlain and Gale 1980) and the laboratory of Begeman at the Max-Planck-Insitut für Chemie in Mainz (Pernicka et al. 1984: Fig. 24). This shows not only that the isotope ratios can be measured to better than 0.1% but also that measurements can be corrected for fractionation to absolute values by reference to the lead isotope standard. This is very important since it allows lead isotope measurements from different laboratories to be used together. As examples of the comparison of lead isotope ratios for the same samples measured in different laboratories, a number of comparisons have been made between our laboratory in Oxford and the U.S. Geological Survey Laboratory in Denver and between the Oxford laboratory and Mainz (Gale 1989: 474). The comparisons showed agreement to be excellent for all three isotope ratios, with measurements agreeing to within 0.1%.

17.4.5. New Developments

Considerable effort has been spent in attempting to use other types of ion sources for which chemical separation techniques are minimized; special mention must be made of the use of an inductively coupled plasma used as the ion source for a quadrupole mass spectrometer (ICPMS; see Date and Gray 1983a, 1983b). At present the sensitivity attained is comparable to that for the thermal ionization magnetic sector mass spectrometer using the silica gel activator, but ordinary ICPMS appears to have much poorer accuracy due to poor peak shape, the inherent instability of the source, and the absence of a multicollector system to average this out. Most archaeological problems, in common with those of isotope geochemistry, need the best achievable accuracy so that at the present stage of development ordinary ICPMS, though excellent for trace chemical analysis, is not suitable for lead isotope analysis in these fields (Russ and Bazan 1987; Hirata and Nesbitt 1995).

An exciting new development has been the successful use, with the latest multicollector mass spectrometers, of a ^{202}Pb–^{205}Pb internal tracer, which for all lead isotope ratios reduces the absolute 2σ error to $\pm 0.002\%$, a reduction in absolute uncertainty by approximately two orders of magnitude (Todt et al. 1984; 1996). Since these artificially produced isotopes of lead are in very short supply and available only in microgram amounts, they cannot be applied routinely to lead isotope provenancing in archaeology. Their use has however enabled much more accurate lead isotope ratios to be established for the NBS/NIST lead isotope standards SRM 981 and 982 (Todt et al. 1996), which are used in normalizing lead isotope ratios measured using TIMS, thus ensuring comparability between laboratories.

The double-spike approach to lead isotope analysis has also been developed using the naturally occurring isotopes ^{204}Pb and ^{207}Pb, which requires at least two mass spectrometer runs for each analysis, one spiked and one unspiked. The gain in precision was initially disappointing (Compston and Oversby 1969), due to the failure to recognize that, with the normal spiking regime, even relatively small but variable blanks cause significant errors in the final results. Hamelin et al. (1985) presented a much more extensive analysis of the methodology and systematics of this method and, using three separate mass spectrometer runs for each sample and a single-collector mass spectrometer, achieved an increase in precision by a factor of 2 to 3 over the usual method of fractionation normalization in terms of repeat measurements of SRM 981. More recently essentially the Hamelin et al. procedure has been adapted to modern multicollector TIMS spectrometers and has allowed an external precision to be achieved for individual analyses of 0.03% (two standard deviations of the population: McCullough and Woodhead 1993; Volker et al. 1993). This technique remains, however, too time consuming for routine application to the large numbers of samples of ores, artifacts, and slags involved in a typical lead isotope provenancing study.

A very important development has been the introduction of mass spectrometers having inductively coupled plasma ion sources followed by magnetic sector mass analysis and multiple ion beam collectors to allow the simultaneous collection and analysis of a number of isotopes, resulting in the so-called MC–ICP–MS technique (Hirata 1996; Halliday et al. 1998). This high-precision mass spectrometric technique was developed by the firm VG Elemental (Walder and Freedman 1992; Walder et al. 1993; Lee and Halliday 1995) and is not to be confused with ordinary ICPMS using a quadrupole mass spectrometer and a single collector that, as briefly reviewed above, cannot provide lead isotope data of a precision necessary for applications to provenancing (Furata 1991). TIMS, though capable of high precision and in many ways the benchmark technique, is well known to be time consuming and to be bedevilled by time-dependent fractionation effects that limit run-to-run accuracy. In contrast, MC–ICP–MS has a fixed bias that is time independent and independent of the chemical properties of the element or admixing of other elements (Walder and Freedman 1992; Halliday et al.

1995; Lee and Halliday 1995). This allows mass fractionation to be corrected out, for Pb, in terms of the fixed isotopic ratio, $^{205}Tl/^{203}Tl$, of additions of an in-house standard of thallium (Walder et al. 1993; Walder and Furuta 1993; Hirata 1996). Though not yet applied to an archaeological study in lead isotope provenancing, the promise is great since, as Hirata (1996) has shown, MC–ICP–MS normalized against an internal Tl standard can attain a reproducibility of lead isotope ratios of the order of $\pm 0.01\%$ using nanogram samples of lead in a very rapid mass spectrometric analytical procedure. Not only is the mass spectrometry faster than TIMS, but the preceding and main time-consuming step of separation and purification can also be simplified and made less time consuming. This new technique has also been used to demonstrate that the isotopic composition of Cu is variable in ores and archaeological artefacts (Gale et al. 1999), which may provide another provenancing method.

17.5. PRINCIPLES AND LIMITATIONS OF LEAD ISOTOPE PROVENANCING

17.5.1. Lead Isotope Fields for Ore Deposits

The basic requirement of lead isotope provenancing is that a number of lead isotope analyses of ores must be determined so as to define the lead isotope fields characteristic of the ore deposits that could, in principle, have supplied the particular metal under consideration. All three possible lead isotope ratios must be used to construct these fields, conveniently depicted in two two-dimensional lead isotope diagrams. Only if *all* the lead isotope ratios of an artifact fall within the fields for an ore source in *both* diagrams could that ore source have provided the metal for the artifact.

The concept of a lead isotope field for a single deposit is useful for the majority of uranium-poor ore deposits in the Mediterranean region. It is less useful for the rare deposits (such as Timna in Israel and Feinan in Jordan; Hauptmann et al. 1992) where present day U/Pb ratios are significant and variable, leading to a wide range of lead isotope compositions for different ore samples within each deposit. Attempts at quantitative definition of lead isotope fields, or the assignment of the numerical probability that the lead isotope composition of a particular artifact derives from a particular isotopic ore field (as attempted in Sayre et al. 1992a), may present statistical difficulties, although this has yet to be demonstrated unequivocally (Baxter and Gale 1998). On the other hand, it remains valid and powerful to use multivariate discriminant analysis to examine the separation of the lead isotope fields of different ore deposits. The extension of the concept of a lead isotope field to an island or a geographical region is fraught with difficulties, as will be seen.

In some cases even the use of multivariate statistical analysis will not completely unscramble partially overlapping lead isotope ore fields. It is then the hope that trace element chemical analyses can be combined with lead isotope data in a multivariate statistical analysis (Doran and Hodson 1975; Pollard 1986; Kaufman and Rousseeuw 1990; Hair et al. 1995) that will group artifacts. Less ambitious, but perhaps more hazardous, is to use trace element data to exclude an artifact from provenance from a particular ore source if the artifact does not contain a metallic element that is present at high concentration in the ore source. If trace element data from different laboratories are to be used, then the data from each laboratory must have good absolute accuracy. For example, a comparison between neutron activation analyses of the same objects by Oxford and Heidelberg shows good agreement within the 20% error bands (Gale 1989).

It is the precise coincidence of lead isotope compositions of particular artifacts with particular ore deposits that is required before it can be concluded that the metal of the artifact came from the named ore deposit. If some ore deposits have overlapping lead isotope compositions, then the conclusions for artifacts plotting in such regions must necessarily be less well defined. Figure 17.7 sets out the quite simple basis of lead isotope analyses as used for provenancing the metal of an ancient artifact. This figure is based on lead isotope data for the ore deposits at Lavrion in Attica, and in the Cycladic islands of Seriphos, Keos and Thera (data in Stos-Gale et al. 1996; Gale 1998). Two examples are given in Figure 17.7 for imaginary lead artifacts having lead isotope compositions plotting at points A and B. The lead isotope data show that artifact A was made of lead metal that was not extracted from any of the four ore deposits whose lead isotope fields appear in Figure 17.7 (the ellipses in this figure do not have any statistical significance but are drawn so as to enclose the data for particular deposits to indicate their extent). Artifact B is consistent only with having been made from metal smelted from the Keos lead ore deposits, since it plots outside Seriphos and Thera fields in both diagrams, outside the Lavrion field in the lower part of the diagram, but inside the Keos field in both diagrams.

It is largely a question of semantics to engage in arguments that lead isotope analyses can only exclude ore sources and can at best establish only that an ore deposit is consistent with being the metal source for particular artifacts. If a set of copper artifacts have lead isotope compositions matching only one copper ore deposit, among the set of copper ore deposits that it is geographically/archaeologically reasonable to consider, the normal application of the scientific method would lead to the conclusion that, *in the present state of knowledge*, those artifacts were made of copper from that deposit.

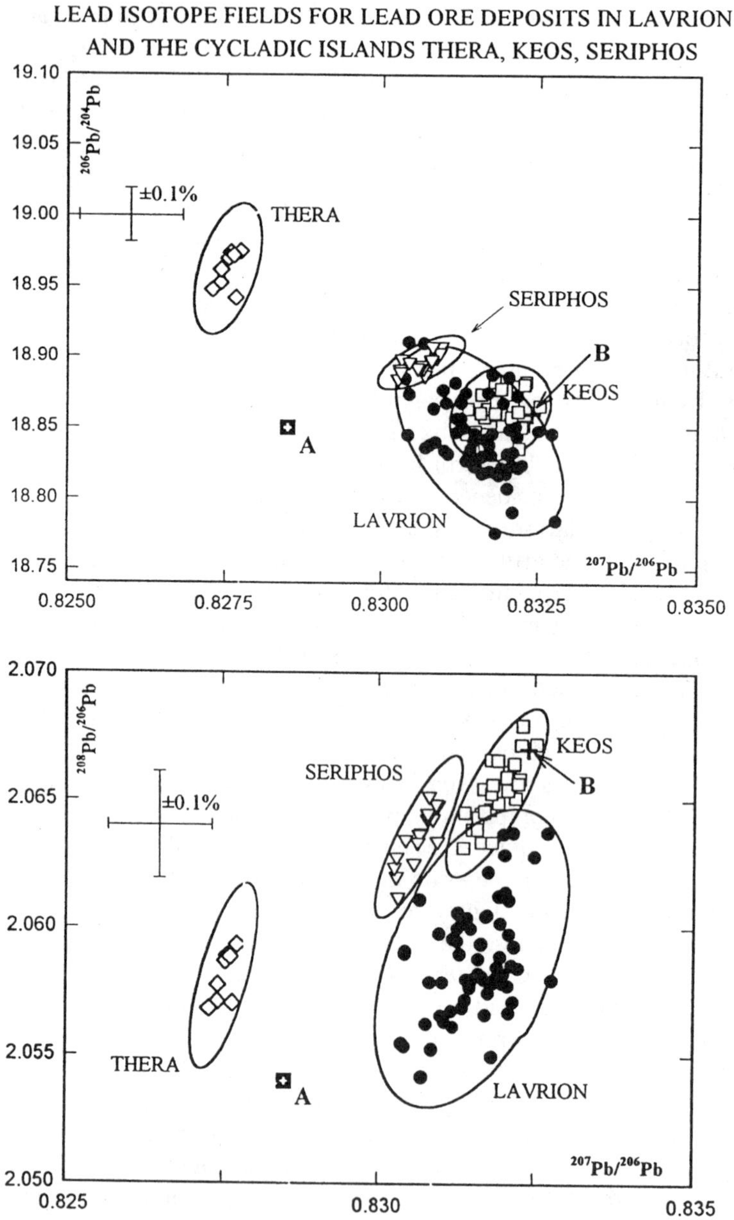

Figure 17.7. Simple illustration of basic principles of lead isotope provenancing, based on lead isotope fields for ore deposits of Lavrion and Cycladic islands of Thera, Keos, and Seriphos, in comparison with lead isotope compositions of two hypothetical lead metal artifacts, A and B.

Naturally such conclusions must rest on a comprehensive lead isotope investigation of all appropriate ore deposits. One must of course avoid the trap of believing that science provides infallible answers to our questions.

17.5.2. Lead Isotope Fractionation in Nature or in Metallurgical Processes

We have seen that ore deposits can usually be defined by a rather restricted range of lead isotope compositions—the lead isotope field. It is vital, for the validity of lead isotope provenancing, that such processes as smelting an ore to produce a metal do not alter the isotopic composition of the lead. Conventional theoretical studies of the mass-dependent fractionation of isotopes prove that, for isotopes as heavy as those of lead, no measurable change (i.e., at the 1 part in 1000 level) in isotopic composition can be brought about by smelting, refining, working, casting, or corrosion processes (Russel and Farquhar 1960: 1–2). Whether the lead occurs as a major element in the antiquity (lead metal, lead pigments, leaded bronzes) or as a minor element (e.g., copper, iron, silver, or zinc artifacts, glasses, glazes), the isotopic composition of that lead will be the same as the isotopic composition of the lead in the ore or mineral from which it derives. Recently this was called into question by Budd et al. (1995a) and Pollard and Heron (1996: 324–326), who suggested that changes in lead isotope composition due to nonequilibrium evaporation during, for example, smelting could produce changes in lead isotope composition great enough, for example, to move the isotopic composition in the smelted metal quite outside the field of isotopic composition of the ores being smelted (Pollard and Heron 1996: Fig. 9.6). Their calculations relied on a thermodynamic model (Mulliken and Harkins 1922) designed to describe evaporation from a liquid where the vapor phase was efficiently removed from the system by condensation on a cold surface. It was not by any means clear that this model could be applied to ancient processes such as smelting, fire refining, and cupellation. Nevertheless, empirical evidence would clearly be welcome on this point.

Such evidence had long been available, in that experiments by Barnes and his colleagues (1978) had given empirical proof that, in a particular case, anthropogenic processes produced no measurable change in lead isotope composition. A large sample of galena was heated with charcoal in a closed crucible at 900°C for 18 hr and then reduced to lead at 900°C in 3 hr in the presence of a weak air blast. Part of this lead metal was reserved for analysis; the remainder was oxidized at 900°C in 1 hr to litharge in a stronger air blast. Some of this litharge was used to make at 1450°C a potassium lead silicate clear glass; some of the litharge was used to make at 950°C a yellow lead antimonate pigment. The lead isotope ratios of the four products were found to be the same as that of the galena to well within the experimental

Table 17.2. Comparison of Lead Isotope Compositions of Galena, Litharge, and Lead

Sample	$^{208}Pb/^{206}Pb$	$^{207}Pb/^{206}Pb$	$^{206}Pb/^{204}Pb$
Argentiferous galena, Rammelsberg: 61E	2.08933	0.85586	18.220
Lead metal smelted from ore 61E	2.09097	0.85610	18.233
Litharge from cupelling lead metal	2.08937	0.85648	18.213

precision of ±0.05%. The work of Barnes et al. is subject to the criticism that it is for circumstances in which the loss of lead is small (<5%), for which even the model of Budd et al. (1995a) would predict but small changes in isotopic composition. Similar confirmation that lead is not measurably fractionated by smelting or cupellation comes from our own work in 1983 with U. Zwicker. As shown in Table 17.2, the lead isotope compositions of argentiferous lead metal smelted by Zwicker from a lead–silver ore from the Rammelsberg mine (Harz) and of litharge resulting from the cupellation of that argentiferous lead are identical within the error of ±0.1% to the lead isotope composition of the ore.

Direct experiments by Pernicka and Bachmann (1983), with lead isotope analyses carried out by Todt, were reported not to have shown any measurable change (no lead isotope data were given) of lead isotopic composition in silver cupelled from argentiferous lead metal that had been smelted from galena from Lavrion. A series of ill-defined metallurgical processes on galenas and lead carried out in Bradford (Budd et al. 1995b) was also claimed to show no measurable change in lead isotope compositions, though again no numerical isotopic data were reported.

Since knowledge of the exact metallurgical processes practiced in ancient times remains scarce, it seems clear that the most rigorous tests of possible lead isotope fractionation should involve the actual remains of ancient metallurgical processes. Recent analyses of ancient lead and copper slags, lead and silver metal, and litharge prove that, in spite of possible considerable losses of lead, the processes of smelting of copper, fire refining, and the cupellation of silver do not affect the lead isotope compositions of the obtained metals (Gale and Stos-Gale 1996 and references therein). All experimental work so far shows that no measurable changes are detected between the lead isotope compositions of the components in the chain: ore–slag–metal and ore–lead metal–litharge–silver. We conclude that there is now overwhelming evidence that anthropogenic metallurgical processes do not measurably alter the isotopic composition of lead and the suggestions of Budd et al. (1995a) to the contrary are incorrect. Based on the same theoretical framework (Mulliken and Harkins 1922), Budd et al. (1995c) suggested that there should be relatively large alterations in the isotopic composition of tin pro-

Table 17.3. Lead Isotope Composition of Weathered and Corroded Objects

Glass				Reference
Less than 0.05% difference observed between interior glass and exterior weathered product for 1600-year water immersed Greek glass and 2600-year old glass from Nimrud, Iraq				Barnes et al. 1978
Lead—Early Cycladic (ca. 2800 B.C.) Lead Boat Models from Naxos				Gale and Stos-Gale 1981
Inv. Mus. No.	Sample	$^{208}Pb/^{206}Pb$	$^{207}Pb/^{206}Pb$	$^{206}Pb/^{204}Pb$
1938.725	Lead metal	2.07663	0.83706	18.735
1938.725	Corrosion	2.07652	0.83697	18.730
1938.726	Lead metal	2.07836	0.83756	18.748
1938.726	Corrosion	2.07924	0.83748	18.750

Copper—Sardinian Copper Oxhide Ingots (ca. 1200–1100 B.C.)				Gale and Stos-Gale 1985	
Site	Sample	$^{208}Pb/^{206}Pb$	$^{207}Pb/^{206}Pb$	$^{206}Pb/^{204}Pb$	Lead Concentration
Serra Ilixi	Copper metal	2.07012	0.84001	18.516	
Serra Ilixi	Sawing dust	2.06846	0.83956	18.524	106 ppm
Lanusei	Copper metal	2.07130	0.84190	18.480	
Lanusei	Sawing dust	2.10382	0.86449	18.034	177 ppm

duced by metallurgical processes. However, both Gale (1997) and Begemann et al. (1999) have, by direct measurement, shown that such alterations of the isotopic composition of tin do not occur at the level predicted by Budd et al.

There is also some direct experimental confirmation that the weathering of glasses over periods of up to 2600 years does not result in altered lead isotope compositions in the weathering product (see Table 17.3) when compared with fresh interior glass (Brill 1969). Our own observations show that the lead isotope composition of the thick corrosion product on the exterior of two Early Cycladic lead boat models from Naxos, dating from about 4800 years ago, is the same as the isotopic composition of the interior uncorroded lead metal (Table 17.3).

Though corrosion also does not alter lead isotope compositions by fractionation, corroded samples of copper objects of low lead content can be dangerous for lead isotope analysis. Corrosion affords the possibility of alteration of the true lead isotope composition of an artifact by exchange of lead, of differing isotopic composition, from the burial environment. An example of this is shown in Table 17.3 for the second Sardinian copper oxhide ingot, from Lanusei, where the sawing dust contains much corrosion and exhibits an isotopic composition characteristic of the Sardinian soil, not the copper of the ingot.

17.5.3. Ore Deposits; Homogeneity of Lead Isotope Composition

In the formation of the majority of metal ore deposits the lead contained in them is separated from the parent uranium and thorium. The "frozen-in" isotopic composition of the lead so separated depends, as we have seen, on the geochemical history of the source materials, specifically on the U/Pb and Th/Pb ratios and the timing of any variations in these ratios, including the time of final separation of the lead. As Barnes (1979) has remarked, it is theoretically possible for each ore deposit in the world to have its own unique isotopic composition, though in many cases the differences between deposits may be so small as to demand the utmost precision and accuracy in analytical techniques in order to differentiate them.

It has sometimes been questioned by archaeometrists whether the isotopic composition of lead in an ore deposit is uniform throughout. Doubt has also been expressed whether the lead in different minerals from the same ore deposit, especially from the primary sulfides as opposed to oxidized minerals, will have the same isotopic composition. An investigation by Barnes et al. (1974) of the Lavrion ore deposit shows that, for the 17 ore samples analyzed, the complete range of variation in isotopic composition is 0.28%, while the isotopic composition is not correlated with geographical position or depth in the ore body, nor with ore type. Further work that confirms these conclusions for the Lavrion ore deposit has been reported by Stos-Gale et al. (1986, 1996).

A number of similar investigations have been carried out by other isotope geochemists. In one example for two deposits in Australia, Gulson and Mizon (1979) have shown that the complete range of lead isotope compositions in the primary sulfide ores is less than 0.3% and that the oxidized ores in the overlying gossan have lead isotope compositions indistinguishable from the deep primary sulfide ores (Fig. 17.8). Many further examples are given by Gulson (1986), in the course of illustrating the value of lead isotope data in mineral exploration.

It is not an easy matter to establish the boundaries of the range of lead isotope compositions for particular ore deposits. Ideally it requires analyses of about 30 to 50 different ore samples from a particular ore deposit to establish its characteristic lead isotope field. Early work by Brill and Wampler (1965b, 1967) identified a so-called Group L of lead isotope compositions that they associated loosely with the Lavrion ore deposit in Attica. Later work in Washington and Oxford showed that the true Lavrion ore deposit

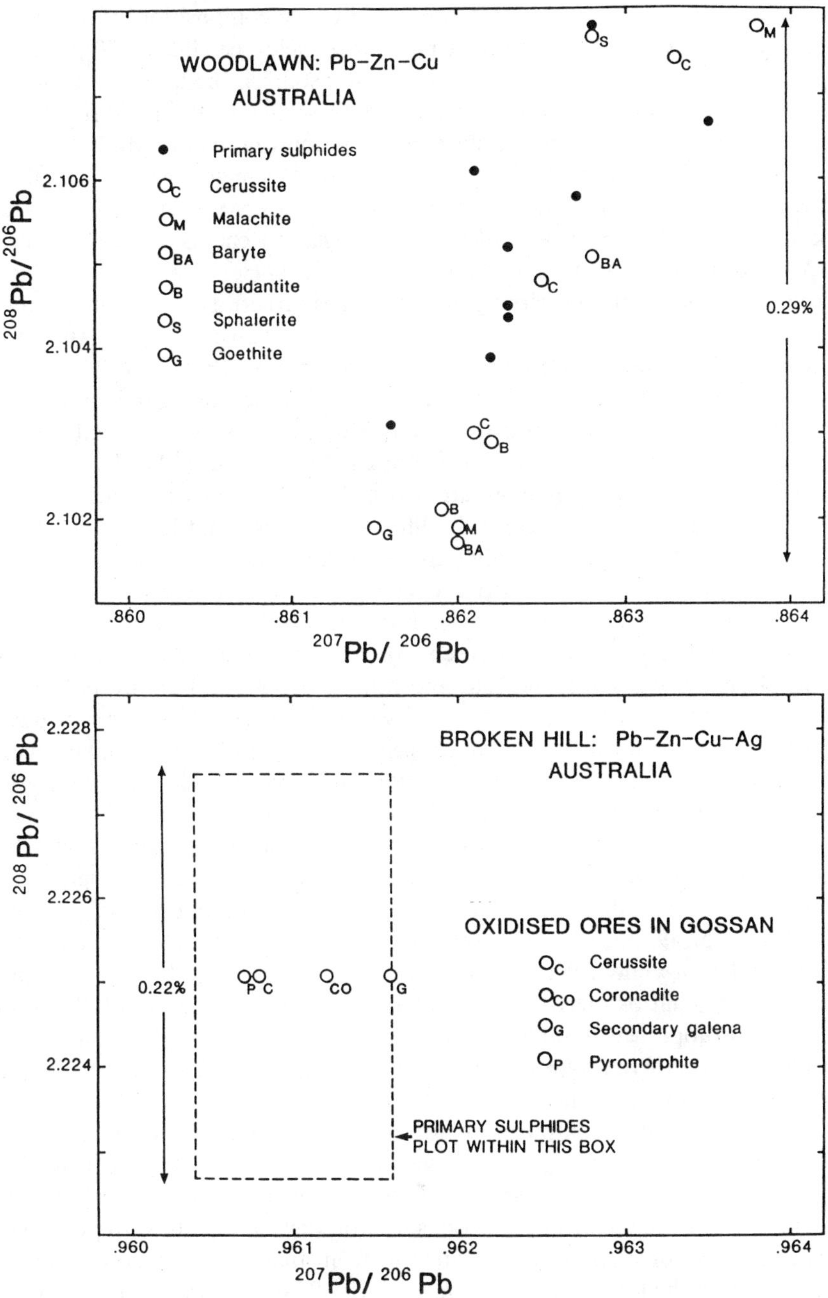

Figure 17.8. Lead isotope compositions for different ore types and associated gossan for two Australian ore deposits. Data from Gulson and Mizon 1979.

exhibited a much more restricted range of isotope compositions [for a comparison between Group L and the true Lavrion field, see Gale (1980: Fig. 8)], while today we have many more analyses to establish the boundaries of the Lavrion field more accurately (Stos-Gale et al. 1986, 1996); in addition there are more recent data from Agrileza and the Soureza valley. Merkel (1985) has recently demonstrated how strongly the trace element pattern of smelted metal is influenced by the initial choice of ore grade charged into the furnace; he has suggested that provenance studies based on chemical analysis must take into account ore beneficiation practiced in ancient times. It was such factors, including the possibility that in Lavrion cerussite (for which only four isotopic analyses exist; Stos-Gale et al. 1986: Table 4) might have been used in ancient times (Conophagos 1980) and that ores exploited by ancient miners are now largely not available, that influenced us in including also analyses of lead and litharge produced in Archaic/Classical industrial sites in our definition of the Lavrion lead isotope field. Agrileza (Ellis Jones 1985) and Soureza (Conophagos 1980) are two such sites, which are situated inland away from the coast, are relatively difficult of access, and had ore-washing facilities, which suggests strongly that they were working the local ores; lead and litharge from these sites moreover have lead isotope compositions in accord with those of the local ores. Litharge was used on a large scale in Lavrion for lining water cisterns, covering ore-washing tables, and even as wall plaster in Agrileza. The lead and litharge from Agrileza are not small samples from a coastal site where lead was in short supply and might have been imported; they are from fourth-century industrial sites equipped with planar ore washeries and set up to exploit on a considerable scale the lead–silver ores of the central Souresa valley region of the Lavrion.

The Lavrion ore deposit is known principally as a lead–zinc–silver deposit, but there is a considerable amount of oxidized copper ore there also, chiefly malachite and azurite and chiefly in the Kamaresa region (Fiedler 1841: 43–45, 49, 55, 61, 89; Cordella 1878: 131–132; Cambresy 1889: 76; Gale and Stos-Gale 1982: 17–18). The data reported by Stos-Gale et al. (1996) show that oxidized copper and arsenic ores from Lavrion fall within the lead isotope field established by analyses of lead ores, providing more evidence that different types of ores from the same ore deposit have the same lead isotope composition.

Another possible complication for metal provenancing using lead isotopes, at least for copper, is the question of fluxing. Copper ores usually contain unwanted gangue, which has, in the smelting process, to be separated as slag from the metal; a flux has to be added to effect this. In principle, lead of an isotopic composition different from that in the metal ore might come in with the flux or the fuel ash. Siliceous fluxes will mostly have a very low lead content; ferruginous fluxes will usually have been taken from the nearest source, which will usually be the gossan of the ore source itself, having the same lead isotope composition as the ore source. Fuel, chiefly charcoal, usually has a low lead content, and in many cases the trees from which it came will have been growing near the ore deposit and will tend to have incorporated lead of the same isotopic composition as that ore deposit. In most cases one might expect the isotope composition of lead in the smelted copper not to be perturbed away from that of lead in the copper ore. The data that we currently have from copper slags from Cyprus and from Thasos show that this expectation is fulfilled for the admittedly rather few analyses made so far.

17.5.4. Remelting of Scrap Copper Alloys: The Mixing Problem

The most serious potential problem for the use of lead isotope analyses for provenancing copper in copper-based alloys is the possible remelting together of metal derived from different ore sources. In our view both archaeologists and archaeometallurgists have been too ready to assume that such recycling is common in prehistoric times, without a serious and thorough examination of the evidence for or against. In reality, this question should be examined, using all available archaeological and other evidence, for each culture and period being studied. For example, the archaeological evidence suggests that this may not have been a very common practice at any time in the Aegean, where it was at all times common practice to bury metal objects with the dead (Renfrew 1972). For Middle Minoan Crete Moody and Ickermann (1985) have shown that metal objects are found not in settlements or palaces but in graves, where they were largely safe from reuse. Further, Branigan (1968) has proved that the reriveting of daggers was a common practice in Early and Middle Minoan Crete, while some 15% of reriveted daggers show clear evidence of having been reriveted on two or more separate occasions. Branigan (1968: 46) comments that "all the re-rivetings are perhaps indicative of the trouble to which Minoans of the Early Bronze Age went in order to avoid scrapping their weapons, or even having them melted down and recast." In fact, it was always easier to repair riveting or rework a damaged edge than to melt down and recast a tool or weapon. Even in the Late Bronze Age the Aegean is unusual in comparison with other parts of Europe in that the quantity of hoarded metal was very small—small in absolute terms, but also small in comparison to the amount of metal deposited in graves (Harding 1984). The quantities of metal deposited in Mycenaean and Minoan graves were much greater than in most of Europe. Neither should one forget that in many cases hoards in the Aegean are not metalworker hoards but are often clearly hoards hidden away in time of danger that the owner hoped to recover in happier times—a hope very often not fulfilled.

Even if a limited amount of smelting and mixing of metal from discarded objects did take place, it may not have caused much confusion in the lead isotope pattern. If the various discarded objects in a founder's hoard were all made from copper that derives from the same ore source, then the mixing together of their metal to make new artifacts will not alter the tell-tale lead isotope composition characteristic of that ore source. This is a quite likely situation since strong economic and technical arguments suggest that at a given time a particular site will tend to receive its copper predominantly from one or perhaps two ore sources. For the Late Minoan II founders' hoard found in the Unexplored Mansion in Crete, this seems to be the case; 85% of objects proved to be made of copper from the Lavrion ore source, 15% from other copper ore sources (Stos-Gale and Gale 1990: 82).

In another archaeological context, Philip (1988) has argued cogently that many hoards in the Near East were put aside, not with the intention of future melting down and recycling, but deliberately to remove metal from circulation to preserve the value of metal. Conspicuous consumption of metal by interment could also raise the prestige of those who could afford this practice concurrently with the aspect of gifts to the gods, simultaneously restricting supply while increasing status and attracting divine favor. At the same time, Philip argued, awkward "counter-gifts" could be avoided, which would sometimes incur unwelcome indebtedness.

17.5.5. Analysis of Lead Isotope Data for Possible Mixing

Pernicka and his colleagues (1984: 592–596) have suggested a valuable method to prove whether or not a given body of data exhibits mixing. It depends on having both lead isotope and trace element data for the metal artifacts in question. Colinearity between three objects in a particular lead isotope–trace element diagram admits the possibility that the middle object might be made of metal mixed from the outer two, but this hypothesis is then negated if in another such diagram these three objects do not lie in a straight line. The method is laborious since all possible sample combinations must be examined, but this can readily be done using a computer. It is also a method that should not be overstressed, since it does not apply in its simple form to a model where scrap metal from many different sources is continually recycled. It should again be stressed, however, that such a model, though often put forward with little effort critically to examine the actual evidence, has to our knowledge never been substantiated as applying to a real archaeological situation.

The patterns of lead isotope compositions that would result if metals were mixed is to a considerable extent constrained by the patterns of the distribution of the lead isotope compositions of ores, which for a specific mineral occurrence is the feature of essential colinearity, which can, for instance, be seen in Figure 17.7. Consequently, the lead isotope compositions of two or more ingredients of a remelted metal will consist of groups of specific numbers that are not random but, to a good approximation, are linearly ordered. For a mixture between just two pieces of metal with different lead isotope compositions, the new isotopic composition of the mixture will fall exactly on a straight line between the isotopic compositions of the two components, the end-member compositions. The proximity of the composition of the mixture to each of the original ratios will be governed by the quantity of lead in each of the original pieces that were mixed. In the extreme case, if 1 g of pure lead were added to 1 g of copper with 0.1% lead, the resulting lead isotope composition of the new alloy will be indistinguishable from that of the lead metal, not the copper, because the quantity of lead from the lead metal will overwhelm the trace of lead in the copper metal.

In practice, more than two broken artifacts might be mixed together, provided that large enough crucibles were available (e.g., the LM II crucibles from Knossos were too small for this to be possible; Popham 1984: plate 206). In such a case the lead isotope composition of the mixture would lie within the polygon, which has as its end members the isotopic compositions of the various artifact fragments being mixed together. The principle will still hold that the lead isotope composition of the mixture cannot lie beyond that of the extreme end-member artifacts from which the mixture was made. We should consider another by-no-means trivial case. If a mixture is made from a number of artifacts that were all made from copper from the same ore deposit, then the mixture will have a lead isotope composition that falls within the isotopic composition of that copper ore deposit.

Thus there are constraints on the lead isotope compositions of metals that result from remelting together broken artifacts. The lead isotope ratios of ancient artifacts cannot be considered in an archaeometallurgical and geochemical vacuum but must always be assessed in the light of what is known about the lead isotope compositions of particular ore deposits. In studying the isotopic compositions of larger groups of artifacts, a helpful feature is that artifacts originating from the same ore deposit will group together in isotopic space. If frequent melting and remixing were practiced of metal originating from different ore deposits, then the new alloys would be expected to show a considerable scatter of lead isotope ratios along the mixing lines between the end members.

It should be mentioned here that Pollard and Heron (1996: 326, 336) endorse the proposal of Budd et al. (1995b), who, basing themselves on the application of the theory of Mulliken and Harkins (1922) to melting processes, proposed that predicted large changes in the isotopic composition of tin might be used as a method of quantifying the degree of recycling of ancient copper/tin alloys. Unfortunately, Gale (1997) and Begemann et al. (1999) have shown that changes in the isotopic composition of tin in ancient tin bronzes are not observed at the levels predicted by Budd et al. (1995b), so that this method is inapplicable in practice. It has now been demonstrated that the application of the Mulliken and Harkins (1922) theory to ancient metallurgical processes, as advocated by Budd et al. (1995b), is unsuccessful for lead (Gale and Stos-Gale 1996), for tin (Gale 1997; Begemann et al., 1999), and for copper (Woodhead et al., 1999; Gale et al. 1999). This is not due to faults in the theory, which was developed purely for the case of non-equilibrium evaporation with the gas phase being removed by condensation on a cold surface, but due to its misapplication to ancient smelting and melting processes, which are not of that nature. We may safely now lay to rest the suggestions in Budd et al. (1995a, 1995b) and Pollard and Heron (1996: 324–326). The review of lead isotope provenance studies presented in Pollard and Heron (1996: 302–344) unfortunately tends to be skewed in a somewhat partial direction, whose foundations have been destroyed by subsequent work (e.g. Stos-Gale et al. 1997); the article as a whole partakes of the character of the curate's egg (*Punch* 1895) and cannot be recommended.

The study by Pernicka and his colleagues of Early Bronze Age (EBA) metals from Troy, the Troas, and Yortan was cautiously interpreted by Muhly as an example of a mixing situation, since when it is all plotted on one diagram the lead isotope data seem to scatter in the way that would be expected from mixing together copper derived from a number of different ore sources (Pernicka et al. 1984: Fig. 31). But the data are from at least three different archaeological sites. If as in Figures 17.9 to 17.11, one plots separately the data for the three sites, the data split into well-defined groups characteristic of different copper ore sources. Moreover, we have carefully examined the trace element and lead isotope data for the Troas samples using Pernicka et al.'s (1984) method and have shown that none can have been formed by mixing metal from some of the others. The data as a whole seem probably not to be a case of the mixing of metals but rather to be a case of the incautious and incorrect plotting together on one diagram of data for at least three (the Troas samples may come from more than one site, are not stratified, and are not firmly dated) separate sites. It ought to have been obvious from the beginning that there was no reason necessarily to expect that these sites should have obtained copper from the same sources or that conflating data from at least three widely separated sites could be interpreted as a case of recycled metals.

There are further important points brought out by the data for Troy, Yortan, and the Troas samples. We have outlined, by hatched lines, regions on the lead isotope diagrams (Figs. 17.9 to 17.11). Though these may not correspond exactly with the isotope fields characteristic of the copper ore

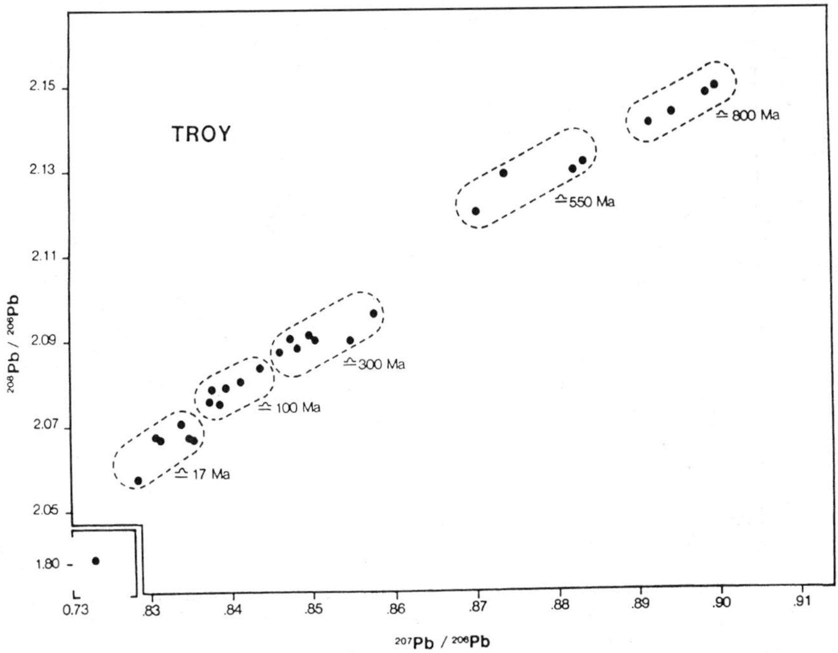

Figure 17.9. Lead isotope compositions for Early Bronze Age copper-based alloy artifacts from Troy (Stos-Gale et al. 1984; Pernicka et al. 1984; Gale et al. 1985a). Lead isotope fields, indicated by broken lines, are approximate; their exact delineation in terms of ore fields must await geographical location of copper ore bodies used by Early Bronze Age Trojans. Approximate model ages (from Cumming–Richards model) are indicated for each grouping of lead isotope data.

deposits appropriate for these objects, they are of about the size characteristic of the maximum range of isotopic compositions yet found for most individual Eastern Mediterranean ore deposits. For the Troas data, which have been shown not to be the result of mixing, model ages (Cumming and Richards 1975) for the four groups are approximately 100, 300, 540, and 845 Ma, respectively. These four groups of artifacts were made of copper that came from ore deposits of completely different geological age, and the model ages give a clue as to which ore deposits to investigate as possible sources. Similar considerations apply to the Early Bronze Age bronzes from Troy itself; for a discussion see Gale et al. (1985a) and Stos-Gale et al. (1984).

It is a very valuable property of the lead isotope provenancing method, not shared by chemical analysis, that the isotopic composition of lead in a metal artifact gives, through a model age, a guide to the geological age of the

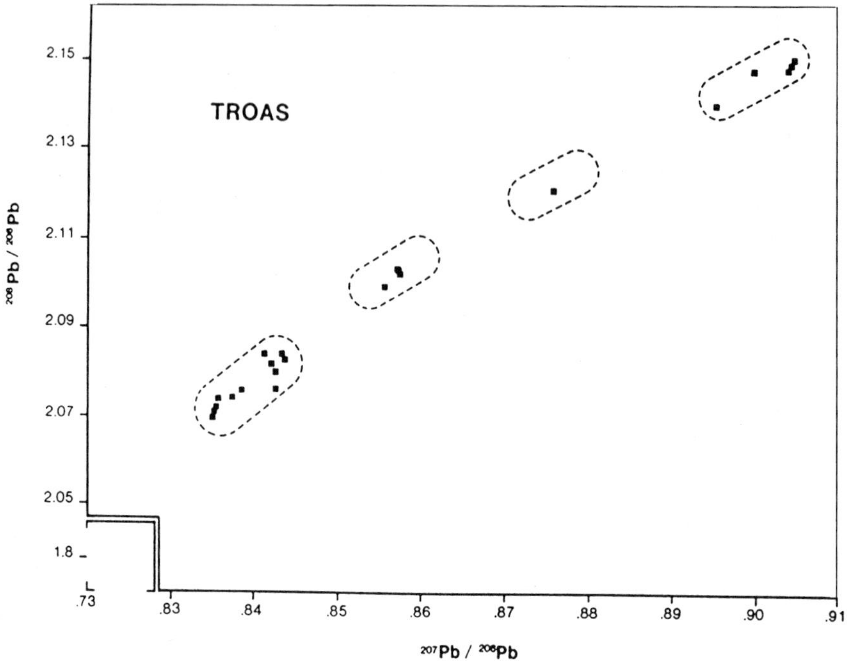

Figure 17.10. Lead isotope compositions for Early Bronze Age copper-based alloy artifacts from the Troas (Pernicka et al. 1984). None of these objects was made by mixture of metal from some of the others, as was proved by using method of Pernicka et al. (1984: 592–596). Lead isotope fields indicated remain tentative at this stage of research.

ore deposit that ultimately supplied its metal. How useful this guide will be in practice depends, of course, on the degree of exactitude with which the geological ages are known of ore deposits in possible source regions and to what extent the ore deposits in question have formation ages that relate to the model ages.

Therefore, it must be emphasized that, in the present poor state of knowledge of the ages of ore deposits in Turkey, the age associations in the case of Turkish ores are tentative. The model age approach can be a valuable guide for further research where geological ages of ore deposits are relatively well established, as in Sardinia and Cyprus. In these two cases the value of the model age approach has been amply demonstrated (Stos-Gale and Gale 1992). However, final associations of artifacts with metal ore deposits must depend on the identity of the actual lead isotope compositions of artifacts and ore deposits.

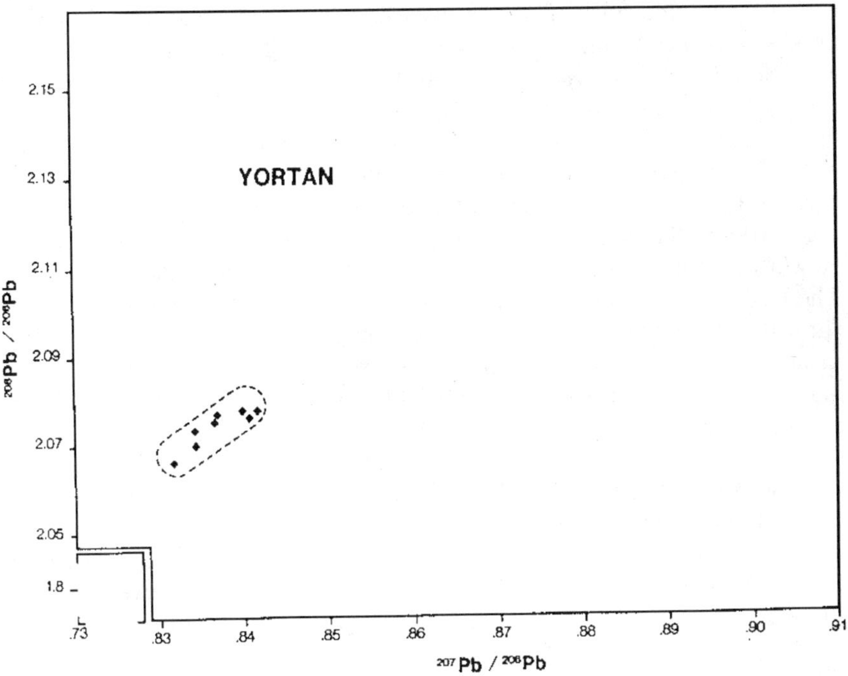

Figure 17.11. Lead isotope compositions for EBII-EBIIIA copper-based artifacts from Yortan cemetery (Pernicka et al. 1984; Gale et al. 1985a).

17.5.6. Metal Ore Deposits with Evolving Lead Isotope Compositions

One caveat, rather unimportant in practice, needs to be mentioned in connection with both the use of model ages and the constancy of lead isotope compositions throughout an ore body. In both cases it is assumed, as is overwhelmingly true, that ore metals such as copper, lead, and silver were separated from uranium and thorium at the time when the ore bodies were formed. For some copper ores the lead concentration can however be very low, as low as 10 ppm. If these low lead contents were associated with appreciable amounts of uranium or thorium, producing high U/Pb ratios, then radioactive decay up to the present day would alter the lead isotopic composition from that established when the ore deposit was formed. For such a sample, simple model ages are not appropriate. Moreover, since the same ore deposit may contain copper ores with widely different U/Pb ratios, this effect could rarely cause cogenetic ores from the same deposit to have rather different lead isotope compositions that may lie on a linear array in a lead isotope ratio plot. Such effects are well known to isotope geochemists, who usually measure uranium, thorium, and lead as well as lead isotope ratios in rock samples and correct the present-day isotope ratios to "initial ratios" using the independently measured age of the rocks (an example for metalliferous sediments from Cyprus, Syria, and Oman is the work of Gale et al. 1981). We emphasize that, though important for the isotope geochemistry of rocks, this effect will only rarely affect the lead isotope geochemistry of older ore deposits; the isotopic signatures for ore deposits younger than about 50 Ma will usually not have been significantly affected by radioactive decay. The U/Pb ratio is so low for lead minerals that lead ore deposits will never show the effect, but it may very rarely be important in copper deposits. For instance, Berthoud's work (1979) shows that copper ores from Afghanistan and Iran have U/Pb ratios between 0.0001 and 10, though chiefly at the lower end of this range. Moreover, the related ore deposits of Timna and Feinan do have high and variable U/Pb ratios and show a corresponding range of lead isotope ratios, from which it is nevertheless possible to derive conclusions about provenance (Gale et al. 1990; Hauptmann et al. 1992). The evolution of lead isotope compositions in lead-poor ore deposits is discussed further, with examples, by Gulson (1986) in a book that deserves study by those applying lead isotope analyses to archaeological science.

17.6. PROVENANCING ANCIENT RAW MATERIALS BY LEAD ISOTOPE ANALYSIS: LEADING PRINCIPLES

17.6.1. Copper Metal and Copper-Based Alloys

There are certain important principles to be observed when using lead isotope analyses for provenancing copper-based metals (Gale and Stos-Gale 1982). The most obvious is that lead isotope analyses of a heavily leaded bronze containing, say, 8 to 15% lead, will determine the provenance of the added lead, not the copper, in the artifact. Clearly the lead in the bronze must come predominantly from the copper ore, with the provisos discussed above about fluxes and fuel, if its isotopic composition is to be of use in provenancing the copper in the bronze. The direct smelting experiments of Tylecote et al. (1977) and the analyses quoted by Percy (1861: 408, 422, 429, 503) of batches of raw smelted copper show that, with a relatively high content of lead in the copper ore, an accidental content up to 4 to 5% of lead in ancient smelted copper metal is quite possible. One example of black copper from Freiberg contained 20.75% lead (Percy 1880: 307). It seems safe to say that copper alloy artifacts containing lead in the range from 50 ppm to 4% are suitable for the provenancing of their copper by lead isotope analysis

(Gale and Stos-Gale 1982). Muhly (1983) questioned whether segregation of lead in copper alloys might make it impossible to determine, by analysis of a small drilled sample, whether one was dealing with an intentionally leaded alloy or not. Our discussion of this point (Gale et al. 1985b: 84–87, 97–98), based largely on casting experiments made by Hughes et al. (1982) and by Staniaszek and Northover (1982), shows that this is not in fact a problem. In fact, Craddock (1976: 97) showed, by analyzing 14 fragments of a large cast statue of a warrior, that throughout the statue there was only a 10% variation of the lead content, which averaged $5.5 \pm 0.4\%$. Samples taken from the left knee and the right little finger of another bronze statue (of Apollo; Craddock 1976: 97) showed lead contents of 20.6 and 20.3%, respectively.

As discussed by Gale and Stos-Gale (1982, 1985), added tin will very rarely bring with it to a bronze sufficient lead to disturb the isotopic composition of the lead associated with the copper metal component. Recent work on both the chemical and lead isotope analysis (Stos-Gale et al., 1998b) of the tin ingots found on the Bronze Age shipwreck of Uluburun (Bass 1987) shows that their lead concentrations fall into two groups and that these groups formed on the basis of lead concentrations are the same as groups formed on the basis of lead isotope compositions. One of these groups has an average lead concentration of 15 ppm; the second group averages 1100 ppm Pb. The contribution of lead to a 10% tin bronze of tin from the first group of ingots is entirely negligible. Those tin ingots from the second group averaging 1100 ppm Pb would, if the copper component had only 0.1% Pb, contribute 12% of the lead in a 10% tin bronze; such a contribution could cause a change away from the lead isotope composition of the copper of the order of 0.2% if the copper is from Cypriot ores and at most 0.1% if the copper is from Lavrion. For a copper metal component carrying with it more than 0.1% Pb the effects are completely negligible. Neither does it seem that artifacts made of arsenical copper will very often have lead isotope compositions that reflect anything other than that of the arsenical copper ore that will usually have been smelted to make them (Gale et al. 1985b; Gale and Stos-Gale 1989: 30–34). There is at present no evidence for the Mediterranean Bronze Age that arsenical copper alloys were ever made by the deliberate use of arsenic minerals per se as an addition to the smelting charge (Gale and Stos-Gale 1989).

It is important that one should not include in an ore data reference bank lead isotope analyses of so-called copper sources that are no more than insignificant and unexploitable occurrences of mineralogical interest only, as are many of the sources shown on the Metallogenetic Map of Greece (Zachos and Maratos, 1973). Attention should be confined to ore deposits, preferably those deposits showing signs of having been worked, and to slag heaps. Dated slags and metal inclusions are ideal materials to define the lead isotope field of a smelting activity, as, for example, the Early Cycladic slag heaps on the Cycladic island of Kythnos (Gale et al. 1985b; Stos-Gale 1989, 1998a) or the Minoan Chrysokamino slag heap on Crete (Stos-Gale 1998a). Ideally one should seek evidence of the date of working from pottery sherds, C-14 and thermoluminescence dating, either in the mines or from slags and slag heaps. Fieldwork is an inescapable part of worthwhile lead isotope studies.

Nevertheless one cannot always expect to find direct evidence of ancient working of an ore deposit; extensive later, even "modern," working may have largely destroyed the earlier evidence, as in Lavrion and Cyprus. In such cases one may only be able to proceed by making, say, lead isotope copper provenance studies on metal artifacts from an archaeological site near the ore deposit. If such studies show that a Late Bronze Age site was using copper from a nearby ore deposit, then that would provide good indirect evidence that the ore deposit in question was worked at least as early as the Late Bronze Age.

In building up an ore lead isotope data bank, the geology of the ore deposits must be taken into account. For instance, most of the mixed sulfide ore deposits in the Aegean are, roughly speaking, pyrite–blende–galena (PBG) deposits, providing predominantly lead, zinc, and silver. Some of these deposits contain minor amounts of chalcopyrite but usually at such a low level and so thoroughly intergrown with the pyrite, blende, and galena that ancient peoples cannot possibly have extracted copper from them. For this reason it is dangerously misleading to construct, as have Pernicka and his colleagues (1984: Figs. 28, 30, 31, 32), a general Aegean field (using lead isotope data almost wholly measured in Oxford and therefore from well-known geological context), based predominantly on PBG deposits and then to compare with it lead isotope analyses of Bronze Age unleaded copper alloy artifacts or even Cypriot copper ores. For copper provenancing purposes copper alloy artifacts must of course be compared only with copper ores that could have been exploited in ancient times; for example, the considerable reserves of oxidized copper ores in parts of the Lavrion ore deposits (e.g., Cordella 1878; Marinos and Petrascheck 1956). It is important also to realize that for many of the PBG deposits used to construct this general Aegean field there is no evidence of ancient working. Further it should be emphasized that within this Aegean field the data for the Lavrion, Siphnian, Kythnian, Thasian, Halkidiki, and other Aegean mines are well-defined subsets. Little value attaches to such an Aegean field save to answer the coarsest of questions; namely, is a given lead artifact consistent or not with having been made from one of the ore deposits used in its construction? On the other hand, uncritical use of such an Aegean field can lead to much confusion. A recent development is the opening of the pages of *Archaeome-*

try to the publication of lists of lead isotope data for ores; five such data sets have been published so far (Stos-Gale et al. 1995a, 1996, 1998a; Rohl 1996; Gale et al. 1997).

17.6.2. Provenancing Other Metals

The classic example of the provenancing of metals using lead isotope analysis is the application to lead metal and silver. The application to lead metal is straightforward, at least for the Bronze Age, where lead was in surplus as a little needed by-product of silver production (Gale and Stos-Gale 1981a, 1981b; Gale et al. 1984). Lead isotope analyses of litharge—lead oxide produced as a by-product of silver extraction—can throw an interesting light on the organization of the silver trade. For example, in the Bronze Age Aegean some of the silver-rich ores from Lavrion were processed in the settlements on Kea and Santorini (Stos-Gale and Gale 1990). For silver metal artifacts one needs to be cautious if the silver is debased to any extent by the addition of copper, in case the added copper brings with it a significant quantity of lead. The most extensive lead isotope analyses of ancient silver are those of the earliest Greek coins (Gale et al. 1980; Hardwick and Stos-Gale 1998; Stos-Gale 1998b) and Predynastic Egyptian silver (Stos-Fertner and Gale 1978; Stos-Gale and Gale 1980). The analyses of silver from the Mycenaean Shaft Graves (Stos-Gale and MacDonald 1991) now need re-evaluation, in view of the considerable augmentation of the lead isotope database of ores during the last decade. Recently an interesting lead isotope study of precoinage silver from the Near Eastern "Hacksilber" hoards has been started (Stos-Gale, in press).

Lead isotope analysis should be of great utility in provenancing iron artifacts where, in the Late Bronze Age and in the Iron Age, the mixing problem does not exist since iron could not be melted. Two such provenance studies have been attempted (Gale et al. 1990; Hirao and Enomoto 1994).

Traces of lead in zinc and tin artifacts may allow these metals to be provenanced and a program is currently in progress to attempt to trace the export of zinc from mediaeval Zawar in India. An important development may be the provenancing of Bronze Age gold, for which no other method exists. Antweiler et al. (1972) have successfully demonstrated that analyses of trace lead in a natural placer gold deposit enabled the location of its primary lode source; it should be possible to extend this to gold artifacts.

17.6.3. Provenancing Minerals and Pigments

In principle, provenance studies can be effected simply by comparative lead isotope analyses of archaeological artifacts and likely ore deposits. Some classes of archaeological artifacts do indeed lend themselves directly to such a simple approach. For instance, it is common to find, in Predynastic Egyptian sites, samples of the lead mineral galena and the copper mineral malachite; galena was also collected by the early cultures of northeastern North America. In Pharaonic Egypt black kohl and eye medications were often of finely powdered galena. Many provenance studies of such mineral samples from archaeological contexts have been made (Brill et al. 1974; Gale and Stos-Fertner 1978; Stos-Gale and Gale 1980; Hassan and Hassan 1981; Gale et al. 1991). Such studies are entirely free of the mixing problem we shall have to discuss for metals; for the most part this is true also for artists' pigments that contain white lead (Keisch 1970) and in more restricted circumstances for some glasses and glazes (e.g., Brill 1968, 1970a, 1970b, 1972, 1976, 1978; Joel et al. 1988; Stos-Gale 1998c).

As an example of direct applications to minerals, we take the provenancing of galena found in Late-Archaic-Early Woodland sites in the Great Lakes area of North America, a study made by Farquhar and Fletcher (1980, 1984). Let us concentrate on the Finlan Site on Lake Ontario, where a number of galena samples were found in the excavations. Comparison with lead isotope data for central and northeastern North American galena ore deposits showed that the Finlan galenas do not come from any of the four local ore deposits (Farquhar and Fletcher 1980: Fig. 1). Instead they originate from the Upper Mississippi Valley deposits, about 1100 km away to the west from the site. Regularities in the lead isotope data within the Upper Mississippi Valley mineral district restrict the source region for the Finlan Site galenas to the cross-hatched region in Figure 17.12, which lies in the southern part of the Upper Mississippi Valley district. This is a good example where lead isotope provenancing information can be derived even in a situation where a large ore deposit region, such as the Mississippi Ore District, is involved that has a large spread of lead isotope compositions.

The lead isotope studies of other mineral pigments on ancient sites in the Old World show contrasting features: In Eneolithic Bulgaria (Gale et al. 1991) the blue/green copper ores were from the nearest mineral occurrences. The interesting feature of these results is the dissimilarity in the origin of pigments and metal excavated in the same context. The azurite and malachite from stratified Eneolithic horizons in the Bulgarian sites is consistent with originating from the excavated Eneolithic mine of Aibunar situated in central Bulgaria (Gale et al. 1991); on the other hand, there is relatively little copper metal from the Varna cemetery and central Bulgarian sites bearing the same isotopic signature (Gale et al. in press). In Eneolithic Serbia no single piece of oxidized copper ore excavated from Eneolithic sites is consistent with derivation from the Eneolithic copper mine of Rudna Glava (Pernicka et al. 1993). In Egypt, while some or all of the predynastic kohl is consistent

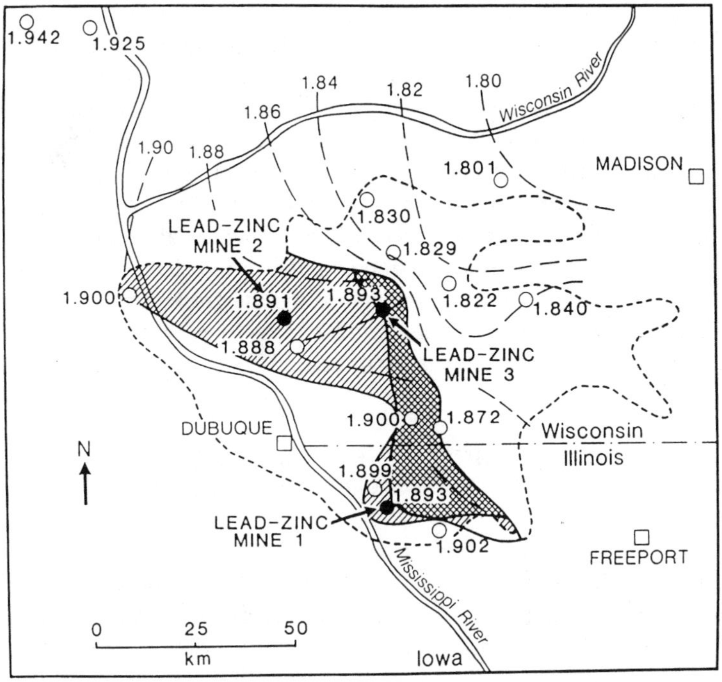

Figure 17.12. Variation in $^{208}Pb/^{206}Pb$ across Upper Mississippi Valley ore district (data from Heyl et al. 1966; Farquhar and Fletcher 1980, 1984). Based on comparative lead isotope data, stratified galenas found at Finlan site can have come only from cross-hatched region.

with the galena from the mines in the Eastern Desert, the lead and silver shows distinctly different lead isotope ratios (Stos-Gale and Gale 1980).

The study of artists' lead-based pigments used in oil paintings, conducted by Keisch (1970) nearly 30 years ago, was never continued on a large scale. Some of Keisch's analyses were repeated on the same samples a year ago in the Isotrace Laboratory in Oxford using a modern multicollector TIMS instrument. It was confirmed that, in spite of the much higher accuracy of the modern measurements, the general pattern of the lead isotope compositions of pigments is the same as that described by Keisch. However, due to the currently available large number of lead isotope analyses of galenas from Europe, it is now possible to begin to relate this pattern to the compositions of ore sources. The preliminary analyses of the data indicate that since mediaeval times much lead white used in pigments was produced using lead ores from large English mines in Derbyshire and the Mendips. The same lead was used for masonry fittings in the Salisbury and Winchester cathedrals.

Brill (e.g., 1968, 1978; Brill et al. 1974; Lilyquist and Brill 1993) conducted lead isotope analyses of numerous ancient Egyptian and Near Eastern glasses. The approach of this group of researchers is directed toward the comparisons of lead isotope patterns in archaeologically and chemically different glasses rather than the direct interpretation of the origin of the lead mineral. This approach can also bring very interesting results and, in view of the present scarcity of lead isotope data for the Near and Middle Eastern ore deposits, very often is the only possible approach. Particularly interesting was a study of Chinese and Japanese glasses. The lead in Chinese glass shows isotopic compositions distinctly different from any European and Near Eastern glasses, while the Japanese glass was successfully compared with the known local ore sources (Brill et al. 1979).

17.7. ARCHAEOLOGICAL CASE STUDIES RESULTING FROM LEAD ISOTOPE ANALYSES OF METALS

The application of lead analysis to studies of the provenance of lead, silver, and copper, together with extensive fieldwork, has made considerable advances in our understanding of the Bronze Age metal exploitation and trade in the Eastern Mediterranean. A number of individual topics have been chosen to illustrate this.

17.7.1. Early Cycladic Metal Sources

The first work in this field (Gale and Stos-Gale 1981b) was an attempt to establish the sources of lead and silver for the Early Cycladic people against the background of the exploration of the mines of Siphnos and other Cycladic lead–silver sources (Wagner et al. 1980; Gale and Stos-Gale 1981b; Wagner and Weisgerber 1985). The Cycladic islands of the central Aegean come into extraordinary prominence in the third millennium B.C.; this was the crucial, formative period for Aegean civilization. The Cyclades were then the home of flourishing cultures with prominent settlements, a rather abundant population, well-developed pottery, and striking achievements in marble sculpture. Production of silver, lead, and copper from their ores was developed early. They were prominent in their use of arsenical copper, while tin bronze appears in this period, in special archaeological circumstances, only at the site of Kastri on Syros. The Cycladic people were the seafarers of the early third millennium B.C. Their cemeteries give evidence of a gradually developing social stratification associated largely with their development of metallurgy (Renfrew 1972). No enquiry into Aegean metallurgy as a whole could afford to omit a study of its Early Cycladic antecedents. We should like to know something about the sources of their metals, their practice of mining and metallurgy, whether metallurgy developed autonomously in the Cyclades, and why arsenical copper is found in use in Early Cycladic times in every investigated Cycladic island but tin bronze occurs in only one site on Syros.

The nearest large sources of lead and silver, well known from Herodotus, are in Lavrion and Siphnos, so for the investigation of Early Cycladic lead and silver sources lead isotope analyses of these ore deposits were essential. Though we now know the shapes of the lead isotope fields for the Siphnian and Lavrion mines rather better than in 1981, the conclusions remain essentially the same and will therefore be summarized here. Eleven of the 21 analyzed Early Cycladic lead and silver artifacts from the islands of Amorgos, Antiparos, Cheiromylos in Despotikon, Naxos, Siphnos, and Syros fall into the Siphnian field, while 7 fall into the Lavrion field. The remaining 3 objects come from other, presently unidentified ore deposits. Later work on Early Cycladic lead from Ayia Irini on Keos shows 4 artifacts of Siphnian lead, 6 of Lavrion lead, and 4 from other sources (Gale et al. 1984). The dominant Early Cycladic sources of lead and silver seem to have been Siphnos and Lavrion. The lead isotope analyses provided the first evidence that the lead–silver mines in both Siphnos and Lavrion were being worked in Early Cycladic II times, about 2800 B.C. In those two deposits there is also a direct evidence of their exploitation at that time (for Siphnos: Gale and Stos-Gale 1981b; Wagner and Weissgerber 1985; for Lavrion: Spitaels 1984). A closer examination of these questions has been given by Gale (1998), in which it is proved that lead deposits in other Cycladic islands, particularly Keos and Thera, were not exploited in Early Cycladic times.

What of the Early Cycladic sources of copper? A systematic survey of the ore deposits and slag heaps in the Cyclades led in 1982 to the discovery of an Early Bronze Age copper smelting slag site of Skouries on the Cycladic island of Kythnos (Gale et al. 1985b; Stos-Gale et al. 1988; Stos-Gale 1989, 1998a); it is therefore the earliest proven copper smelting site yet known in Greece. Chemical analyses of the copper slags, and of prills of copper within them, show that arsenical copper was being produced accidentally here in the Early Bronze Age. Lead isotope analyses of the slags and of stray pieces of copper ore found within the slag heap enable one to construct a characteristic lead isotope field (Fig. 17.13). This field, which is clearly separated from the isotopic field for Lavrion, seems to have structure and probably should be divided into at least two regions. One of these regions appears to fit well with copper ores actually found on Kythnos, with a proviso. This is that we have found copper ores also on Siphnos at the location Aspros Pyrghos and one copper ore sample also at Ayios Sostis; lead isotope analyses of these Siphnian ores overlap with the Kythnian copper ores save that one ore

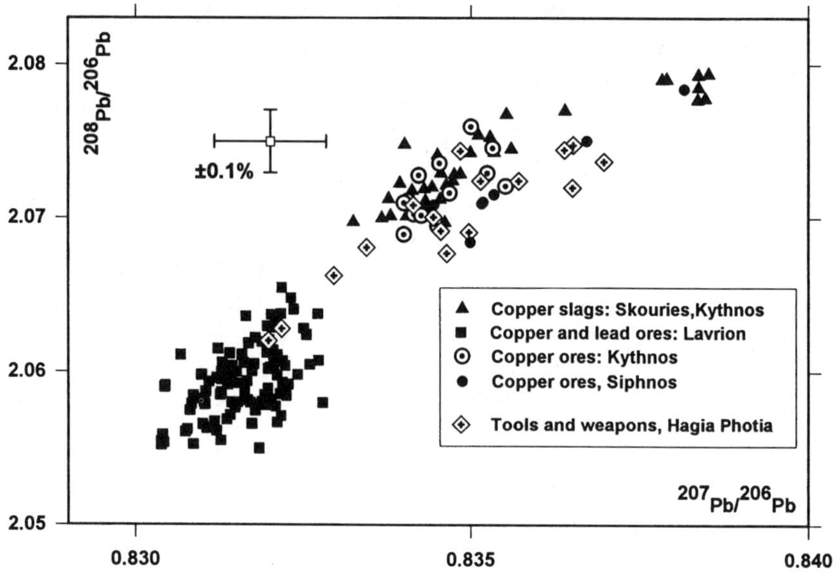

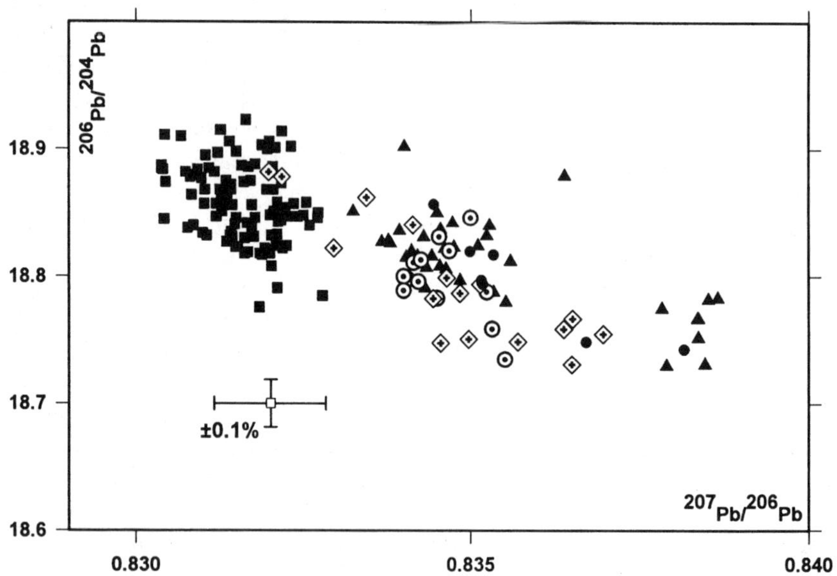

Figure 17.13. Lead isotopic fields for Kythnos and Lavrion, together with lead isotope data for copper ores from Siphnos and for Bronze Age slags from Kythnos.

from Siphnos plots with the upper group of slags from Skouries on Kythnos. It is argued in more detail by Stos-Gale (1998a) that it is probable that copper ores from a number of Cycladic islands were brought to Kythnos and smelted there to produce copper. Lead isotope analyses of Early Cycladic copper objects from the "Kythnos" Hoard (cogently argued by Fitton to be related to Naxos rather than Kythnos) and from Amorgos show that the great majority are consistent with copper that was processed at the ECII copper slag heap on Kythnos (Fig. 17.14). An interesting finding has been that most EMI copper objects from the largest cemetery of Bronze Age Crete at Hagia Photia, the first village east of Siteia on the northern coast of Crete, have lead isotope compositions consistent with copper processed on Kythnos (Fig. 17.15), though two are consistent with Lavrion ores. This is in accord with the pottery and other evidence that shows strong Early Cycladic influences at Hagia Photia, so strong that this may even be the cemetery of a Cycladic colony on Crete (Doumas 1976).

17.7.2. Anatolian Refugees in the Aegean?

A topic of archaeological interest is the question whether there were Anatolian refugees in the Southern Aegean at the end of the Early Cycladic II period or in the Early Cycladic IIIA period, perhaps fleeing from the destructions recorded at Troy, Poliochni on Lemnos, and Thermi on Lesbos at about 2100 B.C. Pottery of Anatolian type has been found in this period at the sites of Manika, Lefkandi, Raphina, Aegina, Kea, Syros, Delos, and Naxos. Moreover, hilltop-fortified sites, clearly standing in uneasy relationship to their truly Cycladic neighbours, have been excavated at Mt. Kynthos on Delos, Panormos on Naxos, and Kastri on Syros (Renfrew 1972; MacGillivray 1984).

The Anatolian links with the Cyclades in ECII/ECIIIA times, as revealed by pottery evidence, have been sought by many scholars on the Western Anatolian coast. However, our evidence from metallurgy points, not in this direction, but toward Troy, Poliochni, and Thermi. The analyses of the metal objects from Kastri on Syros show a large proportion to be of tin bronze, an alloy otherwise unknown in the Early Cycladic culture (Gale and Stos-Gale 1986a). At this period tin bronze was also not common in Anatolia, but it is quite abundant at Troy and at Thermi in Lesbos; at this latter site even an object of tin, a bangle, was excavated. A triangle diagram of Cu–As–Sn (Gale et al. 1985a: Fig. 4) shows also that the other alloy types in use at Troy and Kastri are similar. Finally the lead isotope data (Fig. 17.16) show not only that Troy, in Troy II times, had access to at least five different copper sources, but that the metal objects from Early Cycladic Kastri have lead isotope compositions that are consistent with their copper having come

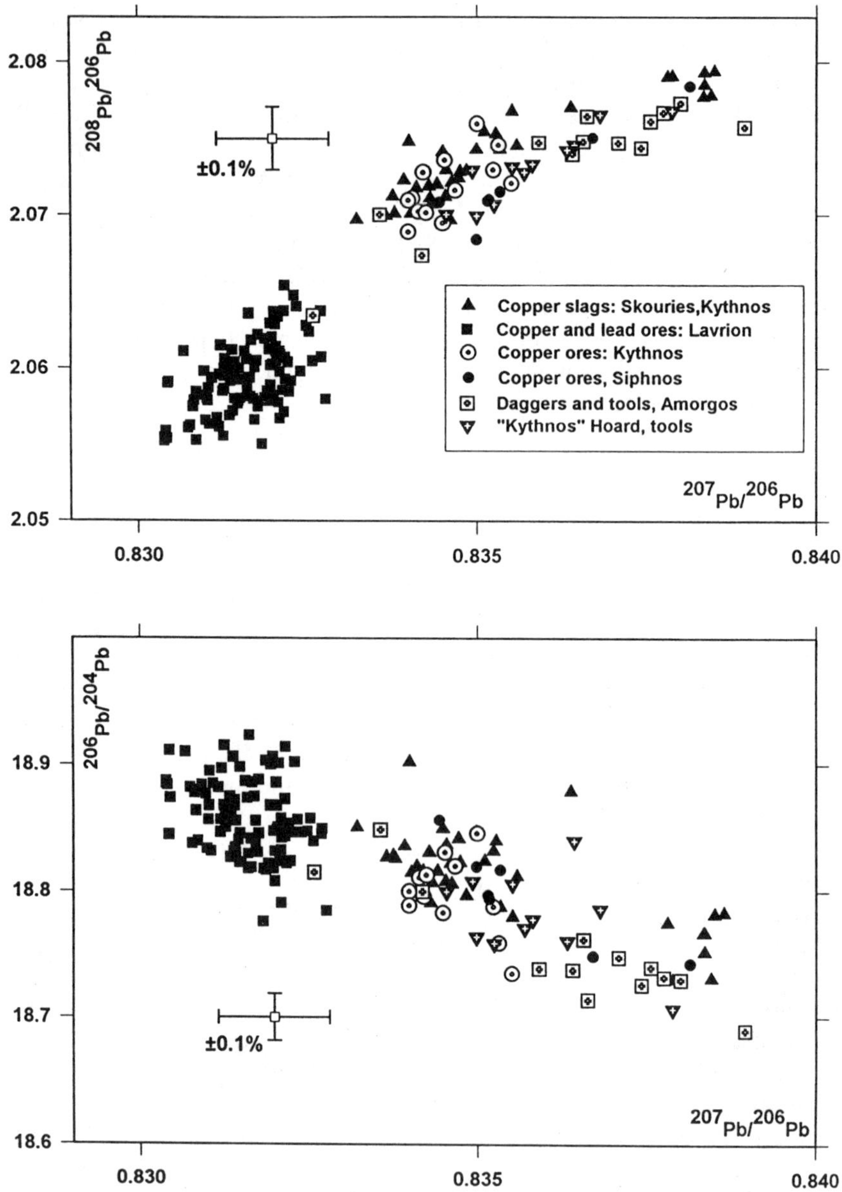

Figure 17.14. Lead isotopic compositions for Early Cycladic copper-based alloy artifacts from Amorgos and Kythnos hoard.

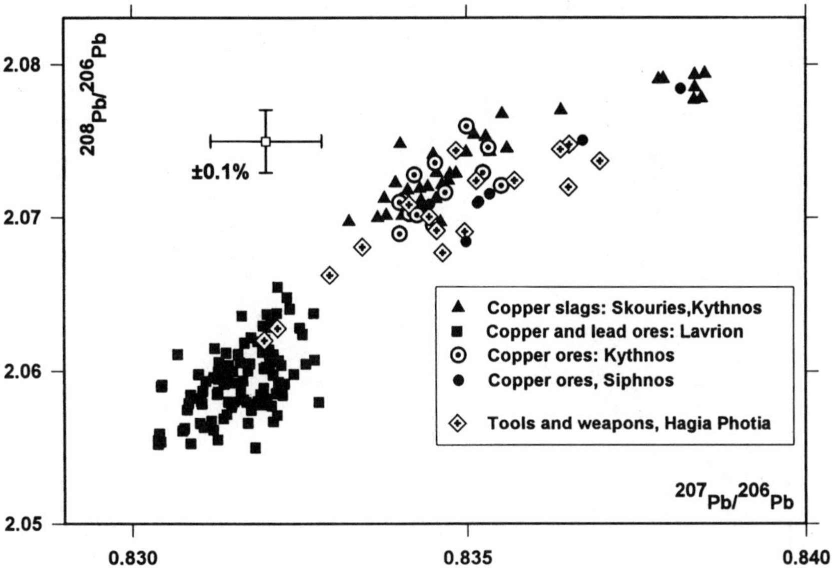

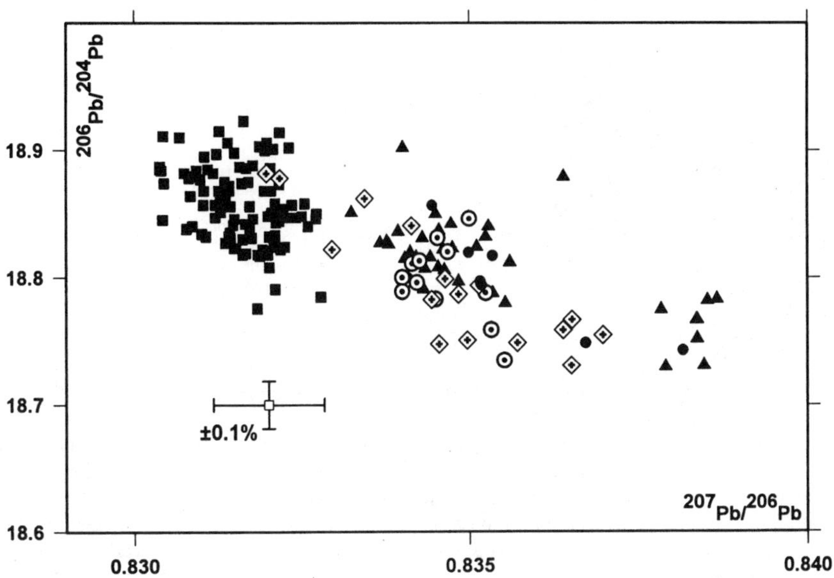

Figure 17.15. Lead isotopic compositions for copper-based alloy artifacts from Early Minoan I (EM I) cemetery of Hagia Photia in eastern Crete.

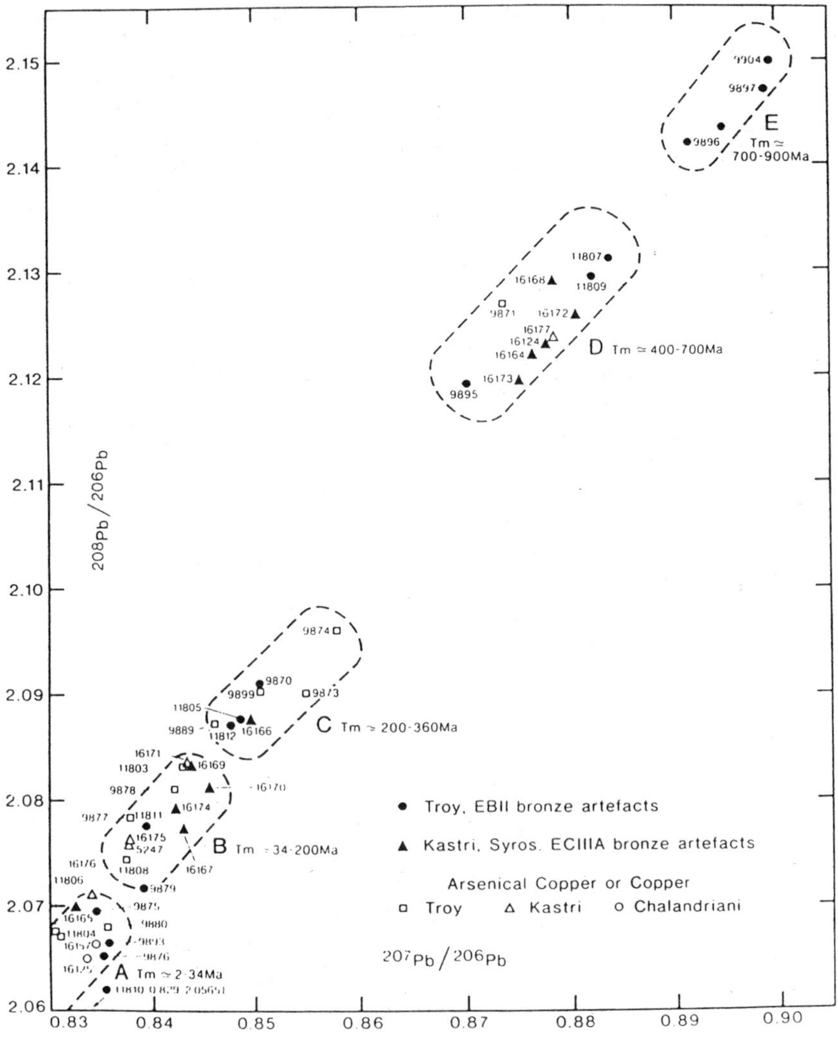

Figure 17.16. Comparison of lead isotope compositions of copper alloy artifacts from Troy and from Kastri on Syros. Approximate lead isotope fields, indicated by broken lines, serve chiefly to emphasise strong grouping of Kastri bronzes with Trojan groups A, B, C, and D. Exact identification of lead isotope fields appropriate to copper ore bodies used to supply copper to Troy groups C, D, and E still await their geographical location; they have not yet been identified as being in Anatolia.

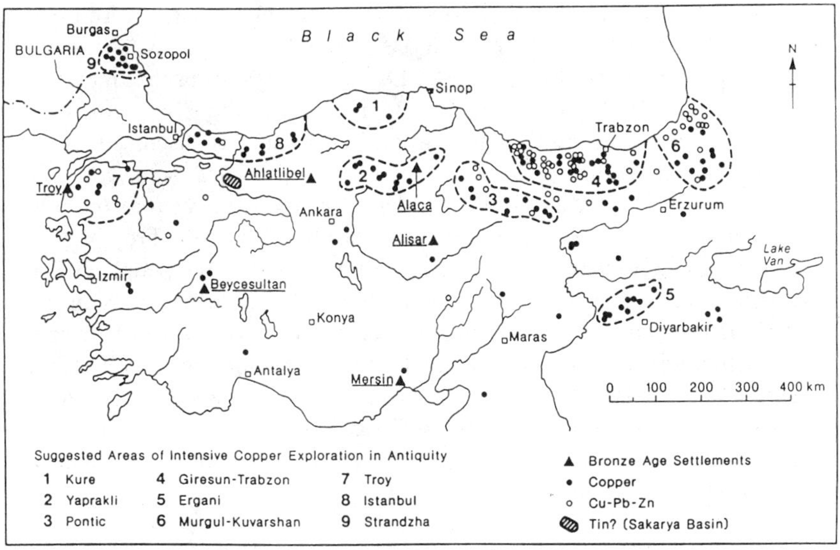

Figure 17.17. Map indicating copper deposit regions in Anatolia and Bulgaria that may have been exploited in antiquity.

from four of the sources in use by Troy (Gale et al. 1985a: see Fig. 29; Gale and Stos-Gale 1986a). [Figure 17.16 shows but one of the two necessary lead isotope diagrams; the other diagram is given by Stos-Gale and Gale (1984) and shows the same groupings.] Work reported by Wagner, Pernicka and others (Wagner et al. 1986; Pernicka et al. 1984) seems to confirm the suggestion that these copper sources are not in the immediate vicinity of Troy. Other potential copper sources exist along the Black Sea coast and elsewhere in modern Turkey (Fig. 17.17), which were investigated by a team from Heidelberg in the 1980s. The relatively few lead isotope analyses published by the German team for these Turkish ore deposits do not identify any of them as the source of copper for the artifacts from Troy and Kastri that have lead isotope compositions plotting in fields C, D, and E of Figure 17.16. Doumas (1988) has recently pointed to Poliochni and Thermi as the oldest protourban centers in Europe and has suggested that Troy itself seems rather provincial in comparison. Could Troy have acted with Poliochni and Thermi in controlling the movement of metals through the Dardanelles, perhaps not only copper but also tin from the Caucasus or beyond in Afghanistan or Uzbekistan? Tin might have accompanied lapis lazuli from the sources for both in Afghanistan. Another factor pointing to Central Asian metal sources is that the lead isotope data for groups D and E in Figure 17.16 correspond with Cambrian or Precambrian model ages. Ore deposits of such ages exist in Afghanistan and Iran, but not in the Mediterranean region, Anatolia, or the Near East. Moreover lead isotope data for the Omani site of Tell Abraq (Weeks 1998) for levels contemporary with the Trojan and Kastri artifacts include data falling into the groups C, D, and E, suggesting that metal from similar or the same deposits came into Oman also in the EBA, which is not inconsistent with a Central Asian origin. This topic is discussed in more detail by Gale (1998b).

The lead isotope and alloy type evidence seems to prove that the hilltop fort at Kastri contained metals coming most probably from Troy and, with the other archaeological evidence, suggests strongly that the people at Kastri were Anatolian refugees who had probably fled the destructions in Troy, Thermi, or Poliochni. This evidence fits very well with ideas put forward recently at the British School at Athens Centenary Conference in Manchester by Doumas (1988), who, on the basis of all the archaeological evidence, suggests that at this time Anatolian refugees came from the North through the straits of Euboea, which fits very well with the distribution of Greek sites producing Anatolian pottery.

17.7.3. Late Bronze Age Copper Oxide Ingots

Later still in time the copper oxhide ingots formed an important part of the international Mediterranean trade in copper. Copper oxhide ingots, of weight from 16 to 30 kg, were a common form in which copper was moved around the larger Mediterranean region in the LBA, from roughly 1600 to 1000 B.C. Buchholz (1959), Bass (1967), and Wachsmann (1987), *in passim*, have given the best archaeological accounts of this important class of traded object. Such copper oxhide ingots have been found from Sardinia and Sicily in the west through Greece, Crete, Turkey, and Cyprus to Syria, the Israeli coast and the Nile Delta, even in Bulgaria, and in Bogazkoy in central Anatolia. A comprehensive distribution map of known findspots was published by Gale (1991). One of the largest bodies of such ingots, more than 350, was excavated underwater, off the coast of Turkey, from the LBA Uluburun shipwreck (Bass 1987), which dates to about 1318 B.C. Large numbers of such ingots, mostly in fragmentary state, have been found on Sardinia (Lo Schiavo, 1990), while the very earliest copper oxhide ingots known, dating from 1550 to 1500 B.C., were excavated at the Minoan site of Hagia Triadha (Paribeni 1903) in Crete. Many scholars have thought that these ingots were all made from Cypriot copper (e.g., Buchholz 1959), though some have thought them to have been made at different places, including Sardinia, perhaps under a central control and definitely for international trade (Bass 1967).

The problem of finding the sources of copper for these ingots using metallurgical or ordinary chemical analyses has proved intractable. As late as 1982 Muhly (1982b: 255) summarized over a decade of work of the Philadelphia group by writing: "Our analytical work on these ingots, involving most of those found in the Eastern Mediterranean as well as of a few examples from Sardinia, has produced no evidence for provenance, no grounds for dividing the known ingots into distinct groups." This problem has at last been solved using lead isotope provenancing. A necessary first step was the acquisition of lead isotope data for the relevant copper deposits around the Mediterranean, particularly for Cyprus and Sardinia, since many copper oxhide ingots have been found on each, and both of these islands possess copper ore deposits. A vital prerequisite to the solution of this problem is of course an answer to the question of whether the lead isotope compositions of the relevant copper ore deposits can be resolved from each other.

Earlier lead isotope analyses of a relatively small number of Cypriot ores suggested that a single overall lead isotope field could be defined for Cypriot ores (e.g., Gale 1991), though the later accumulation of more data already hinted that this field had structure in the sense that the isotopic data for some Cypriot mines could be resolved (Gale and Stos-Gale 1992b). With a further set of detailed lead isotope analyses it was proved beyond doubt that there is large variability of lead isotope composition between different Cypriot ore deposits (Stos-Gale et al. 1997; Gale, 1999) and that some can easily be differentiated on this basis (see Fig. 17.18). For instance, the Sha mine can easily be discriminated from Kalavasos ores and Apliki ores and also from the lead isotope composition of the copper ores from Lavrion in Attica, Greece. This is of considerable archaeological value. For example, there is no analyzed LBA copper-based artifact from Cyprus (see Stos-Gale and Gale 1994), Crete, or Greece that has a lead isotope composition matching the Sha or Kalavasos ore deposits, which tells us something about which Cypriot copper ore sources were *not* exploited in the LBA.

However, can the lead isotope compositions of Cypriot ores be distinguished from the ores in Sardinia? If not, we might only be able to conclude that these oxhide ingots found in Cyprus and Sardinia were made *either* from Cypriot copper ores *or* from Sardinian ores. It was shown conclusively by Gale (1999), *contra* Budd et al. (1995a, 1995c), that both the geologically older Sardinian ores and the much younger ores from North West Sardinia (including the Calabona deposit) have lead isotope compositions that do not overlap at all with any Cypriot copper ores. Lead isotope data for some of these ores were published by Stos-Gale et al. (1995a). New data for the relatively important Calabona deposit were tabulated by Stos-Gale et al. (1998b), and Figure 17.18 shows that these data do not overlap with data for Cypriot ores. Lead isotope data for Sardinian ores (Gale and Stos-Gale 1987; Stos-Gale and Gale 1992; Stos-Gale et al. 1995a) are plotted in Figure 17.19, which shows the large spread of lead isotope data for the Sardinian metal ore deposits, which is to be expected since they range in geological age from the Tertiary to the Cambrian. Comparison in Figure

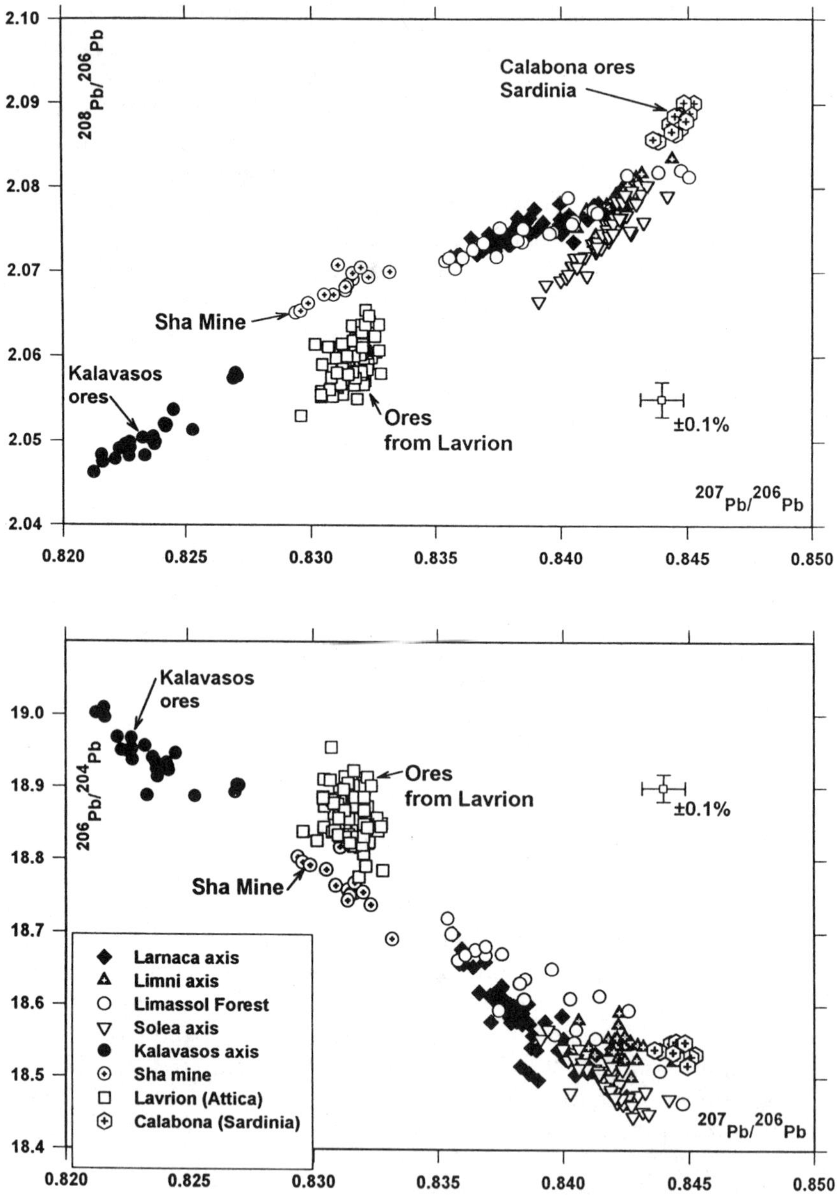

Figure 17.18. Lead isotope data for Cypriot ores and for Lavrion ore deposit (Attica, Greece) and Calabona copper deposit in northwest Sardinia.

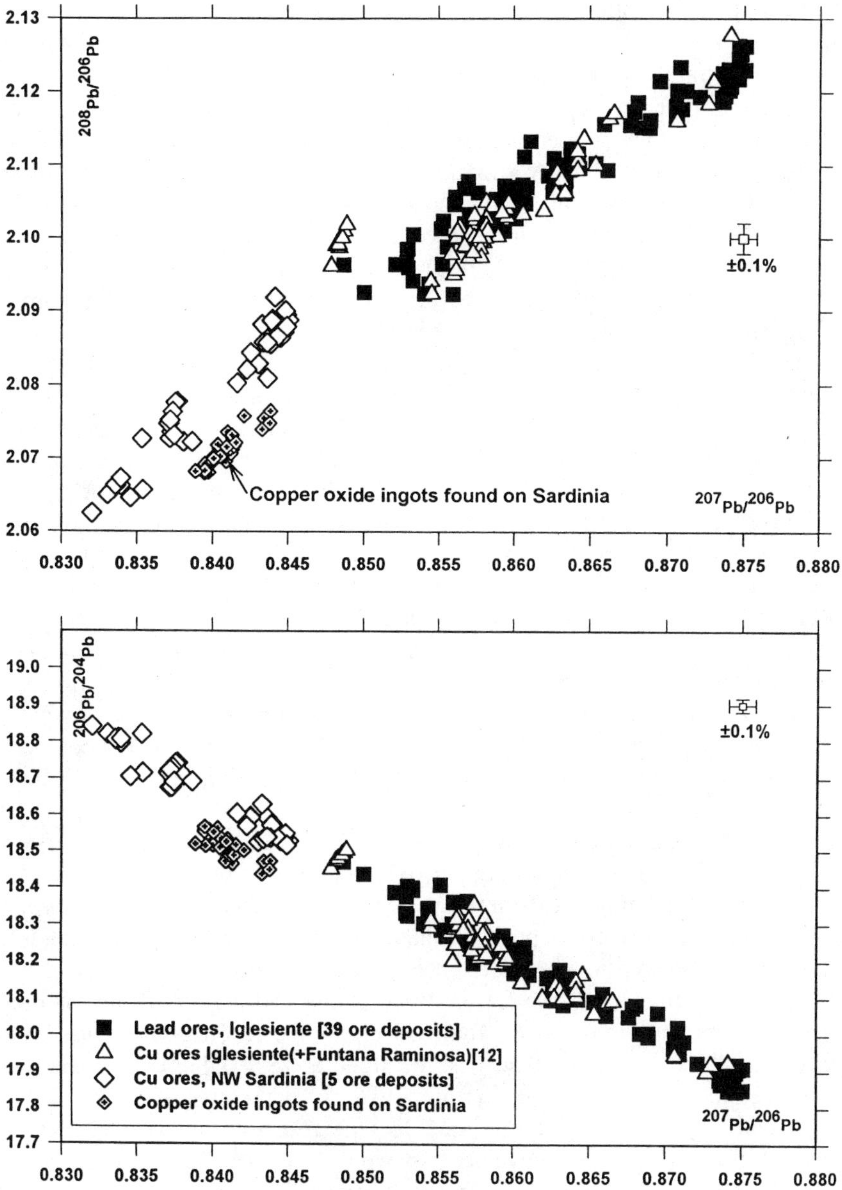

Figure 17.19. Lead isotope data for Sardinian ores and for copper oxhide ingots found on Sardinia.

17.19 of the lead isotope data for Sardinian ores with the isotopic analyses of Sardinian oxhide ingots shows that these ingots are definitely not made from copper coming from Sardinian copper deposits.

Various other overlaps of the lead isotope compositions of Cypriot ore deposits with other copper deposits have been suggested (see Budd et al. 1995c). However, a full investigation of these suggestions proves that the lead isotope compositions of the ores from the Apliki copper deposit in Cyprus have no overlaps with the known isotopic compositions of any other copper deposits in the broad Mediterranean region (Gale, 1999).

We now examine the leading features of the lead isotope data for copper oxhide ingots. A considerable number of lead isotope analyses have been made of such ingots (Gale 1991; Stos-Gale et al. 1997). A notable feature of the lead isotope compositions of those copper oxhide ingots that date later than about 1250 B.C. is that their lead isotope compositions group rather tightly together. At first sight it is surprising that this is so whether the ingots were found in Sardinia, Greece, Crete, Cyprus, the Nile Delta, Turkey, or Bulgaria. It was suggested by Budd et al. (1995c) that this tight grouping cannot be matched by the lead isotope field of any particular copper ore source and that there is even no match between the lead isotope compositions of oxhide ingots found on Cyprus and any Cypriot copper ore source. Figure 17.20 shows this to be incorrect. Copper oxhide ingots excavated from the six different sites on Cyprus of Enkomi, Mathiati, Skouriotissa, Ayios Dhimitrios, Maroni, and Pyla Kokkinokremos were shown to have lead isotope compositions that fully match those of Cypriot copper ores from one region on Cyprus; that of the Apliki mine region in the North East Troodos mountains.

Gale (1999: Fig. 10) has demonstrated the same match with Apliki deposit copper ores for miscellaneous copper oxhide ingots from all over the Mediterranean and beyond. Far from there being no match with the lead isotope compositions of any ore deposit, we have demonstrated that there is such a match with the Cypriot Apliki copper ore deposit for all of the post–1250 B.C. copper oxhide ingots so far analyzed, some 180 ingots in total. Moreover, this match extends to all analyzed ingots found on both Cyprus and Sardinia; all these post–1250 B.C. oxhide ingots were made of copper smelted from the Apliki ore deposit in Cyprus.

In contrast, the lead isotope data for the much earlier copper oxhide ingots from Minoan Crete have been shown not to match any copper ores in Cyprus (Gale and Stos-Gale 1986b). These earliest copper oxhide ingots have lead isotope compositions that do not match any yet analyzed copper ores. The lead isotope compositions of these ingots, dating to 1550 to 1500 B.C., suggest they were made from copper smelted from Precambrian copper deposits, and they do not match the isotopic composition of ores of Cam-

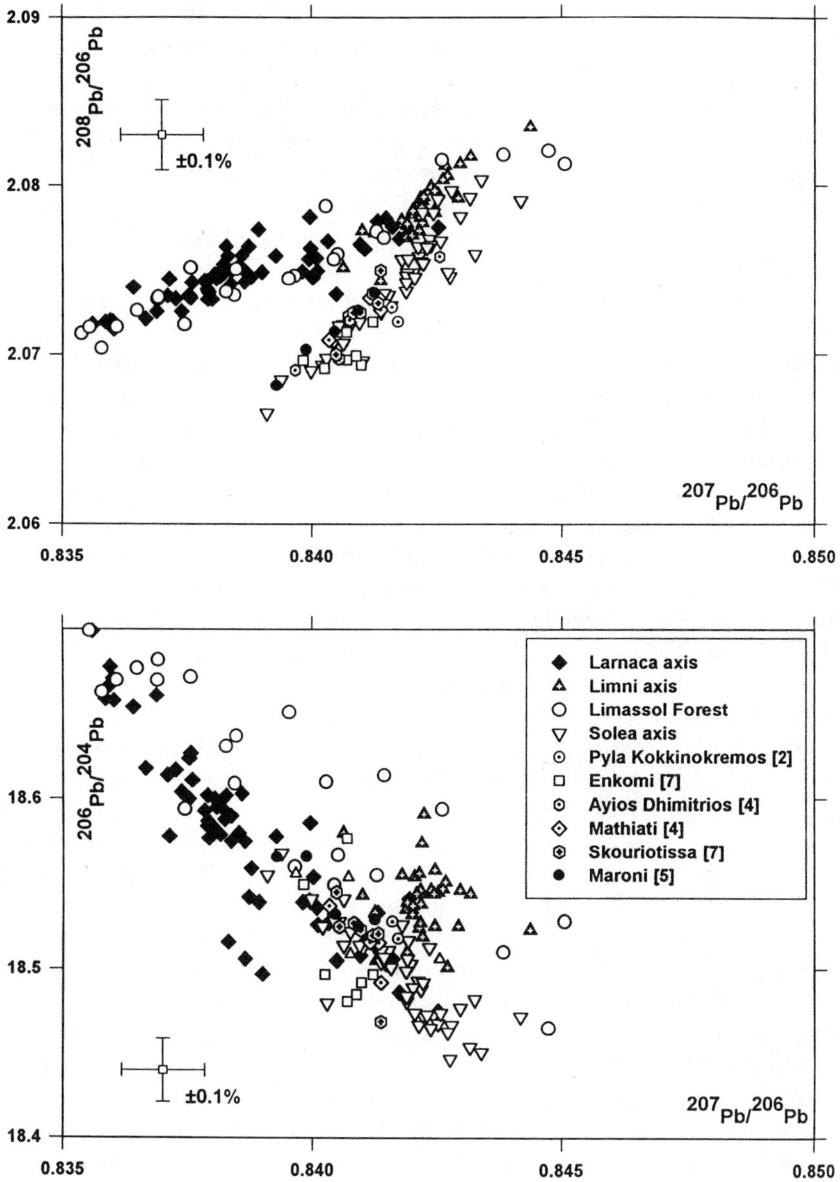

Figure 17.20. Lead isotope data for copper oxhide ingots from six sites on Cyprus, showing overlap with Cypriot ores from Apliki ore deposit. Six sites with oxide ingots are Pyla Kokkinokremos, Enkomi, Ayios Dhimitrios, Mathiati, Skouriotissa, and Maroni.

brian to Silurian age in Sardinia. Copper ores of Precambrian age exist in Afghanistan and other parts of central Asia; we are about to include such ores in our studies.

One of the points of criticism of the use of lead isotopes for provenancing copper oxhide ingots was that these ingots probably incorporated recycled metals, extending even to recycled bronze artifacts (Budd et al. 1995c). In support it was claimed that some copper oxhide ingots contain significant tin contents. This argument was demonstrated to have no basis in fact by Stos-Gale et al. (1998b: Fig. 1).

The postulated need for a mixing model to explain the lead isotope characteristics of copper oxhide ingots (Budd et al. 1995c) was strongly based, as we have seen, on the supposed lack of a match in isotopic composition between that for the oxhide ingots with any feasible copper ore source. Figure 17.21 shows that in fact there is also a match for the lead isotopic compositions of the Cape Gelidonya oxhide ingots with the Apliki region copper ores in Cyprus; the data for the analyzed oxhide ingots are given by Stos-Gale et al. (1998b: Table 2). We have already seen the same match with Apliki region copper ores for the oxhide ingots found on Cyprus and Sardinia. Consequently, we now know the source of copper for the oxhide ingots found on the Cape Gelidonya shipwreck, dating to about 1200 B.C. It has also become clear that the bases of the mixing hypothesis proposed by Budd et al. (1995c) for the copper oxhide ingots have no support from the lead isotope and trace element data.

An important digression involves the plano-convex, or bun, copper ingots also found on the Cape Gelidonya shipwreck. There were well over 30 such ingots from the wreck. There were also 19 flat and thin bars described as "slab ingots" by Bass (1967). Lead isotopic data are published by Stos-Gale et al. (1998b) for 34 of the bun ingots and 6 slab ingots. These data, plotted in Figure 17.22, fall into distinctive groups.

All but one of the bun ingots (which are of very different sizes) fall squarely with the lead isotope compositions of the ores from the Apliki ore deposit of Cyprus. The conclusion can only be that the copper extracted in these mines was cast not only into oxhide-shaped ingots, but also into bun ingots. Indeed, it may well be that the primary product was the bun ingots, which were subsequently fire refined and mostly cast into oxhide ingots. It had been suggested that the oxhide ingots were cast by remelting together many bun ingots that had been made from copper from different sources, having different lead isotope compositions. The lead isotope composition of the resulting copper oxhide ingot would then be a mixture and would have lost the characteristics of the copper sources from which it was made. The evidence from the Cape Gelidonya shipwreck negates that model. Since we now know that by far the majority of the bun ingots on this wreck have the

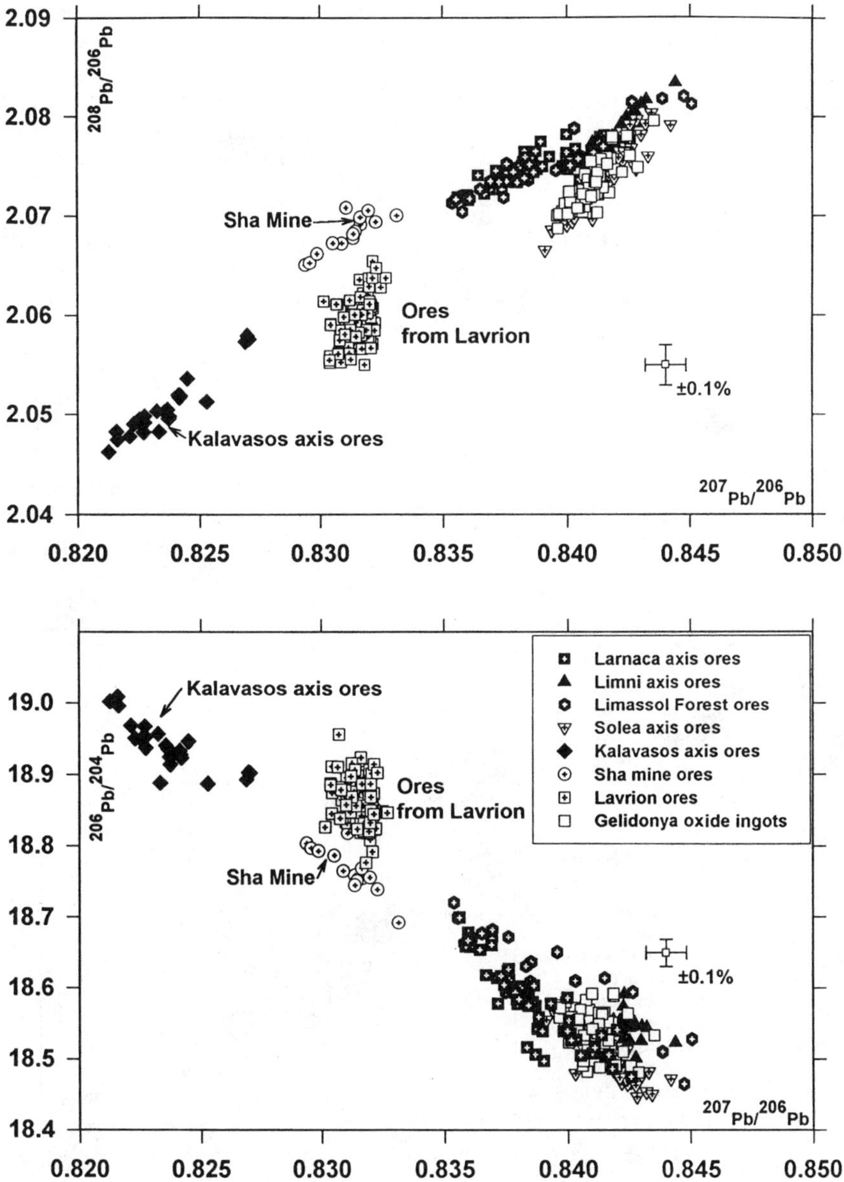

Figure 17.21. Lead isotope data for copper oxhide ingots from Bronze Age shipwreck of Cape Gelidonya (Bass 1967) showing overlap with Cypriot ores from the Apliki ore deposit.

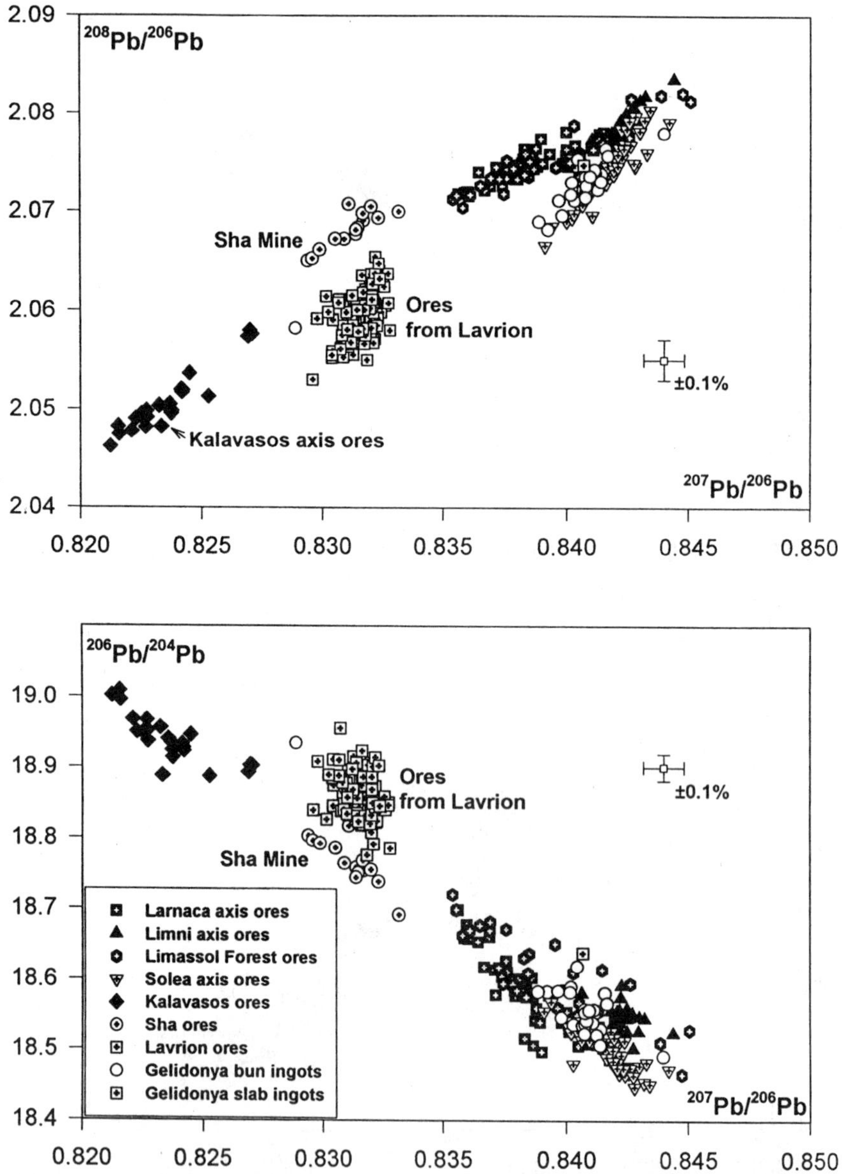

Figure 17.22. Lead isotope data for bun and slab ingots from Bronze Age shipwreck of Cape Gelidonya (Bass 1967). Of 34 analyzed bun ingots, 33 fully overlap with lead isotopic compositions of Cypriot ore deposit of Apliki. Five slab ingots, out of six analyzed, overlap with lead isotope field of Lavrion ore deposit.

isotopic composition linking them to production from the Apliki region copper ores, the lead isotope composition of the oxhide ingot (if made from bun ingots) would also have the isotopic composition characteristic of the Apliki ore deposit. However, the presence of bun ingots on this shipwreck shows that not all of such ingots were cast into oxhide ingots; at least some were traded as bun ingots.

One bun ingot is different and falls with an isotopic group, which we call the "El Amarna" group, because on this Egyptian site we found the largest group of copper-based artifacts having this isotopic composition (Stos-Gale et al. 1995b). This isotopic group is not uncommonly represented among Aegean and Cypriot copper artifacts, but we do not yet know an actual copper deposit characterized by these ratios, nor have we yet found a copper oxhide ingot with the El Amarna lead isotope composition.

It should be mentioned here that all the bun and oxhide ingots are of >99.5% purity copper, containing virtually no tin. In contrast, the six flat rectangular ingots we have analyzed from the Cape Gelidonya shipwreck contain small amounts of tin (2.5% for the one consistent with Cypriot ores; for the other five the tin contents are 0.41, 0.42, 1.4, 3.9, and 7.2%). Obviously these slab or bar ingots, which were all found together in the "Captain's cabin" area of the Gelidonya shipwreck, are of a completely different type: not raw copper, but a rather low-tin bronze alloy. This sort of alloy could in principle, but not necessarily, result from melting together scrap high-tin bronze with pure copper. Note that five of these low-tin slab ingots have lead isotope compositions that fit with Lavrion copper ores, and but one with Cypriot ores.

The idea that different copper ore sources may have been used at different times to supply the copper for oxhide ingots is supported by other data. The earliest copper oxhide ingots, excavated on Crete at Hagia Triadha and dating probably to Late Minoan IA (LMIA) (Georgiou 1979) or about 1550 to 1450 B.C., have a completely different lead isotope composition from all other oxhide ingots so far analyzed (Gale and Stos-Gale 1986b).

For those who may not be acquainted with the location of mines in Cyprus, Stos-Gale et al. (1997: Fig. 2, p. 90) show the locality on Cyprus of the Apliki, Mavrovouni, and Skouriotissa ore deposits, close together to each other, the largest and richest copper deposits in Cyprus, and some of the ore deposits closest to the sea in Cyprus. That all the copper for post–1250 B.C. oxhide ingots came from the Apliki deposit itself may have important implications for our understanding of the organization of the prehistoric copper industry in Cyprus. However, we do not necessarily need to look for some large urban center in the Apliki region that was responsible for organizing and carrying out the manufacture of copper oxhide ingots near the mines. It remains possible, for example, that the operation involved mining the ores and smelting them in the same general Apliki region to produce copper in the form of bun ingots, which was then transported to coastal urban centers such as Enkomi, Kition, Alassa, or Ayios Dhimitrios, for subsequent fire refining and casting into oxhide ingot form. Dikaios (1969) had long ago suggested that *ores* smelted at Enkomi may have been transported there from the Skouriotissa region. On the other hand, the existence of large LBA furnace crucibles in the excavations at Enkomi and Kition, crucibles that contain 30 kg of molten copper in a small fraction of their total volume (Zwicker 1986), suggests the hypothesis that on at least some occasions the copper metal for an oxhide ingot may have been obtained in one smelt.

The application of lead isotope analyses to discovering the sources of copper for the oxhide ingots found all over the Bronze Age Mediterranean has provided evidence about the evolution with time of the copper ore sources used to make these ingots. This is a very positive story. The success of this work has also a negative side. In the course of this work, which depends on approximately a thousand lead isotope analyses of ores and ingots, all of the assumptions underlying the attempts by the Bradford Ancient Metallurgy Research Group to question lead isotope metal provenancing have been vitiated. The subject is now on an even firmer base than before. Finally, we stress that our conclusion that the copper for all post–1250 B.C. copper oxhide ingots was derived from the Cypriot ore deposit of Apliki does not extend to such Bronze Age Cypriot artifacts as the hook tanged weapons. Though some of these Cypriot artifacts were derived from the Apliki region copper ores, most were not, as can readily be seen by a comparison of the lead isotope compositions of these ores with published isotopic data for Cypriot objects (Stos-Gale and Gale 1994). This suggests that copper ores from other Cypriot deposits around the Troodos were being smelted in the Bronze Age and supplied copper for many artifacts, though not for oxhide ingots.

17.7.4. Export of Cypriot Copper to the Western Mediterranean

Another matter vital to Greek prehistory is the question of whether Cyprus was, as is conventionally assumed, the major source of copper for the Mycenaeans and the Minoans. This hypothesis is essentially based on two assumptions and two facts. The first assumption is that the copper-exporting state of Alasia, recorded in Near Eastern tablets, is Cyprus (Hellbing 1979), and the second is that it would have copper available to supply both the eastern Mediterranean and the Aegean. One fact is that copper was certainly being produced on Cyprus from Cypriot ores in the Late Bronze Age, as lead isotope studies have definitely proved. The second fact is that large quantities of Minoan and Mycenaean pottery appear in Cyprus in the fourteenth and thirteenth centuries B.C. (Catling 1980). It has usually been assumed that this pottery and its contents represented the Cypriot end of a trade route that had its motivation in the need for Minoan and Mycenaean procurement of copper in Cyprus.

In fact, lead isotope analyses of Late Minoan II bronzes from the Unexplored Mansion, Knossos, show that the major source of copper for this site was not Cyprus but the ore deposits of Lavrion in Attica, Greece (Stos-Gale and Gale 1984; Stos-Gale and Macdonald 1991). For Late Bronze Age Ayia Irini on Keos the overwhelming copper source again seems to be Lavrion, though one oxhide ingot fragment does indeed come from Cyprus. Finally, for Late Minoan III levels in West Crete at Armenoi, Samonas, and Nerou Kourou, the copper source is again not Cyprus, but most probably Lavrion. We have definite proof that the scanty copper ore sources on Crete itself were not exploited by the Minoans (Stos-Gale, 1993a). So far, then, the evidence is overwhelmingly against Cyprus having been an important source of copper for the Minoans, though more work is needed. Whether Cyprus was an important source of copper for the Mycenaean world on the mainland is a subject that needs further investigation.

17.7.5. Earliest Metallurgy of Copper in the Balkans

The scale and technical development of metallurgy in the Bulgarian Eneolithic completely overshadow what was happening in contemporary Anatolia, Palestine, or the Aegean, while in copper-ore-rich Cyprus there are no more than a handful of tiny metal objects at this time. The earliest Bulgarian copper production is certainly of particular interest, because the copper deposits in Bulgaria are many and of considerable size. The earliest development of mining and metallurgy in Bulgaria has been studied exhaustively by Bulgarian and Russian archaeologists (Chernykh, Raduncheva, Todorova, and others), but the attempts at associating ores from specific mining regions and metal artifacts on the bases of spectral elemental analyses were not entirely successful. The lead isotope provenance study of the Chalcolithic metallurgy in Bulgaria carried out in the last few years by the Oxford team in collaboration with archaeologists and geologists from the Bulgarian Academy of Sciences was based on extensive geological fieldwork, allowing collection of copper minerals from all major occurrences. The lead isotope characterization of Bulgarian ore deposits was also aided by the geologically aimed analyses of Amov et al. (1993). The data for the Bulgarian ore deposits show that they can be readily distinguished from most Greek and Serbian ores (Pernicka et al. 1993; Stos-Gale et al. 1996, 1998a)

Two large analytical projects concerned with the earliest copper artifacts from Serbia and Bulgaria have been completed in recent years (Pernicka et al. 1993, 1997; Gale et al. 1991, in press). Altogether several hundreds of artifacts from nearly 30 sites in Serbia and Bulgaria, including nearly a hundred copper objects from the Varna cemetery (Gale et al. in press), were analyzed for their lead isotope and elemental compositions. Comparisons of their lead isotope compositions with those of the copper ores from southeast Europe show quite clearly that this metallurgy had a relatively local character and relied predominantly on the ore deposits found in the Balkans. Of particular interest is the conclusion that the mines of Rudna Glava and Aibunar, hailed in the archaeological literature as "the Eneolithic copper mines," played a rather small part in the production of copper metal. The main sources of copper for the artifacts produced across the Balkans in the fifth to the fourth millenium B.C. appear to have been the mines of Majdanpek and Rudnik in Serbia and the mines situated in southeast, central, and northwest Bulgaria. No single Balkan Eneolithic copper-based artifact proves to have been made from copper from Rudna Glava, though a small number seem to have been made of copper smelted from the Aibunar ores. This is a striking demonstration of the power of the lead isotope method and of the weakness of the conventional approaches of either archaeology or archaeometallurgy. It is interesting to notice that thorough lead isotope and elemental comparisons between the groups of artifacts led Pernicka and his colleagues to the conclusion that remelting and mixing of metals from different sources were not common in the Chalcolithic Balkans, nor was the use of native copper important.

The lead isotope (LI) patterns (meaning the LI ratios shown by the whole group) presented by the artifacts from Bulgaria are quite different from those of the EBA artifacts from Greece and Cyprus and, as far as we know at present, Anatolia. Only some artifacts from the Eneolithic Bulgarian site of Dolnoslav, which has been identified as a place of cult, might have originated from ore deposits in the southern Rhodope mountains, from where copper was also transported south to Chalcolithic sites located today in Greece.

17.7.6. Lead Isotope Analyses of Artifacts from Other European Countries

So far, apart from studies in the Far East to be mentioned later, the lead isotope provenance studies of metals have chiefly been concerned with the earliest metallurgy of Southern Europe and Anatolia. For Anatolia there was a large project involving extensive fieldwork carried out in the 1980s by a team from Heidelberg, reported in a number of fascicules of the *Jahrbuch des Romisch-Germanischen Zentralmuseums*. It is noteworthy that, in relation to the large amount of ore deposits sampled and the excellence and importance of the geological and archaeological survey work undertaken, relatively few lead isotope analyses of ores have yet been published, with the result that the characterization of the lead isotope fields of the important Anatolian metal ore deposits remains sparse. Another extensive research program in Turkey that involved lead isotope provenancing was led by Yener (Yener et al. 1989a, 1989b, 1991; Yener and Özbal 1987; Yener and Vandiver 1993); it has provided much new evidence of the archaeology and archaeometallurgy of the Bolkardag region, including fascinating evidence for a small tin mine exploited in the EBA, but has again so far apparently added relatively little in the way of published lead isotope data for well-characterized Anatolian ores. Outside Anatolia and Southern Europe, in the recent decade an interesting study of lead isotope provenancing aspects of the early copper and lead metallurgy of the British Isles was initiated (Rohl 1996). This very preliminary study of lead isotope characteristics of ores and artifacts from Britain shows again that the patterns of exploitation and distribution of metals are more complex than expected from purely archaeological research.

Lead isotope provenance studies of post–Bronze Age metals in Europe and the Near East seem to be rare. A review of the contribution of lead isotope analyses to numismatics was published elsewhere (Stos-Gale 1998b), but a few years ago two publications discussed the provenance of copper-based alloys of the Roman period from France and Poland (Beck et al. 1985; Stos-Gale 1993c). In both cases it was concluded that the majority of metals used for the production of the copper alloys originated from the local ore deposits, with only rare imported pieces.

17.7.7. Lead Isotope Provenance Studies in the Far East

A large number of lead isotope analyses of Far Eastern bronzes have been carried out during the past decade in Japan (some of this work was in collaboration with American colleagues). Most of the publications are in Japanese and Chinese, but some of them also have English summaries. The application of lead isotope provenance studies in Japan was, by chance, initiated simultaneously with the beginnings of the Heidelberg/Oxford research on Greek silver coins (Yamasaki et al. 1978, 1979). The lead isotope analyses were carried out on a series of Western and Eastern bronze Han mirrors. The results show that it is possible to distinguish between these groups and also that the Japanese lead sources are very different from the Chinese deposits. The largest project of lead isotope characterization of Far Eastern bronzes was initiated by Mabuchi in 1977. The results of this research have been published in a series of articles in Japanese (Mabuchi and Hirao 1980, 1982a, 1982b, 1983, 1984a, 1984b; Mabuchi et al. 1982, 1983; Nishida 1980,

1982, 1983). The earliest Far Eastern bronze implements studied by lead isotopes are dated to about 300 B.C. The investigated groups of objects included Japanese imitative mirrors, ancient Chinese coins, ancient Korean bronzes, bell-shaped objects found in Japan (*dotaku*), Japanese coins, and copper artifacts of the early historic age (seventh century). A review of the results of this research was published by Mabuchi et al. (1985). This review publishes also over 200 lead isotope data measured for these projects, including a few data for galenas from Japan and China.

The conclusions based on the results of the lead isotope provenance studies of the earliest Japanese bronzes show that bronzes were introduced to Japan from South Korea at first and then from North China. Copper or bronze artifacts containing indigenous Japanese lead appear only in the seventh century. The study of the chemical and isotope composition of chronological groups of *dotaku* revealed very interesting technical features concerning the reuse of metals, so often mentioned in archaeological literature as fact, often without any material evidence. The authors concluded (Mabuchi and Hirao 1982b) that neither random remelting of broken objects nor gradual addition of broken objects occurred during the period of production of these objects. The analytical data indicated that groups of bronze ingots or other objects of specific and perhaps known composition were imported in three different periods for making of *dotaku*.

It should be noted that all so-far published data from the Japanese laboratories are mostly related to medium-to-high lead alloys; therefore the artifacts are compared with sources of galena rather than copper ores. The limited amount of published lead isotope data for these ores indicate that the Japanese galenas are very different in their characteristic ratios from the Chinese ones. The Japanese data groups with $^{208}Pb/^{206}Pb$ ratios of 2.08 to 2.11 ($^{207}Pb/^{206}Pb$: 0.84 to 0.856), while the Chinese minerals range from $^{208}Pb/^{206}Pb$ of 2.96 to 2.22 ($^{207}Pb/^{206}Pb$: 0.848 to 0.893). This broad differentiation allows quite positive conclusions to be drawn from the lead isotope studies of metals and pigments in glasses used in the ancient Far East, though a larger body of data for ores is of course desirable.

17.7.8. Sources of Tin

As a final application we turn to a very brief review of our preliminary work on the subject of the sources of tin for the Bronze Age Mediterranean region. The source(s) of tin ore used to supply the ancient world with tin remains a major unresolved problem in Bronze Age archaeology. The earliest tin bronzes occur in Anatolia, the Troad, Mesopotamia, and the Cyclades, but no tin ores occur in these areas. Nor is there any scientific evidence to link known sources of tin ore with ancient tin metal. If a scientific method could be found to make such a link, then it would be possible to reconstruct a prehistoric trade route in tin ore, or tin metal, and to elucidate interactions between ancient peoples. In his review of work on ancient metallurgy and mining Muhly (1985) wrote, "in certain areas we have yet to see a major breakthrough or significant change in traditional confusion. Foremost in the latter category must be the problem of ancient sources of tin. It is remarkable that, after twenty years of intensive scholarly investigation and fieldwork, we still have no hard evidence regarding the sources of tin being exploited by the numerous and widespread bronze industries of antiquity."

It has been demonstrated that tin bronze can be made not only by additions of tin metal to molten copper but also by directly adding cassiterite under a charcoal cover to molten copper in a crucible, so it is conceivable that either the ore or tin metal may have been traded. However, tin ore has so far not been identified in archaeological sites, while more than 1 tonne of tin metal ingots has been excavated from the Bronze Age shipwreck of Uluburun, off the coast of Turkey (Bass 1987), and tin metal ingots have come from underwater off the coast of Israel, for instance at Hishuley Carmel and Kfar Samir (Galili et al. 1986).

There are various possible ways to approach provenancing tin, among which are lead isotope analyses, tin isotope analyses, trace element analyses, and characterization of intermetallic inclusions by scanning electron microscopy and proton microprobe analysis, all of which we are exploring. Here we restrict our discussion to presenting some preliminary data for the lead isotope approach.

Figure 17.23 shows lead isotope data for 57 tin ingots from the Uluburun Bronze Age shipwreck. It shows a grouping of isotopic compositions into at least two groups. Intriguingly these groups are also characterized by the ingots in each group having quite different concentrations of lead, as indicated on the figure.

Also shown in Figure 17.23 is some of the lead isotope data for tin ingots excavated from the seventeenth-century shipwreck of Aanloop Molengaat, which foundered in the North Sea off the Dutch coast. The importance of these ingots is that the excavators have traced impressed heraldic seals on these ingots to particular tin mines in the Saxon Erzgebirge (Vos 1996). The lack of overlap of the lead isotope compositions of these Dutch ingots with those of the Uluburun ingots would seem to exclude these particular Erzgebirge tin deposits as the source of tin for the ingots aboard the Uluburun ship.

Figure 17.23 shows some lead isotope data for ingots of probable Bronze Age date, excavated from underwater off the coast of Israel. These ingots, from the underwater site Hishuley Carmel, form a discrete isotopic group, which essentially merges or overlaps with one of the isotopic groups for the

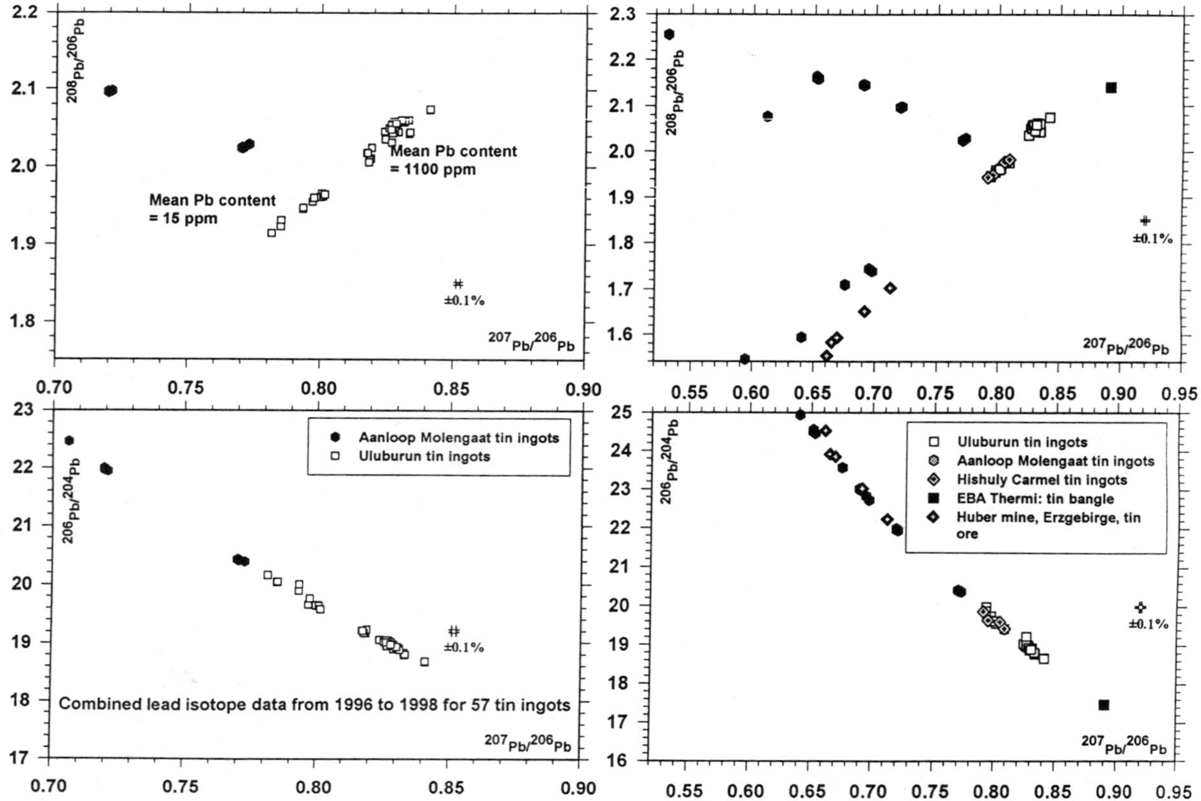

Figure 17.23. Lead isotopic compositions for 57 tin ingots from Bronze Age shipwreck of Uluburun compared with lead isotopic compositions of tin ingots from the seventeenth-century Aanloop Molengaat shipwreck, of tin ingots excavated off coast of Israel and of tin ores from Erzgebirge.

tin ingots from the Uluburun shipwreck. It is noteworthy that these particular Israeli tin ingots have lead concentrations that are very similar to those of the isotopic group of Uluburun tin ingots, which they overlap in isotopic composition. This may indicate that the Bronze Age ship wrecked off the Israeli coast at Hishuley Carmel was carrying metallic tin ingots that derived ultimately from the same tin ore deposits that were the source of one group of the metallic tin ingots carried by the Uluburun ship, which foundered off the Turkish coast.

Figure 17.23 also shows more of the lead isotope data for the tin ingots from the Aanloop Molengaat shipwreck, together with lead isotope data for some cassiterite ores from the Huber mine in the Erzgebirge. This is not an ideal comparison, since we had no cassiterite from the exact tin mines in the Erzgebirge to which Vos (1996) believes the seals on some of these ingots point. Of the Aanloop Molengaat ingots, nine ingot samples fall to the high $^{208}Pb/^{206}Pb$ side of the data array defined by the decay constants and abundance of ^{238}U, ^{235}U, and ^{232}Th along which the other six samples (0.595 to 0.706) lie. It cannot at present be excluded that the individual tin ingots from the Aanloop Molengaat wreck, not all of which bear identifiable stamps, come from a greater range of tin ore sources than has been deduced from those that bear traceable heraldic stamps.

The lead isotope ratios for cassiterite from the Huber Mine (Erzgebirge) are shown in Figure 17.23, in comparison with the data for the Aanloop Molengaat tin ingots. The data are colinear but lie on the low $^{208}Pb/^{206}Pb$ side of six of the Aanloop Molengaat ingots. There is no real basis for claiming a match between these ores and the corresponding tin ingots, bearing in mind the analytical errors. Simple comparison between the two sets of lead isotope data cannot confirm a match between source ore and artifact, any relationship being considered tenuous, if present at all. However, it can be deduced unambiguously that the Huber mine was not the source of tin for the tin ingots from the Uluburun ship, though it would of course be premature to conclude that the Erzgebirge deposits as a whole are excluded. The lead concentration in all cassiterites so far analyzed in Oxford is less than 10 ppm. This project, for which we have also analyzed a number of Cornish tin ingots and cassiterite ores from various other localities, is clearly at an embryonic stage, and so we end our discussion at this point with the remark that lead isotope data for cassiterite ores and tin metal artifacts seem promising.

17.8. CONCLUSIONS

Table 17.4 summarizes some of the more important contributions made by lead isotope analyses to our understanding of metal sources and metal trade

Table 17.4. Some Inferences from Lead Isotope Analyses Bearing on Bronze Age Mediterranean Metal Sources, Trade, and Archaeometallurgy

Lavrion in Attica and Siphnos in the Cyclades were major sources of Early Cycladic lead and silver	Gale and Stos-Gale 1981b; Pernicka et al 1983
ECII/ECIII Kastri on Syros (Cyclades) was a settlement of immigrants who brought metals from outside the Aegean; metals of the same origin were found in Troy	Stos-Gale et al. 1984; Gale et al. 1985a
Copper from the Cyclades (Skouries smelting site on Kythnos) was used predominantly in the EBA Aegean, including Crete	Stos-Gale 1989, 1992, 1993a, 1998a
Copper in the LBA Aegean is predominantly from Lavrion; Cypriot copper was not very commonly used for tools, weapons, etc.	Stos-Gale 1993a; Stos-Gale et al., in press c
Lead and silver in the LBA Aegean is chiefly from Lavrion	Gale et al. 1984
All post–1250 B.C. copper oxhide ingots originate from a single mining region in Cyprus (Apliki)	Stos-Gale et al. 1997
Sixteenth-century B.C. oxhide ingots from Crete (Hagia Triadha) are of non-Aegean origin	Gale and Stos-Gale 1986b
Nuragic copper-based artifacts are made of local Sardinia ores, but all oxhide ingots found in Sardinia were made of Cypriot copper (Apliki mine)	Stos-Gale and Gale 1992; Stos-Gale et al. 1997
Copper and silver from the Aegean was present in fourteenth dynasty Egypt	Stos-Gale et al. 1995b
Predynastic galenas are from the Eastern Desert, Egypt, but silver was imported, possibly from Iran	Stos-Gale and Gale 1980; Stos-Gale, in press; Hassan and Hassan 1981
Ergani Maden mine was not a main copper source in Bronze Age Anatolia	Pernicka et al. 1984; Seeliger et al. 1985; Gale et al. 1985a
Beginnings of copper metallurgy in the Balkans were not based on metals from the mine of Rudna Glava and only to a relatively small extent from the mine of Aibunar.	Gale et al. 1991; Pernicka et al. 1993; Gale et al. in press
In LBA Timna (Israel) both copper and iron objects were made from local ores	Gale et al. 1990

in the Bronze Age Mediterranean. Lead isotope analyses have solved the problem of the source of copper for post–1250 B.C. copper oxhide ingots and have the potential to provide real evidence about the organization of the copper industry within Bronze Age Cyprus. Already the pattern of changing metal sources with time and of the trade in copper, lead, and silver begins to emerge in the Bronze Age Mediterranean. Fifteen years ago nothing of this was known. Lead isotope provenancing is now on a sound footing, with applications to the provenancing of lead, silver, copper, iron, and perhaps tin and gold and with applications to glazes, glasses, and artists' pigments. We hope that this brief review may have given some feeling for the tremendous potential of lead isotope analyses for provenance studies and for archaeology.

Acknowledgments

N. H. Gale acknowledges having been introduced to studies of Mediterranean archaeometallurgy, numismatics, and archaeology by the late Wolfgang Gentner and Lord Colin Renfrew. The research of the Isotrace Laboratory in Oxford would not have been possible without the financial support of the Science Research Council of Great Britain, the Leverhulme Trust, the British Academy, and the Institute of Aegean Prehistory.

REFERENCES*

Akishin, P. A., O. T. Nkitin, and G. M. Panchenkov (1957), *Geokhimya* **5**, 429.

Amov, B. G. (1968), *Isotopenpraxis* **9**, 358.

Amov, B. G. (1983), *Earth Planet. Sci. Lett.* **65**, 61.

Amov, B. G., B. Kolkovski, and R. Dimitrov (1993), *Annuaire de l'Université de Sofia "St. Kliment Ochridski," Faculté de Geologie et Geographie, Livre 1-Geologie* **85**, 73–98.

Antweiler, J. C., B. R. Doe, and M. H. Delevaux (1972), *Econ. Geol.* **67**, 302.

Arden, J., and N. H. Gale (1974), *Anal. Chem.* **46**, 2.

Aston, F. W. (1942), *Mass Spectra and Isotopes*, Arnold, London.

Barnes, H. L. (1979), *Geochemistry of Hydrothermal Ore Deposits*, Wiley-Interscience, New York.

Barnes, J. W. (1988), Ores and Minerals—Introducing Economic Geology, Open University Press, Milton Keynes, Philadelphia.

Barnes, I. L., T. J. Murphy, J. W. Gramlich, and W. R. Shields (1973), *Anal. Chem.* **45**, 1881.

Barnes, I. L., W. R. Shields, T. J. Murphy, and R. H. Brill (1974), in *Archaeological Chemistry*, Vol. 1, American Chemical Society, Washington, DC, p. 1.

Barnes, I. L., J. W. Gramlich, M. G. Diaz, and R. H. Brill (1978), in *Archaeological Chemistry*, Vol. 2, Advances in Chemistry Series 171, G. F. Carter, Ed., American Chemical Society, Washington, DC, pp. 273–279.

*This is also intended as a bibliography for the subject.

Barnes, I. L., T. J. Murphy, and R. H. Brill (1982), in *Technical Examination of the Classical Bronze Horse from the Metropolitan Museum of Art*, K. C. Lefferts, L. J. Majewski, E. Sayre, and P. Meyers, Eds., *J. Am. Insti. Cons.* **21/1**, 32–39.

Barnes, I. L., R. H. Brill, E. C. Joel, and G. V. Piercy (1986), in *Proceedings of the Twenty-Fourth International Archaeometry Symposium*, J. Olin and M. J. Blackman, Eds., Smithsonian Institution Press, Washington DC, pp. 1–12.

Barnes, I. L., W. T. Chase, L. L. Holmes, E. C. Joel, P. Meyers, and E. V. Sayre (1988), in *The Beginning of the Use of Metals and Alloys*, R. Maddin, Ed., MIT Press, Cambridge MA, pp. 296–306.

Bass, G. F. (1967), Cape Gelidonya: A Bronze Age shipwreck, *Trans. Am. Philos. Soc.* **57/8.**

Bass, G. W. (1987), *Nat. Geograph.* **172**(6), 691.

Baxter, M. J., and N. H. Gale (1998), *J. Appl. Stat*, **25**(5), 671.

Beck, F., M. Menu, T. Berthoud, and L.-P. Hurtel (1985), *Recherches Gallo-Romaines I*, Editions de la Rèunion des Musées Nationaux, Paris, pp. 71–139.

Becquerel, H. (1896), *Comptes Rendus* **122**, 420, 501, 559, 689, 762, 1086.

Begemann, F., S. Schmitt-Strecker, and E. Pernicka (1989), in *Old World Archaeometallurgy*, A. Hauptmann, E. Pernicka, and G. A. Wagner, Eds., Deutschen Bergbau-Museums, Bochum, pp. 269–278.

Begemann, F., K. Kallas, S. Schmitt-Strecker, and E. Pernicka (1999), in *Proceedings of the International Symposium on the Beginnings of Metallurgy, Bochum, 1995. Der Anschnitt, Beiheft 9*, A. Hauptmann, Th. Rehren, E. Pernicka, and U. Yalcin, Eds. 277.

Berthoud, T. (1979), Etude par Analyse de Traces et la Modelisation de la Filiation entre Minerai de Cuivre et Objets Archeologiques du Moyen Orient, Thesis, University of Paris.

Boltwood, B. B. (1907), *Am. J. Sci.* **23**(4), 77.

Branigan, K. (1968), *Copper and Bronze Working in Early Bronze Age Crete*, Studies in Mediterranean Archaeology XIX, Lund.

Branigan, L. (1992), *Nature* **296**, 701.

Brill, R. H. (1968), in *Proceedings of the Eighth International Congress on Glass, London*, Society of Glass Technology, Sheffield, England, pp. 47–68.

Brill, R. H. (1969), in *Proceedings of the Fourth International Congress on Glass, Ravenne-Venise*, Society of Glass Technology, Liege, pp. 255–261.

Brill, R. H. (1970a), *J. Glass Studies* **12**, 185.

Brill, R. H. (1970b), *The Impact of the Natural Sciences on Archaeology*, The British Academy, London, pp. 143–164.

Brill, R. H. (1972), *The International Congress on Glass, Versailles, Artistic and Historical Communication*, l'Institut du Verre, Paris, pp. 307–316.

Brill, R. H. (1976), in *Kenchreai, Eastern Port of Corinth*, L. Ibrahim, R. Scranton, and R. H. Brill, Eds., E. J. Brill, Leiden, Netherlands, pp. 225–255.

Brill, R. H. (1978), *Iraq* **40**(Spring), 23.

Brill, R. H., and W. R. Shields (1972), Special Publication No. 8, Royal Numismatic Society, Oxford University Press, Oxford, 279.

Brill, R. H., and J. M. Wampler (1965a), *Am. J. Archaeol.* **69**, 165.

Brill, R. H., and J. M. Wampler (1965b), *Application of Science in Examination of Works of Art* Museum of Fine Arts, Boston, pp. 155–166.

Brill, R. H., and J. M. Wampler (1967), *Am. J. Archaeol.* **71**, 63.

Brill, R. H., W. R. Shields, and G. M. Wampler (1970), *Application of Science in Examination of Works of Art*, Boston Museum of Fine Arts, Boston, pp. 73–83.

Brill, R. H., I. L. Barnes, and B. Adams (1974), *Recent Advances in Science and Technology of Materials*, Vol. 3 Plenum, New York and London, pp. 9–27.

Brill, R. H., K. Yamasaki, I. L. Barnes, K. J. R. Rosman, and M. Diaz. (1979), *Ars Orientalis* **11**, 87.

Buchholz, H. G. (1959), *Praehistorische Zeitschrift* **37**, 1

Budd, P., A. M. Pollard, B. Scaife, and R. G. Thomas (1995a), *Archaometry* **37**(1), 143.

Budd, P., R. Haggerty, A. M. Pollard, B. Scaife, and R. G. Thomas (1995b), *Israel J. Chem.* **35**, 125.

Budd, P., A. M. Pollard, B. Scaife, and R. G. Thomas (1995c), *J. Med. Arch.* **8**, 1.

Cambresy, A. (1889), *Rev. Univers. Mines Sc.* **3**(1), 76

Cameron, A. Z., D. H. Smith, and R. L. Walker (1969), *Anal. Chem.* **41**, 525.

Catanzaro, E. J. (1968), *Earth Planet. Sci. Lett.* **3**, 343.

Catanzaro, E. J., T. J. Murphy, W. R. Shields, and E. L. Garner (1968), *J. Res. Natl. Bur. Stand., Sect.* **72A**, 261.

Catling (1980), *Cyprus and the West 1600–1050 BC*, Ian Sanders Memorial Lecture, Sheffield University Press, Sheffield, England, p. 35.

Chamberlain, V., and N. H. Gale (1980), in *Proceedings of the Fifteenth International Symposium on Archaeometry, Edinburgh*, E. Slater and J. Tate, Eds. National Museum of Antiquities, Edinburgh, pp. 139–155.

Chase, W. T., I. L. Barnes, and E. C. Joel (1990), *Ancient Chinese Bronzes from the A. M. Sackler Collection*, Vol. 2, Harvard University Press, Cambridge, MA, pp. 168–171.

Compston, W., and V. M. Oversby (1969), *J. Geophys. Res.* **74**, 4338.

Conophagos, C. (1980), *Le Laurium Antique*, Ektodike Hellados, Athens.

Cordella, A. (1878), *La Grece, sous le Rapport Geologique et Mineralogique*, Parent, Paris.

Craddock, P. (1976), *J. Archaeol. Sci.* **3**, 93.

Constantinou, G. (1982), in *Early Metallurgy in Cyprus, 4000–500 B.C.*, J. D. Muhly, R. Maddin, and V. Karageorghis, Eds., Pierides Foundation, Nicosia.

Cumming, G. L., and J. R. Richards (1975), *Earth Planet. Sci. Lett.* **28**, 155.

Curie, P., and M. Curie (1898), *Comptes Rendus* **127**, 175, 1215.

Daly, N. R. (1960), *Rev. Scient. Instrum.* **31**, 264.

Date, A. R., and A. L. Gray (1983a), *Spectrochim. Acta* **38B**, 29.

Date, A. R., and A. L. Gray (1983b), *Int. J. Mass Spectrom. Ion Phys.* **48**, 357.

Davis, O. (1935), *Roman Mines in Europe*, Clarendon Press, Oxford.

Dempster, A. J. (1918), *Phys. Rev.* **11**, 316.

Dikaios, P. (1969), *Enkomi*, Verlag Philipp von Zabern, Mainz.

Dodson, M. H. (1963), *J. Sci. Instrum.* **40**, 289

Dietz, M., and E. P. Horwitz (1993), *Mag. Separat. Sci.* **11**, 46

Doe, B. R., and R. Rohrbough (1977), Open File Report 79-661, U.S. Geological Survey, Denver.

Doe, B. R., and R. E. Zartman (1979), in *Geochemistry of Hydrothermal Deposits*, H. L. Barnes, Ed., Wiley, New York, pp. 22–70.

Doran, J. E., and F. R. Hodson (1975), *Mathematics and Computers in Archaeology*, Edinburgh.

Doumas, C. (1976), *Athens Ann. Archaeol.* **9**, 69.

Doumas, C. (1988), *Problems in Greek Prehistory*, in E. B. French and K. A. Wardle, Eds., Bristol Classical Press, Bristol pp. 21–30.

Eiseman, C. J. (1978), *MASCA J.* **1**, 18

Elliott, R. M. (1963), in *Mass Spectrometry*, A. A. McDowell, Ed., McGraw-Hill, New York, pp. 69–103.

Ellis Jones, J. (1985), *Archaeol. Repts.* **31**, 106.

Evans, R. D. (1955), *The Atomic Nucleus*, McGraw-Hill, New York.

Farquhar, R. M., and I. R. Fletcher (1980), *Science* **207**, 640.

Farquhar, R. M., and I. R. Fletcher (1984), *Am. Antiquity* **49**(4), 774.

Farquhar, R. M., and V. Vitali (1987), *MASCA J.* 1.

Faure, G. (1977), *Principles of Isotope Geology*, John Wiley & Sons, New York.

Faure, G. (1986), *Principles of Isotope Geology*, John Wiley & Sons, New York.

Feynman, R. P. (1992), *The Character of Physical Law*, Penguin Books, London.

Fiedler, L. G. (1841), *Reise durch alle Theile des Königreiches Griechenland*, Leipzig.

Furata, N. (1991), *J. Anal. Sci.* **7**, 823.

Gale, N. H. (1978a), *Archaeo-Physika* **10**, 194.

Gale, N. H. (1978b), in *Thera and the Aegean World*, Vol. 1, C. Doumas, Ed., Aris and Phillips, London, p. 529.

Gale, N. H. (1980), in *Thera and the Aegean World*, Vol. 2, C. Doumas, Ed., Aris and Phillips, London, p. 161.

Gale, N. H. (1989), in *Archaeometry*, Y. Maniatis, Ed., Elsevier, Amsterdam, p. 469.

Gale, N. H. (1991), in *Bronze Age Trade in the Mediterranean*, Vol. 40, N. H. Gale, Ed., Paul Aströms Förlag, Studies in Mediterranean Archaeology, p. 197.

Gale, N. H. (1996), *Anal. Chim. Acta* **332**(1), 15.

Gale, N. H. (1997), *Archaeometry* **39**(1), 71.

Gale, N. H. (1998), The role of Kea in metal production and trade in the Late Bronze

Age. In *Kea-Kythnos: History and Archaeology, Meletimata* 27, L. G. Mendoni and A. Mazarakis-Ainian, Eds., Athens & Paris, Diffusion de Bocard, p. 737.

Gale, N. H (in press a), in *Proc. Int. Symposium on Troy and its Treasures, Pushkin Museum Press, Moscow 1996*, M. Treister, Ed.

Gale, N. H. (1999), in *Metals in Antiquity*, S. M. M. Young, P. Budd, R. Ixer, and M. Pollard, Eds., Archaeopress, Oxford, 110.

Gale, N. H., and A. E. Mussett (1973), *Rev. Geophys. Space Phys.* **II**, p. 37.

Gale, N. H., and Z. Stos-Fertner (1978), *MASCA J.* **1**, 19.

Gale, N. H., and Z. A. Stos-Gale (1981a), *Sci. Am.* **244**(6), 176.

Gale, N. H., and Z. A. Stos-Gale (1981b), *Annu. Br. School Athens* **76**, 169.

Gale, N. H., and Z. A. Stos-Gale (1982), *Science* **216**, 11.

Gale, N. H., and Z. A. Stos-Gale (1984), in *Proceedings of the Cycladic Workshop*, J. A. MacGillivray and R. L. N. Barber, Eds., Institute of Archaeology, London, Edinburgh, p. 255.

Gale, N. H., and Z. A. Stos-Gale (1985), Reports of the Department of Antiquities, Cyprus, p. 83.

Gale, N. H., and Z. A. Stos-Gale (1986a), *Pact J.* **15**, 13.

Gale, N. H., and Z. A. Stos-Gale (1986b), *Annu. Br. School Athens* **81**, 81.

Gale, N. H., and Z. A. Stos-Gale (1987), in *Studies in Sardinian Archaeology*, M. Balmuth, Ed., BAR International Series 387, Oxford, p. 135.

Gale, N. H., and Z. A. Stos-Gale (1989), *Miniera y Metalurgia en las Antiguas Civilizaciones Mediterraneas y Europeas*, C. Domergue, Ed., Ministerio de Cultura, Madrid, p. 21.

Gale, N. H., and Z. A. Stos-Gale (1992a), in *New Developments in Archaeological Science. Proceedings of the British Academy 77. A Joint Symposium of the Royal Society and the British Academy*, A. M. Pollard, Ed., Oxford University Press, New York, p. 63.

Gale, N. H., and Z. A. Stos-Gale (1992b), *Archaeometry* **34**(2), 311.

Gale, N. H., and Z. A. Stos-Gale (1996), *Archaeometry* **94**, S. Demirci, A. M. Özer and G. D. Summers, Eds., Tübitak, Ankara, p. 287.

Gale, N. H., W. Gentner, and G. A. Wagner (1980), in *Metallurgy in Numismatics*, Vol. 1, Special Publication No. 13, D. M. Metcalf, Ed., Royal Numismatic Society, London, p. 3.

Gale, N. H., E. T. C. Spooner, and P. J. Potts (1981), *Can. J. Earth Sci.* **18**, 1290.

Gale, N. H., Z. A. Stos-Gale, and J. L. Davis (1984), *Hesperia* **53**(4), 389.

Gale, N. H., Z. A. Stos-Gale, and G. E. Gilmore (1985a), *Anatolian Studies* **35**, 143.

Gale, N. H., A. Papastamataki, Z. A. Stos-Gale, and K. Leonis (1985b), in *Furnaces and Smelting Technology in Antiquity*, P. T. Craddock and M. J. Hughes, Eds., British Museum, London, p. 81.

Gale, N. H., H. G. Bachmann, B. Rothenberg, Z. A. Stos-Gale, and R. Tylecote (1990), in *Ancient Metallurgy of Copper*, B. Rothenberg, Ed., Institute of Archaeology, London, p. 182.

Gale, N. H., H. G. Bachmann, B. Rothenberg and Z. A. Stos-Gale (1991), The Adventitious Production of Iron in the Smelting of Copper in Timna, W. Arabah, *Ancient Metallurgy in Timna*, B. Rothenberg Ed. Thames and Hudson, London. P. 182. ISBN 0 906183 030.

Gale, N. H., Z. A. Stos-Gale, P. Lilov, M. Dimitrov, and T. Todorov (1991), in *Proceedings of the Colloquium "Decouverte du Metal." Saint Germaine-en-Laye. January 1989*, C. Eluere, Ed., Edition Picard, Millenaires, Paris, p. 49.

Gale, N. H., Z. A. Stos-Gale, G., Maliotis, and N. Annetts (1997), *Archaeometry* **39**(1), 37.

Gale, N. H., Z. A. Stos-Gale, A. Raduncheva, I. Ivanov, P. Lilov, T. Todorov, and I. Panayotov (in press), in *Proceedings of the British Museum Conference on Early Metallurgy*, 1995 P. Craddock, Ed., BM Press, London.

Gale, N. H., A. P. Woodhead, Z. A. Stos-Gale, A. Walden and I. Bowen (1999), *Int. J. Mass Spectrometry and Ion Processes* **184**, 1.

Galili, E., N. Shmueli, and M. Artzy (1986), *Int. J. Nautical Archaeol.* **15**(1), 25.

Georgiou, H. (1979), *The Late Minoan Destruction of Crete: Metal Groups and Stratigraphic Considerations*, Monograph IX, University of California, Los Angeles.

Gerstenberger, H., and G. Haase (1997), *Chem. Geol.* **136**, 309.

Göpel, C., G. Manhes, and C. J. Allègre (1985), *Geochim. Cosmochim. Acta* **49**, 1681.

Grögler, N., J. Geiss, M. Grünenfelder, and F. G. Houtermans (1966), *Z. Naturforschung* **21**, 1167.

Gulson, B. L. (1986), *Lead Isotopes in Mineral Exploration*, Elsevier, Oxford.

Gulson, B. L., and K. J. Mizon (1979), *J. Geochem. Explor.* **2**, 299.

Habfast, K. (1998), *Int. J. Mass Spectrom.* **176**, 133.

Hair, J. F., R. E. Anderson, R. L. Tatham, and W. C. Black (1995), *Multivariate Data Analysis*, Prentice-Hall, Englewood Cliffs, NJ.

Hall, C. M., W. Yi, and D. Teagle (1995), *Int. J. Mass Spectrom.* **146/147**, 21.

Halliday, A. N., D.-C. Lee, J. N. Christenson, A. J. Walder, P. A. Freedman, C. E. Jones, C. M. Hall, W. Yi, and D. Teagle (1995), *Int. J. Mass Spectrom. Ion Process.* **146/147**, 21.

Halliday, A. N., D.-C. Lee, J. N. Christensen, M. Rehkamper, W. Yi, X. Luo, C. M. Hall, C. J. Ballentine, T. Pettke, and C. Stirling (1998), *Geochim. Cosmochim. Acta* **62**, 919.

Hamelin, B., G. Manhes, F. Albarède, and C. J. Allègre (1985), *Geochim. Cosmochim. Acta* **49**, 173.

Harding, A. F. (1984), *The Mycenaeans and Europe*, London.

Hardwick, N., and Z. A. Stos-Gale (1998), in *Applications of Scientific Methods for Investigating Coins and Coinage*, A. Oddy and M. Cowell, Eds., British Museum and Royal Numismatic Society, London.

Hart, S. R., N. Shimizu, and D. A. Sverjensky (1981), *Econ. Geol.* **76**, 1873.

Hassan, A. A., and F. A. Hassan (1981), *Archaeometry* **23**, 77.

Hauptmann, A. (1989), in *The Earliest Periods of Copper Metallurgy in Feinan/break, Jordan*, A. Hauptmann, E. Pernicka, and G. A. Wagner, Eds., Old World Archaeometallurgy, Bochum, p. 119.

Hauptmann, A., F. Begemann, E. Heitkemper, E. Pernicka., and S. Schmitt-Strecker (1992), *Archeomaterials* **6**, 1.

Hegde, K. T. M., and J. E. Ericson (1985), in *Furnaces and Smelting Technology in Antiquity*, British Museum Occasional Paper 48, P. T. Craddock and M. J. Hughes, Eds., British Museum, London, p. 59.

Hellbing, L. (1979), *Studies in Mediterranean Archaeology* LVII, Paul Åströms Förlag, Göteborg.

Heyl, A. N., M. H. Delavaux, R. E. Zartmann, and M. R. Brock (1966), *Econ. Geol.* **61**, 933.

Heyl, A. V., J. P. Landis, and R. E. Zartman (1974), *Econ. Geol.* **69**, 992.

Hirao, Y. M., Sasaki, M. Takenaka and T. Uchida (1992), *Anatolian Archaeological Studies* **1**, 165–186.

Hirao, Y. and J. Enomoto (1992), *J. of the Archaeological Society of Nippon* **77**, 124–126.

Hirao, Y., J. Enomoto, and H. Tachikawa (1995), in *Essays on Ancient Anatolia and its Surrounding Civilisations*, H. I. H. Prince Takahito Mikasa, Ed. Harrassowitz Verlag, Wiesbaden, p. 89–114.

Hirao, Y., and J. Enomoto (1994), *Archaeol. Natur. Sci.* **29**, 41.

Hirata, T. (1996), *Analyst* **121**, 1407.

Hirata, T., and R. W. Nesbitt (1995), *Geochim. Cosmochim. Acta* **59**, 2491.

Holmes, A. (1946), *Nature* **157**, 680.

Holmes, A. (1947), *Nature* **159**, 127.

Holmes, A. (1949), *Nature* **163**, 453.

Houtermans, F. G. (1946), *Naturwissenschaften* **33**, 185.

Hughes, N. J., J. P. Northover, and B. E. P. Staniaszek (1982), *Archaeology* **1**, 359.

Jin, Z., W. T. Chase, Y. Hirao, H. Mabuchi, K. Miwa, Y. Hirao, K. Zhao, R. Chen, and J. Hua. (1993–1998), *Wenwu (Cultural Relics)* **447**, 80–89.

Jin, Z., W. T. Chase, Y. Hirao, H. Peng. H. Mabuchi, K. Miwa, and K. Zhan. (1994–1998), *Kaogu (Archaeology)* **323**, 774–747.

Jin, Z., H. Mabuchi, W. T. Chase, D. Chen, K. Miwa, Y. Hirao, and D. Zhao (1995), *Wenwu (Cultural Relics)*, **465**, 80–85.

Jin, Z., G. Zheng, W. T. Chase, Y. Hirao, and Y. Hayakawa (1998), in press.

Joel, E. C., J. S. Olin, M. J. Blackman, and I. L. Barnes (1988), in *Archaeometry '88*, R. M. Farquhar, R. G. V. Hancock, and L. A. Pavlish, Eds., University of Toronto, Toronto, p. 188.

Kaufman, L., and P. J. Rousseeuw (1990), *Finding Groups in Data*, John Wiley & Sons, New York.

Keisch, B. (1970), *Stud. Cons.* **15**, 1.

Kingston, H. M., and L. B. Jassie (1988), *Introduction to Microwave Sample Preparation*, American Chemical Society, Washington, DC.

Köppel, V., and M. Grünenfelder (1979), in *Lectures in Isotope Geology*, E. Jäger and J. C. Hunziker, Eds., Springer Berlin, p. 134.

Köppel, V., and L. Kostelka (1976), *Abstracts of the Fourth European Colloquium on Geochronology*, Amsterdam, p. 58.

Krogh, T. E. (1973), *Geochim. Cosmochim. Acta* **37**, 485.

Lambert, J. B. (1997), *Traces of the Past*, Addison-Wesley, Reading, MA.

Lee, D. C., and A. Halliday (1995), *Int. J. Mass Spectrom.* **146/147**, 35.

Lilyquist, C., and R. H. Brill (1993), *Studies in early Egyptian Glass*, Metropolitan Museum of Art, New York.

Lo Schiavo, F. (1990), *Quaderni* **17**, 16.

Ludwig, K. R. (1997), *Chem. Geol.* **135**, 325.

Mabuchi, H., and Y. Hirao (1980), *Museum* **357**, 36.

Mabuchi, H., and Y. Hirao (1982a), *Museum* **370**, 4.

Mabuchi, H., and Y. Hirao (1982b), *Eokogaku Zasehi* (*Archaeol. Soc. Nippon*) **68**(1), 42.

Mabuchi, H., Y. Hirao, S. Sato, S., N. Midorikawa, and L. Igaki, (1982), *Archaeol. Nat. Sci.* **15**, 23.

Mabuchi, H., Y. Emoto, Y. Hirao, S. Kitada, and K. Kimura (1983), *Sci. Pap. Jpn. Antiq. Art Crafts* **28**, 65.

Mabuchi, H. and Y. Hirao (1990), *J. of the Archaeological Society of Nippon* **75**, p. 385–404.

Mabuchi, H., Y. Emoto, T. Kadokura, Y. Hirao, S. Aoki and K. Miwa (1991), *Science for Conservation* **30**, 1–19.

Mabuchi, H., Y. Hirao, and M. Nishida (1985), *Archaeometry* **27**(2), 131.

MacGillivray, J. A. (1984), in *The Prehistoric Cyclades*, J. A. MacGillivray and R. L. N. Barber, Eds., Edinburgh, p. 70.

Marinos, G. P., and W. E. Petrascheck (1956), *Laurium*, Inst. for Geol. and Mineral Exploration, Athens.

Mattinson, J. M. (1972), *Anal. Chem.* **44**, 1715.

Mayne, K. I. (1960), in *Stable Isotope Geochemistry and Mass Spectrometric Analysis*, A. A. Smales, and L. R. Wager, Eds., Wiley-Interscience, New York, p. 148.

McCulloch, M. T., and J. D. Woodhead (1993), *Geochim. Cosmochim. Acta* **57**, 659.

McDowell, C., Ed. (1963), *Mass Spectrometry*, McGraw-Hill, New York.

McKinney, C. R., J. M. McCrea, S. Epstein, H. H. Allen, and H. C. Urey (1950), *Rev. Sci. Instrum.* **21**, 274.

Merkel, J. F. (1985), *MASCA J.* **3**(5, Archaeometallurgy Suppl.), 164.

Milne, G. W. (1971), *Mass Spectrometry: Techniques and Applications*, Wiley-Interscience, New York.

Molleson, T. I., T. Eldridge, and N. H. Gale (1986), *Oxford J. Archaeol.* **5/2**, 249.

Moody, J. A., and N. E. Ickermann (1985), in *Contributions to Aegean Archaeology*, S. N. C. Wilkie and W. D. E. Coulson, Eds., University of Minnesota Press, Minnesota p. 61

Muhly, J. D. (1982a), in *Acts of the International Archaeological Symposium: "The Relations Between Cyprus and Crete, c. 2000–500* B.C.," Leventis Foundation Nicosia, p. 87.

Muhly, J. D. (1982b), in *Acts of the International Archaeological Symposium: "The Relations Between Cyprus and Crete, c. 2000–500* B.C.," *Nicosia, p. 251.*

Muhly, J. D. (1983), Reports of the Department of Antiquities, Cyprus, Vol. 2, Nicosia, p. 210.

Muhly, J. D. (1985), *Am. J. Archaeol.* **89**, 275.

Mulliken, R. S., and W. D. Harkins (1922), *J. Am. Chem. Soc.* **44**, 37.

Nier, A. O. (1939), *Phys. Rev.* **55**, 153.

Nier, A. O. (1940), *Rev. Sci. Instrum.* **11**, 212.

Nishida, M. (1980), *Museum* **357**, 32.

Nishida, M. (1982), *Museum* **370**, 13.

Nishida, M. (1983), *Museum* **382**, 31.

Paribeni, R. (1903), *Rendiconti dello Accademia dei Lincei XII*, **fascicule 7–8**, p. 1.

Percy, J. (1861), *Metallurgy I—Fuels, Fireclays, Copper, Zinc and Brass*, John Murray London.

Percy, J. (1880), *Metallurgy—Silver and Gold—Part 1*, John Murray London.

Pernicka, E., and H. G. Bachmann (1983), *Erzmetall* **36**, 592.

Pernicka, E., and G. A. Wagner (1988), in *Thasos als Rohstoffquelle für Bunt-und Edelmetalle im Altertum*, G. A. Wagner and G. Weisgerber, Eds., Antike Edel-und Buntmetallgewinnung auf Thasos, Der Anschnitt, Deutschen Bergbau-Museum, Bochum, p. 224.

Pernicka, E., G. A. Wagner, and N. Todt (1983a), Paper MPIH-1983-V37. Max-Planck-Institut für Kernphysik Heidelberg.

Pernicka, E., G. A. Wagner, K. Assimenos, C. Doumas, F. Begemann, and W. Todt (1983b), in *Proceedings of the Twenty-Second Symposium on Archaeometry, Bradford*, University of Bradford, A. Aspinall and S. E. Warren, Eds., p. 292.

Pernicka, E., T. C. Seeliger, G. A. Wagner, F. Begemann, S. Schmitt-Strecker, C. Eibner, O. Oztunali, and I. Baranyi (1984), *Jahrbuch des Romisch-germanischen Zentralmuseums* **31**, 533.

Pernicka, E., F. Begemann, S. Schmitt-Strecker, and A. P. Grimanis (1990), *Oxford J. Archaeol.* **9**(3), 263.

Pernicka, E., F. Begemann, S. Schmitt-Strecker, and G. A. Wagner (1993), *Praehistorische Zeitschrift* **68**(1), 1.

Pernicka, E., F. Begemann, S. Schmitt-Strecker, H. Todorova, and I. Kuleff (1997), *Deutches Archaeologisches Institut: Eurasia Antiqua*, Vol. 3, Philip von Zabern, Mainz am Rhein, p. 41.

Philip, G. (1988), *World Archaeology* **20**(2), 190.

Philip-Horwitz, E., M. L. Dietz, S. Rhoads, C. Felinto, N. H. Gale, and J. Houghton (1994), *Anal. Chim. Acta* **292**, 263.

Pollard, A. M. (1986), in *Greek and Cypriot Pottery*, R. E. Jones, Ed., British School at Athens, Athens, p. 56.

Pollard, A. M., and C. Heron (1996), *Archaeological Chemistry*, Royal Society of Chemistry, London.

Popham, M. R. (1984), *The Minoan Unexplored Mansion at Knossos*, Plates Volume, Thames and Hudson, London.

Punch (1895), London: *The reference to the curate's egg is in a cartoon depicting a nervous young curate dining with his bishop; the curate has been served with a completely bad egg. The curate responds to his bishop's polite enquiry that the egg is good in parts.* See *Universal Dictionary*, Reader's Digest Association, London, (1994), p. 384.

Renfrew, A. C. (1972), *The Emergence of Civilisation*, Methuen London.

Renfrew, A. C., and P. Bahn (1996), *Archaeology: Theories, Methods and Practise*, Thames and Hudson, London, p. 344.

Richards, T. W., and M. E. Lambert (1914), *J. Am. Chem. Soc.* **36**, 1329.

Roddick, J. C., and W. D. Loveridge (1986), Application Report No. 66, Finnigan MAT, Bremen.

Rohl, B. M. (1996), *Archaeometry* **38**(1), 165.

Russ, G. P., and J. M. Bazan (1987), *Spectrochim. Acta Part B* **42**, 49.

Russell, R. D., and R. M. Farquhar (1960), *Lead Isotopes in Geology*, Interscience Publishers, Inc., New York.

Rutherford, E. (1913), *Radioactive Substances and Their Radiations*, Cambridge University Press, London.

Sayre, E. V., K. A. Yener, E. C. Joel, and I. L. Barnes (1992a), *Archaeometry* **34**(1), 73.

Sayre, E. V., P. Jett, and E. C. Joel (1992b), in *Materials Issues in Art and Archaeology III*, Materials Research Society Symposium Proceedings, Vol. 267, P. B. Vandiver, J. R. Druzik, G. S. Wheeler, and I. C. Freestone, Eds., Materials Research Society, Pittsburgh, PA, pp. 225–237.

Seeliger, T. C., E. Pernicka, G. A. Wagner, F. Begemann, S. Schmitt-Strecker, C. Eibner, O. Oztunali, and I. Baranyi (1985), *Jahrbuch des Romisch-Germanischen Zentralmuseums* **32**, 597.

Soddy, F. (1911), *Annu. Repts. Progr. Chem.* **99**, 72.

Soddy, F. (1914), *Chemistry of the Radio-Elements*, Part II, Longmans, London.

Spitaels, P. (1984), *Thorikos VIII*, University of Gent, Gent, p. 151.

Stacey, J. S., and J. D. Kramers (1975), *Earth Planet. Sci. Lett.* **26**, 207.

Staniaszek B. E. P., and P. Northover (1982), in A. Aspinall and S. E. Warren, *Proceedings of the Twenty-Second International Symposium on Archaeometry*, University of Bradford, Bradford, p. 262.

Stanton, R. L., and R. D. Russell (1959), *Econ. Geol.* **54**, 588.

Stos-Fertner, Z. A., and N. H. Gale (1978a), *Proc. 18th Int. Symp. Archaeomet. ArchaeoPhys.* **10**, 299–314.

Stos-Fertner, Z. A., and N. H. Gale (1978b), *Archeo-Physika* **10**, 299.

Stos-Gale, Z. A. (1989), in *Old World Archaeometallurgy, Der Anschnitt Beiheft*, Vol. 9, A. Hauptmann, E. Pernicka, and G. A. Wagner, Eds., Bergbau Museum, Bochum, pp. 279–292.

Stos-Gale, Z. A. (1992), *Oxford J. Archaeol.* **11**(2), 155.

Stos-Gale, Z. A. (1993a), in *Trade and Exchange in Prehistoric Europe*, C. Scarre and F. Healy, Eds., Oxbow Books, Oxford, pp. 115–129.

Stos-Gale, Z. A. (1993b), *J. Eur. Archaeol.* **1/2**, 101.

Stos-Gale, Z. A. (1993c), *Archaeol. Polona* **31**, 149.

Stos-Gale Z. A. (1993d), in *Archeologia delle Attivita Estrattive e Metallurgiche (Proceedings of a Summer School in Certosa di Pontignano-Campiglia Marittima September 1991)*, Quaderni del Dipartimento di Archeologia e Storia delle Arti, Sezione Archeologica—Università di Siena, R. Francovich, Ed., Edizioni All'insegna de Giglio S.A.S. Florence, pp. 593–627.

Stos-Gale, Z. A. (1998a). The role of Kythnos and other Cycladic islands in the origins of Early Minoan metallurgy. In *Kea-Kythnos: History and Archaeology, Meletimata* 27, L. G. Mendoni and A. Mazarakis-Ainian, Eds., Athens & Paris, Diffusion de Bocard, p. 717.

Stos-Gale, Z. A. (1998b). Lead isotope analyses of coins—a review. In *Metallurgy in Numismatics, Vol 4.* Royal Numismatic Society, Special Publication No. 30, A. Oddy and M. Cowell, Eds.. London. p. 348.

Stos-Gale, Z. A. (in press). Lead–isotope analyses of glass, glazes and some metal artefacts. Chapter VI in G. Bass Ed. Serce Limani: An Eleventh-Century Shipwreck, Volume I. G. F. Bass, S. Matthews, J. R. Steffy and F. H. Van Doorninck Eds. Texas A&M University Press, College Station, Texas.

Stos-Gale, Z. A. (in press), in *The Transition from Hacksilber to Coinage*, M. Balmuth, Ed., American Numismatic Society.

Stos-Gale, Z. A. (in press). Trade in Metals in the Bronze Age Mediterranean: an overview of Lead Isotope Data for Provenance Studies. In the *Proceedings of the Conference Supply and Circulation of Metals in Bronze Age Europe* Brimingham June 1997, Ed. C. Pare, Prehistoric Society, London.

Stos-Gale, Z. A., and N. H. Gale (1980), *Rev. d'Archeometr.* **5**(111), 285.

Stos-Gale, Z. A., and N. H. Gale (1982), *J. Field Archaeol.* **9**, 467.

Stos-Gale, Z. A., and N. H. Gale (1984), *Oxford J. Archaeol.* **3**(3), 23.

Stos-Gale, Z. A., and N. H. Gale (1990), in *Thera and the Aegean World*, Vol. 3, A. Hardy, Ed., Thera Foundation, London, pp. 72–92.

Stos-Gale, Z. A., and N. H. Gale (1992), in *Sardinia in the Mediterranean; A Footprint in the Sea. Studies in Sardinian Archaeology presented to Miriam Balmuth*, R. H. Tycot, and T. K. Andrews, Eds., Sheffield Academic, Sheffield, pp. 317–346.

Stos-Gale, Z. A., and N. H. Gale (1994), in *Provenience Studies and Bronze Age Cyprus: Production, Exchange and Politico-Economic Change*, A. B. Knapp and J. F. Cherry, Eds., Prehistory Press, Sheffield, pp. 92–121.

Stos-Gale, Z. A., and C. F. Macdonald (1991), in *Bronze Age Trade in the Mediterranean*, N. H. Gale, Ed., Studies in Mediterranean Archaeology XC, Paul Aströms Förlag, pp. 248–280.

Stos-Gale, Z. A., N. H. Gale, and G.R. Gilmore (1984), *Oxford J. Archaeol.* **3**(3), 23.

Stos-Gale, Z. A., N. H. Gale, and U. Zwicker (1986), Reports of the Department of Antiquities of Cyprus, Nicosia, pp. 123–144.

Stos-Gale, Z. A., N. H. Gale, A. Papastamataki, S. Leonis, O. Hadjianastassiou, and A. MacGillivray (1988), in *Aspects of Ancient Mining and Technology*, J. E. Jones, Ed., Acta of British School at Athens Centenary Conference, Bangor, pp. 1–31.

Stos-Gale, Z. A., N. H. Gale, J. Houghton, and R. Speakman (1995a), *Archaeometry* **37**(2), 407.

Stos-Gale, Z. A., N. H. Gale, and J. Houghton (1995b), *Egypt, the Aegean and the Levant: Interconnections in the 2nd millenium BC*, W. V. Davies and L. Schofield, Eds., British Museum Press, London, pp. 127–135.

Stos-Gale, Z. A., N. H. Gale, and N. Annetts (1996), *Archaeometry* **38**(2), 381.

Stos-Gale, Z. A., G. Maliotis, N. H. Gale, and N. Annetts (1997), *Archaeometry* **39**(1), 83.

Stos-Gale, Z. A., N. H. Gale, N. Annetts, T. Todorov, P. Lilov, A. Radunncheva, and I. Panayotov (1998a), *Archaeometry* **40**(1), 217.

Stos-Gale, Z. A., N. H. Gale, G. Bass, C. Pulak, E. Galili, and J. Sharvit (1998b), in *Proceedings of the Fourth International Conference on the Beginnings of the Use of Metals and Alloys, BUMA IV, Matsue, Japan*, H. Katayama, Ed. The Japan Inst. of Metals, Aoba, Japan 115.

Stos-Gale, Z. A., G. Maliotis, and N. H. Gale (1998c), *A Preliminary Survey of the Cypriot Slag Heaps and Their Contribution to the Reconstruction of Copper Production on Cyprus*, Metallurgica Antiqua, in Honour of H. G. Bachmann and R. Maddin, T. Rehren, A. Hauptmann, and J. Muhly, Eds., Der Anschnitt, Beiheft 8 Deutches Bergbau Museum, Bochum.

Stos-Gale, Z. A., M. Kayafa, and N. H. Gale (in press c), *Opuscula Athenensis*, Lund.

Tatsumoto, M., R. J. Knight, and C. J. Allègre (1973), *Science* **180**, 1279.

Thomson, J. J. (1913), *Rays of Positive Electricity*, Longman, London.

Todt, W., R. A. Cliff, A. Hanser, and A. W. Hofmann (1984), *Terra Cognita* **4**, 209.

Todt, W., R. A. Cliff, A. Hanser, and A. W. Hofmann (1996), in *Earth Processes: Reading the Isotopic Code*, Geophysical Monograph 95, A. R. Basu and S. R. Hart, Eds., American Geophysical Union, Washington, DC, pp. 429–437.

Tuttas, D., and K. Habfast (1982), *High Precision Lead Isotope Ratio Measurements*, Finnigan MAT Application Note 51, Finnigan MAT Gmbh, Bremen.

Tylecote, R. F., H. A. Ghaznavi, and P. J. Boydell (1977), *J. Arch. Sci.* **4**, 305.

Volker, F., M. T. McCulloch, and R. Altherr (1993), *Geophys. Res. Lett.* **20**, 927.

Vos, A. (1996), *Deguwa, Rundbrief* **11**, 22.

Wachsmann, S. (1987), Aegeans in the Theban Tombs, *Orientala Lovaniensis Analecta* **20**, Peeters, Louvain.

Wagner, G. A., and G. Weisgerber (1985), *Silber, Blei End Gold auf Sifnos*, Der Anschnitt, Beiheft 3 Bochum.

Wagner, G. A., E. Pernicka, T. C. Seeliger, I. Lorentz, F. Begemann, S. Schmitt-Strecker, C. Eibner, and O. Oztunali (1986), *Jahrbuch des Romisch-Germanischen Zentralmuseums* **33**, 723.

Wagner, G. A., W. Gentner, H. Groppengieser, and N. H. Gale (1980), in *Scientific Studies in Early Mining and Extractive Metallurgy, British Museum Occasional Paper* 20, P. Craddock, Ed. London, pp. 63–85.

Wagner, G. A., E. Pernicka, M. Vavelidis, I. Baranyi, and Y. Bassiakos (1986), *Archaometallurgische Untersuchungen auf Chalkidiki. Der Anschnitt* **5–6**(38), 166.

Walder A. J., and P. A. Freedman (1992), *J. Anal. Atom. Spectrom.* **7**, 571.

Walder, A. J., and N. Furuta (1993), *Anal. Sci.* **9**, 675.

Walder, A. J., I. Platzner, and P. A. Freedman (1993), *J. Anal. Atom. Spectrom.* **8**, 19.

Webster, R. K. (1960), in *Stable Isotope Geochemistry and Mass Spectrometric Analysis*, A. A. Smales, and L. R. Wager, Eds., Wiley-Interscience, New York, pp. 202–246.

Weeks, L. R. (1999) *Antiquity*. **73**(279), 49.

Woodhead, A. P., N. H. Gale, and Z. A. Stos-Gale (1999), in *Metals in Antiquity*, S. M. M. Young, P. Budd, R. Ixer, and M. Pollard, Eds., Archaeopress, Oxford. 134.

Yamasaki, K., M. Murosumi, S. Nakamura, M. Hinata, and M. Yuasa (1978), *J. Chem. Soc. Jpn. Chem. Ind. Chem.* 1112.

Yamasaki, K., M. Murosumi, S. Nakamura, M. Hinata, M. Yuasa, and M. Watarai (1979), *Proceedings of the International Symposium on the Conservation and Restoration of Cultural Property*, Tokyo National Research Institute of Cultural Properties, Tokyo, pp. 221–234.

Yener, K. A., and H. Özbal (1987), *Antiquity* **61**, 220.

Yener, K. A., and P. J. Vandiver (1993), *Am. J. Arch.* **97**, 207.

Yener, K. A., H. Özbal, E. Kaptan, A. N. Pehlivan, and M. Goodway (1989a), *Science* **244**, 200.

Yener, K. A., H. Özbal, A. Minzoni-Deroche, and B. Aksoy (1989b), *Natl. Geograph. Res.* **5**, 477–494.

Yener, K. A., E. V. Sayre, H. Özbal, E. C. Joel, I. L. Barnes, and R. H. Brill (1991), *J. Archaeol. Sci.* **18**, 541.

Zachos, K. and G. Maratos, (1973), Metallogenic Map of Greece, 1:1,000,000, Institute for Geology and Subsurface Research, Athens.

Zwicker, U. (1986), *Cyprus Association of Geologists and Mining Engineers*, Bulletin 3, Nicosia, pp. 79–91.

Zwicker, U., Z. A. Stos-Gale., and N. H. Gale (1986), in *Proceedings of the Tenth Inernational Congress of Numismatics, London*, J. A. Carradice, Ed., Publication No. 11, Professional Numismatists, London, pp. 117–124, Pl. 12–13.

CHAPTER

18

LUMINESCENCE DATING

SEBASTIANO O. TROJA

Department of Physics and Astronomy, University of Catania, Catania, Italy

RICHARD G. ROBERTS

Department of Earth Sciences, La Trobe University, Melbourne, Australia

18.1. INTRODUCTION

The phenomenon of luminescence in solids has been a subject of scientific interest since the seventeenth century. In a report to the Royal Society of London, Robert Boyle (Boyle 1664) described the shining light, as distinct from black-body emission, produced by a diamond when heated in the dark. Since then, researchers have observed the production of luminescence in many types of mineral and artificial crystals, and many other phenomena in which luminescence is emitted, either through the spontaneous release of energy from a crystal or through stimulation (e.g., thermally, optically, mechanically, or electrically), have been reported. But it was only in the 1940s that Randall and Wilkins (1945) developed the first acceptable model for thermally stimulated luminescence (thermoluminescence, or TL).

After initially using thermoluminescence in radiation dosimetry (for some crystals the TL intensity is proportional to the dose received from nuclear radiation), in the 1950s it was proposed as a means of archaeological dating (Daniels et al. 1953). Many artifacts (such as pottery, ceramic, and brick) have crystalline inclusions of quartz, feldspar, and other minerals that are continuously subjected to low levels of natural radiation present in all environments and that emit thermoluminescence. The first TL dating results were obtained in the 1960s from the mix of luminescent crystals present in ancient pottery (Aitken et al. 1964, 1968; Ichikawa 1965; Mazess and Zimmerman 1966; Ralph and Han 1966).

Modern Analytical Methods in Art and Archaeology, Edited by Enrico Ciliberto and Giuseppe Spoto. Chemical Analysis Series, Vol. 155
ISBN 0-471-29361-X

The dating procedure was subsequently refined for the silt-sized mineral fraction (the "fine-grain" technique) (Zimmerman 1967, 1971) and for 100-μm-diameter quartz crystals (the "quartz inclusion" method) (Fleming 1970, 1979; Langouet et al. 1979, 1980; Hütt and Smirnov 1983). Later applications used zircon crystals (Zimmerman et al. 1974; Sutton and Zimmerman 1976; Templer and Smith 1988; Smith et al. 1991), quartz and feldspar grains greater than 100 μm in diameter (Mejdahl 1983; Mejdahl and Wintle 1984), and calcite crystals (Wintle 1978; Debenham 1983; Debenham and Aitken 1984).

As optical radiation is also often able to stimulate luminescence in these minerals, the bleaching of crystals by sunlight, or by artificial light, was long considered a factor that could seriously affect the stability of the TL signal (e.g., Aitken et al. 1963). But this sensitivity to light was not exploited as a dating method until the 1980s, when the use of optically stimulated luminescence (OSL) was proposed as a means of dating sediments (Huntley et al. 1985). In optical dating the crystal is exposed to light rather than heated to a high temperature, as in TL, and this has allowed new types of samples to be dated, such as those that were previously subjected only to solar illumination (e.g., archaeological and geological sediments). Extensive reviews of sediment dating applications may be found in Roberts (1997) and Prescott and Robertson (1997). Today, the stimulated light emission from crystals is involved in retrospective dosimetry (McKeever et al. 1995) and, increasingly, in the dating of artifacts and Quaternary sediments (Aitken 1985, 1998).

18.2. PHYSICAL BACKGROUND

18.2.1. Mechanism of Luminescence Production

The term *luminescence* defines the phenomenon of the emission of light from a body as the result of an external cause unconnected with the temperature of the body itself and distinct from any process of scattering (Ropp 1991; Kitai 1993; Galanin 1996). Although the process of luminescence is also present in vitreous solids and in molecules (Shiomoya and Yen, 1998), in the case of dating it essentially concerns crystalline solids. For the latter, the physical phenomenon of luminescence is closely linked to the crystalline structure of materials and the impurities present in them.

Electrons in the crystal can occupy, in accordance with the Fermi–Dirac statistic, energy states that are distributed in energy regions that form continuous energy bands (Burns 1990). A characteristic of semiconducting and insulating crystals is that they possess energy bands that, at absolute zero, are completely full of electrons as well as bands that are completely empty. The last of the filled bands (the valence band) is separated from the first empty band (the conduction band) by an energy gap (the "forbidden gap," E_g), which is characteristic of the crystal. At temperatures above absolute zero, electrons can possess enough energy to move into the conduction band (the number depending on the size of the forbidden gap and on the temperature), thereby leaving effective positive charge carriers (holes) in the valence band. In this way, there are free negative charge carriers in the conduction band and free positive charge carriers in the valence band. Such a condition can explain the charge transport phenomena responsible for the electrical, magnetic, and optical properties of pure crystals. However, this scheme corresponds to an ideal crystal.

In reality, each natural or artificial crystal possesses defects such as, for example, vacancies in the periodic structure of the crystal lattice or substitution by foreign atoms different from those expected, as well as interruptions in the periodicity of the crystal lattice. All these defects of the ideal lattice periodicity generate new and different energy levels, even within the forbidden gap, which can be occupied by electrons or by holes. New electrical, magnetic and optical behavior is thus generated.

Although a description of the phenomena founded on the electronic structure of the crystal cannot be stated in rigorous mathematical and physical terms, a qualitative description of the phenomenon is possible. The newly generated energy states possess a characteristic distance from the conduction band (ΔE) and a specific probability that the trapped charge carrier is detrapped (i.e., equivalent to introducing a mean life, τ, for the level). The phenomenon here analyzed essentially concerns insulating crystals with an energy gap so high (some electron volts) that, at room temperature, it does not allow a significant number of electrons to jump to the conduction band, and with energy states in the gap having a very low probability for detrapping of charge carriers (very long mean life energy traps) in the absence of external stimulating processes.

When a crystalline solid of this type is irradiated by nuclear radiation or is subjected to other physical interactions that are able to deliver energy available for the excitation of electrons, the electrons of the valence band can acquire enough energy to occupy free states in the conduction band. They then fall and are trapped at localized energy levels within the forbidden gap. Symmetrically free holes in the valence band are trapped at levels in the gap for holes. The transition by which an electron can then return to the valence band is characteristic of the level, and in general there can be more than a single mechanism (Schulman and Compton 1962; Fowler 1968).

In these crystals the electrons can be trapped in metastable states for a very long time (up to millions of years) until some external cause, in particular heating or electromagnetic radiation of a suitable wavelength, allows the electrons to acquire enough energy to escape from the traps and return to the conduction band. From this band one of the following possibilities will happen: the electrons can fall back into the valence band, fall into the same traps or different traps, or, lastly, fall into a recombination center where they recombine with holes (which in turn are trapped at localized levels for holes). It is these recombination phenomena that can give rise to the emission of light in well-defined wavelength regions that, being correlated to the energy distance between the recombination centers and the conduction band, are characteristic of the crystal.

In this sense, thermoluminescence can be defined as a long-period phosphorescence, a nonstimulated emission phenomenon with a very long mean life, in which temperature has the effect of greatly reducing the permanence of the electrons in the traps (Chen and Kirsh 1981). In the same way, OSL can be considered a phosphorescence in which the transition of electrons from the trap level to the conduction band is stimulated by optical radiation of a suitable wavelength.

Many models of the two stimulated emission mechanisms are present in the literature (a review of the models can be found in McKeever and Chen 1997). Figure 18.1 shows the schematic representation of some excitation and disexcitation mechanisms in a single trap/single recombination center model, as in the Randall and Wilkins (1945) model, as well as a multitrap/multicenter model, such as those of McKeever (1994). In the case of TL, heating at a constant rate empties the traps resulting in the formation of response curve I (luminescence intensity) versus T (temperature), where the presence of one or more convoluted peaks at particular temperatures and with characteristic emission wavelengths depends on the parameters of the traps and recombination centers that are present. In the case of OSL, the application of a constant flow of electromagnetic radiation of a particular wavelength empties the traps resulting in the formation of response curve I versus t (time) showing a typical rapidly decreasing signal, often with a convolution of different exponential decay components.

Figure 18.2 shows the experimental trend of (a) a typical thermoluminescence glow curve (luminescence intensity versus stimulation temperature) and (b) a decay curve of optically stimulated luminescence (luminescence intensity versus stimulation time). TL peaks at high temperatures are normally related to deep electron traps with long-term stability. The area subtended by the curve represents the total contribution of photons due to processes of luminescence in the wavelength region of the detected emission and, for OSL, at the stimulation wavelength used.

18.2.2. Principles of Luminescence Dating

The methods of dating by stimulated luminescence are based on the assumption that the crystalline inclusions, such as quartz (SiO_2), zircon

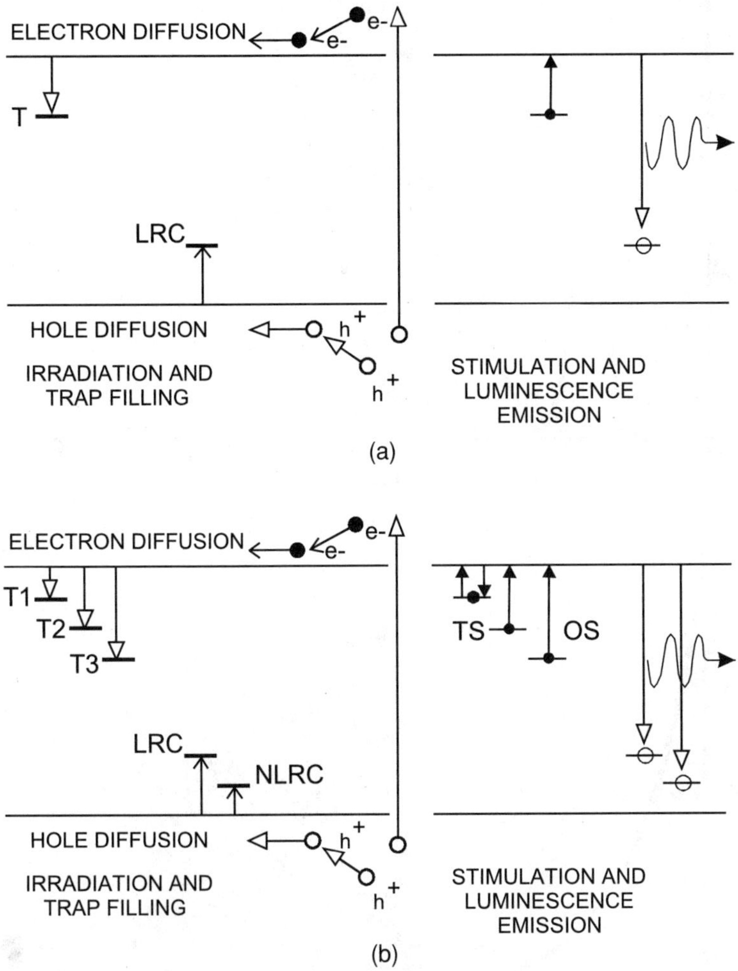

Figure 18.1. Luminescence models: (*a*) Single trap/single luminescence recombination center model for thermally stimulated luminescence and optically stimulated luminescence. (*b*) Multiple trap/multiple recombination center model. It considers three different kinds of electron trap (T1, T2, and T3) and two types of recombination center (nonluminescent recombination center, NLRC, and luminescent recombination center, LRC).

($ZrSiO_4$), calcite ($CaCO_3$), and feldspars (Ca, K, Na, $Al_2Si_2O_8$) (Marfunin 1979; Peckjett 1992), that are present in the sample, are able to accumulate electrons in trap levels and store, for a long time, the dose received from natural radioelements (α, β, and γ radiation from radionuclides belonging to the ^{235}U, ^{238}U, ^{232}Th, ^{40}K, and ^{87}Rb decay chains) (Evans 1955; Lorentz

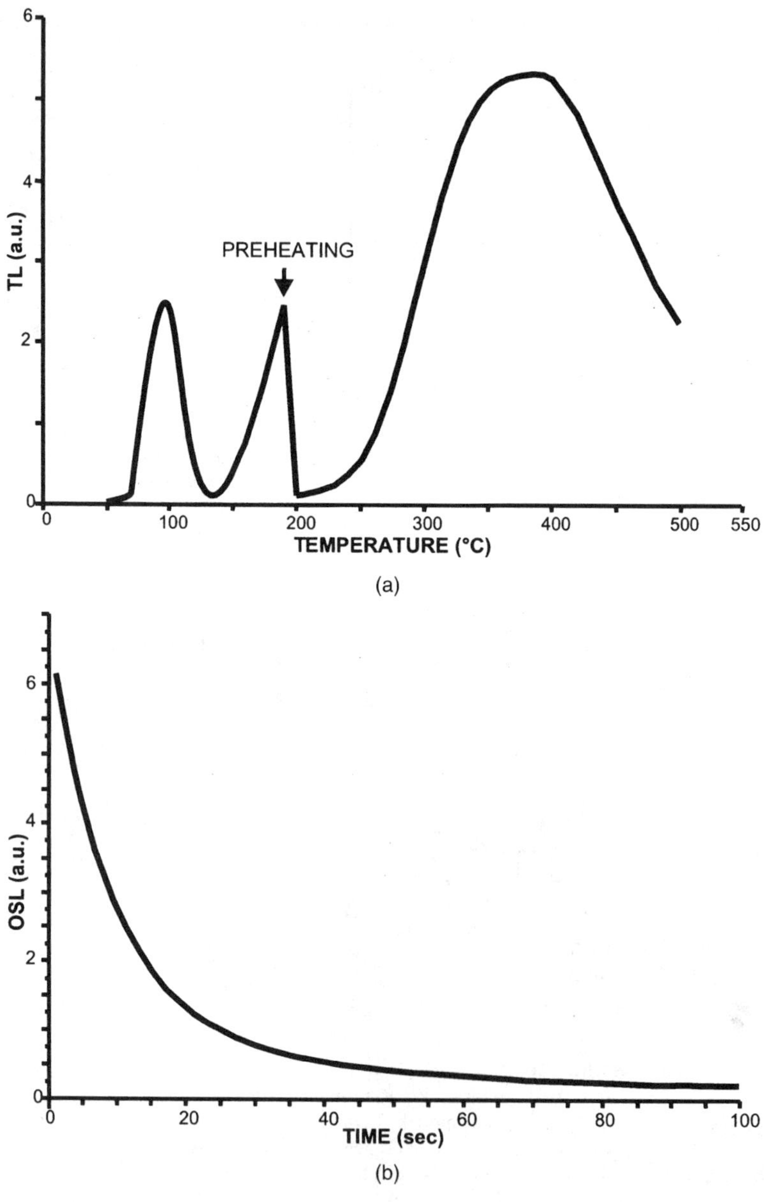

Figure 18.2. (*a*) Typical thermoluminescence glow curve obtained from a coarse-grain quartz sample (100 to 300 μm in diameter) exposed to laboratory irradiation using a calibrated $^{90}Sr/^{90}Y$ β source. Preheating for some seconds reduces the unstable component. The black-body contribution, measured during a second heating cycle, has been subtracted. Infrared absorbing and blue-violet bandpass filters were used during the TL and black-body acquisition. (*b*) Typical optically (green light) stimulated luminescence decay curve obtained from a coarse-grain quartz sample exposed to laboratory irradiation using a calibrated $^{90}Sr/^{90}Y$ β source. Blue-violet filters were used during OSL acquisition.

1983; Aitken 1985). This physical quantity represents the energy absorbed per unit mass and is measured in grays (1 Gy = 1 J/kg). When subjected to thermal or optical stimulation, the crystals are able to release the electrons from the traps, emitting a number of photons proportional to the number of detrapped electrons. For the crystals mentioned above, the response is frequently characterized by a linear relation between the intensity of luminescence and the absorbed dose over a large range of doses.

In the case of crystals that have never been subjected to high-temperature heating or exposed to light, the luminescence emitted is correlated to the total dose absorbed since the geological formation of the crystal. For crystals that have been subjected to heating (such as fired clay, pottery artifacts, or fired stones) or to optical radiation of sufficient duration and/or intensity (such as many archaeological and geological sediments), the heating or optical bleaching event empties the electron traps, thereby zeroing the internal luminescence clock. The crystal loses the information related to the dose accumulated over geological time. As a result of the natural radioelements present in both the matrix and the surrounding environment, the crystals once more start accumulating a dose of radiation, known as the "paleodose" or "archaeodose." The paleodose is accumulated at a rate characteristic of the sample itself and of the environment in which the sample is buried. Therefore, the paleodose is proportional to the age of the artifact or sediment (i.e., the time elapsed since the last heating or bleaching event).

From this it follows that the crystals contained in such samples are able to provide information regarding the date of crystal formation, artifact manufacture, or the time that has passed since their last exposure to high temperature or to light, in accordance with the age equation:

$$t = \frac{P}{\dot{D}} \tag{18.1}$$

where t is the age in years, P indicates the total dose absorbed by the crystal, and $\dot{D}$ is the mean annual dose (Gy/yr). The homogeneity and uniformity in space and time of the exposure to nuclear radiation the intrinsic luminescence sensitivity of the crystals, and determine the datability of the artifacts or sediments. The analysis of the curves of luminescence intensity emitted by crystals during the thermal or optical stimulation allows the determination of the total dose received by the crystals because the luminescence emitted is proportional to the dose absorbed by the sample.

In reference to the luminescence emitted, the age equation can be written:

$$t = \frac{L}{I \times \dot{D}} \tag{18.2}$$

where L is the measured luminescence, and l is the luminescence sensitivity (the luminescence emitted per unit dose absorbed). The annual dose $\dot{D}$ is associated with the natural ionizing irradiation emitted by the radioactive elements contained in the sample, in the sediment that surrounded it, and within the crystals themselves. It also includes the dose, usually small, due to cosmic radiation (Prescott and Stephan 1982; Prescott and Hutton 1988, 1994).

The annual dose can thus be expressed as follows:

$$\dot{D} = \dot{D}_\alpha + \dot{D}_\beta + \dot{D}_\gamma + \dot{D}_{\mathrm{cosm}} \tag{18.3}$$

where $\dot{D}_\alpha$ is the α dose rate component, $\dot{D}_\beta$ the β dose rate component, $\dot{D}_\gamma$ the γ dose rate component, and $\dot{D}_{\mathrm{cosm}}$ the cosmic-ray contribution. On separating the relative components, we obtain:

$$\begin{aligned}\dot{D} = {} & \dot{d}_{\alpha\mathrm{samp}} + \dot{d}_{\beta\mathrm{samp}} + \dot{d}_{\gamma\mathrm{samp}} + \dot{d}_{\alpha\mathrm{sed}} \\ & + \dot{d}_{\beta\mathrm{sed}} + \dot{d}_{\gamma\mathrm{sed}} + \dot{d}_{\mathrm{cosm}} + (\dot{d}_{\alpha\mathrm{crys}} + \dot{d}_{\beta\mathrm{crys}} + \dot{d}_{\gamma\mathrm{crys}})\end{aligned} \tag{18.4}$$

where $\dot{d}_{\alpha\mathrm{samp}}$, $\dot{d}_{\beta\mathrm{samp}}$, and $\dot{d}_{\gamma\mathrm{samp}}$ represent the dose rates due, respectively, to the α, β, and γ radiation emitted by radioactive elements present in the matrix (e.g., pottery) surrounding the crystals; $\dot{d}_{\mathrm{cosm}}$ is the dose rate due to the effect of cosmic radiation, which varies with geographical position and altitude and depth (with values normally ranging between 50 and 150 μGy/yr); $\dot{d}_{\alpha\mathrm{sed}}$, $\dot{d}_{\beta\mathrm{sed}}$, and $\dot{d}_{\gamma\mathrm{sed}}$ represent the components of the dose rate due to the radioactive emissions of the sediment (e.g., burial soil); and $\dot{d}_{\alpha\mathrm{crys}}$, $\dot{d}_{\beta\mathrm{crys}}$, and $\dot{d}_{\gamma\mathrm{crys}}$ are the annual dose components from any radioactive impurities present within the crystals themselves.

The evaluation of the annual dose is usually based mostly on an understanding of the radiation dose received by the crystals from the surrounding material. In some instances the internal dose rate ($\dot{d}_{\mathrm{crys}}$) is significant. An example of this is K feldspars, because of their potassium content, and also zircons, which have a high uranium and thorium content. Experimental determination of all of the significant dose rate components should be performed to identify whether the artifact or sediment had been subjected to a constant or a varying dose rate throughout its entire period of burial (Bell 1976, 1977, 1979b). Ideally, the dose rate should be constant over time, but the evolution of the dose rate through time can be modeled for certain types of time-dependent variation (Olley et al. 1996, 1997).

18.2.3. Some Additional Considerations

It is also necessary to determine the degree to which the luminescence intensity is proportional to the absorbed energy for each sample. Even under

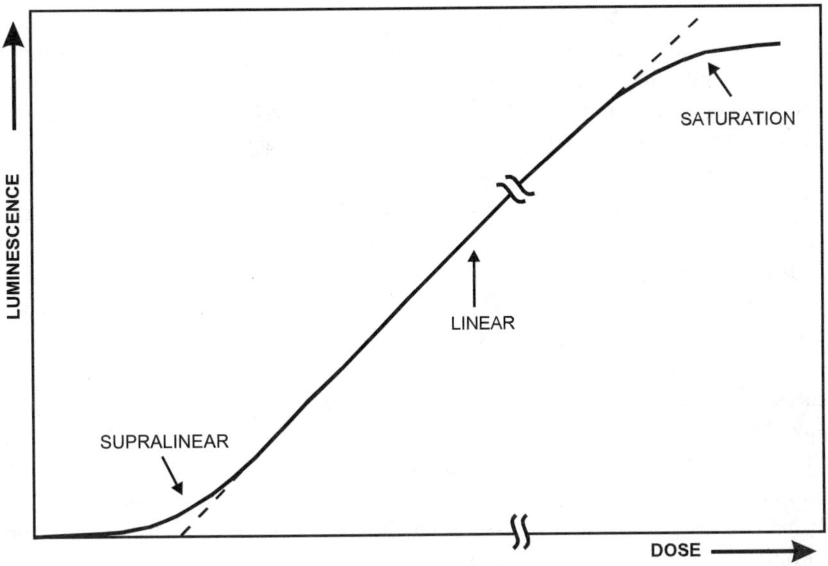

Figure 18.3. Typical response of luminescence intensity plotted against applied dose. The supralinear, linear, and saturated regions are indicated.

optimal conditions, there are variations in the linearity of the response, especially in regions of low and high doses. Figure 18.3 shows the typical behavior of the luminescence intensity versus absorbed dose. In the low-dose region, there is often a nonlinear dose–response behavior, typically supralinear (Chen et al. 1996), connected with various phenomena of sensitivity variation. In the high-dose region, the nonlinearity is present as a result of a natural reduction in the number of free traps available for electron transitions, resulting in a saturation-type behavior. It is therefore necessary to reconstruct the behavior of luminescence growth versus dose for each sample, where both the correction at low doses and the reconstruction in the region near saturation are particularly critical for young and old samples, respectively.

The behavior in the high-dose region leads to the use of a nonlinear fit in the luminescence versus dose relationship for crystals that have received doses close to those of saturation. This constrains the dating range for artifacts. Both the mean life of the levels (which indicates the capacity of the crystal to hold electrons in the traps without effective loss of information regarding the total absorbed dose) and the behavior of the crystal (which, beyond saturation, does not allow variations in the luminescence when the dose is increased) limit the maximum datable age of artifacts and these factors are largely dependent on the sample and on the environment.

Some limitations to dating are related to the presence of different phenomena during all the steps of the process (irradiation and electron accumulation in the traps, stimulation, and recombination). Nonradiative recombination at nonluminescent centers reduces the efficiency of luminescence production (e.g., Chen and Hag-Yahya 1997). On the other hand, the presence of shallow traps with a short mean life allows for competition in trapping, with a quick detrapping of electrons with short-time luminescent emission (normal fading). This effect can induce overcounting in readouts of laboratory-dosed samples immediately after irradiation.

Moreover, the presence of different depth-localized levels allows, in some cases, the tunnel transfer of electrons from the long mean life luminescent traps to different traps, leading to anomalous fading (Visocekas 1979, 1993; Tyler and McKeever 1988; Spooner 1992, 1994; Shepherd 1997). Such fading makes samples unsuitable for dating or requires a correction procedure to be applied to compensate for the loss. Laboratory irradiation and heating can, in turn, modify the behavior of the crystals with sensitization phenomena (Sunta et al. 1994; McKeever et al. 1996). The behavior of the sample during the natural trapping process may then differ from the behavior during the irradiation-readout cycles that are necessary for dating.

18.3. IONIZING RADIATION

The laws regulating the decay processes of the natural radionuclides (^{235}U, ^{238}U, ^{232}Th, ^{40}K and ^{87}Rb), which supply the ionizing dose to the crystals used for dating, are discussed in Chapter 10. The different types of ionizing radiation have influenced the development of dating and preparation methods for fine and coarse grains.

Heavy charged particles (α radiation emitted by the radioactive elements of the U and Th decay chains), light particles (β particles emitted by the U, Th, K, and Rb radioactive chains), and electromagnetic radiation (γ rays emitted by U, Th and K) lose their energy through various interaction mechanisms (Bleurer and Goldsmith 1959; Knoll 1989). In the case of heavy α particles, these interactions lead to the total loss of energy in a very short track length, practically linear, and common to all particles of the same energy (Bowman 1982). In the case of α particles emitted from the U and Th chains, the mean range is about 30 μm. Beta particles, charged but very light, undergo an energy loss characterized by notable variations in their direction, with a specific energy loss per unit path length much lower than that of α particles. It is, however, possible to define an extrapolated range that, for pottery and for the maximum energies of the particles emitted, does not exceed about 2 mm in thickness. Gamma radiation, because of its electromagnetic nature, behaves differently and transfers its energy to those particles that are present in the solid through single interaction mechanisms: photoelectric, Compton, and, at high energies, pair production. After these interactions, the original radiation disappears with an exponential absorption law, making it impossible to define a limiting range.

Some practical consequences are connected with these mechanisms of energy loss. Alpha particles have a range of penetration that, at equal energies, is lower than that of β particles by about two orders of magnitude. Thus, while β and γ (the interaction of the γ radiation has the primary effect of high-energy electron production) lose their energy in large volumes and the luminescence efficiency for exposure to β or γ radiation can be considered equivalent, the α particles produce high ionization density, along very short linear tracks, with luminescence saturation. For equal absorbed doses, the α radiation dose component has a reduced production of luminescence (Singhvi and Aitken 1978). Thus, an efficiency correction coefficient, k_α or a (Aitken 1985, 1998), must be used, and $\dot{D}_\alpha$ in equation (18.3) must be substituted by $k_\alpha \dot{D}_\alpha$, the effective α component.

Evaluation of the rate at which energy is absorbed by a crystal would therefore require the determination of the energy loss rate of the radiation at each of the points of the environment considered. This would make the determination practically impossible, but by adding the "infinite matrix" assumption to environmental homogeneity and uniformity conditions, it is possible to assume that in any environmental volume the energy released is equal to the energy absorbed. The principle of energy conservation indeed suggests that the rate of energy absorbed per unit mass in any volume of an infinite homogeneous matrix (with respect to the particle range) is equal to the rate of energy emission per unit mass.

Dimensions greater than some tens of micrometers constitute an infinite matrix for α particles, and dimensions greater than some millimeters make an infinite matrix for β particles. With regard to γ radiation, although a finite absorption range cannot be defined, it is possible to calculate that, for the γ energies present and for the atomic numbers usually involved in the absorbing matrix, a spherical region of the order of a few tens of centimeters surrounding the absorbing luminescent crystal includes more than 95% of the γ dose. To evaluate the annual dose components, it is sufficient to find the concentrations of the radioactive elements in the decay chains, and the cosmic radiation dose rate, present in the sample/environment system. The various contributions to the annual dose are obtained using concentration/dose conversion factors (Nambi and Aitken 1986; Olley et al. 1996; Adamiec and Aitken 1998).

The chemical-physical characteristics of the elements of the U and Th chains, however, make it necessary to verify whether secular equilibrium exists. The presence of a gaseous element in the radioactive chains suggests

Table 18.1. Annual dose components, quoted in micrograys per year, for 1% of natural K, 50 parts per million (ppm) for natural Rb, 1 ppm of natural Th, and 1ppm of natural U. Values from Adamiec and Aitken (1998).

	α	β	γ
Potassium	—	782	243
Rubidium	—	19	—
Thorium			
Full chain	732	27.3	47.6
Pre-thoron (^{220}Rn)	305	9.1	17.8
Uranium			
Full chain	2780	146	113
Pre-radon (^{220}Rn)	1260	60	4.4

the possibility that physical phenomena, such as the diffusion of gas in solids or its solubility and transport in absorbed water, could cause the total or partial loss of one intermediate element of the chain necessary for the maintenance of secular equilibrium (Aitken 1985, 1998). This appears more probable for radon (^{222}Rn in the ^{238}U chain) because of its mean life (3.82 days), which is comparable to the time required for substantial diffusion, than for the other two gaseous elements, which have very short mean lives (55 sec for thoron and 4 sec for actinon) (Meakins et al. 1978).

Disruption of the radioactive chains could also be caused by natural geochemical processes involving other elements in the chains (Ivanovich and Harmon 1992). High-resolution α or γ spectroscopy (NCRPM 1976, 1985; Martini et al. 1983; Murray et al. 1987; Knoll 1989) is usually used to determine the activity of all (or the majority of) the elements in the chain independently from one other. Table 18.1 shows the concentration/dose conversion factors calculated for α, β, and γ components relative to the content of U, Th, K, and Rb, with and without total radon and thoron escape (Adamiec and Aitken 1998). Due to the very high difference in the dose rates for the two situations, miscalculations in the luminescence ages could arise from the incorrect determination of the radioactive disequilibrium induced by total or partial radon escape, or by the absence of other elements.

Moreover, modifications to the absorbed dose can arise as the result of various phenomena correlated with the interaction of radiation with different substances present (e.g., water). These may modify the conditions of energy transfer, as well as variations in the luminescence efficiency (attenuation of the radiation dose in the crystal itself), and must be taken into consideration (Aitken and Alldred 1972; Aitken 1976). Water present in the environment surrounding the crystal leads to a reduction in the energy per unit mass absorbed by the crystal. This reduction in the effective dose can be evaluated from the mean water content of the sample during its period of burial and the appropriate correction factors for α, β, and γ dose rates (Zimmerman 1971; Aitken 1985) can be introduced.

Also, β radiation energy loss causes a progressive reduction in ionization. Because of the continuous spectrum of β radiation, crystals with dimensions greater than some tens of micrometers show a reduction in the efficiency of luminescence production. Dose attenuation factors directly related to the grain size have, therefore, been calculated, and are used for grains larger than 100 μm in diameter (Bell 1979a; Mejdahl 1979).

18.4. EXPERIMENTAL METHODS

18.4.1. Sample Preparation

The β dose rate contribution, and to an even greater extent the α contribution, from the sediment surrounding the sample (if, as in the case of pottery, this can be distinguished from the sample) can be excluded by removing the surface layer of the sample itself to a depth of about 2 mm; and the γ contribution, except in particular cases, may be determined with very low error by analysis of the external sedimentary matrix (about 30 cm in radius) surrounding the sample (Hubbell 1977). It must be noted, however, that in some cases (for example, sandy sediments) the sample matrix and the sediment matrix are identical. Eq. (18.4) therefore assumes different forms, both for the dose rate components and the correction factors (water, fading, attenuation), depending on the different experimental situations.

Moreover, the luminescent crystalline grains can vary in size from micrometers for clay to millimeters for coarse sand. Comparison of the absorption processes of the various dose rate components and the mean ranges of the radiations thus leads to the development of two distinct methodologies. The first refers to silt-sized (fine) grains with diameters of between 2 to 4 μm and 8 to 10 μm, much less than the α particle mean range, so the α particles, in general, pass through the crystals (Aitken et al. 1964, 1968). For fine grains, the heterogeneous crystalline phases generally present in the samples are not easily separable into discrete mineral species, and so they do not permit optimum selection for TL or OSL stimulation. The second strategy refers to large grains (greater than about 100 μm in diameter) for which the absorption of α radiation occurs only in the external shell of the crystal and does not involve the inner region (Mejdahl 1969). For large grains, moreover, a process of mineralogical selection makes it possible to use mineral-specific TL and OSL stimulation procedures.

Figure 18.4 shows some important phases of the preparation procedure for the selection of the required grain sizes and crystalline phases. A condi-

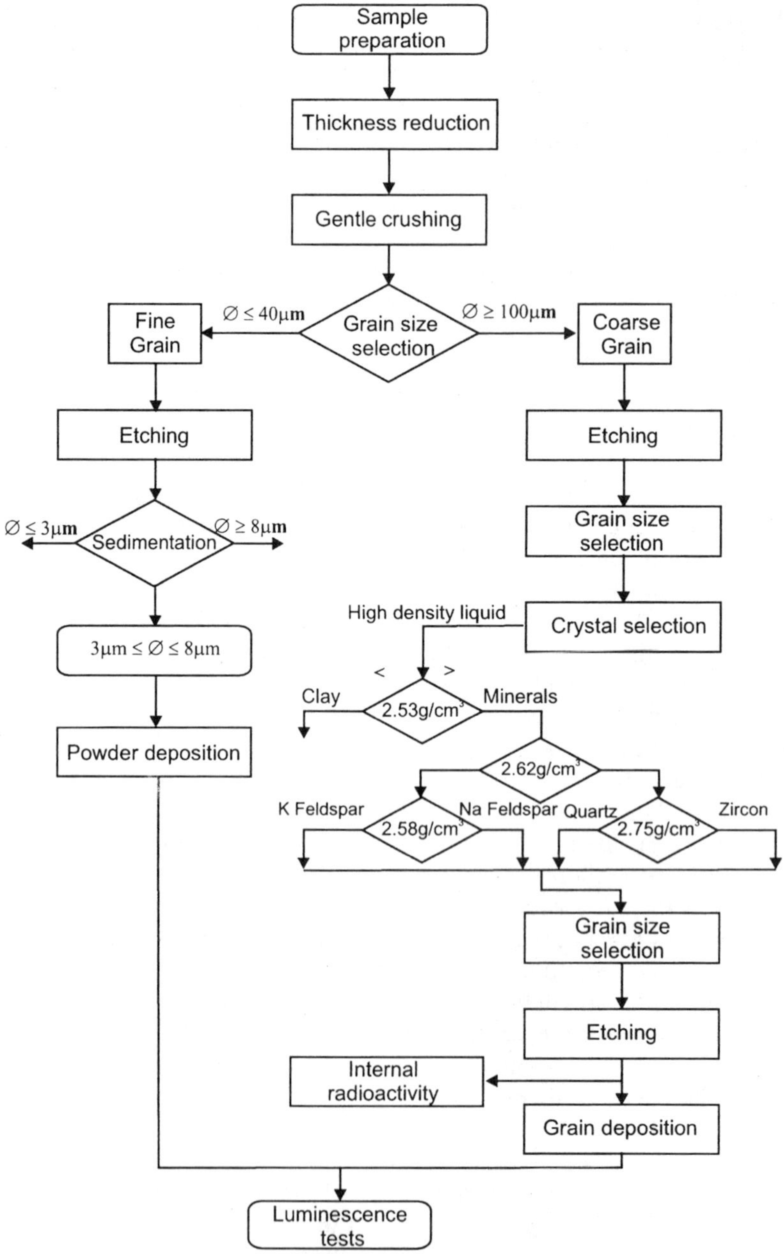

Figure 18.4. Flow chart of sample preparation procedures. Grain size and crystallographic phase separation steps are shown for TL and OSL dating using fine and coarse grains.

tion common to the collection of the samples, and to all of the preparation methods, is that the crystals are not exposed to light (avoiding optical bleaching) nor broken (avoiding reduction in the grain dimensions), and do not undergo mechanical stresses that could induce triboluminescence.

Preliminary measurements related to the determination of some of the annual dose components are common to the different methods. These include *in situ* measurement of the cosmic-ray dose rate component and the *in situ* integral or spectroscopic measurement of the environmental dose rate components (Mejdahl 1970; Murray et al. 1978; Bell 1979b). The collection of samples suitable for laboratory measurements of the dose rate components and the evaluation of correction parameters, and the collection of samples and sediments without humidity loss for water content measurement, are also necessary.

The preparation of fine-grain samples does not present particular difficulties. For the very low quantity of material required for analysis, grams of sample are usually sufficient to prepare the necessary fine-grain phase used for preliminary luminescence tests, the dating of fine artifacts, and authenticity testing (Aitken et al. 1971). After gently crushing the sample and a preliminary sieving to eliminate the coarser phase, the material is washed and etched with H_2O_2 and dilute hydrochloric acid to eliminate the organic and carbonate components. The fine-grain residue of mixed crystals present in an archaeological artifact, or in fine sediment, is then separated by sedimentation (Stokes's law) in liquid, and deposited as a thin layer (almost a monolayer) on aluminum disks for luminescence measurement.

The preparation of the coarse-grain fraction is more complex and, in general, needs tens of grams of raw sample, so that the *coarse-grain* technique is usually applied when the availability of large quantities of sample is not a problem. The α dose rate contribution received by a crystal from the external matrix can be excluded by removing the surface layer by hydrofluoric (HF) acid etching of the crystal to a depth of about 30 μm, without any substantial variation in the β attenuation. This procedure avoids the determination of the coefficient of luminescence efficiency k_α (Aitken and Bowman 1975), essential in the fine-grain procedure.

The separation of the crystalline phases using high-density liquids is only moderately selective. Thus polarized light microscopy, X-ray diffraction, or other morphological and/or crystallographic observations are usually performed to verify the mineral composition and grain size distribution of the prepared sample (Bell and Zimmerman 1978; Goedicke 1984).

18.4.2. Luminescence Measurement

Quartz crystals are often the principal constituent of the fine-grain fraction separated from fired artifacts (i.e., pottery, brick, terracotta) or fine-grain sediment, but they also represent the most commonly used coarse-grain crystals extracted from artifacts. The luminescence emissions from fine-grain samples composed of mixed quartz and feldspar crystals may, however, be dominated by the signals from feldspars, which are more luminescent than quartz. This problem does not occur in coarse-grain dating, because the quartz and feldspar crystals are routinely separated before measurement.

Many quartz TL peaks are reported in the literature, some of them common to all quartz crystals. The temperature of the two peaks useful for dating, 325 and 375°C (Wintle 1997), limits the maximum temperature necessary for TL detection to about 500°C, with a heating rate of up to about 20°C/sec. A third peak occurs at a low temperature, about 110°C, but is detectable only during readout soon after laboratory irradiation because of its high thermal instability. It shows very characteristic behavior in terms of TL sensitivity to the received dose (Fleming 1973), so allowing the development of the "predose" method of dating.

The 375°C peak is stable at room temperature (mean life 10^6 to 10^8 years; Aitken 1985) and is considered to be the most suitable for dating of heated materials, with a strong emission at about 470 nm. This peak is not, however, as light sensitive as the 325°C peak, which is also stable for millions of years. Because of their very different susceptibilities to light (Spooner et al. 1988), the 375°C peak is often called the "hard-to-bleach" TL signal, whereas the 325°C peak is known as the "easy-to-bleach" TL signal.

The OSL emission from quartz has been measured for many excitation wavelengths (Godfrey-Smith et al. 1988), with a strong emission in the near-ultraviolet (Huntley et al. 1991). This emission is associated with the easy-to-bleach TL peak at 325°C (Smith et al. 1986; Spooner 1994; Kaylor et al. 1995; Wintle and Murray 1997). The existence of thermally unstable peaks at lower temperatures, and the equalization of radiative recombination probabilities in the natural and laboratory-irradiated subsamples, makes preheating necessary before OSL readout (Wintle and Murray 2000). No anomalous fading has been reported for quartz, and, in general, quartz crystals do not have a high radioactive content, so the internal dose rate contribution is usually disregarded (Bell 1979b).

Potassium-rich feldspar grains also present two TL peaks at high temperatures (280 and 330°C) and one at a low temperature (150°C) (Mejdahl 1988; Duller 1997; Krbetschek et al. 1997). The two high-temperature peaks have sufficient thermal stability for TL dating, using the blue-violet emissions, but the 150°C peak is unstable. The 330°C peak has a mean life of 10^6 years (Strickertsson 1985), but many authors have reported anomalous fading of this peak (e.g., Wintle 1973).

With regard to OSL, studies have been made at excitation wavelengths of up to 800 nm, with emissions generally below 500 nm and centered at about

400 nm (Krbetschek et al. 1997). The presence of a strong emission with stimulation in the infrared (IR) region led to the development of an infrared stimulated luminescence (IRSL) dating methodology for feldspars (Hütt et al. 1988; Huntley et al. 1991). This allowed the differentiation of feldspar from quartz and the application of different dating techniques for quartz, where green (and/or blue) light appears suitable for optical stimulation. While feldspars still show a strong IR-stimulated emission in the blue-violet region, the presence of emissions associated with unstable traps has suggested preheating at temperatures above 200°C. The presence of a significant internal potassium content makes it necessary also to account for the β dose rate contribution from the crystal itself.

Because of their considerable U and Th contents, zircon crystals are the only crystals that allow "self-dating." This is because the internal contributions to the dose rate are so high, in many cases, that contributions from general variations in the environmental dose rate become negligible. Thus the annual dose calculation must include $\dot{d}_{crys}$ in equation (18.4). These crystals, however, present a notable inhomogeneity in the distribution of the radioactive impurities and nonuniformity within and between the grains, so that, in practice, single-grain dating is usually required (Aitken 1985). TL dating has been feasible on some occasions, but it is often difficult to separate the zircon crystals and they also tend to suffer from anomalous fading. OSL dating of zircons has also been performed in conjunction with TL dating (Smith 1988).

Although proposed some time ago, calcite has allowed effective dating in only a few cases, despite the intense OSL and TL emissions shown by these crystals (Wintle 1978; Debenham and Aitken 1984; Aitken 1985; Ugumori and Ikeya 1980). The variability of the TL and OSL responses has restricted its widespread application, but calcite appears to be a good candidate for future developments and progress has been made in this direction (e.g., Liritzis et al. 1996b).

18.4.3. Measurement Systems

The two terms of the age equation (18.1) refer to different experimental methods: the numerator, regarding the luminescence emissions, is related to solid-state experimental techniques, while the denominator is related to environmental radiation measurement techniques.

18.4.3.1. Luminescence Emissions

The wide application of luminescence dating methods has stimulated the production of many systems suitable for TL and/or OSL measurements (Bøtter-Jensen and Bechmann 1968; Bailiff and Younger 1988; Bøtter-

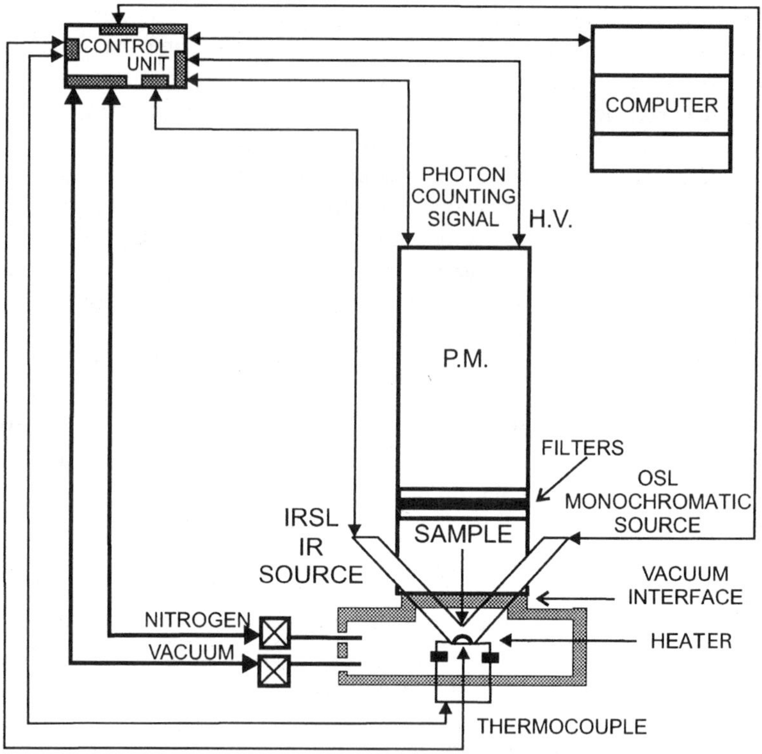

Figure 18.5. Computer-controlled system for measurement of thermally stimulated luminescence and optically stimulated luminescence.

Jensen, 1988, 1997; Hütt and Jaek 1990; Bøtter-Jensen et al. 1991, 1999a,b; Galloway 1991, 1994; Bøtter-Jensen and Duller 1992; Poolton et al. 1994).

Due to the stimulation characteristics common to the crystals most often used in dating, certain performance criteria are required of the standard TL/OSL systems. A temperature range of up to 500 to 700°C, a heating rate of up to 20°C/sec, green and/or blue light (filtered lamp light, light-emitting diodes, or laser beam) for optical stimulation of quartz crystals, and infrared radiation (normally IR light-emitting diodes laser diode) for stimulation of feldspar crystals are found in many of the commercial and custom-made systems. Also fitted are infrared-reflecting and ultraviolet/violet bandpass filters for the light emissions used by fine- or coarse-grain TL and OSL techniques. Figure 18.5 shows the principle scheme of an integrated (TL, IRSL, and OSL) computer-controlled system.

18.4.3.2. Radioactive Elements

With regard to the annual dose components, numerous methods are presently used to determine the radioactivity of sample, sediment, or crystal in thermoluminescence and optical dating. The components produced by α, β, and γ radiation, which result from the nuclear decay of isotopes in the decay chains of ^{238}U, ^{232}Th, ^{235}U, ^{40}K, and ^{87}Rb, and their relative importance with regard to the dating method used (fine or coarse grain), have led to the development and application of different methods of spectroscopic and integral measurement. These generally have the aim of evaluating the α, β, and γ components of the dose rate, either by identifying the concentrations of the radioactive elements (e.g., Meakins et al. 1979; Chowdhary et al. 1982; Mangini et al. 1983; Bøtter-Jensen and Mejdahl 1985; Goedicke 1988; Prescott and Hutton 1988; Guibert and Schvoerer 1991; Hutton and Prescott 1992) or by comparison with standards of known dose rate (e.g., Jensen and Prescott 1983; Bøtter-Jensen and Mejdahl 1988; Prescott and Hutton 1988; Sanderson 1988; Sanzelle et al. 1988; Hutton and Prescott 1992).

The undoubted advantages of high-resolution γ spectroscopy, which allows the activities of individual radioactive elements to be determined and can thus also provide information on the state of radioactive equilibrium of the ^{238}U and ^{232}Th (and, if necessary, ^{235}U) chains, are limited by its high cost. This is even more true when the amount of material available for analysis is reduced to only a few grams for samples, or a few milligrams for crystals, leading to the need for systems with high efficiency, low noise, and long signal accumulation times (Chowdhary et al. 1982; Murray and Aitken 1982; Murray et al. 1987; Guibert and Schvoerer 1991).

Measurements of α and β activity in thick layers are therefore made in many laboratories because of the very small amounts of material needed, the low cost of high-efficiency counters, and the low background levels obtained using normal shields and techniques (Aitken and Bowman 1975; Aitken 1978; Murray 1982; Bøtter-Jensen and Mejdahl 1985, 1988; Sanderson 1988). Different systems are available: scintillation counters for α activity (Aitken and Bowman 1975; Goedicke 1988), and gas counters (Geiger–Muller or proportional counters) for β activity (Bøtter-Jensen and Mejdahl 1988; Sanderson 1988) and α plus β activities (Troja et al. 1995).

Alternatively, uranium and thorium can be measured by neutron activation, and potassium by X-ray fluorescence, or all three elements by inductively coupled plasma mass spectrometry or other methods. However, these methods usually measure only the parent concentrations, so any problems of disequilibrium in the uranium and thorium decay chains are not detected. The U and Th concentrations are usually compared, therefore, with the concentrations of the daughter products in the lower part of the decay chains. These are measured using other methods, and concordance between the parent and daughter concentrations is generally taken to indicate that the decay chains are presently in a state of secular equilibrium.

18.5. PALEODOSE DETERMINATION

18.5.1. Laboratory Irradiation

It is necessary to characterize the dose–response behavior of each sample so that the paleodose can be determined from a direct comparison of the natural luminescence with that induced by a known dose (delivered using calibrated laboratory sources). Different methods are used to define the rate of increase of luminescence intensity with added dose.

$^{90}Sr/^{90}Y$ β sources, with activities of 10 to 100 mCi, are most often used for laboratory irradiation, and the dose delivered to the sample must be calibrated for the grain size and mineralogy of the crystals under investigation. The types of material that hold and surround the crystals during the irradiation also need to be included in the calibration exercise because scattering and backscattering from these materials influence the effective dose absorbed by the crystals (Aitken 1985).

For the fine-grain technique, it is necessary to evaluate the efficiency of luminescence production by α particles, compared to β or γ radiation. This requires access to a calibrated α source. The luminescence obtained after exposure to a known α dose is compared with that induced by a known β (or γ) dose, and the resulting ratio for α efficiency is introduced as a correction factor (k_α) into the dose rate determination for fine grains (Aitken 1985). The α-efficiency factor can be largely ignored for coarse grains because their α-irradiated rinds are mostly removed during sample preparation, by etching in HF acid.

18.5.2. Preheating

After laboratory irradiation, and before measurement of the luminescence signals, it is standard practice to "preheat" the sample by holding it at an elevated temperature for a period of time (Wintle 1997; Aitken 1998). The same preheat is given also to those portions of the sample that have been kept aside to measure the natural TL or OSL signals, and the signals remaining after bleaching. Many different combinations of temperature and time have been tried, the choice being dependent on a number of factors including the mineral species under investigation and the age of the sample.

The preheat is given to empty traps with inadequate thermal stability, to reduce any age shortfall due to anomalous fading (for feldspars), and to equalise the probability of radiative recombination in the natural, bleached, and laboratory-irradiated portions of the sample. The latter may be achieved by means of a preheat "plateau test", using many separate portions of the same sample. The temperature is held constant and the duration is increased, or vice versa, and the paleodose is obtained from the preheat plateau region.

In optical dating, the paleodose can be obtained from a single portion of sample. Sensitivity changes in quartz arising from changes in radiative recombination probability are measured during the dating procedure, and these measurements provide the means to correct the OSL data (Murray and Roberts 1998; Wintle and Murray 2000).

18.5.3. Additive Dose and Regenerative Dose Protocols

The paleodose is usually determined using one or both of two protocols, known as the *additive dose* and *regeneration* procedures. Comprehensive and up-to-date accounts of these and other protocols are given by Aitken (1985, 1998) and Wintle (1997).

The additive dose method is most commonly used for heated materials such as fired pottery and ceramics. With this protocol, the natural TL or OSL signal is measured from a set of subsamples (aliquots) that have not been heated, bleached, or dosed in the laboratory. Other aliquots are given a range of additional laboratory doses (usually from a β source) and are then thermally or optically stimulated, during which the TL and OSL emissions are measured. The luminescence intensities of the natural and laboratory-irradiated aliquots are then plotted against the applied laboratory doses (Fig. 18.6). An initial estimate of the paleodose is obtained by fitting a best-fit line (a "growth curve") to these dose–response data and extrapolating the line to where it intersects the horizontal axis. This represents a first approximation of the dose that has been acquired by the sample since the last heating or bleaching event.

The situation is often more complicated than described above, for three reasons. First, an explicit correction is needed for any supralinearity in the low-dose portion of the growth curve; this correction is essential for young samples but is less important for older samples. The latter are more prone to the second common complication: the tendency for growth curves to saturate at high doses. This requires the data to be fitted with a suitable type of nonlinear growth curve. Third, extrapolation of the best-fit line to the horizontal axis may not be appropriate for unheated materials because the last episode of sunlight exposure may not have been of sufficient intensity or duration to zero the TL or OSL signals. In this case, additional experiments

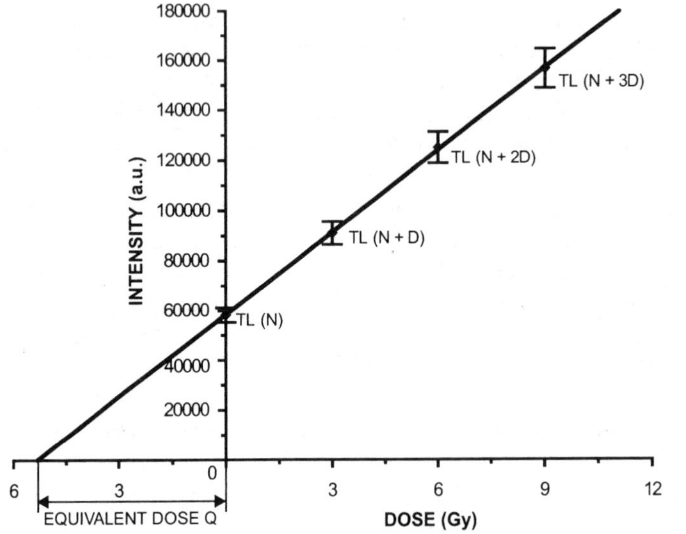

(a)

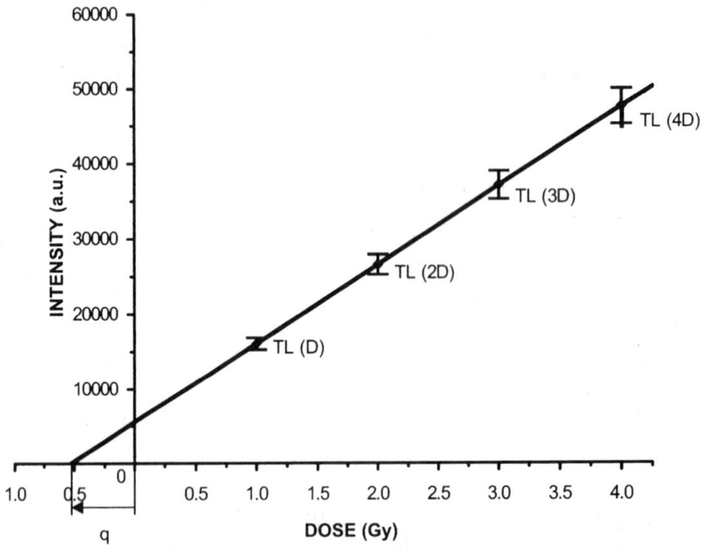

(b)

Figure 18.6. Additive dose method: (*a*) The β equivalent dose (paleodose) determination, where N is the natural luminescence, and $N + D$, $N + 2D$, and $N + 3D$ are subsamples that have received the natural dose and additional β doses in the laboratory. (*b*) Nonlinearity correction calculation in the low-dose region.

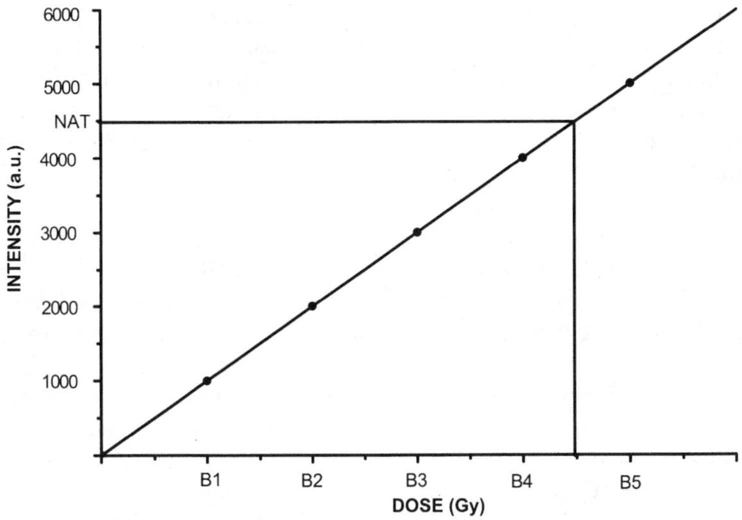

Figure 18.7. Regeneration method. Most subsamples are first emptied of their natural dose (by heating or bleaching) and are then given a range of applied doses that should bracket the natural dose. The paleodose is obtained from the intercept of the regenerated dose–response curve with the natural luminescence intensity (measured from a number of subsamples that have not been heated, bleached, or dosed in the laboratory).

need to be made to estimate the likely magnitude of the past "residual" TL or OSL level; bleaching of the material using filtered sunlight or the examination of a modern sample deposited under similar conditions are two such tests.

The regeneration protocol avoids any complications due to supralinear growth in the low-dose region. It also avoids the problem of extrapolating growth curves when the natural luminescence intensity is close to dose saturation. With this procedure, a small number of natural aliquots are thermally or optically stimulated to measure their natural TL or OSL signals. The remaining aliquots are either heated or bleached to reset the luminescence signals, and then given a range of laboratory doses sufficient to encompass the natural dose. The TL or OSL data obtained from these regenerated aliquots are then fitted with a growth curve, and the paleodose is calculated by matching the natural luminescence intensity with the regeneration dose needed to produce the same luminescence intensity (Fig. 18.7).

However, the regeneration method is not without problems. Large changes in luminescence sensitivity may arise as a result of the initial heating or bleaching step used to construct the regenerated growth curve. This may lead to an incorrect estimate of the paleodose. To recognize such problems, many researchers use the additive dose and regeneration procedures in tandem, to check that both yield the same paleodose. This analysis is done most efficiently by sliding the additive dose curve along the horizontal axis to check that it coincides with the regenerated growth curve (e.g., Huntley et al. 1993; Feathers 1997c; Smith et al. 1997; Folz and Mercier 1999a). If it does, then the two data sets are combined to calculate the paleodose.

The regeneration protocol is also subject to the same uncertainty as the additive dose method about the degree to which the TL or OSL signals were reset by the last heating or bleaching event. But application of the "plateau test" provides one means of checking that the dating signals had been fully zeroed by the last event. This test (Aitken 1985) is routinely performed on samples that are thermally stimulated, a plot being obtained of paleodose versus glow curve temperature. For quartz, the same paleodose should be obtained for temperatures above 250 to 300°C, because only these TL peaks are associated with electrons that have been trapped for geologically long periods of time. Quartz has two useful TL dating signals in this high-temperature region: the easy-to-bleach peak at 325°C and the hard-to-bleach peak at 375°C. If both signals had been fully zeroed by sunlight or by heating in the past, then they should yield the same paleodose. But if the 325°C peak gives a smaller paleodose, then the most likely explanation is that the last heating or bleaching event did not reset the 375°C signal. The plateau test is also carried out on burnt flints, the aim being to check that the flints had been previously heated to at least 450°C (Aitken and Valladas 1992; Mercier et al. 1995a).

18.5.4. Single and Multiple Aliquots

Until recently, several aliquots of each sample were required to determine its paleodose. This restricted the range of possible applications for luminescence dating to samples with abundant crystals. Moreover, to obtain the correct age using multiple aliquots requires that all of them have identical luminescence characteristics as well as identical histories of exposure to light, heat, and environmental radiation. The requirements for sample heterogeneity are almost never met in river-deposited sediments because only some of the water-transported grains are well bleached before burial, while others are partially bleached or retain their geological dose (e.g., Murray et al. 1995; Olley et al. 1998, 1999).

Over the past decade, therefore, optical dating protocols have been developed that require very few, or individual, aliquots for paleodose determination. The additive dose and regeneration protocols are still used, but these have been modified so that the paleodose may be estimated from a single aliquot, or even from a single grain. This permits the degree of sample heterogeneity to be examined directly, from the spread in paleodose values obtained from individual aliquots (e.g., Duller 1994; Li 1994; Murray et al. 1995; Clarke 1996; Olley et al. 1998, 1999; Murray and Olley 1999). Studies such as these have shown that, for many grains, the optical dating signal is often not reset at the time of sediment deposition; under such adverse conditions, less light-sensitive TL signals are even less likely to have been zeroed.

The first single-aliquot protocols for feldspars were proposed by Duller (1991), and several variants of these are now in common usage (e.g., Duller 1995; Galloway 1996). Mejdahl and Bøtter-Jensen (1994) and Tso and Li (1994) developed protocols for quartz crystals and fine grains, respectively, that required very few aliquots, and true single-aliquot additive dose and regeneration methods are now available (Murray et al. 1997; Murray and Roberts 1998; Murray and Mejdahl 1999; Folz and Mercier 1999b; Galbraith et al. 1999; Murray and Wintle 2000). These single-aliquot protocols have recently been extended to single grains of feldspar from geological deposits in Canada (Lamothe et al. 1994; Lamothe and Auclair 1997, 1999) and to individual grains of quartz from archaeological and geological sites in Australia (Murray and Roberts 1997; Roberts et al. 1997, 1998a, 1998b, 1999; Olley et al. 1999; Yoshida et al. 2000).

18.6. APPLICATIONS IN ART AND ARCHAEOLOGY

Luminescence dating methods have been applied to a variety of critical chronological questions in the fields of art and archaeology and at timescales ranging from the last few hundred years to the past several hundred millennia. At the near end of this age range, luminescence methods have been used for authenticity testing of fired works of art presented for sale and for dating of historic and protohistoric pottery found at archaeological sites. At the far end of the age range, luminescence dating of burnt flint tools and unheated artifact-bearing sediments has made some of the most significant and controversial contributions to the chronology of modern human origins and prehistoric artistic activities. We discuss some of these applications below, and further examples may be found in other recent reviews (e.g., Aitken 1997; Feathers 1997a; Roberts 1997).

18.6.1. Heated Objects

18.6.1.1. Authenticity Testing

For works of art that have been fired, in whole or in part, during the process of creation, TL dating has been employed to authenticate or refute their claimed date of manufacture. It is usually possible to establish beyond reasonable doubt that a supposedly ancient object is not, in fact, a modern copy if there is at least a twofold difference in age between the genuine artifact and later copies (Stoneham 1991). But detection of fakes may not be possible if the age gap is too small, such as with nineteenth-century reproductions of eighteenth-century French ceramics (Stoneham 1991) and sixteenth-century German stoneware (Gaimster 1992).

Authenticity testing has been carried out on a large number of fired pottery and ceramics, as well as porcelain and stoneware (Stoneham 1991; Gaimster 1992), the clay cores of bronze casts (Zimmerman et al. 1974; Fleming 1979), and Renaissance terracotta sculptures (Fleming and Stoneham 1973b; Fleming 1979). Fine-grain (Zimmerman 1971) and quartz inclusion (Fleming 1970) methods are most often used to exploit luminescent minerals such as feldspar and quartz, but some objects may require more specialized methods, such as the predose technique (Fleming 1973; Bailiff 1994) or the use of zircon grains (Zimmerman et al. 1974). The focus of current research is the development of TL techniques that demand smaller quantities of sample (Michael et al. 1997) and the application of OSL and IRSL methods that require only a single subsample for dating (Liritzis et al. 1994; Tso and Li 1994).

As art objects have been exposed to a more or less unknown external ionizing dose since their creation, the calculated ages have typical 1σ errors of $\pm 25\%$ (Aitken 1990; Stoneham 1991). But this level of precision is usually adequate to distinguish modern fakes from ancient works of art, and uncertainty in the external dose can be minimized by using zircons (where the dose rate is dictated by the uranium and thorium content of the grains; Zimmerman et al. 1974), sand-sized grains of feldspar (where the internal dose rate from potassium is dominant; Mejdahl 1983), or the subtraction technique for fine grains and quartz inclusions (Fleming and Stoneham 1973a).

Application of these methods has resulted in the identification of a series of forgeries, such as the Hacilar ceramics from Turkey (Aitken 1985), funerary wares allegedly made during the T'ang Dynasty (Fleming 1975), some supposedly Etruscan statuettes (Ralph and Han 1971), numerous bronzes and Hui Hsien style pottery from China (Fleming 1979), and inscribed tiles claimed to date from the final period of Roman occupation of Britain (Fleming 1975). Baked objects (including inscribed tablets) from the infamous site of Glozel in France have also been subject to authenticity testing (Fleming 1979; Aitken 1985). Here, objects considered as modern forgeries by archaeologists gave ancient TL ages, but a final resolution to this conundrum is anticipated (Stoneham 1991).

18.6.1.2. Archaeological Dating

Daniels et al. (1953) were the first to propose that TL methods held the potential for the dating of ancient pottery, and this line of inquiry was pursued by research groups in England, Japan, the United States, and Denmark during the 1960s. By the end of that decade, encouraging signs of success had been reported from TL tests on pot sherds of known ages (based on pottery style, site stratigraphy, and uncalibrated ^{14}C ages) of up to 8000 years (Aitken et al. 1964, 1968; Ichikawa 1965; Mazess and Zimmerman 1966; Ralph and Han 1966, 1971; Mejdahl 1969).

Continued technical refinements (Fleming 1970; Zimmerman 1971) led in two main directions: on the one hand, to the testing of art objects of questionable authenticity (see above) and the validation of ^{14}C ages for controversial archaeological finds, such as those for Yotoco style pottery in Colombia and Jomon pattern pottery in Japan (Fleming 1979); and on the other hand, to TL dating of heated artifacts produced before the advent of pottery and close to or beyond the age limit of ^{14}C dating. Zimmerman and Huxtable (1971) made the first such exploration into the Paleolithic when they obtained TL ages of about 33,000 years for burnt clay lumps associated with human figurines at Dolni Vestonice. They were swiftly followed by applications to other heated artifacts, including burnt flint tools (Göksu et al. 1974) and cooking stones (Huxtable et al. 1976; Huxtable and Aitken 1977).

TL dating of pottery continues to provide valuable chronological frameworks for hitherto poorly dated styles in many regions of the world. Recent examples from the Americas include a 7000-year-old pot sherd from the Brazilian Amazon (Roosevelt et al. 1991), the pre-Hispanic classic and postclassic pottery in western Mexico (Duverger et al. 1993), the St. Croix ceramics from eastern Canada (Godfrey-Smith et al. 1997), and fifteenth- to nineteenth-century brown ware ceramics from the Great Basin (Feathers and Rhode 1998). One of the Great Basin sherds was also dated using OSL, in the wake of earlier such studies of burnt bricks, stones, and ceramics from archaeological sites in Scandinavia (Mejdahl and Bøtter-Jensen 1994, 1997) and the Americas (Feathers 1997a). Optical dating holds several advantages over TL dating of such materials, such as the ease and rapidity of analysis, the improved precision of paleodose estimates, and the requirement for lesser amounts of sample using single-aliquot methods (Murray and Mejdahl 1999).

Pottery found on or near the ground surface is often better suited to luminescence dating than to other numerical-age methods, such as ^{14}C dating, because the object being dated is also the "target" archaeological event, whereas with ^{14}C dating there is the question of association between the artifact and the dated organics (Dunnell and Feathers 1994; Feathers 1997a). There are, however, complicating factors that may hinder luminescence methods. For example, success in TL dating of pot sherds found on the Pacific Islands was strongly dependent on the source of the raw material used to make the pottery (Prescott et al. 1982; Huntley et al. 1983). As a consequence, it may be beneficial to also make mineralogical and petrographical studies of the pottery (e.g., Huntley et al. 1983; Troja et al. 1996).

Another potential problem is radioactive disequilibria in the ^{238}U decay chain, the most common forms being post depositional movement of uranium and radon. The former was found in ceramics from Niger in western Africa (Guibert et al. 1994) and also afflicts pottery from Ban Chiang in Thailand (Carriveau and Harbottle 1983; Aitken et al. 1983). Surface sherds in the Great Basin appear to have absorbed radon gas (Feathers and Rhode 1998), whereas Hong Kong pottery has been shown to suffer from an extreme loss of radon (Huxtable et al. 1982–1983). But provided any disequilibria are detected and the dose rate adjusted accordingly, then the calculated TL ages should not be seriously in error.

In addition to pottery, luminescence dating has been applied to a number of other heated materials of archaeological interest. The sintered remains of vitrified forts in Scotland (Sanderson et al. 1988; Strickertsson et al. 1988) and a Neolithic loam hut in Germany (Buschbeck and Flettner 1994) have been dated by TL, as have quartz grains extracted from volcanic ash deposited on the Aegean island of Yiali (Liritzis et al. 1996a). The latter indicated that this eruption took place in the second millennium B.C. and may have played a role in the decline of the Minoan civilization in Crete. By far the most important archaeological application, however, has been TL dating of burnt flints from western Europe and the eastern Meditteranean to obtain reliable ages for the key sites occupied by Neanderthals and anatomically modern humans.

18.6.1.3. Burnt Flints

Flint heated deliberately or accidentally during tool manufacture has proved to be well suited to TL dating, and has provided the means to investigate the timing of significant archaeological events that lie well beyond the reach of ^{14}C dating. Conversely, the initial results from optical dating of flints suggest that the OSL signal may not have sufficient stability for long-range dating (Poolton et al. 1995).

Burnt flints from Africa, North and South America, Asia, and Europe have been subject to TL dating (see review by Roberts 1997), but those dated from western Europe and the Levant (Figure 18.8) have made the greatest archaeological impact (Valladas et al. 1986, 1987, 1988, 1999; Mercier et al. 1991, 1995a, 1995b). The Levant is situated at the eastern end of the Mediteranean Sea and this region contains some of most important evidence concerning the evolution of modern humans and the demise of the Neanderthals. TL dating of burnt flints from the Neanderthal burial site of Kebara in Israel showed that Neanderthals had occupied this cave about 60,000 years

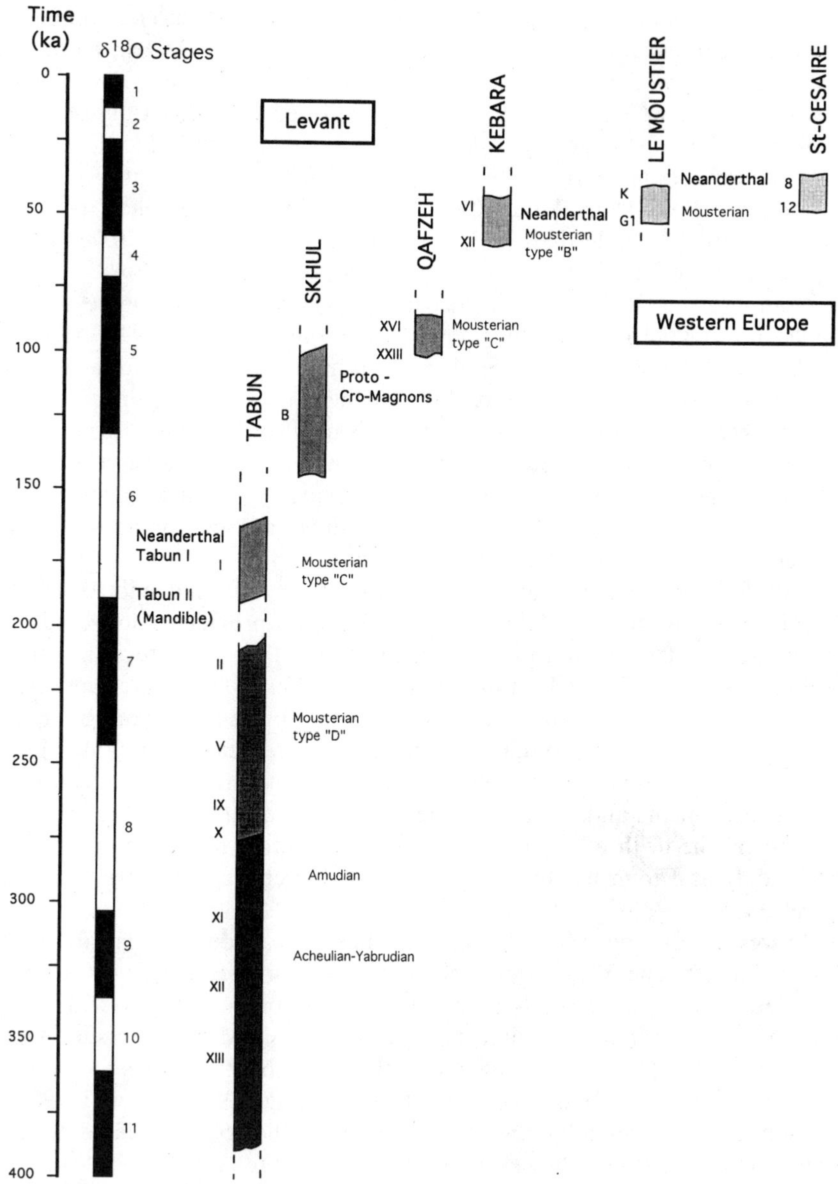

Figure 18.8. Summary diagram of the flint TL chronology constructed for some of the key Paleolithic sites in the Levant and western Europe. The annotations to the left of each bar indicate the archaeological levels, and the oxygen isotope scale is shown on the left-hand side of the diagram, next to the timescale which is marked in thousands of years. (Reprinted, by permission, from Mercier and others, *Quaternary Science Reviews*, Vol. 14, p. 361. Copyright © 1995 Elsevier Science Ltd.)

ago (Valladas et al. 1987)—30,000 years after the first fully modern humans, the Proto-Cro-Magnons, had settled at the nearby site of Qafzeh, based on the flint ages for the skeleton-bearing strata (Valladas et al. 1988). These results, both of which were subsequently confirmed by ESR dating of fossil teeth from the same sedimentary layers, forced a major revision to the theory that modern humans were descended from Neanderthals. Recent TL dating of 19 burnt flints from Amud Cave has confirmed the presence, perhaps intermittent, of Neanderthals in the Levant between 50,000 and 70,000 years ago (Valladas et al. 1999).

Confidence in the flint ages at Kebara and Qafzeh is further increased by application of the isochron technique (Aitken and Valladas 1992; Mercier et al. 1995a). From a plot of paleodose versus internal dose rate for a suite of contemporaneous flints, it is possible to calculate an age that is independent of the external dose rate. The isochron age is unaffected by variations in soil water content and by disequilibria in the uranium series because it is based solely on the internal radioactivity of the flints, which are impermeable (Aitken 1997). For the Kebara and Qafzeh flints, the isochron ages were in agreement with the individual TL age determinations.

Tabun Cave in Israel is the pivotal site for the Paleolithic cultures of the Levant, and here too TL dating of burnt flints has provided a longer chronology for modern human presence in the region than was previously supposed (Mercier et al. 1995b). Flint ages of up to 350,000 years are supported by isochron ages at this site, but they conflict with the much younger ESR and uranium series age estimates for fossil teeth. Mercier et al. (1995b) attributed this difference to the heterogeneous distribution of radionuclides in the cave sediments and to uncertainties in the rate of uranium uptake experienced by the teeth after burial. Taking such factors into account, a new ESR study at Tabun has found that the age discrepancy is greatly reduced (Rink 1998).

In western Europe, TL dating of burnt flints has made a similar archaeological impact. The Mousterian stone tool tradition in France is commonly associated with Neanderthals, and the Châtelperronian industry with anatomically modern humans. Flint dating at the classic Mousterian site of Le Moustier in the Dordogne showed that this rock shelter had been continuously occupied from about 56,000 to 40,000 years ago (Valladas et al. 1986), a result later confirmed by ESR dating of teeth. The youngest dated occurrence of Neanderthals, however, is at Saint-Césaire, where a TL age of $36,000 \pm 2700$ years was obtained for the layer containing Neanderthal remains and Châtelperronian tools (Mercier et al. 1991). This result not only challenged the notion that this industry was the exclusive handiwork of modern humans, but it also raised the prospect that Neanderthals coexisted with modern humans for several millennia in this corner of Europe (Mercier et al. 1991).

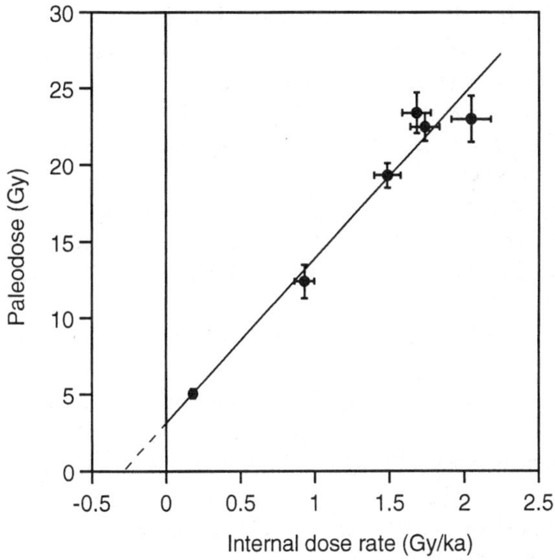

Figure 18.9. Isochron plot for six burnt flints collected from Pedra Pintada rock shelter in Brazil. The flint TL paleodoses are plotted against their internal radioactivities, and a linear least-squares fit is made. The average age of the flints (11,000 $\pm$ 500 years) is given by the slope of the line, and the external dose rate (0.28 $\pm$ 0.03 Gy/ka) is obtained from the intercept on the horizontal axis. (Reprinted, by permission, from Michab and others, *Quaternary Science Reviews*, Vol. 17, p. 1044. Copyright © 1998 Elsevier Science Ltd.)

In both of these European flint dating studies, as with those in the Levant, it was imperative to test several flints for their suitability for TL dating. Only flints that have been heated in antiquity to at least 450°C will yield a reliable age, and the TL plateau test is routinely performed to check that this condition is satisfied for each flint (Aitken and Valladas 1992; Mercier et al. 1995a). About 40% of the flints collected from Le Moustier, Saint-Césaire and Amud Cave had been burnt sufficiently for TL dating.

A similar approach has also been used effectively to date the Paleoindian site of Pedra Pintada in Brazil (Roosevelt et al. 1996; Michab et al. 1998b). Ten burnt lithic artifacts recovered from this rock shelter deposit passed the TL plateau test; four produced ages of 11,900 to 17,500 years and the other six gave an average age of 10,600 $\pm$ 800 years. The latter agreed well with the ages obtained by isochron analysis (Fig. 18.9) of the burnt stones (11,000 $\pm$ 500 years), optical dating of the surrounding quartz sediments (11,500 to 12,000 years), and ^{14}C dating of short-lived plant remains (10,500 to 14,200 years, after calibration). The resulting chronology shows the unequivocal presence of people in the Amazon at or before the time that North America was settled by the Clovis culture (Roosevelt et al. 1996). And TL dating of baked quartz pebbles from the Brazilian site of Pedra Furada is under investigation (Michab et al. 1998a) to test claims of a much earlier colonization of the Americas (Bahn 1993).

In north Africa, six fragments of burnt flint from a large hearth in the lower levels of Sodmein Cave have been TL dated to 118,000 ± 8000 years, indicating that this part of Egypt was occupied during the Middle Paleolithic (Mercier et al. 1999). This date also matches the last interglacial TL and optical ages obtained at the archaeological sites of Bir Tarfawi and Bir Sahara East in the eastern Sahara Desert (Wendorf et al. 1994).

18.6.2. Unheated Materials

18.6.2.1. Anthropogenic Structures

It has long been recognized that the TL signals from materials such as sediments and calcite may be bleached by sunlight (Aitken et al. 1963), and this phenomenon later formed the basis for TL and optical dating of unheated sediments. But only recently have these light-sensitive signals been exploited for luminescence dating of unheated materials used in the construction of ancient monuments and earthen mounds.

Liritzis and colleagues have conducted TL dating studies of ancient monuments in Greece, obtaining ages from the contact surfaces of limestone building blocks (Liritzis 1994; Liritzis et al. 1997; Theocaris et al. 1997). The method is based on the premise that the exterior of each block is bleached by sunlight at the time of monument construction, and that this exposure is sufficient to reset the TL signal in the surface layer. When placed in contact with another block, this surface is concealed from further illumination, and a dating signal begins thereafter to accumulate. By removing successive calcite layers, each 1 mm thick, from the surface of a block from the Mycenean wall, Liritzis (1994) found that only the outermost layer had been bleached during the period of wall construction and that the inner layers retained their geological TL signals. The exterior layer gave a TL age of 3700 ± 450 years, in agreement with archaeological expectations.

This method was then applied to the surface scrapings from a piece of marble extracted from the foundations of the temple of Apollo at Delphi (Liritzis et al. 1997). They showed that sunlight was effective in bleaching the TL signal in this material to a depth of at least 0.5 mm, but modified the technique to accommodate the possibility that the exterior layer had not been bleached completely when the temple was built. Using this "partial bleach" technique, they obtained a TL age of 2420 ± 300 years, which matched the accepted archaeological age of about 2550 years. However, calcite scraped from the interblock surfaces of two limestone pyramids yielded TL ages of 3500 to 5500 years for pyramid construction (Theocaris et al. 1997), much earlier than some archaeological expectations but consistent with the luminescence age for an associated ceramic (Liritzis et al. 1994).

This same approach should be applicable to building blocks (and other stone structures) composed of any light-sensitive luminescent mineral, and if grains of quartz or feldspar are available, then the advantages of optical dating could be enjoyed; the age of emplacement of prehistoric standing stones is a candidate ripe for such investigation. The benefit of using OSL, rather than TL, for dating sun-bleached sediments is well illustrated by the luminescence study of quartz grains extracted from ancient earthen mounds and ridges in Louisiana (Feathers 1997b). Use of the hard-to-bleach TL signal gave ages that reflected the geological age of the landforms on which the earthworks stood, indicating that the grains had not been exposed to sufficient sunlight during construction of these structures. The OSL signal also showed evidence of incomplete bleaching in antiquity, but because this signal had been partially reset (owing to its much greater sensitivity to light), it proved possible to derive a maximum age of about 5500 years for the time of mound building. This estimate is consistent with the optical age of 4000 ± 450 years for an earthen ridge that had been sufficiently bleached, and with calibrated ^{14}C ages of 5300 to 5600 years for charcoal from the construction fill, making it the earliest mound complex so far discovered in North America (Saunders et al. 1997).

18.6.2.2. Archaeosediments

The mound dating study mentioned above is an example of how luminescence dating can be applied to archaeological sediments, or archaeosediments. The earthworks were clearly made by humans in this instance, but there are many other archaeological contexts in which luminescence dating of naturally deposited sediments can provide vital chronological information.

The event of interest when dating artifact-bearing sediments is usually the time elapsed since the most recent episode of sunlight exposure, when the artifacts and the sediments both lay on the ground surface. If the sediments were not sufficiently bleached by sunlight before burial, then the measured paleodose will include a geological contribution and the calculated age will be too large. Much effort has been expended on devising luminescence dating methods to identify and avoid this potential pitfall. Also, the possibility of postdepositional movement of artifacts must be assessed because, unlike the dating of pottery and burnt flint, the artifacts per se are not dated in archaeosediment applications.

On all of the inhabited continents, archaeosediments transported by wind, water, and hillslope processes have been investigated using luminescence

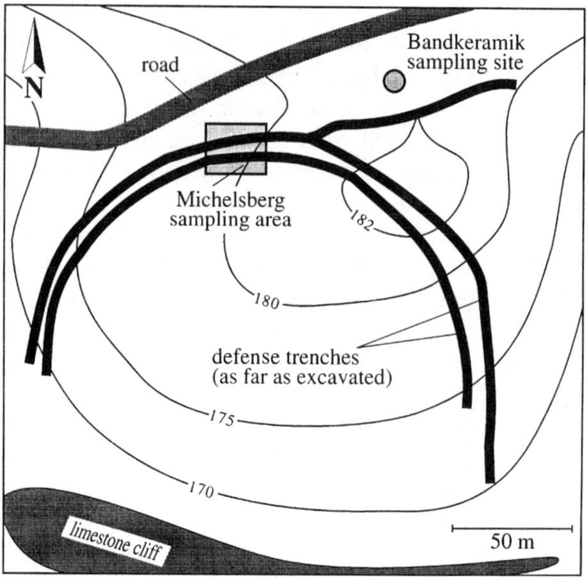

Figure 18.10. Topographic map and ground plan of the Michelsberg defense trenches in southern Germany, showing sampling site locations. Contours are marked in meters above mean sea level. (Reprinted, by permission, from Lang and Wagner, *Archaeometry*, Vol. 38, p. 132. Copyright © 1996 Research Laboratory for Archaeology and the History of Art, Oxford University.)

methods. Remarkably few such studies have been made in the Americas (Feathers 1997a), but two recent examples are those reported above for the Pedra Pintada and mound complex sites (Feathers 1997b; Michab et al. 1998b). Some applications on other continents are discussed below.

Europe. Infrared stimulation of feldspars has been exploited for luminescence dating of German archaeological deposits that may have been poorly bleached during sediment transport. At a Michelsberger Kultur hillslope settlement in southern Germany (Fig. 18.10), two sets of defense trenches are infilled with slopewash (a type of colluvium). Optical dating of the colluvial silt gave IRSL ages of about 4400 years, consistent with the trench being infilled after the abandonment of the settlement in the fourth millennium B.C. (Lang 1994). This result also showed that the IRSL signal had been satisfactorily reset by sunlight in antiquity, despite the short distance of sediment transport by water. Later experimental tests showed that just 30 min of sunlight filtered through cloud and fog was adequate to zero the IRSL signal (Lang and Wagner 1996).

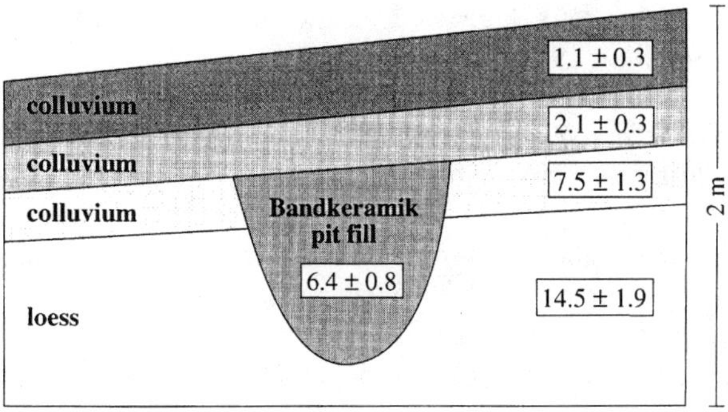

Figure 18.11. Schematic cross section of the Bandkeramik site in southern Germany, showing sediment type, sampling locations, and IRSL ages in thousands of years. (Reprinted, by permission, from Lang and Wagner, *Archaeometry*, Vol. 38, p. 133. Copyright © 1996 Research Laboratory for Archaeology and the History of Art, Oxford University.)

Optical dating was also carried out on silt-sized feldspars from a nearby Early Neolithic site, where a pit had been infilled with sediment and pot sherds of the Mittlere Bandkeramik period (Lang and Wagner 1996). The IRSL age of 6400 ± 800 years for the pit fill (Fig. 18.11), obtained using a narrow detection window to avoid interference from unstable signals, confirmed archaeological expectations (see also Fleming 1979), and may indicate that hillslope erosion increased with the introduction of farming and forestry practices by the Bandkeramik people. At another site in southern Germany, optical dating has been used to identify intensive agriculture during medieval times as the likely cause of accelerated hillslope sedimentation (Lang and Wagner 1997). In both of these studies, dating of the archaeosediments is equivalent to the dating of pottery or burnt flint, insofar as the object being dated is also the target archaeological event.

In another recent study of silt-size sediments derived from ancient anthropogenic soil erosion, Lang et al. (1999) found that only the most light-sensitive IRSL signals gave the correct ages for infilled storage pits at an Iron Age settlement in southern Germany. Experiments showed that the problem was due to the sediments being heterogeneously bleached to differing extents at the time of deposition, as opposed to being partially bleached to the same non-zero level.

In southern Britain, optical dating of fine-grain river sediments has revealed that the onset of valley sedimentation coincided with forest clearance in the Bronze Age and with farming in Roman and Saxon times (Rees-Jones and Tite 1997a). Optical dating of hillslope deposits met with less success, however, with the IRSL and OSL ages for ditch-fill sediments being much greater than archaeological expectations; insufficient bleaching of the sediments before burial was suggested as the likely cause of this discrepancy (Rees-Jones and Tite 1997a).

Africa and Western Asia. Luminescence methods are playing an increasing role in the dating of artifact-bearing sediments associated with Acheulian and later sites in this key region for the evolution of anatomically modern humans. Unlike the above examples from Europe, the African archaeosediments are not the direct result of anthropogenic disturbance and, thus, are not the target archaeological event. It is therefore imperative to show that the artifacts and the sediments are in direct archaeological association, which usually requires that the artifacts are in their primary depositional setting.

In Botswana, the Middle and Late Stone Age levels at White Paintings rock shelter have been dated using TL and OSL techniques for quartz sand (Feathers 1997c). A good indication of sufficient bleaching is given by the concordance in ages obtained from the easy-to-bleach TL and OSL signals, and the hard-to-bleach TL signal. Also, the average luminescence age ($55{,}400 \pm 4700$ years) for the transition from the Middle to the Late Stone Age agrees reasonably well with independent age determinations for this transition at other sites.

More controversial luminescence results have been reported for the Middle Stone Age site of Katanda in Zaire, where carved bone points are thought to provide evidence for fully modern human behavior (Brooks et al. 1995). The quartz river sands were dated using only the hard-to-bleach TL signal because the easy-to-bleach TL and OSL signals appeared to decrease in intensity with added dose. The TL age of $82{,}000 \pm 8000$ years matched the ESR age for hippopotamus teeth collected from the same sedimentary unit, assuming uranium uptake by the tooth soon after burial. Further OSL investigations of the Katanda sediments (Feathers and Migliorini 2000), including the analysis of single aliquots composed of just 20 grains, broadly confirm this surprisingly old age and offer no convincing evidence for insufficient bleaching of the sediments at the time of deposition.

Dating has also been deployed at Acheulian and Middle Stone Age sites discovered in the eastern Sahara Desert in Egypt. Several numerical-age methods (^{14}C, uranium series, ESR, amino acid racemization, and luminescence) were applied to the fossil-lake deposits at Bir Tarfawi and Bir Sahara East, but the chronologies produced by the different methods were far from consistent (Wendorf et al. 1994). Optical dating of wind-blown quartz sand produced the most internally consistent set of ages, and TL dating of burnt and unburnt sediments was also attempted, using the hard-to-bleach TL signal. All three luminescence techniques gave similar ages, and these indicated that the most recent Pleistocene wet episode in Egypt occurred more than 60,000 years ago, with major episodes between 100,000 and 140,000 years ago (Wendorf et al. 1994). This wet period has also been identified at the nearby site of Bir Kiseiba, from optical dating of white river sand that contains Middle Stone Age artifacts (Haynes et al. 1997).

Another site in eastern Egypt, Taramsa 1, has yielded Middle Paleolithic artifacts and the remains of an anatomically modern child (Vermeersch et al. 1998). The surrounding chert extraction pits contained grains of wind-blown quartz sand and these yielded an average optical age of 55,500 $\pm$ 3700 years using the multiple-aliquot additive dose method. This result places modern humans in the Nile valley at a time consistent with the "Out of Africa" model of modern human origins.

On the southern coast of Africa, Feathers and Bush (2000) have applied TL, OSL and IRSL methods to the wind-blown sediments deposited in Die Kelders rock shelter. They obtained ages of 60,000 to 70,000 years from the sand-sized quartz grains (analysed using both the hard-to-bleach TL and easy-to-bleach OSL signals) and the silt-sized feldspars (examined by IRSL). Additive dose and regeneration methods were applied to multiple aliquots (TL, OSL, IRSL) and single aliquots (OSL). The similarity in luminescence ages obtained using these varied methods and signals that bleach at different rates, as well as their agreement with previous ESR age determinations, strongly suggests that insufficient bleaching is not a problem at this Middle Stone Age site (Feathers and Bush 2000). Some other recent archaeological applications in southern Africa are discussed by Wintle (1999).

TL, OSL and IRSL methods have likewise been applied to archaeosediments in the Levant region of western Asia (Porat et al. 1999). Wind-blown sediments correlated with an Acheulian artifact-bearing unit at the open-air site of Holon were dated using multiple-aliquot and single-aliquot methods on sand-size grains of potassium-rich feldspar and quartz. The different methods yielded consistent ages of about 200,000 years, which suggests that, as at Die Kelders, partial bleaching is not a major concern. This conclusion is sustained by the concordance between the luminescence ages and the ESR ages obtained for two teeth from the archaeological bed, assuming a linear rate of uranium uptake (Porat et al. 1999).

Ancient flash-flood sediments deposited along the perimeter of irrigated farmland in southern Yemen have also been the subject of optical dating (Balescu et al. 1998). The IRSL signal from feldspars yielded ages of 1500 to 2100 years for the agricultural silts, in good agreement with an available ^{14}C age. Optical dating has the potential, therefore, to provide reliable chronologies for human-induced sedimentation in arid environments, in the same way as it has been applied to anthropogenic deposits in humid regions of northern Europe.

Eastern Asia. Luminescence dating of archaeological sediments has recently made a controversial contribution to the question of the timing of the human colonization of northeast Asia. Most archaeologists would argue that this region was not occupied earlier than 50,000 years ago, but TL ages of 267,000 ± 24,000 and 366,000 ± 32,000 years have been obtained for the silt-sized sediments that lie above and below the artifact horizon at Diring Yuriakh, situated beside the Lena River (Waters et al. 1997, 1999). Determination of the paleodose proved difficult for these samples, as the additive dose growth curves required substantial extrapolations and as the paleodose values tended to rise with glow curve temperature.

These ages may be too great if the sediments had not been bleached sufficiently before burial. Wind-blown loess from an overlying unit was therefore collected because such material is likely to have received prolonged sunlight exposure before deposition (but see Huntley and Richards 1997 and Lian and Huntley 1999). The loess gave a minimum TL age of 240,000 ± 19,000 years for the artifacts (Waters et al. 1997, 1999); attempts to use the more rapidly bleached IRSL signal for dating were foiled because it was found to be in dose saturation (S. L. Forman, in Gibbons 1997). On the other hand, the TL ages may be too young because of the long-term instability of the TL signal in feldspars, and the existence of magnetically reversed sediments above the artifact layer further suggests that the artifacts were buried more than 780,000 years ago (Huntley and Richards 1997).

Evidence that the Diring artifacts are at least 110,000 years old is presented by Richards (1994), who made an innovative luminescence study of quartzite pebbles collected (at night) from the horizon that includes the artifacts. The pebbles were too thick to allow sunlight to penetrate far beneath their outer surfaces. So he removed grains from the surface of a pebble and from increasing depths (by etching in hydrofluoric acid) to experimentally determine the maximum distance of sunlight penetration; a similar strategy was developed independently by Liritzis (1994) to date the sun-exposed surfaces of limestone building blocks. After several hours of exposure to sunlight, the OSL signal in the surface layer (about 0.25 mm thick) had been reduced to 20% of its original intensity, and layers as deep as 1 to 2 mm had been bleached to some extent (Fig. 18.12). This result showed that the outer rinds of quartzite pebbles, and by analogy quartzite artifacts, may be bleached sufficiently by sunlight to reset the optical dating signal. But similar experiments on a larger number of pebbles are needed to test the generality of this finding.

Richards (1994) also attempted to estimate the time since a pebble buried in the artifact horizon was last exposed to sunlight. On this occasion, he used infrared stimulation and found that the IRSL signal remained constant with added dose, indicating that the sample was in dose saturation. But a minimum paleodose of about 400 Gy was estimated by comparing the "natural" IRSL signal with that measured after bleaching the sample for 3 hr and then giving it a regenerative dose. This paleodose corresponds to a minimum age

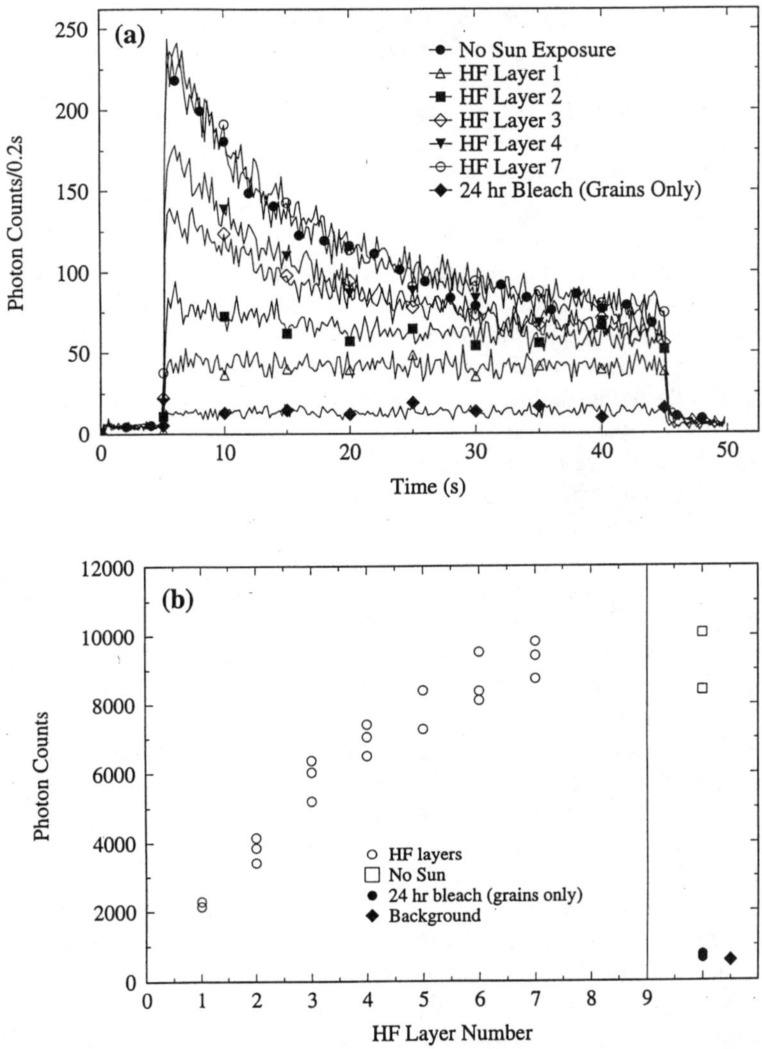

Figure 18.12. (*a*) OSL decay curves for a quartzite pebble from Diring Yuriakh in Siberia, using green light for stimulation. Sand-sized grains were extracted from each hydrofluoric (HF) acid layer after the pebble had been exposed to direct sunlight for 12.5 hr and to cloudy skies for 5.5 hr. The data for no sun exposure are from grains removed before the pebble was exposed to sunlight, and the 24-hr bleach (grains only) data are from grains exposed to a sunlamp for 24 hr. (*b*) Same data as in (*a*) but expressed as the sum of photon counts for the first 10 sec of green light stimulation. (Reprinted, by permission, from Richards, M. A. thesis, Chapter 6, page 7. Copyright © 1994 M. P. Richards.)

of 110,000 years for the archaeological horizon and provides independent evidence of significant antiquity for Diring Yuriakh.

Some geologists have argued, however, that the Diring artifacts are found only in areas where the sedimentary sequence has been disturbed, and that artifacts are absent where the stratigraphy is intact (Kuzmin and Krivonogov 1994). This raises concerns about the association between the sediments and pebbles collected from the undisturbed deposits for luminescence dating and the artifacts. Fortunately, the study by Richards (1994) also offers the means to resolve this and other disputes where the archaeological context is in doubt. He flags the exciting prospect of using optical dating to directly determine burial ages for unheated, sun-bleached artifacts, and thus avoid the problem of their postdepositional displacement. This requires that artifacts are collected at night or under a shroud, using a filtered torch for illumination; see Aitken (1998: 66) for safe-light conditions. If optical dating of individual quartz grains (see below) and quartzite artifacts is done in tandem, then it should be possible to provide a complete account of the stratigraphic integrity of an archaeological deposit.

Australia. Luminescence dating of archaeosediments has seen wide application in Australia, beginning with the dating of sand dunes formed along the eastern shoreline of Lake Mungo in New South Wales. These dunes are famous for entombing the skeletons of the oldest anatomically modern humans yet discovered on this continent. An age of about 33,500 years for human habitation at Lake Mungo was initially obtained from ancient hearths using ^{14}C and TL methods, the latter on baked quartz sand (Huxtable and Aitken 1977; Bell, 1991). Matching TL and ^{14}C ages were obtained for nearby unheated archaeosediments (Readhead 1988), and TL ages of 41,400 to 43,300 years were determined for the sedimentary unit in which a human skeleton had been buried, thereby placing a maximum age on this burial (Oyston 1996; Bowler and Price 1998). The hard-to-bleach and easy-to-bleach TL signals gave comparable ages for the latter deposit, which suggests that it was well bleached at the time of deposition. On the other hand, the overlying dune sands appear to have received insufficient exposure to sunlight before burial, as indicated by the greater ages obtained from the hard-to-bleach signal; the easy-to-bleach signal yields the more reliable TL age estimates in such situations, although even these may be too old. A greater age for this Mungo skeleton has been claimed by Thorne et al. (1999) on the basis of OSL, ESR and uranium series analyses. Optical ages of about 60,000 years were obtained from two samples of dune sand. But these samples were collected a substantial distance from the burial site, which raises questions about stratigraphic association and may explain the discrepancy with the TL chronology (Bowler and Magee 2000).

The potential pitfalls of archaeosediment dating may be illustrated by the recent and high-profile controversy over the antiquity of Jinmium rock shelter in northern Australia. The artifact-bearing deposits at this site were originally dated using the hard-to-bleach TL signal from the quartz sands, with ages of between 116,000 ± 12,000 and 176,000 ± 16,000 years being proposed for initial human occupation (Fullagar et al. 1996). This would have made it two to three times more ancient than any other dated archaeological site in Australia, with broader implications for world prehistory. Also, the rock engravings (pecked cupules) that cover the shelter wall were claimed to be at least 50,000 years old and among the earliest recorded rock art in the world, based on the TL ages of the sediments beneath which these cupules, and an engraved fragment of sandstone, were buried (Fullagar et al. 1996).

A subsequent optical and ^{14}C dating study of the Jinmium deposit (Roberts et al. 1998a, 1999) has shown that the original TL ages were far too old because many of the sand grains were not exposed to sufficient sunlight before burial to reset even the OSL signal. In this situation, the hard-to-bleach TL signal would have been even less well bleached, and the failure of the Jinmium samples to pass the plateau test attests to this problem; the easy-to-bleach TL signal would have given a better indication of the time since sediment burial, albeit still too large because of the inclusion of partially bleached grains in the TL subsamples (Roberts 1997). The correct luminescence age for each sedimentary layer was obtained from optical dating of more than 1000 individual grains, and calculating the age from those grains that had been sufficiently bleached before burial. This single-grain approach indicated that the Jinmium deposit is no older than 10,000 years, and these optical ages agreed with ^{14}C ages for the upper two-thirds of the deposit (Roberts et al. 1998a).

With the redating of Jinmium to the Holocene, the timing of initial human arrival in Australia is based on luminescence ages of 50,000 to 60,000 years for the artifact-bearing quartz sands at the Malakunanja II and Nauwaḷabila I rock shelters in the Northern Territory. TL dating was originally used at Malakunanja II, but these samples passed the plateau test, and agreement between TL and ^{14}C ages for the younger layers at this site lends independent support to the adequacy of bleaching of these sediments before burial (Roberts et al. 1990). Furthermore, the TL ages for the key samples have been confirmed by optical dating of single grains of quartz (Fig. 18.13), and the consistency of the single-grain ages implies that the Malakunanja II sediments, and presumably also the artifacts contained therein, have not experienced significant postdepositional disturbance (Roberts et al. 1998b). At Nauwalabila I, the easy-to-bleach OSL signal was used to date the deepest sediments, and the close match between the ^{14}C, TL, and optical ages from

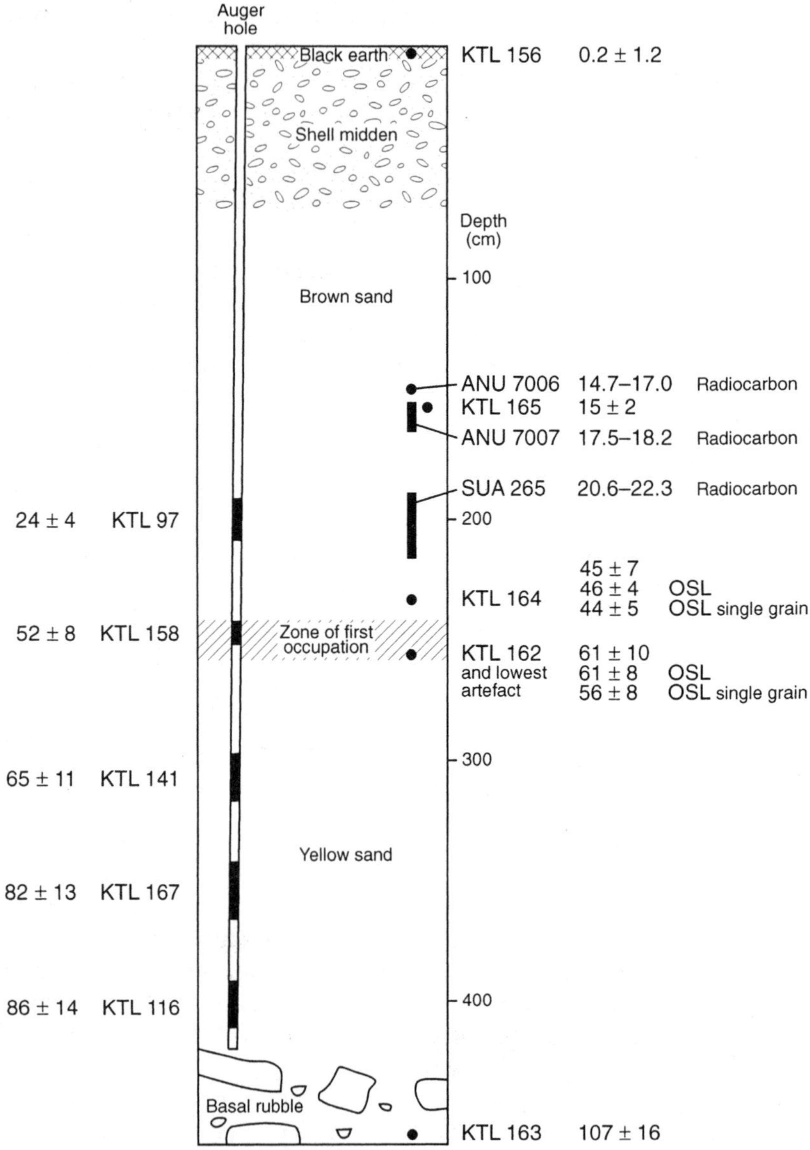

Figure 18.13. Stratigraphic section of the Malakunanja II excavation (southwest face) in northern Australia. Locations, depth intervals, and ages (in thousands of years) are shown for the TL, ^{14}C, and OSL samples. The ^{14}C and OSL ages are labeled to distinguish them from the unmarked TL ages. OSL ages were obtained from multiple-grain single aliquots and from single grains. The ^{14}C ages have been calibrated and are given as 2σ age ranges. (Reprinted, by permission, from Roberts and others, *Ancient TL*, Vol. 16, p. 20. Copyright © 1998 Laboratoire de Physique Corpusculaire, Université Blaise-Pascal.)

the upper occupation levels is good evidence that the sediments had received sufficient exposure to sunlight before burial (Roberts et al. 1994).

Australian archaeosediments are especially well suited to optical dating using subsamples composed of very few grains or single grains of quartz because the OSL signals are more intense (per unit dose) than those emitted by sedimentary quartz from most other regions of the world. The requirement for ever smaller subsamples means that valuable artifacts and fossils stored in museums may be dated if enough sediment has remained attached. This could be useful for sites that have been destroyed or are inaccessible, but the lack of dosimetry data for the original surrounding deposit may introduce some additional uncertainty to the age determination.

The spread in ages from individual aliquots or grains gives an indication of the likelihood of sample contamination by insufficiently bleached grains, as shown by the OSL study at Jinmium, and this approach has also been used to investigate occupation deposits at Ngarrabullgan Cave (David et al. 1997) and Allen's Cave (Murray and Roberts 1997). The sediments in both of these rock shelters had been well-bleached at deposition and gave optical ages in agreement with calibrated ^{14}C ages. At Puritjarra rock shelter, however, contamination of the deposit by insufficiently bleached grains has been proposed as a possible cause of the difference between the ^{14}C and, systematically older, TL ages (Smith et al. 1997). The TL ages were obtained using the easy-to-bleach signal, but as few as 5% contaminant grains would account for the age discrepancy. Single-aliquot and single-grain OSL methods are now being used to identify such material (McCoy et al. 2000).

18.6.2.3. Rock Art

Rock paintings and petroglyphs (pictures produced by engraving, pounding, or pecking) are notoriously difficult to date because of the paucity of suitable material for ^{14}C analysis and other numerical-age methods. However, there have recently been some attempts to date rock art using luminescence methods.

The Uffington White Horse in Oxfordshire is the only prehistoric equine hill figure in Britain and was thought from its style to be Celtic or Saxon in age. This striking figure is 110 m long and was made by cutting a series of ditches and packing them with chalk. Sediments that had been washed down the hillslope and deposited on the figure became buried by chalk added during later repairs. Two samples of buried sediment from the belly of the horse were collected for optical dating: one from below the lowest chalk layer and another from between two chalk layers (Rees-Jones and Tite 1997a, 1997b). Because chalk is low in radioactivity, a significant fraction (about 40%) of the total dose rate was contributed by cosmic rays (Aitken

1998: 139). Ages of about 3000 years were obtained for both samples using the OSL signal from silt-sized quartz, and the IRSL signal from silts of mixed mineralogy yielded a similar age. This places the creation of the figure in the Bronze Age or early Iron Age, making it older than archaeological expectations but consistent with other evidence in the region for intense activity at that time.

Optical dating has also been used to obtain minimum ages for paintings in the rock shelters of northern Australia (Roberts et al. 1997). Many of the spectacular paintings of human figures are partly covered by ancient nests made of sediment gathered by mud-wasps from the margins of streams and pools. The sediment is exposed to sunlight when it is collected by the insects, during their flight to the rock shelter and, finally, during nest construction. Mineral grains concealed inside the nest then begin to accumulate a radiation dose, and this can be used to estimate the time since the nest was built and, hence, give a minimum age for the underlying painting. Roberts et al. (1997) collected fossilized nests at night and, for the largest of these, they extracted quartz grains from the outermost 1 to 3 mm of the nest and from successively deeper layers, in the manner of Liritzis (1994) and Richards (1994). Optical dating of each portion showed that modern sunlight did not penetrate below the surface layer and that the time of nest construction could be estimated using the core of the nest.

Using this approach, figures thought to have been painted most recently (based on the superimposition sequence) yielded minimum ages of 100 to 600 years, and a faded anthropomorphic painting was dated to at least 17,500 $\pm$ 1800 years. This age was confirmed by optical dating of a small nest built on top of the head-dress of this figure. Because of its small size, the entire nest was disaggregated and 47 individual quartz grains were dated (Fig. 18.14). Eleven of the grains examined had paleodoses of less than 8 Gy: this result was expected because grains at and near the surface of the nest are exposed to modern sunlight, which will fully or partially reset the OSL signal. The paleodoses of the remaining grains gave a minimum age of 20,700 $\pm$ 1800 years for the figure, making it the world's oldest dated painting of a human. A great antiquity is also indicated by the condition of equilibrium existing between ^{226}Ra and its parent ^{230}Th in the mud of this nest. If it had been built with a similar initial ^{226}Ra excess as found in the nests constructed during the past 600 years, then it would have taken at least 8000 years for this excess to decay away (Roberts et al. 1997).

Single-grain methods offer the prospect of dating other very small samples. For example, the remains of mud-wasp nests that have been painted over should indicate the maximum age of the painting, and surface grains that are hidden beneath a thick application of paint or charcoal on the wall of a rock shelter should yield an age contemporaneous with the target (artistic) event.

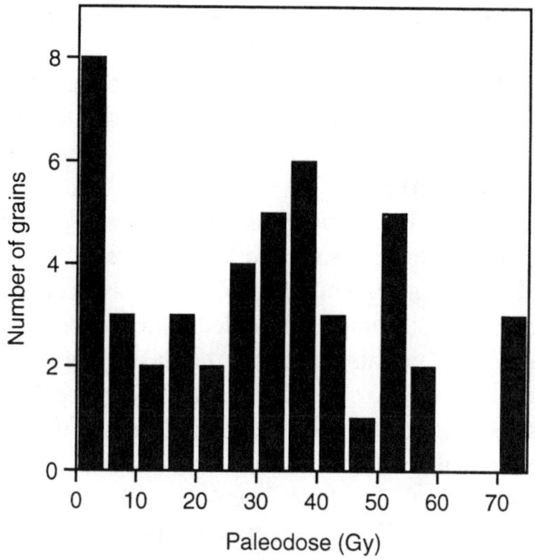

Figure 18.14. Histogram plot of the paleodoses obtained for 47 individual quartz grains extracted from a Pleistocene mud-wasp nest in northern Australia. Grains were recovered from the sun-exposed outside of the nest, as well as from the light-safe inside, and this accounts for the small (less than 8 Gy) paleodoses recorded for 11 grains. The average paleodose for the remaining 36 grains is 39 ± 3 Gy, which corresponds to an age of $20{,}700 \pm 1800$ years. (Data were collected by R. G. Roberts and A. S. Murray.)

The same strategy is applicable, at least in principle, to historic paintings in which the canvas is covered by a ground layer containing quartz grains, as is the case for some seventeenth-century Dutch paintings (Smith 1997); the age obtained would be the time since the most recent application of paint. In practice, however, sample collection would destroy some of the painting and the accurate determination of the paleodose and, in particular, the dose rate would present a formidable challenge.

18.6.3. Other Materials

Other materials of archaeological interest have also been investigated using luminescence methods (Aitken 1985, 1997). Several of the results show signs of promise and, despite being largely ignored in favor of flint and archaeosediment dating, they warrant further investigation.

Bone and tooth enamel have been examined using TL (Jasinska and Niewiadomski 1970; Christodoulides and Fremlin 1971; Driver 1979), but various difficulties, including sample decomposition upon heating, halted further work in this direction. Optical dating is a possibility because the sample can be stimulated at room temperature, but these biological materials will remain subject to the same uncertainties of postdepositional uranium uptake that afflict ESR dating of fossil teeth.

Shells composed wholly or partly of calcite are better suited to TL dating (Johnson 1960; Driver 1979), and calcite has been used to date speleothems in archaeological contexts using TL (e.g., Aitken and Bussell 1982; Debenham 1983; Debenham and Aitken 1984) and ultraviolet stimulation (Ugumori and Ikeya 1980); optical dating by green light may also be possible (Aitken 1998: 140). The luminescence signal is effectively zero when the calcite crystals are formed, but dating beyond 500,000 years is problematic because of signal instability. For best results, the calcite should be clean and must not have been recrystallized, as this will reset the dating signal.

Phytoliths are another promising subject for luminescence studies. Phytoliths are opaline silica bodies produced by plants, and they are commonly found in archaeological deposits. Burnt and unburnt phytoliths from two archaeological sites in Ecuador have been tested for their TL properties by Rowlett and Pearsall (1993). They reported that both types of phytolith emitted TL and responded to laboratory irradiation, and that the 350 to 380°C region of the glow curve gave TL ages of 3000 to 5000 years, in line with independent dating evidence. It is not known if phytoliths also have a light-sensitive signal; none was detected during blue-plus-green or infrared stimulation of phytoliths from New Zealand, even after they had been given a laboratory dose; but further tests are being conducted (Lian 1998).

Metallurgical slags from Roman smelting sites on the Aegean islands of Siphnos and Thasos have also been examined by TL (Elitzsch et al. 1983). The radioactivity was concentrated in the glassy matrix, while the TL was emitted by the quartz and iron silicate inclusions. The latter gave TL ages of 1400 to 5000 years for the Siphnos slags and 1000 to 2000 years for the Thasos slags, both in satisfactory agreement with archaeological expectations. The radioactivity content of most slags is sufficient to permit TL dating (Elitzsch et al. 1983), and optical dating of the quartz component should be feasible.

Archaeological glass is another potential candidate for luminescence dating. A survey of 55 man-made glasses, ranging in age from Roman to medieval, found that the TL signal had a low sensitivity to ionizing radiation (Sanderson et al. 1983). While this presents a serious obstacle to TL dating, larger (e.g., 1 g) samples should yield sufficient luminescence, and signal stability appears to be adequate for dating purposes (Sanderson et al. 1983). The shape of the glow curve may also be useful as a characterization parameter, in the same way as the TL properties of obsidian, a glass of volcanic origin that was widely used by prehistoric people, have been exploited as a means of source identification (e.g., Huntley and Bailey 1978; Leach and Fankhauser 1978; Rendell et al. 1985). The use of TL signals to distinguish between geological materials of different origins is an application that reaches back to the pioneering TL studies of Daniels et al. (1953).

18.7. CONCLUSIONS

Luminescence dating has its origins in TL dating of fired pottery. It later diversified into TL dating of other burnt archaeological materials, most notably flint artifacts, and into authenticity testing of works of art that had been heated during their creation. All of these areas of investigation are still important, and they have been joined by luminescence studies of unheated archaeosediments. The flint and archaeosediment applications have probably made the greatest contribution to archaeology, in subjects as diverse as modern human origins, continental colonizations, and prehistoric rock art.

Technological improvements and archaeological controversies continue to spur the development of luminescence dating methods in the quest to obtain ever more reliable chronologies over a longer span of time. At present, the luminescence age range extends from just a few years, back to almost one million years. The use of smaller samples and individual grains has brought optical dating closer to becoming a nondestructive analytical method. This presents new opportunities for applications in art and archaeology, and the investigation of new materials should further enhance the value of luminescence methods.

REFERENCES

Adamiec, G., and M. J. Aitken (1998), *Ancient TL* **16**, 37.

Aitken, M. J. (1976), *Archaeometry* **18**, 233.

Aitken, M. J. (1978), *J. Coun. Eur. PACT* **2**, 104.

Aitken, M. J. (1985), *Thermoluminescence Dating*, Academic, London.

Aitken, M. J. (1990), *Science-Based Dating in Archaeology*, Longman, London.

Aitken, M. J. (1997), in *Chronometric Dating in Archaeology*, R. E. Taylor and M. J. Aitken, Eds, Plenum, New York, pp. 183–216.

Aitken, M. J. (1998), *An Introduction to Optical Dating: The Dating of Quaternary Sediments by the Use of Photon-Stimulated Luminescence*, Oxford University Press, Oxford.

Aitken, M. J., and J. C. Alldred (1972), *Archaeometry* **14**, 257.

Aitken, M. J., and S. G. E. Bowman (1975), *Archaeometry* **17**, 132.

Aitken, M. J., and G. D. Bussell (1982), *J. Coun. Eur. PACT* **6**, 550.

Aitken, M. J., and H. Valladas (1992), *Philos. Trans. R. Soc. London B* **337**, 139.

Aitken, M. J., M. S. Tite, and J. Reid (1963), *Archaeometry* **6**, 65.

Aitken, M. J., M. S. Tite, and J. Reid (1964), *Nature* **202**, 1032.

Aitken, M. J., D. W. Zimmerman, and S. J. Fleming (1968), *Nature* **219**, 442.

Aitken, M. J., P. R. S. Moorey, and P. J. Ucko (1971), *Archaeometry* **13**, 89.

Aitken, M., J. Huxtable, and A. Murray (1983), *Antiquity* **57**, 217.

Bahn, P. G. (1993), *Nature* **362**, 114.

Bailiff, I. K. (1994), *Radiat. Meas.* **23**, 471.

Bailiff, I. K., and E. J. Younger (1988), *Nucl. Tracks Radiat. Meas.* **14**, 171.

Balescu, S., J-F. Breton, B. Coque-Delhuille, and M. Lamothe (1998), *C. R. Acad. Sci. Paris* **327** (*Earth Planet. Sci.*), 31.

Bell, W. T. (1976), *Archaeometry* **18**, 107.

Bell, W. T. (1977), *Archaeometry* **19**, 99.

Bell, W. T. (1979a), *Ancient TL* **8**, 2.

Bell, W. T. (1979b), *Archaeometry* **21**, 243.

Bell, W. T. (1991), *Archaeometry* **33**, 43.

Bell, W. T., and D. W. Zimmerman (1978), *Archaeometry* **20**, 63.

Bleurer, E., and G. J. Goldsmith (1959), *Experimental Nucleonics*, Rinehart and Company, New York.

Bøtter-Jensen, L. (1988), *Nucl. Tracks Radiat. Meas.* **14**, 177.

Bøtter-Jensen, L. (1997), *Radiat. Meas.* **27**, 749.

Bøtter-Jensen, L., and P. Bechmann (1968), in *Proceedings of the Second International Conference on Luminescence Dosimetry*, Gatlinburg CONF-680920, U.S. National Bureau of Standards, Washington DC, pp. 640–649.

Bøtter-Jensen, L., and G. A. T. Duller (1992), *Nucl. Tracks Radiat. Meas.* **20**, 549.

Bøtter-Jensen, L., and V. Mejdahl (1985), *Nucl. Tracks Radiat. Meas.* **10**, 663.

Bøtter-Jensen, L., and V. Mejdahl (1988), *Nucl. Tracks Radiat. Meas.* **14**, 187.

Bøtter-Jensen, L., C. Ditlefsen, and V. Mejdahl (1991), *Nucl. Tracks Radiat. Meas.* **18**, 257.

Bøtter-Jensen, L., V. Mejdahl, and A. S. Murray (1999), *Quaternary Geochronology* **18**, 303.

Bøtter-Jensen, L., G. A. T. Duller, A. S. Murray, and D. Banerjee (1999), *Radiat. Prot. Dorim.* **84**, 335.

Bowler, J. M., and D. M. Price (1998), *Archaeol. Oceania* **33**, 156.

Bowler, J. M., and J. W. Magee (2000), *J. Human Evol.* (in press).

Bowman, S. G. E. (1982), *J. Coun. Eur. PACT* **6**, 61.

Brooks, A. S., D. M. Helgren, J. S. Cramer, A. Franklin, W. Hornyak, J. M. Keating, R. G. Klein, W. J. Rink, H. Schwarcz, J. N. Leith Smith, K. Stewart, N. E. Todd, J. Verniers, and J. E. Yellen (1995), *Science* **268**, 548.

Boyle, R. (1664), *Experiments and Considerations upon Colours with Observations on a Diamond That Shines in the Dark*, Henry Herringham, London.

Burns, G. (1990), *Solid State Physics*, Academic, Boston.

Buschbeck, H. M., and S. Flettner (1994), *Quat. Sci. Rev.* **13**, 569.

Carriveau, G. W., and G. Harbottle (1983), *Antiquity* **57**, 56.

Chen, R., and A. Hag-Yahya (1997), *Radiat. Meas.* **27**, 205.

Chen, R., and Y. Kirsh (1981), *Analysis of Thermally Stimulated Processes*, Pergamon, London.

Chen, R., G. Fogel, and C. K. Lee (1996), *Radiat. Prot. Dosim.* **65**, 63.

Chowdhary, P., V. Hildebrand, and J. Ney (1982), *J. Coun. Eur. PACT* **6**, 116.

Christodoulides, C., and J. H. Fremlin (1971), *Nature* **232**, 257.

Clarke, M. L. (1996), *Radiat. Meas.* **26**, 611.

Daniels, F., C. A. Boyd, and D. F. Saunders (1953), *Science* **117**, 343.

David, B., R. Roberts, C. Tuniz, R. Jones, and J. Head (1997), *Antiquity* **71**, 183.

Debenham, N. C. (1983), *Nature* **304**, 154.

Debenham, N. C., and M. J. Aitken (1984), *Archaeometry* **26**, 155.

Driver, H. S. T. (1979), *J. Coun. Eur. PACT* **3**, 290.

Duller, G. A. T. (1991), *Nucl. Tracks Radiat. Meas.* **18**, 371.

Duller, G. A. T. (1994), *Quat. Sci. Rev.* **13**, 521.

Duller, G. A. T. (1995), *Radiat. Meas.* **24**, 217.

Duller, G. A. T. (1997), *Radiat. Meas.* **27**, 663.

Dunnell, R. C., and J. K. Feathers (1994), in *Dating in Exposed and Surface Contexts*, C. Beck, Ed., University of New Mexico Press, Albuquerque, pp. 115–137.

Duverger, C., D. Lévine, C. Ney, P. Guibert, M. Schvoerer, and W. G. Mook (1993), *Journal de la Société des Américanistes* **79**, 105.

Elitzsch, C., E. Pernicka, and G. A. Wagner (1983), *J. Coun. Eur. PACT* **9**, 271.

Evans, R. D. (1955), *The Atomic Nucleus*, McGraw-Hill, New York.

Feathers, J. K. (1997a), *J. Archaeol. Meth. Theory* **4**, 1.

Feathers, J. K. (1997b), *Quat. Sci. Rev.* **16**, 333.

Feathers, J. K. (1997c), *Quat. Sci. Rev.* **16**, 321.

Feathers, J. K., and D. A. Bush (2000), *J. Human Evol.* **38**, 91.

Feathers, J. K., and E. Migliorini (2000), *Quat. Sci. Rev.* (in press).

Feathers, J. K., and D. Rhode (1998), *Geoarchaeology* **13**, 287.

Fleming, S. J. (1970), *Archaeometry* **12**, 133.

Fleming, S. J. (1973), *Archaeometry* **15**, 13.

Fleming, S. J. (1975), *Authenticity in Art: The Scientific Detection of Forgery*, Institute of Physics, London.

Fleming, S. J. (1979), *Thermoluminescence Techniques in Archaeology*, Clarendon, Oxford.

Fleming, S. J., and D. Stoneham (1973a), *Archaeometry* **15**, 229.

Fleming, S. J., and D. Stoneham (1973b), *Archaeometry* **15**, 239.

Folz, E., and N. Mercier (1999a), *Quat. Sci. Rev.* **18**, 859.

Folz, E., and N. Mercier (1999b), *Radiat. Meas.* **30**, 477.

Fowler, W. (1968), *Physics of Color Centers*, Academic, New York.

Fullagar, R. L. K., D. M. Price, and L. M. Head (1996), *Antiquity* **70**, 751.

Gaimster, D. R. M. (1992), in *Why Fakes Matter: Essays on Problems of Authenticity*, M. Jones, Ed., British Museum Press, London, pp. 108–115.

Galanin, M. D. (1996), *Luminescence of Molecules and Crystals*, Cambridge International Science, Cambridge.

Galbraith, R. F., R. G. Roberts, G. M. Laslett, H. Yoshida, and J. M. Olley (1999), *Archaeometry* **41**, 339.

Galloway, R. B. (1991), *Nucl. Tracks Radiat. Meas.* **18**, 265.

Galloway, R. B. (1994), *Radiat. Meas.* **23**, 547.

Galloway, R. B. (1996), *Radiat. Meas.* **26**, 103.

Gibbons, A. (1997), *Science* **278**, 220.

Godfrey-Smith, D. I., D. J. Huntley, and W. H. Chen (1988), *Quat. Sci. Rev.* **7**, 373.

Godfrey-Smith, D. I., M. Deal, and I. Kunelius (1997), *Geoarchaeology* **12**, 251.

Goedicke, C. (1984), *Nucl. Tracks Radiat. Meas.* **9**, 87.

Goedicke, C. (1988), *Nucl. Tracks Radiat. Meas.* **14**, 219.

Göksu, H. Y., J. H. Fremlin, H. T. Irwin, and R. Fryxell (1974), *Science* **183**, 651.

Guibert, P., and M. Schvoerer (1991), *Nucl. Tracks Radiat. Meas.* **18**, 231.

Guibert, P., M. Schvoerer, M. P. Etcheverry, B. Szepertyski, and C. Ney (1994), *Quat. Sci. Rev.* **13**, 555.

Haynes, C. V., Jr., T. A. Maxwell, A. El Hawary, K. A. Nicoll, and S. Stokes (1997), *Geoarchaeology* **12**, 819.

Hubbell, J. H. (1977), *Radiat. Res.* **70**, 58.

Huntley, D. J., and D. C. Bailey (1978), *Archaeometry* **20**, 159.

Huntley, D. J., and M. P. Richards (1997), *Ancient TL* **15**, 48.

Huntley, D. J., W. R. Dickinson, and R. Shutler, Jr. (1983), *Archaeol. Oceania* **18**, 106.

Huntley, D. J., D. I. Godfrey-Smith, and M. L. W. Thewalt (1985), *Nature* **313**, 105.

Huntley, D. J., D. I. Godfrey-Smith, and E. H. Haskell (1991), *Nucl. Tracks Radiat. Meas.* **18**, 127.

Huntley, D. J., J. T. Hutton, and J. R. Prescott (1993), *Quat. Sci. Rev.* **12**, 1.

Hütt, G., and I. Jaek (1990), *Radiat. Prot. Dosim.* **34**, 73.

Hütt, G., and A. Smirnov (1983), *J. Coun. Eur. PACT* **9**, 463.

Hütt, G., I. Jaek, and J. Tchonka (1988), *Quat. Sci. Rev.* **7**, 381.

Hutton, J. T., and J. R. Prescott (1992), *Nucl. Tracks Radiat. Meas.* **20**, 367.

Huxtable, J., and M. J. Aitken (1977), *Nature* **265**, 40.

Huxtable, J., M. J. Aitken, J. W. Hedges, and A. C. Renfrew (1976), *Archaeometry* **18**, 5.

Huxtable, J., A. S. Murray, and M. J. Aitken (1982–1983), *J. Hong Kong Archaeol. Soc.* **10**, 84.

Ichikawa, Y. (1965), *Bull. Inst. Chem. Res. Kyoto Univ.* **43**, 1.

Ivanovich, M., and R. S. Harmon, Eds. (1992), *Uranium-Series Disequilibrium: Applications to Earth, Marine, and Environmental Sciences*, Oxford Science, Oxford.

Jasinska, M., and T. Niewiadomski (1970), *Nature* **227**, 1159.

Jensen, H. E., and J. R. Prescott (1983), *J. Coun. Eur. PACT* **9**, 25.

Johnson, N. M. (1960), *J. Sediment. Petrol.* **30**, 305.

Kaylor, R. M., J. Feathers, W. F. Hornyak, and A. D. Franklin (1995), *J. Lumin.* **65**, 1–6.

Kitai, A. H. (1993), *Solid State Luminescence*, Chapman and Hall, London.

Knoll, G. F. (1989), *Radiation Detection and Measurement*, John Wiley & Sons, New York.

Krbetschek, M. R., J. Götze, A. Dietrich, and T. Trautman (1997), *Radiat. Meas.* **27**, 695.

Kuzmin, Y. V., and S. K. Krivonogov (1994), *Geoarchaeology* **9**, 287.

Lamothe, M., and M. Auclair (1997), *Radiat. Meas.* **27**, 107.

Lamothe, M., and M. Auclair (1999), *Earth Planet. Sci. Lett.* **171**, 319.

Lamothe, M., S. Balescu, and M. Auclair (1994), *Radiat. Meas.* **23**, 555.

Lang, A. (1994), *Quat. Sci. Rev.* **13**, 525.

Lang, A., and G. A. Wagner (1996), *Archaeometry* **38**, 129.

Lang, A., and G. A. Wagner (1997), *Quat. Sci. Rev.* **16**, 393.

Lang, A., A. Kadereit, R. H. Behrends, and G. A. Wagner (1999), *Archaeometry* **41**, 397.

Langouet, L., A. Roman, A. Deza, P. Brito, G. Concha, and C. Arsenjo de Roman (1979), *Rev. Archéom.* **3**, 57.

Langouet, L., A. Roman, and R. Gonzales (1980), in *Proceedings of the Sixteenth International Symposium on Archaeometry*, National Museum of Antiquities of Scotland, Edinburgh, pp. 312–320.

Leach, B. F., and B. Fankhauser (1978), *J. R. Soc. New Zeal.* **8**, 331.

Li, S-H. (1994), *Radiat. Meas.* **23**, 563.

Lian, O. B. (1998), Victoria University of Wellington, New Zealand, personal communication, 1998.

Lian, O. B., and D. J. Huntley (1999), *Quat. Sci. Rev.* **18**, 1453.

Liritzis, I. (1994), *C. R. Acad. Sci. Paris* **319** (Ser. II), 603.

Liritzis, I., R. B. Galloway, and P. S. Theocaris (1994), *J. Radioanal. Nucl. Chem. Articles* **188**, 189.

Liritzis, I., C. Michael, and R. B. Galloway (1996a), *Geoarchaeology* **11**, 361.

Liritzis, I., P. Guibert, F. Foti, and M. Schvoerer (1996b), *Nucl. Instrum. Methods Phys. Res. B* **117**, 260.

Liritzis, I., P. Guibert, F. Foti, and M. Schvoerer (1997), *Geoarchaeology* **12**, 479.

Lorentz, A., Ed. (1983), *Proposed Recommended List of Heavy Element Radionuclide Decay Data*, INDC (NDS)-149/NE, IAEA Nuclear Data Section, Vienna.

Mangini, A., E. Pernicka, and G. A. Wagner (1983), *J. Coun. Eur. PACT* **9**, 49.

Marfunin, A. S. (1979), *Spectroscopy, Luminescence and Radiation Centers in Minerals*, Springer-Verlag, Berlin.

Martini, M., G. Spinolo, and G. Dominici (1983), *J. Coun. Eur. PACT* **9**, 9.

Mazess, R. B., and D. W. Zimmerman (1966), *Science* **152**, 347.

McCoy, D. G., J. R. Prescott, and R. J. Nation (2000), *Radiat. Meas.* (in press).

McKeever, S. W. S. (1994), *Radiat. Meas.* **23**, 267.

McKeever, S. W. S., and R. Chen (1997), *Radiat. Meas.* **27**, 625.

McKeever, S. W. S., M. Moscovitch, and P. D. Townsend (1995), *Thermoluminescence Dosimetry Materials: Properties and Uses*, Nuclear Technology, Ashford.

McKeever, S. W. S., L. Bøtter-Jensen, N. Agersnap Larsen, V. Mejdahl, and N. R. J. Poolton (1996), *Radiat. Prot. Dosim.* **65**, 59.

Meakins, R. L., B. L. Dickson, and J. C. Kelly (1978), *J. Coun. Eur. PACT* **2**, 97.

Meakins, R. L., B. L. Dickson, and J. C. Kelly (1979), *Archaeometry* **21**, 79.

Mejdahl, V. (1969), *Archaeometry* **11**, 99.

Mejdahl, V. (1970), *Archaeometry* **12**, 147.

Mejdahl, V. (1979), *Archaeometry* **21**, 61.

Mejdahl, V. (1983), *J. Coun. Eur. PACT* **9**, 351.

Mejdahl, V. (1988), *Quat. Sci. Rev.* **7**, 357.

Mejdahl, V., and L. Bøtter-Jensen (1994), *Quat. Sci. Rev.* **13**, 551.

Mejdahl, V., and L. Bøtter-Jensen (1997), *Radiat. Meas.* **27**, 291.

Mejdahl, V., and A. G. Wintle (1984), in *Thermoluminescence and Thermoluminescent Dosimetry*, Vol. 3, Y. S. Horowitz, Ed., CRC Press, Boca Raton, FL, pp. 133–190.

Mercier, N., H. Valladas, J-L. Joron, J-L. Reyss, F. Lévêque, and B. Vandermeersch (1991), *Nature* **351**, 737.

Mercier, N., H. Valladas, and G. Valladas (1995a), *Quat. Sci. Rev.* **14**, 351.

Mercier, N., H. Valladas, G. Valladas, J-L. Reyss., A. Jelinek, L. Meignen, and J-L. Joron (1995b), *J. Archaeol. Sci.* **22**, 495.

Mercier, N., H. Valladas, L. Froget, J. L. Joron, P. M. Vermeersch, P. Van Peer, and J. Moeyersons (1999), *J. Archaeol. Sci.* **26**, 1339.

Michab, M., H. Valladas, L. Froget, and N. Mercier (1998a), *Ancient TL* **16**, 5.

Michab, M., J. K. Feathers, J-L. Joron, N. Mercier, M. Selo, H. Valladas, G. Valladas, J-L. Reyss, and A. C. Roosevelt (1998b), *Quat. Sci. Rev.* **17**, 1041.

Michael, C. T., N. Zacharias, Y. Maniatis, and D. Dimotikali (1997), *Ancient TL* **15**, 36.

Murray, A. S. (1982), *J. Coun. Eur. PACT* **6**, 53.

Murray, A. S., and M. J. Aitken (1982), *J. Coun. Eur. PACT* **6**, 155.

Murray, A. S., and V. Mejdahl (1999), *Quat. Sci. Rev.* **18**, 223.

Murray, A. S., and J. M. Olley (1999), *GeoRes. Forum* **5**, 121.

Murray, A. S., and R. G. Roberts (1997), *Earth Planet. Sci. Lett.* **152**, 163.

Murray, A. S., and R. G. Roberts (1998), *Radiat. Meas.* **29**, 503.

Murray, A. S., and A. G. Wintle (2000), *Radiat. Meas.* **32**, 57.

Murray, A. S., S. G. E. Bowman, and M. J. Aitken (1978), *J. Coun. Eur. PACT* **2**, 84.

Murray, A. S., R. Marten, A. Johnston, and P. Martin (1987), *J. Radioanal. Nucl. Chem. Articles* **115**, 263.

Murray, A. S., J. M. Olley, and G. G. Caitcheon (1995), *Quat. Sci. Rev.* **14**, 365.

Murray, A. S., R. G. Roberts, and A. G. Wintle (1997), *Radiat. Meas.* **27**, 171.

Nambi, K. S. V., and M. J. Aitken (1986), *Archaeometry* **28**, 202.

NCRPM (1976), Environmental Radiation Measurements, Report 50, National Council on Radiation Protection and Measurements, Bethesda, MD.

NCRPM (1985), A Handbook of Radioactivity Measurement Procedures, Report 58, National Council on Radiation Protection and Measurements, Bethesda, MD.

Olley, J. M., A. Murray, and R. G. Roberts (1996), *Quat. Sci. Rev.* **15**, 751.

Olley, J. M., R. G. Roberts, and A. S. Murray (1997), *Radiat. Meas.* **27**, 433.

Olley, J., G. Caitcheon, and A. Murray (1998), *Quat. Sci. Rev.* **17**, 1033.

Olley, J. M., G. G. Caitcheon, and R. G. Roberts (1999), *Radiat. Meas.* **30**, 207.

Oyston, B. (1996), *Quat. Sci. Rev.* **15**, 739.

Peckjett, A. (1992), *The Colours of Opaque Minerals*, John Wiley & Sons, New York.

Poolton, N. R. J., L. Bøtter-Jensen, A. G. Wintle, J. Jakobsen, F. Jørgensen, and K. L. Knudsen (1994), *Radiat. Meas.* **23**, 529.

Poolton, N. R. J., L. Bøtter-Jensen, and W. J. Rink (1995), *Radiat. Meas.* **24**, 551.

Porat, N., L. P. Zhou, M. Chazan, T. Noy, and L. K. Horwitz (1999), *Quat. Res.* **51**, 328.

Prescott, J. R., and J. T. Hutton (1988), *Nucl. Tracks Radiat. Meas.* **14**, 223.

Prescott, J. R., and J. T. Hutton (1994), *Radiat. Meas.* **23**, 497.

Prescott, J. R., and G. B. Robertson (1997), *Radiat. Meas.* **27**, 893.

Prescott, J. R., and L. G. Stephan (1982), *J. Coun. Eur. PACT* **6**, 17.

Prescott, J. R., G. B. Robertson, and R. C. Green (1982), *Archaeol. Oceania* **17**, 142.

Ralph, E. K., and M. C. Han (1966), *Nature* **210**, 245.

Ralph, E. K., and M. C. Han (1971), in *Science and Archaeology*, R. H. Brill, Ed., MIT Press, Cambridge, pp. 244–250.

Randall, J. T., and M. H. F. Wilkins (1945), *Proc. R. Soc. London A* **184**, 366.

Readhead, M. L. (1988), *Quat. Sci. Rev.* **7**, 257.

Rees-Jones, J., and M. S. Tite (1997a), *Archaeometry* **39**, 177.

Rees-Jones, J., and M. S. Tite (1997b), in *Archaeological Sciences 1995*, A. Sinclair, E. Slater, and J. Gowlett, Eds., Oxbow Monograph 64, Oxbow Books, Oxford, pp. 159–162.

Rendell, H. M., D. C. Steer, P. D. Townsend, and A. G. Wintle (1985), *Nucl. Tracks Radiat. Meas.* **10**, 591.

Richards, M. P. (1994), M. A. Thesis, Simon Fraser University, Vancouver.

Rink, W. J. (1998), McMaster University, Canada, oral presentation at the 63rd Annual Meeting of the Society for American Archaeology, Seattle, March 1998.

Roberts, R. G. (1997), *Radiat. Meas.* **27**, 819.

Roberts, R. G., R. Jones, and M. A. Smith (1990), *Nature* **345**, 153.

Roberts, R. G., R. Jones, N. A. Spooner, M. J. Head, A. S. Murray, and M. A. Smith (1994), *Quat. Sci. Rev.* **13**, 575.

Roberts, R., G. Walsh, A. Murray, J. Olley, R. Jones, M. Morwood, C. Tuniz, E. Lawson, M. Macphail, D. Bowdery, and I. Naumann (1997), *Nature* **387**, 696.

Roberts, R., M. Bird, J. Olley, R. Galbraith, E. Lawson, G. Laslett, H. Yoshida, R. Jones, R. Fullagar, G. Jacobsen, and Q. Hua (1998a), *Nature* **393**, 358.

Roberts, R., H. Yoshida, R. Galbraith, G. Laslett, R. Jones, and M. Smith (1998b), *Ancient TL* **16**, 19.

Roberts, R. G., R. F. Galbraith, J. M. Olley, H. Yoshida, and G. M. Laslett (1999), *Archaeometry* **41**, 365.

Roosevelt, A. C., R. A. Housley, M. Imazio da Silveira, S. Maranca, and R. Johnson (1991), *Science* **254**, 1621.

Roosevelt, A. C., M. Lima da Costa, C. Lopes Machado, M. Michab, N. Mercier, H. Valladas, J. Feathers, W. Barnett, M. Imazio da Silveira, A. Henderson, J. Sliva, B. Chernoff, D. S. Reese, J. A. Holman, N. Toth, and K. Schick (1996), *Science* **272**, 373.

Ropp, R. C. (1991), *Luminescence and Solid State*, Elsevier, Amsterdam.

Rowlett, R. M., and D. M. Pearsall (1993), in *Current Research in Phytolith Analysis: Applications in Archaeology and Paleoecology*, D. M. Pearsall and D. R. Piperno, Eds., MASCA Research Papers in Science and Archaeology 10, University Museum, University of Pennsylvania, Philadelphia, pp. 25–29.

Sanderson, D. C. W. (1988), *Nucl. Tracks Radiat. Meas.* **14**, 203.

Sanderson, D. C. W., S. E. Warren, and J. R. Hunter (1983), *J. Coun. Eur. PACT* **9**, 287.

Sanderson, D. C. W., F. Placido, and J. O. Tate (1988), *Nucl. Tracks Radiat. Meas.* **14**, 307.

Sanzelle, S., H. Erramli, J. Faïn, and D. Miallier (1988), *Nucl. Tracks Radiat. Meas.* **14**, 209.

Saunders, J. W., R. D. Mandel, R. T. Saucier, E. T. Allen, C. T. Hallmark, J. K. Johnson, E. H. Jackson, C. M. Allen, G. L. Stringer, D. S. Frink, J. K. Feathers, S. Williams, K. J. Gremillion, M. F. Vidrine, and R. Jones (1997), *Science* **277**, 1796.

Schulman, J. H., and W. D. Compton (1962), *Color Centers in Solids*, Pergamon, London.

Shepherd, L. A. (1997), *Radiat. Meas.* **27**, 199.

Shiomoya, S., and W. M. Yen (1998), *Phosphor Handbook*, CRC Press, Boca Raton, FL.

Singhvi, A. K., and M. J. Aitken (1978), *Ancient TL* **3**, 2.

Smith, B. L. (1997), La Trobe University, Melbourne, Australia, personal communication, 1997.

Smith, B. W. (1988), *Quat. Sci. Rev.* **7**, 401.

Smith, B. W., M. J. Aitken, E. J. Rhodes, P. D. Robinson, and D. M. Geldard (1986), *Radiat. Prot. Dosim.* **17**, 229.

Smith, B. W., G. C. W. S. Wheeler, E. J. Rhodes, and N. A. Spooner (1991), *Nucl. Tracks Radiat. Meas.* **18**, 273.

Smith, M. A., J. R. Prescott, and M. J. Head (1997), *Quat. Sci. Rev.* **16**, 299.

Spooner, N. A. (1992), *Quat. Sci. Rev.* **11**, 139.

Spooner, N. A. (1994), *Radiat. Meas.* **23**, 625.

Spooner, N. A., J. R. Prescott, and J. T. Hutton (1988), *Quat. Sci. Rev.* **7**, 325.

Stoneham, D. (1991), in *Scientific Dating Methods*, H. Y. Göksu, M. Oberhofer, and D. Regulla, Eds., Kluwer Academic, Dordrecht, pp. 175–192.

Strickertsson, K. (1985), *Nucl. Tracks Radiat. Meas.* **10**, 613.

Strickertsson, K., F. Placido, and J. O. Tate (1988), *Nucl. Tracks Radiat. Meas.* **14**, 317.

Sunta, C. M., E. M. Yoshimura, and E. Okuno (1994), *Radiat. Meas.* **23**, 655.

Sutton, S. R., and D. W. Zimmerman (1976), *Archaeometry* **18**, 125.

Templer, R. H., and B. W. Smith (1988), *Nucl. Tracks Radiat. Meas.* **14**, 329.

Theocaris, P. S., I. Liritzis, and R. B. Galloway (1997), *J. Archaeol. Sci.* **24**, 399.

Thorne, A., R. Grün, G. Mortimer, N. A. Spooner, J. J. Simpson, M. McCulloch, L. Taylor, and D. Curnoe (1999), *J. Human Evol.* **36**, 591.

Troja, S. O., A. Cro, and P. Picouet (1995), *Radiat. Meas.* **24**, 297.

Troja, S. O., A. Cro, A. M. Gueli, V. La Rosa, P. Mazzoleni, A. Pezzino, and M. Romeo (1996), *Archaeometry* **38**, 113.

Tso, M-Y. W., and S-H. Li (1994), *Radiat. Meas.* **23**, 451.

Tyler, S., and S. W. S. McKeever (1988), *Nucl. Tracks Radiat. Meas.* **14**, 149.

Ugumori, T., and M. Ikeya (1980), *Jpn. J. Appl. Phys.* **19**, 459.

Valladas, H., J. M. Geneste, J. L. Joron, and J. P. Chadelle (1986), *Nature* **322**, 452.

Valladas, H., J. L. Joron, G. Valladas, B. Arensburg, O. Bar-Yosef, A. Belfer-Cohen, P. Goldberg, H. Laville, L. Meignen, Y. Rak, E. Tchernov, A. M. Tillier, and B. Vandermeersch (1987), *Nature* **330**, 159.

Valladas, H., J. L. Reyss, J. L. Joron, G. Valladas, O. Bar-Yosef, and B. Vandermeersch (1988), *Nature* **331**, 614.

Valladas, H., N. Mercier, L. Froget, E. Hovers, J. L. Joron, W. H. Kimbel, and Y. Rak (1999), *J. Archaeol. Sci.* **26**, 259.

Vermeersch, P. M., E. Paulissen, S. Stokes, C. Charlier, P. Van Peer, C. Stringer, and W. Lindsay (1998), *Antiquity* **72**, 475.

Visocekas, R. (1979), *J. Coun. Eur. PACT* **3**, 258.

Visocekas, R. (1993), *Nucl. Tracks Radiat. Meas.* **21**, 175.

Waters, M. R., S. L. Forman, and J. M. Pierson (1997), *Science* **275**, 1281.

Waters, M. R., S. L. Forman, and J. M. Pierson (1999), *Quat. Res.* **51**, 195.

Wendorf, F., R. Schild, A. E. Close, H. P. Schwarcz, G. H. Miller, R. Grün, A. Bluszcz, S. Stokes, L. Morawska, J. Huxtable, J. Lundberg, C. L. Hill, and C. McKinney (1994), in *Late Quaternary Chronology and Paleoclimates of the Eastern Mediterranean*, O. Bar-Yosef and R. S. Kra, Eds., Radiocarbon, University of Arizona, Tucson, pp. 147–168.

Wintle, A. G. (1973), *Nature* **245**, 143.

Wintle, A. G. (1978), *Can. J. Earth Sci.* **15**, 1977.

Wintle, A. G. (1997), *Radiat. Meas.* **27**, 769.

Wintle, A. G. (1999), *S. Afr. J. Sci.* **95**, 181.

Wintle, A. G., and A. S. Murray (1997), *Radiat. Meas.* **27**, 611.

Wintle, A. G., and A. S. Murray (2000), *Radiat. Meas.* (in press).

Yoshida, H., R. G. Roberts, J. M. Olley, G. M. Laslett, and R. F. Galbraith (2000), *Radiat. Meas.* (in press).

Zimmerman, D. W. (1967), *Archaeometry* **10**, 26.

Zimmerman, D. W. (1971), *Archaeometry* **13**, 29.

Zimmerman, D. W., and J. Huxtable (1971), *Archaeometry* **13**, 53.

Zimmerman, D. W., M. P. Yuhas, and P. Meyers (1974), *Archaeometry* **16**, 19.

CHAPTER

19

ELECTRON SPIN RESONANCE DATING

RAINER GRÜN

Research School of Earth Sciences, Australian National University, Canberra, Australia

19.1. INTRODUCTION

Electron spin resonance (henceforth ESR) dating was first applied by Ikeya (1975) who dated a speleothem from Akiyoshi Cave, Japan. Since then, a wide range of materials have been studied in applications in geology, geography, and archaeology. ESR can be used for dating of minerals that have been precipitated, such as secondary carbonates or tooth enamel, as well as materials that have been heated in the past such as burnt flint. The method has been comprehensively reviewed by Grün (1989a, 1989b), Ikeya (1993), and Rink (1997).

In archaeology, ESR dating of tooth enamel and carbonates, such as speleothems, spring deposited travertines, and shells, are of particular interest, and these particular applications were recently reviewed by Grün (1997). First ESR results on burnt flint seemed promising (Porat and Schwarcz 1991). However, so far it has not been possible to reach the reliability that is routinely obtained by thermoluminescence (Mercier et al. 1995). Maurer (1980) showed that it is, in principle, possible to obtain ESR age assessments of quartz extracted from ceramics. Unfortunately, ESR is several orders of magnitude less sensitive than thermoluminescence (Aitken 1985, 1997); thus, this ESR dating approach has not been further pursued. Most of recent ESR research focuses on dating of tooth enamel, and only this application is reviewed here.

19.2. BASIC PRINCIPLES OF ESR DATING

Trapped charge dating methods, which comprise of ESR, thermoluminescence [TL; for a recent review see Aitken (1997); see also Chapter 18 this

Modern Analytical Methods in Art and Archaeology, Edited by Enrico Ciliberto and Giuseppe Spoto. Chemical Analysis Series, Vol. 155
ISBN 0-471-29361-X

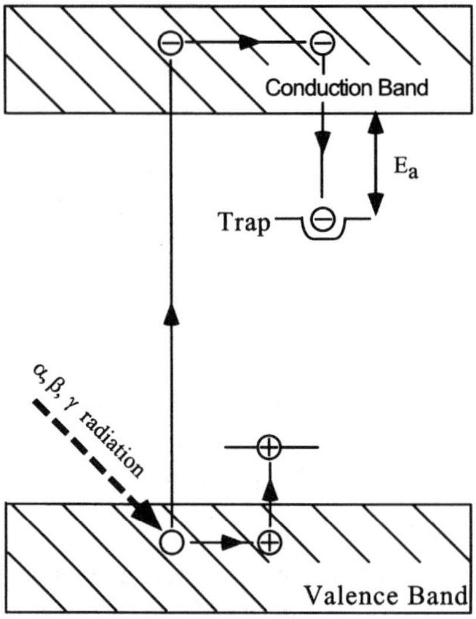

Figure 19.1. Energy band scheme of an insulating mineral (modified from Grün 1996). Radiation emitted from radioactive isotopes ionizes atoms, and electrons are transferred from the ground state (valence band) to the conduction band. After a short time of diffusion most electrons recombine with holes located near the valence band. Some electrons are trapped at charge deficit sites (defects) forming paramagnetic centers that can be detected with ESR (see Fig. 19.2). The number of trapped electron is dependent on the number of traps, the strength of the radiation field (dose rate), and the duration of radiation (= age). E_a is the activation energy of the trap.

book], and optically stimulated luminescence [OSL; Aitken (1998); see also Chapter 18 this book], utilize the fact that many samples are natural dosimeters. Rays emitted from naturally occurring radioactive isotopes ionize atoms (see Fig. 19.1). In this process, negatively charged electrons are knocked off atoms in the ground state (valence band) and transferred to a higher energy state (conduction band), positively charged holes remain near the valence band. After a short time of diffusion, most electrons recombine with holes, returning the mineral back to its original state. However, all natural minerals contain impurities, such as lattice defects, interstitial atoms, and the like, that can trap electrons when they fall back from the conduction band. The trapped electrons and holes form paramagnetic centers that can be detected with ESR, giving rise to a characteristic ESR signal (Fig. 19.2). The number of trapped electrons is dependent on the following factors: the number of traps, the strength (dose rate, $\dot{D}$), and duration of the radiation (age, T). An ESR age is derived from the simple relationship:

$$D_E = \int_0^T D(t)\,dt \qquad (19.1)$$

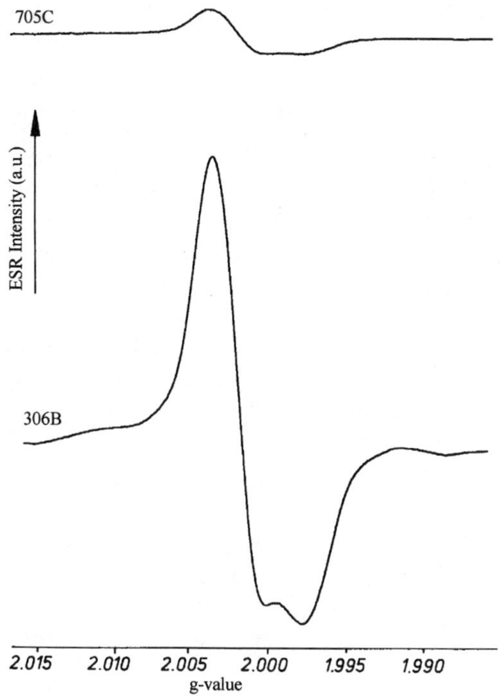

Figure 19.2. ESR signal of a relatively young enamel sample from Le Moustier (top) compared to an old enamel sample from Sterkfontein (bottom). It can be immediately seen that the intensity of the ESR lines is a measure of age (from Grün and Stringer 1991, modified).

where D_E is the equivalent dose. If the dose rate is constant, Eq. (19.1) is reduced to

$$T = \frac{D_E}{\dot{D}} \qquad (19.2)$$

The determination of the D_E value, the actual ESR part of the dating procedure, follows the additive dose method (see Figure 19.3). The dose rate is calculated from the analysis of the radioactive elements (Th, U, and K, other contributions are usually negligible) in the sample and its surroundings. The concentrations of the radioactive elements are converted into dose rates by published tables (e.g., Nambi and Aitken 1986). The determination of the radioactivity that influences the sample is rather complex and has to be carefully evaluated.

To summarize, two parameters have to be determined to estimate an ESR age: the accumulated dose D_E and the dose rate $\dot{D}$. Before presenting dating procedures, I shall briefly describe some basic principles of ESR spectroscopy.

19.2.1. ESR Measurement

Measurements are made with an ESR spectrometer. The sample is placed into a microwave cavity located in a strong external magnetic field. A paramagnetic center (i.e., the trapped electron or hole) has a magnetic moment that is oriented parallel to the magnetic field. In resonance, the magnetic moment is flipped into the opposite direction by absorption of microwave energy, which is conducted to the sample from a microwave generator. The amount of absorbed microwave energy is directly proportional to the number of paramagnetic centers, and, in the end, to the age of the sample.

The ESR spectrometer records the microwave absorption with respect to the magnetic field, resulting in characteristic ESR spectra (e.g., Fig. 19.2). The intensity I of the ESR signal is usually shown as the first derivative. To become independent of specific equipment configurations, the position of an ESR line in a spectrum is described by the g value, which is proportional to the ratio of microwave frequency over magnetic field strength; see X axis in Figure 19.2. The g value is characteristic for a given paramagnetic center and is usually near the value of 2 (which corresponds approximately to the g value for free electrons).

19.2.2. Determination of the Accumulated Dose D_E

For the determination of the D_E value, the ESR response to radioactive irradiation must be known and mathematically expressed. In the laboratory a piece of tooth enamel (between 100 and 500 mg) is separated and thin surface layers of about 50 to 100 μm are removed (to eliminate the volume that was irradiated by alpha rays). The sample is ground and sieved for homogenization and about 10 equal quantities (aliquots) of between 5 and 25 mg are weighed out. These aliquots are irradiated by a calibrated gamma source. Many hospitals and chemistry departments have such gamma sources for sterilization or cancer treatment. During radiation an amount of energy is transferred to the sample, and the unit for this energy dose is Gray (Gy).

The additive dose method is used for the determination of the D_E value of tooth enamel. The signal intensity is plotted versus the calibrated artificial gamma doses. The extrapolation to zero ESR intensity yields the D_E value on the intersection with the X-axis (Fig. 19.3). Detailed simulations of dose response curves (Grün and Brumby 1994) have demonstrated that the data points are best fitted by a single saturating exponential function, weights that are inversely proportional to the square of the measured ESR intensity, and an analytical error calculation that has been suggested by Brumby (1992).

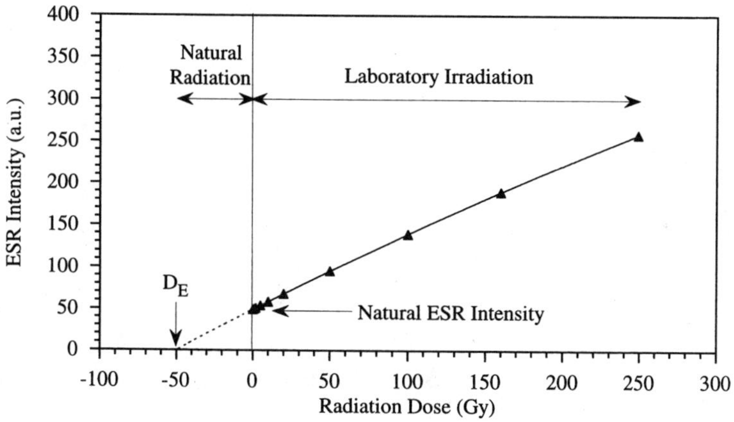

Figure 19.3. Relationship between signal intensity, time, dose rate, and dose (figure modified from Grün 1997). When the sample is formed, it has no ESR intensity. Trapped electrons are created by natural radiation, and their number is dependent on the product of time and dose rate (= dose). When the sample is collected and measured, the ESR signal intensity is equivalent to the number of trapped electrons. Defined laboratory irradiation creates increased ESR signals. The plot of ESR intensity vs. laboratory dose is called the *dose response curve*. The D_E value results from fitting the data points of the dose response with an exponential function and extrapolation to zero ESR intensity.

An ESR signal used for dating should have the following properties:

- When the sample is formed, it does not contain ESR signals.
- The signal intensity grows in proportion to the dose received.
- The signals must have a stability that is at least one order of magnitude higher than the age of the sample.
- The number of traps is constant or changes in a predictable manner. Recrystallization, crystal growth, or phase transitions must not have occurred.
- The signals should not show anomalous fading.
- The ESR signal is not influenced by sample preparation (grinding, exposure to laboratory light).

ESR spectra of tooth enamel are shown in Figure 19.4. On the first glance, the spectrum is dominated by one large signal that consists of two parts, T1 to B1 and B2. These are considered to be two components of the same radical, CO_2^- (Callens et al. 1995). Figure 19.4*b* shows that some spectra contain a variety of additional signals, and these may interfere with the correct estimation of the ESR signal intensity. Traditionally, peak-to-

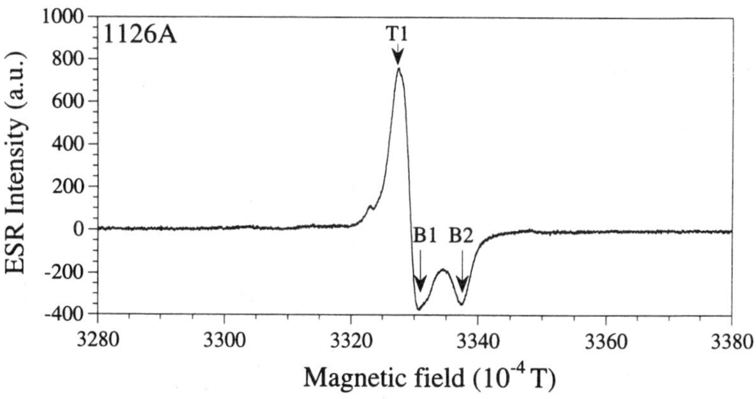

(a)

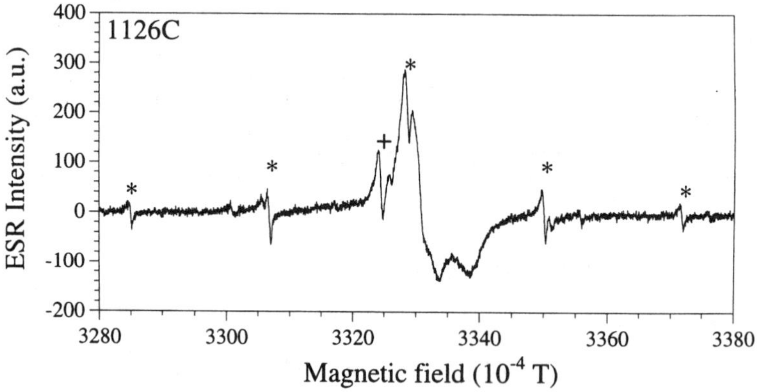

(b)

Figure 19.4. ESR spectra of the natural aliquots of two powdered subsamples of the tooth 1126 [for more details on this sample see Grün et al. (1997a), Oka et al. (1997); figure modified from Grün et al. (1997a)]. The ESR peak marked in the upper spectrum by T1, B1, and B2 are due to CO_2^- (Callens et al. 1995; Vanhaelewyn et al., in press) and show two angular components of this radical. The lower spectrum shows that the ESR signal can be interfered by a range of other ESR peaks. The asterisks in the diagram of sample 1126C indicate the inner five signals of a septet that has been attributed to dimethyl (Bouchez et al. 1988). The signal at $g = 2.0056$ (+) is probably due to a free SO_2^- radical (Barabas 1992; Ikeya 1993).

peak intensities have been used for measuring the signal heights (e.g., the distance between T1–B1 or T1–B2). Grün and Clapp (1996) have shown that this signal intensity can be measured with a precision of better than 1%; D_E values derived from the signal T1–B2 have a precision of better than 4% (Grün 1998a). Further analysis of the ESR spectra using spectrum decon-

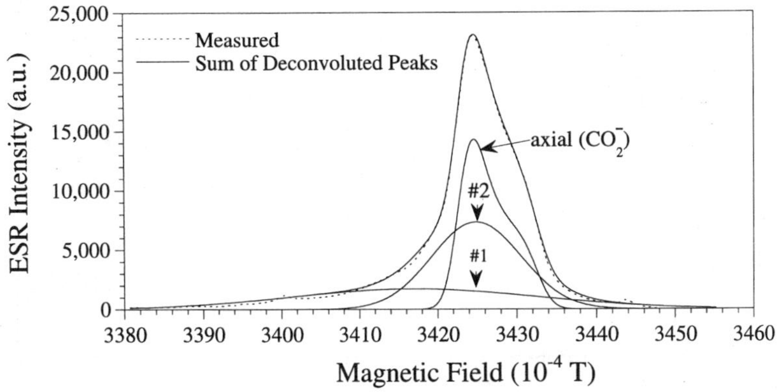

Figure 19.5. Integrated spectrum of tooth enamel and fitting with four Gaussian curves [figure modified from Jonas et al. (1994)]. The axial peak consists of two Gaussian peaks and can be attributed to the CO_2^- radical (see Fig. 19.4*a*). This apparent axial signal is overlapped by two other lines (#1 and #2) which often give different D_E values than the axial peak (Jonas 1997; Grün, 1998c). For reliable D_E estimation it seems therefore necessary to separate these different peaks (by spectrum de-convolution).

volution cast doubt whether the peak-to-peak analysis would result in reliable dose values (Jonas et al. 1994; Jonas 1997). This was due to the fact that the main signal can only be fitted with at least four Gaussian components (Fig. 19.5). Two of these (#1 and #2 in Fig. 19.5) do not belong to the CO_2^- radical, and usually one of these (#2) gives smaller dose values than the components of the CO_2^- radical. However, spectrum deconvolution using Gaussian peaks was criticized by Vanhaelewyn et al. (in press) because the Gaussian peaks had no physical meaning. Vanhaelewyn et al. (in press) carried out a spectrum analysis procedure using maximum-likelihood common factor analysis, resulting in physically meaningful components. They showed that more than 90% of the ESR spectrum is generated by the CO_2^- radical, and dose determinations using this isolated component did not differ much from the peak-to-peak determination. In fact, a comparison between the different methods of spectrum evaluation shows that most dose determinations, whether using peak-to-peak, plateau tests (see Grün and Jonas 1996), spectrum deconvolution using Gaussian or Lorentzian peaks (Grün, 1998c), or maximum-likelihood common factor analysis (Vanhaelewyn et al., in press) are within the experimental errors of the results.

19.2.2.1. Single Aliquot D_E Determinations

To analyze human dental material, it was necessary to develop a method that does not destroy material unnecessarily. Grün (1995) showed that it

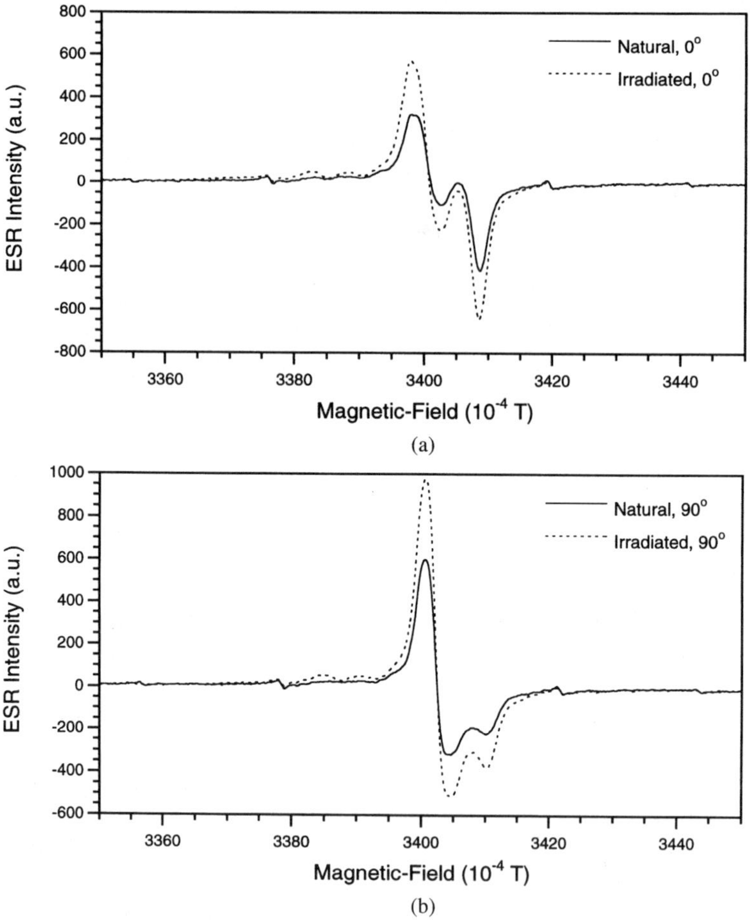

Figure 19.6. Two ESR spectra measured of a single piece of tooth enamel, rotated by 90° in the magnetic field. It can clearly be seen that the two visible components of the CO_2^- radical change in intensity and position in the magnetic field. For reliable dose estimation, the different components have to be separated, and it is best to determine D_E values for a variety of angles.

was possible to determine the dose on single tooth fragments, rather than powdered aliquots. The problem of using single pieces is that the sample has to be removed from the sample holder, irradiated, and replaced into the sample holder in exactly the same position. This is due to anisotropy resulting from the overlapping of the crystal magnetic field with the external magnetic field. The ESR spectrum depends strongly on the relative position of these two fields (see Fig. 19.6). This is, of course, averaged out when using

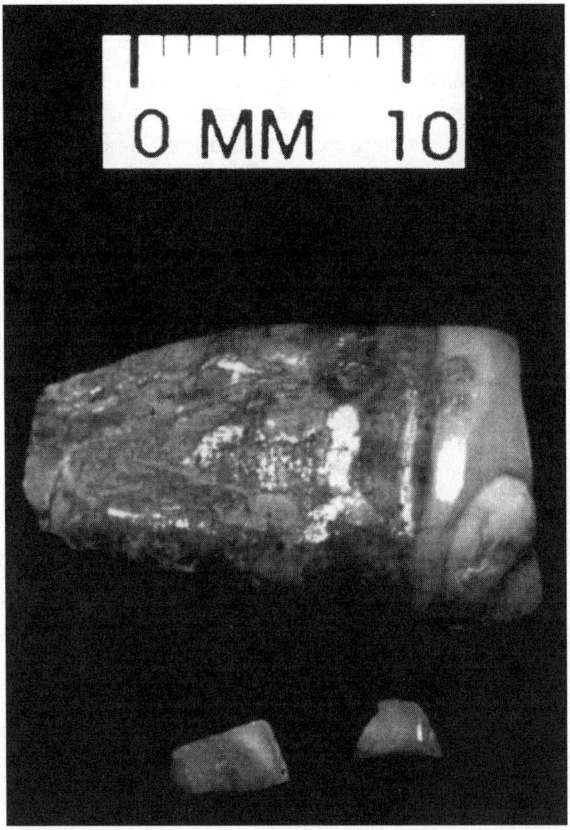

Figure 19.7. Human tooth from Florisbad. The tooth fragments were used for dose determination and the assessment of the age of the Florisbad hominid [see Fig. 19.19, Grün et al. (1996)]. The original, unmodified photo was kindly provided by Dr. J. Brink, National Museum, Bloemfontein, South Africa.

finely powdered samples, as, for example, in X-ray analysis. However, rather than measuring the sample in one position, as suggested by Grün (1995), the sample holder is inserted into a goniometer and measured at 5° to 10° intervals. For the assessment of the D_E value it is necessary to deconvolute the spectrum, which can be done reasonably fast with commercial programs such as Peak Fit (see Grün, 1998). Dose determinations on single fragments were first carried out on a hominid tooth from Florisbad (Grün et al. 1996). Figure 19.7 shows the tooth and the two fragments that were removed. After measurement, the two pieces were reinserted into the tooth and no visible damage remained.

19.2.3. Determination of the Dose Rate $\dot{D}$

The dose rate is calculated from the concentrations of radioactive elements in the tooth and its surroundings plus a component of cosmic rays. In ESR studies, only the U and Th decay chains and the ^{40}K decay are of relevance (a minor contribution comes from ^{87}Rb in the sediment). There are three different ionizing rays that are emitted from the radioactive elements:

- Alpha rays have only a very short range of about 20 μm in tooth enamel because of the large size of the particles. Alpha particles are not as efficient in producing ESR intensity as beta and gamma rays, therefore the alpha efficiency, which is usually in the range of 0.05 to 0.3, has to be determined (see Aitken 1985, 1990).
- Beta rays (electrons) have a range of about 2 mm.
- Gamma rays have a range of about 30 cm.

The concentrations of radioactive elements in the sample are usually very different from its surroundings. Thus, an *internal dose rate* and *external dose rate* have to be assessed independently. Furthermore, it is necessary to estimate the cosmic dose rate, which is about 300 μGy/a at sea level and decreases with depth below ground and is also dependent on altitude as well as geographic latitude (Prescott and Hutton 1988).

The dose rates for the U and Th decay chains and K are shown in Table 19.1. For example, if a sample contains 2 ppm U, 3 ppm Th, and 4% K, an α-efficiency of 0.1 is assumed or measured, the total internal dose rate is generated by alpha and beta particles:

$$
\begin{aligned}
\dot{D}_{\text{internal}} &= k\dot{D}_\alpha + \dot{D}_\beta \\
&= (2 \times 0.1 \times 2781 + 3 \times 0.1 \times 738) \\
&\quad + (2 \times 147 + 3 \times 29 + 4 \times 814)\ \mu\text{Gy/a} = 4415\ \mu\text{Gy/a}
\end{aligned}
$$

If an *external* gamma-*dose rate* of 1300 μGy/a was measured and a *cosmic dose rate* of 150 μGy/a was calculated from the depth of the sample below

Table 19.1. Dose Rates for U, Th, and K

	$\dot{D}_\alpha$	$\dot{D}_\beta$	$\dot{D}_\gamma$	
1 ppm $^{238}U + ^{235}U$	2781	147	114	μGy/a
1 ppm ^{232}Th	738	29	52	μGy/a
1% K		814	242	μGy/a

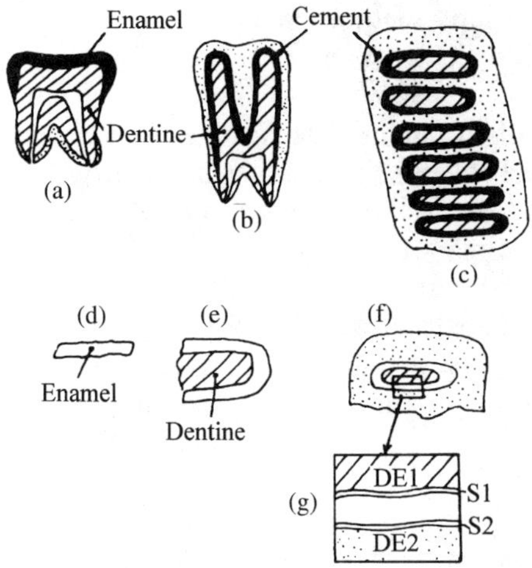

Figure 19.8. Structure of mammalian teeth (from Grün et al. 1987, modified): (*a*) human tooth with low crown, (*b*) high crowned tooth (e.g., horse, bovid), (*c*) elephant tooth, (*d–f*) possible environments of tooth enamel, (*g*) S1 and S2 are >50–200 mm thick enamel layers that are removed to eliminate the effect of the external alpha dose. DE1 and DE2 denote the organic rich dentine/cement layers.

surface, the total dose rate is:

$$\dot{D}_{total} = \dot{D}_{internal} + \dot{D}_{external} = 4415 + 1300 + 150\ \mu Gy/a = 5865\ \mu Gy/a$$

Dose rate calculations become more complicated when disequilibrium in the uranium decay chains or attenuation factors have to be considered (for details see Grün 1989a, 1989b). Typical errors in the estimation of the total dose rate are in the range of 4 to 7%.

19.2.3.1. External Dose Rate

Tooth enamel consists of more than 96% of the mineral hydroxyapatite (Driessens 1980), which is in contact with the more organically rich dentine and cementum (see Fig. 19.8*a–c*). Material in contact with enamel (sediment, dentine, or cementum) irradiates the sample with alpha, beta, and gamma rays, which are emitted from naturally occurring radioactive elements (U, Th, and K). Figure 19.8 shows possible configurations for the enamel layer. The removal of the outer 50 μm of the sample eliminates all the externally alpha-irradiated volume (Fig. 19.8*g*). Since the enamel layer is rather thin, it is normally not possible to remove the outer 2 mm, which received external beta radiation. The externally derived beta dose rate decreases with depth, and therefore attenuation factors have to be considered. Note that there is presently a discussion about the correct assessment of the beta attenuation factor (Brennan et al. 1997; Grün 1998b). In most cases, the results will be affected to less than 3 to 5%. However, if dentine contains high uranium concentrations (>50 ppm) the "one-group" method advocated by Brennan et al. (1997) and experimentally confirmed by Yang (1997) may give considerably older results.

The external beta dose rate has to be calculated separately from the external gamma dose rate because the beta dose rate is generated from the dentine or sediment immediately attached to the enamel layer, whereas the gamma dose rate originates from all sediment that is within a radius of about 30 cm around the sample. The beta dose rate from the sediment is derived from the chemical analysis of U, Th, and K (in cases of Fig. 19.8*d* and *e*). The beta dose rate of the dentine (Fig. 19.8*e* and *f*) originates from its U concentration (dentine is virtually free of Th and K). The dose rate for dentine is affected by uranium disequilibrium and uranium accumulation (see below).

The gamma dose rate can normally not be deduced from laboratory analyses but has to be measured *in situ* with a portable, calibrated gamma spectrometer or TL dosimeters. *In situ* measurements have the advantage to include the present-day water contents. The gamma dose rate from dentine is negligible.

Water absorbs some beta and gamma rays, and its presence in the surrounding sediment has to be considered in the calculation of the beta and gamma dose rate (Bowman 1976; Aitken and Xie 1990). Dentine may contain up to 25% water.

19.2.3.2. Internal Dose Rate

This parameter is mainly generated by alpha and beta rays emitted from elements of the uranium decay chains. An alpha efficiency of 0.15 ± 0.02 has repeatedly been measured (Grün 1985; DeCanniere et al. 1986; Grün and Katzenberger-Apel 1994), and this value is normally used for all samples. However, Chen et al. (1994) measured a higher alpha efficiency of 0.227 ± 0.013 for Chinese samples. Most enamel pieces are rather thin, therefore the internal beta dose rate is not 100% absorbed, and a self-absorption factor has to be calculated (which is 1 minus the external beta attenuation factor). Of course, the same considerations apply here as for the external beta attenuation factor (see above). Quaternary samples, including enamel and dentine, display disequilibrium in the U decay chains, which is actually the basis for uranium-series dating. Uranium-series disequilibrium affects the average dose rates and has to be taken into account mathematically.

19.2.3.3. Problem of Uranium Uptake

Teeth and bones show postdepositional uranium uptake, and this effect further complicates the dose rate determination of tooth enamel. The process of

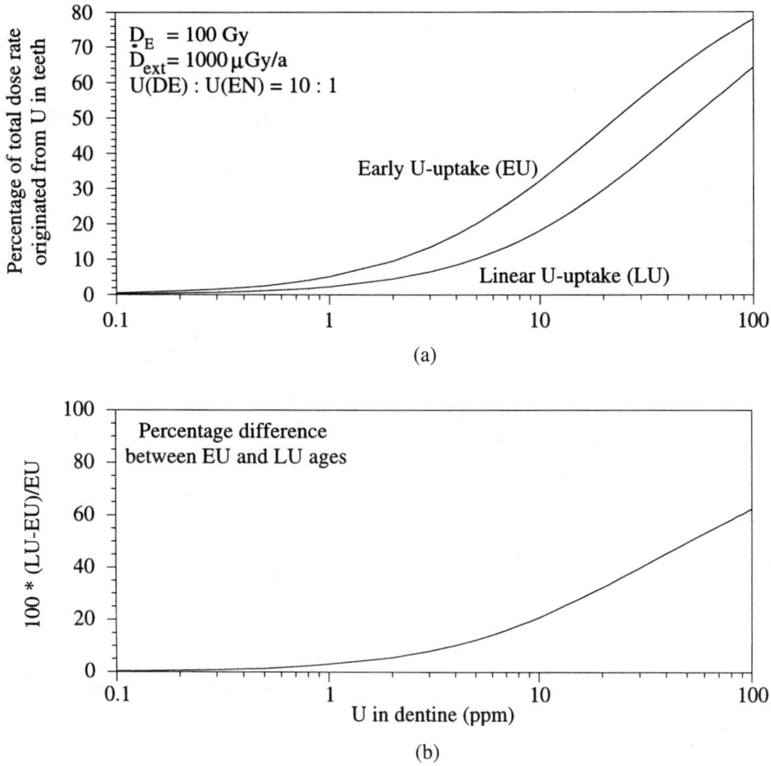

Figure 19.9. Effect of uranium accumulation on the calculation of an ESR age estimate (modified from Grün 1997). The dose rate of enamel and dentine is critically dependent on the uranium uptake model. (*a*) The contribution of the external sediment dose rate to the total dose rate becomes smaller with increasing uranium concentrations in dentine and enamel. Furthermore, the dose rate calculation becomes more dependent on the uranium uptake model. This causes an increasing difference between the early uranium uptake (EU) and linear uranium uptake ages (LU). At high uranium concentrations the LU age is twice the EU age (*b*).

uranium uptake cannot normally be determined exactly. Dentine usually accumulates much more uranium than enamel (by a factor of 10 to 100; Grün and Taylor 1996). For teeth, two models have been suggested (Ikeya 1982): (i) Uranium accumulation shortly after burial of the tooth (early uranium uptake, EU) and (ii) continuous uranium accumulation (linear uranium uptake, LU).

Figure 19.9 shows the influence of the measured uranium concentration on the ESR model ages. As long as the uranium concentrations in the components of a tooth are low (<2 ppm in dentine), the discrepancy between EU and LU ages is less than 10%. With increasing uranium concentrations this intrinsic uncertainty increases very rapidly. In the extreme, the LU model

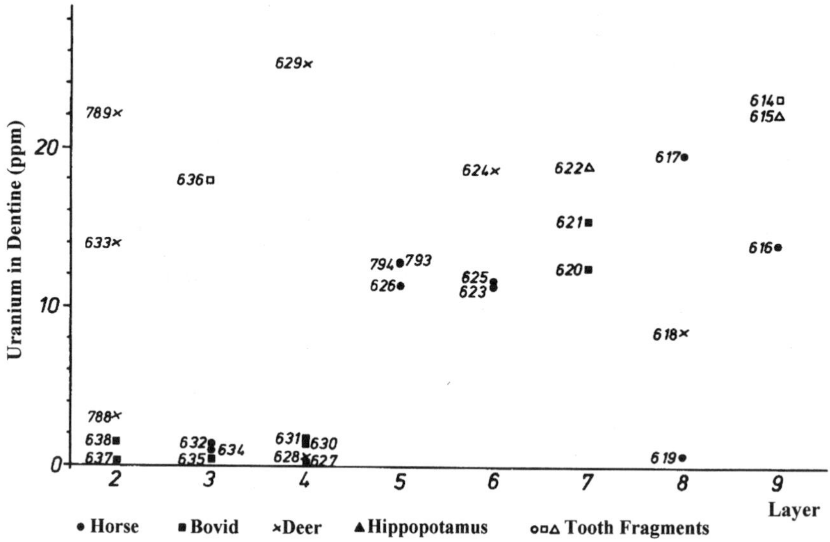

Figure 19.10. Uranium concentrations in dentine of teeth from Pech de l'Azé II (see Fig. 19.11; from Grün et al. 1991, modified).

age is twice the EU age estimate. New developments in ESR and uranium-series dating have shown a strategy to overcome this problem (see below).

If there are no specific indications to justify a particular uranium uptake model, age estimates for both models ought to be given. The most probable age of the sample is somewhere between the two estimates (for a compilation see Grün and Stringer 1991). Most teeth older than the last interglaciation, which is widely accompanied by a more pluvial climate, have accumulated considerable amounts of uranium and therefore most ESR age estimates of such teeth are associated with large uncertainties (>25%). It is *important to note that the EU age estimate is the minimum possible age*. An overall precision of less than 7% can be obtained for teeth with low uranium concentrations (see Grün et al. 1990).

A good example of the relative quality of ESR age estimates is given by the investigation of Pech de l'Aze II (Grün et al. 1991; see Figs. 19.10 and 19.11). In the upper part of the sediment package, the uranium concentrations are relatively low; thus the discrepancies between the EU and LU uptake are small. Taking sedimentological and palaeontological evidence into account, a tight chronology can be established for these layers. Below layer 4, all teeth have considerably higher uranium concentrations, resulting in a large discrepancy between EU and LU age estimates. There is clearly a hiatus between the deposition of layer 4 and layer 5, which corresponds to

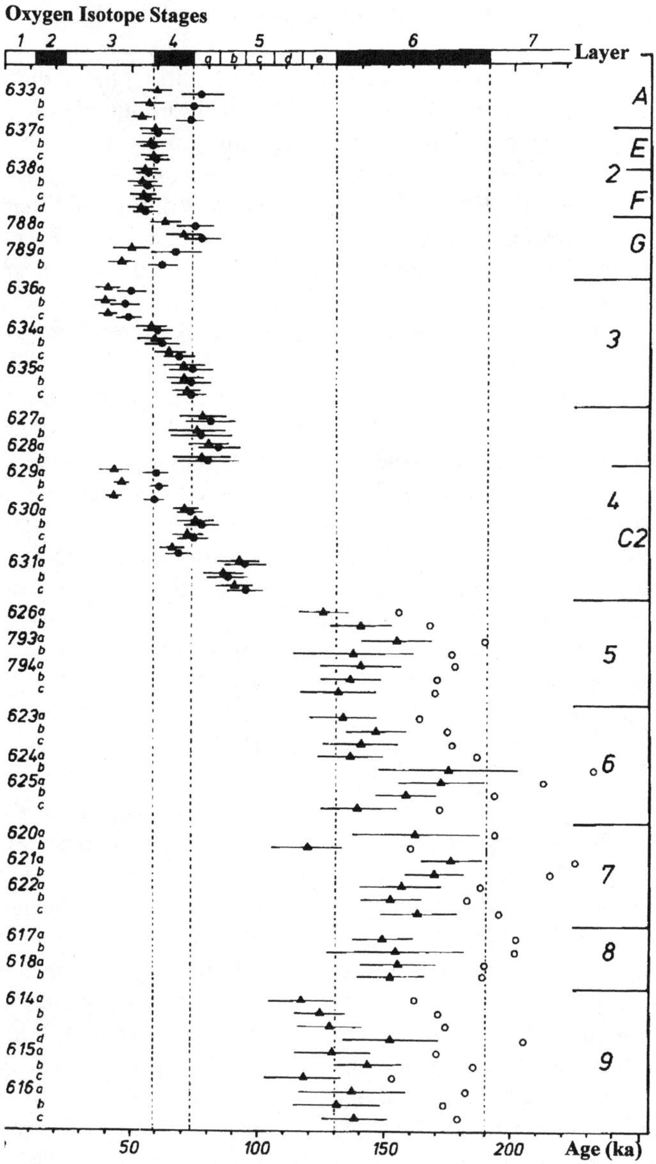

Figure 19.11. ESR age estimates of teeth from Pech de l'Aze II (from Grün et al. 1991, modified). Triangles: EU ages with errors, closed circles: LU ages with errors; open circles: LU age estimates. In layers 2 to 4, the uranium concentration in the teeth is very small, and there is little difference between EU and LU age estimates. For teeth with higher uranium concentrations, sometimes EU age estimates show a better internal consistency with the rest of the data set (e.g., 633, 788); for other samples (e.g., 636, 629) the LU age estimates give a better consistency. The hiatus between layers 4 and 5 corresponds to the last interglacial period (oxygen isotope stage 5), which seems to have caused the increased uranium concentrations in the teeth of the lower layers 5 to 9 (see Fig. 19.10). For more details of uranium accumulation of teeth in layers 5 and 9 see Figures 19.14 and 19.15.

the last interglacial, with a warmer and probably more pluvial climate. This may explain why during this time, water (carrying uranium) was mobilized and the teeth of the lower levels accumulated uranium.

For the upper levels, it is not possible to postulate a general U uptake model. Most larger teeth have low uranium concentrations and therefore the EU and LU age estimates do not differ to a great extent. These teeth should give the most reliable results. For teeth with higher uranium concentrations, sometimes the EU uptake results seem to yield a better internal consistency (e.g., 633, 788), in other cases the LU result seems more appropriate (636, 629).

The ESR results for layers 5 to 9 indicate a deposition somewhere between 100 and 250 ka [representing a time range corresponding to oxygen isotope stages 5 to 7 (Martinson et al. 1987); ka = 1000 years]. If the sediments were of interglacial character (they are not!), the ESR results would not allow the distinction between the last and the penultimate interglaciation (stages 5 and 7). However, the combination of faunal, sedimentological, and archaeological evidence and ESR data results in a clear assignment of layers 5 to 9 to the penultimate glaciation (stage 6). Two teeth from layers 5 and 9 were used for detailed uranium-series analyses to understand how uranium migrated into the teeth (see below).

19.2.3.4. Isochron Dating

Blackwell and Schwarcz (1993) developed an isochron technique for the determination of ESR ages without knowledge of the external dose rate. This is of particular advantage when ESR samples originate from museum collections and it is not possible to measure the gamma-dose rate at the original locality. Figure 19.12 shows the principles: The method only works if the subsamples have varying uranium concentrations, resulting in a range of different combined enamel/dentine dose rates. Several D_E values are determined for a single tooth and are plotted versus the combined enamel and dentine dose rates for the two uranium uptake models. A linear fit can be applied to the two data sets. In an ideal case, the two extrapolations will intersect the Y axis at one point, which corresponds to the total external dose (sediment beta and gamma dose rates and cosmic dose rate). The slopes of the two lines give the EU and LU ESR age estimates for the sample.

This method has been applied in a test on samples from Bau de l'Aubesier (Blackwell and Schwarcz 1993) yielding a reasonable agreement between the isochrone ESR age estimates and independent uranium-series results.

The isochron method actually has the advantage that ESR ages can be determined on teeth being completely independent of the uranium uptake mode: as shown above, in an ideal case the linear fits of the EU and LU data

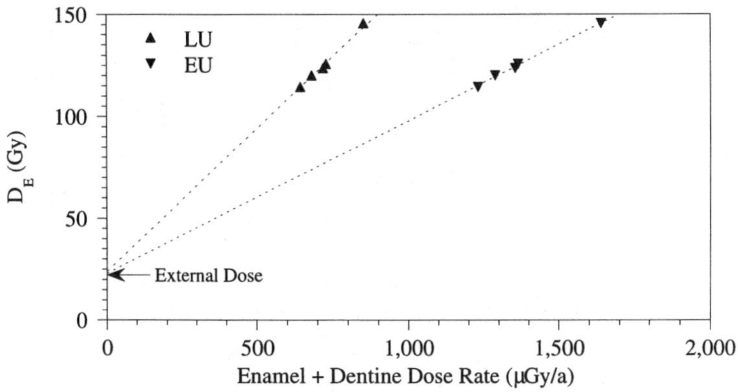

Figure 19.12. Isochron dating (data from Blackwell and Schwarcz 1993; graph modified from Grün 1997). The dose rate, which is dependent on the uranium in the teeth (dentine and enamel), is calculated for the early (EU) and linear (LU) uranium uptake models and plotted versus the measured D_E values. When the D_E values (and uranium dose rates) vary significantly, a linear extrapolation can be carried out. In an ideal case, both lines intersect the Y axis in one point, which yields the external dose. EU and LU age estimates can be calculated from the slopes of the two lines without considering the external dose at all. This method is important when the ESR samples originate from museum collection and no *in situ* measurements of the external dose rate can be carried out. On the other hand, if precise external dose rate values can be determined, ESR ages can be calculated that are independent of the uranium uptake by dividing the external dose by the external dose rate.

intersect the dose axis at the same point. If the total external dose rates can be measured, the age of the sample can simply be derived by dividing the external dose by the external dose rate. However, tests to validate the latter procedure have so far been inconclusive.

19.2.3.5. Combined ESR Uranium-Series Dating: Solving the Uranium Uptake Problem

Grün et al. (1988) proposed a one-parameter equation to describe uranium accumulation in tooth enamel:

$$U(t) = U_m(t/T)^{p+1} \tag{19.3}$$

where $U(t)$ = uranium concentration at time t, U_m = measured uranium concentration, T = age, and p = diffusion parameter. The time dependency of uranium uptake for some given p values are illustrated in Figure 19.13*a*. It can be seen that a p value of $p = -1$ corresponds to a closed system (equal to the EU model); $p = 0$ represents a linear uptake and higher values a supra-linear uranium accumulation (the model does not account for uranium

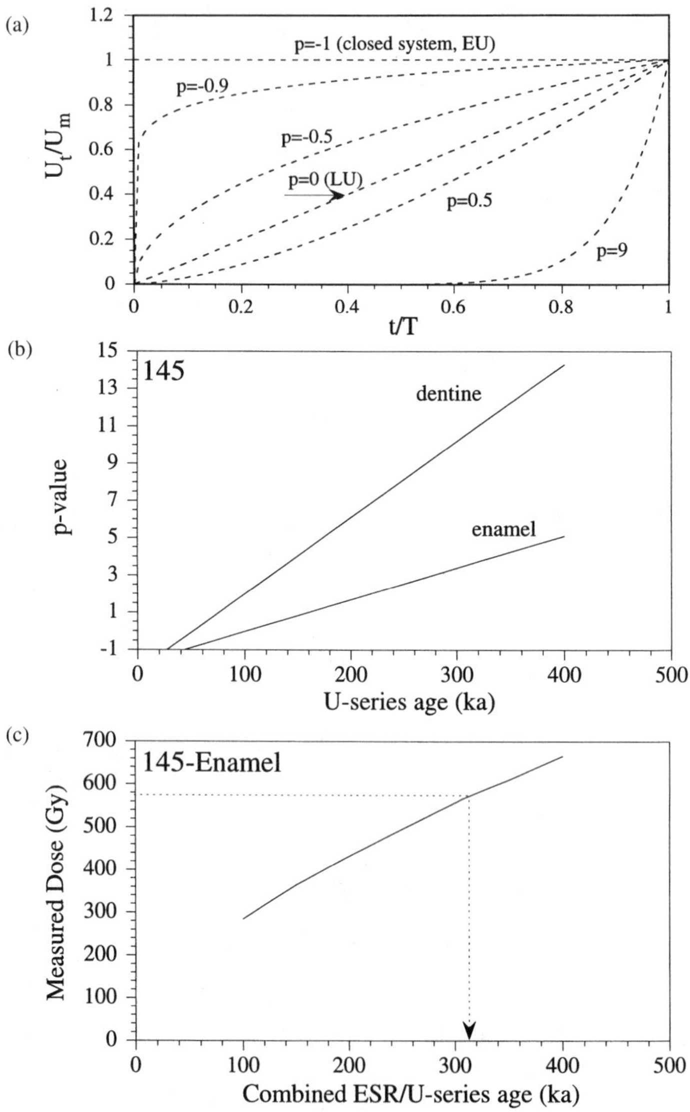

Figure 19.13. (*a*) Time dependency of uranium uptake for different *p* values (= uptake parameter). A closed system (EU) is represented by a *p* value of −1 and the LU model by a *p* value of 0. (U_t = uranium concentration at the time *t*; U_m = measured uranium concentration, T = age of the sample). (*b*) Relationship between *p* values and resulting uranium-series age for measured $^{234}U/^{238}U$ and $^{230}Th/^{234}U$ ratios for sample 145 (Hoxne). Note, that the relationship between *p* value and uranium-series age is nearly linear. (*c*) Plot of the dose that a piece of enamel has received based on the p/t relationship of (*b*) and age determination for sample 145. The *p* values corresponding to the age of sample 145 can be found in (*b*) (all graphs modified from Grün et al. 1988).

leaching). The measured $^{230}Th/^{234}U$ and $^{234}U/^{238}U$ ratios are used to establish the relationship between the p value and a corresponding uranium-series age. Figure 19.13*b* shows that a nearly linear relationship exists between the two parameters. Using this p/t relationship the dose a sample has received at any given time can be calculated. The projection of the measured dose value of the enamel on this function (Fig. 19.13*c*) results in the combined ESR uranium-series age. The associated p values follow from Figure 19.13*b*.

Until recently, only one experimental study had been published (Grün et al. 1988). This reflects the fact that uranium concentrations in enamel are usually very low and alpha spectrometry results have large uncertainties, which seriously compromised any modeling attempts. New advances in uranium-series dating using mass spectrometers (Edwards et al. 1987a, 1987b) allow measurements of uranium-series ratios with great accuracy on relatively small samples with low uranium concentrations. Several mass spectrometric studies have now been carried out, for example, on a series of Israeli paleoanthropological sites (McDermott et al. 1993) and on the site of Hexian (Grün et al. 1998); see below.

To obtain a better understanding of how uranium migrates into the teeth, Grün et al. (1999) carried out a detailed mass spectrometric uranium-series analysis on samples 626 (layer 6) and 616 (layer 9) from Pech de l'Aze II (see above). A series of enamel/dentine pairs were collected from both teeth (see Fig. 19.14), and the results are shown in Figure 19.15. The relative distribution of uranium in dentine and enamel is very similar (Fig. 19.15*a* and *b*) with a partitioning factor of about 50 between dentine and enamel. The apparent uranium-series ages of the dentine/enamel pairs are also similar, however, the dentine age is usually older than the related enamel age. The uranium uptake history of each sample can be calculated from the uranium-series result and the stratigraphic age using the relationship shown in Eq. (19.3) (Fig. 19.15*e*–*h*). The results were summarized as follows:

- Uranium uptake was the dominating process for all subsamples of both teeth.
- The uranium uptake rate of the dentine decreased rapidly during the first 20 ka from about 0.5 to 2 ppm/ka to less than 0.1 ppm/ka.
- The minimum offset between the stratigraphic age of the teeth and uranium-series age estimates was about 35 ka, and the maximum offset about 80 ka for both samples; thus, the uranium-series age estimates of the older sample were generally less affected by uranium uptake than of the younger sample.
- There was a clear relationship between uranium concentration and age between adjacent dentine and enamel samples. The bulk of the uranium

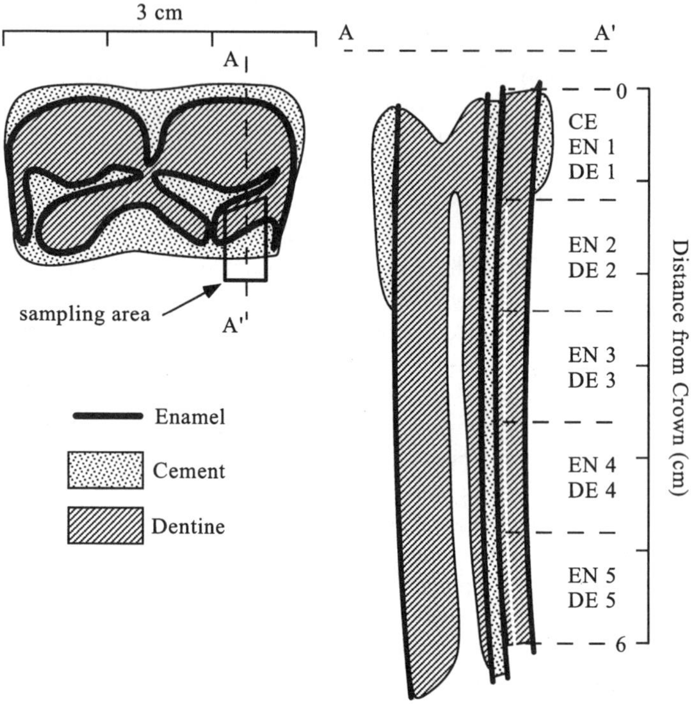

Figure 19.14. Occlusal view (left) and cross section (right) of horse tooth 626 indicating the location of the samples (from Grün et al., 1999, modified).

was absorbed by the organic matter (dentine/cement) earlier than by enamel, which implied that uranium migrated along the dentine layers into the teeth and from there into the enamel.

- The most likely uranium uptake of all components lies between early and linear uranium uptake (see above). However, although it was possible to solve the problem of uranium uptake, it is generally not possible to assume a simple, single parameter uranium uptake model for a given tooth.

Note that the uranium uptake histories apply only to these two teeth and cannot be generalized. Other samples, such as those from Hoxne (Grün et al. 1988), Israeli sites (Grün and McDermott 1994), or Hexian (Grün et al. 1998; see Fig. 19.21) may have experienced very different uranium uptake histories.

To summarize, EU and LU age estimations usually bracket the correct

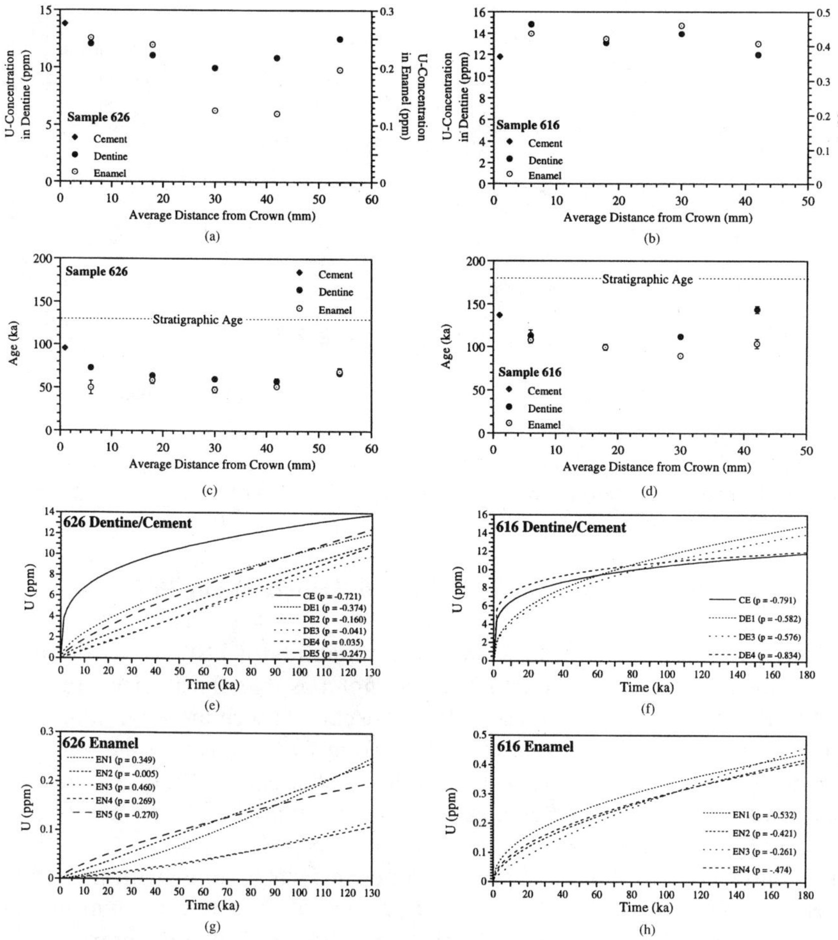

Figure 19.15. A: Uranium concentrations (*a*, *c*) and uranium-series age estimates (*b*, *d*) of the enamel and dentine subsamples. The symbol is larger than the error for those data points that do not show error bars. Uranium uptake for samples 626 (*e*, *g*) and 616 (*f*, *h*). Note that the uranium uptake into enamel is delayed when compared to the adjacent dentine ($p_{\text{dentine}} < p_{\text{enamel}}$). The bulk of the uranium uptake takes place during the first 20,000 years; the older sample (616) shows *p* values closer to early uptake ($p = -1$). Note that the uranium uptake reconstruction applies only to these two teeth; other samples may have experienced very different uranium uptake modes (see Fig. 19.21*c*) (graphs compiled from Grün et al. 1999).

age of the sample. This may involve large uncertainties, if the sample contains significant amounts of uranium (see Fig. 19.9). The explicit uranium uptake history can be estimated by combined ESR and uranium-series analysis.

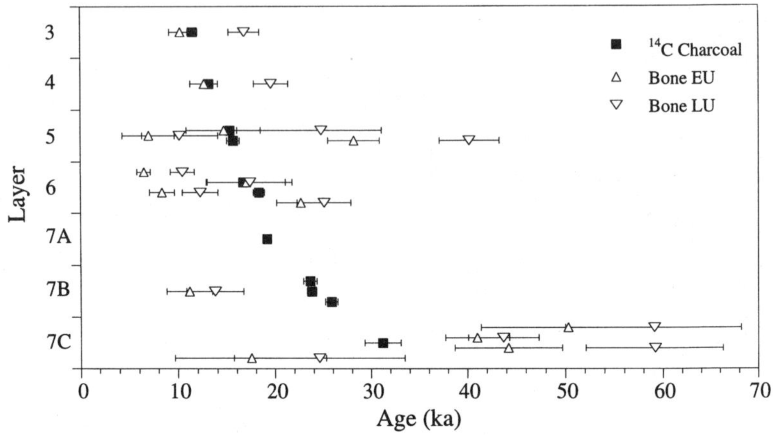

Figure 19.16. ESR dating results from Batadomba cave, Sri Lanka, compared with calibrated radiocarbon results on charcoal (from Abeyratne et al. 1997, modified).

19.2.4. Why Is It Not Possible to Date Bones by ESR?

Grün and Schwarcz (1987) reviewed the possibility of ESR dating of bones and came to the conclusion that it was not possible. Their views resulted from well-known observations: (i) Bones usually absorb more uranium than teeth and (ii) the mineral phase in bones is only somewhere in the region of 40 to 60% (enamel $\approx$ 96%). Along with the alteration of the organic constituents in bone during fossilization, the mineralogic compounds change and the crystal size of hydroxyapatite increases (Hassan et al. 1977). Formation of the new mineral phase (with new defects = traps) with time must generally lead to an age underestimation regardless of the uranium uptake model applied. Abeyratne et al. (1997) attempted to date bone samples from Batadomba Cave, Sri Lanka. Clearly, the ESR results show little resemblance with the ^{14}C chronology of the site (Fig. 19.16).

Oduwole and Sales (1991) observed that the ESR signal begins to fade immediately after irradiation. In older bones this fading component becomes smaller relative to the total signal, and therefore this effect was suggested as a dating technique. The fading has been attributed to collagen, proteins, or surface interactions of small-sized hydroxyapatite crystals (Ostrowski et al. 1974). If the fading is controlled by the latter process, it actually measures the crystallinity of the hydroxyapatite component in the bone. The crystallinity is indeed age dependent but is also influenced by many other processes (see previous paragraph).

19.2.5. Dating Range

The upper dating range of the ESR technique is mainly controlled by thermal stability of the paramagnetic centers. Trapped electrons only have a limited probability of staying in the traps. After some time, which is temperature dependent, the electrons leave the traps and recombine with holes. The thermal stability is expressed by mean life (τ) of a trapped electron at the defect site. The influence of de-trapping on age is negligible only if the mean life is approximately 10 times higher than the age of the sample.

Preliminary annealing experiments by Schwarcz (1985) suggested a mean life in the range of 10 to 100 Ma (Ma = million years). Apart from a very large scattering, an investigation of teeth from Sterkfontein (Schwarcz et al. 1994) showed no particular trend of underestimation in an age range of 1.6 to 2.4 Ma, which corresponds more or less to the expected age (Grine 1988). Therefore, the thermal stability seems sufficient for dating Middle and Late Pleistocene samples.

The lower dating range is determined by the sensitivity of the ESR spectrometer. It is easily possible to detect a signal that is generated by 1 Gy. This may correspond to a few thousand years in a low-dose-rate environment or a few tens of years in a high-dose-rate environment.

19.3. APPLICATION OF ESR DATING

19.3.1. Direct Dating of Hominids: Florisbad

Dating studies on paleoanthropological sites are usually carried out on material associated with human remains, such as the sediment, charcoal, or other fauna rather than the human specimen itself. The reason lies in the fact that most dating techniques are destructive and because the hominid remains are too rare to be sacrificed for dating. This indirect dating approach is in many cases not satisfactory, because:

- The human remains are often buried into the sediments and the association with other materials is uncertain (e.g., Skhul, Qafzeh, etc.).
- Faunal remains or minerals from the sediment are reworked from older deposits (see, e.g., present discussion of the age of the *Homo erectus* remains in Indonesia: Swisher et al. 1996; Grün and Thorne 1997).
- The hominid specimen was discovered at a time when no careful excavations were carried out, and it is impossible to correlate the specimen

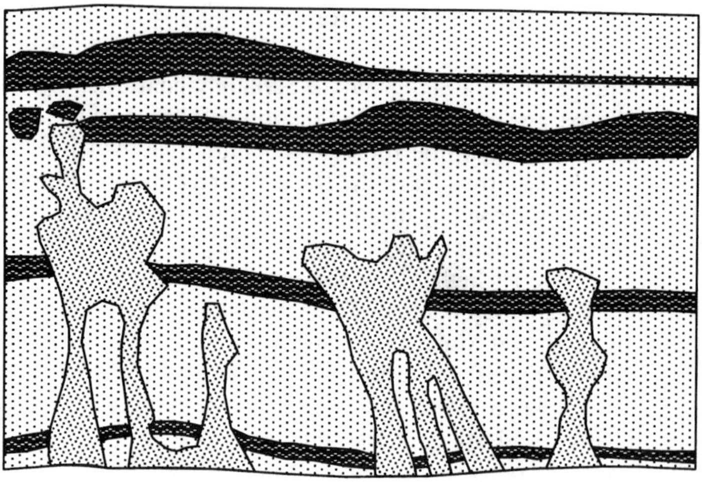

Figure 19.17. Diagrammatic section of sediment layers at Florisbad depicting an east–west section of the original excavations (Dreyer 1938): "*Columns of white sands do not continue to the surface but are terminated at different levels of the stratified deposits . . . Each of these consist of a basal mass of artefacts, bones and teeth etc . . .*" The cranium formed part of a bone assemblage consisting of mostly carnivore prey remains and some stone artifacts found in a spring vent that was capped by the green sand above Peat I. The spring collection is a derived from all spring vents, thus, presents a mixture of specimen with different ages. The animal remains from the springs would have been sampled at various times when carnivores utilized the spring pools as a source for prey (after Brink 1987).

with other datable material (this applies to large number of important human fossils).

The Florisbad site is a good example that often the age of a hominid fossil cannot be derived from any so-called associated materials but can only be obtained by direct analysis of the hominid remains. The Florisbad human cranial fragments were discovered in 1932 (Dreyer 1938). The partial cranium was part of a natural accumulation of mostly carnivore prey remains (Brink 1987, 1988) that were trapped in vertical spring vent structures. The spring and associated white sand bodies intruded horizontal sand and clay layers when the spring was active in a specific locality. When the spring migrated to another area, the vents were sealed by the horizontal deposits, resulting in a succession of ancient spring vents at various depths (see Fig. 19.17).

The horizontal deposits consist of a sequence of sands and clays interbedded with organic horizons, previously and erroneously referred to as "peats." These organic horizons vary in lateral extent (Rubidge and Brink

1985). Peat II has a considerable lateral spread and represents an old land surface on which Middle Stone Age artifacts and bone remains, the products of a short-lived human occupation event, occur in primary sedimentary context (Kuman and Clarke 1986; Brink 1987). From a series of test pits it appears that there is Middle Stone Age material down to bedrock. However, others have described the artifacts associated with the hominid as Faure-smith, an Early Stone Age industry (Beaumont et al. 1978). Previous dating attempts produced infinite radiocarbon age estimates for the human occupation layer, and Peat II and a preliminary uranium-series age estimate in excess of 100,000 years for Peat I (Clarke 1985).

One set of tooth samples was collected *in situ* from the test pit, which spans the whole sequence at Florisbad and the occupation horizon. A second set of samples was selected from the spring collection of Dreyer's original excavation. Additionally, samples for optical dating were collected. Figure 19.18*a* shows the ESR and OSL dating results of the samples from the test pit and the occupational horizon, and Figure 19.18*b* shows the ESR dating attempt for the samples of the spring collection. It is clear that the spring collection does not represent a single, short time span. The provisional ages scatter from about 300,000 years to about 100,000 years. A closer look at the sediments recovered from the spring collection specimen revealed that many sediments were strongly colored (with high U and Th concentrations) indicating that these samples did not originate from spring vents. The repeated analysis of the individual teeth show that the ESR results are well reproducible. The ESR age estimates of the specimen of the test pit also show considerable age spread within layers indicating reworking of material. This is not particularly surprising considering the spring activities and that usually only small bone and teeth fragments could be found in the horizontal layers, indicating remobilization of material. The age spread is smallest for the samples from the human occupation horizon (120 ± 5 ka) where the taphonomy clearly indicates a short-term event and reworking can be excluded.

To tighten the age estimate on the Florisbad hominid, it was decided to analyze the associated tooth directly. Although it cannot be proven that this tooth derived from the same individual as the facial bones, this seems highly probable. Furthermore, it is the only dental sample clearly associated with the other hominid remains found together in a white spring vent that was sealed by the green layer above the first peat (Dreyer 1938). OSL dating analyses were obtained of these two layers, as well as the basal sand layer above the dolerite.

Figure 19.19 shows a schematic diagram of the ESR results of the fragments of the hominid tooth and the occupational horizon with the OSL age estimates of related layers. The dating results support an age for the occu-

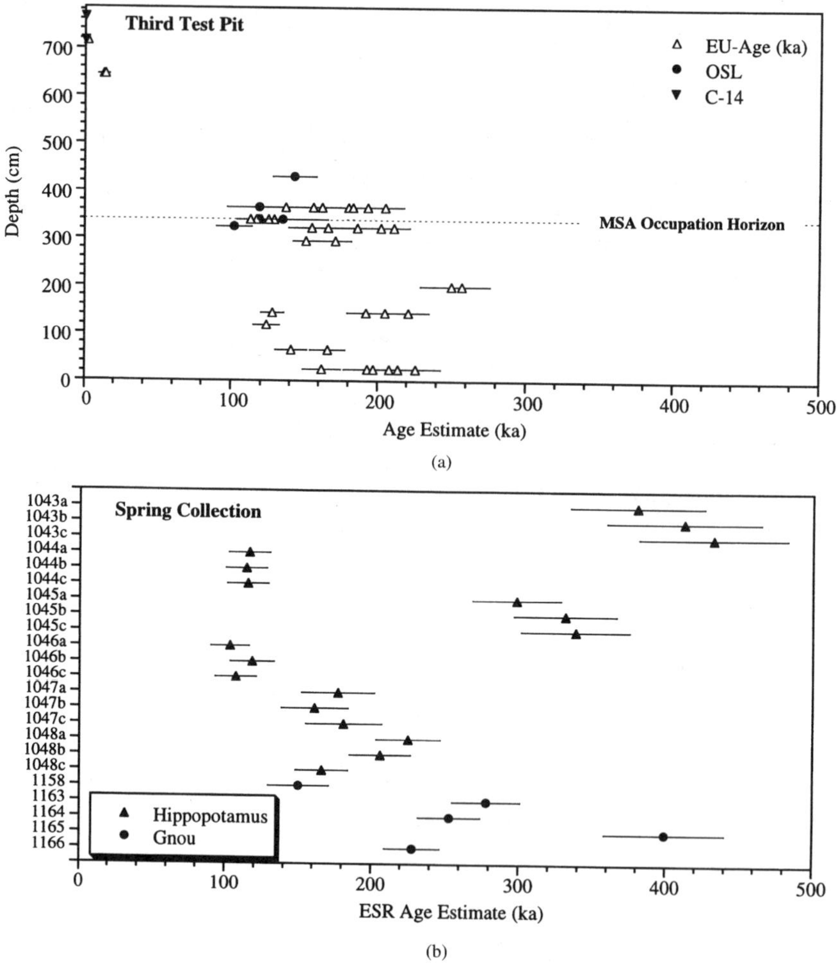

Figure 19.18. (*a*) Age estimates for the test pit (the results of the occupation horizon are projected into the sequence). (*b*) Apparent ESR age estimates of the spring collection samples (both graphs after Grün et al. 1996).

pation horizon corresponding to the last interglacial period of around 125 ka. The results for the hominid are clearly older, most likely coinciding with the previous interglacial (200 to 250 ka) confirming the view of Beaumont et al. (1978) that Peats I and II represent two interglacial periods separated by a full glacial and that the hominid remains could well relate to an Early Stone Age industry. The relatively old age of the hominid fits well with its combination of archaic and modern characteristics, considering that anatomically

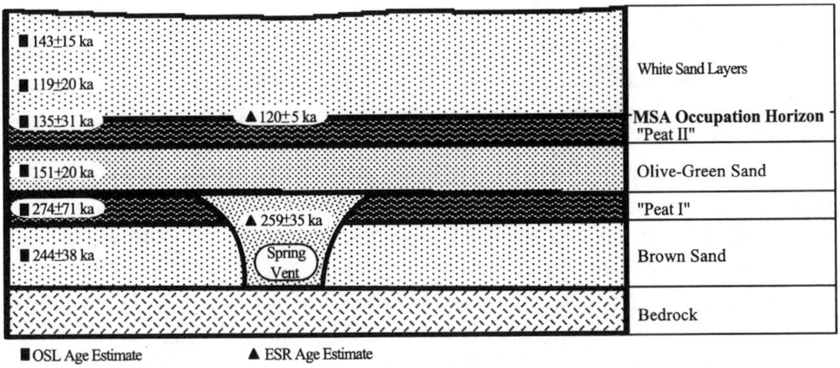

Figure 19.19. Schematic diagram of the lower part of the sediment sequence at Florisbad with a reconstruction of the spring vent yielding the hominid fossils (not to scale) (after Grün et al. 1996). The spring mount dissected the basal sand and Peat I and was covered by the olive green sand. The ESR results yield an age of 259 ± 35 ka for the human tooth and an age of 120 ± 5 ka for the younger human occupation horizon. Although having larger errors (because of saturation problems), the OSL age estimates confirm the large age span between the beginning of sedimentation to the deposition of the occupation horizon.

modern humans are known from several southern African sites from the last interglacial (Grine and Klein 1985; Grün and Stringer 1991; Bräuer 1992). Direct dating of human tooth enamel is presently being carried out on specimens from Skhul, Tabun, Atapuerca, and Lake Mungo.

19.3.2. Human Evolution in China

Longgupo. This cave yielded some human teeth that have been attributed to a precursor of *Homo erectus* (Huang et al. 1995c). Paleomagnetic studies showed that the layers in which the human teeth were found had normal polarity. ESR results on covering layers resulted in ages of between 750 ± 90 ka (EU) and 1020 ± 120 ka (LU). These results could be used to anchor the paleomagnetic stratigraphy, and the normal polarity of the hominid bearing layers could be attributed to the Olduvai event (1.96 to 1.78 Ma).

Yuan Mou. Similar to Longgupo, a few isolated human teeth, attributed to *Homo erectus*, were found in layer 25 of the Yuanmou sequence in Yunnan Province. Paleomagnetic studies found that layer 25 was part of the Matuyama period, however, showing normal polarity. This was attributed to a time period of 1.7 ± 0.1 Ma, just after the Olduvai Event (Qian and Zhou 1991). ESR results of teeth (Huang and Grün 1998) gave ages of between 820 ± 100 ka (EU) and 1300 ± 200 ka (LU), clearly supporting the Early Pleistocene age of the hominid remains.

Zhoukoudian. The site of Peking Man is the most famous paleoanthropological and archaeological site in China. Considerable efforts have been spent to date the site, and in 1985 Wu et al. published a detailed chronology for the site (see Table 19.2). Each of the layers could be attributed to a specific warm or cold climatic cycle of the oxygen isotope record between stage 6 (130 to 190 ka) and 19 (765 to 785 ka). According to this scheme, layer 3 (containing the skull cap HIII) is between 245 and 295 ka; layer 8 to 9 (skull caps LI–LIII) 515 to 460 ka, and layer 11 (skull cap E1) 520 to 555 ka. Teeth from layers 3, 6/7, and 10 were analyzed by ESR (Grün et al. 1997*b*; see Fig. 19.20). The ESR results on their own would imply that the sequence of layers 3 to 10 were deposited between about 300 to 350 ka (layer 3) and about 400 to 550 ka (layer 10), that is, representing a considerably shorter time span as advocated by Wu et al. (1985). However, Grün et al. (1997b) preferred to interpret the ESR results in a manner that the independent age assessment of Wu et al. (1985) could be used to understand the uranium uptake mode of the samples. Therefore, samples from layer 3 would have experienced early uranium uptake, those of layer 6/7 an intermediate uranium uptake between EU and LU, and samples from layer 10 linear uranium uptake.

Alpha spectrometric uranium-series results (Shen 1994) had already implied that the speleothems of layer 2 may be as old as 390 ± 30 ka. The results of this study were reconfirmed by high-resolution thermal ionization mass spectrometry (TIMS) uranium-series analyses that reported ages of 385 ± 25 and 400 ± 12 ka for the base of layer 1 (Cheng and Edwards 1998), 414 ± 13 ka for layer 2 (Shen et al. 1996), and 540^{+80}_{-40} ka for a speleothem from layer 3 (Cheng and Edwards 1998). Some people doubted the exact provenance of these speleothems samples. A compact collected *in situ* at the top of the sedimentary sequence, probably related to layers 1 or 2, yielded ages of about 144 ka at the top and 322 ka at the base (Grün, unpublished data). All uranium-series results clearly imply that hominids found in layer 3 are older than about 400 ka and that it is more likely that the whole sedimentary series of layers 3 to 13 were deposited between about 400 to 500 and 785 ka rather than 240 to 785 ka. Furthermore, it seems doubtful that all isotope stages of these time periods are represented in the sequence. For a discussion about the reassessed significance of the site, the reader is referred to Weiner et al. (1998) and Wuethrich (1998).

Hexian. The site of Hexian produced portions of several human crania. These were attributed to *Homo erectus* and based on some progressive features correlated with the specimen HIII found in layer 3 of Zhoukoudian. The ESR results of Grün et al. (1998) are shown in Figure 19.21*a*. These cover a large time span of 344 ± 48 ka (EU) to 465 ± 94 ka (EU). These

Table 19.2. Depositional Cycles and Ages of Cave Deposits of Zhoukoudian[a,b]

Layer	Thickness (m)	Lithological Description	Depositional Character	Previous Age Estimates (ka)	Loess Sequence[c]	Oxygen Isotope Stages and Boundaries[d]	Magnetic Polarity
1	1.5	Brownish yellow breccia with silt	Colluvium		L2	6 190 ka	Brunhes Magnetic Epoch
2	1.7	Brownish red silty breccia and travertine	Sedimentation by impound water	221 ± 84-E(1) 230^{+30}_{-23}-U(3)	S2	7 245 ka	
3	3.6	Coarse and large breccia with skull cap HIII	Colluvium and collapse of cave walls and roof	256^{+60}_{-40}-U(3) 249 ± 51-U(4) 282 ± 59-E(2)	L3	8 295 ka	
4	6.9	Colored ash and clayey silt	Sheet flow and evidence for the use of fire	292 ± 26-T(8) 306 ± 56-F(6) 316 ± 80-E(2) 320 ± 64-E(2)	S3	9	
5	0.4	Travertine	Chemical precipitation from underground flow			330 ka	
6	7.1	Fine and coarse breccia and huge blocks	Colluvium and collapse of cave walls and roof	355-U(5) 368 ± 85-E(2)	L4	10 375 ka	

Table 19.2. (Continued)

Layer	Thickness (m)	Lithological Description	Depositional Character	Previous Age Estimates (ka)	Loess Sequence[c]	Oxygen Isotope Stages and Boundaries[d]	Magnetic Polarity
7	1.5	Grayish yellow fine silt flow	Sedimentation from underground	385 ± 85-E(2) 396 ± 46-E(2)	S4	11	
						415 ka	
8–9	4.0	Breccia and blocks, skull caps LI–LIII	Colluvium and collapse of cave walls and roof	423 ± 80-E(2)	L5	12	
						460 ka	
10	0.6	Colored ash and silty clay	Sheet flow and evidence for the use of fire	462 ± 45-F(7)	S5a	13	
						520 ka	
11	0.8	Grayish brown breccia with skull cap EI	Disintegration of cave wall	585 ± 105-E(2)	S5bc	14	
						555 ka	
12–1	0.9	Brown coarse sand with fine gravels	Sedimentation from underground flow		S5de	15	
						615 ka	

12–2	0.6	Brown coarse sand and breccia	Disintegration of cave wall	669 ± 84-E(1)	L6	16		
							660 ka	
13–1	0.3	Reddish brown silty clay	Sheet flow		S6	17		
							690 ka	
13–2	3.0	Breccia	Disintegration of cave wall		L7	18		
							765 ka	
13–3	1.5	Brown silty clay	Sheet flow		S7	19		
							785 ka	Brunhes-Matuyama Boundary

[a] From Grün et al. (1997b).

[b] Dating results: U = uranium series; T = thermoluminescence; F = fission track; E = electron spin resonance; (1) preliminary results (Huang et al. 1991); (2) Huang et al. (1993); (3) Zhao et al. (1985); (4) Yuan and Chen (1980); (5) Xia (1981); (6) Guo (1991); (7) Liu et al. (1985); (8) Pei (1985).

[c] Loess sequence from Liu and Ding (1980); Kukla (1987).

[d] Oxygen isotope stage boundaries from Prell et al. (1986); Martinson et al. (1987); Shackleton et al. (1990).

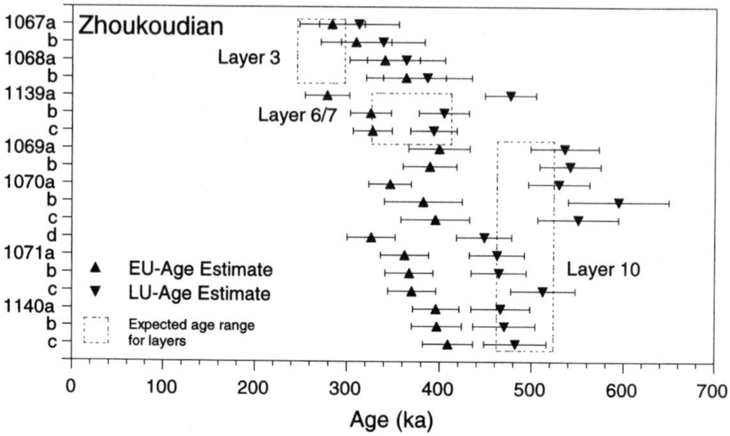

Figure 19.20. ESR age estimates for tooth samples from Zhoukoudian (from Grün et al. 1997, modified). The boxes represent the durations of the oxygen isotope stages that have been assigned to the respective layers (see Table 19.2). No precise depth information was available for the samples.

ages were considerably older than ESR results obtained by Huang et al. (1994; 1995a, 1995b) who obtained ages in the range of 160 to 220 ka (EU) and 250 to 350 ka (EU) and uranium-series results by Chen and Yuan (1988) of 130 to 220 ka (combined $^{231}Pa/^{230}Th$). Grün et al. (1998) also carried out combined uranium-series/ESR age determinations on two samples (Fig. 19.21*b*). While the individual age determinations show a very wide scattering (uranium-series between about 150 and 550 ka; ESR between about 300 and 600 ka), the combination of both methods can explain the very different uranium uptake history that both samples have experienced (Fig. 19.21*c*) and the combination of the uranium-series/ESR data result in a very tight age estimation of 412 $\pm$ 25 ka.

This result was first greeted with scepticism because it implied that the Hexian calvaria were contemporaneous with the morphologically less advanced *Homo erectus* specimen from Zhoukoudian (LI–LIII) rather than more similar HIII. However, when considering the new TIMS uranium-series results on Zhoukoudian, the morphological correlation Hexian–Zhoukoudian HIII is now supported by chronological evidence, but both specimens are considerably older than previously thought.

Yunxian. Layer III of the Middle Pleistocene terrace of the Han River contained two almost complete *Homo erectus* crania that show some features of archaic *Homo sapiens* (Li and Etler 1992). Paleomagnetic measurements suggested an age of 830 to 870 ka. Chen et al. (1997) carried out ESR and

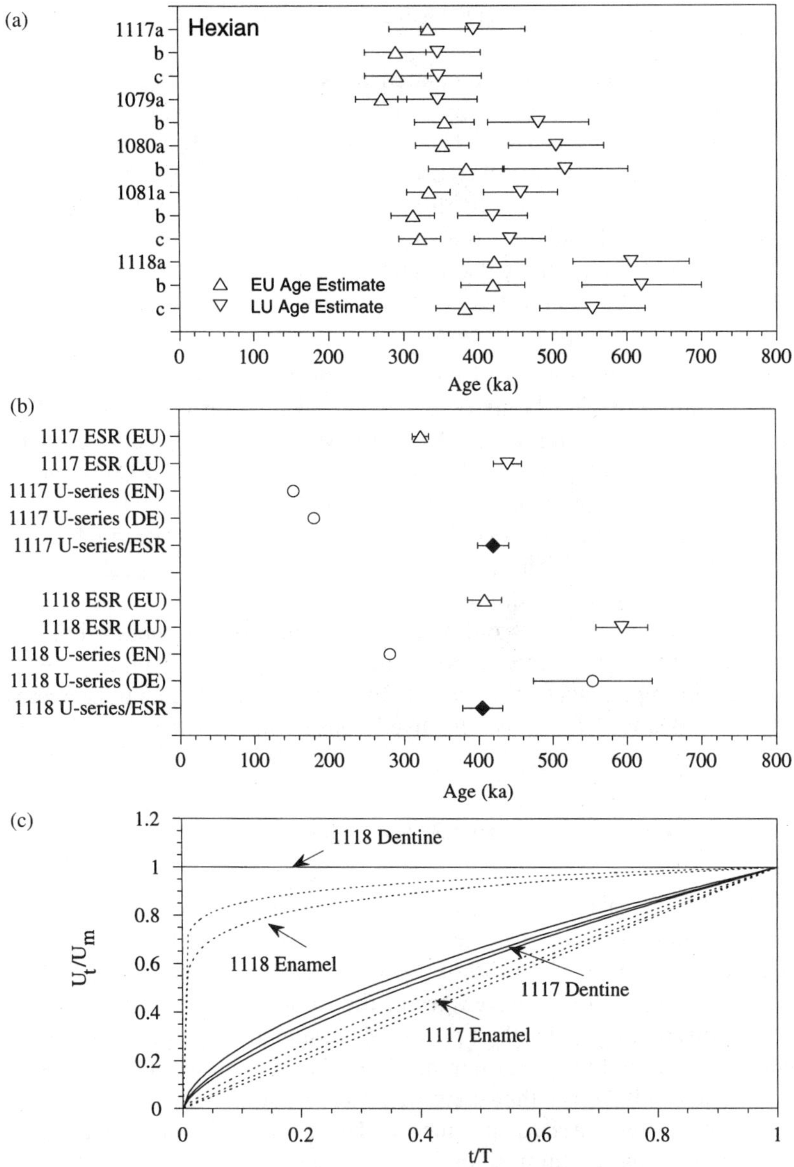

Figure 19.21. Age determination on teeth from Hexian (from Grün et al. 1998, modified). (*a*) ESR age estimates on teeth from Hexian. (*b*) Age estimates for samples 1117 and 1118 (EN = enamel; DE = dentine). Both the ESR and uranium-series age estimates on the two teeth imply that the two teeth have distinctively different ages. The combined uranium-series/ESR age estimates are virtually indistinguishable and give the best age estimate for the site (412 ± 25 ka). (*c*) Modeled uranium uptake histories. Sample 1118 shows uranium uptake that is near to a closed system model whereas the constituents of sample 1118 are closer to a linear uptake model.

uranium-series analyses and calculated ages of 581 ± 93 ka [EU, recalculated by Yang (1997) using one-group beta attenuation values to 599 ± 118 ka] and 1051 ± 150 ka (LU). The uranium-series analyses suggested that enamel experiences slightly delayed uranium uptake whereas minor leaching might have occurred from the dentine. Thus, Chen et al. (1997) preferred the EU age estimates. Similar to the sites of Longgupo and Yuan Mou, ESR seems to underestimate the paleomagnetic reference ages to some extent but confirm, on the other hand, the great antiquity of the sites.

Jinniushan. This site contained large parts of a skeleton of an early *Homo sapiens*. Uranium-series dating of bones had yielded results of about 230 to 300 ka (Chen and Yuan 1988). ESR analysis of several teeth resulted in weighted averages of 185 ± 10 ka (EU) and 278 ± 13 ka (LU), indicating the early occurrence of *Homo sapiens* in China (Chen et al. 1994). Using one-group beta attenuation factors, Yang (1997) obtained a weighted average EU age of 224 ± 13 ka. This particular case illustrates the importance of the correct calculation of beta attenuation factors.

It was assumed that this old age of the Jinnuishan specimen implied a long co-occurrence of *Homo erectus* and *Homo sapiens* in China. However, the newer dating results of the sites of Zhoukoudian and Hexian (see above) imply that advanced *Homo erectus* lived considerably earlier than the specimen found at Jinniushan.

19.3.3. Summary

The ESR results carried out on Chinese paleoanthropological sites confirm that early humans occurred at least 1.5 Ma. The new dating results imply that relatively advanced forms of *Homo erectus*, as found in layers 3 at Zhoukoudian and Hexian have ages in the range of 400 to 500 ka, that is, they are considerably older than previously assumed. A precursor of modern *Homo sapiens* found at Jinnuishan has been dated to between about 200 and 300 ka. The results imply that there was no temporal co-occurrence between the advanced *Homo erectus* specimen at Hexian and Zhoukoudian and the early *Homo sapiens* of Jinnuishan.

ESR seems to lead to age underestimation for the sites of Longgupo, Yuan Mou, and Yunxian. These studies have in common that teeth with lower uranium concentrations yield older apparent ages than those with high uranium concentrations. This points to a problem related to the correct estimation of the uranium dose rate (not related to uranium uptake), which is not yet solved.

19.4. FUTURE DEVELOPMENTS

The main advance in ESR dating of tooth enamel lies in the direct dating of human material using enamel fragments and, possibly, whole teeth. Uranium of enamel fragments can be measured using laser ablation induced coupled plasma–mass spectrometry (ICP–MS), which allows the precise analysis of very small samples (e.g., Eggins et al. 1998). This technique can be used to establish uranium profiles along very thin holes (50 μm diameter) that are nearly invisible. New advances in mass spectrometry, particularly isotopic analyses with new generation multicollector ICP sector mass spectrometers (e.g., Halliday et al. 1998), will significantly reduce laboratory preparation for uranium-series dating. Thus, combined ESR/uranium-series dating could be developed as a routine method within the next few years. ESR measurements can also provide important restraints when hominid fossils are dated nondestructively by gamma spectrometric uranium-series dating (see Simpson and Grün 1998).

Acknowledgment

I wish to thank the numerous colleagues and friends who have provided scientific stimulus and guidance over the years, in particular—apart from the late Gerd Hennig—Henry Schwarcz, Ann Wintle, Ulrich Radtke, Eddie Rhodes, and John Chappell. Samples, advice, collaborative results, and discussions in context of this review were kindly provided by James Brink, Tiemei Chen, Pei-Hua Huang, Henri Laville, Malcolm McCulloch, Frank McDermott, Paul Mellars, Chris Stringer, and Alan Thorne. I thank Anne Gillard and Günther Wagner for proof reading.

REFERENCES

Abeyratne, M., N. Spooner, R. Grün, and J. Head (1997), *Quat. Sci. Rev.* **16**, 243.

Aitken, M. J. (1985), *Thermoluminescence Dating*, Academic, New York.

Aitken, M. J. (1990), *Science-Based Dating in Archaeology*, Longman, London.

Aitken, M. J. (1997), in *Chronometric and Allied Dating in Archaeology*, R. E. Taylor and M. J. Aitken, Eds., Plenum, New York, pp. 183–216.

Aitken, M. J. (1998), *Optical Dating*, Oxford University Press, Oxford.

Aitken, M. J., and J. Xie (1990), *Ancient TL* **8**, 6.

Barabas, M. (1992), *Nucl. Tracks Radiat. Meas.* **20**(3), 453.

Beaumont, P., H. de Villiers, and J. C. Vogel (1978), *S. Afr. J. Sci.* **74**, 409.

Blackwell, B. A., and H. P. Schwarcz (1993), *Appl. Radiat. Isotopes* **44**, 243.

Bouchez, R., R. Cox, A. Herve, E. Lopez-Carranza, J. L. Ma, M. Piboule, G. Poupeau, and P. Rey (1988), *Quat. Sci. Rev.* **7**, 497.

Bowman, S. G. E. (1976), Thermoluminescence Dating, the Evaluation of Radiation Dosage, Ph.D. Thesis, University of Oxford, Oxford.

Bräuer, G. (1992), in *Continuity or Replacement: Controversies in Homo Sapiens Evolution*, G. Bräuer and F. Smith, Eds., Balkema, Rotterdam, pp. 83–98.

Brennan, B. J., W. J. Rink, E. L. McGuirl, H. P. Schwarcz, and W. V. Prestwich (1997), *Radiat. Meas.* **27**, 307.

Brink, J. S. (1987), *Mem. Nas. Mus. Bloemfontein* **24**, 151.

Brink, J. S. (1988), *Palaeoecol. Afr.* **19**, 169.

Brumby, S. (1992), *Nucl. Tracks Radiat. Meas.* **20**, 595.

Callens, F., P. Moens, and R. Verbeeck (1995), *Calcified Tissues Int.* **56**, 543.

Chen, T. M., and S. Yuan (1988), *Archaeometry* **30**, 59.

Chen, T. M., Q. Yang, and E. Wu (1994), *Nature* **368**, 55.

Chen, T. M., Q. Yang, Y. Q. Hu, W. B. Bao, and T. Y. Li (1997), *Quat. Sci. Rev. (Quat. Geochronol.)* **16**, 455.

Cheng, H., and R. L. Edwards (1998), *Chinese Sci. Bull. Suppl.* **43**, 26.

Clarke, R. J. (1985), in *Ancestors: The Hard Evidence*, E. Delson, Ed., Wiley-Liss, New York, pp. 301–305.

DeCanniere, P., R. Debuyst, F. Dejehet, D. Apers, and R. Grün (1986), *Nucl. Tracks Radiat. Meas.* **11**, 211.

Dreyer, T. F. (1938), *Argeol. Navors. Nas. Mus. Bloemfontein* **1**, 65.

Driessens, F. C. M. (1980), *Bull. Soc. Chim. Belg.* **89**, 663.

Edwards, R. L., J. H. Chen, T. L. Ku, and G. J. Wasserburg (1987a), *Science* **236**, 1547.

Edwards, R. L., J. H. Chen, and G. J. Wasserburg (1987b), *Earth Planet. Sci. Lett.* **81**, 175.

Eggins, S. M., L. P. J. Kinsley, and M. Shelley (1998), *Appl. Surf. Sci.* **127–129**, 278.

Grine, F. E. (1988), *Evolutionary History of the "Robust Australopithicines,"* Aldine d'Gruyter, New York.

Grine, F. G., and R. G. Klein (1985), *Anthropology* **8**, 55.

Grün, R. (1985), *Sonderveröffentlichungen des Geologischen Instituts der Universität zu Köln* **59**, 1.

Grün, R. (1989a), *Quat. Int.* **1**, 65.

Grün, R. (1989b), *Die ESR-Altersbestimmungsmethode*, Springer, Berlin-Heidelberg.

Grün, R. (1995), *Ancient TL* **13**, 3.

Grün, R. (1996), *J. Human Evol.* **30**, 227.

Grün, R. (1997), in *Chronometric and Allied Dating in Archaeology*, R. E. Taylor and M. J. Aitken, Eds., Plenum, New York, pp. 217–261.

Grün, R. (1998a), *Radiat. Meas.* **29**, 177.

Grün, R. (1998b), *Radiat. Meas.* **29**, 579.

Grün, R. (1998c), *Ancient TL* **16**, 51.

Grün, R., and S. Brumby (1994), *Radiat. Meas.* **23**, 307.

Grün, R., and R. Clapp (1996), *Ancient TL* **14**, 1.

Grün, R., and M. Jones (1996), *Radiat. Meas.* **26**, 621.

Grün, R., and O. Katzenberger-Apel (1994), *Ancient TL* **12**, 35.

Grün, R., and F. McDermott (1994), *Quat. Geochronol. (QSR)* **13**, 121.

Grün, R., and H. P. Schwarcz (1987), *Ancient TL* **5**, 1.

Grün, R., and C. B. Stringer (1991), *Archaeometry* **33**, 153.

Grün, R., and L. Taylor (1996), *Ancient TL* **14**, 21.

Grün, R., and A. Thorne (1997), *Science* **276**, 1575.

Grün, R., H. P. Schwarcz, and S. Zymela (1987), *Can. J. Earth Sci.* **24**, 1022.

Grün, R., H. P. Schwarcz, and J. M. Chadam (1988), *Nucl. Tracks Radiat. Meas.* **14**, 237.

Grün, R., P. Beaumont, and C. B. Stringer (1990), *Nature* **344**, 537.

Grün, R., P. Mellars, and H. Laville (1991), *Antiquity* **65**, 544.

Grün, R., J. S. Brink, N. A. Spooner, L. Taylor, C. B. Stringer, R. G. Franciscus, and A. S. Murray (1996), *Nature* **382**, 500.

Grün, R., H. Kohno, A. Tani, C. Yamanaka, M. Ikeya, and H. P. Huang (1997a), *Radiat. Meas.* **27**, 425.

Grün, R., P. H. Huang, X. Wu, C. B. Stringer, and A. Thorne (1997b), *J. Human Evol.* **32**, 83.

Grün, R., P. H. Huang, W. Huang, F. McDermott, C. B. Stringer, A. Thorne, and G. Yan (1998), *J. Human Evol.* **34**, 555.

Grün, R., G. Yan, M. McCulloch, and G. Mortimer (1999), *J. Archaeol. Sci.* **26**, 1301.

Guo, S. L. (1991), *Nucl. Tracks Radiat. Meas.* **19**, 719.

Halliday, A. N., D. C. Lee, J. N. Christensen, M. Rehkämper, W. Yi, X. Luo, C. M. Hall, C. J. Ballentine, T. Pettke, and C. Stirling (1998), *Geochim. Cosmochim. Acta* **62**, 919.

Hassan, A. A., J. D. Termine, and C. V. Haynes, Jr. (1977), *Radiocarbon* **19**, 364.

Huang, P., and R. Grün (1998), *Acta Anthrop. Sin.* **17**, 165.

Huang, P. H., S. Z. Jin, R. Y. Liang, Z. J. Lu, L. Z. Zheng, Z. X. Yuan, and B. X. Cai (1991), *Chinese Sci. Bull.* **36**, 1457 (in Chinese).

Huang, P. H., S. Z. Jin, Z. C. Peng, R. Y. Liang, Z. J. Lu, Z. R. Wang, J. B. Chen, and Z. X. Yuan (1993), *Appl. Radiat. Isotopes* **44**, 239.

Huang, P. H., R. Y. Liang, L. Z. Zheng, Y. H. Xu, Y. S. Fang, and D. S. Fang (1994), *Chinese Sci. Bull.* **39**, 1919.

Huang, P. H., L. Z. Zheng, Y. C. Quan, R. Y. Liang, Y. H. Xu, Y. S. Fang, and D. S. Fang (1995a), *Nucl. Techn.* **18**, 491 (in Chinese).

Huang, P. H., L. Z. Zheng, Y. C. Quan, R. Y. Liang, Y. H. Xu, Y. S. Fang, and D. S. Fang (1995b), *Acta Anthrop. Sin.* **14**, 262 (in Chinese).

Huang, W., R. Ciochon, Y. Gu, R. Larick, Q. Fang, H. Schwarcz, C. Yonge, J. de Vos, and W. Rink (1995c), *Nature* **378**, 275.

Ikeya, M. (1975), *Nature* **255**, 48.

Ikeya, M. (1982), *Jpn. J. Appl. Phys.* **21**, L690.

Ikeya, M. (1993), *New Applications of Electron Spin Resonance—Dating, Dosimetry and Microscopy*, World Scientific, Singapore, London, Hong Kong.

Jonas, M. (1997), Electron Spin Resonance Dating and Dosimetry of Tooth Enamel, Ph.D. Thesis, University of Cambridge, Cambridge.

Jonas, M., L. P. Zhou, E. Marseglia, and P. Mellar (1994), *Cambridge Archaeol. J.* **4**, 139.

Kukla, G. (1987), *Quat. Sci. Rev.* **6**, 191.

Kuman, K., and R. J. Clarke (1986), *Geoarchaeology* **1**, 103.

Li, T. Y., and D. A. Etler (1992), *Nature* **357**, 404.

Liu, T. S., and M. L. Ding (1980), *Scientific Papers on Geology for International Exchange*, pp. 77–82.

Liu, S. S., F. Zhang, R. Y. Hu, J. F. Liu, S. L. Guo, S. H. Zhou, W. Meng, B. F. Zhang, and S. F. Sun (1985), in *Multi-disciplinary Study of the Peking Man Site at Zhoukoudian*, R. K. Wu, M. E. Ren, X. M. Zhu, Z. G. Yang, C. K. Hu, Z. C. Kong, Y. Y. Xie, and S. S. Zhao, Eds., Science Press, Beijing (in Chinese), pp. 241–245.

Martinson, D. G., N. G. Pisias, J. D. Hays, J. Imbrie, T. C. Moore, and N. J. Shackleton (1987), *Quat. Res.* **27**, 1.

Maurer, C. A. (1980), Electron Spin Resonance Spectroscopy: A Potential Technique for Dating Ancient Ceramics, Ph.D. Thesis, University of Illinois, Urbana.

McDermott, F., R. Grün, C. B. Stringer, and C. J. Hawkesworth (1993), *Nature* **363**, 252.

Mercier, N., H. Valladas, and G. Valladas (1995), *Quat. Sci. Rev. (Quat. Geochronol.)* **14**, 351.

Nambi, K. S. V., and M. J. Aitken (1986), *Archaeometry* **28**, 202.

Oduwole, A. D., and K. D. Sales (1991), *Nucl. Tracks Radiat. Meas.* **18**, 213.

Oka, T., R. Grün, A. Tani, C. Yamanaka, M. Ikeya, and H. P. Huang (1997), *Radiat. Meas.* **27**, 331.

Ostrowski, K., A. Dziedzic-Goclawska, W. Stachowicz, and J. Michalik (1974), *Ann. N. Y. Acad. Sci.* **238**, 186.

Pei, J. X. (1985), in *Multi-Disciplinary Study of the Peking Man Site at Zhoukoudian*, R. K. Wu, M. E. Ren, X. M. Zhu, Z. G. Yang, C. K. Hu, Z. C. Kong, Y. Y. Xie, and S. S. Zhao, Eds., Science Press, Beijing (in Chinese), pp. 256–260.

Porat, N., and H. P. Schwarcz (1991), *Nucl. Tracks Radiat. Meas.* **18**, 203.

Prell, W. L., J. Imbrie, D. G. Martinson, J. J. Morley, N. G. Pisias, N. J. Shackleton, and H. F. Streeter (1986), *Paleoceanography* **1**, 137.

Prescott, J. R., and J. T. Hutton (1988), *Nucl. Tracks Radiat. Meas.* **14**, 223.

Qian, F., and G. X. Zhou (1991), *Quaternary Geology and Paleoanthropology of Yuanmou, Yunnan, China*, Science Publishing House, Beijing.

Rink, W. J. (1997), *Radiat. Meas.* **27**, 975.

Rubidge, B. S., and J. S. Brink (1985), *Navors. Nas. Mus. Bloemfontein* **5**, 69.

Schwarcz, H. P. (1985), *Nucl. Tracks Radiat. Meas.* **10**, 865.

Schwarcz, H. P., R. Grün, and P. V. Tobias (1994), *J. Human Evol.* **26**, 175.

Shackleton, N. J., A. Berger, and W. R. Peltier (1990), *Trans. R. Soc. Edinburgh: Earth Sci.* **81**, 251.

Shen G. (1994), *L'Anthropologie* **98**, 554.

Shen, G., T. L. Ku, B. Gahleb, and Z. Yuan (1996), *Acta Anthop. Sin.* **15**, 210.

Simpson, J. J., and R. Grün (1998), *Quat. Sci. Rev.* **18**, 1009.

Swisher, C. C., W. J. Rink, S. C. Anton, H. P. Schwarcz, G. H. Curtis, A. Suprijo, and Widiasmoro (1997), *Science* **274**, 1870.

Vanhaelewyn, G., F. Callens, and R. Grün (in press), *Appl. Radiat. Isotopes.*

Weiner, S., Q. Xu, P. Goldberg, J. Liu, and O. Bar-Yosef (1998), *Science* **281**, 251.

Wu, R. K., M. E. Ren, X. M. Zhu, Z. G. Yang, C. K. Hu, Z. C. Kong, Y. Y. Xie, and S. S. Zhao (1985), *Multi-Disciplinary Study of the Peking Man Site at Zhoukoudian*, Science Press, Beijing (in Chinese).

Wuethrich, B. (1998), *Science* **281**, 165.

Xia, M. (1982), *Acta Anthop. Sin.* **1**, 191 (in Chinese).

Yang, Q. (1997), Experimental Determination of Beta Attenuation in Tooth Enamel Layers and Its Implication in ESR Dating, MS.c. Thesis, McMaster University, Hamilton.

Zhao, S. S., M. Xia, Z. H. Zhang, M. L. Liu, S. X. Wang, Q. F. Wu, and Z. B. Ma (1985), in *Multi-Disciplinary Study of the Peking Man Site at Zhoukoudian*, R. K. Wu, M. E. Ren, X. M. Zhu, Z. G. Yang, C. K. Hu, Z. C. Kong, Y. Y. Xie, and S. S. Zhao, Eds., Science Press, Beijing (in Chinese), pp. 246–250.

CHAPTER

20

DATA HANDLING AND STATISTICAL ANALYSIS

MIKE J. BAXTER

Department of Mathematics, Statistics and Operational Research, Nottingham Trent University, Nottingham, United Kingdom

CAITLIN E. BUCK

School of History and Archaeology, University of Wales, Cardiff, Wales, United Kingdom

20.1. INTRODUCTION

The main focus of this chapter is discussion of statistical methods that can be used for investigating structure in tables of data of the kind generated by many of the analytical methods discussed in other chapters. Paradigmatically, suppose that a program of analysis on *n* artifacts gives rise to measurements relating to their chemical composition with respect to *p* oxides/elements. In asking if there is structure in the resultant table of data, we are asking if the *n* artifacts can be divided into chemically distinct subgroups, preferably with archaeological or art historical meaning. If this is possible, then we would also like to display this structure in an informative fashion.

This book is concerned with analytical methods in archaeology and art historical study. In this chapter the term *archaeometry* will be used as "shorthand" for this general area. We believe that archaeometric study should be driven by archaeological and art historical problems—a view that is, perhaps, obvious, but by no means universal. Section 20.2 sets out our views on the kinds of archaeological and art historical problems that necessitate analytical and statistical input.

Given an agreed need for statistical analysis, a variety of approaches are possible. The authors of this chapter have published books on exploratory multivariate methodologies (Baxter 1994a) and Bayesian approaches to in-

Modern Analytical Methods in Art and Archaeology, Edited by Enrico Ciliberto and Giuseppe Spoto. Chemical Analysis Series, Vol. 155
ISBN 0-471-29361-X

terpreting archaeological data (Buck et al. 1996). The latter can be viewed as a specific approach to the statistical modeling of data. In contrast to the exploratory approach, it is more demanding in terms of the assumptions that need to be made about the data but also potentially more rewarding in that formal inferences can be drawn. Exploratory and modeling approaches to data analysis are complementary, and we thought it useful to include discussion of both within the chapter. In archaeometry the exploratory approach has dominated to date; the potential of modeling approaches is something that we hope will attract more attention in the future.

The main exploratory techniques used in archaeometry are illustrated by example in Section 20.3 and discussed more fully in Section 20.5, after a brief review of techniques for preliminary data analysis in Section 20.4. Some practical considerations, that perhaps deserve more attention than they have received in the literature, are addressed in Section 20.6. Section 20.7 provides discussion and illustration of the Bayesian approach, and Section 20.8 concludes the chapter.

While the use of mathematics is unavoidable in a chapter of this kind, we have, for the most part, attempted to discuss and illustrate ideas in what we hope is an accessible fashion. A partial exception to this is the discussion of the uses of Mahalanobis distance in Section 20.5.5 where, for convenience of reference, some relatively inaccessible material is gathered. The books cited earlier, and references in the text, may be consulted for a more detailed and more mathematical treatment of the material covered in this chapter.

It has already been noted that the modeling approach to data analysis has been less exploited in archaeometry than the exploratory approach. For reasons of space this chapter concentrates on the problem defined in the opening paragraph. This means that we have not dealt with one of the great successes of the Bayesian approach in its application to dating problems. The interested reader is referred to Buck et al. (1991, 1992, 1994), Buck and Christen (1998), Christen and Buck (1998), Litton and Buck (1996), Christen (1994a, 1994b), and Zeidler et al. (1998) for a full discussion of such applications.

20.2. PROBLEMS AND STATISTICAL APPROACHES

20.2.1. Nature of the Data

When considering their interpretation, data that arise from the methodologies outlined in the preceding chapters can be considered in two broad categories. The first are data that require very specialist statistical techniques simply to make the methodology usable (e.g., radiocarbon dating and DNA sequencing). The second are data that cannot easily be interpreted without some manipulation or compression but for which standard statistical methodologies exist even if considerable fine tuning is required (e.g., chemical compositional data).

Almost all of the analytical techniques described in this book give rise to data of both types. The specialist statistical techniques required simply in order for a particular methodology to be utilized (e.g., calibration or pattern matching) are outlined in the relevant chapters or other works referenced therein and are not considered here. Here, focus is placed on data of the second type that arise as the final output from almost all the analytical techniques discussed. We look in some detail at the range of statistical techniques utilized to aid in their interpretation, at those most commonly adopted, and at some that we feel deserve further consideration or are likely to prove useful in the future.

In very general terms, analytical data consist of one or more numerical observation from each art historical or archaeological object of interest. Such data, it is usually hoped, characterize the object in some way. They might be as simple as the weight of each object (or some component part), or the percentage of a single chemical element in each object (or a specimen taken from it), or they could be more complex, consisting of a number of observations per object (such as the percentages of 30 chemical elements present in each specimen analyzed). In this chapter we consider a selection of statistical methodologies for representing and interpreting data of these types.

20.2.2. Why Data Are Collected

Such data are commonly collected for two reasons:

1. To allow investigation and characterization of a single object or specimen (such as a possible fraud or an unusual pigment in a painting).
2. To allow investigation and characterization of a group of objects or specimens (such as collections of bronze objects and material from possible copper sources).

Investigations of the first type often require only simple summary statistics (such as a mean and standard deviation) or a simple plot to reveal properties of interest. If the data obtained are to be compared with others (e.g., a library of previous results), then investigation of the second type is usually needed rather than the first.

Data obtained as part of an investigation of the second type require careful selection of appropriate summary statistics, plotting mechanisms, and perhaps more sophisticated numerical analysis to arrive at useful archaeological or art historical conclusions. In the remainder of this chapter focus is primarily placed upon situations in which analytical results from several archaeological or art historical objects are available that are to be investigated and interpreted as one corpus of data.

20.2.3. Form of the Data

Consider, first, a situation in which there is available just one ($p = 1$) numerical observation relating to each of the n specimens analyzed. Such data are described as univariate (or one dimensional) and are commonly reported using summary statistics (such as mean, median, mode, and standard deviation) and/or appropriate graphs (such as histograms and line graphs). Where the data exhibit symmetry and have a single peak (are unimodal), summary statistics alone may be an adequate means for reporting them; but, where they are not symmetric or are multimodal, a graph (such as a histogram) or rather more sophisticated summary statistics will probably be needed.

The next level of complexity (from a data presentation and interpretation perspective) is bivariate (or two dimensional) data for which each of the n specimens gives rise to two ($p = 2$) numerical observations. One example of such data would be the percentages of two chemical elements present in each specimen (e.g., tin and arsenic in bronzes). In some cases the two variables will be considered as independent and interpreted separately, effectively treating the analytical results as two univariate data sets. Commonly, however, if two variables are collected as part of the same analytical work, it is because the two pieces of information are interrelated, and it is not helpful to consider them as independent.

In such situations, the data would normally be plotted (in the first instance) on an X-Y plot so that both variables can be viewed together (one on the X axis and the other on the Y). On such a plot, any obvious structure in the data will be observed as patterning or clustering, which then requires interpretation in the light of our knowledge of the archaeological and art historical setting and the physics and chemistry of the methodology used to produce the data. Once any patterning has been observed, more sophisticated bivariate statistical procedures might then be adopted to allow summarization of the nature of the patterns or quantification of the strength of the relationships they represent (see below).

Any data with more than two observations per specimen are described as multivariate (or multidimensional) and are investigated in much the same way regardless of whether there are 3 ($p = 3$) or 30 ($p = 30$) observations relating to each specimen. In simple terms, each data set consists of p observations from n archaeological or art historical specimens. This gives rise to an $n \times p$ data matrix (n rows and p columns) that can be represented by

X. Individual data values within **X** are referred to as x_{ij} where $i = 1, \ldots, n$ and $j = 1, \ldots, p$. In the remainder of this chapter the terms *cases* or *specimens* will be used to refer to the n rows of **X**, while the p columns (often oxides measured as a percentage or trace elements measured as parts per million) will be referred to as *variables*. Uppercase letters, such as X_j, will be used to denote the jth variable.

Some statistical methodologies require that the data in **X** have specific properties (e.g., that the data are from a multivariate normal distribution). In such situations, investigation of the properties of **X** is needed and adjustments may have to be made so that the data have the appropriate form. Such adjustments are usually undertaken using specific mathematical functions (such as logarithms) and are known as transformations. In the remainder of this chapter any such transformation of **X** will be represented by **Y**. Individual values in **Y** are represented by y_{ij}.

Some of the statistical procedures adopted later in the chapter (such as principal component analysis) require linear combinations of the variables defining **Y** and give rise to output in the form of another matrix **Z**. **Z** summarizes certain important features of **Y**, and the first few columns of **Z** are commonly plotted to aid in the interpretation of **Y** and, indirectly, **X**.

20.2.4. Questions to Be Addressed

The specific archaeological or art historical questions that individual research groups are attempting to tackle are, as highlighted by the preceding chapters of this book, extremely varied. Nonetheless, the types of questions that they are addressing are broadly similar: Where did this object come from? What is this object made from? Could this object have been made at the same time and/or place as those ones? In general, the objectives are to identify similarities and difference between the specimens analyzed. These objectives are usually approached by searching for patterns or groupings in the data in an attempt to establish their properties in both qualitative and quantitative terms.

Having identified initial groupings, it is often useful to use our understanding of these as the basis of further research. Groupings in chemical compositional data from ceramics, for example, might be felt to relate to particular kiln sites, pottery workshops, or clay sources. With such groups established, future research work may include attempts to attribute new specimens to one of the existing groups or to identify new groups not previously recorded. Such problems can, to a certain extent, be tackled by visual inspection of the raw data and use of simple plots, but to quantify the similarities and difference between specimens and groups of specimens somewhat more sophisticated techniques are commonly sought.

In looking for suitable techniques, analysts working on archaeological and art historical data are commonly tackling one of two problems. The first arises in situations where sources or sites of manufacture are known and give rise to samples that can be analyzed alongside those arising from excavation or other research. The second arises in situations where no known production sites have been identified so that the only available data arise from artifacts found during excavation, via field survey or in museum collections.

In investigations of the first type, the objective is to allocate new specimens to one or other of the known sources or sites of manufacture. Statistical techniques for tackling this problem are usually termed *discriminant analysis* as they allow the known groups to be discriminated one from another. Once the statistical properties of the known groups have been determined, samples of unknown origin can be allocated to the group whose statistical properties are most similar to their own. Since production sites have not been identified for very many materials in most archaeological and historical time periods, projects of this type are much less common than those of the second type.

For problems of the second type, a methodology is sought to allow identification of clusters of data points that might (given certain properties such as chemical composition) be related one to another. Analysts hope that any such clusters can then be linked to the provenances from which the artifacts originate. Statistical techniques for achieving this are usually referred to as *cluster analysis* techniques and have been widely used in studies of archaeological and art historical data.

20.2.5. Problem Definition

In this chapter we have, thus far, only considered data handling and analysis in very general terms. When simply defining notation and concepts, this is helpful and appropriate, but it hides much of what is really involved in applying statistical techniques to aid in the interpretation of archaeometric data. In practice, each data set is collected with a view to addressing particular archaeological or art historical issues, which must be carefully defined and the statistical approaches fine tuned accordingly.

Consider, once again, chemical compositional data from pottery specimens. To what are groupings in such data likely to relate? The kiln site at which the pots were made? The source from which the clay (or other raw materials) was obtained? Techniques adopted in the manufacture of the vessels? The problem is that, in fact, any one of these (and many other) factors can have a bearing on the number, nature, and relevance of groupings identified in the data whether they arise from visual inspection or lengthy and rigorous statistical analysis. The statistical analysis should not be divorced from any other parts of the investigation.

It is thus not uncommon for a cyclic approach to data analysis to be needed. For example, a clear and plausible relationship might be established between the data collected and an archaeological or art historical question currently of interest. Suitable statistical techniques could then be adopted and considerable research undertaken only to find that the relationship initially posited is not as expected. It would then be necessary to rethink the relationship between the analytical data and the questions under investigation and try again.

Returning to consider the use of chemical compositional data to aid in the sourcing of pottery. Naively, it could be assumed that the relationship between the chemical composition and the site of production ought to be a simple one and that any groupings in the data will relate directly to production sites of known or unknown location. The reality is, however, nothing this simple. There are factors other than site of production that can produce clusterings in the data. Such factors include:

- The use of tempers mixed with the clay in varying proportions, thus diluting the proportions of some chemical elements and enhancing others.
- The use of clays from more than one source at the same kiln site, thus creating more than one cluster (or overlapping clusters) relating to that site.
- Poor mixing of the raw materials (or large inclusions in the clay) resulting in specimens from different places on the same vessel having different chemical compositions and possibly being allocated to different clusters by the statistical techniques adopted.
- Insufficient care when taking specimens for analysis resulting in contamination (e.g., with glaze or the residue of soil or other material that the vessel contains), thus adding noise to the analytical data that could create phantom groupings in the data.

Such problems are now known in a range of research areas and have been widely documented and discussed (see, e.g., Bishop and Neff 1989). This does not mean that sourcing problems are now easy to address. Identifying the problems does not necessarily bring simple solutions. Indeed, just because the problems are relatively easy to describe does not mean that statistical techniques can be found to take account of them. Most statistical techniques adopted to aid in archaeometric study are descriptive rather than truly interpretative. They allow quantification, display, and summarization of the data but do not necessarily lead to clear or obvious archaeological or art historical interpretations. The interpretation must be made in the light of all the prior information available including production techniques, postproduction usage of the object, storage conditions, laboratory and conservation methods adopted, sampling techniques used to obtain material for analysis, the physics and chemistry of the analytical techniques adopted, and so on.

If there is a desire to use statistical techniques to lead coherently and directly to clear or obvious interpretations (rather than just summaries and descriptions to be considered in the light of the relevant prior information), it is vital to build our understanding of the processes that gave rise to the data into the statistical procedures adopted. In other words, clear theories are needed that relate the data to what we want to learn. Unfortunately, however, it is not sufficient to summarize these theories in words, it is necessary to use them to build mathematical models so that specific properties of the data can be related to specific features of the problem under investigation. With all the prevailing issues discussed above, this is not an easy thing to achieve and, as a consequence, few archaeologists, art historians, or analysts interpreting archaeometric data have adopted model-based approaches. In the remainder of this chapter we will look first at descriptive approaches to interpreting archaeometric data and then move on to discuss the strengths and weaknesses of adopting the model-based alternatives.

20.2.6. Notation and Terminology

Here we collect notation used in the rest of the chapter. Reference can be made back to this section, as appropriate. Many of the analytical techniques described in the book give rise to an $n \times p$ matrix of values $\mathbf{X}$ whose typical element is x_{ij}. The terms *cases* or *specimens* will be used to refer to the n rows, while the p columns (often oxides measured as a percentage or trace elements measured as parts per million) will be referred to as *variables*. Uppercase letters, such as X_j, will be used to denote the jth variable.

Usually the elements of $\mathbf{X}$ are subjected to some form of transformation or pretreatment before a statistical analysis, resulting in the matrix $\mathbf{Y}$, with elements y_{ij}. Several important statistical procedures result in new variables that are linear combinations of the variables defining $\mathbf{Y}$, giving output in the form of a matrix $\mathbf{Z}$, the first few columns of which typically form the basis of the presentation of results.

The mean of X_j is $\bar{x}_j$, and $\bar{\mathbf{x}}$ is the $p \times 1$ vector of means, so

$$\bar{\mathbf{x}}' = (\bar{x}_1 \bar{x}_2 \ldots \bar{x}_p)$$

where the prime notation, $'$, denotes vector or matrix transpose. The estimated covariance for X_i and X_j is defined as

$$s_{ij} = \sum_{k=1}^{n} (x_{ki} - \bar{x}_i)(x_{kj} - \bar{x}_j)/(n - 1)$$

In this notation s_{jj} is the variance of X_j, however, we will use the more usual s_j^2 to denote the variance, so that the square root, s_j, is the estimated standard deviation. In one section it will be necessary to distinguish between the variances of variables in **X**, **Y**, and **Z**, and the notation $v(X_j) = s_j^2$ for the variance of the jth variable in **X** and $v(Z_j)$ for the variance of the jth variable in **Z**, for example, will be used.

The $p \times p$ matrix, **S**, is the estimated *covariance matrix* of **X** and is symmetric, since $s_{ij} = s_{ji}$. In vector notation **S** is defined as

$$\mathbf{S} = \sum_{j=1}^{n} (\mathbf{x}_j - \bar{\mathbf{x}})(\mathbf{x}_j - \bar{\mathbf{x}})'/(n-1)$$

where $\mathbf{x}_j$ is a $p \times 1$ vector of the observations for the jth case. Other results can conveniently be expressed in matrix or vector notation. For example,

$$(\mathbf{x}_i - \mathbf{x}_j)' \mathbf{S}^{-b} (\mathbf{x}_i - \mathbf{x}_j) \qquad (20.1)$$

defines the *squared Euclidean distance* between the ith and jth cases if $b = 0$ and the *squared Mahalanobis distance* if $b = 1$. These are important concepts in the multivariate analysis of data that are discussed in more detail later.

20.3. DATA AND INTRODUCTORY EXAMPLES

20.3.1. Introduction

In this section some common methodologies, discussed in detail later, are applied to some data sets that are used for illustration throughout this chapter. The data sets used in this chapter have all been published in *Archaeometry* (see the full references) and are reproduced, with permission, at the URL http://science.ntu.ac.uk/msor/mjb/artarch.html. In providing illustrations before we discuss the methodologies in detail, it is our intention to give a flavor of the way in which statistical methodology is commonly applied to chemical compositional data in an archaeological context. Readers wholly unfamiliar with the methodologies and terminology used may wish to read the methodological sections first and return to this section.

20.3.2. Lead Isotope Ratio Data

The analysis and interpretation of lead isotope ratio data is discussed by Gale and Stos-Gale (Chapter 17). Since only $p = 3$ ratios are measured, the

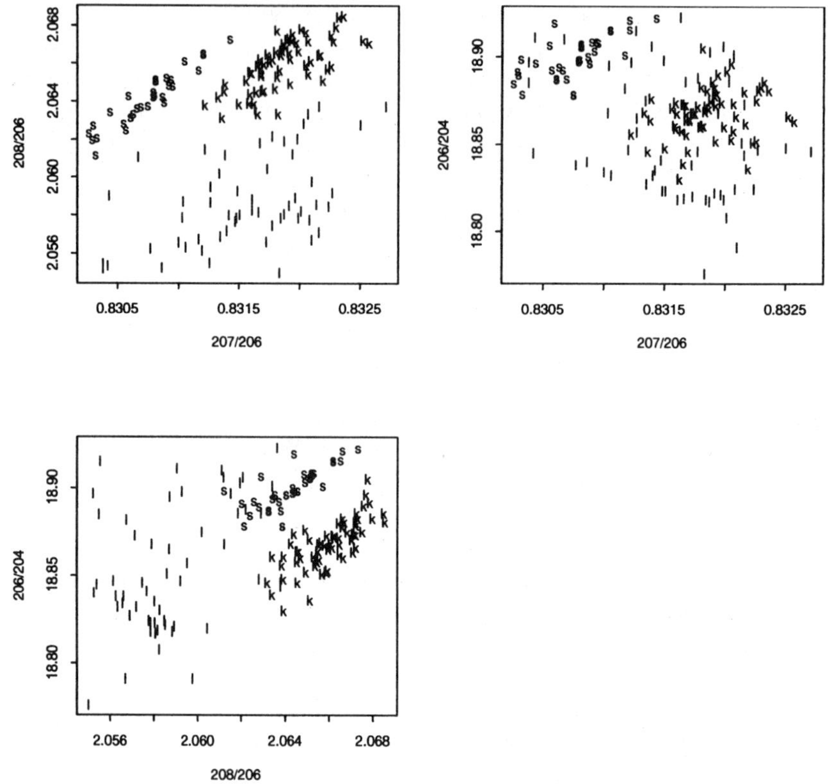

Figure 20.1. Bivariate plots for lead isotope ratio data from Stos-Gale et al. (1996). Axes are labeled with the relevant ratio; labels *k*, *l*, and *s* in the plots refer to the Kea, Lavrion, and Seriphos fields.

structure of lead isotope data sets is relatively uncomplicated compared with data generated by techniques such as neutron activation analysis for which $p > 30$ is typical (Glascock 1992). Despite this apparent simplicity, the statistical analysis of lead isotope data has not been without controversy (Sayre et al. 1992; Budd et al. 1993, 1995, 1996; Scaife et al. 1996; Stos-Gale et al. 1997). The debate is polarized between those who use complex (multivariate) methods to analyze such data and those who maintain that, given the low dimensionality, simple graphical analysis is usually adequate.

For illustrative purposes the data sets for the Kea, Lavrion, and Seriphos fields, published in Stos-Gale et al. (1996) will be used. A common way to present such data is in the form of two-dimensional plots, and Figure 20.1 shows all such bivariate plots, with the source of cases labeled. The plot of the $^{208/206}Pb/^{207/206}Pb$ ratios, in particular, suggests that the three fields are separated, though only just in the case of the Kea and Lavrion fields. Such plots can also be used to delineate the extent of individual fields, identify outliers, and assess whether a field is normally distributed. These are all issues that have arisen in the analysis of lead isotope data that, in the last two cases, apply to more complex kinds of data. The issues raised are considered in more detail later.

20.3.3. Atomic Absorption Spectrophotometry of Romano-British Pottery

Table 20.1 is based on that given on page 156 of Tubb et al. (1980) and shows the composition of 48 specimens of Romano-British pottery, determined by atomic absorption spectrophotometry (AAS). The specimens were found at five separate kiln sites in three different regions of Great Britain, and the regional information is also shown. Tubb et al. (1980) were interested in whether provenance could be related to chemical composition and, if so, which variables were most important.

Visual inspection of Table 20.1 is sufficient to identify some very unusual values (e.g., the low value of TiO_2 for GA4) that are typographical errors (Nickless, pers. comm.). The correct data were used in Tubb et al. (1980), but we have adopted the published data for illustrative purposes, as the detection and effects of unusual data are often of interest in statistical analyses. The aim of our first analysis is to group the data using the chemical information only. This happens to be a "simple" data set to analyze in the sense that the structure in it is fairly obvious; however, it usefully illustrates the utility of the different methodologies that are discussed in more detail later. To anticipate terminology that will be explained later, the data are first standardized (see Section 20.5.2.2) and then specimens are grouped together using the average linkage method of cluster analysis with Euclidean distance as the measure of dissimilarity (see Section 20.5.3.2). The dendrogram resulting from this analysis is shown in Figure 20.2.

If the dendrogram is cut horizontally at the level of about 3, there are three groups in the data, with specimens 4, 35, and 36 isolated as outliers. These three chemical groups correspond to the three regional groups that the data are known to consist of and establish a link between chemistry and provenance.

Cluster analysis has been the main statistical vehicle for compositional data analysis in archaeometry, and numerous published examples exist (Baxter 1994a: 179–181). A limitation of the dendrogram as a method of display is that it is difficult to see the relationships between the groups and assess how genuinely distinct they are. Accordingly, cluster analyses are often accompanied by a principal component plot such as that in Figure 20.3.

Table 20.1. Results of Analyses of Romano-British Pottery[a]

No.	Id.	Reg.	Al_2O_3	Fe_2O_3	MgO	CaO	Na_2O	K_2O	TiO_2	MnO	BaO
1	GA1	1	18.8	9.52	2.00	0.79	0.40	3.20	1.01	0.077	0.015
2	GA2	1	16.9	7.33	1.65	0.84	0.40	3.05	0.99	0.067	0.018
3	GA3	1	18.2	7.64	1.82	0.77	0.40	3.07	0.98	0.087	0.014
4	GA4	1	17.4	7.48	1.71	1.01	0.40	3.16	0.03	0.084	0.017
5	GA5	1	16.9	7.29	1.56	0.76	0.40	3.05	1.00	0.063	0.019
6	GB1	1	17.8	7.24	1.83	0.92	0.43	3.12	0.93	0.061	0.019
7	GB2	1	18.8	7.45	2.06	0.87	0.25	3.26	0.98	0.072	0.017
8	GB3	1	16.5	7.05	1.81	1.73	0.33	3.20	0.95	0.066	0.019
9	GB4	1	18.0	7.42	2.06	1.00	0.28	3.37	0.96	0.072	0.017
10	GB5	1	15.8	7.15	1.62	0.71	0.38	3.25	0.93	0.062	0.017
11	GC1	1	14.6	6.87	1.67	0.76	0.33	3.06	0.91	0.055	0.012
12	GC2	1	13.7	5.83	1.50	0.66	0.13	2.25	0.75	0.034	0.012
13	GC3	1	14.6	6.76	1.63	1.48	0.20	3.02	0.87	0.055	0.016
14	GC4	1	14.8	7.07	1.62	1.44	0.24	3.03	0.86	0.080	0.016
15	GD1	1	17.1	7.79	1.99	0.83	0.46	3.13	0.93	0.090	0.020
16	GD2	1	16.8	7.86	1.86	0.84	0.46	2.93	0.94	0.094	0.020
17	GD3	1	15.8	7.65	1.94	0.81	0.83	3.33	0.96	0.112	0.019
18	GD4	1	18.6	7.85	2.33	0.87	0.38	3.17	0.98	0.081	0.018
19	GD5	1	16.9	7.87	1.83	1.31	0.53	3.09	0.95	0.092	0.023
20	GE1	1	18.9	7.58	2.05	0.83	0.13	3.29	0.98	0.072	0.015
21	GE2	1	18.0	7.50	1.94	0.69	0.12	3.14	0.93	0.035	0.017
22	GE3	1	17.8	7.28	1.92	0.81	0.18	3.15	0.90	0.067	0.017
23	C01	2	14.4	7.00	4.30	0.15	0.51	4.25	0.79	0.160	0.019
24	C02	2	13.8	7.08	3.43	0.12	0.17	4.14	0.77	0.144	0.020
25	C03	2	14.6	7.09	3.88	0.13	0.20	4.36	0.81	0.124	0.019
26	C04	2	11.5	6.37	5.64	0.16	0.14	3.89	0.69	0.087	0.009

27	C05	2	13.8	7.06	5.34	0.20	0.20	4.31	0.71	0.101	0.021
28	C06	2	10.9	6.26	3.47	0.17	0.22	3.40	0.66	0.109	0.010
29	C07	2	10.1	4.26	4.26	0.20	0.18	3.32	0.59	0.149	0.017
30	C08	2	11.6	5.78	5.91	0.18	0.16	3.70	0.65	0.082	0.015
31	C09	2	11.1	5.49	4.52	0.29	0.30	4.03	0.63	0.080	0.016
32	C10	2	13.4	6.92	7.23	0.28	0.20	4.54	0.69	0.163	0.017
33	C11	2	12.4	6.13	5.69	0.22	0.54	4.65	0.70	0.159	0.015
34	C12	2	13.1	6.64	5.51	0.31	0.24	4.89	0.72	0.094	0.017
35	C13	2	12.7	6.69	4.45	0.20	0.22	4.70	0.73	0.394	0.024
36	C14	2	12.5	6.44	3.94	0.22	0.23	0.81	0.75	0.177	0.019
37	G01	2	11.6	5.39	3.77	0.29	0.06	4.51	0.56	0.110	0.015
38	G02	2	11.8	5.44	3.94	0.30	0.04	4.64	0.59	0.085	0.013
39	T11	3	18.3	1.28	0.67	0.03	0.03	1.96	0.65	0.001	0.014
40	T12	3	15.8	2.39	0.63	0.01	0.04	1.94	1.29	0.001	0.014
41	T13	3	18.0	1.50	0.67	0.01	0.06	2.11	0.92	0.001	0.016
42	T14	3	18.0	1.88	0.68	0.01	0.04	2.00	1.11	0.006	0.022
43	T16	3	20.8	1.51	0.72	0.07	0.10	2.37	1.26	0.002	0.016
44	A11	3	17.7	1.12	0.56	0.06	0.06	2.06	0.79	0.001	0.013
45	A12	3	18.3	1.14	0.67	0.06	0.05	2.11	0.89	0.006	0.019
46	A16	3	16.7	0.92	0.53	0.01	0.05	1.76	0.91	0.004	0.013
47	A18	3	14.8	2.74	0.67	0.03	0.05	2.15	1.34	0.003	0.015
48	A26	3	19.1	1.64	0.60	0.10	0.03	1.75	1.04	0.007	0.018

[a] Numbers are percentage metal oxide; sample identifications are those from the original paper; Reg. is region of origin.

From Tubb et al. 1980. Reprinted with permission from *Archaeometry*.

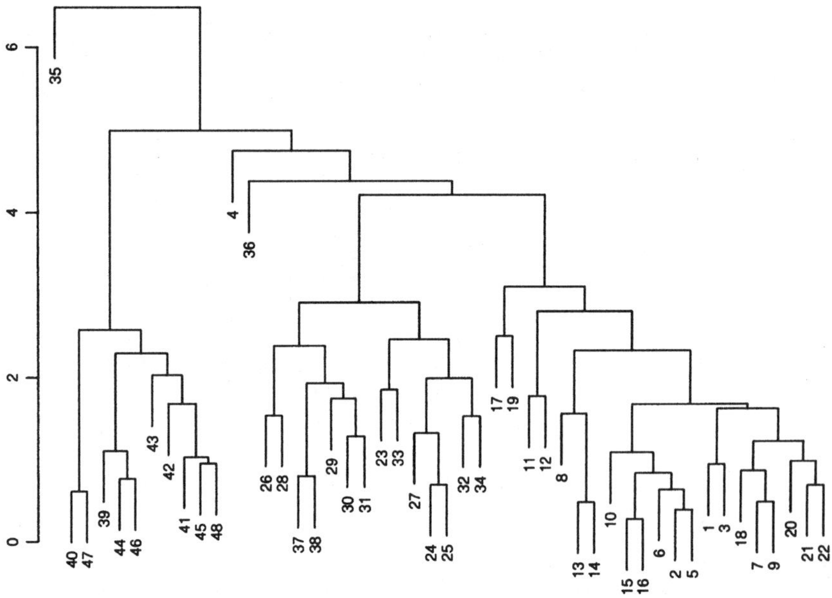

Figure 20.2. Dendrogram of average link cluster analysis using Euclidean distance of standardized Romano-British pottery data.

Using standardized data in this case allows a two-dimensional approximation of the p-dimensional distances between specimens. That there are three reasonably well-separated groups is confirmed, except that specimens 4, 12, 35, and 36 "hover" at the periphery of the plot or lie in-between the three main groups. The group to the left of both the principal components plot and the dendrogram is particularly distinct and compact.

In both the foregoing analyses the regional information is used to interpret the graphs but not to produce them. If discriminant analysis is adopted, the *a priori* regional information is used, and linear combinations of the original variables are sought to display the separation of the presumed groups as well as possible. This results in Figure 20.4 where the separation is even more evident than in Figure 20.3, but unusual data are less obvious.

This section has illustrated the use of the three most commonly used multivariate methods in the analysis of chemical compositional data in archaeometry but has avoided any discussion of the many considerations that arise in applying them. This will be dealt with in Section 20.5 after first discussing some of the simpler approaches to analysis that it is always important to consider.

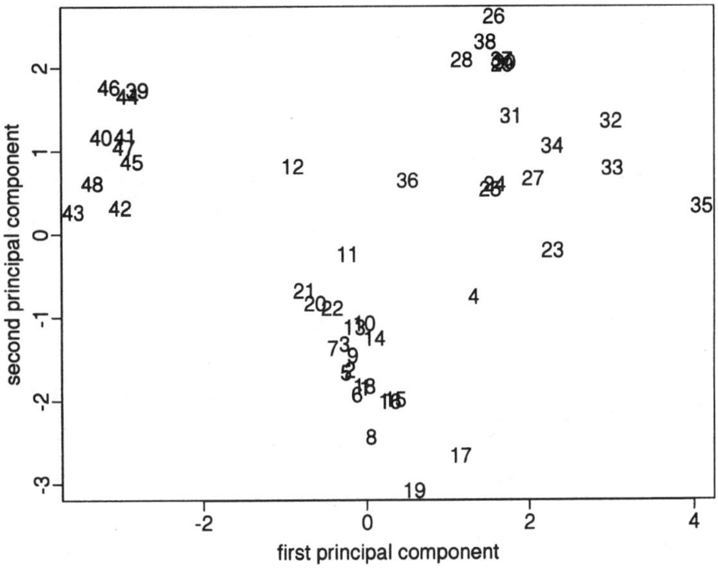

Figure 20.3. Principal component analysis using standardized Romano-British pottery data. Numbers identify pottery specimens.

20.4. PRELIMINARY DATA ANALYSIS

20.4.1. Tabular Inspection

An obvious first stage of analysis is tabular inspection, or "eyeballing" the columns of a data set, for obviously anomalous values. This can be surprisingly effective, even for quite large sets of data, provided they are well presented.

In Table 20.1 cases 1 (high Fe_2O_3), 4 (low TiO_2), 35 (high MnO), and 36 (low K_2O) are clearly unusual relative to the data set as a whole. As previously noted, it is known that some of these unusual values are typographical errors; in practice the reasons for unusual values such as these would be sought (e.g., misrecording, contamination). Consideration could be given to omitting them from further statistical analysis at this stage. An alternative is to "flag" the points as unusual, but retain them and monitor their effect on further analysis.

20.4.2. Univariate Data Inspection and Kernel Density Estimates

An essential part of any program of data analysis should be univariate graphical inspection of the data. This will serve to identify obvious outliers

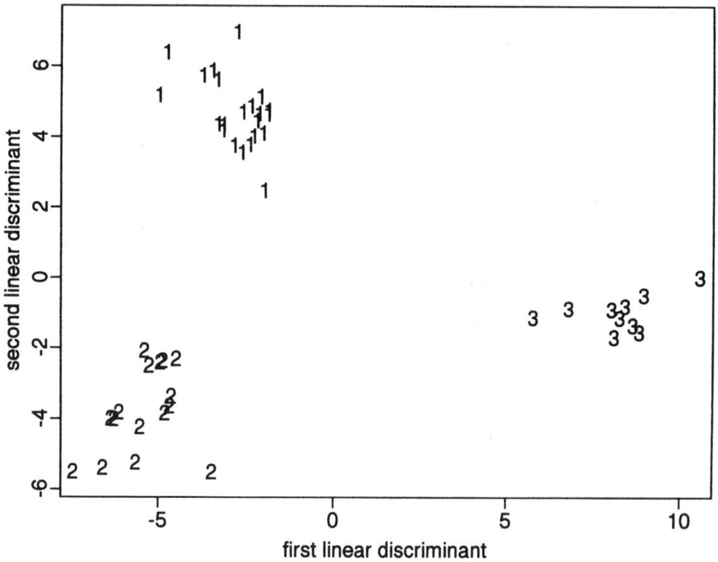

Figure 20.4. Discriminant analysis using Romano-British pottery data. Numbers identify the region of origin of pottery specimens.

or structure in the data, as well as indicating whether data transformation is likely to be beneficial. Histograms and variants of the histogram (e.g., frequency polygons, dot plots) should be familiar to all scientists and are examples of density estimates. The left panel of Figure 20.5 shows an example of a histogram in which the area of a bar is proportional to the frequency of observations in the interval defined by the bar. Usually the bars are chosen to be of equal width so that the height of the bar is also proportional to the frequency. The precise appearance of a histogram depends both on the interval widths and the location of interval boundaries and can sometimes be misleading. Here we shall focus on a perhaps less familiar form of density estimate—kernel density estimates (KDEs)—that are used in other parts of the chapter.

The underlying idea is simple. A "kernel" or probability density function is centered at each data point, plotted on a line graph, and the KDE is obtained by summing the heights of the kernels. This results in a KDE such as those to the right of Figure 20.5. Mathematically, a kernel density estimate can be written as

$$\hat{f}(y) = \frac{1}{n} \sum_{i=1}^{n} K(y - y_i; h)$$

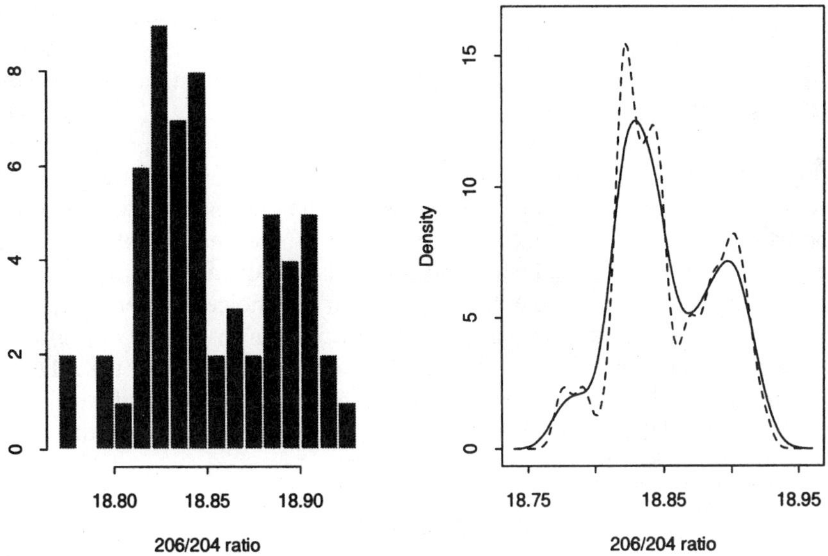

Figure 20.5. A histogram (left panel) and KDEs (right panel) for the Lavrion $^{206}Pb/^{204}Pb$ lead isotope ratio data from Stos-Gale et al. (1996). The solid KDE uses a smoothing parameter, h, determined by the method of Sheather and Jones (1991); the dashed KDE uses a subjectively determined h.

where $K(\cdot)$, the kernel, is a probability density function; the y_i are the data points; and h is a *smoothing* or *bandwidth* parameter that controls the spread of the kernel and the smoothness of the final estimate. In all the examples in this chapter $K(\cdot)$ is taken to be a normal probability density function with standard deviation h.

Choice of h is important and is discussed at length in Wand and Jones (1995). Unless otherwise stated we use the procedure developed by Sheather and Jones (1991). It is sometimes advantageous to allow the value of h to vary in a way that depends on the density in the region of a point, and this leads to an *adaptive* KDE.

The main purpose of this section is to alert readers to the existence of this useful methodology. Silverman's (1986) book helped popularize the technique, Wand and Jones (1995) detail many of the subsequent developments, and Bowman and Azzalini (1997) is an applied text with associated code for the S-Plus software package (see also, Venables and Ripley 1997). Archaeometric applications are to be found in Baxter and Beardah (1995, 1996), Baxter et al. (1997), Baxter and Gale (1998), Baxter (1999a, 1999b), Beardah (1999), Lockyear (1999), and Williams-Thorpe et al. (1996). Beardah and Baxter (1996) gives free access to code for the MATLAB package.

Limited implementation in software packages has no doubt hampered the regular use of KDEs. This situation is rapidly changing, and Simonoff (1996: 28) has argued that histograms should be replaced as the "standard" density estimate in computer packages.

Figure 20.5 shows a histogram and two kernel density estimates for the $^{206}Pb/^{204}Pb$ lead isotope ratio for the Lavrion field, discussed in the previous section. It is apparent from both displays that the ratio is bimodal, violating the assumption of normality sometimes used in the analysis of such data (e.g., Sayre et al. 1992). The advantages of the KDEs are that the appearance is not dependent, as the histogram is, on the choice of interval boundaries; and it is easier to use them for comparative purposes. The appearance of the histogram is affected by the choice of interval width, and the KDE is similarly affected by the choice of h. The dashed line in the second panel uses a subjectively determined choice of h that shows clearly enough the main structure, but possibly undersmooths the data.

20.4.3. Box Plots

Inspection of some form of density estimate will help to reveal the presence of obvious structure (e.g., grouping) in the data. A useful complementary plot is the "box-and-whisker plot" or "box plot," an example of which is shown later in Figure 20.13. Precise details depend on the software package used. Typically a box defined by the interquartile range (or something very similar) is shown that encompasses the central 50% of the data, along with an indication of the position of the median. Whiskers extend to the limits of the data values, and may be broken to identify isolated points that can tentatively be regarded as extreme.

If the data are multimodal, the box plot may be misleading, so it should always be examined in conjunction with a density estimate. If the data are unimodal, lack of symmetry in a box plot may suggest the desirability of some form of data transformation—logarithmic, for instance, if values are strictly positive. Equally useful the box plot, as usually implemented in software packages, provides a quick means of identifying outlying data that will be potentially problematic in more complex statistical analysis. Baxter et al. (1995) use the box plot as their main analytical tool, in a study of compositional variation in colorless Romano-British glass of different types and periods, to demonstrate the greater variability within some types.

20.4.4. Bi- and Trivariate Plotting

For examining correlations between variables, identifying obvious groupings, and for detecting bivariate outliers, plots such as Figure 20.1 are useful.

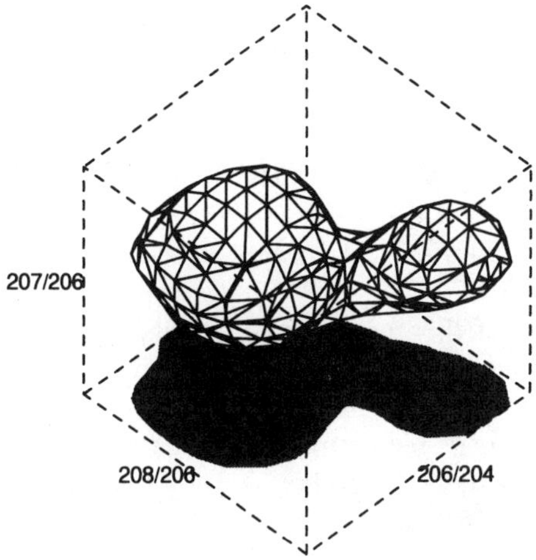

Figure 20.6. Three-dimensional KDE for the Lavrion lead isotope ratio data from Stos-Gale et al. (1996).

Many software packages now enable all possible pairwise plots to be produced in a routine fashion, under the heading of "draftsmen" or "pairs" plots or something similar. The main limitation is that with more than seven or eight variables plots become small and difficult to read, so subsets of variables need to be used.

Methods of trivariate data display are also increasingly common, though their effectiveness is more readily demonstrated at a computer terminal than on the printed page. "Brush-and-spin" plots enable three-dimensional clouds of points to be rotated in real time, to identify useful viewpoints, and subsets of the data—corresponding to a type or provenance, for example—can be highlighted.

Kernel density estimation may also be extended to two and three dimensions. Two-dimensional applications to chemical data are discussed in Baxter et al. (1997) and Beardah and Baxter (1996), where animation is also used. A three-dimensional KDE for the Lavrion lead isotope data is shown in Figure 20.6 and was obtained using the S-Plus software provided with Bowman and Azzalini (1997). It shows, quite effectively, that the data do not appear to have a normal distribution.

Some data, such as lead isotope ratios, are naturally trivariate and graphical analysis is a natural way of viewing them (e.g., Scaife et al. 1996). That three dimensions is a natural limit for viewing data graphically without recourse to "tricks" is not as restricting as it might seem. As will be discussed in the next section, some of the most common multivariate statistical methods used in archaeology work—if successful—by transforming the data to a state where most of the information is contained in two or three dimensions.

20.5. STANDARD METHODS OF MULTIVARIATE ANALYSIS

20.5.1. Concepts of Distance

The prime purpose of this section is to discuss those multivariate methods most commonly used in the exploration and presentation of chemical compositional data in archaeology. Before doing this it will be helpful to discuss the idea of the "distance" between cases in a data matrix, which is crucial for the understanding of these methods.

A definition of the distance between cases was given in Eq. (20.1). For the case $b = 0$ the special case of *squared Euclidean distance* is obtained. This is just the extension to p dimensions of the concept of distance as measured in everyday two- and three-dimensional space and can also be written more simply as

$$d_{ij}^2 = \sum_{k=1}^{p} (y_{ik} - y_{jk})^2$$

Some graphical methods of data analysis are based on principal component analysis (PCA) and attempt to produce data displays in a low dimension that reproduce the Euclidean distances between cases as closely as possible. Other common methods, including many approaches to cluster analysis, are dependent on the use of Euclidean distance.

Euclidean distance is not ideal for all purposes. To see why, consider the upper left diagram in Figure 20.1. It can be seen that the lead isotope ratios $^{208}Pb/^{206}Pb$ and $^{207}Pb/^{206}Pb$ for the Seriphos field are highly correlated, resulting in a strongly linear pattern that suggests a group distinct from the other two fields. The Euclidean distance between many of the cases in the Seriphos field and cases in the Kea field (i.e., straight-line distance) is, however, smaller than distances between many of the cases within the Seriphos field. This can adversely affect the performance of some common multivariate methods. Similarly, some of the cases to the right of the Seriphos field appear to be as close to the centroid of the Kea field as to that of their own field.

The problem is that Euclidean distance does not allow for correlation between variables; Mahalanobis distance, the square root of Eq. (20.1) with $b = 1$, does. A simple expression for squared Mahalanobis distance, similar

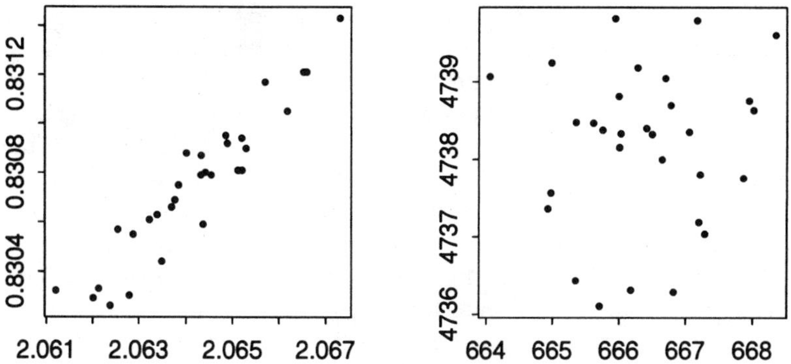

Figure 20.7. Left panel shows a bivariate plot of the lead isotope ratios $^{208}Pb/^{206}Pb$ (horizontal axis) and $^{207}Pb/^{206}Pb$ using data from Stos-Gale et al. (1996). The strongly linear pattern, reflecting a high correlation, is removed on transforming to Mahalanobis distance in the second panel.

to that for squared Euclidean distance, is not readily obtained, however, the effect of using it is easily demonstrated. Figure 20.7 shows the data for the Seriphos lead isotope ratios $^{208}Pb/^{206}Pb$ and $^{207}Pb/^{206}Pb$ before and after transformation to Mahalanobis distance. The effect is to convert a highly linear scatter to an uncorrelated, spherical scatter. The importance of Mahalanobis distance is discussed further and more technically in Section 20.5.5.

20.5.2. Principal Component Analysis

20.5.2.1. Basic Idea Behind PCA

Principal component analysis is most often used to obtain graphs, such as that in Figure 20.3, that show the relationship between cases. The method is discussed in most textbooks on multivariate analysis, two of which (Jolliffe 1986; Jackson 1991) are devoted entirely to the method. Baxter (1994a) and Shennan (1997) provide accounts written for an archaeological readership.

New variables of the form

$$Z_i = a_{1i}Y_1 + a_{2i}Y_2 + \cdots + a_{pi}Y_p$$

are defined, for $i = 1, 2, \ldots, p$, with the property that they are uncorrelated. Z_1 has maximum variance, Z_2 has second maximum variance subject to the lack of correlation, and so on.

The total variance in the data is $\sum_{j=1}^{p} v(Y_j) = \sum_{j=1}^{p} v(Z_j)$. The hope is that these sums will be approximated by $\sum_{j=1}^{q} v(Z_j)$ for some value of q (much) less than p. Where this hope is realized, we can expect that plots based on $Z_1, Z_2, \ldots, Z_q$ will reveal structure that exists in the full p-dimensional data but is hard to perceive. In Figure 20.3 a plot for $q = 2$ is sufficient to show that there are three groups in the data. An alternative way of looking at PCA is that it approximates, in a low-dimensional space, the Euclidean distances between cases. The quality of the approximation is measured by the percentage of variation accounted for by the leading components. In Figure 20.3 this is 67% for the first two components and 79% for the first three. Even for large p the first two components often account for over 50% of the variance, and four components are often sufficient to account for 80% (Baxter 1995). Even with values as low as 50%, component plots such as Figure 20.3 can be informative, though it is advisable to look at components other than the first two. It is increasingly common for three-dimensional plots based on the first three components to be used (e.g., Burton and Simon 1993, Castellano et al. 1996, Rotonno et al. 1997) though these can be difficult to read on the printed page and bivariate plots are probably to be preferred (Neff, n.d.). If the third and higher-order components are of interest, as is sometimes the case (e.g., Williams-Thorpe et al. 1984), the draftsman plot is an effective device for examining all pairwise plots of potential interest.

20.5.2.2. Scale Dependence of PCA

It is well known that PCA is scale dependent (Mardia et al. 1979, Section 8.2.3), and analyses will tend to be dominated by those variables for which $v(Y_j)$ is relatively large. Thus, the way in which y_{ij} is defined is important. Let

$$y_{ij} = (x_{ij} - \bar{x}_j)/s_j^a.$$

If $a = 0$, the variables are *centered* to have mean zero and variance s_j^2; if $a = 1$, the variables are *standardized* to have mean zero and unit variance.

Centering alone is usually inappropriate because a PCA will be predictably dominated by a few variables with large variances. Accordingly, standardization is common and is often the default in computer packages. Another common approach is to *transform* the data to logarithms (to base 10) and then use the transformed data in either centered or standardized form (Bishop and Neff 1989). There is no consensus about which is the 'best' approach, nor should one be expected.

Standardization gives all variables (potentially) equal weight in an analysis, but distances between two cases will depend on the entire sample (through the influence of s_j), and not just the two cases. Some consider this a drawback that is avoided by using transformed but unstandardized data.

The log transformation will tend to equalize variances, but it is still possible for a subset of variables to dominate an analysis. Another reason for advocating transformation is that it will render some variables, and particularly trace elements, more normally distributed. This is discussed in Section 20.6.3.

Researchers at the Missouri University Research Reactor (MURR), one of the leading laboratories generating neutron activation analysis (NAA) data for archaeological analysis, have observed that experience "has shown that standardization and log-transformation lead to equivalent results" (Neff and Glascock, p. 3, 1992; Neff, n.d.). Baxter (1994c, 1995) reached similar, if more qualified, conclusions. If data are standardized, with or without logarithmic transformation, then, in general, similar results are obtained unless there are obvious outliers in the data. The principal component plots for unstandardized logged data can appear very different, but often lead to identical substantive conclusions.

20.5.2.3. Biplots

A recent development in the presentation of PCA results in archaeometry is the increased use of biplots (Baxter 1992, 1994a; Neff 1994, n.d.), in which information on the relationship between variables is simultaneously presented and interpreted with information on the relationship between cases. Gower and Hand (1996) provide an extended treatment of the methodology. A biplot is shown in Figure 20.8 for a PCA of a small set of standardized medieval glass compositional data given in Cox and Gillies (1986). It can be seen that there are two compact groups, to the right and bottom left of the plot and a more dispersed one to the top left. The arrowed lines, or vectors, from the origin are associated with the variables and are labeled with the metal (oxide).

An acute angle between vectors is indicative of a high positive correlation (e.g., Al and Fe have a correlation of 0.95); an oblique angle is indicative of high negative correlation (e.g., −0.95 for Al and Cu); an approximate right angle indicates a low correlation (e.g., 0.04 for Al and Na). The plot as a whole provides a visual approximation of the correlation structure in the data if standardized data are used (or the covariance structure if unstandardized data are used). The length of a vector is an indication of how well the associated variable is approximated and how much it contributes to the analysis. For equally scaled axes, the arrows should lie approximately on a circle; this is so here apart from Pb, which is being poorly represented. The relative positions of the vectors and groups is indicative of the groups' chemical compositions. Thus, it can be anticipated that the bottom left group will have relatively high levels of Al and Fe and low levels of Cu; the group to

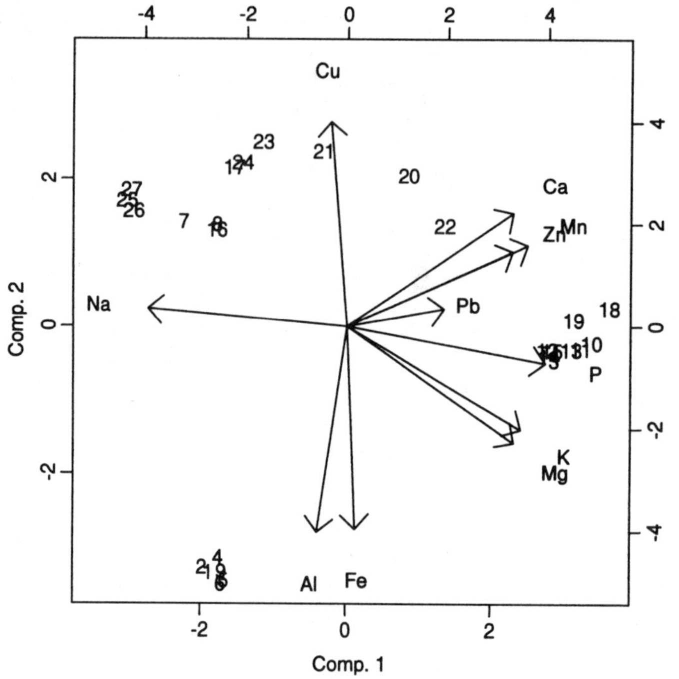

Figure 20.8. Biplot of the compositions of 27 specimens of medieval window glass from data given in Cox and Gillies (1986).

the right will have relatively high levels of the oxides associated with the vectors to the right and low levels of Na.

It is possible to define different kinds of biplot, and mathematical details are given in the references cited. That in Figure 20.8 is the default produced in the S-Plus package (Venables and Ripley 1997). It is possible for the plot to be rather crowded if there are many cases and variables; however, they have been used with 20/30 variables (e.g., Hegmon et al. 1997; Neff et al. 1996).

20.5.3. Cluster Analysis

20.5.3.1. Basic Idea behind Cluster Analysis

Cluster analysis has been the most widely used multivariate statistical technique in archaeometry. Pollard (1986), Baxter (1994a), and Shennan (1997) provide accounts for an archaeological readership, while Everitt (1993) is an accessible general introduction.

Cluster analysis is a generic term for a wide range of specific techniques. The essential components of many methods are (a) a measure of (dis)similarity between cases in a data matrix is defined and (b) an algorithm for grouping cases on the basis of their (dis)similarity is specified. Given the many ways in which (a) and (b) can be done there are a very large number of ways in which a cluster analysis might be carried out. The following review is highly selective and concentrates on the methods that have found most favor in practice.

The vast majority of published cluster analyses in the archaeometric literature are *hierarchic agglomerative* in nature. Each case is initially regarded as a single cluster, and cases are successively amalgamated until a single group remains. Clustering algorithms differ in the criterion used for merging clusters, once cluster formation is begun. These methods may be contrasted with *partitioning* methods in which cases are iteratively allocated and reallocated to clusters to optimize some criterion of success.

The issues concerning standardization and transformation in cluster analysis are identical to those that apply in PCA. Usually, after log transformation or standardization, Euclidean distance or its square is chosen as the measure of dissimilarity between cases. This and the choice of clustering algorithm is, we suspect, often a function of the defaults in available software.

20.5.3.2. Specific Methods of Cluster Analysis

A particularly simple clustering algorithm is the *single linkage* method in which the measure of dissimilarity between two clusters is the smallest distance between two cases, one from each cluster. At any stage the closest two clusters, according to this criterion, are merged. Single linkage is often avoided because of the problem of "chaining," whereby quite distinct clusters can be linked because of the effect of a few intermediate cases. The *complete linkage* method, in which cluster dissimilarity is the largest distance between two cases, one from each cluster, has found more favor (e.g., Mirti et al. 1993; Rotonno et al. 1997; Troja et al. 1996). It has a tendency to produce small and compact clusters.

The two methods just described take no account of group structure in the definition of cluster dissimilarity. In the *average linkage* method cluster dissimilarity is defined as the average distance between all pairs of points, one from each cluster. This, along with *Ward's method*, is the most commonly used method in practice. In this latter method the variability within the jth cluster is defined as

$$S_j = \sum_{i=1}^{n_j} \sum_{k} (y_{ik} - \bar{y}_k)^2$$

where n_j is the number of cases in cluster j and $\bar{y}_k$ is the mean of the kth variable in the cluster. If $S = \sum_{j=1}^{J} S_j$, where J is the number of clusters at a given stage, is the total variability, those two clusters are merged that give the least increase in S.

In Ward's method, for fixed J, it may be possible to reduce S by moving cases from one cluster to another. That is, Ward's method does not necessarily optimize the criterion being used to assess the "success" of the clustering. This idea lies at the heart of *k-means* or *relocation* methods, whereby some initial partition of the data—which may be determined by a hierarchical method—is iteratively modified to try and optimize some criterion such as S. This approach is used in Argyropoulos (1995), Castellano et al. (1996), and Taylor and Robinson (1996a, 1996b), for example.

20.5.3.3. Interpreting Dendrograms

Since the use of relocation can, according to the criterion used, only improve the quality of a clustering, it is surprising that it is not used more frequently. One reason may be that results are not then as conveniently represented, since the dendrogram that results from a hierarchical analysis and which is the usual mode of presentation is no longer suitable. Figure 20.9 shows two

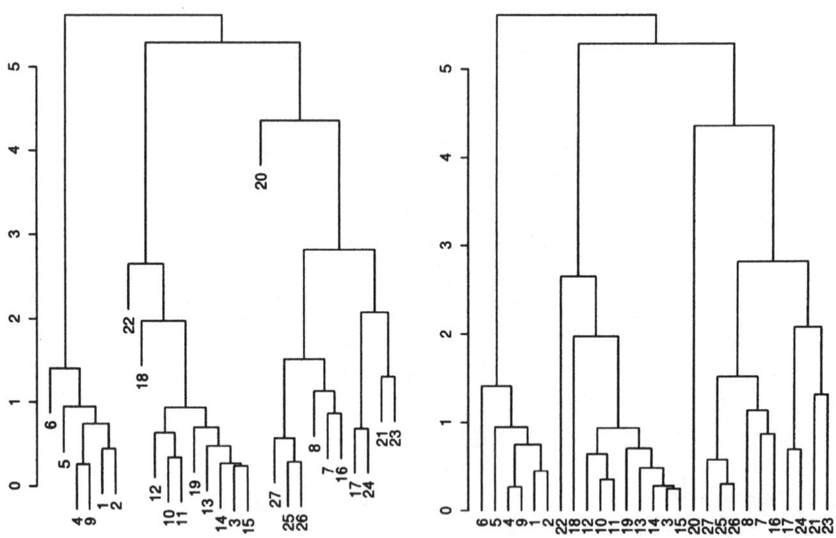

Figure 20.9. Two alternative styles of dendrogram presentation for a cluster analysis of the compositions of 27 specimens of medieval window glass from data given in Cox and Gillies (1986). The average linkage method for standardized data has been used with Euclidean distance as the dissimilarity measure.

different styles of dendrogram presentation for the York medieval glass data; that to the left is the style favored in the S-Plus package, while that to the right is the style more often to be found in the literature. The vertical axis shows the dissimilarity level at which cases or clusters merge. Cases that merge at a low level (e.g., 4 and 9) show a high level of similarity.

Interpretation of dendrograms is usually subjective; in the present case there seems to be three fairly clear clusters and a possible outlier (20). The subjectivity is compounded by the fact that the appearance depends on the choice of style, any scaling of the axis, and the distance measure used. Squared Euclidean, as opposed to Euclidean, distance will often result in dendrograms showing apparently clearer clustering. In summary, the appearance and apparent interpretability of a dendrogram can be misleading and needs to be checked. This is particularly so, since it is well known that clustering methods can impose structure on a data set even when such structure does not exist (Baxter 1994a: 166; Neff, n.d.).

A simple and common way of confirming the interpretation of a dendrogram is to examine principal component plots on which the points are labeled on the basis of the cluster to which they have been allocated. More "objective" methods of determining the number of clusters exist but are not widely used and not necessarily effective (Baxter 1994a: 161–165).

20.5.3.4. Problems with Correlated Data

Archaeometric compositional data can be highly correlated, and it is well known that this can cause problems for cluster analysis. Figure 20.1 showed that the lead isotope field for Seriphos was distinct from that for Kea, and this was evident using the $^{208}Pb/^{206}Pb$ and $^{207}Pb/^{206}Pb$ ratios. The correlation between ratios is 0.94 for Seriphos, 0.69 for Kea, and 0.77 for the combined fields. Using only these two ratios, an average linkage cluster analysis was undertaken using standardized data and Euclidean distance. The resulting dendrogram in Figure 20.10 is labeled according to the field from which cases come. Such a dendrogram would often be interpreted as showing two main clusters in the data, but at this level it can be seen that the two fields are not separated. (At the 5 cluster level there is separation into three clusters from Seriphos and two from Kea). Figure 20.11 plots the raw data, using as labels the clusters to which cases are assigned at the two-cluster level. It can be seen that the clustering does not respect the evident grouping, with "spherical" rather than elongated clusters being produced that ignore the gap between the fields.

Although this is an artificial example, it illustrates a problem that causes some researchers to regard a cluster analysis as little more than an initial exploratory method, the results of which require modification (e.g., Slane

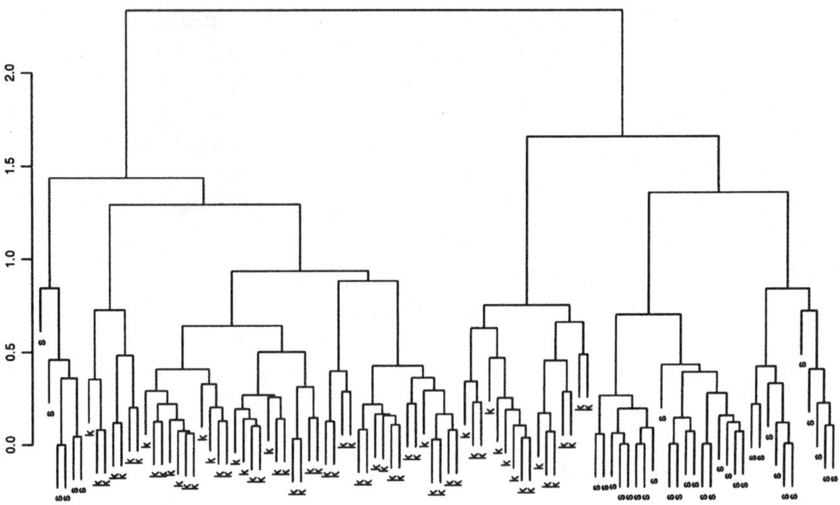

Figure 20.10. Dendrogram is for an average link cluster analysis based on lead isotope ratio data for the combined Seriphos and Kea fields, using the $^{208}Pb/^{206}Pb$ and $^{207}Pb/^{206}Pb$ ratios (data from Stos-Gale et al. 1996).

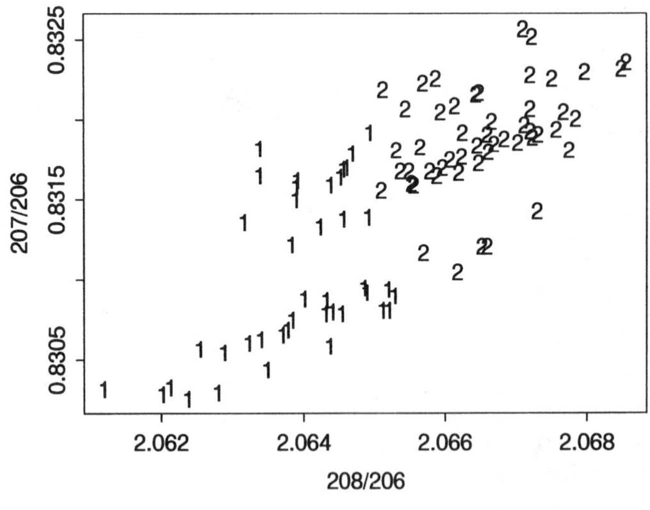

Figure 20.11. Plot is based on lead isotope ratio data for the combined Seriphos and Kea fields, using the $^{208}Pb/^{206}Pb$ and $^{207}Pb/^{206}Pb$ ratios (data from Stos-Gale et al. 1996). Labels indicate to which of the two main clusters, determined from Figure 22.10, cases belong. At this level the cluster analysis fails to distinguish between the upper Kea field and the lower Seriphos field.

et al. 1994; Steponaitis et al. 1996). This typically involves a form of relocation methodology, using Mahalanobis distance, and will be discussed in Section 20.5.5.

Baxter (1994a: 167–170) discusses the issues more fully. It is noted there that Solomon (1971) has suggested that if $r_{ij} < 0.5$ for most (i, j) pairs, there is not a practical problem, whereas Harbottle (1976: 53–54) suggests the same if, for $p = 20$, $r_{ij} > 0.8$ in fewer than 20 instances. A suggestion occasionally advanced is that cluster analysis be carried out on the uncorrelated principal components, standardized to have equal variance. This seems to have been little applied to archaeometric data and awaits practical evaluation, but there are reasons for believing it will not be a satisfactory general solution to the problems posed by correlated data (Baxter 1994a: 169).

20.5.4. Discriminant Analysis

20.5.4.1. Basic Idea of Discriminant Analysis

Whereas PCA and cluster analysis are used to discover if there are groups within a set of data, discriminant analysis uses prior assumptions about the existence of groups. These might correspond to the known provenances of specimens, for example, or a typological classification, or to groups determined statistically by a cluster analysis. One use of discriminant analysis is to establish or confirm whether such externally or statistically defined groups are chemically distinct. A second use is to establish rules for allocating specimens, of unknown provenance or type, to a group. There are numerous ways of undertaking a discriminant analysis and several books are devoted to this and related matters (e.g., McLachlan 1992; Hand 1997); Baxter (1994a) includes an account written for an archaeological audience. Archaeometric practice is dominated by one particular approach, Fisher's linear discriminant analysis (LDA), and the account here concentrates on this. Alternative approaches are discussed in Hand (1981), Anderson (1984), and Bishop (1995).

Superficially, LDA looks a lot like PCA. Suppose that there are g groups in the data and that $p > g$. In LDA $(g - 1)$ new variables are defined that are linear combinations of the original variables and that are constructed to be uncorrelated. The first variable is defined to maximize the separation between groups; the second is similarly defined, subject to its having no correlation with the first variable. For $g > 2$ it is usual to present a plot of the first two discriminant functions to demonstrate separation, as in Figure 20.4.

The texts cited may be consulted for mathematical details, which are most simply understood in the two-group case. In this instance a single discriminant function

$$Z = a_1 Y_1 + a_2 Y_2 + \cdots + a_p Y_p$$

is defined. For any choice of the a_i the mean and variance of Z can be calculated for each group, it being assumed that the groups are sampled from populations with equal covariance matrices. In LDA, loosely speaking, the coefficients are determined to separate the group means, on the Z scale, as much as possible, while retaining the compactness of the groups. More formally, the ratio of the between-group to within-group variation is maximized. It can be shown (e.g., Lachenbruch 1975: 10) that the difference between the group means on the Z scale is then just the Mahalanobis distance between the group mean vectors on the original scale. The idea of maximizing between- to within-group variation extends to $g > 2$ (Lachenbruch 1975: 66).

The "success" or "error rate" of an LDA can be judged in various ways; Hand (1997) provides a discussion of this and related problems. The most common approach in archaeometry is the use of the *resubstitution* method in which the cases used in the LDA are subsequently classified into the group that they are nearest to, in terms of Mahalanobis distance from the group centroid. This tends to produce an optimistic assessment because a case affects the results that lead to its later classification. Methodologies variously termed "leave-one-out," "jack-knifing," or "cross-validation" in the literature effectively assign cases to a group on the basis of an LDA carried out on the $(n - 1)$ specimens omitting that case. Differences in results for resubstitution and cross validation can be large; Knutsson et al. (1988) report an example where use of the latter method compared to the former increases the error rate by 16%, and Baxter (1994b) reports an example where the difference is 20%.

20.5.4.2. Assumptions of LDA

Fisher's (1936) original derivation of LDA made no assumptions about the distribution of the variables within groups. In particular, when used as a purely descriptive tool to investigate or demonstrate group separation, it is *not* assumed that variables have a (multivariate) normal distribution. What is assumed is that distinct groups have a common covariance matrix. Although tests of this assumption exist, they are rarely reported in the archaeometric literature, even though in some applications the assumption is often patently false. The justification for using LDA is then essentially pragmatic; it is believed to be robust to departures from the assumptions that underpin it, and the effective use of techniques such as *quadratic discriminant analysis* (see below), that avoid the assumption, require larger sample sizes than are commonly available. Baxter (1994a: 197–200) discusses this in more detail.

The descriptive use of LDA is probably its most common application in archaeometry but, as noted in Section 20.2.4, one of its potential attractions is its use in establishing the provenance of specimens not included in the original analysis. Once a group is established, whether on *a priori* grounds or as a result of statistical analysis, the Mahalanobis distance of a new specimen from a group centroid can be established (see Section 20.5.5). This, in turn, can be converted to a probability that can be used to assess whether or not the specimen could plausibly belong to the group. This approach does require the assumption of normality. This use of LDA in archaeometry was pioneered at the Brookhaven National Laboratory (Sayre 1975; Bieber et al. 1976) and has been restated more recently, in the context of lead isotope ratio analysis, by Sayre et al. (1992, 1995). Precise details of calculations depend on the use, or otherwise, of the equal covariance matrix assumption, as well as normality, and a useful discussion of the issues involved is given by Leese (1992).

The use of Mahalanobis distance results in an estimate of the *absolute* probability that a specimen belongs to a group. Some computer packages present *relative* rather than absolute probabilities of group membership. If such packages are used, it needs to be remembered that there is an implicit assumption that specimens belong to one of the groups represented in the sample used for LDA.

20.5.4.3. Quadratic Discriminant Analysis

Given $g = 2$, in a successful LDA values of Z for the first group should be quite distinct from those for the second group. For $g > 2$ the most usual way of presenting results is as a plot of Z_1 against Z_2. The allocation of a specimen to a group depends on its position in relation to the linear boundaries defined by the LDA. It is possible to generalize LDA by including quadratic and cross-product terms in the discriminant function, and this gives rise to the technique of quadratic discriminant analysis (QDA). This is equivalent to relaxing the equal covariance assumption and treating groups as if they are sampled from populations with different covariance matrices.

The more general QDA approach has been discussed in the archaeometric literature (e.g., Leese 1988, 1992) but not much used. One reason may be the more demanding sample size requirements, discussed in Section 20.6.4.

20.5.4.4. Stepwise Discriminant Analysis

Some software packages (e.g., SPSS, BMDP) have options that allow the use of stepwise discriminant analysis in which a subset of the variables are selected for the final discriminant functions. Klecka (1980) provides an account geared toward the use of the SPSS package as it then was. These options should be used with caution. There are many different criteria that can be used for selection that do not, in general, result in the same discriminant functions. There is also no guarantee that a selected subset will be the "best" one available, however "best" is defined.

Baxter (1994b) provides a fuller discussion. Examples are given where different selection criteria produce completely disjoint subsets of variables that perform at about the same level. As judged by the error rate, several examples were provided where different, and smaller, subsets of variables out performed those identified by a variety of variable selection procedures. The main message here is that one cannot assume that a selected subset is "optimal" in any useful sense of the term, so that interpretation of results that goes beyond graphical display should be undertaken cautiously.

20.5.5. More on Mahalanobis Distance

20.5.5.1. Theory

Mahalanobis distance is important in certain approaches to data handling and has been used in archaeometry for some time (Sayre 1975). This subsection collects together a number of results for reference and supplements the informal discussion given in Section 20.5.1.

Let $\mathbf{z}_i$ and $\mathbf{z}_j$ be two $p \times 1$ vectors and $\mathbf{S}_a$ a suitably defined estimate of a covariance matrix. Mahalanobis distance (MD) is defined as

$$D^2 = (\mathbf{z}_i - \mathbf{z}_j)' \mathbf{S}_a^{-1} (\mathbf{z}_i - \mathbf{z}_j)$$

If $\mathbf{z}_i$ and $\mathbf{z}_j$ are the *mean* vectors for two distinct groups, with $\mathbf{S}_a$ their pooled covariance matrix, then the MD between group centroids is obtained. If $\mathbf{z}_i = \mathbf{x}_i$ and $\mathbf{z}_j = \mathbf{x}_j$, and $\mathbf{S}_a = \mathbf{S}$ for the data matrix $\mathbf{X}$ consisting of a single group, then the MD between the two cases i and j is obtained. If $\mathbf{S}_a = \mathbf{S}$, for the data matrix $\mathbf{X}$ whose mean vector is $\mathbf{z}_j$, and if $\mathbf{z}_i$ is a case that may or may not belong to the group, then the MD between that case and the group centroid is obtained.

In this last case let D_i^2 be the MD when $\mathbf{z}_i$ is *not* a member of $\mathbf{X}$. Using results given in Mardia et al. (1979: 77), it can be shown that if $\mathbf{X}$ has a multivariate normal distribution $nD_i^2/(n+1)$ is a $T^2(p, n-1)$ variable, where this last quantity is Hotelling's T statistic with p and $(n-1)$ degrees of freedom. From Theorem 3.5.2 of Mardia et al. (1979) it then follows that

$$\frac{(n-p)nD_i^2}{(n^2-1)p} \sim F_{p,n-p}$$

the F distribution with p and $(n-p)$ degrees of freedom (see also Krzanowski 1988: 227). When n is large in relation to p, then, approximately, $D_i^2 \sim pF_{p,n-p}$. Since, under these conditions, it is also the case that $F_{p,n-p} \sim \chi_p^2/p$, where χ_p^2 is the chi-square distribution with p degrees of freedom (Johnson and Kotz 1970: 75), it follows that $D_i^2 \sim \chi_p^2$ for large n. This large sample approximation is used in some approaches to data handling in archaeometry (e.g., Beier and Mommsen 1994).

Using the above results, an MD value may be converted into a value from a standard probability distribution, the probability of which may be determined. If the probability is sufficiently great (conventionally greater than 0.05), then the case may plausibly be regarded as a member of the group.

For the situation where $\mathbf{z}_i$ is a member of $\mathbf{X}$, matters are a little more complicated since case i will affect the computation of $\bar{\mathbf{x}}$ and $\mathbf{S}$ and, in cases of doubt, lead to an overoptimistic assessment of the probability of group membership, especially for small samples. This problem may be overcome using the *leave-one-out* method in which case i is omitted from the group and treated as external to the remaining $(n-1)$ specimens. If $D_{(i)}^2$ is the leave-one-out estimate of MD,

$$\frac{(n-p-1)(n-1)D_{(i)}^2}{n(n-2)p} \sim F_{p,n-p-1}$$

(e.g., Campbell 1985; Penny 1996). As

$$D_{(i)}^2 = \frac{(n-2)n^2}{(n-1)^3} \frac{D_i^2}{1 - nD_i^2/(n-1)^2}$$

(Atkinson and Mulira 1993; Penny 1996), computations can be based on calculations of MD for the full data set (Leese and Main 1994). A "brute-force" methodology in which each case is omitted in turn to calculate $D_{(i)}^2$ is also feasible with modern computing power, if less elegant.

20.5.5.2. An Example

An example, based on the AAS data for Romano-British pottery of Table 20.1, usefully brings together the different techniques discussed here and in earlier sections. Tabular inspection of the data suggested a number of possible outliers, only one of which was clearly detected by cluster analyses. There was a clear indication of three groups in the data, corresponding to specimens from kiln sites from three distinct regions (Figs. 20.2 and 20.3). The discriminant analysis of Figure 20.4 appears to confirm this conclusion.

Table 20.2. Mahalanobis Distance and Probability Calculations for the Group Defined by Region 1 for the Romano-British Pottery[a]

No.	Id.	D_i^2	$D_{(i)}^2$	p_i	D_i^2	$D_{(i)}^2$	p_i
1	GA1	18.0	179.6	.00	13.4	41.8	.00
2	GA2	3.1	3.8	.98	1.3	1.5	.87
3	GA3	11.5	28.1	.17	0.6	0.6	.97
4	GA4	19.8	2006.6	.00	1.0	1.1	.91
5	GA5	6.8	10.7	.72	3.3	4.2	.51
6	GB1	6.3	9.6	.77	0.2	0.2	1.00
7	GB2	3.3	4.1	.97	2.0	2.3	.76
8	GB3	10.6	23.7	.25	7.9	13.5	.06
9	GB4	3.2	3.9	.98	2.9	3.6	.59
10	GB5	7.1	11.5	.67	5.5	7.9	.22
11	GC1	8.4	15.2	.51	2.0	2.4	.75
12	GC2	16.8	108.9	.00	16.3	92.2	.00
13	GC3	6.3	9.5	.77	4.3	5.8	.36
14	GC4	9.0	17.1	.43	3.8	4.9	.44
15	GD1	4.2	5.5	.94	0.8	0.9	.95
16	GD2	6.7	10.5	.72	2.4	2.8	.69
17	GD3	13.1	39.9	.07	1.6	1.8	.83
18	GD4	8.8	16.5	.46	7.9	13.6	.06
19	GD5	7.0	11.3	.68	3.2	3.9	.54
20	GE1	6.6	10.2	.74	1.8	2.0	.80
21	GE2	9.0	16.9	.44	1.1	1.2	.91
22	GE3	3.5	4.4	.97	0.8	0.9	.95

[a] D_i^2 and $D_{(i)}^2$ are Mahalanobis distances from the group centroid, the second quantity being calculated on a "leave-one-out" basis; p_i is the estimated probability that the specimen could come from the group, based on $D_{(i)}^2$. The first block of results is based on all nine variables, the second on four variables only.

Many published statistical analyses of archaeometric data do not advance beyond this point. The PCA, however, suggests the presence of outliers in the data that the cluster and discriminant analyses do not detect and it is of interest to take matters further.

In Table 20.1 there are 22 cases from the first region. Treating this as a single group and calculating the MD of specimens from other groups to its centroid confirms that they are highly unlikely to be members of the group. Table 20.2 shows the values of D_i^2 and $D_{(i)}^2$ for the group; the latter are, characteristically, greater than the former. In this instance the probability calculations are very similar whether one assumes the validity of the large sample approximation or uses results for $D_{(i)}^2$, and only the latter are shown as p_i. The probability calculations suggest that specimens GA1, GA4, and GC2 may not be plausible members of the group. It was noted earlier that the first two of these have obvious univariate outliers for some variables, which may be misprints, whereas GC2 appears to be a genuine multivariate outlier. It may be noted that GC2 differs from other specimens from the same region primarily in having lower values for each variable used. This might be attributable to differential SiO_2 dilution—a variable not recorded. That is, it is essentially similar to other cases but for a higher proportion of SiO_2, which consequently reduces the proportions of other variables.

The approach just illustrated forms the basis of the group refinement or evaluative techniques used in some laboratories (Sayre 1975; Bishop and Neff 1989; Glascock 1992). After initial group definition, cases may be added to or rejected from a group on the basis of probability calculations. New calculations are then done for this "refined" group, and this proceeds in an iterative manner until stability is achieved.

In Tubb et al. (1980) it was concluded that only four or five variables were really needed to differentiate the three groups, and it is of some interest to repeat calculations using just Fe_2O_3, MgO, CaO, and K_2O. These calculations are shown in the final three columns of Table 20.2. Specimen GA1 is still indicated as an implausible group member, with the extreme value for Fe_2O_3 being responsible for this; GC2 is also suggested as an implausible member and is extreme with respect to all four variables used. Most interestingly, GA4 now sits quite happily within the group because TiO_2, on which it was outlying, is no longer included in the analysis. These results indicate how important outliers and variable selection can be in affecting the outcome of a statistical analysis of compositional data, and this is discussed further in Section 20.6.

Studies that have used MD for the purposes discussed above include Blackman et al. (1993), Lizee et al. (1995), Slane et al. (1994), and Steponaitis et al. (1996). All these involve the use of NAA with ceramic materials. The work by Slane et al. (1994) presents statistical results in particular detail and includes probability calculations based on both D_i^2 and $D_{(i)}^2$.

If interest centers on assessing the similarity of two groups, of size n_1 and n_2, similar results are available. Specifically,

$$\frac{(n_1 + n_2 - p - 1)n_1 n_2 D^2}{(n_1 + n_2 - 2)(n_1 + n_2)p} \sim F_{p,n-p-1}$$

(Glascock 1992; Arnold et al., 1999). In this case, as well as the normality assumption, the calculation of $\mathbf{S}_a$ assumes that the two groups are sampled from populations with equal covariance matrices.

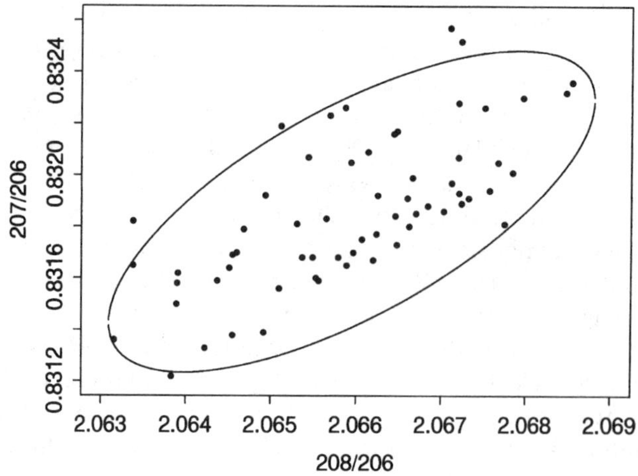

Figure 20.12. A 90% confidence ellipse shown on a bivariate plot of the $^{208}Pb/^{206}Pb$ and $^{207}Pb/^{206}Pb$ lead isotope ratios for the Kea field, using data from Stos-Gale et al. (1996).

20.5.5.3. Confidence Ellipsoids

The laboratories that have made most use of MD have also regularly used confidence ellipsoids, based on MD, on bivariate plots ($p = 2$), to estimate the space occupied by a group and compare the extent to which groups overlap. Such plots can be based on pairs of the original variables or on pairs of linear combinations of the variables such as principal components (e.g., Glascock 1992; Hegmon et al. 1997; Lizee et al. 1995; Slane et al. 1994; Steponaitis et al. 1996; Triadan et al. 1997). The methodology has influenced the way in which archaeological lead isotope ratio data are often presented (e.g., Sayre et al. 1992; Stos-Gale and Gale 1994) in a manner that has attracted some controversy (Scaife et al. 1996; Scaife 1998). Figure 20.12 shows a 90% confidence interval for the Kea field, based on the data in Stos-Gale et al. (1996). That there are points just outside the ellipse is not a cause for concern; this is to be expected since the ellipse is not designed to enclose all points. There is, in fact, evidence that the normality assumption used here is not valid (Baxter and Gale 1998; Baxter 1999a), and Beardah (1999) has investigated an alternative way of representing the extent of this field using KDEs.

Given the mean and covariance matrix of a group **X**, pairs of values for which $D_i^2 = c$, where c is a constant, define elliptical contours of constant probability. The chosen value of c depends on the probability level required and whether the χ^2 or F distribution is used. For the 90% confidence ellipsoid in Figure 20.12 the former was used since, with $n = 62$ and $p = 3$, this large sample approximation gives almost identical results to the use of F.

The controversy alluded to concerns the fact that the construction of confidence ellipsoids requires the assumption of (bivariate) normality, and this is rarely rigorously checked for. In the field of lead isotope studies there is mounting evidence that the assumption is often incorrect (Baxter and Gale 1998; Baxter 1999a, 1999b; Scaife 1998). We are unaware of similar studies for other applications; it is probable, however, that chemical groups defined by statistical procedures are more likely to be normal than groups defined by archaeological criteria (e.g., type). This is because the statistical procedures used are often designed to identify (chemically) homogeneous groups, and these are more likely to approximate to normality than potentially non-homogeneous groups defined by other criteria. Two-dimensional kernel density estimates (Beardah, 1999; Scaife 1998) provide an alternative to confidence ellipsoids that do not require the normality assumption, but are more demanding of data. Sample size questions, and the normality assumption, are discussed below and in Section 20.6.

20.5.5.4. Sample Size Requirements

An unstated assumption so far has been that MD is readily, and reliably, calculable. In fact, this is a nontrivial problem. The calculation requires, for the case of a single group, that $\mathbf{S}$ be estimable, which in turn requires that $n > p$. For stable estimation of $\mathbf{S}$ it is desirable that n be much greater than p. Harbottle (1976) suggests $n > 3p$ as a minimum requirement and $n > 5p$ for preference. Techniques such as NAA and ICPS can routinely generate information for 20+ variables. This implies that for a data set consisting of several groups the sample size needed to make effective use of methods based on MD is demanding. One approach to the problems is to base MD calculations on principal components; for example, Slane et al. (1994) base calculations on 28 variables when group sizes are large enough and on the first eight principal components when they are not.

20.5.6. Other Methods

Here brief note is made of other approaches that have sometimes been used. *Correspondence analysis* can be thought of as a form of PCA suited to categorical data, but which may be applied to any data matrix of non-negative numbers, possibly after transformation, where it can be thought of as a form of nonlinear PCA. It has been widely used in the general archaeological literature in recent years (e.g., Madsen 1988; Müller and Zimmermann 1997) and has occasionally been applied to archaeometric data.

Greenacre (1984, 1993) discusses the technique in detail, with the latter the more accessible account, and Underhill and Peisach (1985) describe, in detail, an algorithm for carrying out correspondence analysis that is applied to chemical compositional data. These researchers have all been based in South African universities, and it is from this source that the few archaeometric applications of correspondence analysis have mostly come (e.g., Bollong et al. 1997).

An attraction of correspondence analysis sometimes used to "sell" the technique is that it results naturally in the joint representation of the rows and columns of a data matrix, and their joint interpretation. That biplots (Section 20.5.2.3) provide a similar facility in PCA has not always been fully appreciated.

The term *compositional data* is sometimes used by statisticians to refer to data for which $\sum_j x_{ij} = c$ for some constant c and for all i (Aitchison 1986). Typically, $c = 1$ or $c = 100\%$. Such data can cause problems for standard methods of statistical analysis because the constraint induces correlations among the variables. Aitchison (1986) discusses a variety of approaches to the analysis of such data, among them the use of the log-ratio transformation $y_{ij} = \log(x_{ij}/x_{ip})$ for $j = 1, 2, \ldots, p-1$. In their original form the p-variate data are constrained to lie within a $p-1$ subspace [more exactly, the data define a $p-1$ simplex embedded within p-dimensional real space (Aitchison 1986: 27)], and it is this constraint that may cause problems for standard statistical techniques. The effect of the transformation is to produce unconstrained $p-1$ dimensional data for which, it is hoped, standard methodology is applicable.

Some archaeometric data can arise in this form, when the analytical method determines the major and minor oxides that sum to approximately 100%, and Baxter (1989) suggested that Aitchison's approach might be useful in such circumstances. Later experience led to this suggestion being strongly qualified (Tangri and Wright 1993; Baxter 1993) because of the undue influence that transformation of the minor oxides (i.e., with values typically less than 1%) can have on subsequent analysis. Omission of the dominant major oxide from analysis (often SiO_2 for ceramic and glass compositions) is often a preferable approach. Comparisons of different data treatments for compositional data are given in Baxter (1989, 1994c).

It is worth noting that the transformation given above has sometimes proved of value, even when the data are not fully compositional in the sense used in the previous two paragraphs. Suppose, in a temporary modification of notation, that the x_{ij} represent the composition of a clay source used in ceramic manufacture. This may be *diluted* in manufacture, by the addition of a pure temper, for example, so that what is measured is $a_i x_{ij}$. If the a_i vary greatly for ceramics made from the same clay, statistical analysis may create the impression that there are more groups or outliers in a sample (in terms of clay source) than really exist. The log-ratio transformation can eliminate this dilution effect, because the a_i cancel out, and it has occasionally been used to good effect for this reason (Leese et al. 1989; Stopford et al. 1991). The transformation also underpins the "best relative fit" method described by Sayre (1975) and exploited in Taylor and Robinson (1996a), for example, where, after statistical analysis, groups or outliers that are disparate solely because of dilution effects may be adjusted to "fit" a group.

Beier and Mommsen (1994) discuss approaches to handling dilution effects in some detail, noting that they can arise for a range of reasons, including systematic measurement error. Baxter et al. (1990) examine the relationship between different forms of log-ratio analysis, correspondence analysis and PCA and discuss the circumstances under which they give rise to similar results.

20.6. PROBLEMS AND PRACTICAL CONSIDERATIONS

20.6.1. Introduction

In this section a number of issues pertinent to the practical application and interpretation of exploratory multivariate methods in archaeometry are discussed. Most of these have received attention in the statistical literature, but their importance in the specific context of archaeometric problems has not been systematically studied, and we believe that future research could usefully examine them in more detail.

20.6.2. Outliers

Barnett and Lewis (1994: 7) define an *outlier* in a set of data to be an "observation (or subset of observations) which appears to be inconsistent with the remainder of that set of data." Many archaeometric data sets are investigated in the expectation that there are several groups in a set of data, so an outlier would be a case that is chemically distinct from the groups in the data.

Although chemical outliers may also prove to be of archaeological interest, they are more often regarded as a nuisance because of their potentially adverse effects on the statistical analyses discussed in Section 20.5. If the main purpose of an analysis is to define, and distinguish between, groups in a set of data, then the omission of outliers will often be sensible.

To omit outliers they must first be identified. Much of the statistical literature on this topic is not suited to typical archaeometric data since the assumed model, against which outliers are judged, is that there is a *single* subset of at least 50% of the data that has a multivariate normal distribution (e.g., Rousseeuw and van Zomeren 1990; Hadi 1992, 1994; Atkinson and Mulira 1993).

For the multiple group problem what is ideally needed is a method that jointly determines both groups and outliers. The filtering procedure of Beier and Mommsen (1994) results in the sequential determination of groups, with outliers being those cases not assigned to a group. The group evaluative procedures described in Section 20.5.5.2 usually begin with the simultaneous determination of groups that are subsequently modified; in the process, cases that do not apparently belong to any group will be identified as outliers. Both procedures, which are similar in spirit, use Mahalanobis distance as a measure of the similarity between a case and a group. The main limitation here is that large group sizes are needed for Mahalanobis distance to be used effectively.

In practice much outlier detection is carried out informally. Eight of 10 studies in the 1996 and 1997 volumes of the journal *Archaeometry* using multivariate statistics note the existence of outliers detected using clustering algorithms; the remaining two works identify outliers using PCA. Cluster analysis is not a panacea. For the data of Table 20.1, case GC2 is a multivariate outlier (Sections 20.3.3 and 20.5.5.2). It is suggested as such by PCA and the group evaluative methodology of Section 20.5.5.2 but not by any of the common forms of cluster analysis, including the single linkage methodology recommended by Beier and Mommsen (1994: 294). Similarly, two outliers in the Lavrion lead isotope data set that can be identified visually using brush-and-spin plots, or by the more formal methodology of Caroni and Prescott (1992), are not suggested by cluster analysis. They are, however, revealed on the third (and final) component of a PCA of the correlation matrix, as the box plot in Figure 20.13 shows.

We would not recommend any single method for multivariate outlier detection and find the use of several techniques is often helpful. Cluster analysis should be supplemented by inspection of the principal components, and the latter is often more useful. If sample size permits, and if it is reasonable to assume that groups have a normal distribution, group evaluative algorithms or that of Beier and Mommsen (1994) result in outliers being determined at the same time as groups are formed. Where there is a reasonable expectation that there is a single main group in a sample, the methodologies of Atkinson and Mulira (1993) or Leese and Main (1994), or more formal tests such as those in Caroni and Prescott (1992) or Penny (1996), might be used. All these procedures use Mahalanobis distance in some form.

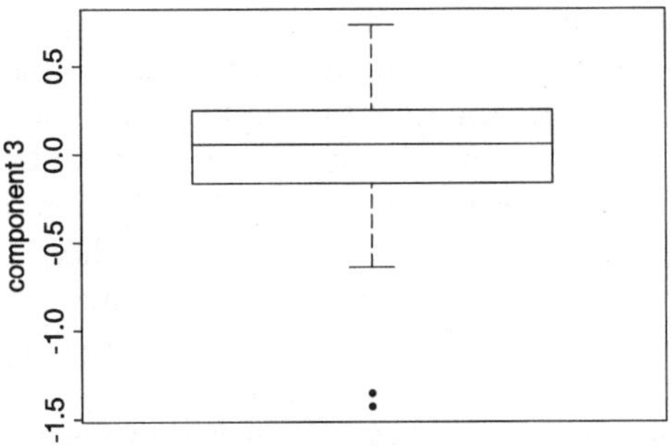

Figure 20.13. Box plot of the scores on the third principal component from a PCA of the Lavrion lead isotope ratio data (data from Stos-Gale et al. 1996). Two outliers, not evident from cluster analyses, are isolated at the bottom of the plot.

20.6.3. Multivariate Normality

Contrary to what is sometimes suggested, multivariate normality (within groups) is *not* a requirement for exploratory methods such as cluster analysis, PCA, or discriminant analysis. It is needed in those approaches to data handling that use probability calculations based on Mahalanobis distance (e.g., Slane et al. 1994; Sayre et al. 1995).

The practice of logarithmically transforming trace element data suggested in Sayre (1975) and Bieber et al. (1976), discussed in Pollard (1986), and widely emulated is often advocated as an aid to achieving normality. The extent to which normality really is obtained is rarely tested, so it is usually an article of faith. Transformation can improve the performance of exploratory methodologies, but this is not necessarily because normality has been obtained. It may simply be that the graphically distorting effects of skewed distributions (where numbers can differ by orders of magnitude) have been avoided.

Baxter and Gale (1998) and Baxter (1999a, 1999b) have recently reviewed methods for testing uni- and multivariate normality and applied these latter to (conveniently low-dimensional) lead isotope ratio data. The conclusion that most data sets examined were non-normal contradicts an assumption that underpins some methodologies for processing lead isotope data (Sayre et al. 1992, 1995). A problem in the past has been the small sample sizes used and/or the use of suboptimal methods for testing normality. It would be of interest to extend these studies to other kinds of data set and problems for which p is greater.

Also requiring further study is the seriousness of departures from normality for those procedures that assume it. A subtle and interesting point arises in connection with group evaluative methodology. The Mahalanobis distance calculations assume normality so that cases that do not conform with this assumption (e.g., apparent outliers) may be rejected from a group. It would thus not be surprising if groups determined in this way did have a normal distribution, but this might simply be a reflection of the assumptions used in the analysis rather than the underlying "reality." This kind of concern underpins criticisms that have been made of some applications of group evaluative methodology (e.g., Budd et al. 1993; Burton and Simon 1996).

It would be nice to see more, and more rigorous, testing of the assumption of normality when it is used. The main practical problem is, as ever, sample size, a discussion of which follows.

20.6.4. Sample Size and Variable Selection

In this, and previous sections, frequent reference has been made to sample size considerations. "What should my sample size be?" is a frequently asked and often unanswerable question, at least in precise terms. In practice, pragmatic considerations, such as the specimens to hand, size of a research budget, and cost of an analytical technique, often determine what is done. Whether this is a good basis for *archaeological* research is a legitimate concern, and we are acquainted with archaeologists who are suspicious of many published archaeometric analyses for this and related reasons. There is the basis for a debate that needs to be aired more in the archaeometric literature; here, however, we will confine ourselves to more technical matters.

The reason that sample size questions are difficult to answer is that they depend intimately on the techniques to be used, the purpose of an analysis, and the structure of the sample, which is unknown in advance of data collection. To illustrate with an apparently simple problem, in the analysis of lead isotope ratio data, with $p = 3$, there is some agreement that 20 geologically well-selected ore samples from a deposit constitute an "agreeable minimum" for estimating the lead isotope field for the deposit (Pollard and Heron 1996: 328). This may be so if the field is multivariate normal; if one wishes to test this, much larger samples may be needed. Results in Baxter (1999a, 1999b) suggest that the data from the Lavrion field are non-normal, and KDEs based on selected linear combinations of the data, and Figure 20.6, indicate that it is bimodal. Unpublished work, based on repeatedly subsampling from the $n = 59$ cases for different sample sizes, suggests that 40+ observations may be needed to detect the non-normality with reasonable confidence.

For certain kinds of analysis *lower* bounds on the sample size can be specified. For example, in linear discriminant analysis $n > p$ is a requirement, and for quadratic discriminant analysis a similar criterion must be satisfied for each group. The minimum sample size permits an analysis, but not necessarily a satisfactory one, as was discussed in Section 20.5.5.4, where it was noted that $n > 3p$ or even $n > 5p$ is desirable if a technique requires an estimate of the covariance matrix for a group. For certain kinds of analysis, such as the assignment of specimens to groups in discriminant analysis, cross validation (Section 22.5.4.1) or *leave-one-out* (Section 22.5.5.1) methods partially alleviate the problem, in the sense that an individual does not affect the estimate that subsequently influences the assignment. Nevertheless, if n is small in relation to p, individual cases (other than those "left out") may still have an undue influence.

Sample size requirements for techniques such as PCA and cluster analysis have received little attention. Investigating this problem is difficult because of the dependence on the unknown structure. If there are distinct and widely separated groups within a data set, it is likely that this can be revealed with fairly small sample sizes, but also that multivariate methods may not be needed to show this. Where group separation is less obvious, but exists and is archaeologically interpretable, larger sample sizes than those often used may be needed.

For certain methods the sample size requirements are clearly related to the number of variables used (as opposed to measured), and an obvious way of reducing requirements is to be selective about which to use. This kind of point was made in Bieber et al. (1976) but has not been systematically studied in an archaeometric context. Variable selection has been studied quite extensively in discriminant analysis (Section 20.5.4.4) and applied in archaeometry. Variable selection for PCA has received some attention (e.g., Krzanowski 1987) but has not been applied to archaeometric problems. Krzanowski (1988: 516) notes that variable selection is an issue for several reasons. These include the possibility that irrelevant variables can mask structure in the data and that the more variables the greater the sample size needed for a reliable analysis.

Questions of sample size and variable selection are thus intimately entwined and not easily answered. This is an area we believe is deserving of further study, particularly in view of the numbers of variables that can routinely be measured by modern analytical techniques. Many existing approaches involve some form of automatic selection. These do not necessarily guarantee selection of those variables that contribute most to an archaeometric or archaeological *interpretation*, and this is a particular problem that needs addressing.

20.6.5. Missing Data

Problems may sometimes occur with missing data. These can be resolved in a number of ways depending on the scale of the problem and the techniques that are to be used. The most draconian approach is to omit cases and/or variables for which missing values occur. If very few cases are affected, then omitting them from an analysis is likely to have little effect. If there are a lot of missing values occurring mostly within a small number of variables, it is less obvious what to do. If the data are missing in a random fashion, then it is likely that little will be lost by not using them. If, however, the data are "missing" nonrandomly (e.g., values are below the level of detection), it is possible that they are carrying information about structure in the data. In these situations it is worth looking to see if other variables appear to carry similar information [e.g., do the cases for which data are missing all have very high/low values for some other variable(s)]. If they do, then omitting the variable with missing values may not matter.

Ideally, however, one would like to use as much of the information collected as possible. If a cluster analysis is intended, one approach is to measure the dissimilarity between pairs of cases as the *average* distance across those variables for which there is no missing data,

$$\bar{d}_{ij}^2 = \sum_k (y_{ik} - y_{jk})^2 / m$$

where $m \leq n$ is the number of variables for which the pair can be compared, and the summation is over these m variables (Sayre 1975; Bieber et al. 1976; Glascock 1992). This approach requires the use of clustering software that allows the direct input of a set of (dis)similarities.

For PCA analyses it is possible to calculate covariances or correlations between pairs of variables for those cases where there are no missing values of the variables, and submit this to a software package that allows direct input of a covariance or correlation matrix. We have not seen any applications of this idea in the archaeological literature, and there are problems with it. One is that the data need to be missing randomly, so that the estimated covariances are not badly biased. Another is that the covariance or correlation matrix calculated in this way may not possess the mathematical properties to allow a PCA to be carried out.

It is also possible to model or estimate missing data. A simple example is when data, missing because values are below the level of detection, are replaced with a value corresponding to the level of detection, to one just below, or to a value half-way between zero and the detection limit. A related practice, necessary if a logarithmic transformation is to be used, is to increment zero values by some small amount. This needs to be done with caution and the effects checked since the effect of transformation can sometimes be to create artificial groups or outliers. A more sophisticated modeling approach is described by Sayre (1975) who notes that, in calculating the

Mahalanobis distance of a specimen from a group centroid, missing data may be replaced by values that minimize the Mahalanobis distance. A recent text (Schafer 1997) is devoted entirely to the analysis of incomplete multivariate data and gives access to S-Plus software for this purpose. The methodologies described have yet to be applied to archaeometric data and might form the basis of future applied research.

20.7. STATISTICAL MODELING

20.7.1. Background

Before moving to consider the use of statistical modeling as an aid to the interpretation of analytical data for art and archaeology, some discussion of the term *model* is required. For around 25 years archaeologists have been using the term *model* to refer to a construct used to aid in relating observations in the field or laboratory to the past populations they are trying to understand (for many examples of this see Clarke 1972). As such, most modern archaeological data interpretation could be seen as model based.

For mathematicians and statisticians, however, the term *model* is a very specific one. In stating that an approach is model based, mathematicians imply that the observed data and the features that we are trying to learn about can be explicitly linked via coherent mathematical and/or statistical functions that can then be computed to provide results upon which direct inferences are based. Such models are not excluded from the definition of model commonly used by archaeologists; they are just one particular manifestation. Researchers working with archaeological data are increasingly aware that "investment in statistical modeling and analytical techniques can yield dividends" (Fieller and Flenley 1988: 81), but it is still not widely undertaken.

While considering use of terms, it might be useful to note that *statistical models* are a specialist subset of *mathematical models*. Statistical models are special in that they allow us, via the use of probability distributions, to represent stochastic elements in the relationships between data and the features we are trying to learn about. Since there is variability involved in most natural and human processes (e.g., decay of organic materials or the movement of groups of people around a landscape), statistical models are usually the most useful for representing our ideas about the processes that gave rise to the archaeological data observed today.

In the remainder of this chapter the term *model* will be used to refer to a statistical model. Before looking at specific models that might be used, it seems appropriate to outline the general approach normally adopted for model-based data interpretation. The approach is cyclic and can be divided into steps as follows:

1. Specify the problem, laying down clearly all the features to be contained within the model and define carefully what the aim of the modeling exercise is.
2. Using all the points laid out in step 1 and, being careful not to include ideas that might unnecessarily complicate the picture, set up an appropriate statistical model.
3. After selecting a particular statistical approach for undertaking the investigation, formulate the statistical model appropriately.
4. Obtain the mathematical solution of the model defined in step 3.
5. Interpret the results obtained in step 4.
6. Compare the interpretations with other knowledge. Do they make sense in the light of what is already known? If so, we can consider making inferences on the basis of the results obtained. If not, is there something we overlooked in the original problem definition? If so, return to step 1. Or, does the nature of the model selected need refining? If so, return to step 2.

With this framework in mind, it can be seen that much of what archaeologists term modeling is covered by step 1 of the statistical modeling cycle. Expressing ideas in a sufficiently coherent and explicit manner to allow building of a statistical model is not always straightforward, but for some fields of archaeological investigation it seems an appropriate next step.

It is our opinion that in the case of the collection of analytical data for investigating the past, the first step of the modeling cycle has already been achieved. Indeed, most sourcing studies have at their core a very clear statement of the relationship between the observed data and the processes that gave rise to them. As outlined throughout this chapter, however, there are (in addition to the basic relationships between data and the source) many other issues that should be taken into account if a model-based approach is to be truly useful. If such information cannot be included when adopting the model-based approach, there may be little gain in interpretative power over and above what can be gained using an exploratory approach since it will still be necessary to "factor in" extra information when we come to make inferences on the basis of the results.

Until recently, this was reason enough for few archaeologists and art historians to consider using model-based approaches; they were seen as vastly more complex and time consuming than exploratory methods and to hold few interpretative advantages. In the last decade, however, major advances in a statistical framework known as the Bayesian paradigm has led to an opportunity to include data, statistical model, and relevant *a priori* information all in the same analysis. Since *a priori* evidence is often seen as a vital component of modern archaeological and art historical investigations, the Bayesian approach seems to offer a model-based statistical framework that is relevant to the way archaeologists and art historians work.

20.7.2. Bayesian Approach

The Bayesian approach has its origins in the work of Thomas Bayes (1763). At its core is a single theorem that is both simple and intuitive and allows a formal means for combining

1. a statistical model of the underlying processes that gave rise to the data (known as the likelihood), and
2. a probabilistic representation of any prior knowledge available about the parameters of the model (known as the prior), to arrive at
3. a probability assessment about the likely values of the parameters of the model, based on both the data and the prior information (known as the posterior).

We do not propose to discuss the theorem in any detail here; for this see Buck et al. (1996) in which its application to a wide range of archaeological problems is also discussed in some detail. Here we will look at the pros and cons of adopting the Bayesian approach and outline the considerations needed for its adoption in sourcing studies.

Those who prefer a more traditional approach to statistical modeling (classical or frequentist statisticians) see the inclusion of subjective prior information as a serious drawback of the Bayesian approach. Classical statistical methods (it is held) provide the ability to arrive at objective conclusions based solely on the current data and not on subjectively assessed prior information. We argue that prior information exists regardless of which statistical framework is adopted. It is better that the people who investigate a specific data set be explicit about their prior knowledge and conditional assumptions and include them coherently in their interpretation than that they behave as if they had no prior experiences that have a bearing on the problem. If we are explicit about our prior knowledge, other researchers can counter what we say, reanalyze the data with their own prior information, and compare the outcomes.

Clearly the arguments about which statistical framework to adopt are complex, and there is insufficient space to pursue them further here. Readers particularly interested in the arguments might like to consult Howson and Urbach (1993) who provide a philosophical discussion of Bayesian reasoning, or Barnett (1982) who gives a balanced and clear view of both classical and Bayesian statistical methods.

20.7.3. Selecting Suitable Models

Since model-based approaches are always specific to the type of problem currently being addressed, it is necessary to focus on just one to illustrate the applicability of the methodology. Many of the analytical methodologies discussed in this book were developed as a result of a desire on the part of archaeologists to identify the source (or provenance) of the artifacts under investigation (for an outline of the problems see Section 20.2.4), and we now turn to focus on these.

Formulating such problems in mathematical terms, suppose that there exist k likely sources for a particular artifact or material type. Now, consider just one chemical compositional vector X_m from an archaeological or art historical sample known to originate from a particular source, m. X_m is then a single vector selected from

$$f_m(X_m|\theta_m) \tag{20.2}$$

where f_m represents a mathematical function specific to source m and θ_m represents a vector of parameters that defines the chemical composition of source m. The vertical bar (|) is a standard symbol meaning "given," which indicates that the values of the compositional vector obtained for a particular specimen from source m are dependent upon the general chemical compositional properties of material from source m.

For sourcing studies to work, a mechanism is needed for characterizing material from the sources in question and then assigning specimens to the closest match. Using the basic definition of source given in Eq. (20.2), Buck (1994), Buck and Litton (1996), and Buck et al. (1996) develop models for a Bayesian approach to both discriminant and cluster analyses. The details of the methodologies they describe are complex and will not be repeated here; rather, what follows is an outline of the nature of the approaches they suggest.

In the case of discriminant analyses, where material from possible sources has already been identified and analyzed, what is required is a mechanism for characterizing each source and then attributing newly analyzed specimens to one or other of the preexisting groups. First, a model is required that will represent the properties of data from the same source. As discussed in Section 20.6.3, it is not uncommon for analysts to assume that samples from the same source exhibit multivariate normality or that a simple transformation will induce this. There are researchers who feel that this is not valid, and more research is needed if model-based approaches are to be widely used, but for the purposes of illustration a model must be selected and the multivariate normal is convenient. Note, however, that following the necessary research, any suitable model could be substituted.

With a suitable model selected, it is necessary to think about the practicalities of allocating new specimens appropriately. It is not reasonable to assume, for example, that each source gives rise to the same quantity of specimens. Consequently, a means is needed for representing the probability that specimens come from each of the possible sources. Let ϕ_m represent the probability that a specimen derives from source m where $m = 1, \ldots, k$ and k is fixed *a priori*. Then, in very basic terms, everything is available to allow reformulation of Eq. (20.2) in terms suitable for adoption in a Bayesian sourcing study. The data (or some transformation of them) from each source are modeled as

$$Y_m | \mu_m, \Omega_m \sim N(\mu_m, \Omega_m)$$

Modeling the probability of samples coming from each source by the vector of *mixing probabilities*, ϕ, allows us the flexibility to include not just prior information about the nature and characteristics of each source but also to express any information available about which sources are more likely to have produced most specimens. Initially, it might be felt that all sources are likely to have produced equal numbers of specimens, but as experience improves this is unlikely to be the case and the prior on ϕ can be adjusted accordingly.

To make this model definition really useful for the problem at hand, one further parameter is needed, **z**. This is the classification parameter that indicates to which source each specimen in the analysis has been assigned. Of course, the specimens of known origin will have fixed z_j's, but those of unknown origin will obtain their z_j's only as a result of the Bayesian analysis. With the model constructed as outlined, the probability that specimen i comes from source m is proportional to

$$\phi_m \times p(Y_i | \mu_m, \Omega_m, z_j = m)$$

We then assign sample i to the source for which this function returns the highest value.

Consider now what is needed to extend the framework for use in situations where no known source material exists, and we are hoping to use data from specimens of unknown origin to establish likely provenance related clusterings. In such studies we wish to partition the data into k provenances where the value taken by k is not known *a priori*. Statistically, this is a much more complex problem than the discriminant analysis outlined above, but the basic approach recommended by Buck (1994) and Buck et al. (1996) can be explained in a fairly intuitive manner. The same basic model is adopted, but, since no sources are known *a priori*, it is necessary to allow for the possibility that each specimen is from a unique provenance and consequently that k could be as large as n. This means that initially each specimen must be modeled with its own μ_j and Ω_j, where $j = 1, \ldots n$. In practice, however, the likely number of provenances will be less than n so that *a posteriori* some samples will share the same μ_j and Ω_j. Such samples are said to form a cluster with the parameters of cluster c being μ_c and Ω_c. Samples from c will all have $z_j = c$ and hence will be seen as coming from the same provenance.

20.7.4. Nature of the Prior Information

Sources of potential prior information have been identified at various points both in this chapter and throughout the book. To take account of the most important of these, researchers involved in provenancing studies (regardless of which interpretative methodology they adopt) might like to ask themselves the following questions.

- Are any provenances known *a priori*? Can their chemical compositional properties be characterized and used to help in the interpretation of new data that arise?
- Even if specific provenances are not known, is anything known about the likely properties of data from sources in general? If so, can this be used to gain insights into the characteristics (perhaps the properties of the covariance structure) of likely sources in the present study?
- Even though the exact number of provenances is commonly not known, is there any prior information about the likely number? If so *a priori* information is available about the form of k. This can be incorporated into a Bayesian study or be used to aid in the interpretation of a dendrogram from cluster analysis or a PCA plot.
- Is there any *a priori* knowledge about which specimens are from the same provenance even though nothing is known about the source itself? For example, are there several specimens taken from the same object that are undisputedly from the same provenance? If so, in a Bayesian analysis, the prior on $\mathbf{z}$ should reflect this and, in an exploratory analysis it will be necessary to check carefully to see where these specimens appear on the final plots.

20.7.5. Implementation

For details of the implementation of the Bayesian methodologies readers are referred to Buck (1994), Buck and Litton (1996) and Buck et al. (1996). It is useful to outline the nature of the approach they advocate here, however, as it provides insight into why the methodologies are only now becoming available. As noted earlier, Bayes's theorem was published in 1763 and allows intuitive statement of even quite complex problems. Given this, it is unfortunate that the nature of the statistical functions required to represent even simple real problems within the framework meant that for many years the mathematics proved intractable. In the early 1980s, however, statisticians with access to powerful computers began to realize that even complex functions could be utilized for models and priors if posterior distributions were simulated rather than obtained analytically (Geman and Geman 1984; Gilks et al. 1996). As a consequence, most modern implementations of the Bayesian paradigm use Markov chain Monte Carlo (MCMC) methods, which involve simulating from the posterior in an iterative fashion until the distributions obtained are stable.

This means that the results of all such analyses are samples from the posterior distributions of the parameters of the model. In some cases such samples relate directly to what we want to learn about but in others, due to the way the problem has been formulated or parameterized, much work is needed to translate numerical results into archaeological or art historical ones. Discriminant and cluster analyses are a case in point. The results obtained are simulations of the classification vectors and some work (e.g., tabulation or graphical representation) is needed before these can be interpreted in terms of the archaeological or art historical questions posed.

20.7.6. An Example

To provide an example of the use of the Bayesian approach to cluster analysis, we reconsider the standardized medieval glass data from York (Cox and Gillies 1986) that were discussed in Section 20.5. These data were collected in an attempt to learn more about the origins of the glass used in York in medieval times. The expectation was that groupings within the data would allow identification of samples that had been produced using similar production techniques and/or from similar raw materials.

The nature of glass production is such that, when one looks for prior information, useful hints as to what might be reasonable are found rather than detailed specific information. This means that clear statements of the values for $\boldsymbol{\Omega}$ or k are not available. Rather, suggestions are available about what might be reasonable. When using prior information of this type, it must be recognized that the posterior information may not be robust. Nonetheless, it is preferable to explicitly include it and then investigate the sensitivity of the results obtained.

In the case of the prior for $\mathbf{\Omega}$ (following suggestions in Buck, 1994), we select a matrix with zeros as the off-diagonal elements and with reasonable values to reflect likely within-cluster variances on the diagonals. In the case of the standardized glass data we know that the variances on each element for the entire data set is one (this is one of the effects of standardizing the data) so a reasonable "ball-park" prior for the diagonals of $\mathbf{\Omega}$ is 1.0. We actually expect the within-cluster variance to be less than this and in using such limited prior information, we are careful to give it relatively little importance, allowing the information in the data to carry much more weight. The mechanism for doing this relates to the details of how $\mathbf{\Omega}$ is parameterized and is discussed in detail in Buck (1994) and outlined in Buck et al. (1996). In simple terms, it is assumed that the prior information about $\mathbf{\Omega}$ arises from a very small number of imaginary samples.

A similar approach can be taken when selecting a suitable prior for k. *A priori* the exact number of groups is not known, so we cannot provide a point estimate of k. Reasonable estimates for the range of likely values are, however, possible. Following the recommendations of Buck (1994), we selected a prior for k with the form shown in Figure 20.14*a*. This form allows for up to 10 groups *a priori*, but suggests that 1 to 6 groups are likely with the most likely being about 3. The mechanism used to arrive at this form of prior on k is extremely flexible and allows great variety in the form of possible prior distributions.

Adopting the methodology of Müller et al. (1996) as implemented by Buck (1994), the York medieval glass data were analyzed with the priors outlined above. The approach used is an iterative, sample-based one and produces results in the form of probabilities of association between each of the samples for a large number of simulations. Interpreting the results of such an analysis could prove extremely difficult unless a graphical summary can be obtained. Fortunately, it is possible to adopt the dendrograms already widely used for displaying the results of exploratory cluster analysis. Essentially, all that is required to produce a dendrogram is a pairwise table of similarities between samples, and this is exactly what is obtained by computing pairwise probabilities of association between samples. We simply have a Bayesian measure of similarity between samples. As a result, Figure 20.15 shows a dendrogram of *a posteriori* association between the samples in the York medieval glass data when the prior information is as outlined above.

Interpreting any posterior results without some sensitivity analysis is difficult (and potentially dangerous as it may lead to very ill-informed infer-

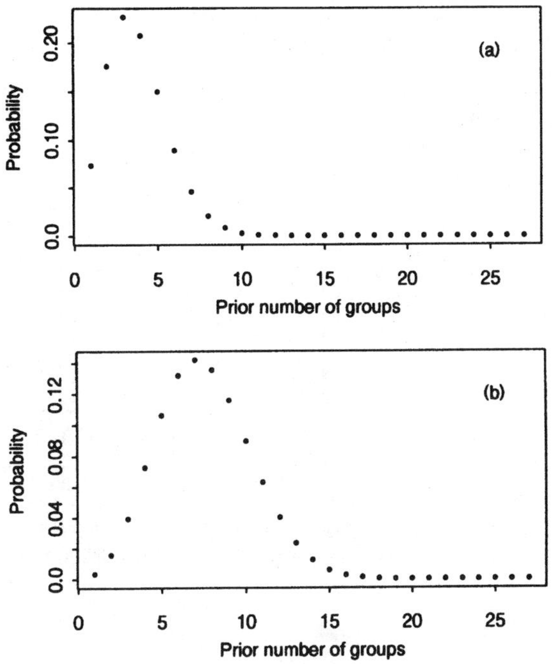

Figure 20.14. Prior distributions for k: (a) is the form used to indicate the belief that k is likely to lie in the range one to six and that the most likely number is around three and (b) is the form used as part of our sensitivity analysis (see below).

ences) and so, in Figures 20.16 and 20.17, plots are provided of the results of similar analyses undertaken with slightly different prior assumptions. To obtain Figure 20.16, we adopted the same prior information used to obtain the results in Figure 20.15, but this time the diagonals of $\mathbf{\Omega}$ were all 0.5 rather than 1.0. In other words, it was assumed that the within-cluster variance was half what it was before. As a result, *a priori*, the clusters were assumed to have smaller volumes. Note, from Figure 20.16, that this has the effect of increasing our *a posteriori* estimate of the number of clusters. To obtain Figure 20.17 we adopted the same prior information used to obtain the results in Figure 20.15, but this time the prior on k had the form shown in Figure 20.14b.

Taking these results in turn, Figure 20.15 suggests that there are four main groups one of which is small and is formed around sample 20, which was identified as an outlier in Section 20.5. Figure 20.16 shows the same basic groupings, but, because our prior on the within-cluster variances is much smaller, some samples are allocated to a group with lower probability

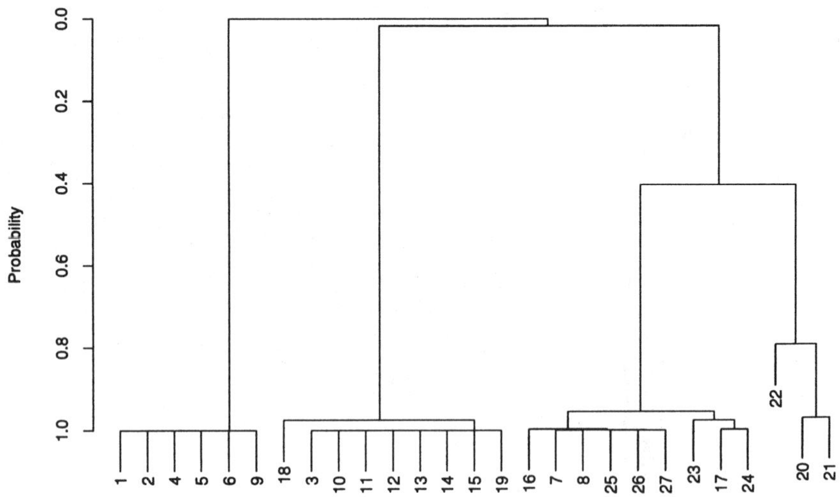

Figure 20.15. Dendrogram of *a posteriori* association between the samples in the York medieval glass data when the prior for k is that shown in Figure 20.14a and the prior for $\mathbf{\Omega}$ is the identity matrix.

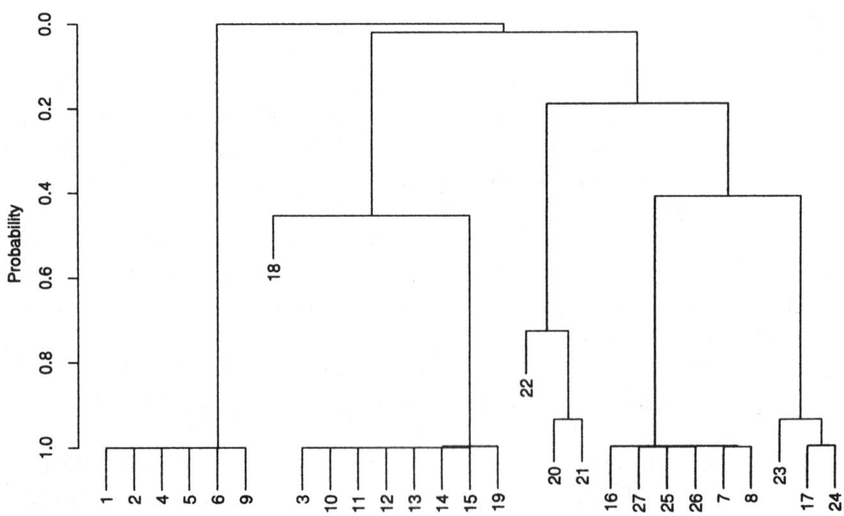

Figure 20.16. Dendrogram of *a posteriori* association between the samples in the York medieval glass data when the prior for k is that shown in Figure 20.14a and the prior for $\mathbf{\Omega}$ has zeros on the off-diagonals and 0.5 on the diagonals.

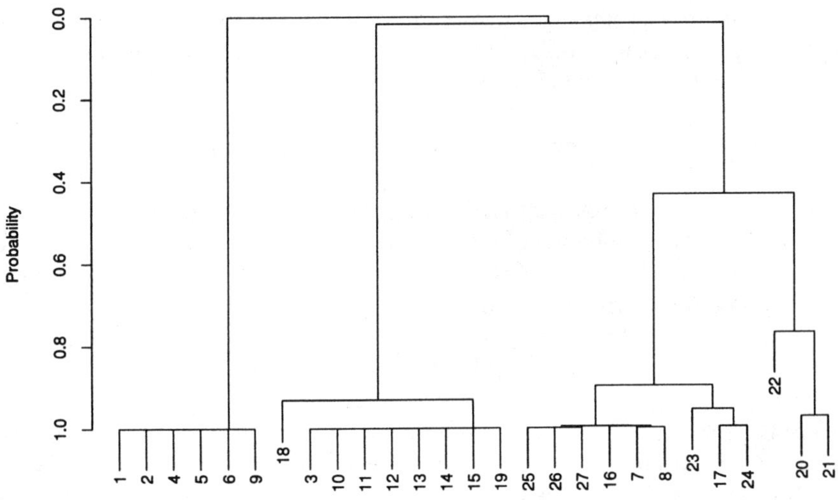

Figure 20.17. Dendrogram of *a posteriori* association between the samples in the York medieval glass data when the prior for k is that shown in Figure 20.14b and the prior for $\mathbf{\Omega}$ is the identity matrix.

than in Figure 20.15. This is entirely to be expected since our prior is that groups should have smaller variability, and so any sample that increases the within-group variance beyond that given by the prior will be allocated to the group with low probability. This is a clear demonstration of the importance of our prior information about $\mathbf{\Omega}$. Figure 20.17 also shows the same basic groupings as the other two plots, but here some samples are allocated to groups with slightly less certainty than in Figure 20.15.

Comparing the Bayesian results with those from the exploratory methods used in Section 20.5, consider first the average linkage dendrograms in Figure 20.9. These show very similar groupings to those obtained using the Bayesian methodology, with the notable exception that samples 20, 21, and 22 are grouped together in a single small cluster in the Bayesian dendrograms and are not in a separate group in the average linkage dendrograms. Some insight into the reason for this can be obtained by inspection of the biplot in Figure 20.8. This shows two compact groups (one to the right and the other the bottom left of the biplot) that can also be identified in the Bayesian dendrograms. The more dispersed group to the top left of the biplot can be seen to be associated with the more dispersed group on the Bayesian dendrograms too. What is most interesting, perhaps, is that the three samples (20, 21, and 22) that consistently appear in a group on their own in the Bayesian dendrograms clearly lie in an area of the biplot between two of the previously noted groups and might reasonably be seen as forming a very dispersed group of their own. In other words, although the methodologies are quite different, the biplot and the Bayesian dendrograms would lead to broadly the same conclusions.

20.7.7. Some Problems That Must Be Addressed

For a model-based approach to be really useful, it is necessary to be able to tackle not just the modeling of the problem under study but also problems of the type outlined in Section 20.6 and others that arise specifically because we have chosen a Bayesian approach.

20.7.7.1. Outliers

Since the Bayesian methodology outlined above results in probabilistic classification of specimens to sources or clusters, outliers should be relatively readily detected. In the case of Bayesian cluster analysis, such specimens are likely to appear in a cluster all of their own or in a small cluster with other samples that are also outliers. Indeed, in our example it was noted that outliers can form small groups of their own. In the case of Bayesian discriminant analysis, outliers are likely to have low probability of membership to several sources that have properties similar to their own, but no clear association with any single group. In some cases the presence of outliers can distort model-based analyses just as they can exploratory ones (see Section 20.6.2), and there can be advantages to removing the possible outliers and then running the analysis again.

20.7.7.2. Multivariate Normality

Although the multivariate normal distribution was used as an illustration in Section 20.7.2, it should be noted that there is no necessity to assume multivariate normality when undertaking Bayesian (or any other model-based) analysis. In theory, researchers are free to adopt any model that is felt to adequately reflect the properties of the problem under study. In practice, some models are statistically easier to implement than others, and this is partly why the multivariate normal is widely used. It has properties that mean that the prior is fairly easily defined and that the computations that lead to the posterior are not intractable.

As the popularity of the Bayesian approach grows, model definition and selection are increasingly seen as cornerstones to its success. As a consequence much research is being invested in this area (see, e.g., studies in Gilks et al. 1996), and the restrictions regarding practicable models are reducing all the time.

20.7.7.3. Sample Size

The potential problems associated with analysis of small sample sizes (as outlined in Section 20.6.4) are particularly pertinent when undertaking model-based analyses. For stable and reliable posterior estimates of parameter values, it is desirable that n be much greater than p. The York medieval glass data (Cox and Gillies 1986) consist of the concentrations of 11 elements in 27 samples and hence contains only barely sufficient samples to make a model-based analysis sensible. As a result, we must be very careful not to place great faith in the posterior probability distributions obtained and certainly should not use the results of a Bayesian analysis alone to inform our interpretation. In general, we suggest that exploratory methods be adopted for all initial investigations (especially when the numbers of samples are small) and that the results of these be used to inform prior opinion if, or when, more data become available.

20.7.7.4. Missing Data

When adopting a Bayesian approach to the analysis of chemical compositional data, random missing data points are much less of a problem than when using exploratory methods. Provided that the numbers of missing data values are not large relative to the total available data, a missing X_{ij} can be treated as a parameter and simulated in the same way as other values we wish to learn about. It is necessary to provide an initial estimate of such values (usually using one of the approaches outlined in Section 20.6.5), but then at each iteration of the sampling scheme a value is resimulated based on the properties of the source or cluster to which specimen X_i is presently allocated. Other types of missing data, including those arising because elements or oxides are present at or below the detection limit of the analytical equipment employed, are usually handled in the ways described in Section 20.6.5, but might in the future be modeled into the statistical framework. Further research into handling missing data when using model-based methods is certainly reasonable.

20.7.7.5. Elicitation and Definition of Prior Information

The nature, quality, and quantity of prior information available for our example Bayesian investigation was extremely limited. Indeed, purists might suggest that without more reliable prior information about $\mathbf{\Omega}$ we should not really be undertaking a Bayesian analysis at all. The present authors feel that if new methodologies are to be more widely adopted they must be illustrated. We hope that by illustrating the potential for such methods, other researchers will try model-based approaches and that, as a result, more helpful prior information will begin to become available that can be used to inform future research. The variability in the chemical compositions of glasses from the same provenance or site of manufacture is complex, presently little understood, and still undergoing research (see, e.g., Henderson, 1988, and Wedepohl, 1997). As far as we are aware, this is the first application of Bayesian cluster analysis to archaeological glass data, and we are optimistic about the results, but further research is certainly needed.

20.8. DISCUSSION—THE FUTURE

20.8.1. Introduction

Predicting future developments is a foolhardy task, however, the one certainty is that data handling in archaeometry in the future will be influenced by developments in both computer hardware and software. Many methods that have been developed in the statistical literature, of apparent or potential relevance to archaeometric problems, have not been exploited. This is almost certainly because they are not widely available in commercial and user-friendly software packages. In these concluding sections attention is drawn to some of these methodologies and what may be required for their effective use.

20.8.2. Exploratory Methods

Increased computing power promoted the use of multivariate statistical methodology in archaeometry from the early 1970s (Sayre 1975; Bieber et al. 1976; Harbottle 1976). In some ways there has been surprisingly little development since then; the work initiated in the works cited remains influential (Glascock 1992), and many statistical applications in archaeometry are recognizably similar to the early work in the way in which methods of cluster analysis and PCA are exploited, for example. This, no doubt, testifies to the utility of the techniques, and they are likely to remain widely used in the foreseeable future.

A subtle change that has occurred is in the way some techniques are viewed by some research groups. Specifically, cluster analysis—the most widely used multivariate method in archaeometry (Baxter 1994a)—occupies a much less central position in some approaches to data analysis than was the case a few years ago (e.g., Glascock 1992; Beier and Mommsen 1994; Neff, n.d.). This has arisen from a more critical appreciation of the limitations of cluster analysis as applied to archaeometric data.

Principal component analysis is a particular example of *projection pursuit* methodology (Jones and Sibson 1987; Nason 1995). In such methods combinations of variables are sought that optimize some index of "interestingness," such as cluster structure. Principal component analysis seeks linear combinations with maximum variance, but this does not necessarily equate with interesting structure, and its popularity stems at least partly from its tractability and widespread availability. This position may change if and when projection pursuit software becomes more readily accessible to the nonspecialist.

The wider archaeological literature has seen a number of attempts at using *neural networks* for classification problems (Durham et al. 1996; Gibson 1996; Bell and Croson 1998). This is a computer-intensive methodology that is attracting increasing attention in the statistical literature (Ripley 1996). Neural networks are a form of nonlinear methodology in contrast to the linear methods of PCA and LDA, and the same is true of *classification trees* (Clark and Pregibon 1993), which can be used as a nonlinear form of discriminant analysis and is available in the S-Plus package. There is no guarantee that such methods will improve much on those currently in use, but they merit investigation.

20.8.3. Model-Based Methods

Model-based methods of cluster analysis exist that are not available in the software packages most used in archaeometry (e.g., McLachlan and Basford 1988; Banfield and Raftery 1993). Such methods depend on specific assumptions about the stochastic nature of the phenomena being studied, and this was discussed in the context of Bayesian methodology in Section 20.7. As with other model-based approaches, their potential for archaeometric use has not been evaluated.

Another potentially attractive methodology is that of *graphical modeling* (Whittaker 1990), though the relatively early archaeological application of Scott et al. (1991) has not been emulated. The attraction of the methodology is that it can be used for mixed-mode data analysis so that, for example, the analysis of quantitative geochemical data can be integrated with qualitative information. The drawback is that the methodology may be difficult to implement with large numbers of variables.

Statisticians are rather keen on model-based methods, so why have they not been more widely used in archaeometry? One reason is that relatively few statisticians are involved in archaeometric study, and the methods are more complex and less readily accessible than exploratory methods. They are also based on assumptions that may not be satisfied by archaeometric data. Finally, models for multivariate data can be very demanding in terms of sample size because of the number of parameters that need to be estimated. To explore the potential of such methods may thus require the use of larger samples than have often been available in the past, and may also require more active collaboration with statisticians.

20.8.4. Software Considerations

Published statistical analyses in archaeometry are usually accomplished using a commercially available software package such as SPSS or SAS or software written in-house. The justification for the latter has been that archaeometric data present problems of analysis not readily handled within standard packages.

For the "bread-and-butter" techniques discussed in Section 20.5, this may have been the case but is no longer true. Packages such as S-Plus (Venables and Ripley 1997), used for many of the analyses and graphics in this chapter, come with many powerful and sophisticated functions. The package, additionally, has programming facilities that enable the calculations one might wish to do with archaeometric data, and the graphics, to be readily produced. Software developed in-house can have the effect of "fossilizing" an approach to data handling, as it can be time-consuming to modify. Additionally, such software can be idiosyncratic and may not be easily portable, so that a great deal of "reinventing the wheel" has taken place in laboratories around the world. We think that nowadays there is a very strong case for using commercial software such as S-Plus, in place of in-house packages, because of its quality and flexibility because it makes it much easier to share ideas and analytical strategies and because it makes it easier to explore newly developed or more complex methodologies.

If we are to turn to model-based approaches, such as the Bayesian one outlined in Section 20.7, however, specialist software will be required. Such methodologies are currently under development and are by no means standard. Statisticians increasingly have access to (or have written their own) suitable software that is not in the public domain. As a result we strongly encourage archaeometrists to collaborate with statisticians to benefit from state-of-the-art methodologies.

Acknowledgments

The authors would like to thank the following for their invaluable comments on earlier drafts of this chapter: Geoff Boden, Jaume Buxeda i Garrigos, J. Andrés Christen, Hilary Cool, Ron Hancock, Caroline Jackson, and Hector Neff. We would also like to extend additional thanks to Hector Neff for access to as yet unpublished papers.

REFERENCES

Aitchison, J. A. (1986), *The Statistical Analysis of Compositional Data*, Chapman and Hall, London.

Anderson, T. W. (1984), *An Introduction to Multivariate Statistical Analysis*, John Wiley & Sons, New York.

Argyropoulos, V. (1995), *Archaeometry* **37**, 271.

Arnold, D. E., H. A. Neff, R. L. Bishop, and M. D. Glascock (1999), in *Material Meanings: Critical Approaches to Interpreting Material Culture*, E. S. Chilton, Ed., University of Utah Press, Salt Lake City, p. 61.

Atkinson, A. C., and H. M. Mulira (1993), *Statist. Comput.* **3**, 27.

Banfield, J. J., and A. E. Raftery (1993), *Biometrics* **49**, 803.

Barnett, V. B. (1982), *Comparative Statistical Inference*, 2nd ed., John Wiley & Sons, New York.

Barnett, V., and T. Lewis (1994), *Outliers in Statistical Data*, 3rd ed., John Wiley & Sons, New York.

Baxter, M. J. (1989), *Archaeometry* **31**, 45.

Baxter, M. J. (1992), in *Computer Applications and Quantitative Methods in Archaeology 1991*, G. Lock and J. Moffett, Eds., BAR International Series 577, Oxford, p. 141.

Baxter, M. J. (1993), *Archaeometry* **35**, 112.

Baxter, M. J. (1994a), *Exploratory Multivariate Analysis in Archaeology*, Edinburgh University Press, Edinburgh.

Baxter, M. J. (1994b), *J. Archaeol. Sci.* **21**, 659.

Baxter, M. J. (1994c), *Arch. Calc.* **5**, 23.

Baxter, M. J. (1995), *Appl. Statist.* **44**, 513.

Baxter, M. J. (1999a), *J. Archaeol. Sci.* **26**, 117.

Baxter, M. J. (1999b), *Computer Applications and Quantitative Methods in Archaeology 1997*, L. Dingwall et al., Eds., BAR International Series 750, Oxford, p. 107.

Baxter, M. J., and C. C. Beardah (1995), in *Computer Applications and Quantitative Methods in Archaeology 1994*, J. Huggett and N. Ryan, Eds., BAR International Series 600, Oxford, p. 63.

Baxter, M. J., and C. C. Beardah (1996), *Arch. Calc.* **7**, 397.

Baxter, M. J., and N. H. Gale (1998), *J. Appl. Statist.* **25**, 671.

Baxter, M. J., H. E. M. Cool, and M. P. Heyworth (1990), *J. Appl. Statist.* **17**, 229.

Baxter, M. J., H. E. M. Cool, M. P. Heyworth, and C. M. Jackson (1995), *Archaeometry* **37**, 129.

Baxter, M. J., C. C. Beardah, and R. V. S. Wright (1997), *J. Archaeol. Sci.* **24**, 347.

Bayes, T. R. (1763), *Philos. Trans. R. Soc.* **53**, 370.

Beardah, C. C. (1999), *Computer Applications and Quantitative Methods in Archaeology 1997*, L. Dingwall et al., Eds., BAR International Series 750, Oxford, p. 107.

Beardah, C. C., and M. J. Baxter (1996), *Internet Archaeol.* **1**, http://intarch.ac.uk/journal/issue1/beardah.

Beier, T., and H. Mommsen (1994), *Archaeometry* **36**, 287.

Bell, S., and C. Croson (1998), *Archaeometry* **40**, 139.

Bieber, A. M., D. W. Brooks, G. Harbottle, and E. V. Sayre (1976), *Archaeometry* **18**, 59.

Bishop, C. M. (1995), *Neural Networks for Pattern Recognition*, Clarendon, Oxford.

Bishop, R. L., and H. Neff (1989), in *Archaeological Chemistry IV*, R. O. Allen, Ed., Advances in Chemistry Series 220, American Chemical Society, Washington, DC, p. 57.

Blackman, M. J., G. J. Stein, and P. B. Vandiver (1993), *Am. Antiq.* **58**, 60.

Bollong, C. A., L. Jacobson, M. Peisach, C. A. Pineda, and C. G. Sampson (1997), *J. Archaeol. Sci.* **24**, 319.

Bowman, A. W., and A. Azzalini (1997), *Applied Smoothing Techniques for Data Analysis*, Clarendon, Oxford.

Buck, C. E. (1994), Towards Bayesian Archaeology, Ph.D. thesis, University of Nottingham, Nottingham, United Kingdom.

Buck, C. E., and J. A. Christen (1998), *J. Archaeol. Sci.*, **25**, 303.

Buck, C. E., and C. D. Litton (1996), in *Valencia 5*, J. M. Bernardo, J. O. Berger, A. P. Dawid, and A. F. M. Smith, Eds., Clarendon, Oxford, p. 499.

Buck, C. E., J. B. Kenworthy, C. D. Litton, and A. F. M. Smith (1991), *Antiquity* **65**, 808.

Buck, C. E., C. D. Litton, and A. F. M. Smith (1992), *J. Archaeol. Sci.* **19**, 497.

Buck, C. E., C. D. Litton, and S. J. Shennan (1994), *Germania* **2**, 427.

Buck, C. E., W. G. Cavanagh, and C. D. Litton (1996), *The Bayesian Approach to Archaeological Data Interpretation*, John Wiley & Sons, Chichester.

Budd, P., D. Gale, A. M. Pollard, R. G. Thomas, and P. A. Williams (1993), *Archaeometry* **35**, 241.

Budd, P., A. M. Pollard, B. Scaife, and R. G. Thomas (1995), *J. Mediterranean Archaeol.* **8**, 1.

Budd, P., R. Haggerty, A. M. Pollard, B. Scaife, and R. G. Thomas (1996), *Antiquity* **70**, 168.

Burton, J. H., and A. W. Simon (1993), *Am. Antiq.* **58**, 45.

Burton, J. H., and A. W. Simon (1996), *Am. Antiq.* **61**, 405.

Campbell, N. A. (1985), *Appl. Statist.* **34**, 235.

Caroni, C., and P. Prescott (1992), *Appl. Statist.* **41**, 355.

Castellano, A., A. D'Innocenzo, C. Pagliara, and F. Raho (1996), *Archaeometry* **38**, 59.

Christen, J. A. (1994a), Bayesian Interpretation of Radiocarbon Results, Ph.D. thesis, University of Nottingham, Nottingham, United Kingdom.

Christen, J. A. (1994b), *Appl. Statist.* **43**, 489.

Christen, J. A., and C. E. Buck (1998), *Appl. Statist.* **47**, 543.

Clark, L. A., and D. Pregibon (1993), in *Statistical Models in S*, J. M. Chambers and T. J. Hastie, Eds., Chapman and Hall, London, p. 377.

Clarke, D. L. (1972), *Models in Archaeology*, Methuen, London.

Cox, G. A., and K. J. S. Gillies (1986), *Archaeometry* **28**, 57.

Durham, P., P. Lewis, and S. J. Shennan (1996), *Analecta Praehistorica Leidensia* **28-I**, 235.

Everitt, B. S. (1993), *Cluster Analysis*, 3rd ed., Edward Arnold, London.

Fieller, N. R. J., and E. Flenley (1988), in *Computer Applications and Quantitative Methods in Archaeology 1987*, C. L. N. Ruggles and S. P. Q. Rahtz, Eds., BAR International Series 393, Oxford, p. 79.

Fisher, R. A. (1936), *Ann. Eugen.* **7**, 376.

Geman, S., and D. Geman (1984), *IEEE Trans. Pattern Anal. Mach. Intell.* **6**, 721.

Gibson, P. M. G. (1996), *Analecta Praehistorica Leidensia* **28-I**, 229.

Gilks, W. R., S. Richardson, and D. J. Spiegelhalter (1996), *Markov Chain Monte Carlo in Practice*, Chapman and Hall, London.

Glascock, M. D. (1992), in *Chemical Characterization of Ceramic Pastes in Archaeology*, H. Neff, Ed., Prehistory Press, Madison, WI, p. 11.

Gower, J. C., and D. J. Hand (1996), *Biplots*, Chapman and Hall, London.

Greenacre, M. J. (1984), *Theory and Applications of Correspondence Analysis*, Academic, London.

Greenacre, M. J. (1993), *Correspondence Analysis in Practice*, Academic, London.

Hadi, A. S. (1992), *J. R. Statist. Soc. B* **54**, 761.

Hadi, A. S. (1994), *J. R. Statist. Soc. B* **56**, 393.

Hand, D. J. (1981), *Discrimination and Classification*, John Wiley & Sons, New York.

Hand, D. J. (1997), *Construction and Assessment of Classification Rules*, John Wiley & Sons, New York.

Harbottle, G. (1976), in *Radiochemistry*, Vol. 3, G. W. A. Newton, Ed., Chemical Society, London, p. 33.

Hegmon, M., J. R. Allison, H. Neff, and M. D. Glascock (1997), *Am. Antiq.* **62**, 449.

Henderson, J. (1988), *Archaeometry* **30**, 77.

Howson, C., and P. Urbach (1993), *Scientific Reasoning: The Bayesian Approach*, 2nd ed., Open Court, IL.

Jackson, J. E. (1991), *A User's Guide to Principal Components*, John Wiley & Sons, New York.

Johnson, N. L., and S. Kotz (1970), *Distributions in Statistics—Continuous Univariate Distributions*, Vol. 2, John Wiley & Sons, New York.

Jolliffe, I. T. (1986), *Principal Component Analysis*, Springer-Verlag, New York.

Jones, M. C., and R. Sibson (1987), *J. R. Statist. Soc. A* **150**, 1.

Klecka, W. R. (1980), *Discriminant Analysis*, Sage, Newbury Park, CA.

Knutsson, K., B. Dahlquist, and H. Knutsson (1988), in *Industrie Lithiques: Aspectes Archaeologique*, S. Beyries, Ed., BAR International Series 411(i), Oxford, p. 253.

Krzanowski, W. J. (1987), *Appl. Statist.* **36**, 22.

Krzanowski, W. J. (1988), *Principles of Multivariate Analysis*, Clarendon, Oxford.

Lachenbruch, P. A. (1975), *Discriminant Analysis*, Hafner, New York.

Leese, M. N. (1988), in *Classical Marbles: Geochemistry, Technology, Trade*, N. Herz and M. Waelkens, Eds., Kluwer Academic, Dordrecht, p. 347.

Leese, M. N. (1992), *Archaeometry* **34**, 318.

Leese, M. N., and P. L. Main (1994), *Archaeometry* **36**, 307.

Leese, M. N., M. J. Hughes, and J. Stopford (1989), in *Computer Applications and Quantitative Methods in Archaeology 1989*, S. P. Q. Rahtz and J. Richards, Eds., BAR International Series 548, Oxford, p. 241.

Litton, C. D., and C. E. Buck (1996), in *Markov Chain Monte Carlo in Practice*, S. Richardson, W. Gilks, and D. Spiegelhalter, Eds., Chapman and Hall, London.

Lizee, J. M., H. Neff, and M. D. Glascock (1995), *Am. Antiq.* **60**, 515.

Lockyear, K. (1999), *Computer Applications and Quantitative Methods in Archaeology 1997*, L. Dingwall et al., Eds., BAR International Series 750, Oxford, p. 85.

Madsen, T., Ed. (1988), *Multivariate Archaeology*, Aarhus University Press, Aarhus.

Mardia, K. V., J. T. Kent, and J. M. Bibby (1979), *Multivariate Analysis*, Academic, London.

McLachlan, G. J. (1992), *Discriminant Analysis and Statistical Pattern Recognition*, John Wiley & Sons, New York.

McLachlan, G. J., and K. E. Basford (1988), *Mixture Models*, Marcel Dekker, New York.

Mirti, P., A. Casoli, and L. Appolonia (1993), *Archaeometry* **356**, 225.

Müller, J., and A. Zimmermann, Eds. (1997), *Archäologie und Korrespondenzanalyse*, Verlag Marie Leidorf GmbH, Espelkamp.

Müller, P., A. Erkanli, and M. West (1996), *Biometrika* **83**, 67.

Nason, G. (1995), *Appl. Statist.* **44**, 411.

Neff, H. (1994), *Archaeometry* **36**, 115.

Neff, H. (n.d.), in *Ceramic Source Determination by Neutron Activation Analysis in the American Southwest*, D. Glowacki and H. Neff, Eds., in preparation.

Neff, H., and M. D. Glascock (1992), unpublished manuscript, Missouri University Research Reactor.

Neff, H., M. D. Glascock, R. L. Bishop, and M. J. Blackman (1996), *Am. Antiq.* **61**, 389.

Penny, K. I. (1996), *Appl. Statist.* **45**, 73.

Pollard, A. M. (1986), in *Greek and Cypriot Pottery: A Review of Scientific Studies*, R. E. Jones, Ed., British School at Athens Fitch Laboratory Occasional Paper 1, Athens, p. 56.

Pollard, A. M., and C. Heron (1996), *Archaeological Chemistry*, Royal Society of Chemistry, Cambridge.

Ripley, B. D. (1996), *Pattern Recognition and Neural Networks*, Cambridge University Press, Cambridge.

Rotonno, T., L. Sabbatini, and M. Corrente (1997), *Archaeometry* **39**, 343.

Rousseeuw, P. J., and B. C. van Zomeren (1990), *J. Am. Statist. Assoc.* **85**, 633.

Sayre, E. V. (1975), unpublished manuscript, Brookhaven National Laboratory.

Sayre, E. V., K. A. Yener, E. C. Joel, and I. L. Barnes (1992), *Archaeometry* **34**, 73.

Sayre, E. V., K. A. Yener, and E. C. Joel (1995), *J. Mediterranean Archaeol.* **8**, 45.

Scaife, B. (1998), Lead Isotope Analysis in Archaeology, Ph.D. Thesis, University of Bradford, Bradford, United Kingdom.

Scaife, B., P. Budd, J. G. McDonnell, A. M. Pollard, and R. G. Thomas (1996), in *Archaeometry 94: Proceedings of the 29th International Symposium on Archaeometry, Ankara, 9–14 May 1994*, S. Demirci, A. M. Ozer, and G. D. Summers, Eds., Tubitak, Ankara, p. 301.

Schafer, J. L. (1997), *Analysis of Incomplete Multivariate Data*, Chapman and Hall, London.

Scott, A., J. Whittaker, M. Green, and S. Hillson (1991), in *Computer Applications and Quantitative Methods in Archaeology 1990*, K. Lockyear and S. P. Q. Rahtz, Eds., BAR International Series 565, Oxford, p. 111.

Sheather, S. J., and M. C. Jones (1991), *J. R. Statist. Soc. B* **53**, 683.

Shennan, S. (1997), *Quantifying Archaeology*, 2nd ed., Edinburgh University Press, Edinburgh.

Silverman, B. W. (1986), *Density Estimation*, Chapman and Hall, London.

Simonoff, J. S. (1996), *Smoothing Methods in Statistics*, Springer-Verlag, New York.

Slane, K. W., J. M. Elam, M. D. Glascock, and H. Neff (1994), *J. Archaeol. Sci.* **21**, 51.

Solomon, H. (1971), in *Mathematics in the Archaeological and Historical Sciences*, F. R. Hodson, D. G. Kendall, and P. Tautu, Eds., Edinburgh University Press, Edinburgh, p. 62.

Steponaitis, V. P., M. J. Blackman, and H. Neff (1996), *Am. Antiq.* **61**, 555.

Stopford, J., M. J. Hughes, and M. N. Leese (1991), *Oxford J. Archaeol.* **10**, 349.

Stos-Gale, Z. A., and N. H. Gale (1994), in *Provenience Studies and Bronze Age Cyprus*, A. B. Knapp and J. F. Cherry, Eds., Prehistory Press, Madison, WI, p. 92.

Stos-Gale, Z. A., N. H. Gale, and N. Annetts (1996), *Archaeometry* **38**, 381.

Stos-Gale, Z. A., G. Maliotis, N. H. Gale, and N. Annetts (1997), *Archaeometry* **39**, 83.

Tangri, D., and R. V. S. Wright (1993), *Archaeometry* **35**, 103.

Taylor, R. J., and V. J. Robinson (1996a), *Archaeometry* **38**, 231.

Taylor, R. J., and V. J. Robinson (1996b), *Archaeometry* **38**, 245.

Triadan, D., H. Neff, and M. D. Glascock (1997), *J. Archaeol. Sci.* **24**, 997.

Troja, S. O., A. Cro, A. M. Gueli, V. La Rosa, P. Mazzoleni, A. Pezzino, and M. Romeo (1996), *Archaeometry* **38**, 113.

Tubb, A., A. J. Parker, and G. Nickless (1980), *Archaeometry* **22**, 153.

Underhill, L. G., and M. Peisach (1985), *J. Trace Microprobe Tech.* **3**, 41.

Venables, W. N., and B. D. Ripley (1997), *Modern Applied Statistics with S-Plus*, 2nd ed., Springer-Verlag, New York.

Wand, M. P., and M. C. Jones (1995), *Kernel Smoothing*, Chapman and Hall, London.

Wedepohl, K. H. (1997), *Glastech. Ber. (GST)* **70**, 246.

Whittaker, J. (1990), *Graphical Models*, John Wiley & Sons, New York.

Williams-Thorpe, O., S. E. Warren, and J. O. Nandris (1984), *J. Archaeol. Sci.* **11**, 183.

Williams-Thorpe, O., M. C. Jones, A. G. Tindle, and R. S. Thorpe (1996), *Archaeometry* **38**, 15.

Zeidler, J. A., C. E. Buck, and C. D. Litton (1998), *Latin Am. Antiq.* **9**, 135.

INDEX*

*Boldface numbers indicate pages on which a general discussion of the term appears.